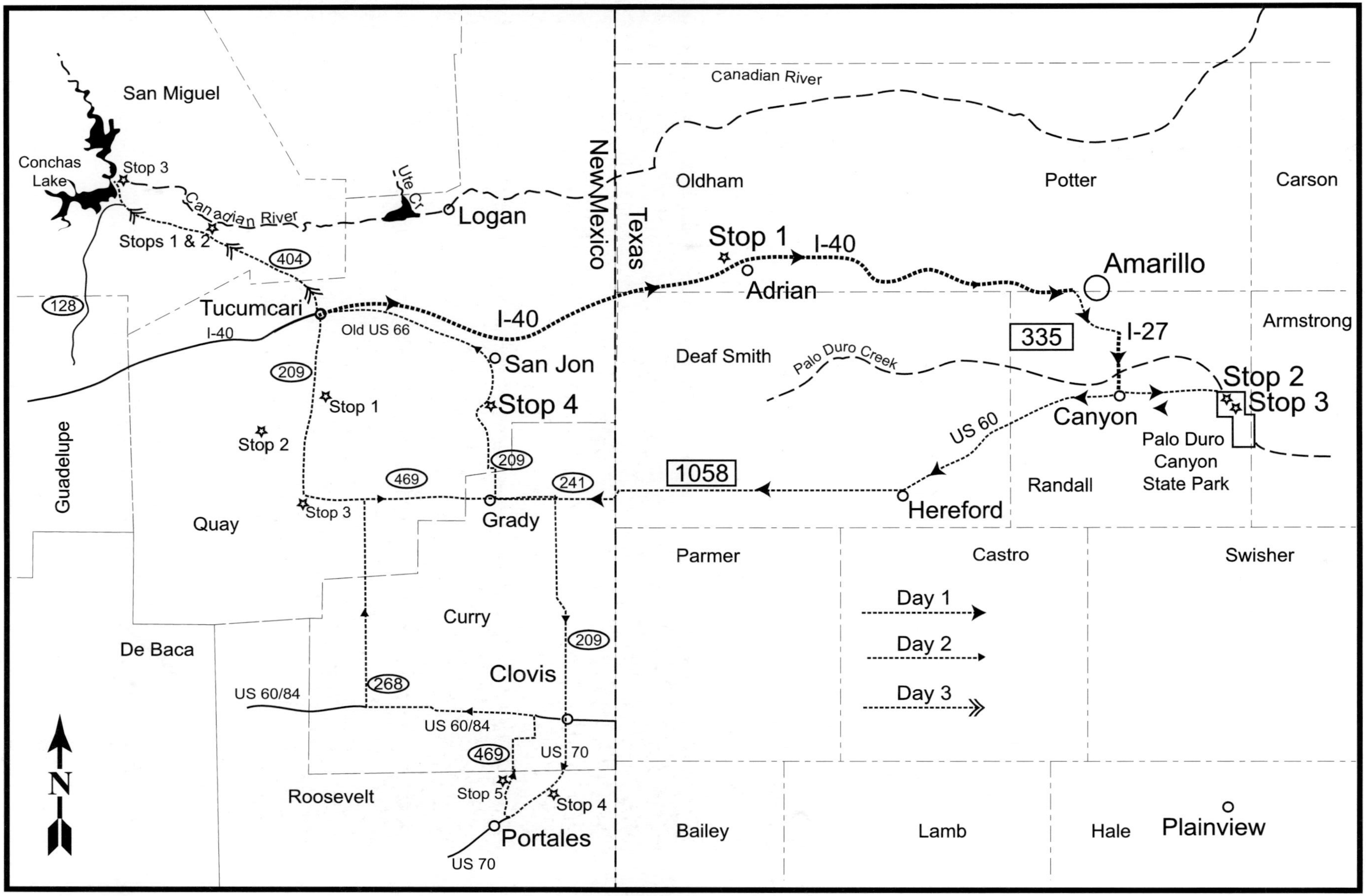
San Miguel
Conchas Lake
Stop 3
Canadian River
Ute Cr
Logan
Stops 1 & 2
404
128
Tucumcari
I-40
Old US 66
I-40
San Jon
209
Stop 1
Stop 2
Stop 4
Guadelupe
209
469
241
Stop 3
Grady
Quay
Curry
209
De Baca
Clovis
268
US 60/84
US 60/84
469
US 70
Roosevelt
Stop 5
Stop 4
Portales
US 70
New Mexico
Texas
Canadian River
Oldham
Potter
Carson
Stop 1
I-40
Adrian
Amarillo
Armstrong
335
I-27
Deaf Smith
Palo Duro Creek
Stop 2
Stop 3
Canyon
US 60
Palo Duro Canyon State Park
1058
Hereford
Randall
Parmer
Castro
Swisher
Day 1
Day 2
Day 3
Bailey
Lamb
Hale
Plainview
N

Geology of the Llano Estacado

Editors
Spencer G. Lucas
Dana S. Ulmer-Scholle

New Mexico Geological Society
Fifty-second Annual Field Conference
September 26-29, 2001

CONTENTS

Dedication ... vi
President's Message ... vii
Editor's Message ... vii
Committees ... ix
Field Conference Schedule and Credits ... x

ROAD LOGS

First-day road log, from Tucumcari to the edge of the Llano Estacado at Gruhlkey, Texas, Palo Duro Canyon, Texas, and San Jon Hill, New Mexico ... *S. G. Lucas, A. P. Hunt, B. S. Kues, A. B. Heckert, and V. T. McLemore* 1

Minipapers:

Paleontology and age of the Upper Triassic Trujillo Formation, east-central New Mexico and West Texas ... *A. P. Hunt* 3
Neogene land-mammal "ages" in the Texas Panhandle ... *S. G. Lucas and G. S. Morgan* 8
Abandon the term Dockum! ... *S. G. Lucas* 12
The San Jon site, Quay County, New Mexico ... *V. T. Holliday and E. Johnson* 17
Late Albian marine invertebrate fauna from the basal Mesa Rica Sandstone at San Jon Hill, New Mexico ... *B. S. Kues* 20
The Frio uplift: A Paleozoic-Mesozoic control on sedimentation patterns in east-central New Mexico ... *A. P. Hunt and S. G. Lucas* 21
Paleocurrents in the Trujillo Formation (Chinle Group; Upper Triassic), east-central New Mexico, and the myth of the Dockum lake ... *A. P. Hunt, S. G. Lucas, and A. B. Heckert* 22

Second-day road log, from Tucumcari to Mesa Redonda, Pyramid Mountain, Ragland, and Blackwater Draw, New Mexico ... *S. G. Lucas, A. P. Hunt, A. B. Heckert, B. S. Kues, and V. T. McLemore* 25

Minipapers:

Oasis State Park ... *V. T. McLemore* 34
The shovel-tusked Gomphothere *Amebelodon* (Mammalia: Proboscidea) from the Miocene Ogallala Formation at the Blackwater Draw site, Roosevelt County, New Mexico ... *G. S. Morgan and S. G. Lucas* 37
Laguna Colorado: A forgotten Holocene lake in Quay County, New Mexico ... *A. P. Hunt* 38

Third-day road log, from Tucumcari to Conchas Dam, New Mexico ... *S. G. Lucas, A. P. Hunt, A. B. Heckert, B. S. Kues, and V. T. McLemore* 41

Minipapers:

The Laramide Tucumcari structural zone, east-central New Mexico ... *A. P. Hunt and S. G. Lucas* 41
The type area of *Bison bison* (Linnaeus, 1758), Canadian River Valley, east-central New Mexico ... *A. P. Hunt and S. G. Lucas* 44
Conchas and Ute Reservoirs and water issues in eastern New Mexico ... *A. P. Hunt* 48
The first vertebrate track (*Brachychirotherium*) from the upper Carnian Garita Creek Formation, east-central New Mexico ... *A. P. Hunt and S. G. Lucas* 51
Conchas Lake State Park ... *V. T. McLemore* 52

Road log references ... 54

ARTICLES

Geophysics and Volcanology

Magnetostratigraphic correlation and dating of West Texas and New Mexico Late Permian strata ... M. Steiner 59
Raton-Clayton and Ocate volcanic fields ... J. C. Aubele and L. S. Crumpler 69

Stratigraphy, Paleontology and Sedimentology

Notes on Upper Permian and Lower Triassic strata and structures in east-central and central New Mexico..... *R. D. Maier* 77

Triassic stratigraphy, biostratigraphy and correlation in east-central New Mexico.................... *S. G. Lucas, A. B. Heckert and A. P. Hunt* 85

The Moenkopi Formation in east-central New Mexico: Stratigraphy and vertebrate fauna............ *J. A. Boy, R. R. Schoch and S. G. Lucas* 103

The bennettitalean leaf "*Zamites*" *powellii* from the Middle Triassic Moenkopi Formation, east-central New Mexico.......... *S. G. Lucas* 111

New occurrence of *Trilophosaurus* (Reptilia: Archosauromorpha) from the Upper Triassic of West Texas and its biochronological significance..........................*A. B. Heckert, S. G. Lucas, R. Kahle and K. Zeigler* 115

The vertebrate fauna, biostratigraphy and biochronology of the type Revueltian land-vertebrate faunachron, Bull Canyon Formation (Upper Triassic), east-central New Mexico.................................*A. P. Hunt* 123

Lacustrine depositional environments of the Upper Triassic Redonda Formation, east-central New Mexico.......... *P. M. Hester and S. G. Lucas* 153

A giant phytosaur (Reptilia: Archosauria) skull from the Redonda Formation (Upper Triassic: Apachean) of east-central New Mexico..*A. B. Heckert, S. G. Lucas, A. P. Hunt and J. D. Harris* 169

Tetrapod footprint ichnofauna of the Upper Triassic Redonda Formation, Chinle Group, Quay County, New Mexico.......... *S. G. Lucas, A. P. Hunt and M. G. Lockley* 177

The Late Triassic sauropod track record comes into focus: Old legacies and new paradigms...................... *M. G. Lockley, J. L. Wright, A. P. Hunt and S. G. Lucas* 181

Geology of the Ima NW quadrangle, east-central New Mexico............................ *S. G. Lucas, G. Weadock, K. K. Kietzke, A. P. Hunt and B. S. Kues* 191

Jurassic strata in east-central New Mexico and their regional significance........................*S. G. Lucas and L. A. Woodward* 203

A brachiosaurid dinosaur (Saurischia: Sauropoda) from the Upper Jurassic of east-central New Mexico.......... *A. P. Hunt* 213

Cretaceous stratigraphy and biostratigraphy, east-central New Mexico.................. *S. G. Lucas, B. S. Kues and A. P. Hunt* 215

Albian-Cenomanian depositional cycles transgressed from Chihuahua trough to Western Interior.................. *R. W. Scott, J. M. Holbrook, M. J. Evetts and F. E. Oboh-Ikuenobe* 221

Nearshore fauna of the Tucumcari Formation (Lower Cretaceous, Albian), Quay County, New Mexico............*B. S. Kues and S. G. Lucas* 229

Stratigraphy and biostratigraphy of an Upper Cretaceous outlier near Roy, Harding County, New Mexico........................ *S. G. Lucas and A. B. Heckert* 251

Paleontological resources of Lake Meredith National Recreation Area and Alibates Flint Quarries National Monument, West Texas.......... *A. P. Hunt and V. L Santucci* 257

A review of Pleistocene vertebrate faunas from northeastern New Mexico...........*G. S. Morgan, S. G. Lucas, P. L. Sealey and A. P. Hunt* 265

Late Pleistocene mammoths (*Mammuthus columbi*) from Mesa Redonda, Quay County, east-central New Mexico.......... *G. S. Morgan, S. G. Lucas and M. E. Gordon* 285

A new late Neogene paleofauna from the Kiowa/Rita Blanca National Grassland, northeastern New Mexico................. *B. A. Schumacher* 293

Hydrogeology

Quantification of spatially varying hydrogeologic properties for a predictive model of groundwater flow in the Ogallala aquifer, northeastern Texas Panhandle...........................*A. R. Dutton, R. E. Mace and R. C. Reedy* 297

Geophysical and hydrochemical analysis of the White River alluvial aquifer, Crosby County, Texas............... *B. J. Hibbs* 309

The Endee Pipeline leak—implications for treating groundwater contamination beneath the Great Plains, eastern New Mexico.......... *W. J. Stone* 315

Long-term monitoring of the geochemistry of surface water and stream-sediment samples from the upper Pecos River, from the southern Pecos Wilderness to Brantley Dam, north of Carlsbad, eastern New Mexico: Year nine..*V. T. McLemore, L. A. Brandvold and E. A. Carey* 321

Economic Geology

A geologic membrane—microbial metabolism mechanism for the origin of the sedimentary copper deposits in the Pastura district, Guadalupe County, New Mexico.....................................*V. W. Lueth and T. M. Whitworth* 333

DEDICATION

Frederick Dale Trauger inherited his interest in water. Fred was born in California's San Joaquin Valley, where his father was Chairman of the Lindsay-Strathmore Irrigation District. Geology was a natural course of study, and Fred received his B.A. from Fresno State University in 1939. He then enrolled in the graduate program at Stanford University. During this time Fred also worked with the Strategic Metals Branch of the U. S. Geological Survey as a geologic field assistant evaluating mercury and manganese deposits in Nevada and California.

At the outbreak of World War II, Fred enlisted in the Army Air Corps and was trained as a photo interpreter. He served in that capacity in North Africa and Italy until 1945. With his return to civilian life, Fred worked briefly with the Bureau of Reclamation conducting hydrologic studies in California before joining the Ground-Water Branch of the U. S. Geological Survey in the Oregon District. In 1952, Fred was transferred to the New Mexico district office in Albuquerque.

Fred Trauger joined the New Mexico Geological Society in 1956; he became Secretary of the Executive Committee in 1964 and President in 1966. Through Fred's efforts, the NMGS became fiscally solvent. Between 1966 and 1975, he was the driving force in the Society's Publications Committee. Fred set up a marketing strategy for the sale of guidebooks, upgraded their quality, and helped establish guidelines to improve the technical presentation. In 1972 he was made an Honorary Member of the NMGS.

Frederick D. Trauger, at Palo Duro Canyon, Texas, December 2000

Participation in the annual field conference was always a highlight for Fred. For many years he was largely responsible for planning the routes and preparation of the road logs. He was editor and co-editor, respectively, of the 1967 and 1972 guidebooks; and in the 1965-1987 period, he was the author or coauthor of many guidebook technical papers and road logs. Particularly memorable was the 1972 Field Conference jointly organized with Vincent Kelley. Fred wound up ably leading Day One, and Vin then took over with characteristic vigor. With the efforts of many persons, including Fred's, the New Mexico Geological Society's annual field conference has become a widely recognized professional event that attracts persons from across the nation and abroad.

Besides being a hydrogeologist of note, Fred's expertise in Mesozoic and Cenozoic stratigraphy and structure is widely recognized—particularly in southwestern and eastern New Mexico. On more than one outcrop, Fred has gone toe-to-toe with others on topics such as Exeter and Entrada Sandstone terminology, and Triassic stratigraphic concepts. His lengthy bibliography includes a number of ground water investigations in California, Oregon, and New Mexico. His report "Water resources and general geology of Grant County, New Mexico" (NMBMMR Hydrologic Report 2, 1972) remains a classic for southwestern New Mexico. He also belongs to the small group of recognized authorities on the ground-water resources of eastern New Mexico and the Southern Great Plains. He is senior author (with F. X. Bushman) of New Mexico State Engineer Technical Report 30 (1964) on "Geology and ground water in the vicinity of Tucumcari…," so the 2001 NMGS Field Conference is certainly on Fred's professional home turf.

Following his retirement from the Water Resources Division of the U. S. Geological Survey, Fred founded the consulting firm of Geohydrology Associates, Inc. with T. E. (Tim) Kelly in 1975. The firm specializes in water supplies for municipal and industrial purposes.

As President until his retirement, Fred consulted on projects throughout the country, but his area of special interest was always eastern and southwestern New Mexico.

Fred is also a recognized authority on turquoise. He has visited most of the turquoise-producing areas in the Southwest and made a detailed study of the Carico Lake Mine near Austin, Nevada. The late Stuart A. Northrop and Fred combined to present a series of lectures on the occurrence and quality of turquoise, with lectures being presented as far away as New York City.

In addition to his love of geology, Fred is an over-achiever in many other areas. When his late beloved wife Jeanne wanted to grow roses, he not only planted roses but also became an expert who won a number of ribbons in rose shows. Fred has become a recognized authority and has judged rose shows throughout the Southwest. He later developed the same love of orchids. Fred is a knowledgeable arborist. The Trauger yard in northeast Albuquerque contains a number of exotic trees, which Fred nurtures, including giant redwood, and various species of pine and cedar found only in California and Arizona. And cats . . ..while most people are happy with one or two, the Traugers have had as many as nine. Besides innumerable friends, Fred also collects movie videos, and currently has over 1,100 films.

In 1996, friends and associates from throughout the country joined in celebrating Fred's 80th birthday. An Irish toast was made: "May Fred live to see 100 years, then be given one extra year to repent." The way he is going, it looks like he will get there! And the New Mexico Geological Society will be one of the principal benefactors. By the way, Fred mixes a mean martini, and Happy Hour always begins at 1630 hrs!

Tim Kelly and John Hawley

PRESIDENT'S MESSAGE

Tucumcari Tonight! Anyone who has traveled the major thoroughfares of New Mexico has encountered the billboards recommending that they make this town on old Route 66 their destination for the evening. The 52nd New Mexico Geological Society Fall Field Conference succumbs to the advertising as we base the 2001 conference out of this hub of eastern New Mexico. It has been 16 years since the Society's last visit to the area. Please take time to thank Spencer Lucas (New Mexico Museum of Natural History) and Adrian Hunt (Mesa Technical College), this year's conference leaders, for their efforts on behalf of the Society.

I am pleased Dana Ullmer-Scholle (NMIMT) has agreed to serve as managing editor for the Society. This guidebook represents her first effort on behalf of the Society, although she has served in such a capacity for other geoscience groups. Although retired, the former managing editor, George Austin, continues in that role for our forthcoming special publications. We expect to see the first volume of the Geology of New Mexico edited by Greg Mack (NMSU) and Katherine Giles (NMSU) very soon. A second volume concentrating on geologic resources that will follow in a year or two.

The pre-press version of the updated Geologic Highway Map of New Mexico will be available about the time of this conference. The drafting and cartographic staff at the New Mexico Bureau of Geology & Mineral Resources has also assisted in the completion of the map. The Bureau continues to be a significant contributor to the continuing success of the NMGS under the direction of Dr. Peter Scholle, State Geologist and Director. Please note the Bureau changed its name as of June 15 this year. Frank Kottlowski spearheaded this project for many years.

Speaking of Frank, we will all miss his presence at the Conference this year and for the years to come. Frank Kottlowski (Honorary member 1967) passed away this spring after a prolonged illness. Frank was a tireless supporter of the Society and to those who studied the geology of this great state. His family asked his friends to honor him by continuing to support the Kottlowski Fellowship in addition to a new endowment in his honor established by the New Mexico Bureau of Geology and Mineral Resources. Frank Kottlowski's legacy lives on in the many geologists and friends touched by his generosity and kindness, this one included.

The Society will miss another one of its hardest workers and greatest supporters. Norma Meeks retired from the New Mexico Bureau of Geology & Mineral Resources after the NMGS Spring Meeting. Linda Ulbricht has agreed to assume some of Norma's duties at the Bureau and serve as registration chair for the Society.

Brian Brister (NMBMMR) and Gary Smith (UNM) co-chaired the Spring Meeting in Socorro with much success. We encourage Society members from all universities and occupations to volunteer to help with the Spring Meeting.

The Society continues to support geoscience education in the state by increasing our scholarship awards again this year. The Robert and Beverly Wellnitz and Lucille Pipkin Scholarships have allowed the Society to take support of geoscience education to new heights and directions. The Robert Root Special Projects Fund and Reprint Fund (established by Fred Trauger) provide the means for the Society to pursue special publications (Geology of New Mexico), develop infrastructure (the new publication storage facility built in conjunction with the Bureau) and reprint our publications. I encourage all members to consider the Society in their estate financial plans.

White Sands will be the focus of next year's field conference. The Society has never specifically visited this part of New Mexico. Access to the missile range affords the participants the opportunity to see "up close" those rocks only viewed from a distance. Kate Giles and I are co-chairs and welcome your participation.

The Society's success comes from the many talented individuals working for the benefit of the geologic community. Please find time to volunteer and continue the traditions that make the NMGS a model for all other professional societies. Enjoy the results of the hard work of this year's conference leaders and please consider how you might contribute to the Society in the future.

Virgil W. Lueth

EDITOR'S MESSAGE

The history of geological studies in New Mexico began on 22 September 1853 when Swiss geologist Jules Marcou (1824-1898) climbed a hill about 20 km southwest of what is now Tucumcari. Dubbed by Marcou "Pyramid Mountain," this hill exposes red beds at its base, overlain by cliffs of yellow sandstone and capped by a thin shale interval from which poured shells of marine bivalves. Marcou identified these bivalves as the oyster-like *Gryphaea* (they actually pertain to a closely related genus, subsequently named *Texigryphaea*), and assigned them a Jurassic age. He thus concluded that the High Plains surface here, and throughout the region, is underlain by Jurassic strata. But the bivalves and the rocks that contain them are actually Cretaceous in age, and Marcou's mistake engendered the first scientific debate about the geology of New Mexico, one that was not resolved during his lifetime.

This year, nearly 150 years after Marcou's visit, the attendees of the 52nd Field Conference of the New Mexico Geological Society will visit Pyramid Mountain. And, although the mountain has not changed much since Marcou, our understanding of the rocks exposed on its flanks has, and is a major focus of the field conference and this guidebook.

It has long been obvious that much of what we know of the Mesozoic geological history of New Mexico comes to us from exposures on the Colorado Plateau in the northwestern portion of the state. A plethora of correlation charts and paleogeographic maps of the Mesozoic cover the Four Corners and mystically are truncated somewhere around the Plateau's southeastern edge, west and northwest of Albuquerque. Yet, the Mesozoic depositional systems knew no such boundaries—the vast Triassic, Jurassic and Cretaceous basins of the North American Western Interior extended eastward, out onto what are now the prairies of West Texas and the Oklahoma panhandle, and in some cases beyond. Understanding their Mesozoic history—the tectonics, sediments and extinct biotas—requires geographically broad study that encompasses the outcrop and subsurface in east-central New Mexico.

Fortunately, in Guadalupe and Quay Counties and adjoining areas, the Pecos and Canadian Rivers have sculpted canyons, escarpments and mesas that well expose a Mesozoic section about one km thick. As at Pyramid Mountain, the section begins with nonmarine red beds of Triassic age, which are loaded with fossils of amphibians and reptiles, including some of the earliest dinosaurs. Bold cliffs of Jurassic eolianites follow, mantled by relatively thin sections of Jurassic claystones and Early Cretaceous marine shale—the latter are the source beds of Marcou's "*Gryphaea*" fossils. Cliff-forming Lower Cretaceous sandstones usually cap the Mesozoic section, and they are locally overlain by erosional outliers of Upper Cretaceous marine strata.

This is the third time the New Mexico Geological Society has visited the Mesozoic section in east-central New Mexico. The first time was in 1972, led by the late Vincent Kelley, one of the giants of New Mexico geology, and Fred Trauger, to whom this book is appropriately dedicated. The second time was in 1985, and I led that field conference (I gotta stop doing this!).

This year, Adrian Hunt, of the Mesalands Dinosaur Museum, is co-chairman of the conference, which is headquartered in the fabled city of Tucumcari. Andrew Heckert, Barry Kues and Virginia McLemore were the other dedicated members of the roadlogging committee. Their hard work is praiseworthy, and I am also grateful to the following organizations and to members of their staffs for providing diverse support of this field conference and guidebook: Blackwater Draw Museum, Blackwater Locality No. 1, Mesa Technical College, New Mexico Bureau of Mines and Mineral Resources, New Mexico Museum of Natural History and Palo Duro Canyon State Park. Finally, I owe special thanks to Matt Celeskey and Daniel Weismann for help with guidebook graphics; Joanne Dickenson for access to Blackwater Draw archives and collections; and Dana Ulmer-Scholle for her skills as managing editor.

Spencer G. Lucas

COMMITTEES

2001 EXECUTIVE COMMITTEE

Virgil W. Lueth., President....New Mexico Bureau of Geology & Mineral Resources
Maureen Wilks, Vice-President....New Mexico Bureau of Geology & Mineral Resources
Gary Stephens, Treasurer....U. S. Bureau of Land Management
Ray Kenny, Secretary....New Mexico Highlands University
Katherine Giles, Past President....New Mexico State University

FIELD CONFERENCE

Adrian P. Hunt, Co-Chair....Mesa Technical College
Spencer G. Lucas, Co-Chair....New Mexico Museum of Natural History

GUIDEBOOK

Spencer G. Lucas, Editor....New Mexico Museum of Natural History
Dana S. Ulmer-Scholle, Managing Editor....New Mexico Insitute of Mining and Technology

ROAD LOGS

Spencer G. Lucas, Chair....New Mexico Museum of Natural History
Adrian P. Hunt....Mesa Technical College
Andrew B. Heckert....University of New Mexico
Barry S. Kues....University of New Mexico
Virginia T. McLemore....New Mexico Bureau of Geology & Mineral Resources

REGISTRATION

Maureen Wilks....New Mexico Bureau of Geology & Mineral Resources

LOCAL ARRANGEMENTS

Adrian P. Hunt....Mesa Technical College

PUBLICATIONS COMMITTEE

James M. Baker, Chair....New Mexico Bureau of Geology & Mineral Resources
Richard M. Chamberlin....New Mexico Bureau of Geology & Mineral Resources

PUBLICATION SALES

Norma J. Meeks....New Mexico Bureau of Geology & Mineral Resources
Linda S. Ulbricht....New Mexico Bureau of Geology & Mineral Resources

2001
FIELD CONFERENCE SCHEDULE

Wednesday, September 26, 2001—Registration Day

6:00-9:00 PM	Registration and ice-breaker party, Holiday Inn
6:00-8:00 PM	Free admission to Mesalands Dinosaur Museum, Tucumcari

Thursday, September 27, 2001—First Day, round trip from Tucumcari to West Texas and return

6:00-7:15 AM	Breakfast (not provided)
7:00-7:30 AM	Registration, Mesalands Dinosaur Museum, Tucumcari
7:30 AM	Buses leave Mesalands Dinosaur Museum
5:30 PM	Leave last stop
6:00 PM	Annual barbecue, Caprock Park, San Jon

Friday, September 28, 2001-- Second Day, round trip from Tucumcari to Blackwater Draw and return

6:00-7:15 AM	Breakfast (not provided)
7:00-7:30 AM	Registration, Mesalands Dinosaur Museum, Tucumcari
7:30 AM	Caravan leaves Mesalands Dinosaur Museum
5:30 PM	Caravan returns to Tucumcari
7:00 PM	Annual banquet (provided), Holiday Inn, Tucumcari

Saturday, September 29, 2001—Third Day, from Tucumcari to Conchas Dam

6:00-7:15 AM	Breakfast (not provided)
7:30 AM	Caravan leaves Mesalands Dinosaur Museum
1:00 PM	Field conference ends at Conchas Dam

CREDITS

Front Cover: Watercolor of photograph of Pyramid Mountain in Quay County by Spencer G. Lucas and Dana S. Ulmer-Scholle.

Inside Front Cover: Satellite image from the new NASA TERRA satellite (using the new sensor called MODIS) provided by Thomas K. Budge, Earth Data Analysis Center, University of New Mexico.

Inside Back Cover: Stratigraphic chart of east-central New Mexico and west Texas, by Spencer G. Lucas.

Frontespiece: Drawing of Tucumcari Mountain by Dana Ulmer-Scholle after a photograph by Wayne Lambert.

Logistical Assistance: Mesa Technical College, New Mexico Museum of Natural History, New Mexico Bureau of Geology.

Photography and Figures: Spencer G. Lucas, Andrew B. Heckert, New Mexico Museum of Natural History, New Mexico Bureau of Geology.

Formatting and Production: Dana Ulmer-Scholle.

Printer: McNaughton & Gunn, Inc., Saline, Michigan.

FIRST-DAY ROAD LOG, FROM TUCUMCARI TO THE EDGE OF THE LLANO ESTACADO AT GRUHLKEY, TEXAS, PALO DURO CANYON, TEXAS, AND SAN JON HILL, NEW MEXICO

SPENCER G. LUCAS, ADRIAN P. HUNT, BARRY S. KUES, ANDREW B. HECKERT and VIRGINIA T. MCLEMORE

THURSDAY, SEPTEMBER 27, 2001

Assembly point: Mesalands Dinosaur Museum, 222 East Laughlin, Tucumcari
Departure time: 7:30 AM
Distance: 286.9 miles
Stops: 4

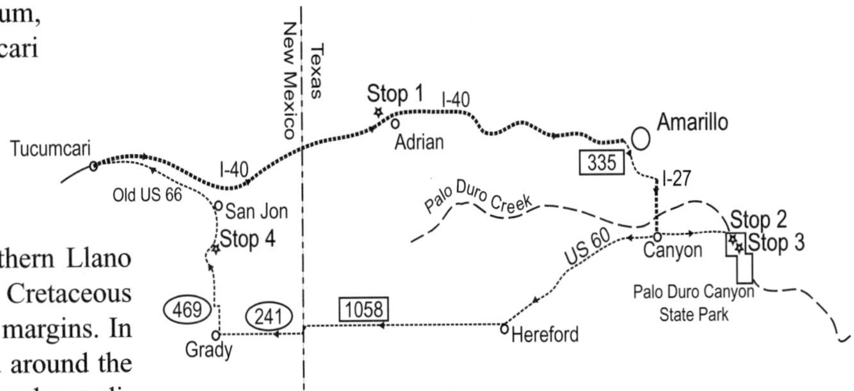

SUMMARY

Today's trip takes us across part of the northern Llano Estacado to examine Permian, Triassic, Jurassic, Cretaceous and Neogene sedimentary rocks exposed at its margins. In so doing, we review the entire section exposed around the staked plains, focusing on problems of stratigraphy, sedimentation, biostratigraphy and magnetostratigraphy.

At STOP 1, just as we ascend the Llano Estacado, we examine the Neogene Ogallala Group and discuss its stratigraphy, sedimentology and geohydrology. We then "scoot" over to Palo Duro Canyon, south of Amarillo, the "Grand Canyon of Texas," on the eastern edge of the Llano Estacado. STOPS 2 and 3 feature the canyon's spectacular scenery due to Neogene erosion into Permian and Triassic red beds. These red beds, and problems regarding their age and correlation, are the focus of the stops.

Palo Duro Canyon is the high point and the endpoint of the Texas portion of this field conference. The trip continues back to New Mexico to STOP 4 at "San Jon Hill," where the eroded northern face of the Llano Estacado exposes a classic Jurassic-Cretaceous section. Here, we collect fossils and discuss regional Jurassic and Cretaceous sedimentation.

Mileage

0.0 Assemble at the Mesalands Dinosaur Museum, 222 East Laughlin, Tucumcari. Mesalands Dinosaur Museum opened in May 2000 and is a part of Mesa Technical College, a state-funded community college. The Museum includes extensive exhibits on the paleontology and geology of eastern New Mexico in a 10,000 ft^2 exhibit hall. The exhibits are designed as a trek through time and reflect the geological sequence exposed in this area—Triassic, Jurassic, Cretaceous and Neogene—that we will be examining on the field conference. Distinctive emphases of the museum include: (1) Triassic paleontology; (2) bronze replicas of paleontological and zoological specimens, including complete skeletons; (3) individual exhibits on the main vertebrate trace fossil groups including coprolites, footprints, skin impressions and eggs; and (4) hands-on exhibits that include dinosaur bones, eggs and even coprolites. **Turn left out of museum parking lot onto S. Adams and immediately turn left (west) onto East Laughlin. 0.1**

0.1 Stop sign. **Turn left** (south) on First Street and proceed south. **0.1**

0.2 Continue straight through traffic light with Tucumcari Boulevard (Old Highway 66) and continue south out of town. Settlement in the Tucumcari area began with the small Hispanic community of Tierra Blanca, about 3 mi to the north. This community became Liberty in the 1860s, after Fort Bascom was established in 1863. Liberty was beyond the five-mile limit within which soldiers could not consume alcohol—needless to say, Liberty became a boisterous community. In 1901, it became widely known that the Rock Island Railroad would pass through the valley just north of Tucumcari Mountain. Land speculators quickly surveyed a town site before the railroad arrived. Half the population of Liberty moved north, and a community was estab-

FIGURE 1.1. Main street, Tucumcari, date unknown (but certainly prior to the arrival of I-40!). Courtesy Museum of New Mexico (negative 11596).

lished, initially and unofficially as Ragtown and then Six Shooter Siding. From 1901-1902, the post office was named Douglas, but this was changed to the current Tucumcari in 1902 (Julyan, 1996; Geyler, 1998).

Tucumcari (Fig. 1.1) is named for the prominent butte of Tucumcari Mountain. As early as 1840 this landmark was associated with the name Tucumcari in its modern spelling (Kues, 1985), and perhaps the earliest version is a 1777 burial record of a Comanche woman captured at Cuchuncari (Julyan, 1996). The name is obviously Native American, but there are disagreements as to its etymology (Pearce, 1965; Julyan, 1998). The most elegant explanation is that is a variant of *tukamukaru,* a Comanche word meaning “to lie in wait for someone or something to approach” (Pearce, 1965). This is consistent with Comanche occupation of the area and with the obvious utility of Tucumcari Mountain as a lookout post. **1.4**

1.6 Pass under bridge of Interstate 40. **Move into left lane to turn left and enter eastbound Interstate 40. 0.5**

2.1 Bridge over canal. Tucumcari Mountain at 2:30. The stratigraphic section exposed at Tucumcari Mountain is (in ascending order): Upper Triassic Bull Canyon and Redonda formations, Middle Jurassic Entrada Sandstone and Summerville Formation, Upper Jurassic Morrison Formation, and Lower Cretaceous Tucumcari Formation and Mesa Rica Sandstone (Fig. 1.2). The Tucumcari area receives water via a canal from Conchas Lake reservoir. The roots of this irrigation project have a long history dating back to 1925 when the “Canadian River Development Association” was formed by H. B. Jones, James L. Briscoe and Arch Hurley. These gentlemen spent an enormous effort, including tens of trips to Washington, to further their goal of providing abundant irrigation water to the area. Finally, in 1936, Congress passed the Flood Control Act that appropriated funding for Conchas Dam. The dam was finished in 1939, and the Arch Hurley Conservancy District (AHCD) was completed in 1954. AHCD utilizes 40 mi of canals and 350 mi of laterals and ditches to provide approximately 1.33 acre feet to over 42,000 acres of land (Geyler, 1998). **0.6**

2.7 Exit 333. **1.3**

4.0 Roadcut developed in the Upper Triassic Bull Canyon Formation immediately before exit 335. The Bull Canyon Formation is a relatively soft, mudstone-rich formation that overlies the sandstone- and conglomerate-dominated Trujillo Formation. As we drive east, I-40 is essentially lying on the contact between the Trujillo and the Bull Canyon formations. Thus, the Trujillo underlies the flat plain between the Canadian River to the north and I-40. South of I-40, badlands of the Bull Canyon

FIGURE 1.2. Tucumcari Mountain, with stratigraphic units numbered: 1 = Bull Canyon Formation, 2 = Redonda Formation, 3 = Entrada Sandstone, 4 = Summerville and Morrison formations, 5 = Tucumcari Formation, 6 = Mesa Rica Sandstone, 7 = Ogallala Group.

Formation will be intermittently visible. **2.0**

6.0 Mesa Redonda at 3:00; in the distance the edge of the Llano Estacado is between 1:00-2:00. **1.7**

7.7 Exit 339. **2.6**

10.3 Pass under overpass. Extensive badlands to the southeast are developed in the Upper Triassic Bull Canyon Formation. These badlands yield numerous Triassic vertebrate fossils, and an inset Pleistocene fluvial deposit also contains bones of sloth (*Paramylodon*), horse (*Equus*), camel (*Camelops*), bison (*Bison*) and mammoth (*Mammuthus*). This is the Rancholabrean (mid-late Pleistocene) Badlands Ranch local fauna of Morgan et al. (2000). **1.0**

11.3 Substation on right. **0.7**

12.0 Low roadcuts in tributary of Barranca Creek are developed in the Upper Triassic Trujillo Formation (Cuervo Sandstone of Kelley, 1972). Indeterminate phytosaurs and the aetosaur *Typothorax coccinarum* have been found in this area (see accompanying minipaper). *Typothorax coccinarum* is an index taxon of the Revueltian land-vertebrate faunachron and indicates an early-mid Norian (Late Triassic: 210-218 Ma) age (Lucas, 1998). **0.6**

PALEONTOLOGY AND AGE OF THE UPPER TRIASSIC TRUJILLO FORMATION, EAST-CENTRAL NEW MEXICO AND WEST TEXAS

Adrian P. Hunt
Mesalands Dinosaur Museum, Mesa Technical College,
911 South Tenth Street, Tucumcari, NM 88401

The Trujillo Formation of the Chinle Group (Upper Triassic) is as much as 70 m of sandstone and conglomerate with minor mudrock that crops out widely in West Texas and northeastern New Mexico. This unit is unfossiliferous except for two areas, Randall County, Texas and Quay County, New Mexico. MDM refers to the Mesalands Dinosaur Museum.

The putative angiosperm *Sanmiguelia* and other fossil plants have been recovered from the Trujillo Formation in Sunday Canyon, Randall County (Ash, 1976, 1980; Cornet, 1986). Dunay (1972), Dunay and Fisher (1979) and Cornet (1986) described pollen from the Trujillo in Sunday Canyon. The only vertebrate fossils reported from the Trujillo in this area are coelacanth and *Semionotus*-like scales (Olsen, 1984).

Hunt (1991) described fossils from the Trujillo Formation in Revuelto and Barranca Creeks in Quay County, New Mexico. The Revuelto Creek specimens include the bivalve *Unio* sp., an indeterminate phytosaur, the aetosaur *Typothorax coccinarum*, an indeterminate rauisuchian and vertebrate coprolites. The Barranca Creek assemblage includes an indeterminate phytosaur and *Typothorax coccinarum*.

During the last five years, the Mesalands Dinosaur Museum has been collecting specimens from the Trujillo Formation from a variety of localities in central Quay County. Generally, the exposures of the Trujillo Formation in east-central New Mexico are not ideal for fossil collecting. In Quay County, the Trujillo underlies the plain that extends from approximately I-40 north to the Canadian River and is thus poorly exposed. Northwest of Mesa Rica in San Miguel County, the Trujillo forms the steep edges of the escarpments. However, some drainages in east-central New Mexico provide low outcrops of the Trujillo Formation that are ideal for fossil prospecting. Several of these outcrops are associated with the Quay anticline.

Large mass mortality beds of unionid bivalves have been found at two localities. These contain thousands of specimens, most of which are articulated. These specimens are found in pale reddish orange (10 R 5/4) to moderate reddish orange (10 R 4/6), matrix-supported, intraformational conglomerates (e.g., MDM 212). The majority of clasts in these beds are rounded calcrete pebbles up to 6 cm in diameter. Taxa include *Unio arizonensis*, *Unio* sp. and *Antediplodon dockumensis*. The amphibian *Apachesaurus gregorii* is represented by a single vertebra (MDM 213).

By far the most common fossils pertain to phytosaurs. The most important specimens are one side of a partial crested rostrum that extends from just anterior to the external nares to near the tip (MDM 214), the tip of a rostrum (MDM 215) and a posterior fragment of a squamosal (MDM 216). The squamosal indicates the presence of a phytosaur with rod-like squamosal processes such as *Pseudopalatus* or *Nicrosaurus*.

The stagonolepidid *Typothorax coccinarum* is represented in the fauna by a large fragment of a paramedian osteoderm (MDM 217) and by small fragments (e.g., MDM 218). These specimens have the diagnostic features of an ornament of rounded pits, an anterior bar and also a ventral bar. One specimen is a portion of a skeleton of an aetosaur with articulated osteoderms and vertebrae (MDM 219), but it is too poorly preserved to be identified.

Other reptile specimens include caudal vertebrae of a rauisuchian (e.g., MDM 220) and half a centrum that is similar to the herrerasaurid *Chindesaurus* (MDM 221). Vertebrate coprolites are also common (e.g., MDM 222, 223). Fragments of fossil wood are abundant, and a few specimens of the tree fungus *Polyporites* have been collected (e.g., MDM 224).

The presence of *Typothorax coccinarum* indicates that the Trujillo Formation belongs to the Revueltian land-vertebrate faunachron, which is of early-mid Norian age (Lucas and Hunt, 1993; Lucas, 1998). This is consistent with the presence in the Trujillo Formation of Texas of *Sanmiguelia* of Ash's (1980) *Sanmiguelia* floral zone, which is also Norian in age (Lucas, 1993). Further support for this age is the presence of *Pseudopalatus/Nicrosaurus*-grade phytosaurs, which are only known from Norian strata (Hunt, 1994). Pollen data from Sunday Canyon have been interpreted to indicate a late Carnian age for the Trujillo (Dunay, 1972; Dunay and Fisher, 1979; Cornet, 1986), but the preponderance of the data indicates that the Trujillo Formation is Norian in age.

12.6 Badlands of Barranca Creek exposed at 2:00-3:00. Llano Estacado in the distance. *Barranca* means gorge or ravine in Spanish (Julyan, 1996). The former community of Barranca was several miles to the south of here.

Drake (1892) was the first to apply formal stratigraphic nomenclature to the Triassic rocks in east-central New Mexico. He measured a series of stratigraphic sections in eastern New Mexico and brought Cummins' (1890) name Dockum beds into Quay County. One of his measured sections is at the head of east Fossil Creek. Marcou (1858) showed a Fossil Creek as a north-south flowing stream northeast of Mesa Redonda (Big Tucumcari Mountain of Marcou). The geographic location of Fossil Creek identifies it as Barranca Creek (cf. Kues, 1985, fig. 11), even though Marcou shows the location of the headwaters incorrectly. By inference, East Fossil Creek must be Revuelto Creek, which joins Barranca Creek near I-40. Note that there are no other areas east of Fossil Creek (Barranca Creek) where Drake (1892, pl. 5) could have measured his thick section of "Dockum beds" (approximately 181 ft estimated from his plate) in Quay County. Drake (1892, plate 5) indicated he measured 81 ft of lower beds (Garita Creek Formation), 25 ft of central beds (Trujillo Formation) and 75 ft of upper beds (Bull Canyon Formation), but these thicknesses are much too thin, and the entire section was probably measured in the Bull Canyon and upper? Trujillo formations. **0.4**

13.0 Exit 343. **3.2**

16.2 Cross bridge over Revuelto Creek. On the right the Upper Triassic Trujillo Formation is overlain by Pleistocene sediments that include fragmentary mammoth teeth and bones. Tributaries of Revuelto Creek, Barranca Creek and Plaza Larga contain Pleistocene terraces that have yielded several vertebrate specimens, including at least two partial mammoth tusks and horse teeth (see Morgan et al., this guidebook). **1.0**

17.2 Upper Triassic badlands of Revuelto Creek at 3:00. This creek is named for Monte Revuelto, a topographic feature that is not on modern maps, but it is the point of the Llano Estacado that extends farthest north between Tucumcari and the state line. The headwaters of what are now called San Jon and Revuelto Creeks are in this area. The ornithischian dinosaur *Revueltosaurus callenderi* Hunt, 1989 and the Revueltian land-vertebrate faunachron are named for this drainage. This name is doubly appropriate because: (1) the holotype of *Revueltosaurus* was found near the creek; and (2) *revuelta* means revolution in Spanish, and the Late Triassic, including the Revueltian (early-mid Norian), saw a great evolutionary turnover of life on land, when the sprawling reptiles and amphibians characteristic of the Paleozoic were replaced by "modern" groups with upright gaits, such as dinosaurs, pterosaurs and mammals. **2.8**

20.0 Note blowouts developed in the Bull Canyon Formation along the road for next few tenths of a mile. **2.2**

22.2 Pass under overpass. Little Rattler mine (sec. 11, T11N, R33E) is located a few miles north of here. The mine produced 59 tons of ore containing 41 lbs U_3O_8 and 44 lbs V_2O_5 and was hosted by Chinle Group strata (McLemore and North, 1985). Numerous small, sedimentary copper-uranium (redbed copper) deposits are present north and south of I-40 in the San Jon area. **2.9**

25.1 San Jon exit (exit 356). **1.7**

26.8 Weigh station. **3.3**

30.1 Exit 361 to Bard, a tiny village that was never more than a trading point along the railroad. Bard dates from the time of the railroad (1908), although it is several miles to the north of the Chicago, Rock Island and Pacific line. Julyan (1996) gives two versions of the origin of the name--it was either named after a place in Texas or it was derived from the Bar-D Ranch. Given its proximity to Endee, which is named after the ND Ranch, the latter explanation seems more likely. Currently, Bard and Endee share a volunteer fire department. **2.6**

32.7 Pass under overpass. **2.4**

35.1 Mile marker 366. Note Triassic roadcuts are cross-bedded, channelized Trujillo Formation sandstones and conglomerates for next 1.5 mi. **3.6**

38.7 Exit 369 to Endee, essentially a vanished settlement that was named for the brand of the ND Ranch. Endee was originally founded in about 1885 by John and George Day. The settlement of Endee has had three main iterations. It started several miles south of here, had a second and more prosperous existence on US-66 and is finally represented by the abandoned tourist facilities on I-40. The original settlement consisted of picket houses with sod roofs (Julyan, 1998). **3.0**

41.7 Cross Trujillo Creek; exposures are Triassic redbeds and Quaternary alluvium and dune sands. **0.4**

42.1 Port of entry on left. **0.4**

42.5 Texas-New Mexico state line. **0.2**

42.7 Exit 0 to Glenrio. Glenrio represents one of many small towns that were killed off by I-40. An attempt to move one of the gas stations to the I-40 exit ended in failure. Glenrio started as the settlement of Rock Island when the Chicago, Rock Island and Pacific railroad came through this area in 1909. The name was changed in 1915 to avoid confusion in delivering mail to the town and to the CRI & P depot here. The name derives from the Celtic *glen* and the Spanish *rio*. Julyan (1996) states that both names are inappropriate, but views of the Llano Estacado off to the east and south leave you no doubt that you are in a valley, and the perennial streams nearby were important to early settlers. The town prospered until World War II, and, indeed, some scenes in the 1940 movie "The Grapes of Wrath" were filmed here. After the construction of I-40, however, all that remains are a post office and two residences. **0.3**

43.0 Pass under overpass. Low bluffs on the right are developed in the lower part of the Upper Triassic Bull Canyon Formation. **1.6**

44.6 We have been in Deaf ("DEEF" to you native English speakers!) Smith County; Oldham County line. Deaf Smith County was named for Erastus "Deaf" Smith, a famous scout of the Texas Revolution. Oldham County derives its name from William Simpson Oldham, pioneer Texas lawyer and Confederate Senator. **1.0**

45.6 Redbeds of Bull Canyon Formation sediments immediately to the right of the road. The Bull Canyon Formation is the youngest Upper Triassic unit in West Texas. The overlying Mesozoic section has been truncated to the east. South of San Jon, the Jurassic and Cretaceous overlie Upper Triassic strata. Farther east in the escarpment south of Endee, the Redonda Formation is the highest Mesozoic unit, but by the Texas line Cenozoic erosion has removed the Redonda Formation. **0.7**

46.3 Old railroad bed of the Chicago, Rock Island, and Burlington Pacific on right. Ermine Cowles Case of the University of Michigan followed this route when he was searching for Upper Triassic fossils here in 1912 (Case, 1914). **1.2**

47.5 Mile marker 5. Low bluff to the right and low hills in distance to left are lower Bull Canyon Formation. Note that Triassic stratigraphy here can be read from the topography. Thus, the sandstone-dominated Trujillo Formation underlies the nearly flat surface we are driving on, and slopes and low hills immediately above it are developed in the mudstone-dominated Bull Canyon Formation.

Gould (1907) published one of the earliest and still one of the most accurate geologic maps of this part of Texas (Fig. 1.3). You can trace the Texas portion of our route on it, which begins just south of the Chicago, Rock Island and Pacific railway going east nearly to Amarillo, then south to Canyon, east to Palo Duro Canyon, then returning west through Hereford, south of Garcia Lake and into New Mexico. **2.1**

49.6 Cross a tributary of Mujares (probably should be Mujeres, "women") Creek. Lower Bull Canyon Formation crops out to right in creek. Note the beveling down of the Ogallala surface toward the east. **3.2**

52.8 Crest low ridge. Edge of Llano Estacado that we will cross is visible ahead. Roadcuts, such as they are, are developed in the Bull Canyon Formation. **1.1**

53.9 Cross creek; note Bull Canyon Formation exposures in creek to right. **1.0**

54.9 Picnic area on right. **1.6**

56.5 Mile marker 14. Roadcuts on the left and right side of the road for the next 0.8 mi are in the Bull Canyon Formation. We are now climbing the escarpment of the Llano Estacado. **0.8**

57.3 Exit 15, Ivy Road. **0.9**

58.2 Ogallala Group roadcuts left and right. The Ogallala in usually tripartite in this area with a basal conglomerate, a medial sandy interval and an upper massive calcrete. Formation of the calcrete ceased about 2.5-3 million years ago. Vertebrate fossils date the Ogallala Group (or Formation) as primarily of Miocene age (Wood et al., 1941). **0.4**

58.6 Crest hill; we are now on the Llano Estacado. Our first stop is the abandoned pit on the left. The Llano Estacado extends from the Pecos River on the west to Palo Duro Canyon in Texas on the east and southward to Hobbs, New Mexico. Also known as the Staked or Stockaded Plains, it covers an area of about 32,000 mi^2 (Reeves, 1972; Hawley, 1993; Holliday, 1995). The origin of this name is disputed by New Mexican and Texan historians and geologists (e.g., Rathjen, 1974; Spearing, 1991; Metz, 1994; Julyan, 1996). There are two possible translations of the Spanish and three possible explanations. The term could be translated "staked plains" from *estaca* or stake/picket and this could denote: (1) stakes were used to mark routes across the featureless plains or to tie up horses; or (2) the frequent yucca plants looked like stakes. *Estacada* literally means stockade or palisade and could refer to the fortress-like aspect of the edge of the caprock. **0.6**

59.2 Sign for exit 18, Gruhlkey. W. H. Gruhlkey was a roadmaster of the Rock Island Railway and owned the land here where a small railroad station was built. Prepare to take exit. **1.0**

60.2 **Exit Interstate 40 at exit 18**, Gruhlkey. **0.1**

60.3 **Turn left** to go east on frontage road. **0.1**

60.4 **Turn left** to cross bridge over Interstate 40. **0.2**

60.6 Stop sign. **Turn left** to go west on frontage road. **Do not get back on Interstate 40. 1.1**

61.7 **STOP 1. Pull off on small road to left** to parking area for the City of Adrian type 3 dump site. Permit #352. Cross road to north and pass through fence. Note, no dumping allowed!

The Neogene Ogallala Formation (Group) contains the Ogallala (High Plains) aquifer, the major water source for domestic and agricultural use on the southern High Plains of New Mexico-Texas. The Ogallala Formation (or Group) is present east of the Rocky Mountains over an area that extends from South Dakota to West Texas (Fig. 1.4). It consists of alluvial sediments that partly fill paleovalleys and widespread eolian sediments that cap paleouplands, deposited under mostly semiarid to subhumid climatic conditions (Gustavson, 1996). Fossil mammals from the Texas Panhandle indicate a Clarendonian-Hemphillian (middle Miocene-early Pliocene) age (see accompanying minipaper).

Here, at Gruhlkey, we can see typical fluvial valley fill sands and gravels and eolian sands below the Caprock calcrete (Fig. 1.5). This calcrete (caliche) is a distinctive, approximately 6-ft thick bed of erosion-resistant, calcium-carbonate rich rock that supports the rim of the Caprock Escarpment. Early workers thought the calcrete represented deposition by a vast lake on the High Plains (e.g., Elias, 1931). However, its pedogenic origin was demonstrated in the 1940s (e.g., Smith, 1940; Bretz and Horberg, 1949a, b). Gustavson (1996) has most recently argued that the Ogallala Caprock calcrete is time transgressive, representing pedogenic carbonate that locally accumulated at various times during the Miocene-Pleistocene interval. Indeed, the Caprock calcrete represents a long period of landscape stability, rather than a particular period of increased aridity (Gustavson, 1996).

After stop retrace route to Interstate 40. 1.0

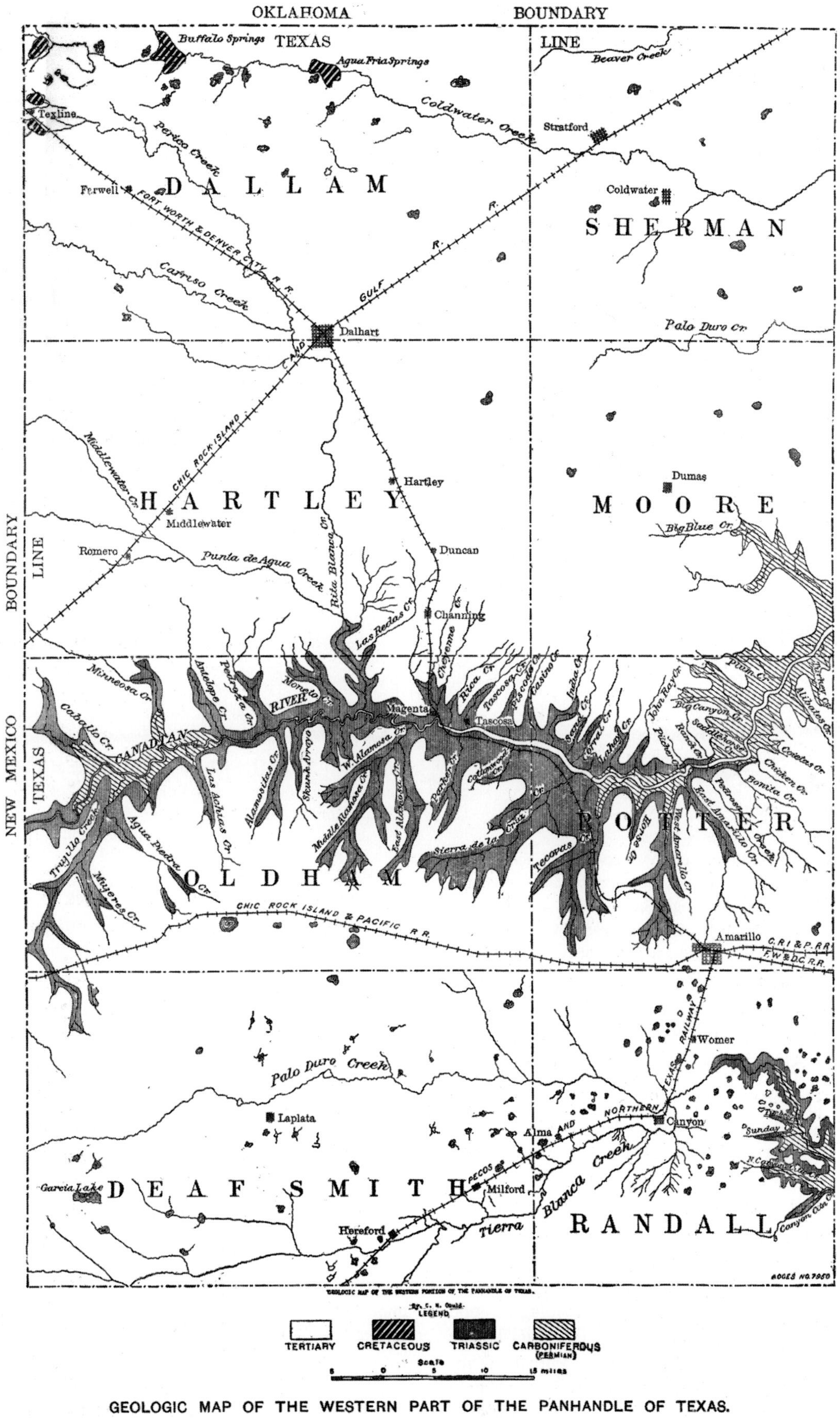

FIGURE 1.3. Gould's (1907) geologic map of the western part of the Texas Panhandle.

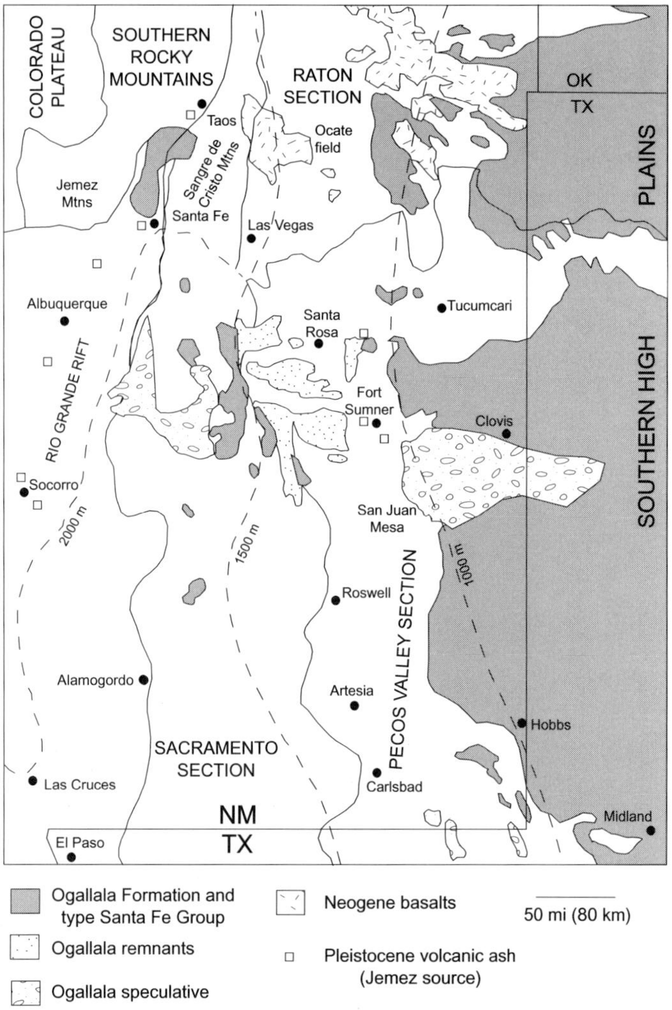

FIGURE 1.4. Map of northeastern New Mexico and adjacent areas showing distribution of Ogallala deposits, Neogene basaltic volcanics, Pleistocene volcanic ashes, major physiographic elements and structural features. The heavy dashed lines indicate magnitude of vertical crustal movement during the past 10 million years (after Hawley, 1984).

FIGURE 1.5. The Ogallala Caprock calcrete at Stop 1. A, World-famous dinosaur paleontologist Adrian Hunt in front of the calcrete. B, Close-up of the calcrete.

NEOGENE LAND-MAMMAL "AGES" IN THE TEXAS PANHANDLE

Spencer G. Lucas and Gary S. Morgan
New Mexico Museum of Natural History, 1801 Mountain Road N.W., Albuquerque, NM 87104

Perhaps the most important contribution vertebrate paleontologists have made to the subdivision of geologic time is the succession of land-mammal "ages" (lmas) Wood et al. (1941) proposed for Cenozoic time in North America. Three of the North America lmas were originally based on fossil mammal assemblages found in the Texas Panhandle – Clarendonian, Hemphillian and Blancan (Fig. 1.6). The first two were coined for fossil mammal assemblages from the Ogallala Formation, whereas the Blancan type assemblage comes from a localized lacustrine deposit termed the Blanco Formation.

An aggregation of mammalian fossils with a stratigraphic distribution in a body of sedimentary rock is an assemblage zone. Vertebrate paleontologists have long referred to such an assemblage zone as a fauna or local fauna. The geologic time equivalent to such a fauna is a lma. This means that lmas are biochronological units—intervals of geologic time recognized by distinctive fossils (Tedford, 1970). Lmas are not the ages of formal stratigraphic terminology. Such ages are intervals of geologic time equivalent to stages, which are bodies of strata. Each age is thus based directly on strata in a specific area; these strata are the stratotype of the stage/age. Lmas lack such stratotypes, so to call them ages is to use that term in a different sense than is usual, which is why the term "age" should be placed in quotation marks. Nevertheless, each lma is based on a type assemblage of mammal fossils, and the lma name is taken from a geographic feature near (or at) that type assemblage (Fig. 1.6).

Wood et al. (1941, p. 12) based the Clarendonian on "the Clarendon local fauna (and member?) near Clarendon, Donley County, Panhandle of Texas." Cope (1893) first reported fossil mammals from this area. The Clarendonian lma as currently understood spans part of late Miocene time, ~ 9-11.5 Ma (Woodburne and Swisher, 1995). The first appearance of the cat-like nimravid *Barbourofelis* marks the beginning of the lma, and the dispersal of

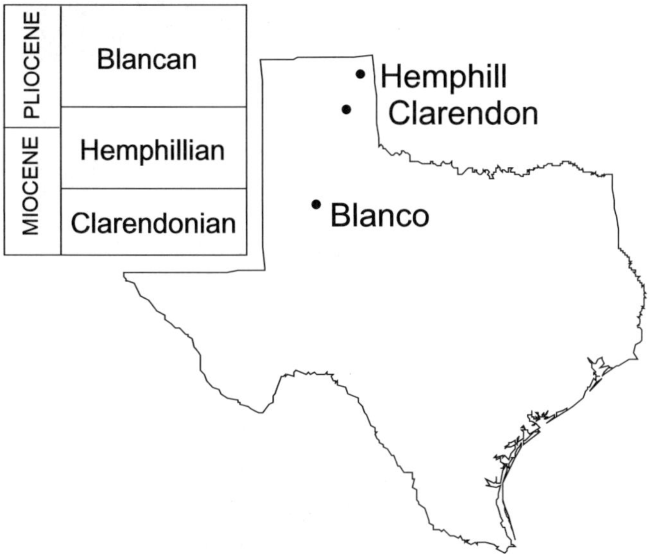

FIGURE 1.6. Location of type fossil assemblages of Texas Neogene land-mammal "ages."

hipparionine horses from North America to Eurasia (the famous "*Hipparion* datum") took place close to the beginning of the Clarendonian. The North American Clarendonian fauna includes an ochotonid, leporids, diverse rodents, chiropterans, erinaceoid and soricoid insectivorans, diverse carnivorans, artiodactyls (especially camels and pronghorn antelopes), horses (especially hipparionines), rhinoceroses (especially aceratheres) and gomphotheres. Mammal assemblages of Clarendonian age are distributed across a broad area in the western United States, and are especially abundant in Nebraska and South Dakota. In New Mexico, the upper part of the Tesuque Formation in the Española basin contains mammals of Clarendonian age (Tedford, 1981).

Wood et al. (1941, p. 12) based the Hemphillian lma on "the Hemphill Member of the Ogallala, which includes both the Hemphill local fauna from the Coffee Ranch quarry and the Higgins local fauna, Hemphill County, Panhandle of Texas." However, there is no recognized "Hemphill Member," and the "Higgins local fauna" is actually from Lipscomb County, Texas (Gustavson et al., 1990). The first discoveries of Hemphillian mammals in the Texas Panhandle were in the 1920s (e.g., Matthew and Stirton, 1930). As now construed, the Hemphillian lma encompasses part of late Miocene-early Pliocene time, ~ 4.5-9 Ma (Woodburne and Swisher, 1995). The immigration of some cricetid rodents from Eurasia to North America and an immigration of ground sloths from South America mark the beginning of the Hemphillian. Hemphillian mammal assemblages are widely distributed in the United States, from Washington and California to Florida and North Carolina. In New Mexico, the best documented Hemphillian mammal assemblage is from the Chamita Formation in the Española basin (MacFadden, 1977; Tedford, 1981). The North American Hemphillian fauna includes sloths, lagomorphs, diverse rodents, soricid and talpid insectivores, diverse carnivores, diverse artiodactyls (especially camels and pronghorn antelopes), horses (especially *Dinohippus* and hipparionines), gomphotheres and the last North American rhinocerotids.

W. F. Cummins first collected fossil mammals in the Blanco Canyon area of Texas in 1890 (Cope, 1893). Wood et al. (1941, p. 12) defined the Blancan lma "based on the local fauna at 'Mt. Blanco' and the adjoining draws, near the 'old rock house,' north of Crawfish Draw, Crosby County, Panhandle of Texas." The type assemblage comes from the Blanco Formation, a localized basin accumulation about 60 m thick of predominantly lacustrine origin (Evans and Meade, 1945). The Blancan lma is now construed to represent much of Pliocene time, about 1.8-4.5 Ma (Woodburne and Swisher, 1995). Blancan mammal faunas are found over a broad region, from Washington/California to Florida. In New Mexico, several Blancan assemblages are found in the upper part of the Santa Fe Group of the Rio Grande rift basins (Morgan and Lucas, 2001). The Blancan begins with the first appearances of several immigrant taxa of rodents, carnivores and true cervids from Eurasia to North America. The North American Blancan fauna includes a few sloths, diverse lagomorphs and rodents (especially sciurids and cricetids), chiropterans, soricid and talpid insectivorans, diverse carnivorans, camels, horses (especially the last hipparionines and first "true" *Equus*) and most of the last gomphotheres.

62.7 Yield sign. Continue straight for 100 ft then **turn right** to cross bridge over Interstate 40. **0.3**

63.0 Stop sign. **Turn left to enter Interstate 40** eastbound. **0.2**

63.2 **Sharp left onto Interstate 40** (leave Old Highway 66). **Do not go straight**. Welcome back to the Llano Estacado. You are now driving across a landscape that is almost unchanged for the past 2 million years! **2.3**

65.5 Large sink on right reflects dissolution at depth. **1.3**

66.8 **Exit 22 to Adrian.** Named for Adrian Cullen, an early farmer in the area, Adrian was founded in 1900 when the Rock Island Railway picked the site as a station. The population in 1990 was 220, including one "E. Presley." **1.0**

67.8 Exit 23. **3.9**

71.7 Large playa on right. **1.6**

73.3 Exit 28 to Landergin; grain elevator to left. Brothers Pat and John Landergin established ranch headquarters here in 1906 that later grew into a small community. **3.0**

76.3 Picnic area on right. **3.4**

79.7 Exit 35 to Vega, the seat of Oldham County. The name is Spanish for "plain" or "meadow." A town since 1903, the 1990 population was 840. Vega is located within the boundaries of the enormous XIT Ranch that extended 200 mi north to south. **1.3**

81.0 Exit 36, town of Vega to left. **1.7**

82.7 Exit 37 (last exit to Vega). **2.7**

85.4 Note trees to left, good indicators of which way the wind blows. **1.5**

86.9 Exit 42 to Everett Road. **6.7**

93.6 Exit 49 to Wildorado. Named for nearby Wildorado Creek, a community was established here in 1900 as a shipping point on the Chicago, Rock Island and Pacific Railway. 1990 population was 180. **2.7**

96.3 Potter County Line. Amarillo is the seat of Potter County, which was named after Robert Potter, a signatory of the Texas Declaration of Independence. **1.9**

98.2 Parking area to right. **0.8**

99.0 Exit 54 to Adkisson Road. **1.8**

100.8 Windmills on right are part of USDA experimental station. **1.1**

101.9 Exit 57 to Bushland. Bushland was established in 1908 as a station on the Chicago, Rock Island and Pacific Railway. It was named for William Henry Bush of Chicago, who donated the land for a townsite. 1990 population was 130. **3.1**

105.0 Exit 60 to Arnot Road. **1.5**

106.5 Cadillac Ranch to right. **0.5**

107.0 Exit 62A to Hope Road and Helium Road. **0.5**

107.5 Exit 62B to Business 40 Amarillo Boulevard. **1.7**

109.2 **Take exit 64, Texas loop highway 335 south to Palo Duro**, leave Interstate 40 here at Amarillo city limit. Amarillo is the metropolis of the Texas Panhandle (1998 population = 171,000) and the largest city on I-40 between Albuquerque and Oklahoma City. When the Fort Worth and Denver City Railway began building across the Texas Panhandle in 1887, a group of merchants chose a site to build stores that became Amarillo. By 1890, the fledgling town was one of the world's busiest cattle-shipping points. Today, its principal income derives from petroleum and cattle. **0.3**

109.5 **Turn right** on Soncy Road, Texas Loop Highway 335. **0.1**

109.6 Proceed through traffic light at Westgate. **0.1**

109.7 Enter Randall County. The county takes its name from Horace Randal, a Confederate general who was killed at the battle of Jenkins' Ferry, Arkansas, in 1864. In 1876, a clerical error doubled the l's in the name! **0.5**

110.2 Proceed through traffic light at West 34th Street. **1.0**

111.2 Leave Amarillo after crossing through traffic light at West 45th Street. **2.6**

113.8 Road curves sharply to left—stay on main highway. **1.3**

115.1 Cross traffic light at Coulter Road. **0.5**

115.6 Interstate 27 bridge. **Just before bridge, turn right at Amarillo city limit to get on Interstate 27 south. 0.2**

115.8 Get into left lane onto Interstate 27 south. **0.7**

116.5 Exit 115. **1.8**

118.3 Exit 113, McCormick Road. **1.2**

119.5 Exit 112, Texas Farm Road 2219. **1.0**

120.5 Exit 111, Rockwell Road. **1.4**

121.9 Exit 110. **Leave I-27 on South US-87, to Canyon, Texas. 0.9**

122.8 Pass under overpass. Entering greater Canyon, Texas. Note deep incision in Ogallala Formation here. **1.0**

123.8 Cross Palo Duro Creek. **0.1**

123.9 Canyon city limit. Canyon, originally Canyon City, was founded in 1878 and named for Palo Duro Canyon to the east. The site was originally the headquarters of the T Anchor Ranch. The Pecos & Northern Railroad made Canyon a shipping point in 1898 (Metz, 1994). Canyon is home to West Texas A & M University and the Panhandle Plains Historical Museum. 1990 population was 11,365. **Get into left lane. 0.8**

124.7 Street light at North 3rd Avenue. Continue straight. **0.3**

125.0 **Turn left on Texas 217/4th Avenue to Palo Duro Canyon. 0.1**

125.1 Panhandle Plains Historical Museum on left. This is not only the largest historical museum in Texas, but it also includes impressive paleontological displays as well as a large exhibit on the petroleum industry, part of which is visible outdoors. Continue straight on Texas 217 to Palo Duro Canyon (12 mi). **1.4**

126.5 Cross Tierra Blanca Creek (leaving Canyon). **0.8**

127.3 Junction with Interstate 27 (go under overpass). Continue straight. **1.4**

128.7 Road bends left. **0.5**

129.2 Junction with Texas Farm Road 1541. Stop sign and light. Continue straight ahead. **6.3**

135.5 Crest low ridge. Breaks of Palo Duro Canyon visible ahead. The Spanish words *palo duro* ("hardwood") refer to the hardwood shrubs and trees in the canyon. **0.8**

136.3 Ogallala on right in headcut of Sunday Canyon, a Palo Duro tributary. **0.2**

136.5 Ogallala Group deposits over red beds of Triassic Tecovas Formation on right in canyon. **0.9**

137.4 Enter Palo Duro Canyon State Park. **0.1**

137.5 Stop to enter park and pay here. **0.7**

138.2 **STOP 2.** Scenic overlook on right, **turn right into parking lot of visitor center**.

From this vantage point (Fig. 1.7), it is easy to understand why Palo Duro Canyon is called the "Grand Canyon of Texas." The canyon is devel-

FIGURE 1.7. Overview of part of Palo Duro Canyon from Stop 2. The cliff-forming sandstone in the foreground is part of the Trujillo Formation of the Chinle Group.

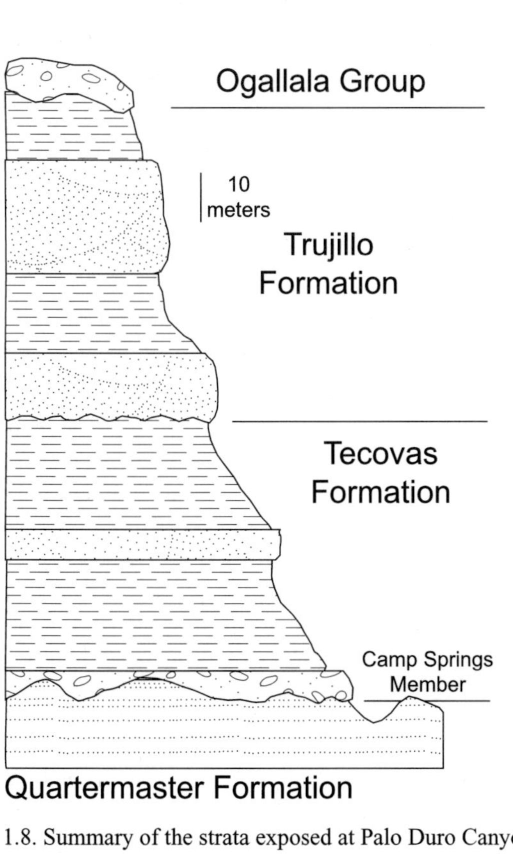

FIGURE 1.8. Summary of the strata exposed at Palo Duro Canyon.

oped around the Prairie Dog Town Fork of the Red River; it is 120 mi long, 1.5-20 mi wide, has a depth of 800-1000 ft and an elevation of 3500 ft at the rim. Palo Duro Canyon State Park opened on 4 July 1934 and encompasses ~16,402 acres of the northernmost portion of the canyon. The Civilian Conservation Corps built most of the park buildings and roads during the 1930s.

Here, we can summarize the stratigraphic section exposed in the park (Fig. 1.8) and examine its uppermost part. That section is (in descending order):

1. Neogene Ogallala Formation (Group), less than 100 ft thick, consisting of a lower interval of reddish brown sandstone and conglomerate overlain by the Caprock calcrete. Note the excellent outcrop here in the lower parking lot next to the visitor center.

2. Upper Triassic Trujillo Formation of Chinle Group (see accompanying minipaper), about 150 ft thick, of cliff-forming sandstone and conglomerate with lesser interbeds of mudstone. These beds are the youngest Triassic strata exposed in the canyon, and form most of the cliffs that rim the canyon (Fig. 1.7).

3. Upper Triassic Tecovas Formation of Chinle Group, up to 200 ft of red-bed mudstone and minor, trough-crossbedded sandstone. A basal, 1-6-ft thick

conglomerate bed, the Camp Springs Member, is locally present.

4. Upper Permian Quartermaster Formation, about 60 ft of which is exposed in the park, consists of red beds of cyclically bedded siltstone, sandstone and gypsum.

After stop return to main road and turn right to continue into park. 0.1

ABANDON THE TERM DOCKUM!

Spencer G. Lucas
New Mexico Museum of Natural History, 1801 Mountain Road NW, Albuquerque, NM 87104

The oldest name for an Upper Triassic stratigraphic unit in the American Southwest is the term "Dockum Beds" of Cummins (1890, p. 189-190), introduced as follows:

> A few miles before reaching Dockum, situated in the western edge of Dickens County [Texas], I came upon a bed of conglomerate sandstone and red clay, resting unconformably upon the clays and sandstones of the upper Permian, entirely unlike anything I have heretofore seen in Texas. This formation lies along the foot of the Staked Plains in a narrow belt. Because of its extensive occurrence in the vicinity of Dockum, I gave the formation the name of Dockum Beds, but will not for the present attempt to determine their correlation. I have found everywhere on the beds of the Permian belt pieces of conglomerate and large pebbles of white quartz that did not belong to the Trinity sands of the Cretaceous which were supposed to overlie the Permian to the westward, and it is a matter of interest to know where this drift came from. The fragments of conglomerate increased in size as we traveled westward until we came upon the beds of that material in the vicinity of Dockum, and the question was solved as to the origin of the fragments of conglomerate and the quartz pebbles.
>
> In the conglomerates are many silicified trunks of trees, some of them of great length. In the red clay above the conglomerate are fossil remains of large reptiles, whose species I was unable to determine in the field. In the upper sandstones were many casts of a Unio that I have provisionally called Unio documensis [sic]. In most places that fossil occurs only as casts, and in one place only did I find specimens of both valves, and they were so badly encrusted with carbonate of lime that the peculiar markings of the shell could not be seen. The sandstone was everywhere full of scales of mica, some of the scales being one-sixteenth of an inch square. The whole thickness of this formation in the vicinity is about 150 feet. These beds extend under the Staked Plains. I traced them up Blanco Canyon to the falls of White River, where they pass out of sight under the beds of the overlying strata.

Cummins' (1890) description makes it clear what strata pertain to his "Dockum Beds:" they are the units subsequently named Camp Springs by Beede and Christner (1926) and Tecovas and Trujillo by Gould (1907) (Fig. 1.9). However, few modern (post-World War II) workers seem to have concerned themselves with what strata composed the original Dockum of Cummins (1890). This perhaps explains the wide variety of usages of the term Dockum, many of which included strata in the Dockum that are clearly not equivalent to strata of Cummins' original Dockum.

Dockum as used by most workers is thus an imprecise and confusing stratigraphic term. Examples include:

1. Various workers (e.g., Darton, 1928; Gorman and Robeck, 1946; Lehman, 1994) have used Dockum to refer to all Upper Triassic strata in east-central New Mexico. This greatly expands Cummins' (1890) original concept to include, among other strata, a thick post-Trujillo section (Bull Canyon and Redonda formations) in east-central New Mexico. Furthermore it also raises the question, usually not answered, of what are the western and northern limits of Dockum strata? When the question is answered, it is often to assert that there is a distinct Dockum basin of Upper Triassic deposition in eastern New Mexico-West Texas, a concept that has been refuted.

2. The term Dockum has been applied to Triassic strata in western Kansas and southeastern Colorado (e.g., Parker, 1934; Merriam, 1963) that are much older than the Dockum; indeed these strata are not of Late Triassic age (Lucas, 1993).

3. Most of the Upper Triassic section in the Dry Cimarron Valley of northeastern New Mexico-western Oklahoma is stratigraphically above the type Dockum, but it has been assigned to the Dockum (e.g., Baldwin and Muehlberger, 1959).

4. In central New Mexico (Socorro and Lincoln Counties), the term Dockum has been applied in various ways to the Triassic section (e.g., Bates et al., 1947; Wilpolt and Wanek, 1951; Myers et al., 1986), including strata both older and younger than the type Dockum section.

5. In West Texas, the term Dockum has been used in various ways, to refer to all or part of the Triassic section (e.g., Adams, 1929; Adkins, 1932, Reeside et al., 1957).

Furthermore, in all the above areas, the term Chinle has also been applied to some or all of the Triassic strata.

Previously, I have tried to retain the term Dockum, using it as a formation in the Chinle Group and restricting its use to West Texas (e.g., Lucas, 1993; Lucas et al., 1994). But now I believe it would be much simpler if the term Dockum were abandoned altogether. The NACSN (1983) certainly allows for this by stating (Article 20a) that "widespread misuse in diverse ways which compound confusion" may justify abandonment of a stratigraphic name. Dockum has been so misused, and it should die a merciful death.

All Upper Triassic strata from the Colorado Plateau to West Texas represent one integrated depositional system that produced a single, complex lithosome, so what should these strata be named? On the Colorado Plateau they are the Chinle Formation of most authors. The Chinle had a well-defined type section when Gregory (1916, 1917) named it, one that encompasses almost all of the Upper Triassic lithosome of the American Southwest. Chinle has long been used in a precise and consistent manner in a voluminous literature and a myriad of geologic maps. Further-

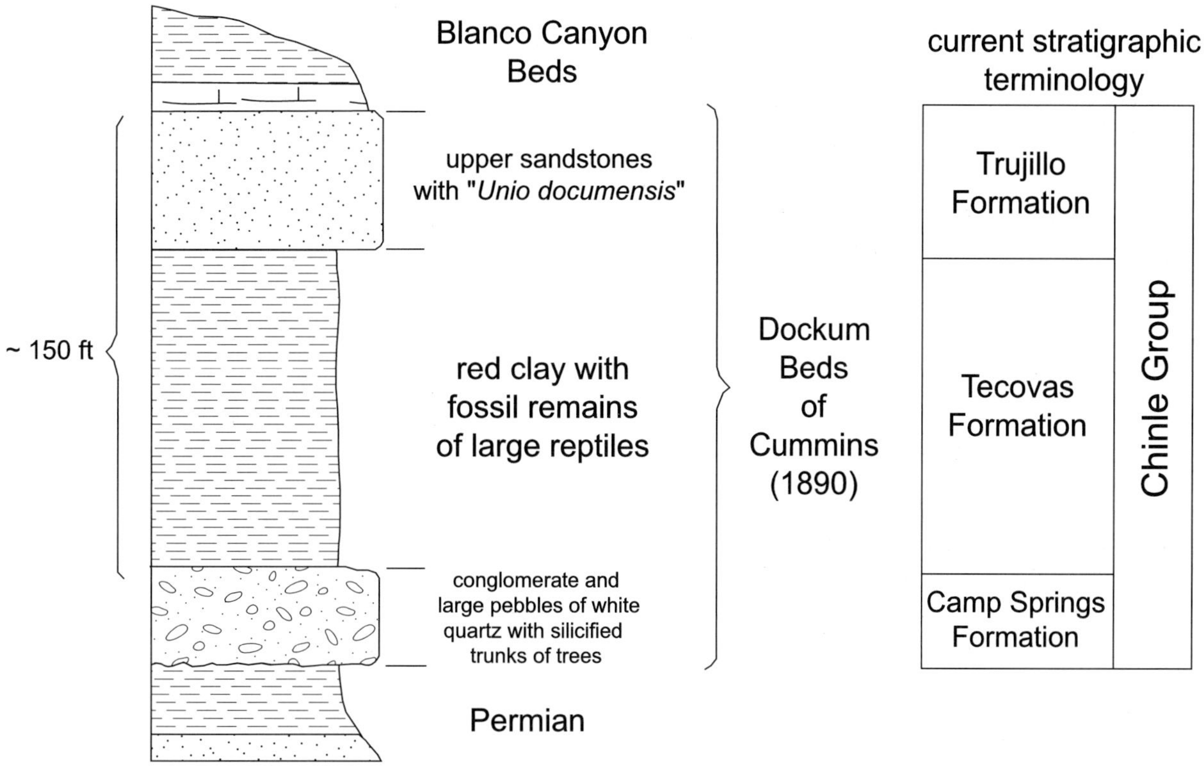

FIGURE 1.9. Stratigraphic interpretation of Cummins' (1890) original description of the "Dockum Beds."

more, the term has been used in eastern New Mexico and West Texas by several workers, dating back to at least the 1930s.

Therefore, I prefer to assign all Upper Triassic nonmarine strata in the American Southwest to the Chinle Group (Lucas, 1993). Dockum admittedly has priority over Chinle (as do the names Dolores and Popo Agie, for Upper Triassic deposits in Colorado and Wyoming, respectively: Lucas, 1993), but Chinle is a much more widely used name with a type section that represents more of the Chinle Group than the original Dockum section of Cummins. I have no doubt that the term Dockum will remain in use in the parochial literature of Texas geology, but for those interested in a precise and consistent nomenclature for the Upper Triassic strata of the American Southwest, all Upper Triassic strata in West Texas can be referred to as Chinle Group divided into formations and members as appropriate (Fig. 1.9).

138.3 Turn right to go down hill. **0.8**

139.1 Begin descent into Palo Duro Canyon. Roadcut to left is in the Ogallala Formation (Group). **0.2**

139.3 Contact between the Ogallala and underlying Triassic Trujillo Formation at green band on left. **0.2**

139.5 Good Trujillo Formation outcrop on right. **0.2**

139.7 Unconformable contact of Upper Triassic Trujillo Formation sandstones and conglomerates over Upper Triassic Tecovas Formation mudstones on right. **0.3**

140.0 Contact of the Upper Triassic Tecovas Formation (redbeds) over the Upper Triassic Camp Springs Conglomerate (white conglomeratic sandstone) at 9:00. **0.2**

140.2 Unconformable contact of the Upper Triassic Camp Springs Conglomerate (white conglomeratic sandstone) over the Upper Permian Quartermaster Formation (orange siltstones and sandstones) on right. **0.1**

140.3 **STOP 3.** Turn right into Pioneer Amphitheater. Here, we can examine an excellent exposure of the Permian-Triassic disconformity (Fig. 1.10) and discuss the following points:

1. The Quartermaster Formation is the stratigraphically highest Permian unit in the Texas Panhandle, and it is the equivalent (senior synonym) of the Dewey Lake Formation of the Midland basin to the south (Lucas and Anderson, 1994). The only biostratigraphically significant fossils from the Quartermaster Formation are three molluscan taxa from Briscoe County, Texas—*Naticopsis transversus* (Beede), *Schizodus oklahomaensis* Beede and *Myalina acutirostratus* Newell and Burma—that suggest a Permian age (Roth et al., 1941). Magnetostratigraphy indicates a Permian age for the Quartermaster Formation—it preserves the mixed polarity characteristic of the Illawara zone of the latest Permian (Molina et al., 1989, 2000). Two ash beds in the Quartermaster Formation at Cap-

FIGURE 1.10. Outcrop of the Permo-Triassic boundary at Stop 3--Permian Quartermaster Formation (Q) beneath Upper Triassic Tecovas Formation (T).

rock Canyon State Park, east of here, have been radioisotopically dated (remember that the Permian-Triassic boundary is well calibrated at ~ 251 Ma: Bowring et al., 1998). K/Ar ages are 261 $\pm$ 9 Ma and 251 $\pm$ 4 Ma (Fracasso and Kolker, 1985; Kolker and Fracasso, 1985). Ar/Ar ages, only published in an abstract (Renne et al., 1996), are 251.2 $\pm$ 0.6 Ma and 249.4 $\pm$ 0.7 Ma, and the latter date was not an ideal plateau age. On face value, these dates suggest that the uppermost Quartermaster Formation may be of earliest Triassic age. Indeed, Schiel (1987, 1988) suggested this based on lithologic similarity of the Quartermaster and Early Triassic portion of the Moenkopi Formation on the Colorado Plateau. We are inclined, nevertheless, to trust the ages given by the bivalves and the magnetostratigraphy, and consider the entire Quartermaster Formation to be Late Permian.

2. There is a substantial unconformity between the Quartermaster Formation and the base of the overlying Chinle Group, representing a hiatus of nearly 25 million years. The age of the Chinle Group base is well constrained biostratigraphically, and Palo Duro Canyon is a critical locality in this constraint. Thus, here, in the basal Camp Springs Member of the Tecovas Formation, a skull of the phytosaur *Paleorhinus* was collected (Fig. 1.11). *Paleorhinus* also is found in the marine Opponitzer Schichten of Austria, strata of late Carnian (Tuvalian) age (Hunt and Lucas, 1991). Other *Paleorhinus*

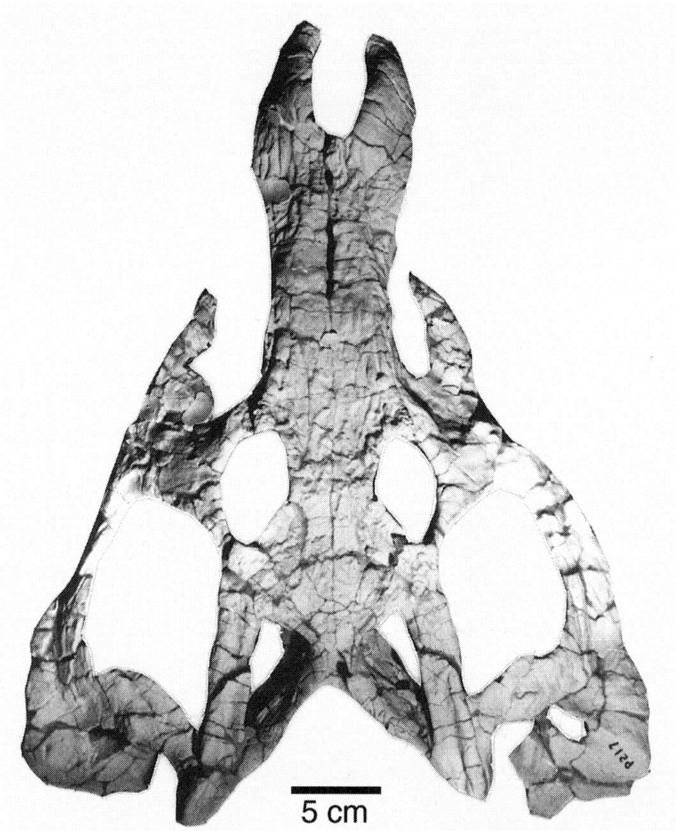

FIGURE 1.11. Dorsal view of skull of the phytosaur *Paleorhinus* found in the Camp Springs Member of the Tecovas Formation at Palo Duro Canyon.

records are in the lowermost Chinle Group in Texas, Arizona and Wyoming, and in Morocco, India and Germany—all strata of late Carnian (Otischalkian) age based on their vertebrate fossil assemblages, palynoflora and other data (Fig. 1.12) (Hunt and Lucas, 1991; Lucas, 1998). The best estimate of a late Carnian date is about 228 Ma, based largely on an Ar/Ar age from Argentina (Rogers et al., 1993). Therefore, the Quartermaster-Chinle hiatus is a gap from ~251 Ma (or slightly older) to ~228 Ma.

3. McGowen et al. (1979, 1983), Granata (1981) and Johns and Granata (1987) interpreted Tecovas Formation mudstones as lacustrine beds, and Trujillo Formation sandstones/conglomerates as easterly flowing rivers and river deltas that flowed into the lake. Indeed, they reconstructed a Late Triassic lake basin that covered much of West Texas, their "Dockum lacustrine basin." However, this idea has not withstood critical scrutiny.

Tecovas mudstones encase small fluvial channels, contain nodular calcretes and other evidence of pedogenesis and yield fossils of terrestrial plants and animals. These mudstones thus closely resemble Chinle Group mud-

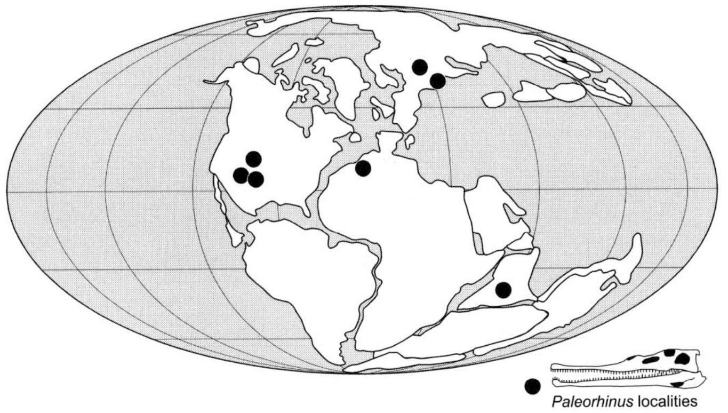

FIGURE 1.12. Late Triassic Pangea showing the distribution of *Paleorhinus* localities.

stones of the Colorado Plateau long interpreted as the deposits of floodplains, stable interfluves and small, isolated ponds (e.g., Dubiel, 1989). Furthermore, paleocurrents measured in Upper Triassic strata on the High Plains of eastern New Mexico and West Texas are generally directed to the northwest (Cazeau, 1960; Kiatta, 1960; Lupe, 1988; May, 1988; DeLuca and Erickson, 1989; May and Lehman, 1989). Riggs et al. (1996) also used age groups of detrital zircons found across the entire Chinle basin to establish paleoflow from the Dockum in the southeast across the Chinle to offshore deposits in Nevada. Yet, McGowen et al. (1979, 1983) combined a few easterly paleocurrent indicators with the generally northwest-southeast orientation of Triassic sand bodies in the subsurface to reconstruct a regional Late Triassic paleoslope down to the east. The orientations of the sand bodies, however, are bidirectional flow indicators consistent with paleoflow to the northwest. Chinle Group sedimentation in West Texas thus was by rivers flowing primarily to the northwest and their associated floodplains (Lucas and Anderson, 1992). No evidence of extensive lakes exists in Chinle Group mudrocks in West Texas.

After stop retrace route to park entrance. 2.9

143.2 Park exit. Retrace route back to Canyon. **8.3**

151.5 Stop sign and light at Texas Farm Road 1541. Continue straight. **1.9**

153.4 Pass under I-27 overpass at eastern edge of Canyon. **2.2**

155.6 **Turn right onto Highway 87** at light at 23rd Street. **0.3**

155.9 **Get into right lane** at 2nd Avenue. Entrance to West Texas State University to right. **0.3**

156.2 Junction with US-60, **go right onto on ramp to Hereford. 0.3**

156.5 Bear right to Hereford on West US-60 (the Woodie Guthrie Memorial Highway). **2.4**

158.9 Breaks at 2:30 are head of a tributary to Palo Duro Canyon. **7.4**

166.3 Enter Umbarger. Umbarger was founded with a two-room schoolhouse in 1902. It grew in 1915 when the Panhandle and Santa Fe Railroad arrived. During World War Two, about 7000 Italian prisoners of war were housed in Hereford, and they carved a wooden reproduction of Da Vinci's "The Last Supper" in the Umbarger church as well as painting a ten-foot canvas of the Assumption (Metz, 1994). 1990 population = 327. **0.2**

167.5 Road to left leads to Buffalo Lake. **1.4**

168.9 Picnic area on right. **1.9**

170.8 Buffalo Lake at 9:00. **0.8**

171.6 Re-entering Deaf Smith County. **1.8**

173.4 Enter Dawn, Texas. The story goes that in 1887, founder J. T. Parrish referred to the town as either "dawn of a new country" or "dawn of civilization." We invite you to look around and draw your own conclusions! 1990 population = 94. **1.8**

175.2 Large gravel pit on left. **6.3**

181.5 Hereford Municipal airport to right. **2.2**

183.7 Junction with Texas Farm Road 2943 to Dimmitt; entering greater Hereford. Dimmitt was founded in 1890, and is the seat of Castro County; 1990 population = 4408. **1.1**

184.8 Hereford city limits. Hereford, the seat of Deaf Smith County, was named in the 1890s for the large herds of Hereford cattle that were raised in this area. The local groundwater is rich in fluorine and iodine, so Hereford is known as the "town without a toothache." In 1881, one of the first barbed wire fences in Texas was erected here to prevent cattle from the T Anchor Ranch from drifting south (Metz, 1994). Local feedlots process 3 million cattle a year. Hereford is also home to the National Cowgirl Hall of Fame and Western Heritage Center. 1990 population was 14,745. **1.8**

186.6 Junction with Texas Farm Road 1259 (Main Street); continue straight. **0.5**

187.1 **Turn right** on North US-385 (25 Mile Avenue). **Immediately get into left lane. 0.7**

187.8 **Turn left on Texas Farm Road 1058** toward Grady at its junction with 211. **0.8**

188.6 Leaving Hereford. **5.2**

193.8 Junction with Texas Farm Road 1057. **0.5**

194.3 Pass through playa lake. **1.0**

195.3 Westway. **5.8**

201.1 Pass over small ridge of Ogallala Group. Note playa lake to right. **1.3**

202.4 Aromatic stockyard to left. **2.6**

205.0 Texas Farm Road 2298 to Friona to left. Established as "Frio" in 1898 as a shipping station on the Pecos Valley and North Texas Railway, the town took its name from nearby Frio Draw. 1990 population was 3600. **2.3**

207.3 Note Ogallala Group deposits in trench on right in large playa basin. **2.0**

209.3 Playa lake on right. **1.5**

210.8 Village of Bootleg. Formerly Bootleg Corner, the town dates back to the early 1900s. It was either named for a moonshine still, or more likely for a "bootleg school." Such schools were small buildings built by a land syndicate to convince prospective buyers that there was a school near land they were buying (clever folks, those Texans!). **0.2**

211.0 Junction with Texas 214 to Adrian, continue straight. **6.5**

217.5 Mailbox for Garcia Lake. Low Chinle (Bull Canyon Formation) redbeds in slopes under Ogallala Formation deposits at 3:00. **4.3**

221.8 Playa on left. **0.5**

222.3 Cross through playa. **1.7**

224.0 Enter New Mexico. Note asphalt change—road is now NM Highway 241. **3.7**

227.7 Enter Bellview, a tiny (population 15) community dating from 1905. Several settlements formed in this area around 1905, each with a different post office. Rosedale was on the present site of Bellview and it took its current name in 1918, although the Baptist church still retains the name Rosedale. The village may have received its name from a local school named Liberty Bell (Julyan, 1996) or it may be a variant of *belle vue,* French for a beautiful view, as a humorous allusion to the lack thereof. **6.0**

233.7 Enter Broadview, population 50, founded about 1925. Broadview was first named Boney Curve, for a prominent local family and the fact that it lies on a conspicuous bend in the highway. When the post office was established in 1931, the community settled on the name Broadview. Broadview lies on a slight rise, so it does allow a broad view of the high plains (Julyan, 1996). **0.1**

233.8 NM-241 ends at junction with NM-209. **Bear right on NM-209** after stop sign. **4.4**

238.2 Playa lake on left. **1.0**

239.2 Enter Grady, one of New Mexico's smallest incorporated towns (population 110 in 1990). Grady was founded in 1907 in eager anticipation of the arrival of the AT & SF railroad. It is still waiting! The town was named after a local lady, likely the first postmistress and owner of most of the original town site. Grady still retains a school and is a force in small school high school athletics. It survives as a farming and ranching community. The town has lost more than half of its population since 1940. Be aware of speed traps at speed limit changes. **0.5**

239.7 **Turn right on NM-469** to go north toward San Jon. **3.0**

242.7 Quay County line. **0.9**

243.6 Slow down and stay on main road through hard curve to left. **1.0**

244.6 Playa on right. **1.0**

245.6 Slow through hard curve to right. Roads are on section lines with mile or mile- multiple spacing. **1.0**

246.6 Junction with NM-231; continue straight. **0.9**

247.5 Abandoned schoolhouse on left at Wheatland is made from sandstone from the Upper Triassic Redonda Formation quarried off the caprock ahead (Fig. 1.13). Slabby calcarenites such as these formed on the shorelines of the Redonda lake. They are more commonly used for paving than in wall construction in this area. Wheatland was settled and named in about 1915 as a result of a ballot when three school districts were consolidated. It was named for the main crop that formerly was cultivated in this area (Julyan, 1996). It has been a near ghost town since 1965, when the post office closed. **1.1**

248.6 Junction with NM-275, continue straight. **2.0**

250.6 Sharp bend to left for offset between principal meridians. **0.7**

251.3 Road bends hard right. **0.8**

252.1 Edge of caprock and Ogallala Formation deposits visible at 2:00. **1.2**

253.3 Road bends hard right just before TV translator. View to north of Canadian country. **0.2**

FIGURE 1.13. The abandoned schoolhouse at Wheatland built from slabs of Triassic sandstone.

253.5 Sign reads "Caprock, elevation 4900." San Jon archeological site is east of here (see accompanying minipaper). **0.2**

THE SAN JON SITE, QUAY COUNTY, NEW MEXICO

Vance T. Holliday[1] and Eileen Johnson[2]
[1]Department of Geography, University of Wisconsin, Madison WI 53706; [2]Museum of Texas Tech University, Box 43191, Lubbock, TX 79409

The San Jon site is an important archaeological and Quaternary stratigraphic locality on the northwestern edge of the Southern High Plains (Hill et al., 1995; Holliday, 1997). The site was discovered in 1940. The first systematic research was an archaeological excavation in 1941 by Frank H.H. Roberts of the Smithsonian Institution (SI) (Roberts, 1942) with accompanying geological studies by Sheldon Judson, a graduate student at Harvard University working under noted geomorphologist Kirk Bryan (Judson, 1953). Additional stratigraphic work was carried out by Jerry Harbour of the University of Arizona in the 1960s (Harbour, 1975). The authors began continuing archaeological and geoarchaeological work in 1995.

The sediments, soils and artifacts are exposed in an eroded "playa" basin, one of thousands of small depressions that dot the surface of the High Plains (Gustavson et al., 1995; Holliday et al., 1996). The basin fill was dissected in the late Holocene when the prominent High Plains escarpment was breached, and a system of deep arroyos formed as tributaries to the breach (Fig. 1.14). The playa fill containing the archaeological material is preserved in several peninsulas or promontories isolated between arroyos (Fig.1.14).

The San Jon basin has a complex history of filling. The regional surface of the High Plains is underlain by the Blackwater Draw Formation, composed of Pleistocene eolian deposits and buried soils (Reeves, 1976; Holliday 1989, 1990). Below are Neogene eolian and alluvial sediments of the Ogallala Formation. The High Plains escarpment is formed by the famous "Ogallala Caprock Caliche," developed in the upper Ogallala Formation. Bedded pink and buff sands of the Ogallala are exposed low on the arroyo walls in the site. The well-developed, deep reddish brown soils and buried soils of the Blackwater Draw Formation are exposed in the upper arroyo walls below the playa fills.

The archaeological material is associated with late Quaternary sediment that filled a playa cut into Pleistocene deposits that filled an older, larger basin (Fig.1.15). The older Pleistocene basin was inset against the Blackwater Draw Formation (Hill et al., 1995; Holliday et al., 1996). Its fill covered an area that probably was over 700 m in diameter. The younger basin with the archaeological deposits probably was about 360 m in diameter. The late Quaternary sediments, up to 10 m thick, almost completely filled the younger basin before canyon entrenchment. The following geoarchaeological discussion focuses on the late Quaternary sediments in the smaller, younger basin.

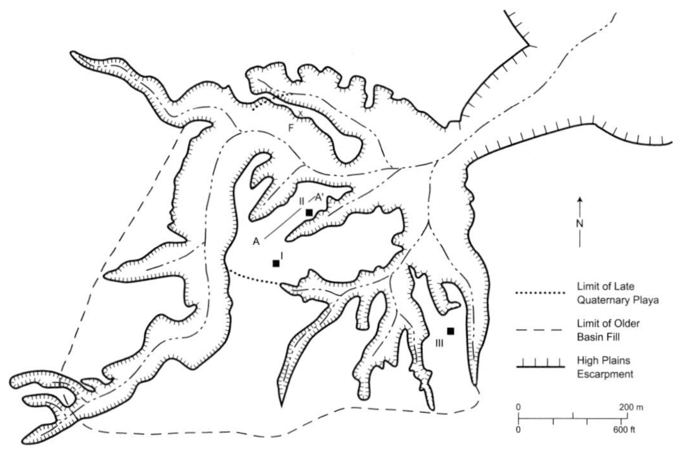

FIGURE 1.14. Map of the San Jon archeological site.

Two basic lithologies were identified in the playa fill. Most of the archaeologically significant sediments are a light gray (sometimes greenish gray) to dark gray, silty loam mud, essentially identical to the Randall Clay soil series (a vertisol) mapped in most playas. Toward the basin margins, especially to the west, southwest and south, there are layers of reddish brown, somewhat sandier loam and silty loam interbedded with the darker muds. Some of these sandier, redder layers are bedded, and other redder deposits are massive, homogeneous and have a higher content of fines. The redder facies of basin fill also exhibits pedogenic modification (Bw and Bt soil horizons), indicative of landscape stability and weathering prior to burial. The fill in the center of thc basin essentially is a homogeneous accumulation of the gray muds, but toward the basin margin in the area of most of the archaeological excavations (Area II, Fig.1.14) this deposit splits into four layers (1m-4m: Fig.1.15). There also are four layers of the redder, sandy sediment (1s-4s: Fig.1.15). The gray mud probably is lacustrine and palustrine fill. The coarser, redder deposits include some alluvium from slopewash, and a high content of eolian sediment. The oldest fill in the sequence, stratum 1m, began accumulating >12,000 rcyr BP (radiocarbon years before present), based on dating of the upper 1m. Otherwise the age of the basin or beginning of filling is not known. Filling continued episodically during the late Quaternary. The youngest basin fill is dated to ~1900 rcyr BP. Erosion began sometime after that date.

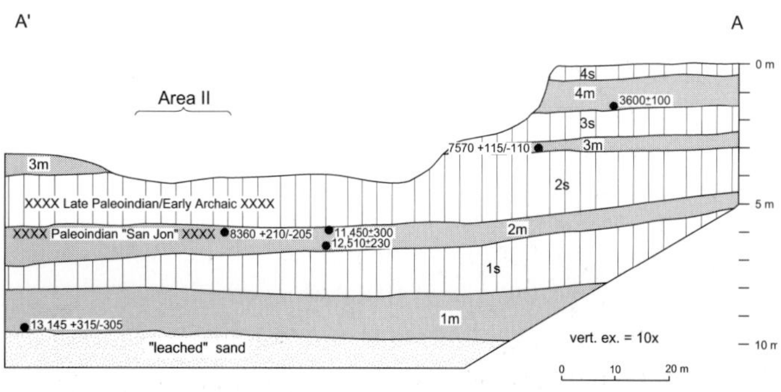

FIGURE 1.15. Stratigraphic cross section of the San Jon archeological site (see location on Figure 1.14).

A number of occupation layers were found at and around the San Jon basin. The best known is near the top of stratum 2m in Area II (Fig.1.15). The SI crew tested a bone bed in 1941 where at least five *Bison antiquus* were represented (Hill et al., 1995). A notable characteristic of some of the bone is that lower limb units were recovered articulated and in an upright position in the mud (Roberts 1942, pl. 3.1; Hill et al., 1995, fig. 7). This situation suggested that animals were mired in mud, died and subsequently were butchered. Data are not available to indicate whether the animals died naturally and then were scavenged or were purposely killed by hunters. A single projectile point was recovered from this feature: an unfluted lanceolate projectile point named "San Jon" by Roberts (1942, p. 8), subsequently identified as a reworked Firstview point (Hill et al., 1995; Knudson, 1995). The bone bed dates to ~8400 rcyr BP.

Above the bone bed, in the sandy sediment of stratum 2s, is an occupation surface discovered by the senior author in 1993. The zone has been the object of intense archaeological investigation by the junior author. The occupation zone consists of considerable debris from stone tool manufacture, as well as broken or discarded stone tools, charcoal and burned rock. The artifacts include several side-notched projectile points similar in morphology to varieties from the central and eastern United States considered to be very Late Paleoindian or Early Archaic. This occupation is dated at ~8000 rcyr BP.

Roberts and the SI crew also discovered and tested a bone bed at the southeastern edge of the basin, in Area III. The bone probably was in stratum 4m (≤3600 rcyr BP), though direct stratigraphic correlations with other areas of the site are impossible because of the intervening arroyos. Portions of seven *Bison bison* were recovered (Hill et al., 1995).

Beginning in 1996, archaeological crews from Texas Tech University have conducted extensive archaeological surveys throughout the large San Jon basin. A number of occupation areas, both buried and at the surface, were found.

The San Jon site has proven to be significant for several reasons. Playa basins undoubtedly were attractive to Native American groups since they first arrived in the region >11,000 rcyr BP. San Jon, however, is one of the few playas where an intact archaeological record has been found (probably because it is one of the rare playas where the fill is exposed). More broadly, it is one of the few sites in the region in any setting with a documented record of multiple, *in situ* occupations spanning the Paleoindian, Archaic and Late Prehistoric traditions. The exposures and the well-stratified and well-dated fill also provide an opportunity to study late Quaternary environmental changes. Samples were collected and are being analyzed (as of the winter of 2000-2001) for stable-carbon isotopes and phytoliths.

253.7 Road to Caprock Amphitheater to right. Begin descent from Llano Estacado. **0.2**

253.9 Ogallala Group roadcuts on left. **0.1**

254.0 Roadcuts of the Cretaceous Pajarito and Mesa Rica formations on left. **0.2**

254.2 Roadcuts of Lower Cretaceous Tucumcari Formation on left. **Prepare to turn into pull out on right. 0.1**

254.3 **STOP 4** at well exposed section of Tucumcari Formation on both sides of highway (Fig. 1.16). The Tucumcari Formation is a marine unit containing a moderately diverse late Albian (late Early Cretaceous) invertebrate fauna (see Kues and Lucas, this volume) that is widely exposed in Quay and eastern Guadalupe Counties. It consists mainly of dark gray shale and mudstone, with thin siltstone, sandstone and shell conglomerate horizons, representing open bay to middle shoreface environments (Scott, 1974).

The Tucumcari Formation was deposited along the western shoreline of the late Albian seaway that extended north across Texas, Oklahoma, Kansas, and eastern New Mexico, linking up briefly with a northern seaway in Colorado. Regression following Tucumcari deposition, resulting in the deposition of the largely nonmarine overlying Mesa Rica Sandstone, severed this link, which was subsequently reestablished, for a longer time, with onset of the Cenomanian Greenhorn transgression (Fig. 1.17). The Tucumcari fauna, together with a similar but less diverse fauna from the Glencairn Formation of northeastern New Mexico, western Oklahoma and southeastern Colorado (see Kues and Lucas, 1987), constitutes the "western endemic center" of the Southern Western Interior biogeographic province of Scott (1986). The Tucumcari fauna is thus distinct from that of the "eastern endemic center" (Kiowa Formation of south-central Kansas), and

FIGURE 1.16. Outcrop of the Tucumcari Formation at Stop 4.

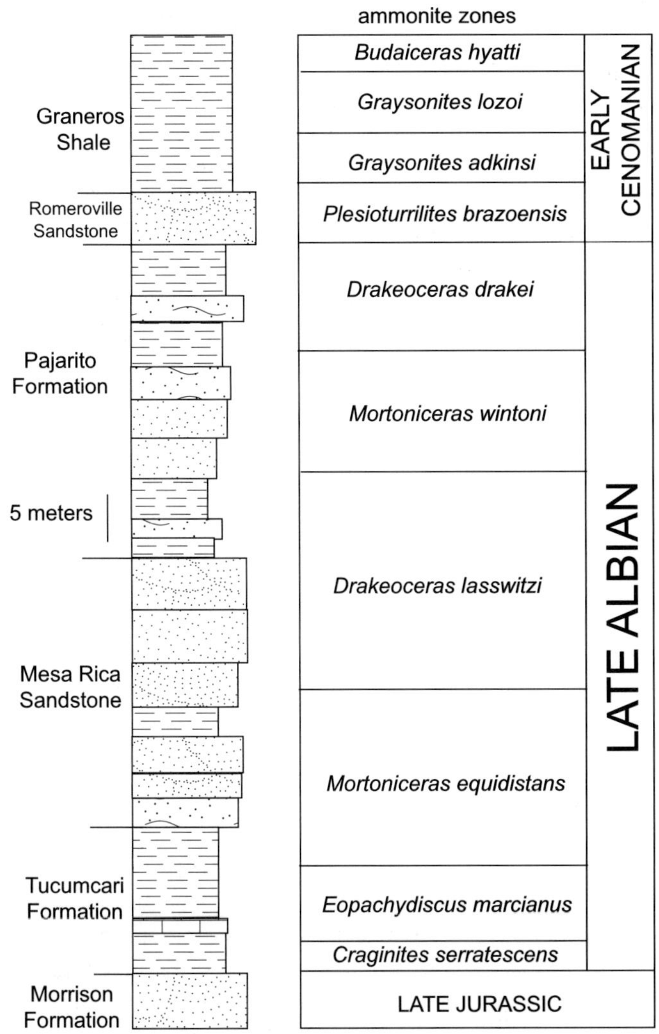

FIGURE 1.17. Summary of the Lower Cretaceous section in east-central New Mexico.

the subtropical Caribbean Province (extreme southern New Mexico, extending into northern Mexico and eastward across Texas).

Here, at San Jon hill, the Tucumcari disconformably overlies the Brushy Basin Member of the Upper Jurassic Morrison Formation, and is about 25 m thick. Only the lower few meters are exposed on the south side of the highway, but the entire Tucumcari Formation is exposed in the tall roadcut on the north side (Fig. 1.16), where it underlies 5-6 m of massive, orange-brown sandstone with late Albian marine fossils representing the basal Mesa Rica Sandstone. The San Jon hill section is probably the most intensively studied exposure of the Tucumcari Formation. Brand and Maddox (1972), Scott (1974) and Lucas and Kisucky (1988) described the stratigraphy of the section, the latter authors establishing it as a primary reference section for the Tucumcari Formation. Scott (1974) studied its sedimentology and faunal associations, and Kues (1997) reported on a distinctive fauna in the basal part of the formation.

Many benthic marine invertebrate fossils typical of the Tucumcari Formation can be observed at this locality. On the low southern roadcut, the lowest 1.5 m contain faunal elements that predate the earliest Tucumcari strata at most other localities in the region, especially the exogyrid oyster *Ceratostreon texanum*. This species is common and characteristic of older Albian strata across Texas and southern Oklahoma, and its presence here is an unusually late occurrence. Two other recently described species, the small ribbed bivalve *Plicatula quayensis* and a small round oyster, *Gyrostrea hinchada*, occur with *Ceratostreon texanum* in the basal Tucumcari Formation. A little higher on the south road cut, beginning about 3 m above the base of the Tucumcari, a more typical fauna appears, including the large oyster *Texigryphaea pitcheri* (formerly known as *T. tucumcarii*, but found to be a synonym of *T. pitcheri*, originally described from Oklahoma: Kues, 1989), abundant specimens of the small, curved, costate oyster *Peilinia levicostata* (identified as *Lopha quadriplicata* in earlier work on the Tucumcari fauna), *Plicatula incongrua*, and the pectinid bivalve *Neithea occidentalis*.

Across the road, in sandier horizons along the western end of the northern road cut, large numbers of *T. pitcheri* have weathered free on the slopes, together with a variety of other bivalves (especially *Peilinia* and *Neithea*) and the high-spired gastropod *Turritella seriatimgranulata*.

Farther upsection, the Tucumcari grades into the basal, massive, orange-brown sandstone of the Mesa Rica Sandstone. These beds contain a nearshore, bivalve-dominated fauna that includes many of the taxa found below in the Tucumcari Formation (see accompanying minipaper). This lower marine facies of the Mesa Rica is absent in most exposures of the formation, including at the type section on Mesa Rica, to the north (Lucas and Kisucky, 1988). Above the basal marine unit, the overlying 15-20 m of the Mesa Rica is regionally a cliff forming, mainly orange-to-brown sandstone deposited in a south- to southeasterly- prograding fluvio-deltaic complex (e.g., Holbrook and Wright, 1992), and lacks marine fossils.

After stop continue north on highway. 0.1

LATE ALBIAN MARINE INVERTEBRATE FAUNA FROM THE BASAL MESA RICA SANDSTONE AT SAN JON HILL, NEW MEXICO

Barry S. Kues

Department of Earth and Planetary Sciences, University of New Mexico, Albuquerque, NM 87131

The basal Mesa Rica Sandstone at San Jon hill is a light brown to orange-brown, 3+ m-thick, cliff-forming, fine- to medium-grained quartz sandstone unit, locally containing small to occasionally large (30 mm) chert pebbles. This massive unit is overlain by about 2 m of softer, slope-forming, brown-to-gray, mottled sandstone. The lower massive unit is locally highly bioturbated, with sparse body fossils, but near the top is a thin horizon that is abundantly fossiliferous, consisting mainly of bivalve steinkerns and some shell material, which continue into the overlying, less well indurated sandstone. These fossils are largely the same taxa as occur in the underlying Tucumcari Formation, and, following Scott (1974), this fauna lived in a mid- to upper shoreface environment, above wave base and close to the shoreline, in shallow marine conditions marked by an unstable sandy substrate. The environment and fauna reflect a shallowing of the Tucumcari sea, and therefore these sandstones were included in the Tucumcari Formation by Scott (1970a, 1974), whereas Griggs and Read (1959), Brand and Maddox (1972) and Lucas and Kisucky (1988) considered them to be better placed within the Mesa Rica Sandstone on lithological grounds. The ammonite *Mortoniceras equidistans* occurs in the basal Mesa Rica at other localities (Kues et al., 1985: Cobban, 1985: Lucas and Hunt, 2000), and the *M. equidistans* range zone extends down into the upper part of the Tucumcari Formation, indicating that there is little or no time missing at the contact of the Tucumcari and Mesa Rica formations. The upper Tucumcari-basal Mesa Rica strata are approximately correlative with the upper Duck Creek Formation of the classic eastern Texas Albian sequence (Kues, 1997).

The fauna of the basal Mesa Rica is generally poorly preserved, consisting of steinkerns of many mollusc taxa, together with shell fragments to occasional complete valves of some bivalves, especially ostreid and gryphaeid taxa. Bivalves dominate this fauna, comprising both the great majority of identifiable taxa, and most of the specimens collected. *Scabrotrigonia emoryi, Protocardia texana, Peilinia levicostata*, and *Pleuriocardia kansasense* are the most abundant bivalves in the basal Mesa Rica collections; *Flaventia belviderensis, Homomya* sp., *Neithea* sp., *Plicatula incongrua*, *Ludbrookia* sp. and *Phelopteria* sp. are uncommon to rare. Gastropods, including *Turritella seriatimgranulata*, are uncommon and typically too poorly preserved to allow accurate identification. Moderate amounts of ostreid shell debris are present, but few specimens of *Texigryphaea* were observed, in contrast to their great abundance in some horizons of the underlying Tucumcari Formation. *Serpula* tubes are sparsely present as epizoans on bivalve shells. Discussion and illustrations of most of these taxa may be found in Kues and Lucas (this volume). In general, this fauna resembles that of the mid-shoreface *Scabrotrigonia-Turritella* association described by Scott (1974), but with few specimens of *Turritella* and considerable numbers of *Peilinia levicostata* (formally *Lopha quadriplicata*), which Scott found to be most abundant in slightly deeper water, together with *Texigryphaea,* in biostromal communities. In the basal Mesa Rica Sandstone, *P. levicostata* is far more abundant than *Texigryphaea*.

Two bivalve genera occur in the basal Mesa Rica Sandstone that have not been reported previously from New Mexico Albian sequences. The first is a small species of exogyrine oyster, with height and length about equal at 28 mm and an articulated width of 15 mm (Fig. 1.18A-B). The left valve is inflated, subcircular to vaguely subtriangular in marginal outline, with a narrow, closely coiled beak; the right valve is flat to slightly concave, and bears a strongly-coiled, depressed beak. Wide, moderately lamellate growth increments characterize the valves; radial ornamentation is lacking. On the best preserved specimen, a very large, shallowly depressed attachment scar some 18 mm in maximum diameter occupies the entire posterodorsal area of the left valve, nearly obliterating the beak. This species is quite different from *Ceratostreon texanum* from the lower Tucumcari Formation, and is unlike any other oyster reported from the Tucumcari Formation, from the Albian sequence at Cerro de Cristo Rey (Böse, 1910), or from the Albian sequences in Texas and Kansas (Stanton 1947; Scott 1970b).

The morphology of the left and right valves of these Mesa Rica specimens closely resembles that of the Cenomanian species *Rhynchostreon levis* (Stephenson), which is known from

FIGURE 1.18. The bivalves *Rhynchostreon* and *Anomia* in the basal Mesa Rica Sandstone at San Jon hill. A, B, *Rhynchostreon* aff. *R. levis* Stephenson: A, left valve showing large attachment pit near beak, UNM 13,129, x 1.25; B, right valve view, same specimen, x 2. C, D, *Anomia* sp.: C, left valve, UNM 13,130, x 2; D, left valve of a small, relatively inflated specimen, UNM 13,132, x 2.

middle to upper Cenomanian strata in Texas (Woodbine Formation: Stephenson, 1952), New Mexico (intertongued Dakota-Mancos sequence: Cobban, 1977), the Black Mesa region of Arizona (upper Dakota Sandstone: Kirkland, 1996), and equivalent strata in Utah and Montana (Cobban, 1977). The Mesa Rica specimens differ from Cenomanian specimens of *R. levis* in having a pronounced attachment scar (Cenomanian specimens lack significant attachment scars), and in possessing a more strongly coiled right valve. It is doubtful that the Mesa Rica specimens are conspecific with the Cenomanian species *R. levis*, but they appear to be a closely related predecessor of that species, and are here referred to *R.* aff. *R. levis*.

Small specimens of the thin-shelled pterioid bivalve *Anomia* also occur in the basal Mesa Rica Sandstone at San Jon hill, and are reported here for the first time from the Albian of New Mexico. Only left valves are preserved, rarely completely and are generally associated with their underlying steinkern. These valves are subcircular in shape, moderately inflated and possess a slightly convex umbo that ends in a small beak that does not extend to the hingeline. The valves lack radial striae and comarginal wrinkles such as characterize some species of Cretaceous *Anomia*; ornamentation is limited to fine growth lamellae. A typical left valve is about 14 mm long, 12.5 mm high, and 3 mm wide (Fig. 1.18C). Some smaller specimens (e.g., Fig. 1.18D) are more inflated, but this may represent ecological variability (correlated with shape of substrate) rather than having taxonomic significance.

Several Cenomanian to Turonian species of *Anomia* have been described from the Texas-Southern Western Interior region (e.g., Stephenson, 1952; Hasenmueller and Hattin, 1990; Kirkland, 1996), but Albian reports of *Anomia* are rare. Hill (1893) described *A. texana* from the Trinity Group of Texas, and Scott (1970b) described an unnamed species from the Kiowa Formation of Kansas. Species of *Anomia* tend to display high intraspecific variability and are separated on the basis of rather subtle morphological features (Hasenmueller and Hattin, 1990). Preservation of the Mesa Rica specimens is not sufficient to allow identification to species, but these specimens do not differ significantly from the Kiowa specimens.

Anomia lived attached by its right valve to hard substrates, including the shells of other organisms. Turonian species described by Hasenmueller and Hattin (1990) were inferred to have been attached to the shells of the ammonoid *Baculites*, but there is no indication what the host substrate of the Mesa Rica specimens might have been. The survival of thin *Anomia* valves in the relatively high energy conditions in which the basal Mesa Rica Sandstone was deposited is somewhat surprising; most other elements of the Mesa Rica bivalve fauna have strong, thick shells, and even these were typically abraded and broken.

254.4 Upper Jurassic Morrison Formation/Middle Jurassic Entrada Sandstone contact. **0.1**

254.5 Approximate position of Entrada Sandstone/Upper Triassic Redonda Formation contact. **0.4**

254.9 Prior to a highway widening project an outstanding cross-section of a Gilbert delta in the Upper Triassic Redonda Formation was exposed in the roadcut to the left (Hester, 1988). The quarry that produced the sandstone for the Wheatland School was in this vicinity. **0.3**

255.2 At 9:00, the Redonda Formation dips north in the Bonita fault zone, which means we are passing over the northwestern edge of the buried Frio uplift (see accompanying minipaper). **7.4**

THE FRIO UPLIFT: A PALEOZOIC-MESOZOIC CONTROL ON SEDIMENTATION PATTERNS IN EAST-CENTRAL NEW MEXICO

Adrian P. Hunt[1] and Spencer G. Lucas[2]
[1]Mesalands Dinosaur Museum, Mesa Technical College, 911 South Tenth Street, Tucumcari, NM 88401; [2]New Mexico Museum of Natural History, 1801 Mountain Road NW, Albuquerque, NM 87104

The Frio uplift is a structural positive element in eastern New Mexico that affected sedimentation patterns from the late Paleozoic until at least the middle Mesozoic. Here, we briefly review its influence from the Desmoinesian (Pennsylvanian) through the Kimmeridgian (Jurassic).

The Frio uplift is a northeast-trending structure that is located approximately along the southeastern Quay County line in east-central New Mexico (Broadhead and King, 1988, fig. 1). It separates the Tucumcari basin to the west from the Palo Duro basin to the east. During the Atokan and Desmoinesian (Pennsylvanian), a north/northeast-trending strip of coarse sandstone delimited the western edge of the emergent Frio uplift and separated it from marine limestones to the west (Broadhead and King, 1988, figs. 14a, b). Marine limestones of the overlying Missourian strata onlapped the Frio uplift and are absent over it (Broadhead and King, 1988, figs. 18a, b, 22a, b).

Early Permian (Wolfcampian) deposition began with the Hueco Formation. Throughout Hueco deposition, stratigraphic units thinned over the emergent Frio uplift (Broadhead and King, 1988, figs. 24a, 26a). There are also facies changes across the uplift during this time. In the lower unit of the Hueco, marine limestones grade into dolostones over the uplift, and in the upper Hueco, coarse sandstones change to marine limestones from northwest to southeast (Broadhead and King, 1988, figs. 24b, 26b). The local fine-grained nature of the Yeso Formation and the absence of thick clastic wedges indicates that the Frio uplift was tectonically inactive by the Leonardian (Early Permian). Similarly, isopachs of the Guadalupian Artesia Group show no influence of a Frio uplift (Broadhead, 1984, fig. 7)

During the Triassic, the Frio uplift was resurgent again. Isopachs of the Middle Triassic Anton Chico Member of the Moenkopi Formation and Upper Triassic Tecolotito Member of the Santa Rosa Formation by Broadhead (1984, fig. 8)--his lower sandstone unit of the Santa Rosa Sandstone--clearly demonstrate thinning over the Frio uplift. The uppermost, Tres Lagunas Member of the Santa Rosa Formation--Broadhead's (1984) upper sandstone member of the Santa Rosa Sandstone--actually pinches out over the Frio uplift (Broadhead, 1984, fig. 9). This explains the absence of the Tres Lagunas Member or any homotaxial equivalent in the Chinle Group in West Texas (Lucas et al., 1994). Between the deposition of the Tecolotito and Tres Lagunas Members in the late Carnian, ispoachs indicate that the geographic expression of the uplift was expanded.

In Norian time, there is a marked thinning of the Trujillo Formation--Broadhead's (1984) Cuervo Sandstone Member of the Chinle Formation--over the Frio uplift (Broadhead, 1984, fig. 10). The isopachs of the Trujillo Formation are more complex, but suggest that the Frio uplift had expanded in size in the early Norian, with two large areas of zero thickness in northern Curry County and southeastern Quay County (Broadhead, 1984, fig. 10). The uppermost unit of the Chinle Group in this area is the late Norian-Rhaetian Redonda Formation, which represents a large lacustrine basin (Hester, 1988). The only subaerial shoreline, deltaic and fluvial facies of the Redonda are in southeastern Quay County, indicating that the Frio uplift was still influencing sedimentation.

The Upper Jurassic sequence exposed in southeastern Quay County (Apache Canyon and San Jon Hill) exhibits a very thin Upper Jurassic Morrison Formation (< 10 m) and the absence of the Middle Jurassic Summerville Formation. This indicates onlap of these units on the Frio uplift during the Callovian-Kimmeridgian. Thinning isopachs of the Cretaceous Mesa Rica Sandstone (Mateer, 1985, fig. 3) could also be interpreted to suggest that the Frio uplift was a positive element as late as the Albian. In conclusion, it is clear that the Frio uplift had a significant impact on sedimentation patterns during the Pennsylvanian (Desmoinesian) to Permian (Wolfcampian), and the Triassic (Anisian) to Jurassic (Kimmeridgian), with a period of quiescence between the two.

262.6 Enter San Jon. Like Tucumcari, San Jon owes its origins to the railroad. The town was named in 1906 by the first postmaster, although there had been dwellings in the area since 1902. "San Jon" is meaningless in Spanish, and Julyan (1996) cited the conventional explanation that the word is a corruption of the Spanish *zanjon* or deep gully. However, elsewhere San Jon is a corruption of San Juan, so the origin of the name is problematic. The town incorporated in 1946; its 1990 population of 277 was down 33% from its height in 1960. Be aware of speed traps at speed limit changes. **0.4**

263.0 **Turn left at stop sign and continue west on old U.S. 66**. Tucumcari Mountain ahead in distance. **3.2**

266.2 Road crosses below overpass. **1.2**

267.4 Cross tributary to Revuelto Creek with exposures of Bull Canyon Formation in badlands to left and right. **0.7**

268.1 Reddish, Triassic-derived alluvium in low roadcut to left. **2.0**

270.1 Crest low ridge. Mesa Redonda at 9:00-9:30, Bulldog Mesa at 10:00, Tucumcari Mountain at 10:30. Note extensive badlands of Bull Canyon Formation in the Revuelto Creek drainage exposed in front of Mesa Redonda. **0.3**

270.4 Note outcrops of Upper Triassic Trujillo Formation in bottom of Revuelto Creek to left. **0.5**

270.9 Roadcut on right is red mudstone under uppermost Trujillo Formation sandstone. Note paleoflow indicators (trough crossbeds) are to west-northwest in these Triassic sandstones (see accompanying minipaper). **0.3**

PALEOCURRENTS IN THE TRUJILLO FORMATION (CHINLE GROUP; UPPER TRIASSIC), EAST-CENTRAL NEW MEXICO, AND THE MYTH OF THE DOCKUM LAKE

Adrian P. Hunt[1], Spencer G. Lucas[2] and Andrew B. Heckert[3]

[1]Mesalands Dinosaur Museum, Mesa Technical College, 911 South Tenth Street, Tucumcari, NM 88401; [2]New Mexico Museum of Natural History and Science, 1801 Mountain Road NW, Albuquerque, NM 87104-1375; [3]Department of Earth and Planetary Sciences, University of New Mexico, Albuquerque, NM 87131

During the late 1970s and 1980s, geologists from the Bureau of Economic Geology in Austin, Texas published a series of studies that suggested the Upper Triassic strata of the Chinle Group in eastern New Mexico and West Texas were deposited in large lakes fed by extensive river deltas (McGowan, 1979, 1980; McGowan et al., 1979, 1983; Granata, 1981; Johns and Granata, 1987; Johns, 1989). These studies were based primarily on subsurface data, but they also included limited outcrop studies. Virtually all strata of the lower Chinle (Trujillo Formation and below) were interpreted as part of a vast lake system. Mudrocks were interpreted as lacustrine, and sandstones as deltaic.

As a local test of this model, we measured 62 paleocurrents from crossbeds in the Trujillo Formation in Quay County, New Mexico, near the confluence of Revuelto and Plaza Larga Creeks (secs. 28-29 T11N, R33E). These measurements indicate a dominantly northwesterly paleoflow (Fig. 1.19), which is inconsistent with the model of McGowen and others. Similar paleocurrents were obtained from a smaller sample by Deluca and Eriksson (1989) from the Trujillo Formation in northeastern New Mexico. Clearly, Trujillo rivers in east-central and northeastern New Mexico were not flowing into lakes in West Texas.

Lower in the Upper Triassic section, Riggs et al. (1996) used age groups of detrital zircons to demonstrate through-flow of a fluvial system from highlands south and east of the Dockum across the Colorado Plateau and into an ocean in the vicinity of modern-day Nevada. Furthermore, several authors have reported paleoflow directions in West Texas from near the posited depocenter of the lacustrine basin, and these too indicate that rivers were flowing to the north and northwest (e.g., Cazeau, 1960;

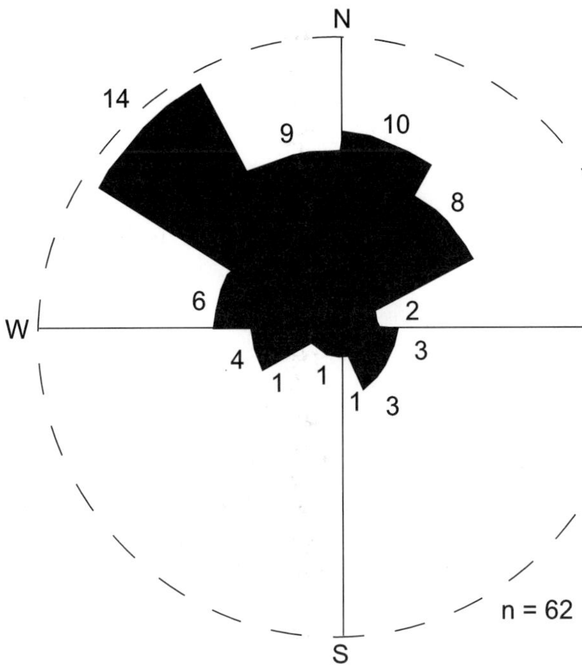

FIGURE 1.19. Rose diagram of Trujillo Formation crossbed azimuths near the confluence of Revuelto and Plaza Larga Creeks, Quay County.

Kiatta, 1960; Asquith and Cramer, 1975; Frelier, 1987; Lupe, 1988; May, 1988). Much of the case for the "Dockum Lake" was based on analysis of the orientation of sandstone bodies in the subsurface of West Texas. Newell (1993, fig. 9) was able to demonstrate that if these are plotted on a rose diagram they are consistent with a northwest flow direction.

Johns (1989) argued that east-west sections of subsurface lithofacies fence diagrams demonstrated that interbedded sandstone-mudrock "deltaic facies" could be seen to laterally grade into "lacustrine" mudstone. Newell (1993) noted that these could just as easily be interpreted as levee deposits grading into floodplain mudrocks associated with northward flowing streams. Indeed, detailed outcrop study of the Tecovas and Trujillo formations in West Texas has demonstrated that they are dominantly fluvial in origin (Asquith and Cramer, 1975; Frelier, 1987; May, 1988). Thus, the Dockum lake model is clearly contradicted by all paleocurrent data and by recent sedimentological studies.

271.2 Old Highway 66 gasoline station. **0.2**

271.4 Excellent Trujillo Formation sandstones in roadcut on right. **0.3**

271.7 Cross Revuelto Creek bridge; more Trujillo sandstones to left. **0.6**

272.3 Bridge over Plaza Larga Creek. **1.4**

273.7 Trees on left mark an abandoned picnic area along Old Highway 66. **2.3**

276.0 Cross creek with Trujillo Formation outcrops here. These outcrops are locally fossiliferous. The Trujillo Formation in this part of Quay County yields vertebrate fossils as well as a large concentration of unionid bivalves. **2.2**

278.2 Pass beneath overpass. **2.0**

280.2 Junction with NM 278. Continue straight. **0.3**

280.5 Junction with NM 286. Continue straight. **3.2**

283.7 Junction with Tucumcari Boulevard. Excellent view of Tucumcari Mountain, just to south. **At yield sign turn right to go under Interstate 40** and enter Tucumcari. **1.4**

285.1 Tucumcari Lake and Tucumcari historical marker on right. The origin and age of Tucumcari Lake are the subject of some debate (Dobrovolny and Townsend, 1946; Trauger and Bushman, 1964; Love, 1985). The lake probably formed as a result of dissolution of Permian evaporites and it is at least Quaternary in age. **1.7**

286.8 **Turn right** at light at First Street (NM-104). **0.1**

286.9 **Turn right** at Laughlin and stop at Mesalands Dinosaur Museum.

End of First-Day Road log.

At Palo Duro Canyon in West Texas, Upper Triassic strata of the Chinle Group rest with profound disconformity on Upper Permian strata of the Quartermaster Formation. A white bed of pebbly sandstone marks this hiatus of approximately 25 million years.

SECOND-DAY ROAD LOG, FROM TUCUMCARI TO MESA REDONDA, PYRAMID MOUNTAIN, RAGLAND, AND BLACKWATER DRAW, NEW MEXICO

SPENCER G. LUCAS, ADRIAN P. HUNT, ANDREW B. HECKERT, BARRY S. KUES, and VIRGINIA T. MCLEMORE

FRIDAY, SEPTEMBER 28, 2001

Assembly point: Mesalands Dinosaur Museum, 222 East Laughlin, Tucumcari
Departure time: 7:30 AM
Distance: 229.5 miles
Stops: 5

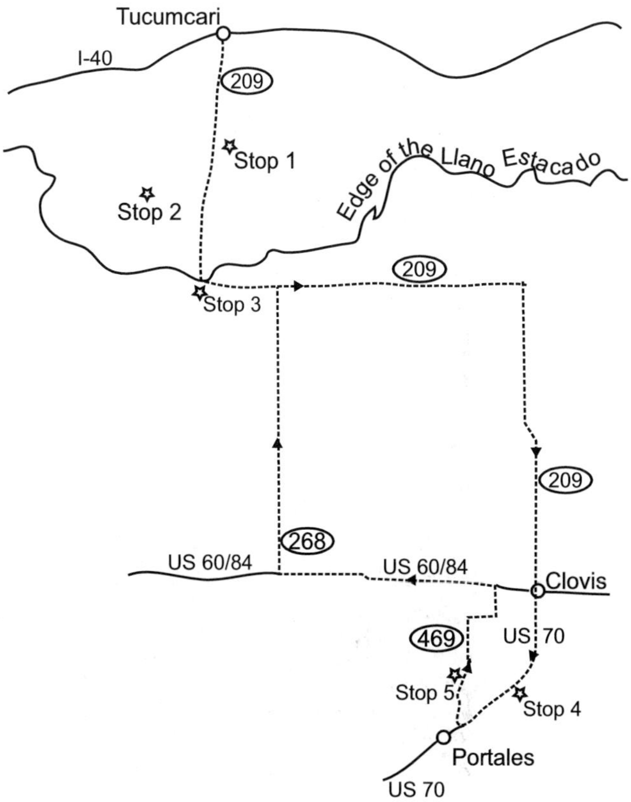

SUMMARY

Today's trip begins by examining Mesozoic strata exposed around erosional outliers of the northern edge of the Llano Estacado in Quay County, New Mexico. STOP 1 at Mesa Redonda is at the type section of the Redonda Formation, the youngest Upper Triassic strata beneath the Llano Estacado. Here, we discuss the base of the Redonda Formation, a possible regional unconformity, and the depositional environment of the formation.

STOP 2 is at Pyramid Mountain, where, in September 1853, Jules Marcou undertook the first detailed scientific observations made by a trained geologist in New Mexico. Here, we examine the Triassic-Cretaceous section. STOP 3 at Ragland affords an excellent exposure of the Neogene Ogallala Formation. STOPS 4 and 5 are at the Blackwater Draw Museum and Blackwater Locality No. 1, respectively. Here, we examine the type Clovis locality, its stratigraphy, deposition and dating.

Mileage

0.0 Assemble at the Mesalands Dinosaur Museum, 222 East Laughlin, Tucumcari. **Turn left out of museum parking lot onto S. Adams and immediately turn left (west) onto East Laughlin. 0.1**

0.1 Stop sign. **Turn left** (south) on First and proceed south. **0.1**

0.2 Go straight through traffic light with Tucumcari Boulevard and continue south out of town. **1.4**

1.6 Pass under bridge of Interstate 40. Tucumcari Mountain at 10:00. **Continue south (straight) on NM-209. 0.2**

1.8 Brown sandstone on both sides of the road is Lower Cretaceous Mesa Rica Sandstone; road surface is on Upper Jurassic Brushy Basin Member of the Morrison Formation.

The Mesa Rica Sandstone that is close to road level forms the highest cliff on Tucumcari Mountain about 1.3 mi to the E, so there is obviously a significant structure between the road and Tucumcari Mountain. Dobrovolny et al. (1946) mapped a westward dip of 11° in the Bull Canyon Formation about 1 mi east of the highway. Therefore, the structure must be northerly in strike and located about a third of the way from the western boundary of sec. 36 T11N, R30E. Dobrovolny et al. (1946) mapped no fault, so by implication they must have assumed that this was the limb of a monocline. Trauger et al. (1972) noted a "sharp flexure" here that dropped the Chinle Group about 500 ft. This structure continues under the city of Tucumcari because there are outcrops of the Morrison Formation, not mapped by Dobrovolny et al. (1946), in the western portion of the city near Tucumcari Elementary School, whereas the Chinle Group is exposed in the eastern area near Tucumcari Lake. Given the gentle dips of all folds in the area (Dobrovolny et al., 1946) and the presence of the southwest-northeast trending Bonita fault southeast of Tucumcari (Stearns, 1972), it is most likely that the structure on the western side of Tucumcari Mountain is a fault. **0.4**

2.2 Historical Marker on left for the Llano Estacado; good view of western side of Tucumcari Mountain. **0.3**

2.5 Green mudrocks on right are Upper Jurassic Brushy Basin Member of the Morrison Formation. The majority of the Morrison Formation in this area consists of green claystone with minor channel-form sandstone. In Quay County, dinosaur bones are relatively common in the Morrison, but the vast majority are isolated and fragmentary. **0.3**

2.8 Bulldog Mesa at 10:00. Most of the mesa is red beds of the upper part of the Chinle Group (Bull Canyon and Redonda formations); cliff of Middle Jurassic Entrada Sandstone is visible near its top. These units are distinctly lower on Bulldog Mesa than on either Tucumcari Mountain or Mesa Redonda as a result of a large Laramide fault that underlies Plaza Larga Creek between Bulldog and Redonda mesas (Dolliver, 1985). On right, sandstones of the Lower Cretaceous Mesa Rica Formation cap cuestas in the Tucumcari structural zone. **1.4**

4.2 Crest hill, Lower Cretaceous Mesa Rica Sandstone faulted on right of road. Mesa Redonda visible from 10:00-11:00. **1.2**

5.4 Vegetation-covered Quaternary dunes(?) visible at 3:00. **0.6**

6.0 Edge of Llano Estacado in distance at 9:00-10:00. Hills at 2:30 are Lower Cretaceous Tucumcari Formation and Mesa Rica Sandstone over Upper Jurassic Morrison Formation. The latter has yielded bones of sauropod dinosaurs in this area. **1.3**

7.3 Crest low ridge. From 1:00 to 2:00 are Saddleback Mesa, Circle S Mesa, Pyramid Mountain and edge of Mesa Quemado. **1.2**

8.5 Road to Briscoe Ranch to right. Mesa to right is informally called Section 29 Mesa and includes Triassic through Cretaceous strata of the Redonda through Mesa Rica formations. **1.0**

9.5 Quay County Road QR-54 to right, which skirts the southern edge of Mesa Quemado. Kues and Kietzke (1985) provided a log of this road, to reach Pyramid Mountain, but the road has since been abandoned and is now impassible. **0.3**

9.8 Cross tributary of Plaza Larga Creek. **0.7**

10.5 **Turn left** on unpaved QR-53, stop to open gate; now entering private land owned by Larry and Virginia Wright. **0.5**

11.0 Road forks—**turn right on lesser road** and cross cattleguard and gate. **0.4**

11.4 Good overview of Mesa Redonda ahead. Stratigraphic succession is (in ascending order) Upper Triassic Chinle Group, Middle Jurassic Entrada Sandstone, Upper Jurassic Morrison Formation, Lower Cretaceous Tucumcari Formation and Lower Cretaceous Mesa Rica Sandstone. **1.6**

13.0 Type section of the Upper Triassic Redonda Formation (Chinle Group) overlain by Middle Jurassic Entrada Sandstone at 12:00. **0.3**

13.3 **Stay left** as road enters from the right. **0.1**

13.4 **STOP 1.** This is the type section of the Redonda Formation (Fig. 2.1) and also affords us a broad overview of many of the erosional outliers of the northern edge of the Llano Estacado in Quay County (Fig. 2.2). The Redonda Formation was named by Dobrovolny et al. (1946) for the widespread and distinctive lithosome that represents the youngest Triassic strata of the Tucumcari basin in east-central New Mexico (Lucas and Hunt, 1989). The Redonda is as much as 92 m thick and rests unconformably on the early to middle Norian Bull Canyon Formation. The top of the formation is an erosional unconformity at the base of the overlying Middle Jurassic Entrada Sandstone or the Neogene Ogallala Formation.

Here, the Redonda Formation consists of mudstone, fine-grained sandstone, limestone, siltstone and a single bed of conglomerate (Fig. 2.1). The laterally perisistent beds of sandstone and siltstone, the moderate reddish brown color and the general lack of smecitic mudstone distinguish the Redonda Formation from the underlying Upper Triassic Bull Canyon Formation. The base of the Redonda Formation here is a thin (~1-6 m), mottled, burrowed-to-structureless, calcareous siltstone that is laterally continuous for several mi. Hester (1988) interpreted this to represent lacustrine precipitation of carbonate mud and subsequent bioturbation of a soupy, unconsolidated substrate. Overlying Redonda strata are massive mudstones and sandstones that form laterally continuous repetitive beds, followed by ripple-laminated micrites. The upper interval of the formation is mostly massive mudstones and sandstones.

The Redonda Formation pertains to the youngest of three Chinle Group depositional sequences,

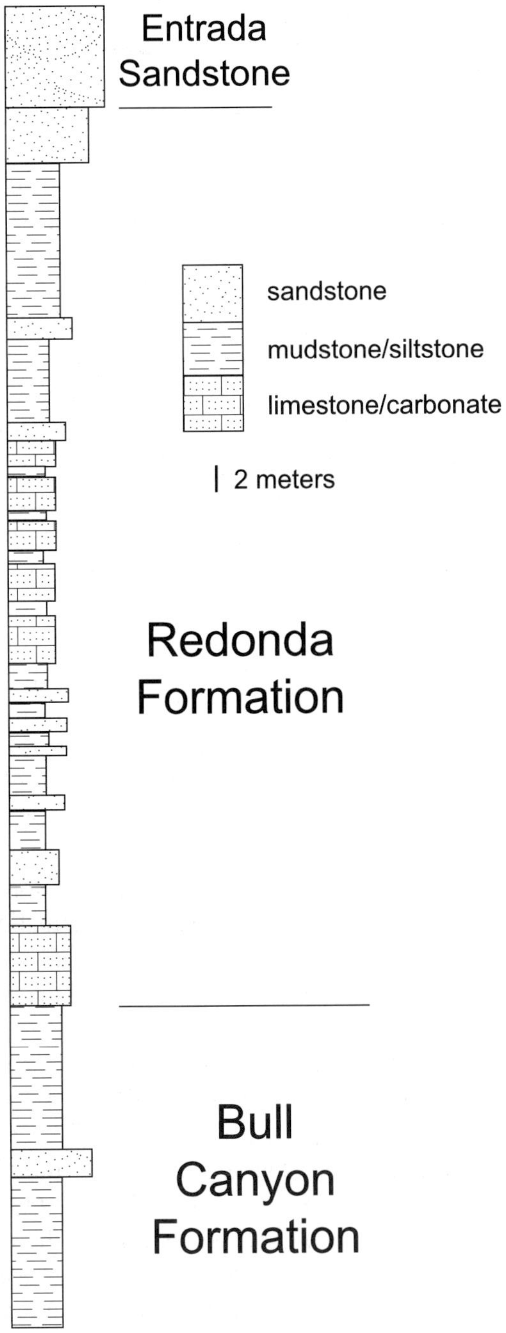

FIGURE 2.1. Stratigraphic section of the Upper Triassic Redonda Formation (type section) at Stop 1 (modified from Hester, 1988).

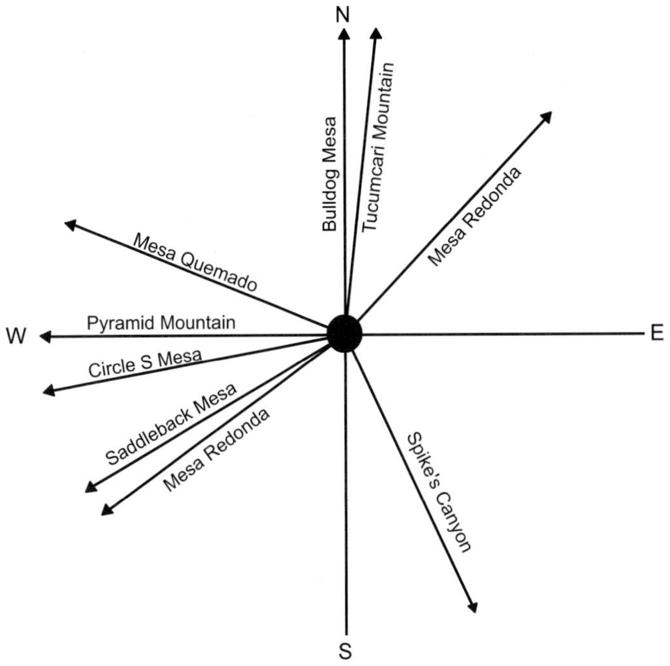

FIGURE 2.2. Vista diagram at Stop 1.

the late Norian-Rhaetian Rock Point sequence of Lucas (1991, 1993), and is correlative with the Rock Point Formation on the Colorado Plateau, the Bell Springs Formation in Wyoming, and the Travesser and Sloan Canyon formations in the Dry Cimarron Valley of northeastern New Mexico, westernmost Oklahoma and southeastern Colorado. Most of these units contain significant, laterally extensive lacustrine deposits whose origin is related to fluctuating intrabasinal hydrologic gradients punctuated by seasonal, monsoonal paleoclimatic circulation patterns (cf. Dubiel et al., 1991).

Long term climatic influence over Redonda Formation sedimentation and facies distribution may be indicated by the laterally-continuous, cyclic, ripple-laminated micrites that comprise the middle portion of the formation (Fig. 2.1). Two orders of cyclicity are expressed by these strata that may correspond with the 21 Ka precessional and the 100 Ka eccentricity wavelengths of the Milankovitch spectrum. Note that four, well-defined 100 Ka cycles appear to modulate and form a single, 400 Ka eccentricity cycle, and this hierarchy of periodicities is similar to that expressed by transgressive-regressive lacustrine cycles ("van Houten cycles") of the Lockatong and Passaic formations of the Newark basin as defined by Olsen (1984, 1986) and Olsen and Kent (1996). However, such an interpretation is equivocal at best, because the maximum thickness of the cyclic interval is only ~50 m, and hence, not amenable to Fourier and the other component analyses needed to establish a cyclostratigraphic framework.

After stop turn around and retrace route to NM-209. 0.1

13.5 Bear right. **0.3**

13.8 Bulldog Mesa and Tucumcari Mountain at 12:00. **2.0**

15.8 Cross cattleguard, gate, **turn left. 0.5**

16.3 **Turn left** onto NM-209 southbound. **1.1**

17.4 Cross tributary to Plaza Larga Creek. **0.6**

18.0 Type section of Upper Triassic Redonda Formation discussed at Stop 1 was measured on point at 9:00. **0.9**

18.9 Historical marker for the notorious outlaw Black Jack Ketchum on left, Quay County road 50.4. This year is the centenary of Black Jack's execution in Clayton. The hanging was botched, and Black Jack's head was ripped from his body. The whole event was captured on film, and the gory event can be revisited at the Herzstein Museum in Clayton. **0.6**

19.5 Milepost 71. **0.3**

19.8 Cross creek. **1.7**

21.5 **Turn right** on Quay County road 48, house on right. Supplemental roadlog 3 of the 1985 NMGS Guidebook (Kues and Kietzke, 1985) ends here. **0.1**

21.6 House on right. **0.7**

22.3 Road makes sharp turn to right. **1.0**

23.3 Road makes sharp turn to left. **0.3**

23.6 Ranch houses on right. **0.7**

24.3 Abandoned ranch house on right. **0.5**

24.8 Road turns hard left. Cross cattleguard. **0.4**

25.2 Cross creek to ranch house (on left), pass house. Private land from the ranch house onward belongs to Aaron S. Parker and is managed by Rick Thompson. **0.1**

25.3 **Stay on road to right. 0.2**

25.5 Road veers left, crosses cattleguard. **0.2**

25.7 Cross cattleguard. **0.2**

25.9 Holding pens on right, **bear left. 0.1**

26.0 Stock pond. **Turn left toward gate. 0.1**

26.1 Pass through gate. **0.2**

26.3 Road enters from right. **Go left. 0.4**

26.7 Road forks, **stay left**. Circle S Mesa dead ahead, Pyramid Mountain at 1:00. **0.5**

27.2 Road to right, **continue straight. 0.3**

27.5 Cattle tank to left. **0.4**

27.9 Gate. **0.3**

28.2 Road forks, note rare, solar-powered well ahead (Fig. 2.3), **go left** (south). **0.3**

28.5 Road forks, **continue right. 0.6**

29.1 Corrals, pass through gate, road turns sharply to north. **0.7**

29.8 Cattle feeder on right, begin descent to Adobe Walls creek. This is not to be confused with the site of the Battle of Adobe Walls, which is near Stinnett in Hutchinson County, Texas. **0.2**

30.0 Road bends to right. Windmill at 8:00. **0.2**

30.2 Cross Adobe Walls creek. **0.2**

FIGURE 2.3. Solar-powered well at mile 28.2.

30.4 Road enters from left, **bear right. 0.5**

30.9 Road forks, **stay left**. Note Laramide fault between Pyramid Mountain and Circle S Mesa. **0.1**

31.0 Cattle feeder and stock tank on left. **0.1**

31.1 Road forks, **stay left**. **0.3**

31.4 Stock pond. **0.2**

31.6 Gate. **0.3**

31.9 Tank on left. **0.4**

32.3 Good view of north face of Pyramid Mountain (Fig. 2.4) displays red beds of Upper Triassic Redonda Formation of Chinle Group, overlain by yellow sandstone cliffs of Jurassic Entrada and Morrison formations capped by dome of Lower Cretaceous Tucumcari and Mesa Rica formations. **0.1**

32.4 Road curves left. **0.6**

33.0 Stock tank. **STOP 2** at north face of Pyramid Mountain (Fig. 2.4). Pyramid Mountain occupies an important place in the history of New Mexico geology. Here, on 22 September 1853, Jules Marcou, geologist of the Whipple Expedition, produced the first stratigraphic section of a sedimentary sequence in New Mexico (Fig. 2.5) and collected bivalve fossils that would later be described as *Gryphaea* (now *Texigryphaea) tucumcarii*, one of the first new species of fossil described from New Mexico (Fig. 2.6).

Marcou recognized that the basal red beds exposed here are Upper Triassic, and assigned all the overlying strata to the Jurassic, based on his belief that *G. tucumcarii* was closely related to European Jurassic gryphaeids (Kues, 1985). We now know that the thick, yellowish, cliff-forming sandstones along the face of Pyramid Mountain are

FIGURE 2.4. Two photos of Pyramid Mountain. A, From the north. B, From the west. Stratigraphic units are: 1 = Triassic Redonda Formation, 2 = Jurassic Entrada Sandstone and Morrison Formation, 3 = Cretaceous Tucumcari Formation and Mesa Rica Sandstone.

indeed Jurassic (Entrada and Morrison Formations), but that the fossiliferous shales (Tucumcari Formation) and the thick sandstone cap of the peak (Mesa Rica Sandstone) are of Early Cretaceous age.

Marcou's identification of Jurassic strata in New Mexico was his most controversial conclusion (DeFord, 1972; Kues, 1985) because Marcou assigned all of the strata defending the Llano Estacado and its erosional outliers a Jurassic age. But, the fossils Marcou collected are not true *Gryphaea* and instead belong to a closely related Early Cretaceous taxon, now called *Texigryphaea*. Thus, the shale strata at Pyramid Mountain are actually of Cretaceous age, and Marcou's error engendered a lengthy debate that was never resolved in his lifetime (DeFord, 1972; Lurie, 1974; Kues, 1985). Marcou's assignment of the underlying rocks to the Triassic lacked a paleontological basis; instead, stratigraphic position and similarity to the nonmarine Keuper (Upper Triassic) strata of Europe pro-

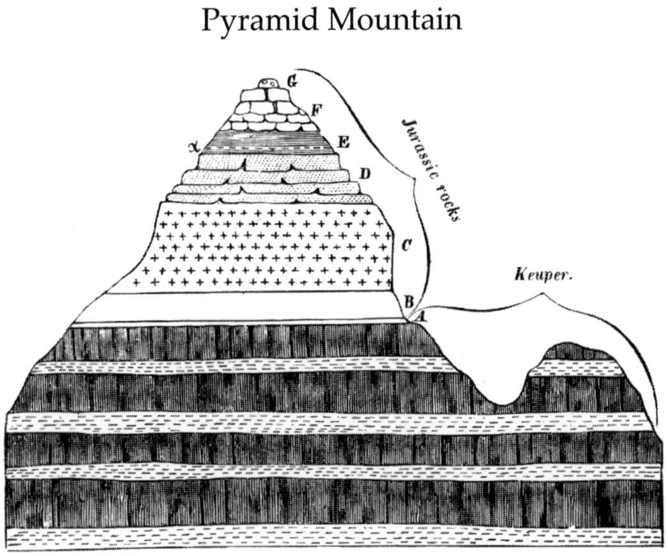

FIGURE 2.5. Marcou's (1858) drawing of Pyramid Mountain.

vided the principal basis for Marcou's determination. It is possible to directly relate the strata in Marcou's drawing of Pyramid Mountain (Fig. 2.5) to a measured section (Fig. 2.7).

After stop, begin retracing route to NM-209. 0.6

33.6 **Turn right**. (Do not go on fork straight ahead). **0.7**
34.3 Gate. **1.3**
35.6 **Stay left** on main road. **0.2**
35.8 Cross Adobe Walls creek. **0.2**
36.0 Road forks, **go left. 0.2**
36.2 Road forks, **go right. 0.7**
36.9 Gate. **0.6**
37.5 **Go left, then go right. 0.5**
38.0 Gate. **0.6**
38.6 Road to left, **continue to right. 0.6**
39.2 Road to left, **continue right. 0.3**
39.5 Road to left, **continue right. 0.2**
39.7 Road to left, **go right. 0.2**
39.9 Gate, then road curves to right. **0.3**
40.2 Cattleguard. **0.2**
40.4 Cattleguard. **0.4**
40.8 Ranch house.**0.4**
41.2 Cattleguard. **Turn right. 1.5**
42.7 Road makes sharp turn to right. **1.0**
43.7 Road makes sharp turn to left. **0.8**
44.5 Stop sign. **Turn right** and go south on NM-209. **1.0**
45.5 Quay Cemetery to right. **0.3**
45.8 Cross creek. **0.1**
45.9 Enter greater Quay. Edge of the Llano Estacado visible ahead. **0.6**
46.5 "Downtown" Quay. Virtually a ghost town now,

FIGURE 2.6. Copy of Marcou's (1858) plate 4, showing specimens designated as *Gryphaea tucumcarii* (fig. 3) and *Ostrea marshi* (now *Lopha subovata*: fig. 4), collected from the Tucumcari Formation at Pyramid Mountain.

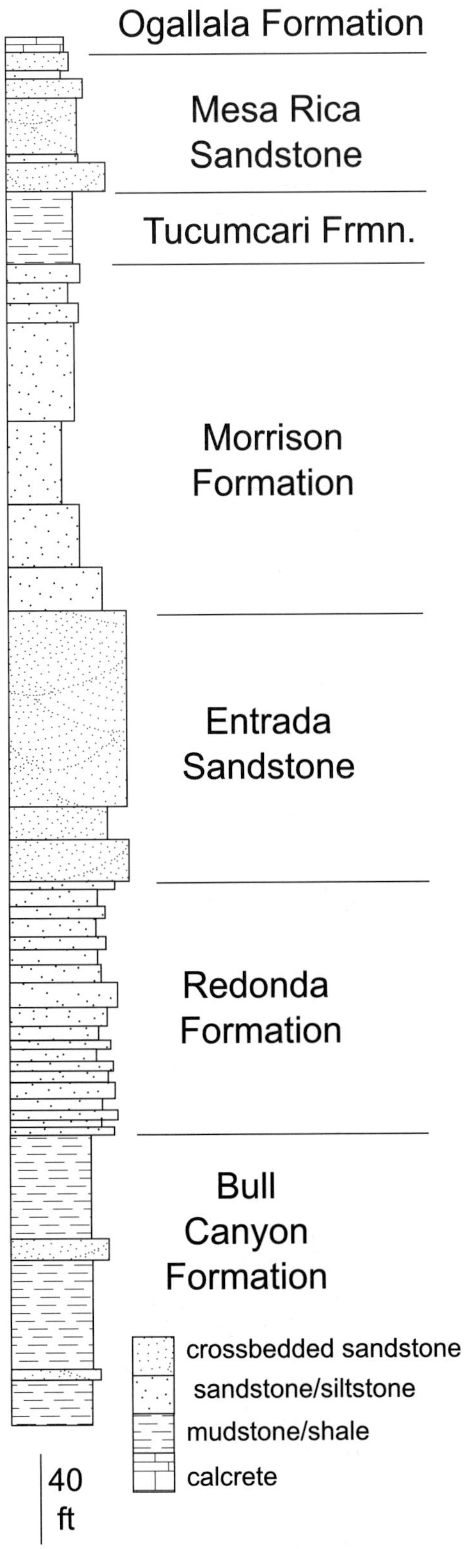

FIGURE 2.7. Measured stratigraphic section at Pyramid Mountain.

Quay was settled in 1902 and named (as is the county) for Pennsylvania Senator Matthew Quay, a supporter of New Mexico statehood. **0.2**

46.7 Leave Quay. **1.0**

47.7 Red shale pit to left, QR-45 to right. **3.2**

FIGURE 2.8. Ogallala Formation overlying Triassic Redonda Formation on edge of Llano Estacado at mile 50.9.

50.9 Neogene Ogallala Formation overlies Upper Triassic Redonda Formation on left. (Fig. 2.8). The Jurassic and Cretaceous strata that topped Tucumcari Mountain, Mesa Redonda and Pyramid Mountain are no longer present here, having been removed by Cenozoic erosion prior to Ogallala deposition. However, 14 mi to the east, the entire sequence is preserved in the Bonita fault zone from near Apache Canyon to the San Jon Hill area (Stop 4 of Day 1). **0.4**

51.3 More of the same. **1.5**

52.8 Begin ascent of escarpment of Llano Estacado. **0.7**

53.5 Picnic table to left. **0.2**

53.7 Roadcut of the Upper Triassic Redonda Formation on the right. **0.1**

53.8 Neogene Ogallala Formation overlies Upper Triassic Redonda Formation on right. **Prepare to stop. 0.2.**

54.0 **Pull over on right side of the road. STOP 3.** The Neogene Ogallala Formation section here at Ragland (Figs. 2.9-2.10) has been described by Evans et al. (1949), Dolliver (1984) and Gustavson (1990, 1996). Here, the Ogallala is about 25 m thick and rests on the Upper Triassic Redonda Formation. Its lower 1.5 m are a conglomerate of well-rounded clasts of volcanic rocks, quartzite and other fine-grained metamorphics. Clasts are imbricated, and the horizontally bedded conglomerate consists of at least three, fining-upward sequences.

The basal conglomerate is overlain by pebbly sandstone and siliceous conglomerate with low angle crossbeds. A 2.4-m-thick covered interval follows below about 1.4 m of pebbly sand and gravel. These beds contain calcrete nodules and mounds, and the amount of calcrete increases upward. Above them are 6 m of pinkish-gray, fine to very fine-

FIGURE 2.9. Ogallala Formation outcrop at Stop 3.

grained sand overlain by a 2.2-m-thick massive, pedogenic calcrete, the "Caprock caliche." This complexly brecciated and recemented carbonate is a stage VI pedogenic calcrete.

Obviously, the lower part of the Ogallala section here represents fluvial deposition of coarse sands and gravels in a major channel system. In contrast, fine sands higher in the section appear to be of eolian origin—note their fine grain size, frosting of some grains and high percentage of silt- and clay-sized particles. The lack of sedimentary structures in these eolian sands is consistent with their deposition in a sand sheet, not in dunes. The pervasive development of calcic solis and calcretes also dominates the upper part of the section.

Reworked basalt clasts in the basal conglomerate can only have come from the Ocate or Raton basalt field to the north. These fields are about 8 Ma (late Miocene) at oldest (Størmer, 1972; Mehnert and O'Neil, 1991), so this is a maximum age of the Ogallala Formation at Ragland.

After stop, **continue south on NM-209 into Ragland. 0.2**

54.2 Enter Ragland as road bends to east, and climb atop Llano Estacado. Continue on NM-209. **6.1**

60.3 Junction with NM-268 to Melrose. Continue straight. **1.9**

62.2 Cross swale. **0.1**

62.3 Junction with NM-210 to Forrest. Continue straight. **2.9**

65.2 Playa to right. **3.4**

68.6 Cross large playa lake deposits. The High Plains are

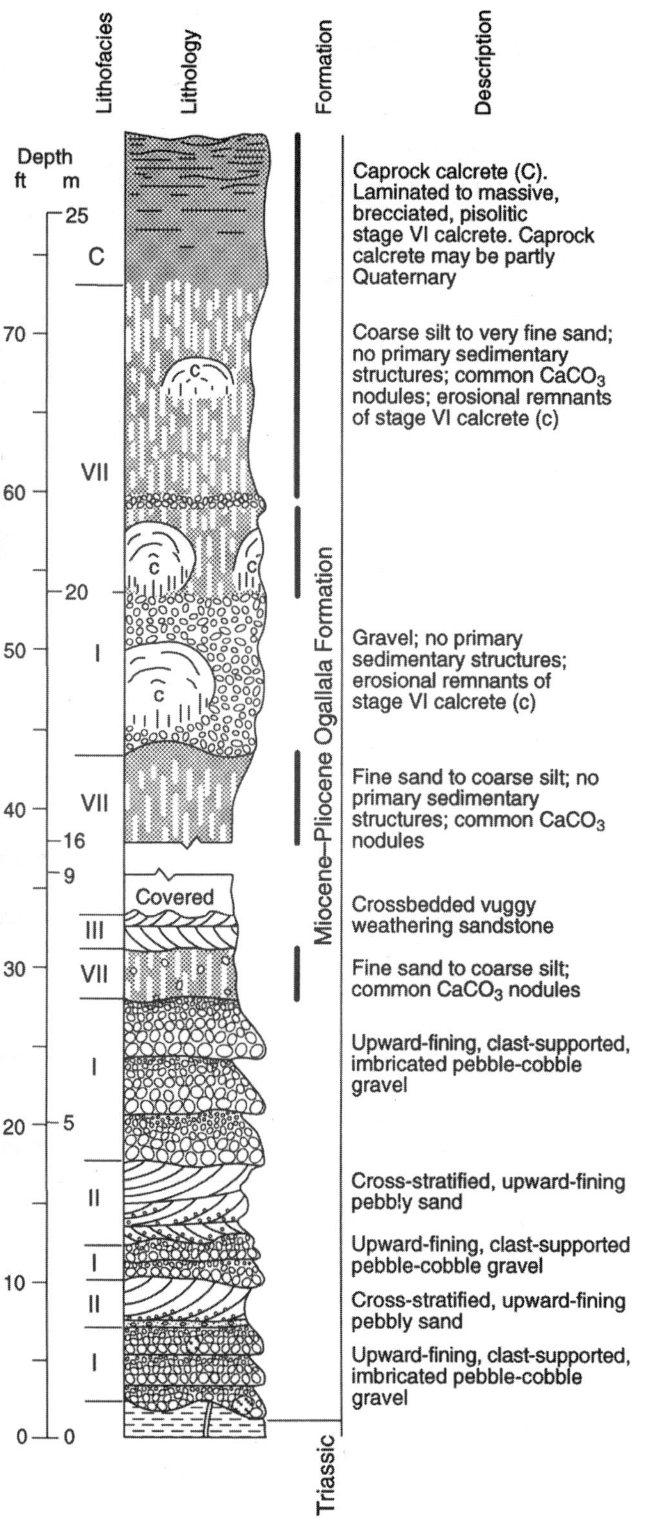

FIGURE 2.10. Measured section of Ogallala Formation at Stop 3 (from Gustavson, 1996).

dotted with circular playa lakes. These depressions are caused either by subsidence or deflation **1.3**

69.9 Playa to left. **1.5**

71.4 Junction with NM-278 to Norton. Continue straight. **2.5**

73.9 Mile marker 40. Very large playa to left. **0.6**

74.5 Enter Curry County, named for George Curry, Territorial Governor in 1909, when the county was formed. **0.9**

75.4 Playa to right. **3.0**

78.4 Enter Grady. **0.2**

78.6 Junction with NM-469 to San Jon. Continue straight. **0.5**

79.1 Leave Grady. **5.0**

84.1 Enter Broadview. **0.4**

84.5 Junction with NM-241 to Bellview. **Continue on NM-209** as it bends to the right (south). **0.7**

85.2 Leave Broadview. **5.4**

90.6 Junction with NM-19. Continue straight. **4.2**

94.8 Cross Frio Draw. **3.1**

97.9 Junction with NM-289. **Stay left on NM-209 to Clovis. 2.9**

100.8 Junction with NM-288 to NMSU Agricultural Science Center. Continue straight. **1.0**

101.8 Curry County Road 22 to left. **1.0**

102.8 Curry County Road 21 to right and left. **1.0**

103.8 Junction with County Road 20. Continue straight. This area, from north of Clovis south to Portales, is by far New Mexico's richest agricultural area. Curry and Roosevelt Counties account for about 25% of the irrigated and 60% of the dryland farming acres in the state, providing mainly wheat, corn and sorghum. Curry County also ranks among the top three counties in New Mexico in total cattle population. Clovis has the largest grain storage capacity in the state and manufactures more livestock feed than any city west of Kansas City. **1.0**

104.8 Curry County Road 19. **0.5**

105.3 Cross Running Water Draw. **0.5**

105.8 Ned Houk Park to left. **1.1**

106.9 Junction with County Road 17. Continue straight. **1.0**

107.9 Junction with County Road 16. Continue straight into greater Clovis. **2.0**

109.9 Junction with NM-77 to left; continue straight. **1.1**

111.0 Clovis city limits. Clovis (population 32,000 in 1998), named after a local ranching family, began in 1906 as Riley Switch. The name was changed to Clovis in 1907 when the Atchison, Topeka and Santa Fe Railroad was completed through the town. Agriculture quickly became the dominant economic activity, especially with the beginning of deep-well pumping for irrigation in the 1940s. Establishment of Cannon Air Force Base nearby in 1942 also bolstered the city's growth. **1.0**

112.0 Junction with NM-245 (Llano Estacado Boulevard). **Continue straight** through several traffic lights. **1.0**

113.0 Continue through traffic light (21st Street) south on NM-209. **1.5**

114.5 Junction US-60/84 and US-70. **Continue straight and cross bridge over railroad.** At this point, NM-209 becomes US 70 west. **2.5**

117.0 Leaving Clovis. **2.2**

119.2 Note large assemblage of fossil tractors in field to right, index fossils of the late Holocene. **1.9**

121.1 Roosevelt County line. **0.9**

122.0 Large dune on left. Note dunal topography here for the next few mi. The dunes in this area are part of the west-east-trending Muleshoe dunes that extend in a strip about 110 mi long and less than 10 mi wide (Reeves, 1972, 1973). **1.9**

123.9 Sign says Blackwater Draw Museum 2 mi. **2.0**

125.9 **Turn left** into parking lot for Blackwater Draw Museum. **0.1**

126.0 **STOP 4.** Blackwater Draw Museum, showcasing the archeology, paleontology, and geology of the famous Blackwater Locality No. 1. Opened to the public in 1969, the Museum is part of Eastern New Mexico University in Portales. Its exhibits detail the Blackwater Draw Paleoindian site, first discovered in 1932, which is the type locality of the Clovis culture, where human tools (Clovis points: Fig. 2.11) were found associated with extinct late Pleistocene mammals, such as mammoth, camels, horse and bison. The site was incorporated into the National Register of Historic Places in 1982. Humans occupied it as early as about 11,500 years ago. Indeed, there are multiple levels of human occupation (i.e., the site is "stratified") from the late Pleistocene into the Late Prehistoric Period (Fig. 2.12), one of the nearly unique and extremely important features of Blackwater Locality No. 1.

After stop return to US-70. 0.1

126.1 **Turn left** onto US-70. **0.7**

126.8 Junction with NM-202. Continue straight. **1.0**

127.8 Entering another dune field. **2.5**

130.3 Sign says turn right in 1 mile to go to Blackwater Draw site. **1.0**

131.3 Note Portales city limit sign, then **turn right at junction with NM-467.** Portales is older than the more populous Clovis. Springs once issued from seven caves that early Spanish settlers called Las Portales because the overhanging cliffs resembled porches ("portales") of Spanish adobe houses (Julyan, 1996). Las Portales became an important water stop on the Fort Summer Trail. Billy the Kid

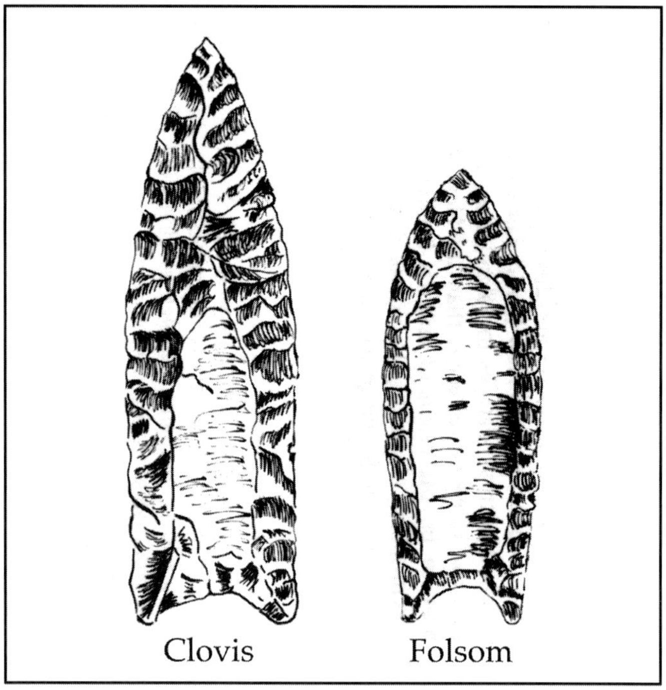

Figure 2.11. Clovis and Folsom points (after Agogino and Eaton, 1972).

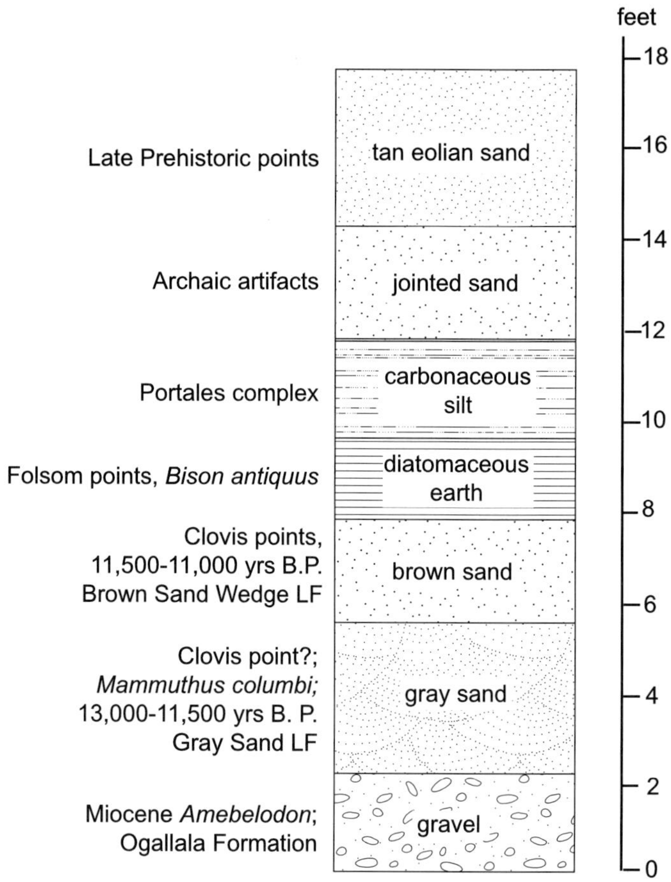

FIGURE 2.12. The classic stratigraphy at Blackwater locality No. 1.

camped at the springs, and it is said that he dreamed of building a ranch at Las Portales. Josh Morrison built a store in the late 1800s, and soon afterward the Pecos Valley and Northeastern Railroad built a construction camp nearby. A post office was established in 1899, and the town's name was shortened to Portales. Today the springs are dry because of pumping for irrigation. Portales is home to Eastern New Mexico University, established in 1934. Agriculture, especially dairy production, is a major industry for both the Portales and Clovis areas. **2.0**

133.3 Enter another vegetated dune field. **1.9**

135.2 Junction with turn to Oasis State park and historical marker (see accompanying minipaper). Continue straight. **1.2**

OASIS STATE PARK

Virginia T. McLemore
New Mexico Bureau of Geology and Mineral Resources, New Mexico Tech, Socorro, NM 87801

Oasis State Park lies 18 mi southwest of Clovis and 7 mi north of Portales via US-60 and NM-467. It was established in 1961 to preserve a true oasis in the sandy desert of the Llano Estacado or "staked plains" of the Great Plains physiographic province. The surrounding area is flat, treeless, featureless and relatively dry. The summers are hot, the winters are cold, and the wind seems to blow constantly. In contrast, the park offers shade trees and a pond, as well as the conveniences such as water, showers, electric hookups and dump stations expected of picnic areas and modern campgrounds (Fig. 2.13). Many of the facilities are accessible to people with disabilities. In addition to picnicking and camping, fishing, hiking and bird watching are popular activities. The pond is stocked with catfish and trout. Trails weave up, down and around the sand dunes. A ballfield lies near the center of the park.

The area is known to have been occupied by man about 11,500 yrs ago as evidenced by stone tools and projectile points found at nearby Blackwater Locality No. 1 (Agogino and Egan, 1972, Holliday, 1995, Haynes, 1995). At the Blackwater Draw archaeological site, projectile points were found with fossil remains of late Pleistocene fauna, including mammoth, camel, horse, bison, sabertooth tiger and dire wolf. A large pond at the site attracted the game animals and hunters, including man, from approximately 11,500 to 6000 yrs ago (Hester, 1972). Evidence indicates that the winters were warmer than today, and the summers were cooler; permanent streams and rivers were common. The oldest artifacts are of the Clovis or Llano culture; they are spear points, cutting tools and a butchered mammoth. The Clovis culture was dominant in this area from about 11,000 to 9000 B.C. It was followed by the Folsom culture from about 10,000 to 8000 B.C. Other, lesser-known cultures followed, including Portales and Archaic (Agogino and Egan, 1972; Hester, 1972; Haynes, 1995). About 7000 yrs ago, the pond dried up following widespread change in the climate, but humans continued to occupy the area as evidenced by camp fires, hand-dug wells and butchering sites.

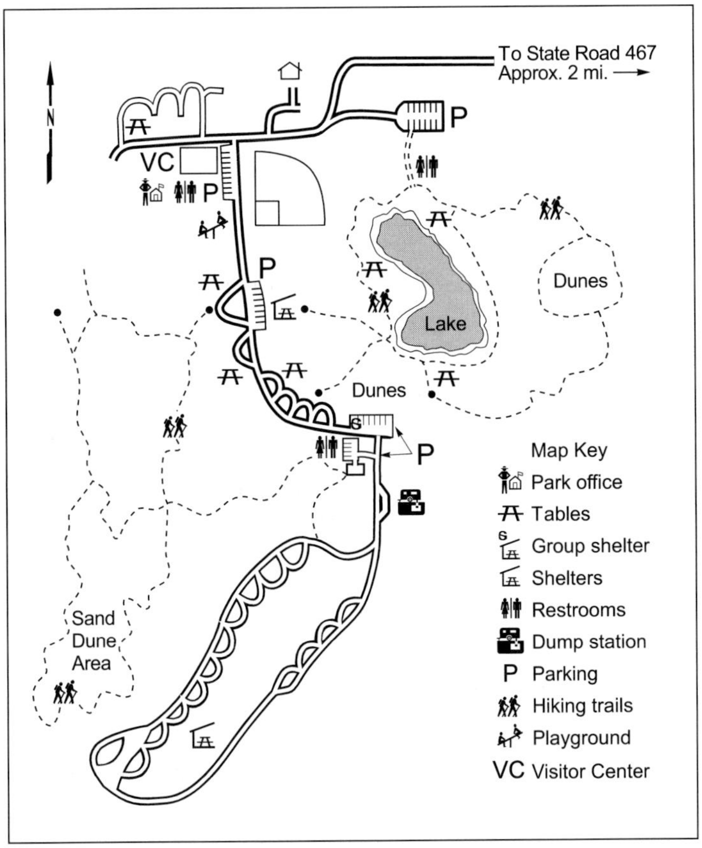

FIGURE 2.13. Map of Oasis State Park.

Finally, the site was abandoned about 4950 yrs ago as the water completely dried up (Hester, 1972).

Later, Comanche, Cheyenne and Kiowa roamed these plains; subsequently, settlers migrated into the area and lived in sod huts and depended on buffalo chips for heat. Numerous cattle drives went through the area in the 1800s. Early settlers came and dug shallow wells to tap the high water table; the stagnant water quickly turned dark, hence the name Blackwater Draw (Julyan, 1996). In 1902, a homesteader, Will Taylor, settled at the present location of Oasis State Park and planted cottonwood, chinaberry, locust, elm and cedar trees. Some of these trees still stand within the state park. Before the area became a state park, it was known as Taylor's Grove, a popular picnicking site (Young, 1984). The natural artesian springs that attracted Taylor to the area have long since dried up as a result of heavy irrigation. Now water is pumped from a well to feed a four-acre pond. The pond also serves as a stopover for migratory waterfowl.

In 1932, Walter Burns, a New Mexico State Highway Department engineer, discovered bones in a gravel pit being quarried for construction of US-70. Burns contacted archaeologists at the University of New Mexico, and excavations began at the quarry now known as Blackwater Locality No. 1 north of the state park. Studies were also conducted by the University of Pennsylvania in 1932-1937 (Katz, 1997). Research over the next 30-40 years slowly pieced together the prehistory of the area (Haynes, 1995; Holliday, 1995; Katz, 1997).

Oasis State Park is in the Portales Valley, an inset in the Llano Estacado. The Llano is formed on the Ogallala Formation. The Ogallala Formation consists of eolian (wind-blown) sand and silt and fluvial (stream or river) and lacustrine (lake) sand, silt, clay, and gravel derived from the Rocky Mountains between mid-Miocene to early Pliocene time, about 3–12 Ma (Reeves, 1972; Hawley, 1984, 1993; Gustavson et al., 1991). These surficial deposits were reworked and deposited by water and wind. In many places a caprock of hard, impermeable calcium carbonate, locally called caliche, formed near the surface in arid conditions during the early Pliocene or later. This caprock forms the rim of the Llano Estacado and is about 10–40 ft thick in the Clovis-Portales area (Galloway, 1956, 1972). It consists of fractured and permeable deposits of gravel, sand, silt and clay that are cemented by calcium carbonate (McGrath and Hawley, 1985). It is used locally for crushed and decorative stone. The caprock is a calcrete and is formed primarily by the process of downward percolation of surface waters depositing calcium carbonate in the upper soil horizons within the Ogallala Formation and perhaps, even in the overlying Blackwater Draw Formation and recent eolian deposits (Bachman, 1976; McGrath and Hawley, 1985). Important aquifers (reservoirs of ground water) occur in the Ogallala Formation, and they supply much of the water used for livestock, irrigated farms and recreation areas such as the state park. Much of this water originally was derived from the Sangre de Cristo Mountains before the Pecos River captured most of the water now migrating from those mountains. The only recharge or new water to the Ogallala aquifer in this area of New Mexico-Texas is from rain along the Llano Estacado. Pumping for irrigation and municipal use is depleting the aquifer much faster than it is being recharged!

The Pleistocene Blackwater Draw Formation overlies the Ogallala Formation in parts of Curry and Roosevelt Counties and consists of thin, silty to clayey soils and eolian sands. The Blackwater Draw Formation was formed by episodic cycles of eolian deposition during dry periods, followed by subhumid to semiarid conditions forming more cemented soils. Numerous cycles occurred for at least 1.6 million years to form the Blackwater Draw Formation. This unit is as thick as 81 ft and extends from the Pecos River valley eastward to Amarillo and Lubbock, Texas (Holliday, 1989). The grain size of the soils decreases from the southwest (predominantly sand) to the northeast (predominantly silt), indicating that the source was the Pecos River valley (Holliday, 1989). Three ash units occur in the formation and have been dated as 1.61 Ma (Guaje or Otowi Ash), 1.59 Ma (Cerro Toledo Ash) and 1.22 Ma (Tshirege Ash). These ash beds were formed by ash erupted and blown by the wind from the Jemez volcanic field (Izett et al., 1972; Gardner and Goff, 1996; Spell et al., 1996). The Brazos River flowed through the Portales area earlier and deposited the gravels that localized the pond used by prehistoric man at Blackwater Locality No. 1 (Taylor and Pitt, 1972; Hester, 1972; Holliday, 1995). During the Pleistocene, the Pecos River captured the Brazos River, and the springs were formed. The springs fed the pond at the Blackwater Draw archaeological site.

Recent sand dunes locally overlie the Blackwater Draw Formation. At the state park, these sand dunes are white because the sand is composed of medium-to fine-grained, white to clear

FIGURE 2.14. Photograph of north bank stratigraphy at Blackwater locality No. 1 taken in December 1962 or January 1963. Strata are: 1 = spoil, 2 = "jointed sand," 3 = "carbonaceous silt," 4 = laminated zone, 5 = white sand lens, 6 = Folsom and Midland zone, 7 = Clovis gray sand, 8 = gravel, 9 = red clay lens. Courtesy of Joanne Dickenson and the site archives of Blackwater Draw locality no. 1.

quartz, gypsum, calcite and feldspar crystals. Most of the dunes are crescent-shaped in plan view because they are formed by constant wind on a flat desert floor; these are called barchan dunes. Parabolic dunes, which are also crescent shaped, are locally formed by wind from the opposite direction.

Acknowledgments—Special thanks to the state park personnel and Joanne Dickenson, curator of the Blackwater Draw Locality No. 1, for discussions and information on the history of the park and Blackwater Draw Locality No. 1. Frank Kottlowski, Bruce Harrison and Dave Love reviewed an earlier version of this manuscript, and their comments are appreciated. The New Mexico Bureau of Geology and Mineral Resources Cartography Department drafted the figure.

136.4 **Turn left** to Blackwater Draw site. Blackwater Draw Locality No. 1 historical marker to right; drive through gate to the parking area at the building that covers the site. **0.4**

136.8 **STOP 5.** Blackwater Draw locality no. 1 is the type locality of the Clovis culture, long considered to be the oldest New World culture (older sites are now known). The site was originally a gravel pit, and in 1932 mammoth bones were discovered in association with large, fluted projectile points (Katz, 1997), later termed "Clovis points" (Fig. 2.11). Excavations for artifacts and fossils began from 1932-1937, then restarted in 1948-1956 and have continued on and off since. These excavations and related geological studies (e.g., Haynes, 1995; Holliday, 1997) establish that the Clovis site was a natural depression occupied initially by a late Pleistocene lake.

FIGURE 2.15. Pedestalled mammoth bed and north wall of excavation of Blackwater Locality No. 1. Photo taken 15 December 1962, courtesy of Joanne Dickensen and the site archives of Blackwater Draw locality no. 1.

This basin is inset in the Pleistocene Blackwater Draw Formation and underlying gravels of the Ogallala Formation, which are latest Miocene (6-7 Ma) in age (see accompanying minipaper).

The paleobasin here contains a stratified fill (Figs. 2.12, 2.14). The oldest fill is spring-deposited sands about 11,000-13,000 years old that grade toward the basin margin into sand and rubble eroded from the valley walls. These sands have yielded the Clovis points, associated mammoths (Fig. 2.15) and other mammals. Indeed, remains of at least eight mammoths and two bone beds of *Bison antiquus* are present in these sands. Although long believed to be a mammoth kill site, the mammoths here were actually scavenged (Saunders and Daeschler, 1994). However, the bison do appear to have been killed and butchered by the Clovis hunters.

Above the Clovis stratigraphic level is diatomaceous earth interbedded with mud and sand that dates back to 10,000-10,800 years ago. It yields Folsom tools and other *Bison* bone beds. Other Folsom features occur in eolian sands on the uplands northwest of the ancient basin. Overlying strata, the "carbonaceous silt" (pond and marsh deposits), contain Late Paleoindian artifacts and *Bison* bones dating to about 10,500-8,500 years ago. The section at the site (Figs. 2.12, 2.14) thus exposes multiple levels of occupation.

After stop return to highway. 0.4

THE SHOVEL-TUSKED GOMPHOTHERE *AMEBELODON* (MAMMALIA: PROBOSCIDEA) FROM THE MIOCENE OGALLALA FORMATION AT THE BLACKWATER DRAW SITE, ROOSEVELT COUNTY, NEW MEXICO

Gary S. Morgan and Spencer G. Lucas
New Mexico Museum of Natural History, 1801 Mountain Road NW, Albuquerque, NM 87104

Blackwater Draw, in Roosevelt County, east-central New Mexico, is the type Clovis site, and one of the best known localities in North America that documents the association of Paleoindians with a diverse assemblage of extinct Pleistocene megafauna (Hester, 1972; Lundelius, 1972; Haynes, 1995). The Clovis level at Blackwater Draw, also called the "Gray Sand," is latest Pleistocene in age. However, a much older, late Miocene mammal, the shovel-tusked proboscidean *Amebelodon*, was recovered from a lower stratigraphic level at Blackwater Draw during gravel mining operations in the early 1960s.

The *Amebelodon* fossil consists of a right dentary with m2 and partially erupted m3 (Fig. 2.16). The mandible was discovered in December 1962 or January 1963 during gravel mining operations in the North Pit at Blackwater Draw by J. D. Murray, a dredge operator for the Sanders Sand and Gravel Company. The *Amebelodon* jaw was derived from gravels of the Miocene Ogallala Formation, which unconformably underlies the late Pleistocene Clovis or Gray Sand level. Unfortunately, the two photographs reprinted here (Fig. 2.16) and another photograph in Katz (1997, p. 2) are the only tangible evidence now available of this fossil. The jaw was retained in J. D. Murray's private collection for many years, but after his death the fossil disappeared, and attempts to locate it have been unsuccessful (Joanne Dickenson, oral commun., 2000).

Even though the Miocene proboscidean mandible from Blackwater Draw cannot be located, the photos of the specimen are sufficient to reveal several aspects of its morphology that are diagnostic of the genus *Amebelodon*. In Figure 2.16 a lateral view of the right dentary shows that the m2 is tetralophodont (i.e., it has four lophids). The m3 is not fully erupted but has four unworn lophids visible. The tetralophodont condition of the m2 is characteristic of advanced species of *Amebelodon* (subgenus *Konobelodon*), whereas more primitive species of *Amebelodon* (subgenus *Amebelodon*) have only three lophids on m2 (Lambert, 1990). One other proboscidean known from New Mexico, the

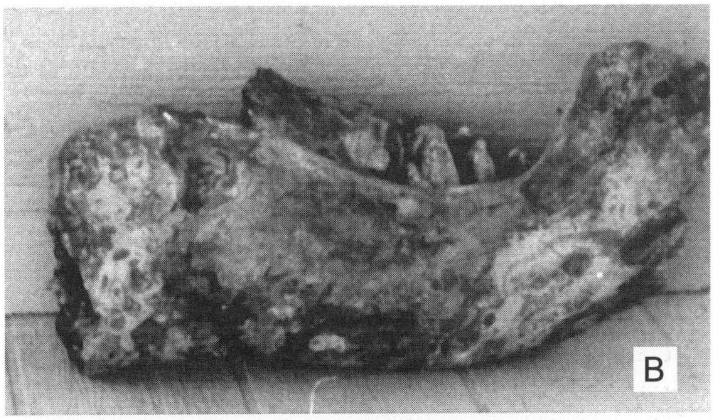

FIGURE 2.16. *Amebelodon* mandible from the Miocene (Hemphillian) Ogallala Formation at Blackwater Draw, Roosevelt County, New Mexico. A. Lateral view of unprepared right dentary, just after it was found in gravels in the North Pit at Blackwater Draw. Hand and foot belong to J. D. Murray, the discoverer of the fossil. B. Medial view of prepared right dentary. Photograph 2.16A courtesy of Joanne Dickenson and the site archives of Blackwater Draw locality no. 1. Photograph 2.16B courtesy of Lienke Katz.

Pliocene gomphothere *Stegomastodon*, also has a tetralophodont m2. However, *Stegomastodon* has a very short mandibular symphysis and lacks lower tusks, whereas the Blackwater Draw fossil has a robust mandibular symphysis and would have possessed a pair of large, shovel-shaped lower tusks.

Although the gomphothere jaw from Blackwater Draw is broken anterior to m2, enough of the horizontal ramus is preserved (Fig. 2.16) to reveal two other important morphological characters. The horizontal ramus increases in depth gradually from below m3 anteriorly, reaching its greatest depth just anterior to the anterior edge of the toothrow. At the anterior edge of the toothrow, the dorsal margin of the horizontal ramus curves noticeably dorsally, and the ventral edge curves somewhat ventrally, which in combination greatly increases the depth of the dentary. The strong dorsal curvature and deepening of the horizontal ramus just anterior to the toothrow are very characteristic of *Amebelodon*, probably reflecting the strengthening of the mandibular symphysis necessary to support the massive shovel tusks of this genus. Figure 2.16 also reveals the presence of a narrow, deep groove between the right and left dentaries just anterior to the toothrow, another feature typical of *Amebelodon*.

The morphological characters of the teeth and dentary evident in Figure 2.16 confirm the presence of the shovel-tusked gomphothere *Amebelodon* from the Ogallala Formation at Blackwater Draw. Furthermore, the presence of a tetralophodont m2 indicates that this dentary is referable to the subgenus *Konobelodon*. *Amebelodon* (*Konobelodon*) has a restricted stratigraphic range in North American late Miocene faunas (Lambert, 1990), where it is known from the late early Hemphillian (about 7 Ma) of Florida and the early late Hemphillian (about 6 Ma) of Kansas. The youngest occurrence of *Amebelodon* is in faunas equivalent in age to the Coffee Ranch local fauna in the Texas Panhandle, the type fauna of the Hemphillian land-mammal "age" (Schultz, 1977). *Amebelodon* is unknown after the early late Hemphillian (Lambert, 1990), so its presence establishes an age of 6 Ma or older for the Ogallala Formation at Blackwater Draw.

The primary importance of the Blackwater Draw *Amebelodon* is that the restricted biostratigraphic range of this genus establishes a fairly narrow age interval (latest Miocene, between 6 and 7 Ma, possibly somewhat older) for the Ogallala Formation in eastern New Mexico. The only previously reported fossils from the Ogallala Formation in New Mexico are a mammalian footprint fauna from San Juan Mesa in Chaves County (Williamson and Lucas, 1996). The San Juan Mesa footprint fauna, which includes a large cat, a large camel, a smaller artiodactyl, and perhaps a dog, does not provide any further information on the age of the Ogallala Formation in New Mexico (Williamson and Lucas, 1996).

137.2 **Turn left** onto NM-467. **1.8**

139.0 Curry County line. **2.1**

141.1 Large stockyard to left. **1.9**

143.0 Junction with road to left to Cannon Air Force Base; continue on NM-467 through right bend in road. **3.0**

146.0 Sharp left bend in road. **2.7**

148.7 Cross railroad. **0.3**

149.0 Stop sign. Junction with US-60/84, **turn left. 2.9**

151.9 Exit for Cannon Air Force Base/Junction NM-311 to right. Continue straight. **9.9**

161.8 Don't miss St. Vrain (dating back to 1907) on right. **3.4**

165.2 Junction with NM-224. Continue straight. **4.1**

169.3 Junction with NM-267; entering Melrose (population 660 in 1990). Melrose began in 1882 as a village named Brownhorn; the Santa Fe railroad came through in 1906 and selected the town as a site for a roundhouse. **0.6**

169.9 **Turn right** at junction with NM-268. **0.1**

170.0 **Turn right** to continue north on NM-268. **0.3**

170.3 Downtown Melrose; most stores are empty and boarded up. **0.2**

170.5 **Veer left** to continue on NM-268. **0.3**

170.8 Road turns sharply left. **0.2**

171.0 Road turns sharply right. **10.0**

181.0 Junction with NM-89. Continue straight. **3.9**

184.9 Junction with NM-288. Continue straight. **4.1**

189.0 Junction with NM-312 at Quay County line. Continue straight. **7.1**

196.1 Junction with NM-210. Continue straight. **2.0**

198.1 Stop sign. **Turn left** at junction with NM-209 to retrace our route this morning. **5.9**

204.0 Enter Ragland. Continue north on NM-209 towards Tucumcari. **0.3**

204.3 Stop 3 of this morning; descend from the Llano Estacado. **7.4**

211.7 Quay. **3.0**

LAGUNA COLORADO: A FORGOTTEN HOLOCENE LAKE IN QUAY COUNTY, NEW MEXICO

Adrian P. Hunt

Mesalands Dinosaur Museum, Mesa Technical College, 911 South Tenth Street, Tucumcari, NM 88401

On 5 October 1841, the main body of an ill-fated Texas-Santa Fe expedition of General Hugh McLeod surrendered to the Mexicans on the shores of the only substantial lake in east-central New Mexico at that time--Laguna Colorado (Carroll, 1951). Laguna Colorado (or Laguna Colarada) was a well-documented lake in the nineteenth century. Josiah Gregg stopped there in 1839 (Gregg, 1844). Subsequently, several exploratory expeditions camped at the lake and mapped its location (Simpson, 1850; Marcy, 1852; Marcou, 1858). None of these expeditions recorded the presence of any other lake in east-central New Mexico; notably there is no mention of Tucumcari Lake. Capt. R. B. Marcy and Lieut. J. H. Simpson were members of the same expedition and published separate maps of the area (Simpson, 1850; Marcy, 1852). Simpson's (1850) map (reproduced by Kues, 1985) shows Laguna Colorado as an elongate, northeast-trending lake drained by a northeast-flowing drainage.

Carroll (1951) located Laguna Colorado approximately 4 mi southeast of the village of Newkirk in Guadalupe County. and Kues (1985) gave a similar location for the surrender of the 1841 expedition. Marcy (1852) and Simpson (1850) show Laguna Colorado as located just south of the 35th Parallel at about 104° 15' W. Marcou (1858) shows the lake at about the same longitude, but just north of the 35th Parallel. Carroll (1951) estimated that the 1841 camp was about 0.5 mi north of the common corner of townships 26 and 27 E.

How long did Laguna Colorado exist? Obviously, the lake was in existence in 1839 (Gregg, 1844) and probably long before that time. In May 1858, a survey along the range line between Range 26 East and 27 East had to triangulate across a lake a distance of 6.07 chains. A later resurvey in February and March of 1881 recorded no lake. However, at the point where the previous survey had been forced to triangulate, they recorded a northeast-flowing arroyo (Carroll, 1951). Thus, it appears that between 1858 and 1881, Laguna Colorado ceased to exist, or at least it became much smaller. Local residents still remembered the lake as late as 1935 (Carroll, 1951).

What became of the lake? Carroll (1951) hypothesized that the area was overgrazed by sheep, resulting in increased erosion. Subsequently, Bull Canyon Creek cut a deep canyon across the plain into the eastern end of the holding basin of Laguna Colorado and drained it. Laguna Colorado was obviously a significant geographic feature in east-central New Mexico at least in the nineteenth century and maybe for a significant time prior to that.

214.7 Tucumcari Mountain in distance, straight ahead. **12.9**

227.6 Tucumcari city limits.**1.7**

229.3 Traffic light at Tucumcari Boulevard. Continue straight. **0.2**

229.5 Turn right at Laughlin and stop at museum.

End of Second-day Road Log.

Coprolites such as these are fossilized feces of Late Triassic amphibians and reptiles. They are among the most common fossils found in Chinle Group strata in east-central New Mexico.

THIRD-DAY ROAD LOG, FROM TUCUMCARI TO CONCHAS DAM, NEW MEXICO

SPENCER G. LUCAS, ADRIAN P. HUNT, ANDREW B. HECKERT, BARRY S. KUES and VIRGINIA T. MCLEMORE

SATURDAY, SEPTEMBER 29, 2001

Assembly point: Mesalands Dinosaur Museum, 222 East Laughlin, Tucumcari
Departure time: 8:00 AM
Distance: 34.2 miles
Stops: 3

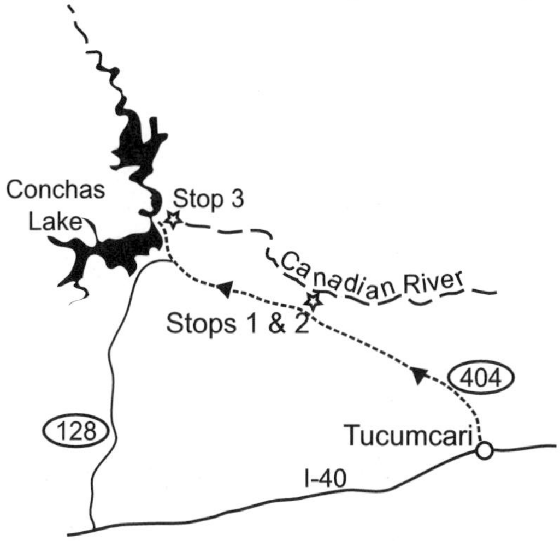

SUMMARY

This trip explores the geology of the Canadian River drainage from Tucumcari north to Conchas Dam. Here, we contrast differences between the beginning and the end of Late Triassic deposition in the vast Chinle basin, and the continuity of Middle Jurassic eolian deposition across the American Southwest.

At STOP 1 we examine an unusual lithofacies of the Upper Triassic Redonda Formation. At STOP 2, a roadcut fortuitously created a three-dimensional exposure of the Middle Jurassic Entrada Sandstone, which facilitates discussion of Middle Jurassic wind patterns and paleogeography. STOP 3 ends the fieldtrip at Conchas Dam, where the onset of Chinle Group deposition in the western United States is featured.

Mileage

0.0 Assemble at the Mesalands Dinosaur Museum, 222 East Laughlin, Tucumcari. **Turn left out of museum parking lot onto S. Adams and immediately turn left (west) onto East Laughlin. 0.1**

0.1 Stop sign. **Turn right** (north) on First Street (NM-104) and proceed north. **0.3**

0.4 **Move into center lane** and proceed through traffic light at Main Street, continue straight. **0.2**

0.6 Cross bridge over railroad tracks. **0.4**

1.0 Leave Tucumcari. **1.3**

2.3 Crest ridge in dune field, Liberty Mesa at 9:00-11:00; note that to the north the Canadian escarpment is visible in the distance. **0.6**

2.9 Red-bed Chinle mudstones apparent beneath alluvium to right in roadcut are probably part of the Bull Canyon Formation. **0.1**

3.0 Cross Pajarito Creek, namesake of the Lower Cretaceous Pajarito Formation. The late Albian index ammonite *Mortoniceras equidistans* was found just south of here in the basal Mesa Rica Sandstone (Lucas and Hunt, 2000). **1.9**

4.9 Low exposures of Upper Triassic Chinle Group red-bed mudstones (probably Bull Canyon Formation) on right of road. Note Laramide folds and faults in this area disrupt the Mesozoic (mostly Lower Cretaceous) strata (see accompanying mini-paper). **0.4**

THE LARAMIDE TUCUMCARI STRUCTURAL ZONE, EAST-CENTRAL NEW MEXICO

Adrian P. Hunt[1] and Spencer G. Lucas[2]

[1]Mesalands Dinosaur Museum, Mesa Technical College, 911 South Tenth Street, Tucumcari, NM 88401; [2]New Mexico Museum of Natural History, 1801 Mountain Road NW, Albuquerque, NM 87104-1375

Northeastern New Mexico is characterized by an apparent lack of structural complexity and a sedimentary cover with very low regional dips. The only major structure to appear on large scale geologic maps (e.g., Dane and Bachman, 1965; Clemons, 1982) is the Bonita fault zone in southeastern Quay County and northern Roosevelt County (Stearns, 1972). The purpose of this paper is to demonstrate that east-central New Mexico actually has a more complex Laramide structural history.

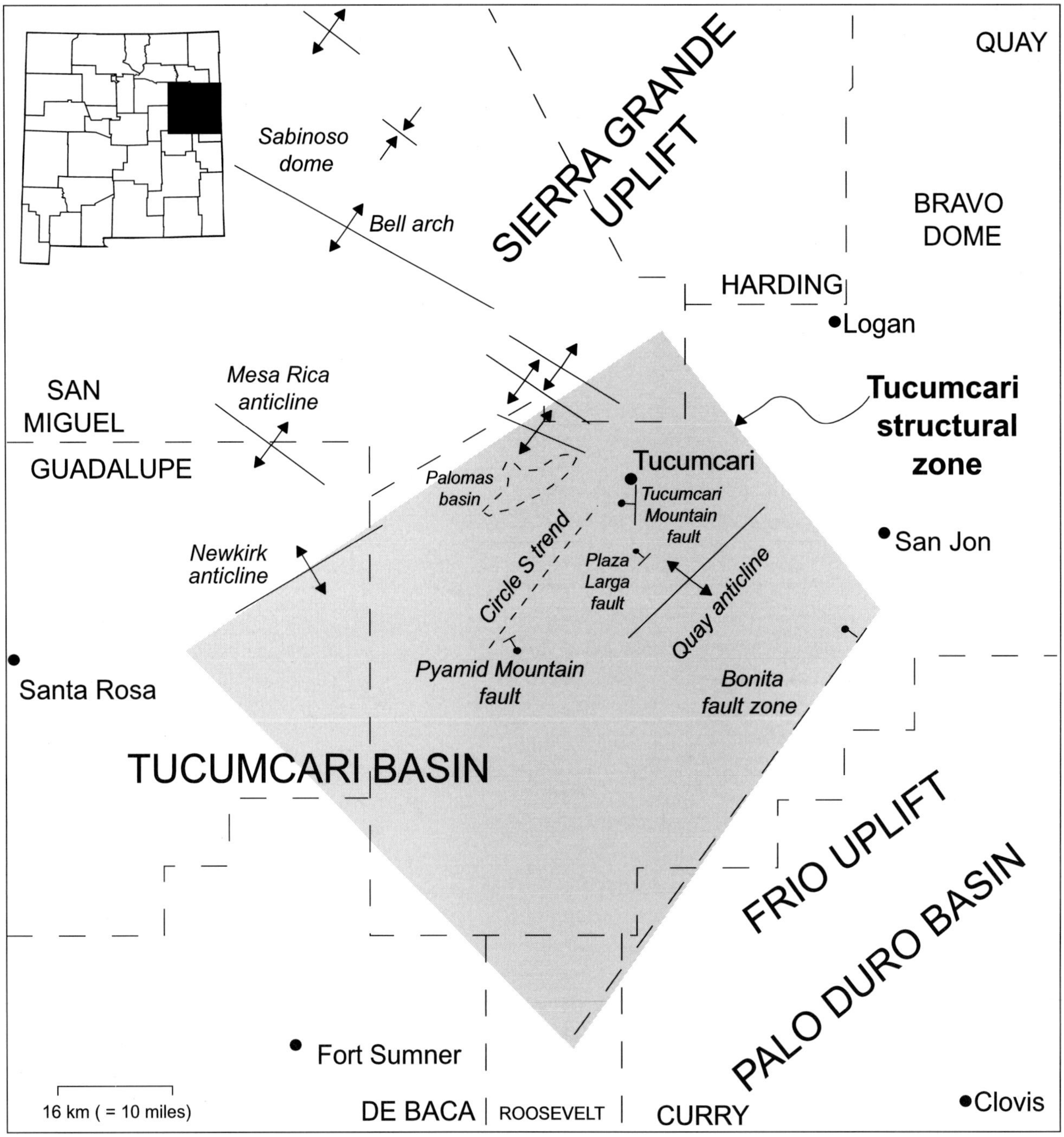

FIGURE 3.1. Simplified structural map of east-central New Mexico showing the Tucumcari structural zone and related features (data from Dobrovolny et al., 1946; Berkstresser and Mourant, 1966; Stearns, 1972; and original observations).

Stearns (1972, fig. 1) provided a structural sketch of east-central New Mexico that implied that there were very few structures in the area. However, the little detailed mapping that has been done in the area demonstrates greater structural complexity (Fig. 3.1; Dobrovolny et al., 1946; Wanek, 1962). From Tucumcari to the southwest, there is a northeast-trending series of structures that we term the Tucumcari structural zone. This zone is bounded on the northwest by the Newkirk anticline and on the southeast by the Bonita fault zone (Fig. 3.1). Most structural features within this zone trend northeast, whereas northwest-trending structures terminate the Tucumcari structural zone to the north/northwest. Neogene strata cover the northeastern and southwestern ends of the Tucumcari structural zone, so its full extent is unclear.

The majority of structures in the Tucumcari structural zone are gentle folds that trend northeast and include, from northwest to southeast: (1) the Newkirk anticline of Stearns (1972); (2) the Palomas basin, which we name for the structural basin mapped by Dobrovolny et al. (1946) that extends to the northeast from Palomas; and (3) a northeast-plunging anticline, originally mapped by Dobrovolny et al. (1946) as extending from Mesa Redonda to the area where Revuelto Creek crosses I-40, that we name the Quay anticline. There is also a northeast trend of small structural basins extending from Circle S Mesa in the southwest to the Five Mile Park area west of Tucumcari (Dobrovolny et al., 1946) that we term the Circle S trend. This trend probably represents a series of dissolution features aligned along a fault zone. These structures, particularly the Quay anticline, appear to influence drainage patterns in the area, with most drainages in the southern two-thirds of Quay County trending to the northeast (e.g., Plaza Larga, Barranca Creek, Trujillo Creek).

The Bonita fault zone is the best studied structure in east-central New Mexico (e.g., Stearns, 1972; Lucas et al., 2000), but there are also several faults in the area that have been overlooked. The most prominent of these is on the western side of Tucumcari Mountain, where the Mesozoic sequence drops approximately 150 m within 0.4 km. This structure continues under the city of Tucumcari, because there are outcrops of the Morrison Formation, not mapped by Dobrovolny et al. (1946), in the western part of the city near Tucumcari Elementary School, whereas the Chinle Group is exposed in the eastern part of the city near Tucumcari Lake. Therefore, the structure must be northerly in strike and located about a third of a mile from the western boundary of T11N, R30E. Dobrovolny et al. (1946) mapped no fault, so by implication they must have assumed that this was the limb of a monocline. Trauger et al. (1972) noted a "sharp flexure" that dropped the Chinle Group about 150 m. Given the gentle dips of all folds in the area (Dobrovolny et al., 1946), and the presence of the southwest-northeast-trending Bonita fault southeast of Tucumcari (Stearns, 1972), it is likely that the structure on the western side of Tucumcari Mountain is a fault.

Dolliver (1985) demonstrated the presence of a fault underlying Plaza Larga, north of Mesa Redonda. Smaller, unmapped faults occur in the area, such as between Circle S Mesa and Pyramid Mountain, where the drag on the fault gives the illusion of an angular unconformity between the Redonda Formation and the Entrada Sandstone.

The most precise age for the Tucumcari structural zone can be derived from the Bonita fault zone. These faults cut through the Upper Cretaceous Bridge Creek Member of the Greenhorn Limestone, which is as young as earliest Turonian (Lucas et al., 2000), but they do not disrupt the Miocene Ogallala Group (Formation). Elsewhere in east-central New Mexico, there is an unconformity between the late Albian Pajarito Formation and the Ogallala Group that is never deformed, and in these areas the structures can be less precisely dated. However, it is clear that all the structures in the Tucumcari structural zone postdate the Turonian Late Cretaceous and predate the Neogene. This age, together with their northeasterly trend, suggests that they are most likely related to the Laramide orogeny.

Are these structures Laramide reactivations of older features? Broadhead and King (1988, fig. 3) produced the most accurate structural map of the Precambrian basement in this area, which shows a number of northeast-trending faults between northwestern Curry County and Tucumcari that form a series of horsts and grabens. Activity on most of these faults had ceased by Early Permian time (Broadhead, 1984).

To the southeast of the Tucumcari structural zone is the southwest-northeast-trending Frio uplift that separates the Tucumcari basin from the Palo Duro basin, and to the northwest is the Sierra Grande uplift that also trends southwest-northeast (Broadhead, 1988, fig. 1). During the Laramide orogeny, the southeastern-dipping Pecos shelf formed in southeastern and east-central New Mexico (Kelley, 1971). It appears likely that the Laramide buckling in the Tucumcari structural zone was a result of this southeastern tilting affecting an area between two northeast-trending uplifts that had healed faults with a northeast trend. Broadhead (1984; Broadhead and King, 1988) and McKallip (1985) documented the reactivation of Paleozoic faults during the Laramide. Kelley (1971) mapped a number of northeast-trending buckles (folds and/or faults) across the Pecos slope in southeastern New Mexico that are presumably a crustal response to the same stress pattern in a more unconfined terrane.

5.3 To left, the cap of Liberty Mesa is Lower Cretaceous Mesa Rica Sandstone overlying Lower Cretaceous Tucumcari Formation and Upper Jurassic Morrison Formation. **0.7**

6.0 San Miguel County line. San Miguel County was named by the Republic of Mexico in 1844 for San Miguel del Bado, one of the oldest communities in the area. It was the county seat until 1864, when the seat was moved to Las Vegas (Julyan, 1996). **0.3**

6.3 Historical marker for Fort Bascom (1863-1870), built to protect the area from Kiowa and Comanche Indians and to encourage settlement of the Canadian River valley. The post was first known as Camp Easton, after Major Langdon C. Easton of the Quatermaster's Department. The name was changed in January 1864 in honor of Captain George N. Bascom, who was killed at the Battle of Valverde, near Socorro. Fort Bascom was located in a horseshoe bend of the Canadian River, but the garrison of 88 men proved ineffectual, and the post was closed in 1870 and the troops transferred to Fort Union. A civilian settlement sprang up with a post office, but this was subsequently abandoned after changing its name to Johnson (Julyan, 1996).

Note Carpenter Mesa to right at 2:00 with prominent white sandstone near its top. This sandstone is similar to the Upper Jurassic Jackpile Member of the Morrison Formation (Holbrook et al., 1987) and is overlain by the Lower Cretaceous Tucumcari Formation and Mesa Rica Sandstone. The Brushy Basin Member of the Morrison Formation contains dinosaur fossils in this valley, including a complete sauropod ("brontosaur") femur. **2.2**

8.5 Cross canal. Note the low hills to the north of Mesa Rica Sandstone, which are much lower topographically than the same beds on the Canadian escarpment to the north. Again, this displacement reflects

FIGURE 3.2. The flank of Liberty Mesa from mile 8.9. Sandstone cliffs of the Cretaceous Mesa Rica sandstone cap the mesa above slopes developed primarily in the Upper Jurassic Morrison Formation.

Laramide structure. **0.4**

8.9 Road descends from eolian sand-covered bench, note red Chinle badlands ahead and Lower Cretaceous Mesa Rica Sandstone overlying Tucumcari Formation and Upper Jurassic Morrison Formation to left on mesa flanks, from 9:00-11:00 (Fig. 3.2). **0.3**

9.2 Cross swale. Note Upper Triassic Redonda Formation red-beds to left.

Francisco Vazquez de Coronado was dreaming of fame and fortune as he led his soldiers across eastern New Mexico in the Spring of 1541. A plains Indian, whom the Spaniards called El Turco (the Turk), was leading Coronado to the kingdom of Quivira. In Quivira even ordinary folk used silver spoons and golden bowls.

Coronado did achieve fame, but not for getting rich, rather for *not* getting rich. Coronado had a knack for *not* finding cities of gold. The previous year Coronado had found the mythical seven cities of Cibola, which turned out to be adobe pueblos of the Zunis. Quivira was no more promising -- it turned out to be villages of the Wichitas in Kansas built of sticks, not gold.

A minor part of Coronado's fame rests on his encounter with the North American bison (or buffalo). Coronado's men first encountered bison in the Canadian River Valley of eastern New Mexico, and they continued to see them as they journeyed into Kansas. It is difficult to imagine the impact of the first sight of these magnificent animals. Capt. Hernando de Alvarado, one of Coronado's assistants, had seen bison the year before and he described them as "the most monstrous beasts we have ever seen or heard about." They were so numerous that he "did not know to what to compare them unless it be to the fish of the sea." Coronado's encounter with the bison in 1541 was not forgotten. Ninety years later, in 1651, Sr. Hernandez, who had never actually seen a bison, described this animal based on descriptions of the "cattle of the Quivira region" written by various members of Coronado's expedition (see accompanying minipaper).

Coronado met Apaches in 1541 who were hunting bison. Bison provided not only meat but also many of the other basics of life such as clothing. The Spaniards may have been acknowledging the importance of the bison when they called them "cibolo," which is from the same word source as "cibola," which they used for the mythical golden cities. **0.3**

THE TYPE AREA OF *BISON BISON* (LINNAEUS, 1758), CANADIAN RIVER VALLEY, EAST-CENTRAL NEW MEXICO

Adrian P. Hunt[1] and Spencer G. Lucas[2]

[1]Mesalands Dinosaur Museum, Mesa Technical College, 911 South Tenth Street, Tucumcari, NM 88401; [2]New Mexico Museum of Natural History, 1801 Mountain Road NW, Albuquerque, NM 87104

Linnaeus (1758) recognized two species of bison in his tenth edition of *Systema Naturae* — the American bison as *Bos bison* and the European wisent as *Bos bonasus* — although he had previously listed them in the sixth edition (Thomas, 1911; McDonald, 1981). Because the American bison was described in 1651 by Hernandez (Fig. 3.3), and the European species was described in 1693 by Ray, the type species of the genus is the American *B. bison*. Subsequently, Bojanus (1827) placed extinct bison in the genus *Urus*, Smith (1827) utilized *Bison* as a subgenus and Knight (1849) accorded *Bison* generic rank (McDonald, 1981). Now the species *bison* and *bonasus* are placed within the genus *Bison* together with other species, and with *B. bison* as the type species.

According to the *International Code of Zoological Nomenclature*, every nominal taxon has, "actually or potentially," a name-bearing type specimen. In the case of *Bison bison* there is no actual type specimen because Linnaeus (1758) named the taxon based on a description published by Hernandez (1651), who in turn had never actually seen a bison! Hernandez (1651) based his description on the writings of members of Francisco Vazquez de Coronado's expedition, which passed through eastern New Mexico in the Spring of 1541.

Hernandez (1651) explicitly stated that his description of the American bison was based on descriptions of "cattle of the Quivira region" by members of Coronado's expedition even though other eyewitness accounts of these animals were available to him (Reed, 1952, 1955; Brand, 1967; McDonald, 1981). Given the lack of a type specimen of *Bison bison*, it is important to locate

FIGURE 3.3. Dr. Francisco Henandez' 1651 illustration of the Mexican bull, "*Taurus mexicanus*," which Linnaeus (1758) used as the basis for his *Bos bison*, now *Bison bison* (after McDonald, 1981).

the type locality for this species, from which a neotype specimen might be collected. The type locality is the area where bison were encountered and described by Coronado's expedition and thus formed the basis of Hernandez' (1651) description and Linnaeus' (1758) naming of the species. Four authors have discussed the search for the type area of *B. bison:*

1. Thomas (1911) identified Mexico as the type area for *B. bison,* with the recognition that seventeenth century Mexico included much of the southwestern United States.

2. Reed (1952, 1955) identified the southern or southeastern United States as the type area because this was where Europeans first saw bison.

3. Hershkovitz (1957) reasoned that because Hernandez (1651) had explicitly based his description on animals of the Quivira region, that the type area should be in central Kansas, which is where the Quivira villages were located.

4. McDonald (1981) noted that Quivira could not be considered the type locality. The main force of Coronado's expedition first saw bison in the Canadian River Valley of east-central New Mexico and then encountered them until the force was split into two units in West Texas. One unit headed north to Quivira, where they reported seeing bison until they were within three or four days of the Quivira villages. In the next month in the Quivira region they did not see, or at least did not report seeing, any bison. Therefore, McDonald (1981, p. 62) concluded "considering the Coronado expedition's experience with bison, and accepting the reports of members of Hernandez' information on *B. bison*, the type locality is here designated as the Canadian River Valley of eastern New Mexico."

Given the geography of the Canadian River area, it is clear that the type area of *Bison bison* can be further restricted to eastern San Miguel County or Quay County, New Mexico. Due to the distribution of open range land, it is most likely that bison would be encountered in the plains of Quay County as opposed to the canyon country to the west. Therefore, in conclusion, it is most likely that the type area for *B. bison* is in Quay County, New Mexico.

9.5 Road descends to cross creek. Note well-exposed Redonda Formation calcretes and siltstones. Canadian River and associated bosque visible at 2:00, with Carpenter Mesa in the distance behind them. In this area, there are a series of northwest-southeast-trending anticlines with low dips along the northwestern boundary of the Tucumcari structural zone.

The Canadian River has a couple of very unusual characteristics. Its headwaters are in the Sangre de Cristo Mountains of Colorado and New Mexico from whence it flows almost due south, but at Conchas Lake it takes an abrupt 90° turn to the east. Also, on the Harding and Mora County boundary, the Canadian is entrenched in a 150-m deep canyon, whereas here the river lies in a broad floodplain. The strange bend in the river is the result of the capture of east-flowing tributaries of the Canadian by the Pecos River during the middle Pleistocene, and the incision occurred mainly in the late Pliocene-early Pleistocene as a result of epeirogenic uplift and climatic changes (Dolliver, 1984). **0.1**

9.6 Cross Road Creek tributary. **0.5**

10.1 Red-beds adjacent to the road are Upper Triassic Bull Canyon Formation. Look left to see unconformable contact between the Bull Canyon Formation and the overlying Redonda Formation. Road drops into the Bull Canyon Formation. **0.3**

10.4 Cross another tributary of Road Creek. Badlands for next mile on left and right are developed in Bull Canyon Formation. **1.0**

11.4 Crest alluvium-capped ridge. **0.6**

12.0 At 3:00, note salmon-colored, ribbed cliffs on point of mesa. These are strata of the Jurassic Summerville Formation (= type Bell Ranch Formation). The terms Bell Ranch Formation (Griggs and Read, 1959) and "lower member" of the Morrison Formation (e.g., Wanek, 1962; Trauger and Bushman, 1964) were long used to refer to strata in east-central New Mexico properly called Summerville Formation (Anderson and Lucas, 1992; Lucas and Anderson, 1998). These strata occupy the same stratigraphic position and are remarkably similar to Summerville strata farther west. Thus, Summerville strata in east-central New Mexico are as much as 30 m thick and consist of thinly and repetitively bedded, fine-grained sandstone, gypsum, siltstone and sandy mudstone. **0.8**

12.8 At 10:00, Middle Jurassic Entrada Sandstone crops out as tan sandstone overlying Upper Triassic Redonda Formation. **0.6**

13.4 Ledges of ripple-bedded Upper Triassic Redonda Formation on left. **0.9**

14.3 Mesa Rica at 9:30-12:00 to left of highway. Mesa Rica (Spanish for "rich mesa") is said to have been named for the treasure of a group of '49er's returning from California who were massacred by Indians, or possibly for a gold-bearing burro train that met a similar fate. In either case, the loot was hidden on the mesa and has yet to be found (Julyan, 1996). **1.2**

15.5 Cross Johnson Creek. Note gentle fold in Mesa Rica Sandstone on mesa to north, another Laramide structure. **2.3**

17.8 For the next 0.4 mi, excellent red-bed exposures, with prominent eolian sandstones on right, just below road are Upper Triassic Redonda Formation. **0.7**

18.3 **STOP 1. Pull over and park on right side of road.** This stop affords us a close examination of an unusual lithofacies of the Redonda Formation--massive, laterally continuous sandstone--and our discussion is based largely on Hester (1988). This sandstone lithofacies (Fig. 3.4) is composed of very fine to medium grained, moderately sorted sand with minor amounts of coarse sand and occasional wood fragments. Vertical and sub- horizontal *Scoyenia gracilis* burrows are sometimes present in the otherwise structureless sandstones.

These sandstones occur as laterally continuous, 40 cm to 13 m thick, single bed units, although average thickness is 1.5 m. Some sandstone beds persist at the same stratigraphic level for as much as 24 km. Lower contacts are irregular due to load casts, but spheroidal weathering accentuates the irregularity. Locally, small (2-3 cm wide, 10-100 cm long) sandstone dikes may taper downward and branch or be of consistent thickness as they project down into underlying mudstones. Upper contacts are sharp. Burrows on the upper surface are filled with overlying mudstone.

Lack of traction structures within these sandstones could be attributed to: (1) destruction of primary bedforms by organisms, (2) homogeneity of sand, or (3) deposition where currents or waves did not affect sediments. The well-preserved, distinctive character of the burrows, rather than a totally bioturbated fabric, rules out the first possibility. Although not abundant, the occurrence of some coarser material indicates transport energy sufficient to carry coarse sand. Wood fragments might float until they became waterlogged and sunk. Once deposition took place,

FIGURE 3.4. Hammer rests on massive sandstone lithofacies of the Upper Triassic Redonda Formation at Stop 1.

this coarser material was not reworked into traction structures, suggesting that once deposited, massive sandstones were not influenced by moving water.

The overall environment of deposition of the Redonda Formation was lacustrine (Hester, 1988). Coarse materials can be carried into deep lake waters by river plumes (under-,inter-, and overflows), turbidity, or slumping processes. If episodic turbidity currents deposited these sands, some traction structures would be expected. Slumps would generate structures indicative of rapid deposition such as dewatering structures, but homogeniety of the sand and fine grain size might obscure such bedforms. Density flows from rivers transport fine sand and silt into a basin on a continuous basis. This massive sandstone lithofacies of the Redonda Formation thus contains evidence of continuous deposition (bioturbation), but lack of other structures make the process that transported these sands difficult to identify.

After stop continue west on highway. 0.9

19.2 Old road to abandoned rest area comes in from left. This area is known locally as "The Narrows." **0.3**

19.5 **STOP 2.** Pull off on right after crossing small creek. Middle Jurassic Entrada Sandstone well-exposed in road cuts with bouldery alluvium above (Fig. 3.5). Note the Upper Triassic Redonda Formation-Middle Jurassic Entrada Sandstone contact on right side of the road. The uppermost Redonda bed is moderate reddish brown, silty, very fine-grained calcareous litharenite, whereas the basal Entrada bed is moderate orange-pink, fine-grained calcareous subarkose.

FIGURE 3.5. The Middle Jurassic Entrada Sandstone at Stop 2. A, Overview of Entrada roadcut. B, Close-up of eolian foresets.

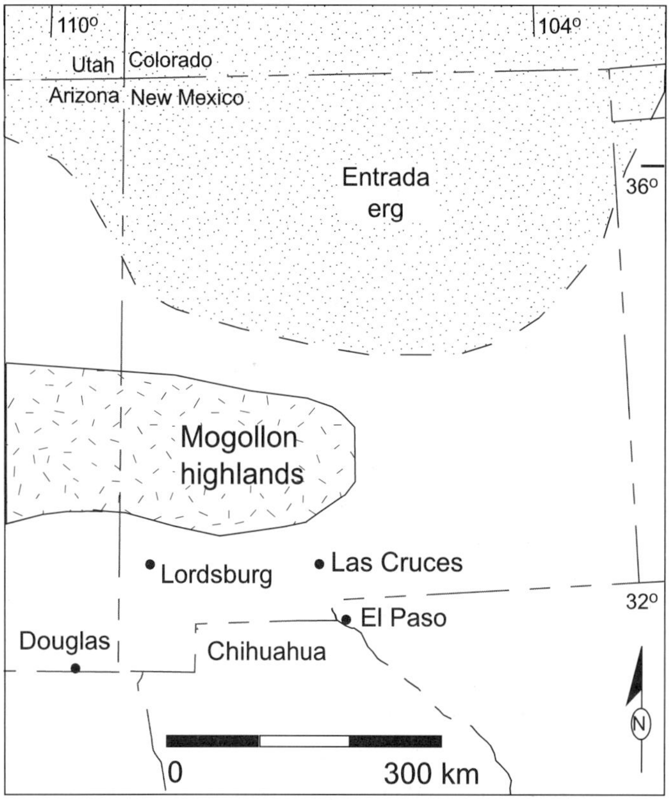

FIGURE 3.6. Simplified paleogeography of New Mexico during Entrada deposition.

The Entrada Sandstone accumulated in a vast sand sea (erg) in the American Southwest during the Middle Jurassic (Bathonian-Callovian) (e.g., Kocurek and Dott, 1983; Blakey et al., 1988; Lucas and Anderson, 1997). This erg covered parts of eastern Utah, southern Wyoming, northeastern Arizona, much of Colorado, northern New Mexico and western Oklahoma (Fig. 3.6). Eolian deposition of the Entrada is inferred from its fine- to medium-grained, well-sorted, subrounded sand grains, which typically display large scale trough crossbeds. Crossbed azimuths indicate regional paleocurrent (wind) direction to the south and southwest (Poole, 1962; Blakey et al., 1988).

This Entrada outcrop here is significant for three reasons:

1. It preserves only dunal and interdunal facies of the upper member of the Entrada Sandstone, the Slick Rock Member. These are characteristic deposits of the vast Entrada erg.

2. Paleocurrent azimuths (foreset dips) here are to the east (Fig. 3.7), which is somewhat anomalous for the regional paleowind patterns.

3. The absence of the lower, Dewey Bridge Member here, and across eastern New Mexico, means we are landward of the Carmel transgression. The Dewey Bridge Member is the paralic facies of the Carmel seaway, which transgressed into central Utah during the Bajocian-Bathonian. In New Mexico, the easternmost outcrops of the Dewey Bridge Member are at Placitas and in the Hagan basin of north-central New Mexico (Lucas et al., 1995; Anderson and Lucas, 2000). This is thus the easternmost extent of the paralic facies of the Carmel seaway

After stop continue west on highway. 0.7

20.2 Crest rise; note Mesa Rica is ahead south of the highway. **1.1**

21.3 Currently driving on Middle Jurassic Entrada Sandstone bench. Note at 10:00 that Summerville Formation forms a ribbed slope overlying the Entrada. **0.2**

21.5 Descend into Upper Triassic Redonda Formation. Note contact of Bull Canyon and Redonda forma-

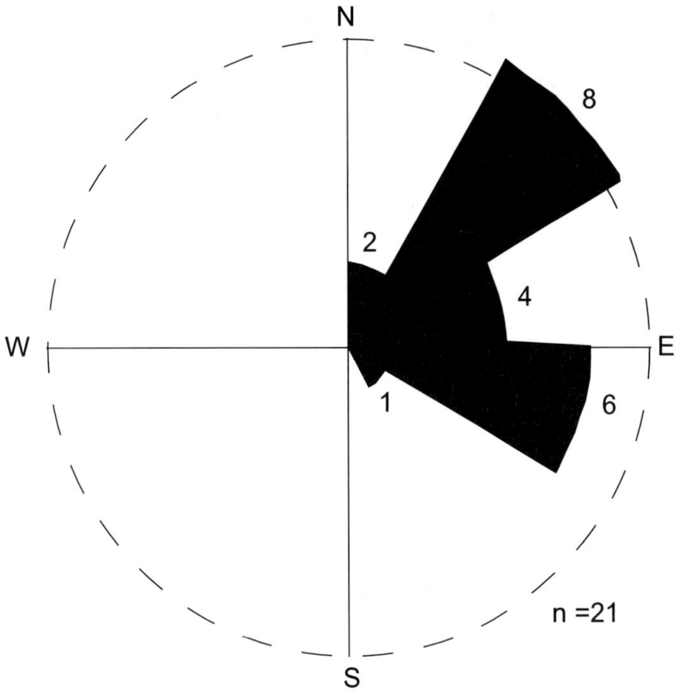

FIGURE 3.7 Rose diagram of crossbed azimuths in Entrada Sandstone at Stop 2.

tions on right. The calcrete that is well exposed on the left side of the road is the basal unit of the Redonda Formation. Note the differences between the two units in the outcrops to the right of the road as we descend the hill. The Bull Canyon Formation has a more brick-red color, and contains more mudstones and lenticular sandstones. **0.8**

22.3 Cross Alamosa Creek. **0.9**

23.2 Crest ridge; to the north (right) the lower breaks are developed in the Upper Triassic Bull Canyon Formation. Beyond, cliffs of Carpenter Mesa are red-ribbed Upper Triassic Redonda Formation overlain by salmon-colored Middle Jurassic Entrada Sandstone and Summerville Formation under a slope of Jurassic Morrison Formation capped by the Cretaceous Mesa Rica Sandstone. **1.5**

24.7 Cross Corralitos Creek; badlands up creek and along the road for the next mile are Upper Triassic Bull Canyon Formation. **2.5**

27.2 Crest low ridge. Note low badlands developed in Upper Triassic Bull Canyon Formation to right in front of Jurassic-Cretaceous section on Carpenter Mesa. Outcrops in this area have yielded fragmentary remains of Triassic reptiles, including numerous teeth of phytosaurs and dinosaurs. **1.3**

28.5 Cross creek. Sandstone exposed in the floor of the creek bed may be the top of the Upper Triassic Trujillo Formation. **0.9**

29.4 Butte at 9:30 is made up of red-bed Upper Triassic Bull Canyon Formation. **0.9**

30.3 Enter greater Conchas Dam metropolitan area. **0.6**

30.9 Junction with NM-433. **Turn right to Conchas Dam State Park. 0.3**

CONCHAS AND UTE RESERVOIRS AND WATER ISSUES IN EASTERN NEW MEXICO

Adrian P. Hunt

Mesalands Dinosaur Museum, Mesa Technical College, 911 South Tenth Street, Tucumcari, NM 88401

The history of dammed reservoirs in eastern New Mexico dates back to 1912 to the design of the first Tucumcari Irrigation Project, which would, if completed, have involved the construction of two dams on Pajarito Creek. In 1925, the Canadian River Development Association was formed to push for a reliable source of irrigation water in east-central New Mexico. After many years of lobbying, Congress passed the Flood Control Act in 1936 that authorized and appropriated funds for the construction of Conchas Dam. In 1937, the New Mexico Legislature approved the formation of the Arch Hurley Conservancy District. The Corps of Engineers finished Conchas Dam in 1939, which was the flood control portion of the project, and by 1954 the Bureau of Reclamation had completed the irrigation component of the project (Geyler, 1998). The total capacity of Conchas Dam is 318,000 acre-feet at spillway crest elevation, but about 68,000 is inactive storage below the invert elevation of Conchas Canal (Mutz, 1998).

The Arch Hurley Conservancy District utilizes approximately 40 mi of main canals and 350 mi of smaller ditches and laterals to deliver 1.33 acre feet of water to 42,000 acres of land. However, while yields from irrigated lands are greatly elevated, the irrigation is used to produce the same crops grown on a "dryland" basis on adjacent acreage, so the full potential of the project has yet to be realized (Geyler, 1998).

The Entrada Sandstone aquifer west of Tucumcari has remained essentially stable since 1943 (Wilson, 1998). However, this aquifer has been recharged by leakage from the Arch Hurley canal, and proposals to line the canal fully could adversely effect this aquifer, which provides all of the water for the city of Tucumcari.

Ute Dam and Reservoir were authorized by the New Mexico State Legislature in 1957, and the initial stage of construction was completed in 1962, with a total capacity at spillway crest elevation of about 110,000 acre-feet. In 1975-1981, the New Mexico State Legislature authorized funding for additional construction to increase the reservoir capacity. By 1984, the expansion was completed, and the capacity increased to 272,000 acre-feet. This increase in size resulted in a lawsuit by the states of Texas and Oklahoma over the terms of the Canadian River Compact of 1950. This suit was settled by the United States Supreme Court, utilizing a Special Master, in 1993 (Mutz, 1998).

Two multi-million dollar projects will result from the construction of Ute Dam. The first is due to a problem at Lake Meredith northeast of Amarillo, which provides a major portion of that urban area's municipal and industrial water. Lake Meredith has chloride levels as high as 350-450 mg/l as a result of the inflow of natural brines into the Canadian River in a six-mile stretch from Ute Reservoir to Revuelto Creek. The brine originates from the dissolution of Permian salts that flow upward from an artesian brine aquifer in the lower Chinle Group. Many factors influence the location and rates of inflow into the Canadian River, but there is strong evidence that the enlargement of Ute Reservoir increased the rate of groundwater brine inflow. Two projects aim to alleviate the chloride problem. First, there will be a large well field in an undeveloped portion of the Ogallala aquifer in Roberts County, Texas that will provide low chloride waters to be blended with those from Lake Meredith. Second, there is the Lake Meredith Salinity Control Project, which involves the drilling of 11 wells near Logan to depths of 120 to 260 ft. Brine from these wells will be transferred by a buried pipeline south of the river to a facility where the water will be injected 3100 feet down into Paleozoic strata. This project is a cooperative effort of the Bureau of Reclamation, the Canadian River Municipal Water Authority and the Texas Water Development Board and will cost about $9.5 million (Wilson and Esparza, 1998).

Wilson and Esparza (1998) detail the Lake Meredith Salinity Control Project and demonstrate that it is utilizing unconventional stratigraphic nomenclature. First, they interpret the Triassic sequence in the Ute Dam area to be Trujillo Sandstone of the Santa Rosa Formation overlying the Tecovas Sandstone of the Santa Rosa Formation. In fact, the bedrock around the dam is more properly considered the Trujillo Formation of the Chinle Group. The Trujillo has never been a member of the Santa Rosa Formation. The Tecovas Formation is not present in New Mexico, and has never been considered a member of the Santa Rosa Formation. The mudstone-dominated Garita Creek Formation underlies the Trujillo Formation, and it is likely that the "Tecovas" is actually the lower Trujillo Formation. Wilson and Esparza (1998, fig. 4) show the Trujillo separated from the "Tecovas" by a "confining shale." It is likely that this is just one of many mudstone intervals within the Trujillo Formation. Also, Wilson and Esparza (1998) note that the brine will be injected into the Pennsylvanian Sangre de Cristo Formation. This unit is not present in the subsurface of the Tucumcari basin (Broadhead and King, 1988), and it is likely that the unit in question is the Permian Abo Formation.

A second major project involving Ute Reservoir is the eastern New Mexico Water Supply Project that has been known by a number of names over 40 years and whose aim is to deliver water from Ute Reservoir to communities in east-central and southeastern New Mexico (Martinez, 1998). This is especially important given the alarming (but predicted) depletion of the Ogallala aquifer, which has a lifetime of from 33 years in Curry County to 40 years in Union County (Wilson, 1998). In 1983, the City of Tucumcari contracted with the Interstate Stream Commission for 3400 acre-feet. Subsequently, in 1987, the Ute Water Commission was formed and involved county commissions from Quay, Curry and Roosevelt counties as well as major and minor communities in these counties. The commission is actually a joint powers agreement entity allowing a sharing of authority among local jurisdictions and the Eastern Plains Council of Governments. In 1987, the United States Congress appropriated $300,000 for an engineering study of the Quay County portion of a proposed pipeline that would go south from Ute Lake to Roosevelt County. Ute Lake's total yield is estimated to be 24,000 acre-feet. Current estimates by Smith Engineering put the total cost of the project, involving a pipeline from Ute Lake as far south as Elida, and including the daunting prospect of raising water up the Llano Estacado, at $212 million with 90% paid by state and federal sources.

At the time of writing (January 2001) the Ute pipeline project is mired in acrimony. Logan threatens lawsuits because of the economic impact on tourism of lowered lake levels. Quay County is considering its own pipeline at an estimated cost of $8-10 million, principally because a shared pipeline to Clovis and Portales would be too large (and expensive) to accommodate peak demand in these urban areas. Tucumcari currently reserves rights to one quarter of the water reserved by the commission, but under the full plan this would cost this small community $2.5 million per year for 20 years. Two things are clear: (1) the Ogallala aquifer is near the end of its life, and the High Plains communities are going to need to find their water somewhere else, so Tucumcari may face a depleted aquifer; and (2) if by 2006, the Ute Water Commission does not implement a plan, other entities such as the City of Albuquerque will most assuredly buy the Ute water rights.

31.2 Lodge Road intersection. **Continue straight. 0.1**

31.3 Road forks, **stay right**, going north by dam. **0.5**

31.8 Parking area and information sign to right; note sandstones of Upper Triassic Trujillo Formation over mudstones of Upper Triassic Garita Creek Formation in slope to right. An incomplete phytosaur jaw was collected from these Garita Creek outcrops. **0.2**

32.0 Trujillo roadcuts to right. Dam to left. **0.8**

32.8 Intersection with road to Conchas recreation area, **continue straight. 0.8**

33.6 Drive along crest of dam, **turn hard right and proceed into parking lot. 0.1**

33.7 **STOP 3.** Conchas Dam (Fig. 3.8). For information on Conchas Dam and the State Park see the accompanying minipaper. The bedrock exposed at Conchas Dam is the lowest strata of the Upper Triassic Chinle Group, the Santa Rosa Formation (Fig. 3.9). In east-central New Mexico, the Santa Rosa Formation consists of three members. The basal Tecolotito Member is up to 104 ft thick and is mostly grayish orange, trough-crossbedded and ripple-laminated quartzarenite. The overlying Los Esteros Member is as much as 134 ft thick and consists of yellow and reddish brown mudstone with minor quartzarenite. The Tres Lagunas Member is up to 140 ft thick and is composed of orange and yellowish brown, crossbedded quartzarenite.

Conchas Dam is anchored in the sandstone-dom-

FIGURE 3.8. Construction of Conchas Dam in 1939 (courtesy of Musem of New Mexico, negative 59160).

inated Santa Rosa Formation, and when the dam was built these strata were informally termed the "canyon sandstone" (e.g., Matthes, 1936). Spiegel (1972) correlated this sandstone interval to the Trujillo Formation of West Texas, which is in the middle of the Chinle Group, but other workers (e.g., Griggs and Hendrickson, 1951; Wanek, 1962; Dane and Bachman, 1965; Kelley, 1972; Lucas et al., 1985a) have assigned it to the Santa Rosa Formation. Recently Fritz (1991, p. 110), in a regional study of the Santa Rosa Formation, concluded "it is not possible to determine whether or not the 'canyon sandstone' represents a member of the Santa Rosa Formation."

Actually, all three members of the Santa Rosa Formation named by Lucas and Hunt (1987) make up the "canyon sandstone." A detailed measured section on the north bank of the river, at the point just downstream of Conchas Dam, makes this clear (Fig. 3.9):

1. The top of the section, the strata on which we are standing, are 10 ft or more of red-bed mudstones, siltstones and ripple-laminated sandstones at the base of the Upper Triassic Garita Creek Formation. Tetrapod footprints are present in these strata (see accompanying minipaper).

2. The underlying cliff is grayish orange and pale yellowish brown sandstones that are trough-cross-bedded and ripple-laminated quartzarenites above a 7-ft-thick calcrete-pebble conglomerate. These strata are ~30 ft thick and are readily assigned to the Tres Laguna Member of the Santa Rosa Formation.

3. These strata overlie ~40 ft of red-bed mudstone and pale reddish brown, very fine grained, ripple-laminated sandstone that belong to the Los

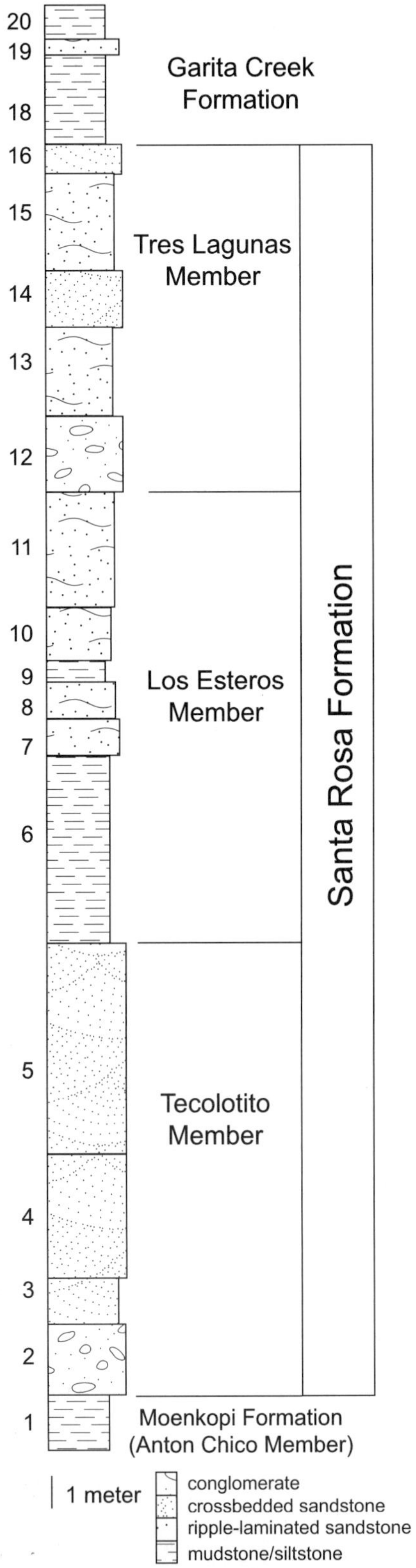

FIGURE 3.9. Measured stratigraphic section of Santa Rosa Formation at Conchas Dam. See Lucas et al. (this guidebook) for a more detailed description of this section.

Esteros Member of the Santa Rosa Formation.

4. Underlying sandstones are pale yellowish brown, trough-crossbedded quartzarenites above a distinctive extraformational conglomerate (Fig. 3.10). Clasts in this conglomerate are extrabasinal lithics, Paleozoic limestones and quartzites up to 6 inches in diameter. This is the extrabasinal conglomerate characteristic of the base of the Chinle Group throughout the western United States (Lucas, 1993). This basal sandstone and conglomerate interval, ~40 ft thick, is thus the Tecolotito Member of the Santa Rosa Formation.

5. Just above river level, the base of the exposed section is pale reddish brown mudstones and siltstones of the Anton Chico Member of the Moenkopi Formation.

After stop proceed out of parking lot. 0.1

FIGURE 3.10. Basal, extraformational conglomerate of Tecolotito Member of Santa Rosa Formation at Conchas Dam.

THE FIRST VERTEBRATE TRACK (*BRACHYCHIROTHERIUM*) FROM THE UPPER CARNIAN GARITA CREEK FORMATION, EAST-CENTRAL NEW MEXICO

Adrian P. Hunt[1] and Spencer G. Lucas[2]

[1]Mesalands Dinosaur Museum, Mesa Technical College, 911 South Tenth Street, Tucumcari, NM 88401; [2]New Mexico Museum of Natural History, 1801 Mountain Road NW, Albuqeruqe, NM 87104-1375

Late Triassic vertebrate tracks in eastern New Mexico are restricted to the Norian Bull Canyon and Redonda formations (Hunt et al., 1993). Here, we report on the first late Carnian tracks, from the Garita Creek Formation. The Garita Creek has otherwise yielded a vertebrate body-fossil assemblage that is of Adamanian (late Carnian) age (Hunt et al., 1989; Hunt and Lucas, 1993).

The tracks are preserved in a 0.4-m thick bed of ripple-laminated and mudcracked, pale reddish brown, fine-grained micaceous litharenite sandstone about 2.5 m above the top of the Santa Rosa Formation. A single *in situ* slab preserves a partial trackway of a quadrupedal tetrapod. One pedal and manual pair of tracks are preserved well enough to be identified (Fig. 3.11). The tracks are preserved in convex hyporelief. The pes impression is 130 mm long and 108 mm wide. There is a well-developed posteromedially-directed posterior extension of the track, giving an over-

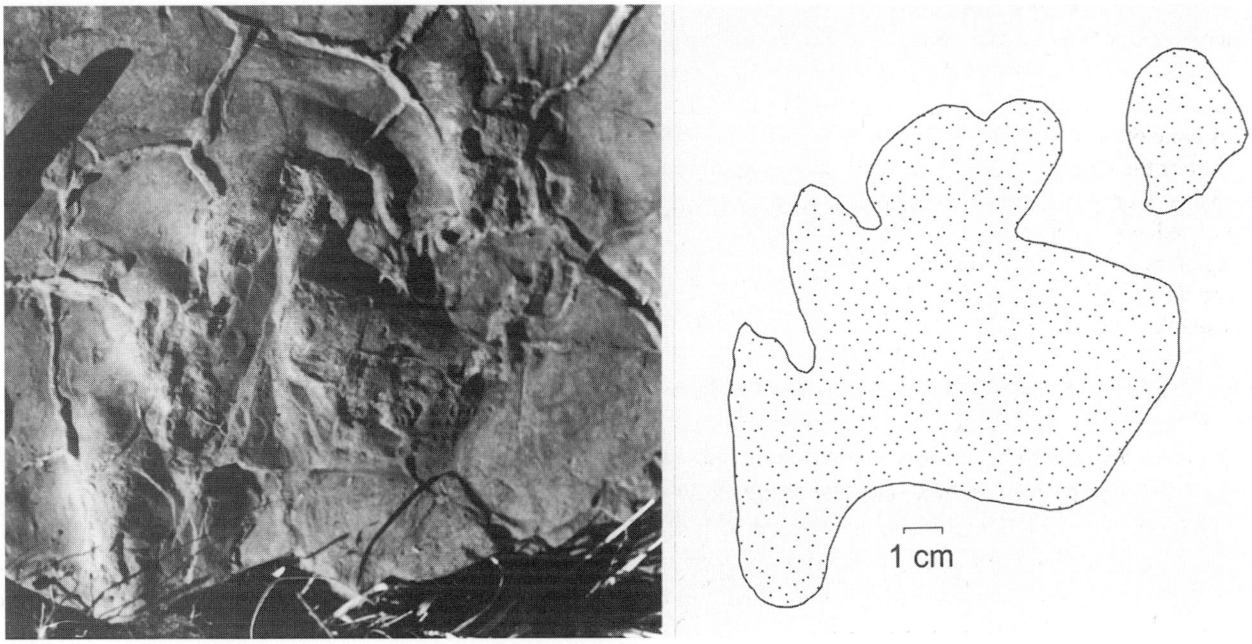

FIGURE 3.11. Track of *Brachychirotherium* in Garita Creek Formation at Conchas Dam.

all L-shape to the pes impression. There is a distinct posterolateral digit impression that is oriented anterolaterally. Anterolaterally, there are the impressions of an additional three, rounded digits. The manus impression is ovoid in shape, with a maximum length and width of 37 mm and 32 mm, respectively. It is located anterior to the midline of the pes impression. There is a second, very poorly preserved pes impression on the slab with a pace of 31 cm. The trackway is trending 242°.

The vertebrate ichnofauna of the Chinle Group of western North America has been well-studied (e.g., Conrad et al., 1987; Hunt et al., 1993; Lockley and Hunt, 1999). There is only one known large (pedal length >100 mm) quadrupedal tracktype with a marked disparity between the size of the pes and manus impressions, and that is *Brachychirotherium*. Further, the lateral direction of the pes digit impressions distinguishes the Garita Creek track from other known Late Triassic tracks that have large pes impressions and small manus imprints, such as *Apatopus* and *Atreipus* (e.g., Baird, 1957; Olsen and Baird, 1986). The narrow gait of the Garita Creek specimens is also consistent with that of *Brachychirotherium* (e.g., Lockley et al., 1987, fig. 5A). Thus, we consider it most parsimonious to assign the Garita Creek track to *Brachychirotherium*. *Brachychirotherium* is conventionally considered to be the track of a derived crurotarsan such as an aetosaur.

This occurrence is the first of a tetrapod track from the Garita Creek Formation and the first from the late Carnian of eastern New Mexico. Indeed, the majority of the tracks known from the Chinle Group are from the uppermost Norian (Apachean) strata in Colorado, Utah, Arizona and New Mexico (Hunt and Lucas, 1992). Only three late Carnian tetrapod tracksites are known from Adamanian strata of western North America, compared to more than 60 from the Apachean (Santucci et al., 1994; Lockley and Hunt, 1999).

CONCHAS LAKE STATE PARK

Virginia T. McLemore

New Mexico Bureau of Geology and Mineral Resources, New Mexico Tech, Socorro, NM 87801

Conchas Lake State Park (Fig. 3.12) was established in 1955 and named after the Conchas River, one of the tributaries of the Canadian River. Conchas is Spanish for shells and was applied to a group of Indians living in the area when seventeenth century Spanish explorers arrived. The word conchas also may be a corrupted name that is confused with the Spanish word conchos, a term also used to describe the Native American tribes in northern New Mexico (Julyan, 1996).

The dam that formed Conchas Lake was the seventeenth dam in the country built by the U. S. Army Corps of Engineers (Young, 1984). It is one of the state's oldest dams, completed in 1939 to control floods, store water for irrigation and local supplies, and assist in local economic recovery from the Depression (Welsh, 1985). More than 3000 people were employed to build the dam. The structure is a concrete and earthen dam that is 235 ft high and 1250 ft long. The lake has a reservoir capacity of 528,951 acre feet of water and sediment. It extends 14 mi upstream of the Canadian River and 11 mi upstream of the Conchas River. The total cost of land purchases, construction and maintenance amounted to nearly $15.8 million (Young, 1984; Welsh, 1985). Water from the lake is used to irrigate 41,400 acres of farmland of the Tucumcari Irrigation Project and travels through 300 mi of canals and laterals.

Conchas Lake State Park ranges in elevation from 4150 to 4250 ft above sea level and occupies 1160 acres of land and 9000 acres of water (surface). Only three sites are administered by the State Parks Department (Northside, Central, and South); most of the land surrounding Conchas Lake is private (Fig. 3.12). The U. S. Army Corps of Engineers administers a day-use site and a visitor's center (open on weekdays). All of the water is open to the public. Sandy beaches are separated by rocky beaches and offer a diversity of opportunities appreciated by both swimmers and fisherman. Coves abound and provide solitude, even during busy holiday weekends.

Recreational activities include hiking, picnicking, camping, swimming, fishing, boating, sailing, scuba diving and water skiing. Camp and picnic sites are along the lake as well as along hills and mesas above water level. Launching ramps for boats and a marina are available. The State Park Office is located at the Northside Recreation Area. The State Department of Game and Fish stocks the lake with walleye pike, small- and large-mouth bass, catfish, crappie, bluegill and northern pike. A playground, golf course, lodge, seaplane landing area, paved airport and full restroom facilities with showers, water and electrical hookups are available. Scuba diving is popular. Cannon Air Force Base also has a small recreation area for military personnel.

Arrowheads, stone tools and pottery shards found throughout the area remind us that prehistoric man probably inhabited the area long ago. Arrowheads found in now extinct bison indicate that Folsom man, named for the town in Union County, lived north of Conchas Lake about 8000–10,000 years ago. Petroglyphs are found in sandstones along some of the bays and inlets of Conchas Lake and also testify to the presence of early inhabitants. Most of these early inhabitants were primarily hunters and gathers. A relic petroglyph in sandstone rests in a building near the Army Corps of Engineers administration building, north of the spillway. When the Spanish explored the area in the 1500s, the Ute, Comanche and Apache Indians called the area covered by Ute Lake and Logan their hunting grounds (Scott, 1986). The first Spaniard in the immediate area was probably Don Antonio de Valverde, who traveled along the Canadian River to Kansas in 1719, before Conchas Lake was formed (Scott, 1986).

Later, Spanish explorers settled in the area that was once part of the Pablo Montoya or Baca land grant. In 1867, John Watts acquired title to the Baca land grant and transferred it to Wilson Waddington in 1872. Waddington conveyed usage of the land to the Consolidated Land, Cattle and Wool Growing Co. This was the beginning of the Bell Ranch.

Most of the rocks surrounding Conchas Lake, including those that form the bedrock of the dam, belong to the Upper Triassic Chinle Group. The Chinle Group consists of alternating layers of red-brown to maroon to gray mudstone, siltstone and sandstone that were deposited in continental fluvial and lacustrine environ-

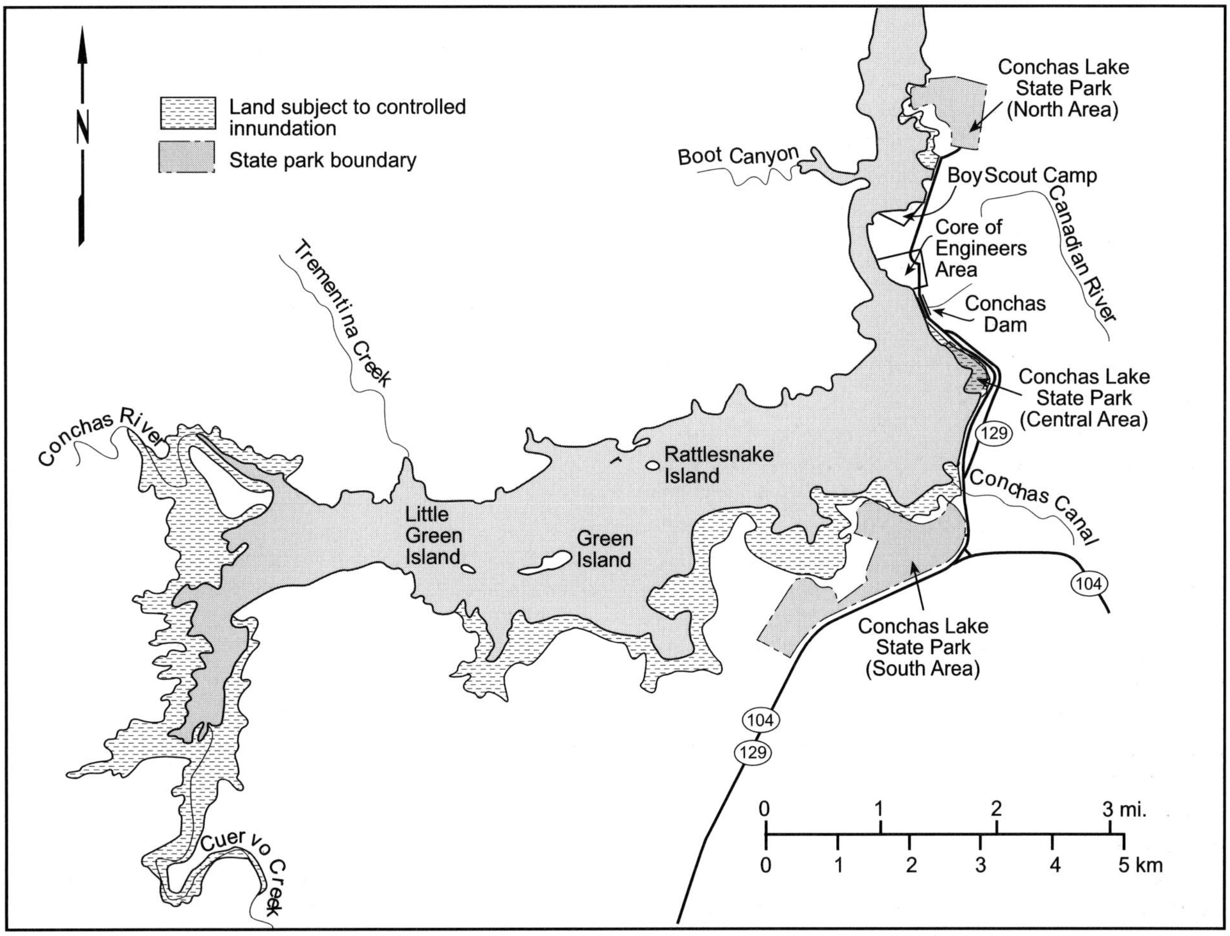

FIGURE 3.12. Map of Conchas Lake.

ments about 205-225 Ma. The Chinle Group is divided into five formations in eastern New Mexico (in ascending order): Santa Rosa, Garita Creek, Trujillo, Bull Canyon and Redonda formations (Lucas, 1995). The medium- to light-gray, massive to crossbedded sandstone exposed at the dam is the Santa Rosa Formation. The crossbeds in the sandstone are consistent with fluvial deposition. Mudcracks indicate that sometimes the sandstones were exposed to drying. The red color is produced by the oxidation of iron within the minerals forming the sandstone; this oxidation is common to seasonally arid environments such as existed at the time the Chinle was deposited. Sections of the core drilled and recovered during construction of the dam are on display near the Army Corps of Engineers administration building, north of the spillway.

Channel deposits of gravel and sand derived from the glaciated terrain in the Sangre de Cristo Mountains during the Pleistocene are found along the Canadian River above and below the dam (Spiegel, 1972). The Canadian River formed during the Pleistocene after the Wisconsinan glaciation (Dolliver, 1984). The headwaters are in the Rocky Mountains in Las Animas County, Colorado, northwest of Raton, New Mexico. The river flows southeastward from the headwaters to Raton, where it flows south to Conchas Lake State Park. At Conchas Lake the river flows due east to Ute Lake State Park and eastward into West Texas and Oklahoma. For most of the river's course in New Mexico, it is a sinuous meander belt that rarely exceeds 750 ft in width (Kessler, 1972). Floods since 1938 have eroded and degraded the river rather than deposit new stream deposits (Kessler, 1972).

33.8 **Turn right** at yield sign and cross dam. **0.4**

34.2 **Turn right** into picnic area for lunch. Note at the parking lot the large cylinders of Triassic sandstone from the coring done while constructing Conchas Dam.

End of Third-day Road Log.

ROAD-LOG REFERENCES

Adams, J. E., 1929, Triassic of West Texas: American Association of Petroleum Geologists Bulletin, v. 12, p. 1045-1055.

Adkins, W. S., 1932, The Mesozoic systems in Texas: University of Texas, Bulletin 3232, p. 239-518.

Agogino, G. A., and Egan, G. N., 1972, Indians of eastern New Mexico: New Mexico Geological Society, 23rd Field Conference, Guidebook, p. 137-140.

Anderson, O. J., and Lucas, S. G., 1992, The Middle Jurassic Summerville Formation, northern New Mexico: New Mexico Geology, v. 14, p. 79-92.

Anderson, O. J., and Lucas, S. G., 2000, Jurassic stratigraphy at Placitas, New Mexico and its regional significance: Geological Society of America, Abstracts with Programs, v. 32, no. 7, p. A-458.

Ash, S. R., 1976, Occurrence of the controversial plant *Sanmiguelia* in the Upper Triassic of Texas: Journal of Paleontology, v. 50, p. 799-804.

Ash, S. R., 1980, Upper Triassic floral zones of North America, *in* Dilcher, D. L., and Taylor, T. N., eds., Biostratigraphy of Fossil Plants: Stroudsburg, Dowden, Hutchinson and Ross, p. 153-170.

Asquith, G. B., and Cramer, S. L., 1975, Transverse braid-bars in the Upper Triassic Trujillo Sandstone of the Texas Panhandle: Journal of Geology, v. 83, p. 657-661.

Bachman, G. O., 1976, Cenozoic deposits of southeastern New Mexico and an outline of the history of evaporite dissolution: U. S. Geological Survey, Journal of Research, v. 4, p. 135-149.

Baird, D., 1957, Triassic reptile footprint faunules from Milford, New Jersey: Museum of Comparative Zoology, Bulletin 117, p. 449-520.

Baldwin, B., and Muehlberger, W. R., 1959, Geologic studies of Union County, New Mexico: New Mexico Bureau of Mines and Mineral Resources, Bulletin 63, 171 p.

Bates, R. L., Wilpolt, R. H., MacAlpin, A. J., and Vorbe, G., 1947, Geology of the Gran Quivira quadrangle, New Mexico: New Mexico Bureau of Mines and Mineral Resources, Bulletin 26, 57 p.

Beede, J. W., and Christner, D. D., 1926, The geology of Foard County: University of Texas, Bulletin 2607, p. 18-53.

Berkstresser, C. F. Jr., and Mourant, W. A., 1966, Ground-water resources and geology of Quay County, New Mexico: Socorro, New Mexico Bureau of Mines and Mineral Resources Ground-water Report 9, 115 p.

Blakey, R. C., Peterson, F., and Kocurek, G., 1988, Synthesis of late Paleozoic and Mesozoic eolian deposits of the Western Interior of the United States: Sedimentary Geology, v. 56, p. 3-125.

Bojanus, L. H., 1827, De uro nostrate eiusque sceleto commentario. Scripsit et *Bovis primigenii* sceleto auxit: Nova Acta Leopoldina, v. 13, p. 413-467.

Böse, E., 1910, Monografia geologica y paleontologica del Cerro de Muleros cerca de Ciudad Juarez, Estado de Chihuahua, y descripcion de la fauna Cretacea de la Encantada, Placer de Guadalupe, Estado de Chihuahua: Instituto Geologico de Mexico, Boletin no. 25, 193 p.

Bowring, S. A., Erwin, D. H., Jin, Y. G., Martin, M. W., Davidek, K., and Wang, W., 1998, Geochronology of the end-Permian mass extinction: Science, v. 280, p. 1039-1045.

Brand, J. P. and Maddox, R. B., 1972, Pre-Dakota Cretaceous formations in northwestern Texas and northeastern New Mexico: New Mexico Geological Society, 23rd Field Conference, Guidebook, p. 98-104.

Bretz, J. H., and Horberg, C. L., 1949a, The Ogallala Formation west of the Llano Estacado: Journal of Geology, v. 57, p. 477-490.

Bretz, J. H., and Horberg, C. L., 1949b, Caliche in southeast New Mexico: Journal of Geology, v. 57, p. 491-511.

Broadhead, R. F., 1984, Subsurface petroleum geology of Santa Rosa Sandstone (Triassic), northeast New Mexico: New Mexico Bureau of Mines and Mineral Resources, Circular 193, 22 p.

Broadhead, R. F., and King, W. E., 1988, Petroleum geology of Pennsylvanian and Lower Permian strata, Tucumcari basin, east-central New Mexico: New Mexico Bureau of Mines and Mineral Resources Bulletin 119, 75 p.

Carroll, H. B., 1951, The Texan Santa Fe trail: Panhandle-Plains Historical Review, v. 24, p. 1-201.

Case, E. C., 1914, The red beds between Wichita Falls, Texas and Las Vegas, New Mexico in relation to their vertebrate fauna: Journal of Geology, v. 22, p. 243-259.

Cazeau, C.J., 1960, Upper Triassic deposits of west Texas and northeastern New Mexico [Ph. D. dissertation]: Chapel Hill, University of North Carolina, 94 p.

Clemons, R. E., 1982, New Mexico highway geology map: Socorro, New Mexico Geological Society, 1 sheet, scale 1:1,000,000.

Cobban, W.A., 1977, Characteristic marine molluscan fossils from the Dakota Sandstone and intertongued Mancos Shale, west-central New Mexico: U.S. Geological Survey, Professional Paper 1009, 30 p.

Cobban, W.B., 1985, The Lower Cretaceous ammonite *Schloenbachia leonenis* Conrad var. *equidistans* Cragin: U.S. Geological Survey, Bulletin 1641-A, 4 p.

Conrad, K., Lockley, M. G., and Prince, N. K., 1987, Triassic and Jurassic vertebrate-dominated trace fossil assemblages of the Cimarron Valley region: implications for paleoecology and biostratigraphy: New Mexico Geological Society, 38th Field Conference, Guidebook, p. 127-138.

Cope, E. D., 1893, A preliminary report on the vertebrate paleontology of the Llano Estacado: Geological Survey of Texas, Fourth Annual Report, p. 1-137.

Cornet, B., 1986, The leaf venation and reproductive structures of a Late Triassic angiosperm, *Sanmiguelia lewisii*: Evolutionary Theory, v. 7, p. 231-309.

Cummins, W. F., 1890, The Permian of Texas and its overlying beds: Texas Geological Survey, First Annual Report, p. 183-197.

Dane, C. H., and Bachman, G. O., 1965, Geologic map of New Mexico: Washington, D. C., U. S. Geological Survey, 2 sheets, scale 1:500,000.

Darton, N. H., 1928, "Red beds" and associated formations in New Mexico, with an outline of the geology of the state: U. S. Geological Survey, Bulletin 794, 372 p.

DeFord, R. K., 1972, Marcou in east-central New Mexico: New Mexico Geological Society, 23rd Field Conference, Guidebook, p. 65-71.

DeLuca, J. L., and Ericksson, K. A., 1989, Controls on synchronous ephemeral and perennial-river sedimentation in the middle sandstone member of the Triassic Chinle Formation, northeastern New Mexico: Sedimentary Geology, v. 61, p. 155-175.

Dobrovolny, E., and Townsend, R. C., 1946, An eolian dam near Tucumcari: Economic Geology, v. 42, p. 492-497.

Dobrovolny, E., Bates, R. L., and Summerson, C. H., 1946, Geology of northwestern Quay County, New Mexico: U. S. Geological Survey, Oil and Gas Investigations Map OM-62, 2 sheets.

Dolliver, P. N., 1984, Cenozoic evolution of the Canadian River basin: Baylor Geological Studies, Bulletin 42, 96 p.

Dolliver, P. N., 1985, The Plio-Pleistocene Canadian breaks of New Mexico: A profile: New Mexico Geological Society, 36th Field Conference, Guidebook, p. 315-318.

Drake, N. F., 1892, Stratigraphy of the Triassic of west Texas: Geological Survey of Texas, Third Annual Report, p. 227-247.

Dubiel, R. F., 1989, Depositional and climatic setting of the Upper Triassic Chinle Formation, Colorado Plateau, *in* Lucas, S. G., and Hunt, A. P., eds., Dawn of the Age of Dinosaurs in the American Southwest: Albuquerque, New Mexico Museum of Natural History, p. 171-187.

Dubiel, R. F., Parrish, J. T., Parrish, M. J., and Good, S. C., 1991, The Pangaean megamonsoon – evidence from the Upper Triassic Chinle Formation, Colorado Plateau: Palaios, v. 6, p. 347-370.

Dunay, R. E., 1972, The palynology of the Triassic Dockum Group of Texas, and its application to stratigraphic problems of the Dockum Group [Ph. D. dissertation]: College Station, Pennsylvania State University, 382 p.

Dunay, R. E., and Fisher, M J., 1979, Palynology of the Dockum Group (Upper Triassic), Texas, U.S.A.: Review of Paleobotany and Palynology, v. 28, p. 61-92.

Elias, M. K., 1931, The geology of Wallace County, Kansas: State Geological Survey of Kansas, Bulletin 18, 254 p.

Evans, G. L., and Meade, G. E., 1945, Quaternary of the Texas High Plains: University of Texas, Publication 4401, p. 485-507.

Evans, G. L., Read, C. B., Kelley, V. C., Wood, G. H., Smith, C. T., Northrop, R. C., Strain, W. S., and DeFord, R. K., 1949, Cenozoic geology of the Llano Estacado and Rio Grande valley: West Texas Geological Society and New Mexico Geological Society, 2nd Field Conference, Guidebook, p. 23-24.

Finch, W. I., and Wright, J. C., 1983, Measured stratigraphic sections of uranium-bearing Upper Triassic rocks in eastern New Mexico, west Texas, and the Oklahoma Panhandle with brief discussion of stratigraphic problems: U.S. Geological Survey, Open-file Report 83-701, 118 p.

Fracasso, M. A., and Kolker, A., 1985, Late Permian volcanic ash beds in the Quartermaster and Dewey Lake formations, Texas Panhandle: West Texas Geological Society Bulletin, v. 24, no. 6, p. 5-10.

Frelier, A. P., 1987, Sedimentology, fluvial paleohydrology, and paleogeomorphology of the Dockum Formation (Triassic), west Texas [M. S. thesis]: Lubbock, Texas Tech University, 198 p.

Fritz, T., 1991, Depositional history of the mid-Late Triassic Santa Rosa Formation, eastern New Mexico [M.S. thesis]: Lubbock, Texas Tech University, 120 p.

Gage, J. E., and Asquith, G. B., 1977, Sedimentology of Mesa Rica Sandstone in Tucumcari basin, New Mexico: New Mexico Bureau of Mines and Mineral Resources, Circular 157, 14 p.

Galloway, S. E., 1956, Geology and groundwater resources of the Portales Valley area, Roosevelt and Curry Counties, New Mexico [M. S. thesis]: Albuquerque, University of New Mexico, 167 p.

Galloway, S. E., 1972, Urban geology of Clovis: New Mexico Geological Society, Guidebook 23, p. 212-213.

Gardner, J. N., and Goff, F., 1996, Geology of the northern Valles caldera and Toledo embayment, New Mexico: New Mexico Geological Society, 47th Field Conference, Guidebook, p. 225-230.

Geyler, J., 1998, Persistence: Brief development history of the Tucumcari Irrigation Project (Arch Hurley Conservancy District), *in* Ortega Klett, C. T., ed., Water Issues of Eastern New Mexico: Las Cruces, New Mexico Water Resources Institute Report 304, p. 47-50.

Gorman, J. M., and Robeck, R. C., 1946, Geology and asphalt deposits of north-central Guadalupe County, New Mexico: U. S. Geological Survey, Oil and Gas Investigations Preliminary Map OM-44.

Gould, C. N., 1907, The geology and water resources of the western portion of the Panhandle of Texas: U.S. Geological Survey, Water-supply Paper 191, 170 p.

Granata, G. E., 1981, Regional sedimentation of the Late Triassic Dockum Group, west Texas and eastern New Mexico [M.A. thesis]: Austin, University of Texas, 199 p.

Gregg, J., 1844, Commerce of the Prairies: Norman, University of Oklahoma Press [1954 edition edited by M. L. Moorhead], 469 p.

Gregory, H. E., 1916, The Navajo country, a geographic and hydrographic reconnaissance of parts of Arizona, New Mexico, and Utah: U. S. Geological Water-supply Paper 380, 219 p.

Gregory, H. E., 1917, Geology of the Navajo country - a reconnaissance of parts of Arizona, New Mexico, and Utah: U. S. Geological Survey, Professional Paper 93, 161 p.

Gregory, J. T., 1972, Vertebrate faunas of the Dockum Group, Triassic, eastern New Mexico and west Texas: New Mexico Geological Society, 23rd Field Conference, Guidebook, p. 120-123.

Griggs, R. L., and Hendrickson, G. E., 1951, Geology and ground-water resources of San Miguel County, New Mexico: New Mexico Bureau of Mines and Mineral Resources, Ground-water Report 2, 121 p.

Griggs, R. L., and Read, C. B., 1959, Revisions in stratigraphic nomenclature in Tucumcari-Sabinoso area, northeastern New Mexico: American Association of Petroleum Geologists Bulletin, v. 43, p. 2003-2007.

Gustavson, T. C., 1991, Age and deveopment of the Ogallala "Caprock" calcrete, southern High Plains of northwestern Texas and eastern New Mexico: Geological Society of America, Abstracts with Programs, v. 5, p. A63.

Gustavson, T. C., 1996, Fluvial and eolian depositional systems, paleosols, and paleoclimate of the upper Cenozoic Ogallala and Blackwater Draw formations, southern High Plains, Texas and New Mexico: Bureau of Economic Geology, Report of Investigations 239, 62 p.

Gustavson, T. C., Holliday, V. T., and Hovorka, S. D., 1995, Origin and development of playa basins, sources of recharge to the Ogallala Aquifer, Southern High Plains, Texas and New Mexico: Austin: Bureau of Economic Geology, Report of Investigations 229, 44 p.

Gustavson, T. C., Holliday, V. T., and Schultz, G. E., 1990, Introduction, *in* Gustavson C., ed., Tertiary and Quaternary Stratigraphy and Vertebrate Paleontology of Parts of Northwestern Texas and Eastern New Mexico: Austin, Bureau of Economic Geology Guidebook 24, p. 1-21.

Gustavson, T. C., Baumgardner, R. W., Jr., Caron, S. C., Holiday, V. T., Mehnert, H. H., O'Neill, J. M., and Reeves, C. C., Jr., 1991, Quaternary geology of the southern Great Plains and an adjacent segment of the rolling plains, *in* Morrison, R. B., ed., Quaternary Nonglacial Geology; Conterminous U. S.: Boulder, Geological Society of America, Vol. K-2, The Geology of North America, p. 477-501.

Harbour, J. 1975. General stratigraphy, *in* Wendorf, F., and Hester, J. J., eds., Pleistocene Environments of the Southern High Plains: Taos, Fort Burgwin Research Center Publication 9, p. 33-55.

Hasenmueller, W.A., and Hattin, D.E., 1990, New species of the bivalve *Anomia* from lower and middle Turonian parts of the Greenhorn Limestone, central Kansas: Journal of Paleontology, v. 64, p. 104-110.

Hawley, J. W., 1984, The Ogallala Formation in eastern New Mexico: Proceedings of the Ogallala Aquifer Symposium II: Texas Tech University, Water Resources Center, p. 157–176.

Hawley, J. W., 1993, The Ogallala and Gatuña formations in southeastern New Mexico region, a progress report: New Mexico Geological Society, 44th Field Conference, Guidebook, p. 261-269.

Haynes, C. V., 1995, Geochronology of paleoenvironmental change, Clovis type site, Blackwater Draw, New Mexico: Geoarchaeology, v. 10, p. 317-388.

Hernandez, F., 1651, Rerum medicarum Novae Hispaniae thesaurus; seu, Plantarum animalium mineralium Mexicanorum historia: Rome, Vitalis Mascardi.

Hershkovitz, P., 1957, The type locality of *Bison bison* Linnaeus: Proceedings of the Biological Society of Washington, v. 70, p. 31-32.

Hester, J. J., 1972, Blackwater Draw Locality No. 1. A stratified, early man site in eastern New Mexico: Fort Burgwin Research Center, Publication 8, 239 p.

Hester, P. M., 1988, Depositional environments of a Late Triassic lake, east-central New Mexico [M.S. thesis]: Albuquerque, University of New Mexico, 152 p.

Hill, M. G., Holliday, V. T., and Stanford, D. J., 1995, A further evaluation of the San Jon site, New Mexico: Plains Anthropologist, v. 40, p. 369-390

Hill, R.T., 1893, Paleontology of the Cretaceous formations of Texas--the invertebrate paleontology of the Trinity Division: Proceedings of the Biological Society of Washington, vol. 8, p. 9-40.

Holbrook, J. M., and Wright, R., 1992, Depositional history of Lower Cretaceous strata in northeastern New Mexico: Implications for regional tectonics and depositional sequences: Geological Society of America Bulletin, v. 104, p. 802-813.

Holbrook, J. M., Wright, R., and Kietzke, K. K., 1987, Stratigraphic relationships at the Jurassic-Cretaceous boundary in east-central New Mexico: New Mexico Geological Society, 38th Field Conference, Guidebook, p. 161-165.

Holliday, V. T., 1989. The Blackwater Draw Formation (Quaternary): A 1.4-plus m.y. record of eolian sedimentation and soil formation on the Southern High Plains: Geological Society of America Bulletin, v. 101, p. 1598-1607.

Holliday, V. T., 1990, Soils and landscape evolution of eolian plains: The Southern High Plains of Texas and New Mexico: Geomorphology, v. 3, p. 489-515.

Holliday, V. T., 1995, Stratigraphy and paleoenvironments of late Quaternary valley fills on the southern High Plains: Geological Society of America, Memoir 186, 136 p.

Holliday, V. T., 1997, Paleoindian Geoarchaeology of the Southern High Plains. Austin, University of Texas Press, 297 p.

Hunt, A.P., 1989, A new ornithischian dinosaur from the Bull Canyon Formation (Upper Triassic) of east-central New Mexico, *in* Lucas, S. G., and Hunt, A. P., eds., Dawn of the Age of Dinosaurs in the American Southwest: Albuquerque, New Mexico Museum of Natural History, p. 355-358.

Hunt, A. P., 1991, The first tetrapod faunas from the Trujillo Formation (Late Triassic) of east-central New Mexico and their biochronological and paleoecological significance: New Mexico Geology, v. 13, p. 93.

Hunt, A P., 1994, Vertebrate paleontology and biostratigraphy of the Bull Canyon Formation (Chinle Group: Norian), east-central New Mexico with revisions of the families Metoposauridae (Amphibia: Temnospondyli) and Parasuchidae (Reptilia: Archosauria) [Ph.D. dissertation]: Albuquerque, University of New Mexico, 403 p.

Hunt, A. P., and Lucas, S. G., 1991, The *Paleorhinus* biochron and the correlation of the nonmarine Upper Triassic of Pangaea: Palaeontology, v. 34, p. 487-501

Hunt, A. P., and Lucas, S. G., 1992, Stratigraphic distribution and age of vertebrate tracks in the Chinle Group (Upper Triassic), western North America: Geological Society of America, Abstracts with Programs, v. 24, no. 6, p. 19.

Hunt, A. P., and Lucas, S. G., 1993, Triassic vertebrate paleontology and biochronology of New Mexico: New Mexico Museum of Natural History and Science, Bulletin 2, p. 49-60.

Hunt, A. P., Lockley, M. G., and Lucas, S. G., 1993, Vertebrate and invertebrate tracks and trackways from Upper Triassic strata of the Tucumcari Basin, east-central New Mexico: New Mexico Museum of Natural History and Science, Bulletin 3, p. 199-201.

Hunt, A. P., Lucas, S. G., and Sealey, P. L., 1989, Paleontology and vertebrate biochronology of the Upper Triassic Garita Creek Formation, east-central New Mexico: New Mexico Journal of Science, v. 29, p. 61-68.

Izett, G. A., Wilcox, R. E., and Borchart, G. A., 1972, Correlation of a volcanic ash bed in Pleistocene deposits near Mt. Blanco, Texas with the Guaje pumice bed of the Jemez Mountains, New Mexico: Quaternary Research, v. 2, p. 554-578.

Johns, D. A., 1989, Lithogenetic stratigraphy of the Triassic Dockum Formation, Palo Duro Basin: Bureau of Economic Geology, Report of Investigations 182, 71 p.

Johns, D. A., and Granata, G. E., 1987, Regional tectonic setting and depositional trends of the Dockum Group, West Texas and eastern New Mexico: Journal of the Arizona-Nevada Academy of Science, v. 22, p. 53-72.

Judson, S., 1953. Geology of the San Jon Site, eastern New Mexico: Smithsonian Miscellaneous Collections 121, 70 p.

Julyan, R., 1996, Place Names of New Mexico. Albuquerque, University of New Mexico Press, 385 p.

Katz, L., 1997, The History of Blackwater Draw. Portales, Eastern New Mexico University, 30 p.

Kelley, V. C., 1972, Triassic rocks of the Santa Rosa area: New Mexico Geological Society, 23rd Field Conference, Guidebook, p. 84-90.

Kesser, L. G., II, 1972, Channel geometry, development, and variation, South Canadian River, eastern New Mexico and West Texas: New Mexico Geological Society, Guidebook 23, p. 165-167.

Kiatta, H. W., 1960, A provenance study of the Triassic deposits of northwestern Texas [M.A. thesis]: Lubbock, Texas Tech University, 63 p.

Kirkland, J.I., 1996, Paleontology of the Greenhorn cyclothem (Cretaceous: late Cenomanian to middle Turonian) at Black Mesa, northeastern Arizona: New Mexico Museum of Natural History and Science, Bulletin 9, 131 p.

Knight, C., 1849, Sketches in Natural History: History of the Mammalia: London, C. Cox.

Knudson, R., 1995. The San Jon points and Paleoindian typology: Plains Anthropologist, v. 40, p. 391-397.

Kocurek, G., and Dott, R. H., Jr., 1983, Jurassic paleogeography and paleoclimate of the central and southern Rocky Mountain region, *in* Reynolds, M. W., and Dolly, E. D., eds., Mesozoic Paleogeography of the West-central United States: Denver, Society of Economic Paleontologists and Mineralogists, Rocky Mountain Section, p. 101-118.

Kolker, A., and Fracasso, M. A., 1985, K-Ar age of a volcanic ash bed in the Quartermaster and Dewey Lake formations (Late Permian); Isochron/West, no. 42, p. 17-19.

Kues, B. S., 1985, Stratigraphy of the Tucumcari area: A historical account: New Mexico Geological Society, 36th Field Conference, Guidebook, p. 119-140.

Kues, B. S., 1989, Taxonomy and variability of three *Texigryphaea* (Bivalvia) species from their Lower Cretaceous (Albian) type localities in New Mexico and Oklahoma: Journal of Paleontology, v. 63, p. 454-483.

Kues, B.S., 1997, New bivalve taxa from the Tucumcari Formation (Cretaceous, Albian) New Mexico, and the biostratigraphic significance of the basal Tucumcari fauna: Journal of Paleontology, vol. 71, p. 820-839.

Kues, B. S., and Kietzke, K. K.,1985, Supplemental road log 3, Mesa Quemado and Pyramid Mountain: New Mexico Geological Society, 36th Field Conference, Guidebook, p. 80-85.

Kues, B. S., and Lucas, S. G., 1987, Cretaceous stratigraphy and paleontology in the Dry Cimarron valley, New Mexico, Colorado and Oklahoma: New Mexico Geological Society, 38th Field Conference, Guidebook, p. 167-198.

Kues, B. S., Lucas, S. G., Kietzke, K. K., and Mateer, N. J., 1985, Synopsis of Tucumcari Shale, Mesa Rica Sandstone and Pajarito Shale paleontology, Cretaceous of east-central New Mexico: New Mexico Geological Society, 36th Field Conference, Guidebook, p. 261-281.

Lambert, W. D., 1990, Rediagnosis of the genus *Amebelodon* (Mammalia, Proboscidea, Gomphotheriidae), with a new subgenus and species, *Amebelodon* (*Konobelodon*) *britti*: Journal of Paleontology, v. 64, p. 1032-1040.

Lehman, T. M., 1994, The saga of the Dockum Group and the case of the Texas/New Mexico boundary fault: New Mexico Bureau of Mines and Mineral Resources Bulletin 150, p. 37-52.

Linnaeus, C., 1758, Systema naturae per regna tria naturae, secundum classes, ordines, genera, species cum characteribus, differentiis, synonymis, locis: Stockholm, Laurentii Salvii.

Lockley, M. G., and Hunt, A. P., 1999, Dinosaur Tracks and Other Fossil Footprints of the Western United States. New York, Columbia University Press (revised paperback edition), 338 p.

Love, D. W., 1985, Geomorphic history of part of Tucumcari Lake drainage basin, New Mexico: New Mexico Geological Society, 36th Field Conference, Guidebook, p. 319-324.

Lucas, S. G., 1991, Correlation of Triassic strata of the Colorado Plateau and southern High Plains, New Mexico: New Mexico Bureau of Mines and Mineral Resources, Bulletin 137, p. 47-56.

Lucas, S. G., 1993, The Chinle Group: revised stratigraphy and biochronology of Upper Triassic nonmarine strata in the western United States: Museum of Northern Arizona, Bulletin 59, p. 27-50.

Lucas, S. G., 1995, Triassic stratigraphy and chronology in New Mexico: New Mexico Geology, v. 17, p. 8-13, 17.

Lucas, S. G., 1998, Global Triassic tetrapod biostratigraphy and biochronology: Palaeogeography, Palaeoclimatology, Palaeoecology, v. 143, p. 347-384.

Lucas, S. G., and Anderson, O. J., 1992, Triassic stratigraphy and correlation, west Texas and eastern New Mexico, *in* Cromwell, D. W., Moussa, M. T., and Mazzullo, L. J., eds., Transactions Southwest Section AAPG, Midland, Texas, 1992: Midland, West Texas Geological Society, p. 201-207.

Lucas, S. G., and Anderson, O. J., 1994, Ochoan (Late Permian) stratigraphy and chronology, southeastern New Mexico and west Texas: New Mexico Bureau of Mines and Mineral Resources, Bulletin 150, p. 29-36.

Lucas, S. G., and Anderson, O. J., 1997, The Jurassic San Rafael Group, Four Corners region: New Mexico Geological Society, 48th Field Conference, Guidebook, p. 115-132.

Lucas, S. G., and Anderson, O. J., 1998, Jurassic stratigraphy and correlation in New Mexico: New Mexico Geology, v. 20, p. 97-104.

Lucas, S.G., and Hunt, A. P., 1989, Revised Triassic stratigraphy, Tucumcari basin, east-central New Mexico, *in* Lucas, S. G., and Hunt, A. P., eds., Dawn of the Age of Dinosaurs in the American Southwest: Albuquerque, New Mexico Museum of Natural History, p. 150-170.

Lucas, S. G., and Hunt, A. P., 1993, Tetrapod biochronology of the Chinle Group (Upper Triassic), western United States: New Mexico Museum of Natural History and Science, Bulletin 3, p. 327-329.

Lucas, S.G., and Hunt, A.P., 2000, The Lower Cretaceous ammonite *Mortoniceras equidistaus* (Cragin) from the Mesa Rica Sandstone, Quay County, New Mexico: New Mexico Museum of Natural History and Science, Bulletin 16, p. 97-100.

Lucas, S.G., and Kisucky, M. J., 1988, Type and reference section of the Tucumcari, Mesa Rica and Pajarito formations, Cretaceous of east-central New Mexico: New Mexico Geology, v. 10, p. 82-89.

Lucas, S. G., Anderson, O. J., and Hunt, A. P., 1994, Triassic stratigraphy and correlations, southern High Plains of New Mexico-Texas: New Mexico Bureau of Mines and Mineral Resources, Bulletin 150, p. 105-126.

Lucas, S. G., Anderson, O. J., and Pigman, C., 1995, Jurassic stratigraphy in the Hagan basin, north-central New Mexico: New Mexico Geological Society, 46th Field Conference, Guidebook, p. 247-255.

Lucas, S. G., Hunt, A. P., and Morales, M., 1985, Stratigraphic nomenclature and correlation of Triassic rocks of east-central New Mexico: A preliminary report: New Mexico Geological Society, 36th Field Conference, Guidebook, p. 171-184.

Lucas, S. G., Estep, J. W., Pence, R., and Hunt, A. P., 2000, Upper Cretaceous outlier in Quay County, New Mexico: Stratigraphy and biostratigraphy: New Mexico Museum of Natural History and Science, Bulletin 16, p. 121-127.

Lundelius, E. L. Jr., 1972, Vertebrate remains from the Gray Sand; *in* Hester, J. J., ed., Blackwater Draw Locality No. 1. A stratified early man site in eastern New Mexico: Fort Burgwin Research Center, Publication 8, p. 148-163.

Lurie, E., 1974, Marcou, *in* Gillespie, C. C., ed. Dictionary of Scientific Biography: New York, Charles Scribner and Sons, v. 9, p. 99-101.

MacFadden, B. J., 1977, Magnetic polarity stratigraphy of the Chamita Formation stratotype (Mio-Pliocene) of north-central New Mexico: American Journal of Science, v. 277, p. 769-800.

Marcou, J., 1858, Geology of North America, With Reports on the Prairies of Arkansas and Texas, the Rocky Mountains of New Mexico, and the Sierra Nevada of California, Originally Made for the United States Government: Zurich, Zurcher and Furrer, 144 p.

Marcy, R. B., 1852, Map of the country between the frontiers of Arkansas and New Mexico, embracing the section explored in 1849, 50, 51 & 52 accompanying Explorations of the Red River of Louisiana in the Year, 1852: 32nd Congress, 2nd Session, Senate Executive Document 54.

Martinez, E., 1998, Keynote Address, *in* Ortega Klett, C. T., ed., Water Issues of Eastern New Mexico: Las Cruces, New Mexico Water Resources Institute Report 304, p. 1-4.

Mateer , N. J., 1985, Pre-Graneros Cretaceous stratigraphy of northeastern New Mexico: New Mexico Geological Society, 36th Field Conference, Guidebook, p. 243-246.

Matthes, G. H., 1936, Problems at Conchas Dam: Civil Engineering, v. 6, p. 437-441.

Matthew, W. D., and Stirton, R. A., 1930, Equidae from the Pliocene of Texas: University of California Publications, Bulletin of the Department of Geological Sciences, v. 19, p. 349-396.

May, B. S., 1988, Depositional environments, sedimentology, and stratigraphy of the Dockum Group (Triassic) in the Texas Panhandle [M. S. thesis]: Lubbock, Texas Tech University, 180 p.

May, B. S., and Lehman, T. M., 1989, Sedimentology of The Dockum Group (Triassic), Texas Panhandle: Geological Society of America, Abstracts with Programs, v. 21, p. 34.

McDonald, J. N., 1981, North American Bison: Their Classification and Evolution. Berkeley, University of California Press, 315 p.

McGowan, J. H., 1979, Depositional facies and uranium occurrences, Triassic (Dockum Group), Texas Panhandle: American Association of Petroleum Geologists Bulletin, v. 63, p. 835.

McGowan, J. H., 1980, Triassic lacustrine basins, Texas, *in* Galloway, W. E., ed., Depositional and Groundwater Flow Systems in the Exploration for Uranium: Austin, University of Texas Bureau of Economic Geology, p. 80-117.

McGowan, J. H., Granata, G. E., and Seni, S. J., 1979, Depositional framework of the lower Dockum Group (Triassic), Texas Panhandle: Bureau of Economic Geology, Report of Investigations 97, 60 p.

McGowan, J. H., Granata, G. E., and Seni, S. J., 1983, Depositional setting of the Triassic Dockum Group, Texas Panhandle and eastern New Mexico, *in* Reynolds, M. W., and Dolly, E. D., eds., Mesozoic Paleogeography of the West-central United States: Denver, Society of Economic Paleontologists and Mineralogists, Rocky Mountain Section, p. 13-38.

McGrath, D. A., and Hawley, J. W., 1985, Calcretes of the Santa Rosa area: New Mexico Geological Society, 36th Field Conference Guidebook, p. 4–6.

McKallip, C. Jr., 1985, History of the Newkirk oil field: New Mexico Geological Society, 36th Field Conference, Guidebook, p. 36-37.

McLemore, V. T., and North, R. M., 1985, Copper and uranium mineralization in east-central New Mexico: New Mexico Geological Society, 36th Field Conference Guidebook, p. 289-300.

Mehnert, H. H., and O'Neil, M. J., 1991, Late Cenozoic vulcanism, uplift, and erosion in the Ocate volcanic field, north-central New Mexico; in Morrisson, R. B., ed., Quaternary nonglacial geology: Conterminous United States: Boulder, Geological Society of America, p. 486-491.

Metz, L. C., 1994, Roadside History of Texas. Missoula, Mountain Press, 448 p.

Molina-Garza, R. S., Geissman, J. W., and Lucas, S. G., 2000, Paleomagnetism and magnetostratigraphy of uppermost Permian strata, southeast New Mexico: Correlation of the Permian-Triassic boundary in nonmarine environments: Geophysical Journal International, v. 141, p. 778-786.

Molina-Garza, R. S., Geissman, J. W., and Van der Voo, R., 1989, Paleomagnetism of the Dewey Lake Formation (Late Permian), northwest Texas: End of the Kiaman Superchron in North America: Journal of Gephysical Research, v. 94, p. 17881-17888.

Morgan, G. S., and Lucas, S. G., 2001, Mammalian biochronology of Blancan and Irvingtonian (Pliocene and early Pleistocene) faunas from New Mexico: Bulletin of the American Museum of Natural History, in press.

Morgan, G. S., Lucas, S. G., Sealey, P. L., McDonald, H. G., and Gordon, M. E., 2000, Pleistocene (Rancholabrean) vertebrate and invertebrate fossils from Badlands Ranch, Quay County, New Mexico: New Mexico Museum of Natural History and Science, Bulletin 16, p. 241-251.

Mutz, P. B., 1998, Reservoir storage development in the Canadian River Basin and related litigation, *in* Ortega Klett, C. T., ed., Water Issues of Eastern New Mexico: Las Cruces, New Mexico Water Resources Institute Report 304, p. 11-16.

Myers, D. A., Sharps, J. A., and McKay, E. J., 1986, Geologic map of the Becker SW and Cerro Montoso quadrangles, Socorro County, New Mexico: U. S. Geological Survey, Miscellaneous Investigations Series I-1567, scale 1:24,000.

Newell, A. J., 1993, Depositional environment of the Late Triassic Bull Canyon Formation (New Mexico): implications for "Dockum Formation" paleogeography: New Mexico Museum of Natural History and Science, Bulletin 3, p. 359-368

Nielsen, R. L., and Dungan, M. A., 1985, The petrology and geochemistry of the Ocate volcanic field, north-central New Mexico: Geological Society of America Bulletin, v. 96, p. 296-312.

Olsen, P. E., 1984, Comparative paleolimnology of the Newark Supergroup: a study of ecosystem evolution [Ph. D. dissertation]: New Haven, Yale University, 563 p.

Olsen, P. E., 1986, A 40-million year lake record of early Mesozoic orbital climatic forcing: Science, v. 234, p. 842-848.

Olsen, P. E., and Baird, D., 1986, The ichnogenus *Atreipus* and its significance for Triassic biostratigraphy, *in* Padian, K., ed., The Beginning of the Age of Dinosaurs: Faunal Change Across the Triassic-Jurassic Boundary: Cambridge, Cambridge University Press, p. 61-87.

Olsen, P. E. and Kent, D. V., 1996, Milankovitch climate forcing in the tropics of Pangea during the Late Triassic: Palaeogeography, Palaeoclimatology, Palaeoecology, v. 122, p. 1-26.

Parker, B. H., 1934, Geology of Two Buttes Dome in southeastern Colorado: American Association of Petroleum Geologists Bulletin, v. 18, p. 1544-1546.

Pearce, T. M., 1965, Place names of New Mexico. Albuquerque, University of New Mexico Press, 187 p.

Poole, F. G., 1962, Wind direction in late Paleozoic to middle Mesozoic time on the Colorado Plateau: U. S. Geological Survey, Professional Paper 450-D, p. 147-155.

Reed, E. K., 1952, The myth of Montezuma's bison and the type locality of the species: Journal of Mammalogy, v. 33, p. 390-392

Reed, E. K., 1955, Bison beyond the Pecos: Texas Journal of Science, v. 7, p. 130-135.

Reeside, J. B., Jr., Applin, P. L., Colbert, E. H., Hadley, H. D., Kummel, B., Lewis, P. J., Love, J. D., Maldanado-Koerdell, M., McKee, E. D., McLaughlin, D. B., Muller, S. W., Rodgers, J., Sanders, J., Silberling, N. J., and Waage, K., 1957, Correlation of the Triassic formations of North America exclusive of Canada: Geological Society of America Bulletin, v. 68, p. 1451-1514.

Reeves, C. C., Jr., 1972, Tertiary-Quaternary stratigraphy and geomorphology of west Texas and southeastern New Mexico: New Mexico Geological Society, 23rd Field Conference, Guidebook, p. 108-117.

Reeves, C. C., Jr., 1976, Quaternary stratigraphy and geologic history of southern High Plains, Texas and New Mexico; in Mahaney, W. C., ed., Quaternary Stratigraphy of North America: Stroudsburg, Dowden, Hutchinson and Ross, Inc., p. 213-234.

Renne, P. R., Steiner, M. B., Sharp, W. D., Ludwig, K. R., and Fanning, C. M., 1996, $^{40}Ar/^{39}Ar$ and U/Pb dating of latest Permian tephras in the Midland basin, Texas: Eos, v. 77, no. 46, p. F794.

Riggs, N. R., Lehman, T. M., Gehrets, G. E., and Dickinson, W. R., 1996, Detrital zircon link between headwaters and terminus of the Upper Triassic Chinle-Dockum paleo-river system: Science, v. 273, p. 97-100.

Roberts, F.H.H., 1942. Archaeological and geological investigations in the San Jon district, eastern New Mexico: Smithsonian Miscellaneous Collections, v. 103(4), 30 p.

Roth, R. I., Newell, N. D., and Burma, B. H., 1941, Permian pelecypods in the lower Quartermaster Formation: Journal of Paleontology, v. 15, p. 312-317.

Santucci, V. L., Hunt, A. P., and Lockley, M. G., 1994, Late Triassic vertebrate tracks from Petrified Forest National Park, in Santucci, V. L. ,ed., The Paleontological Resources of the National Parks: Denver, National Park Service Technical Report NPS/NRPO/NRTR-95/16, p. 109-112.

Saunders, J. J., and Daeschler, E. B., 1994, Descriptive analyses and taphonomical observations of culturally-modified mammoths excavated at "The Gravel Pit," near Clovis, New Mexico in 1936: Proceedings of the Academy of Natural Sciences Philadelphia, v. 145, p. 1-28.

Schiel, K. M., 1987, Investigations of the Dewey Lake Formation, Delaware basin, New Mexico: El Paso Geological Society, 18th Field Conference, Guidebook, p. 149-156.

Schiel, K. M., 1988, The Dewey Lake Formation: End Stage deposit of a peripheral foreland basin [M. S. thesis]: El Paso, University of Texas, 181 p.

Schultz, G. E., 1977, Guidebook field conference on Late Cenozoic Biostratigraphy of the Texas Panhandle and Adjacent Oklahoma, August 4-6, 1977: West Texas State University (Canyon), Kilgore Reseach Center, Departments of Geology and Anthropology, Special Publication 1, 160 p.

Scott, G. R., 1986, Historic trails map of the Raton and Springer 30' x 60' quadrangles, New Mexico and Colorado: U.S. Geological Survey, Miscellaneous Map I-1641, 10 p.

Scott, R.W., 1970a, Stratigraphy and sedimentary environments of Lower Cretaceous rocks, southern Western Interior: American Association of Petroleum Geologists Bulletin, v. 54, p. 1225-1244.

Scott, R.W., 1970b, Paleoecology and paleontology of the Lower Cretaceous Kiowa Formation, Kansas: The University of Kansas Paleontological Contributions, Article 52 (Cretaceous 1), 94 p.

Scott, R.W., 1974, Bay and shoreface benthic communities in the Lower Cretaceous: Lethaia, v. 7, p. 315-330.

Simpson, J. H.,1850, Report from the Secretary of War, in compliance with a resolution of the Senate, the report and map of the route from Fort Smith, Arkansas to Santa Fe, New Mexico, made by Lieutenant Simpson: 31st Congress, 1st Session, Senate Executive Document 12, 25 p.

Smith, C. H., 1827, Supplement to the order Ruminantia; in G. Cuvier, The Animal Kingdom arranged in Conformity with its Organization: London, George B. Whitaker.

Smith, H. T. U., 1940, Geological studies in south-wetern Kansas: State Geological Survey of Kansas, Bulletin 34, 212 p.

Spell, T. L., Kyle, P. R., and Baker, J.,1996, Geochronology and geochemistry of the Cerro Toledo Rhyolite: New Mexico Geological Society, 47th Field Conference Guidebook, p. 263-268.

Spiegel, Z., 1972, Cenozoic geology of the Canadian River valley, New Mexico: New Mexico Geological Society 23rd Field Conference, Guidebook, p. 118-119.

Stanton, T.W., 1947, Studies of some Comanche pelecypods and gastropods: U.S. Geological Survey, Professional Paper 211, 256 p.

Stearns, D. W., 1972, Structural interpretation of the fractures associated with the Bonita fault: New Mexico Geological Society, 23rd Field Conference, Guidebook, p. 161-167.

Stephenson, L.W., 1952, Larger invertebrate fossils of the Woodbine Formation (Cenomanian) of Texas: U.S. Geological Survey, Professional Paper 242, 226 p.

Størmer, J. C., Jr., 1972, Ages and nature of volcanic rocks of northeastern New Mexico: Geological Society of America Bulletin, v. 83, p. 2443-2448.

Taylor, R. G., and Pitt, W. D., 1972, Urban geology of Portales: New Mexico Geological Society, 23rd Field Conference Guidebook, p. 214-215.

Tedford, R. H., 1970, Principles and practices of mammalian geochronology in North America: Proceedings of the North American Paleontological Convention, v. 1, p. 666-703.

Tedford, R. H., 1981, Mammalian biochronology of the late Cenozoic basins of New Mexico: Geological Society of America Bulletin, v. 92, p. 1008-1022.

Thomas, O., 1911, The mammals of the tenth edition of Linnaeus; an attempt to fix the types of the genera and the exact bases and localities of the species: Proceedings of the Zoological Society of London, v. 1911, p. 120-158.

Trauger, F. D., and Bushman, F. X., 1964, Geology and groundwater in the vicinity of Tucumcari, Quay County, New Mexico: New Mexico State Engineer, Technical Report 30, 178 p.

Trauger, F. D., Mankin, C. J., and Brand, J. P., 1972, Road log of Tucumcari, Mosquero, and San Jon country: New Mexico Geological Society, 23rd Field Conference, Guidebook, p. 12-45.

Wanek, A. A., 1962, Reconnaissance geologic map of parts of Harding, San Miguel, and Mora counties, New Mexico: Washington, U. S. Geological Survey Oil and Gas Investigations Map OM 209, 1 sheet.

Welsh, M., 1985, A Mission in the Desert, Albuquerque District: Washington, D.C., U.S. Government Printing Office, Region 8, 262 p.

Williamson, T. E., and Lucas, S. G., 1996, Mammal footprints from the Miocene-Pliocene Ogallala Formation, eastern New Mexico: New Mexico Geology, v. 18, p. 1-5.

Wilpolt, R. H., and Wanek, A. A.,1951, Geology of the region from Socorro and San Antonio east to Chupadera Mesa, Socorro County, New Mexico: U. S. Geological Survey, Oil and Gas Investigations Map OM-121, scale 1:62,500.

Wilson, C. R., 1998, Overview of the hydrology of northeastern New Mexico; in Ortega Klett, C. T., ed., Water Issues of Eastern New Mexico: Las Cruces, New Mexico Water Resources Institute Report 304, p. 37-46.

Wilson, L., and Esparza, L. E., 1998, Lake Meredith Salinity Control Project; in Ortega Klett, C. T., ed., Water Issues of Eastern New Mexico: Las Cruces, New Mexico Water Resources Institute Report 304, p. 131-141.

Wood, H. E., Chaney, R. W., Clark, J., Colbert, E. H., Jepsen, G. L., Reeside, J. B., Jr., and Stock, C., 1941, Nomenclature and correlation of the North American continental Tertiary: Geological Society of America Bulletin, v. 52, p. 1-48.

Woodburne, M. O., and Swisher, C. C. III, 1995, Land mammal high-resolution geochronology, intercontinental overland dispersals, sea level, climate, and vicariance: SEPM Special Publication 54, p. 335-364.

Young, J. V., 1984, The state parks of New Mexico. Albuquerque, University of New Mexico Press, 160 p.

MAGNETOSTRATIGRAPHIC CORRELATION AND DATING OF WEST TEXAS AND NEW MEXICO LATE PERMIAN STRATA

MAUREEN STEINER
Dept. Geology & Geophysics, University of Wyoming, Laramie, WY 82071, magnetic@uwyo.edu

Abstract.—Ochoan formations in west Texas-New Mexico have been investigated magnetostratigraphically for the purpose of obtaining age information. The locally-named Dewey Lake Formation has been studied in two areas of northwest Texas and in southeastern New Mexico. The subjacent Alibates and Rustler formations also are being investigated in these areas. Data from the Magenta Dolomite Member of the Rustler Formation near Carlsbad, NM, indicate three magnetic polarity intervals, contained within 11 m of dolomitic strata. Only preliminary Magenta data are available at present; nevertheless, four stratigraphic sections in the member consistently indicate the three polarity intervals. Additionally and surprisingly, reliable paleomagnetic data have been obtained from the Castile Formation. The reconnaissance data show a relatively strong remanent magnetization which is stable to thermal demagnetization procedures. A probable reversed polarity Late Permian magnetization was obtained. Within the Dewey Lake Formation of west Texas, magnetostratigraphy shows that relative thickness of enclosed volcanic ash beds is not a reliable correlation tool; ash beds of similar thickness located 120 km apart lie in different magnetic polarity intervals. Magnetostratigraphy has shown that the 0.6 m ash layer at Dickens, TX, corresponds to a 0.02 m ash layer (the upper of two) at Caprock Canyons State Park. Magnetostratigraphy also demonstrates that fully one-half of the Dewey Lake Formation present in northwest Texas (Caprock Canyons) has been removed by pre-Late Triassic erosion in the more southerly localities. Late Permian, rather than Triassic, paleopoles are obtained from the Dewey Lake beds. The data of this study suggest a correlation of the Magenta Dolomite with the lower exposed portion of the Alibates strata of northwest Texas (Caprock Canyons State Park). A Late and Middle Permian magnetostratigraphic time scale is derived by combining all available magnetostratigraphic studies. The Dewey Lake-Alibates/Rustler magnetostratigraphic sequences correlate well with the global magnetic polarity pattern and suggest that the Alibates/Rustler strata may be latest Capitanian in age and that the Dewey Lake (Quartermaster) strata were deposited in the Wujiapingian and early Changsingian Stages of the Late Permian.

DEWEY LAKE (QUARTERMASTER) FORMATION

Introduction

The youngest Permian strata in North America are the red beds of the Permian Basin of west Texas and southeastern New Mexico. Numerous names have been applied to these strata , including Quartermaster, Pierce Canyon Redbeds, and Dewey Lake and Alibates (stratal relationships were reviewed by Lucas and Anderson, 1994). The age of these red bed strata is unknown due to an almost total absence of fossils, and has been speculated to be both Late Permian and Triassic. A rarely-occuring molluscan fauna suggested a Permian age (Roth et al.,1941), but Schiel (1987) favored a Triassic age. The red beds are commonly unconformably overlain by Upper Triassic (upper Carnian) sandstones of the basal Chinle Group.

For the purpose of age-dating, an intensive magnetostratigraphic study of both the Dewey Lake and the underlying, anhydrite-rich Alibates exposures was undertaken at Caprock Canyons State Park, near Quitaque, Texas (approximately 107 km NE of Lubbock, TX: Fig. 1). The Dewey Lake and Alibates are distinguished by the presence of bedded anhydrite within the Alibates and its absence in the Dewey Lake strata. The total section exposed and studied at Caprock Canyons (Steiner et al., in prep.) is nearly 140 m thick, 55 m of Alibates beds and 94 m of Dewey Lake strata. The Dewey Lake beds contain two volcanic ash beds, one about 4 m above the formation base defined by the absence of bedded anhydrite and a second about 20 m higher than the first. The lower ash bed is relatively thick, varying between 20 and 28 cm thickness over lateral exposures of 12 km. In sharp contrast, the higher ash bed is only 2 mm thick.

Another thick exposure of volcanic ash was observed (D. Cheney, pers. comm., 1997) in the Dewey Lake Formation near Dickens, Texas (90 km due east of Lubbock, TX). The Dickens exposure is approximately 120 km south-southeast of the Caprock Canyons exposure (Fig. 1), and the ash bed is 0.57 m thick. The thickness of exposure therefore greatly physically resembles the lower ash bed at Caprock Canyons. The need for rudimentary age control for paleoclimatic studies in the Early and Middle Permian exposures farther east of Dickens prompted a magnetostratigraphic investigation of the Dickens exposures of the Dewey Lake Formation. The intent was to establish a correlation of this ash bed with the one it resembles at Caprock Canyons, which is being dated radiometrically (Renne et al., 1996).

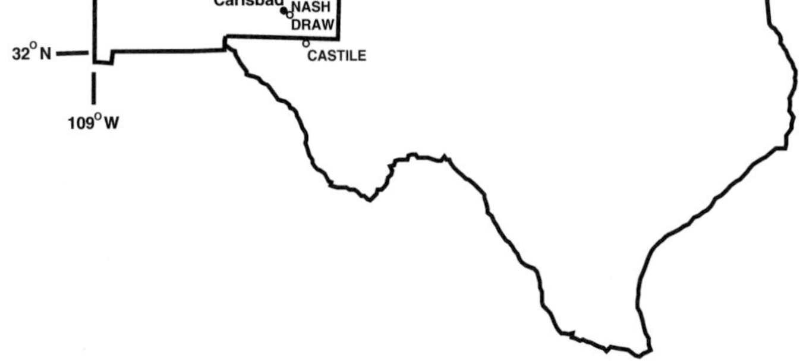

FIGURE 1. Location sampled sections discussed in the text. City locations are indicated by black dots and sampling localities by open symbols.

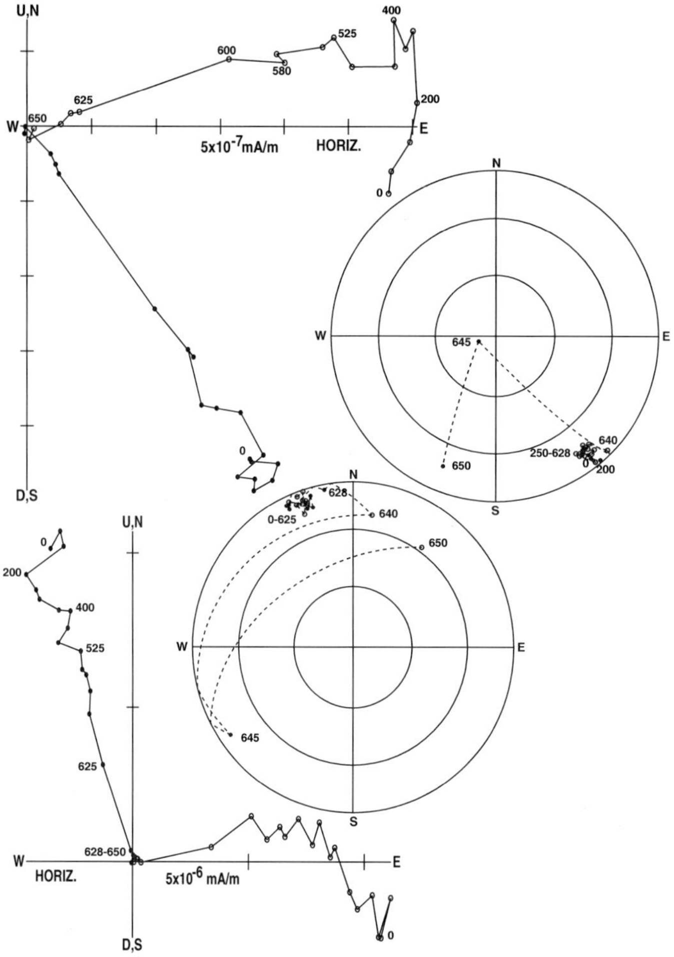

FIGURE 2. Typical demagnetization responses of the Dewey Lake samples from the Dickens locality. Equal area stereographic plots show lower (upper) hemisphere directions as solid (open) circles. Orthogonal axes plots show the horizontal projection of the magnetic vector (declination) as solid symbols and the vertical projection (inclination) as open symbols.

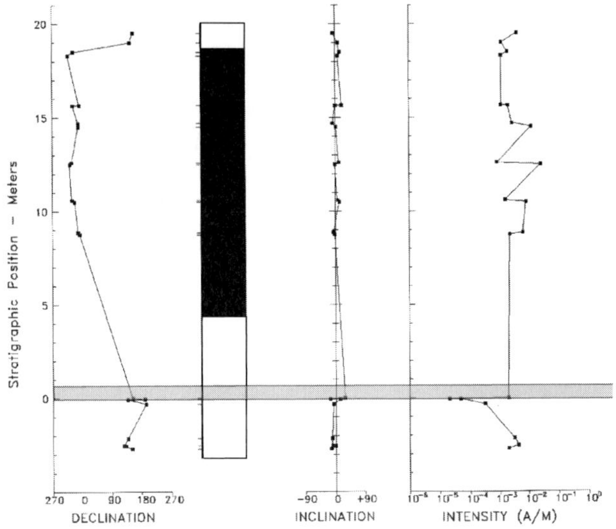

FIGURE 3. The magnetostratigraphy at Dickens; black indicates normal polarity. Grey band represents the volcanic ash bed.

Magnetostratigraphy

Twenty-two oriented paleomagnetic samples were collected in a reconnaissance sampling of Dewey Lake exposures adjacent to US Hwy 82, at the rest stop approximately 4 km east of Dickens, TX. Two partial sections were sampled just south of the highway. At one exposure, the volcanic ash bed forms the top of the exposure, and 3 m of strata below the ash bed were sampled. Just east of this exposure, a gully exposes the section from the ash bed to the top of the formation. The formation is overlain unconformably by conglomeratic sands of the basal Camp Springs Formation of the Chinle Group. The reconnaissance sampling was preformed under stringent time constraints, so it was guided by an assumption that this ash bed corresponded to the lower of the two at Caprock Canyons. That assumption meant that the ash layer should be encompassed by a thick normal polarity interval, and therefore sampling aimed for the top of that polarity interval. Sampling in the gully began at about 8 m above the ash bed and continued in reconnaissance fashion to the top of the formation. Samples were collected with a battery-powered electric drill fitted with a diamond bit and using forced-air as a bit coolant.

The paleomagnetic samples were measured with a cryogenic magnetometer and progressively thermally demagnetized in 17 steps from 150° to 650° C. Like the Dewey Lake samples of the Caprock Canyons area, the magnetizations are remarkably uncomplicated by later secondary magnetizations, a rare phenomenon in paleomagnetic studies. Most samples decay univectorially to the origin of orthogonal axes plots after removal of a component between 150° and 300° C that is steeply inclined. Demagnetization response is illustrated by two examples (Fig. 2), one sample from below, and one from above the ash bed. The low temperature component from all samples comprises a quite scattered population of generally northerly directed, steeply inclined directions; the component is far steeper than the presently observed and theoretical axial dipole directions for this site, and its origin presently is unclear.

With such relatively uncomplicated magnetizations, the magnetostratigraphic interpretation is straightforward. The Dewey Lake magnetostratigraphy at Dickens (Fig. 3) consists of two intervals of reversed polarity separated by one of normal polarity. The volcanic ash bed lies within reversed polarity, unexpected because its thickness. Unfortunately, the sampling coverage is inadequate to fully define the reversed polarity interval containing the ash bed, although further samples will be taken. Despite the sampling gap, correlation of the Dewey Lake strata at Dickens and contained ash layers between Dickens and Caprock Canyons is clear and unambiguous, as will be shown.

RUSTLER FORMATION, MAGENTA MEMBER

The Magenta Member is one of five members of the Rustler Formation of extreme southeastern New Mexico and southwest-

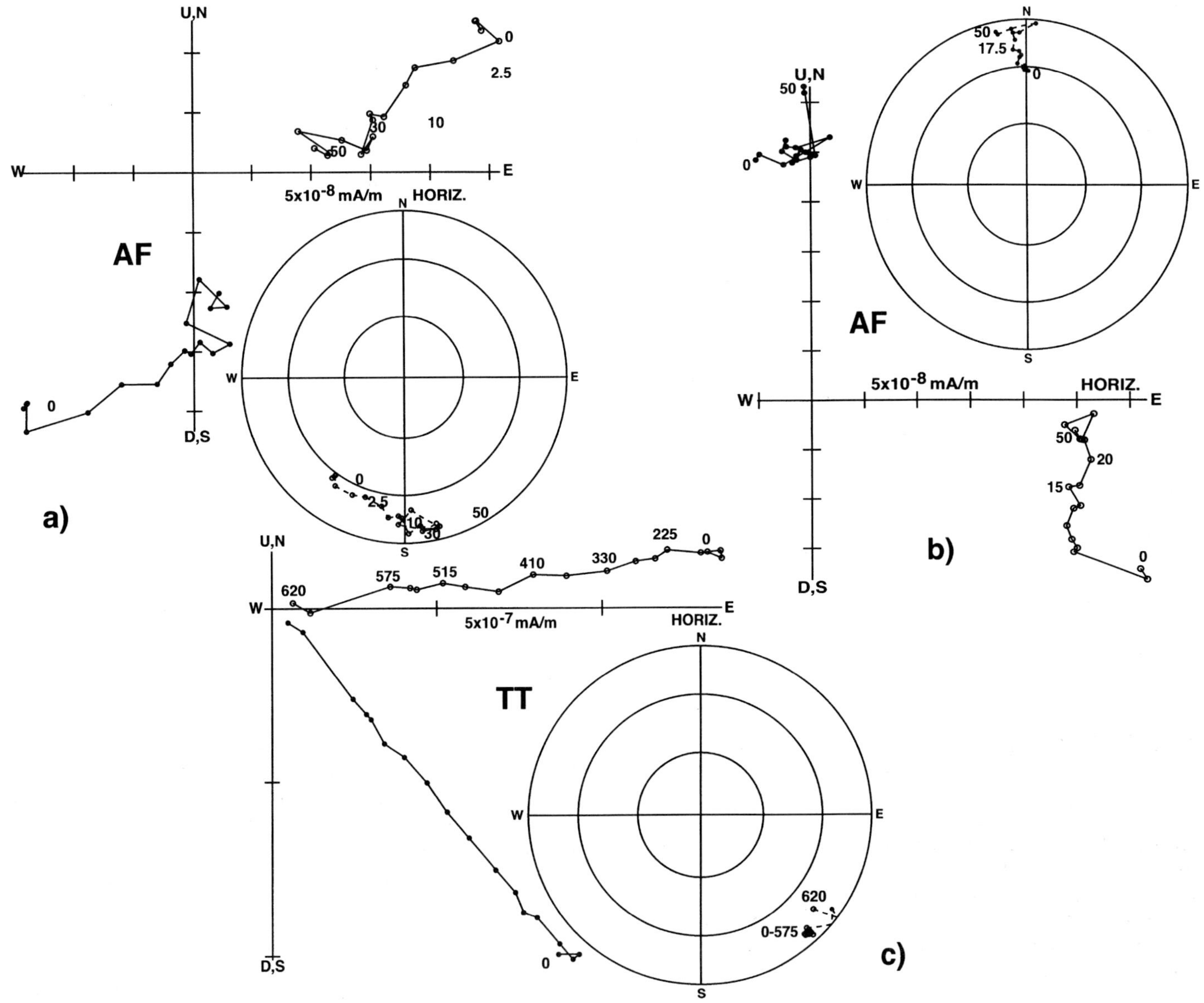

FIGURE 4. a) and b) Alternating field (AF) and c) thermal (TT) demagnetization of Magenta Dolomite samples from the Rustler Formation. Plots and symbols as in Fig. 2.

ern Texas. Pink dolomite characterizes the member; five collections were made from stream-cut exposures on the west side of Nash Draw (Fig 1); all from exposures accessed from NM Highway 31. One nearly complete section was sampled near an abandoned mine at the more northerly end of NM 31, near its intersection with US 62-180. This sampling spanned 11 m (of a probable 15 m thickness) at 0.5 to 1 m increments, yielding twenty four samples. Four other partial sections of the Magenta Member were sampled farther south on NM 31, in exposures created by a deeply incised creek that drains eastward from NM 31 out into Nash Draw. Most of these collections spanned 2.5 m, and one spanned 3.3 m of the Magenta Member, resulting in another 55 samples from the member.

The magnetic polarity of about 2/3 of the Magenta Member samples is readily apparent from the initial measurements (natural remanent magnetizations, or NRMs). As yet, only pilot samples have been demagnetized. Both alternating field (AF) and thermal demagnetization have been tested. AF demagnetization was successful in removing overprint magnetizations in some samples (Fig. 4a-b), but not in others. Thermal demagnetization was more successful than AF (Fig. 4c), and thermal demagnetization of the rest of the samples in progress.

However, four of the five sections are sufficiently free of overprinting magnetizations that their NRM directions indicate a polarity sequence; moreover, the four sequences suggest the same polarity succession, reversed polarity followed by a thin normal polarity interval, followed again by reversed polarity (Fig. 5). Data from the Magenta Member are reported here in the preliminary stage of their investigation because of a suggested correlation with the Alibates "Beds" of the Caprock Canyons sequence in Texas, and the constraint that that correlation provides on the age of the superjacent Dewey Lake strata.

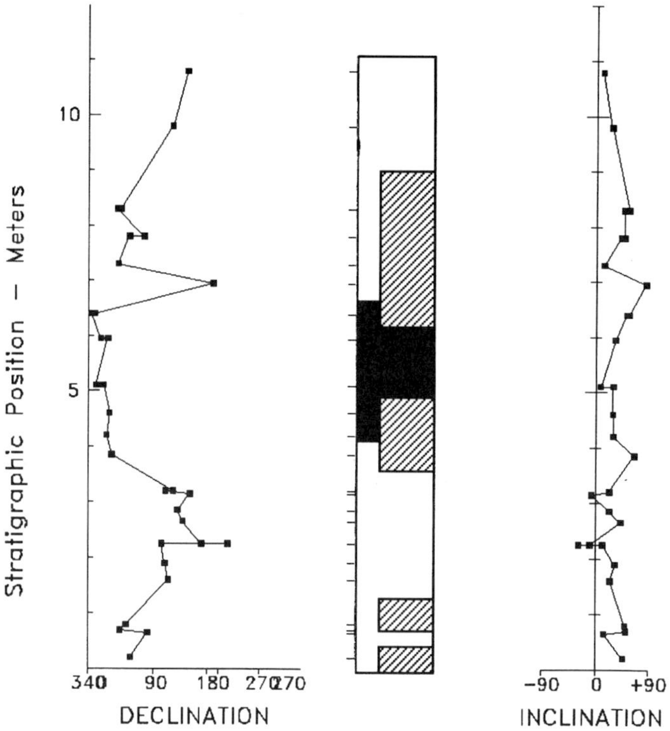

FIGURE 5. Preliminary Magenta Member magnetostratigraphy from the natural remanent magnetizations (NRMs).

CASTILE FORMATION

The Castile gypsum consists of alternate layers of white gypsum/anhydrite and dark layers. The dark color suggested the possibility that a detrital component may reside in these layers, and thus that magnetic components could be present. To test this possibility, a sample was taken from a block sample already in the possession of the author. The block consisted of thin (approximately 2 mm) alternating layers of black and pure white anhydrite. The insitu vertical direction of the block was known, but the azimuthal orientation was unknown.

Quite surprisingly, this sample was magnetic at the level of many chemical sedimentary rocks such as limestones and dolomites; initial magnetization values were about 1.5 x 10^{-7} mA/m. More significant than the unexpectedly strong magnetization, was the stability of the magnetization to demagnetization techniques. Both AF and thermal demagnetization were tested on samples from this block. Low-field AF demagnetization (1 to 7 mT) had little effect in changing the magnetization direction, but the intensity decayed dramatically (Fig. 6) until it was quickly at or below the resolution of a cryogenic magnetometer (5x10^{-9} mA/m). This suggests the presence of some amount of very "soft" (unstable) component of magnetization, such as equant magnetite of >1 micron size or an iron sulfide

In contrast, the magnetization direction remained unchanged by thermal demagnetization to 500° C. Moreover, the magnetic intensity decayed fairly linearly to a near-zero intensity by 500° C (Fig. 6). This behavior suggests either fine-grained magnetite, maghemite, or hematite; obviously, more detailed work on more samples is required to determine the actual magnetic carriers. The stability of the Castile sample to thermal demagnetization was a surprise, partly because of the unstable behavior in response to AF demagnetization. Moreover, predominantly gypsiferous sedimentary rocks incorporated in red beds usually have unstable magnetizations; additionally, sedimentary rocks possessing weak initial intensities not infrequently are unstable above about 300° C.

The stability of the Castile sample to 500° C rules out magnetic carriers in the iron sulfide solid solution series; that stability suggests magnetite or maghemite as a magnetic carrier, thus possible evidence of a detrital origin of the carrier. Further support of a detrital carrier is the found in the fact that the magnetic direction is similar to other Late Permian magnetizations for North America. Because the block was azimuthally unoriented, the compass direction of the remanence is unknown. But, the vertical orientation of the magnetic vector was known, thus the magnetization is referenced to the horizontal planes of the parallel dark and light laminations. A locus of all possible paleopoles that could correspond to this magnetization is generated by the distance of the site to the pole, information determined from the inclination component (the vertical component) of the magnetic vector obtained. This results in a circular track (Fig. 7) which is bracketed by two other circles representing the 95% confidence limits on the determination of the magnetic direction itself.

At 95% confidence, the Castile locus of poles statistically overlap with the slightly younger Alibates paleopole of northwest Texas (A in Fig. 7). Hence, if the Castile sample block and its magnetic direction were rotated about 120-130° in a counterclockwise sense, the Castile pole would agree well with the next younger Late Permian paleopole known from North America. Obviously, this Castile "pole" is from one sample only; the comparison is merely undertaken to show that the Castile Formation probably preserves a stable Late Permian magnetization.

DISCUSSION

Figure 7 also displays the paleopoles from the Dewey Lake strata at Dickens (DD) and at Caprock Canyons (CC). The CC data suite was divided into the Alibates portion (A), and a lower (DL) and upper (DU) portions of the Dewey Lake beds, and poles calculated for each (Steiner et al., in prep.). The reason for the division within the Dewey Lake strata is a significant facies change that occurs at 26 m, a change from very evenly bedded siltstones and claystones to medium grained fluvial sandstones and thick eolian sandstones. This lithologic change is accompanied by shift in magnetic direction and a degradation of the quality of magnetization preservation; the lower 26 m display extremely linear demagnetization trends to the origin of orthogonal axes plots above 300° C demagnetization, clearly indicating that no secondary magnetizations influence the magnetic direction above that treatment step. On the other hand, the upper sandstone samples display considerably less linear trends and consistently have higher errors associated with their characteristic directions. Further, the mean direction of the strata above 26 m is displaced from that of the lower strata (Fig. 7). Several possibilities could explain the difference (Steiner and Renne, in prep.): 1. The

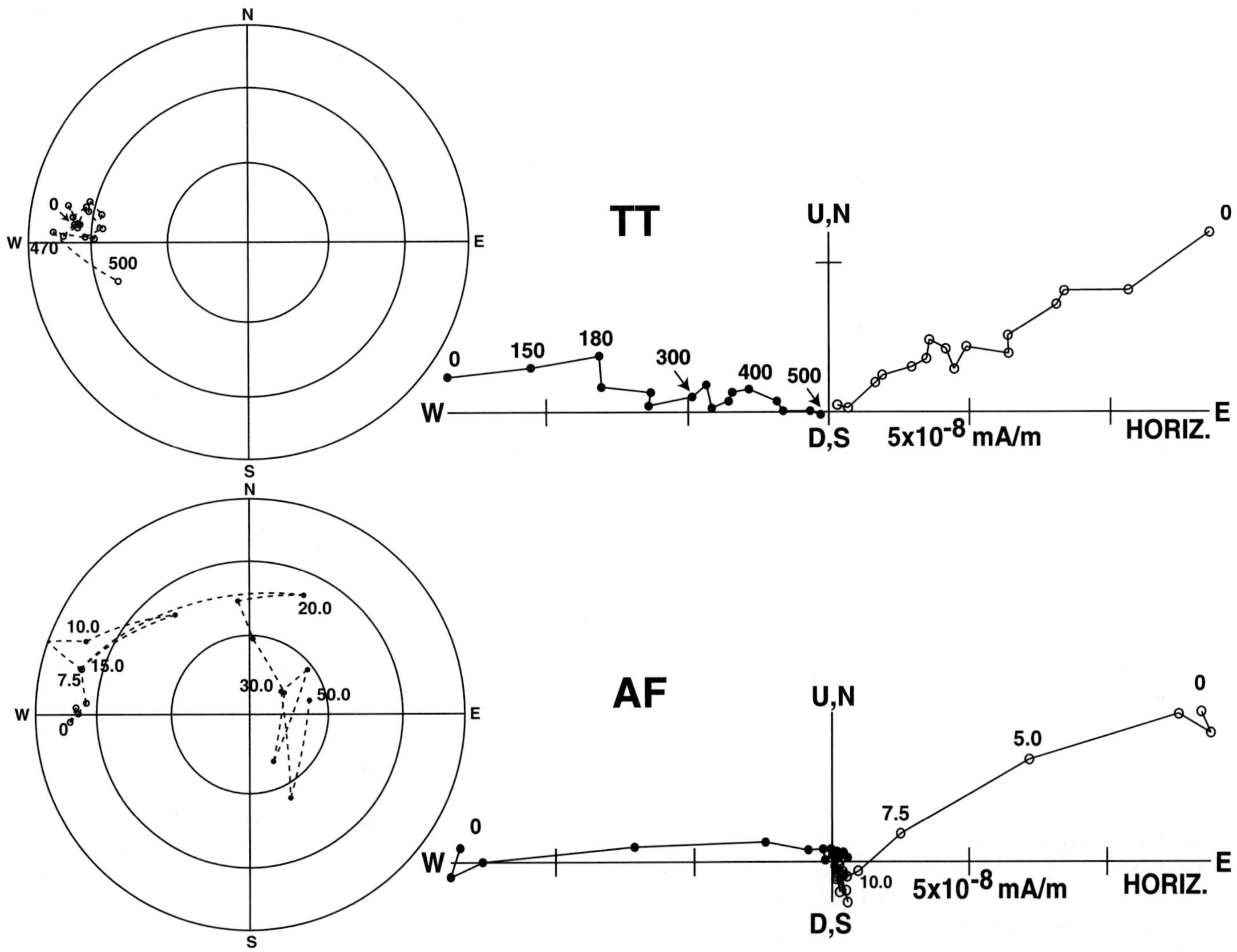

FIGURE 6. Castile reconnaissance sample demagnetization by alternating fields (AF) and thermal treatment (TT). Plots and symbols as in Figure 2.

porosity of the sandstones may have allowed additional magnetization in the sandstones during the time represented by the unconformity between the Dewey Lake and the overlying Chinle Group. For that reason, the paleopole for cratonic North America from the Late Triassic Chinle Group (C) is shown in Figure 7. The upper Dewey Lake strata appear biased towards the Late Triassic pole, possibly representing an unremoved overprint associated with the unconformity. 2. The lithologic change could represent a change in formation, hence somewhat younger strata (perhaps Early or Middle Triassic). In fact, the strata above 26 m in part physically resemble the Middle Triassic Moenkopi Formation (Anton Chico Member) of the area between Albuquerque and Santa Rosa, New Mexico. However, the absence of an unconformity or any change in age of the strata is demonstrated by a comparison of the Dewey Lake paleopoles between Dickens and Caprock Canyons.

The fine-grained, evenly bedded lithology of the Dewey Lake strata at Dickens is identical to that from the lower 26 m at Caprock Canyons (Fig. 8). The Dickens Dewey Lake paleopole (DD in Fig. 7) is statistically indistinguishable from that of the CC strata below 26 m (DL in Fig. 7). This means that the Dickens and lower Caprock Canyons strata are closely similar in age.

This paleopole agreement supports the stratigraphic correlation of the magnetic polarity sequences and correlation among the volcanic ash beds between the two sites, shown in Figure 8. The volcanic ash at Dickens has reversed polarity beds below and above it, suggesting that it correlates with the higher of the two ash beds at CC. The relative lengths of the overlying normal polarity interval at Dickens and CC support that correlation. Moreover, the relative position of the ash bed within the reversed polarity interval at Dickens matches that at CC, providing yet a third similarity in support of this correlation, making the proposed correlation between Dickens and Caprock Canyons quite probable.

The magnetostratigraphic and paleopole correlations between the Dewey Lake Formation at Dickens and Caprock Canyons also address the stratigraphic questions raised by the lithology and pole position change at the Caprock Canyons site. The similarity of the Dickens pole from magnetic polarity intervals Rc3 to Rd4 to that from the just older strata of Nb2 and the lower 3/4 of Rb2 polarity intervals at CC proves that the lithology change is simply

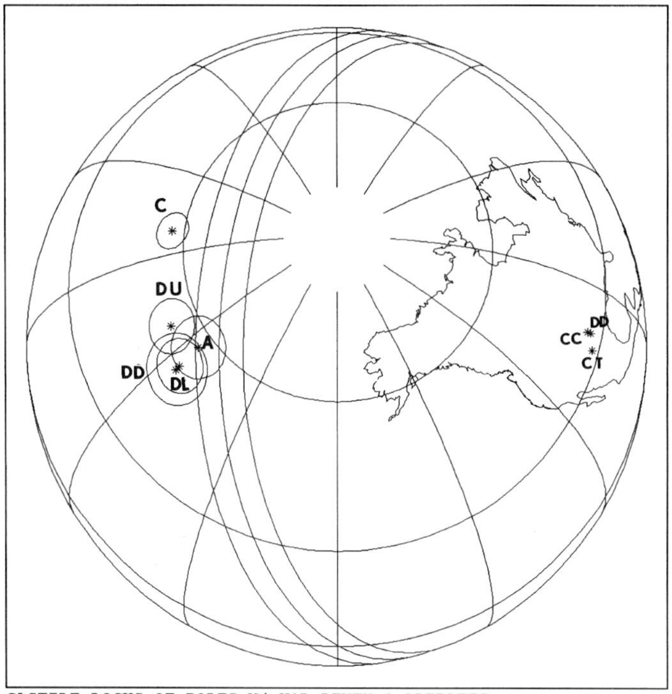

FIGURE 7. Paleomagnetic pole positions. The Dewey Lake Formation is represented by Caprock Canyons and Dickens localities; the Caprock Canyons sequence has been divided into the Alibates (A), lower Dewey Lake (DL) and upper (DU) Dewey Lake. The cratonic Chinle Group pole (C) is shown for reference. A locus of poles and their 95% confidence limits are shown for the Castile sample. Site locations are shown by asterisks (CT = Castile; CC = Caprock Canyons; DD = Dickens).

that; the same fine-grained, evenly laminated siltstone-mudstone facies at Dickens gives an identical paleopole to that of the fine-grained, evenly laminated but slightly older strata at CC, and not the same paleopole as magnetostratigraphically coeval fluvial and eolian sands comprising the upper 1/4 of Rc3-Nc3-Rd4 at CC. Because the magnetostratigraphy and ash correlations prove that these different lithologies represent the same time interval, this pole difference is strong evidence that the coarser and more porous lithology has influenced the characteristic direction of the strata, probably by being more susceptible to fluid penetration and overprinting by later magnetizations. The magnetostratigraphic match between the two sections and the display of identical paleopoles from the same lithology negates the possibility of an unconformity formational and age change in the Caprock sequence.

Figure 8 also shows a magnetostratigraphic sequence from the Dewey Lake beds at Nash Draw (Molina-Garza et al., 2000). Here the Dewey Lake beds rest on the Rustler Formation, whereas at Caprock Canyons, they lie on the Alibates 'Formation'. The proposed correlation of the Nash Draw Dewey Lake strata with the Dickens and CR exposures was guided by the relative position of the Nash Draw and CR magnetographies with respect to the base of the Dewey Lake beds at each locality.

The proposed magnetostratigraphic correlations of Figure 8 also were influenced by the apparent polarity sequence within the Magenta Member of the Rustler Formation. As shown in the figure, the Magenta magnetostratigraphic sequence closely resembles that of the lower beds of the Alibates Formation exposed at Caprock Canyons, in the display a polarity sequence containing a very short normal polarity interval enclosed by dominantly reversed polarity. Both that moderately distinctive polarity sequence and its position relative to the base of the Dewey Lake Formation at each locality suggest that the Rustler Formation correlates approximately to the Alibates Formation. The proposed correlation can be tested because it predicts that the upper member of the Rustler should contain entirely reversed polarity.

The experimentation on a single sample of the Castile Formation was a test to determine the possibility of magnetic remanence carriers in the formation. Similar tests were carried out on samples of the Salado Formation with negative results (Steiner, unpublished data). The single Castile sample suggests reversed polarity at that time. While not at all definitive, that suggestion is in agreement with global magnetostratigraphy (Fig. 9).

Figure 9a shows all magnetostratigraphic studies determined globally from Middle and Late Permian strata. The data shown are the magnetostratigraphic interpretations and ages quoted by the authors. Among the sequences, the Iranian section (Gallet et al., 2000) possesses the best combination of good marine faunal control and good magnetic characteristics indicative of a reliable magnetization record. The Pakistan sequence also is well-dated biostratigraphically, and has a similar magnetostratigraphic signature, but possesses slightly less pristine magnetizations. Using the assumption that the Iranian data present the best representation of geomagnetic field behavior for the Late Permian (Wujiapingian and Changsingian Stages), all other sequences have been adjusted to match this sequence in general polarity signature in Figure 9b.

The best Dewey Lake paleopoles definitely demonstrate that the Dewey Lake strata are Permian and not Triassic in age. Comparison of the Dewey Lake magnetostratigraphy (Caprock Canyons) with the Iranian and Pakistani sections (Fig. 9b) suggests that the age of the Dewey Lake strata is probably Wujiapingian and early Changsingian. The three reversed polarity intervals in the Dewey Lake magnetostratigraphy tentatively are inferred to correspond to the lower three reversed polarity intervals of the Iranian sequence. This interpretation implies that the upper Changsingian is missing in Texas.

One other conclusion follows from acceptance of the Iranian sequence as representative of Late Permian magnetic polarity. The lower part of the North American Ochoan Stage then becomes equivalent to the uppermost portion of the global Capitanian Stage. This implies that the top of the Capitanian Stage stratotype (as represented by the Tansill and Reef Trail formations) is older than the top of the global Capitanian Stage; that is, the uppermost North American Capitanian strata are several million years older than the oldest Wujiapingian or Djulfian strata of the Tethys. This is a seemingly inescapable conclusion at this point.

CONCLUSIONS

Five conclusions devolve from the above data and discussions.

1) Ash beds can dramatically change thickness along strike over distances of 100 km.

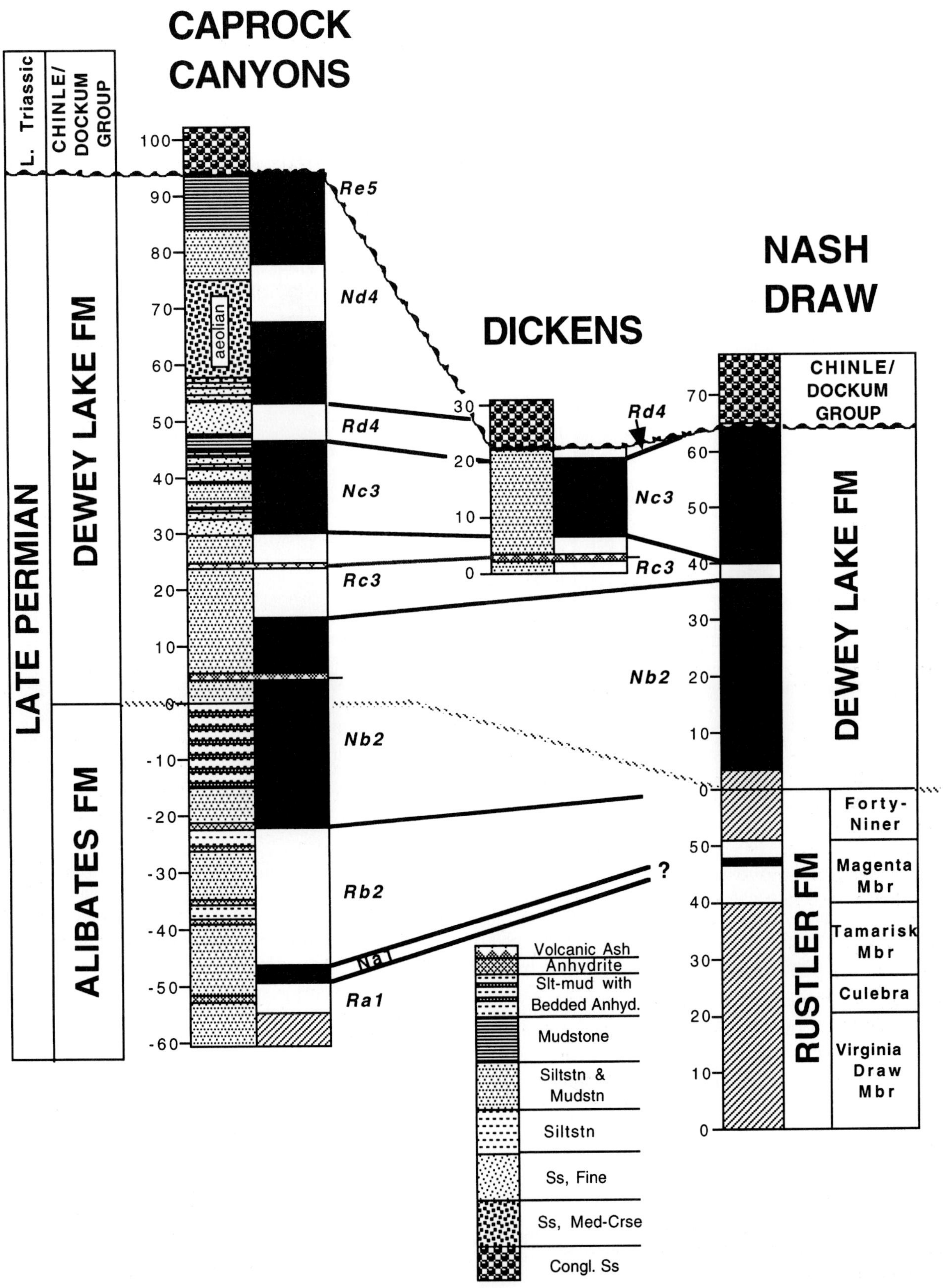

FIGURE 8. Correlation of Dewey Lake, Alibates and Rustler Formations from Caprock Canyons, Dickens and Nash Draw. Correlations are based on a combination of polarity interval boundaries, ash bed occurrences and the contact of the Dewey Lake Formation with the underlying formations in the various localities. Combinations of the letters R and N with other letters and numbers are used to designate individual polarity intervals for discussion. Thickness scales are in meters.

GLOBAL MIDDLE & LATE PERMIAN MAGNETOSTRATIGRAPHY

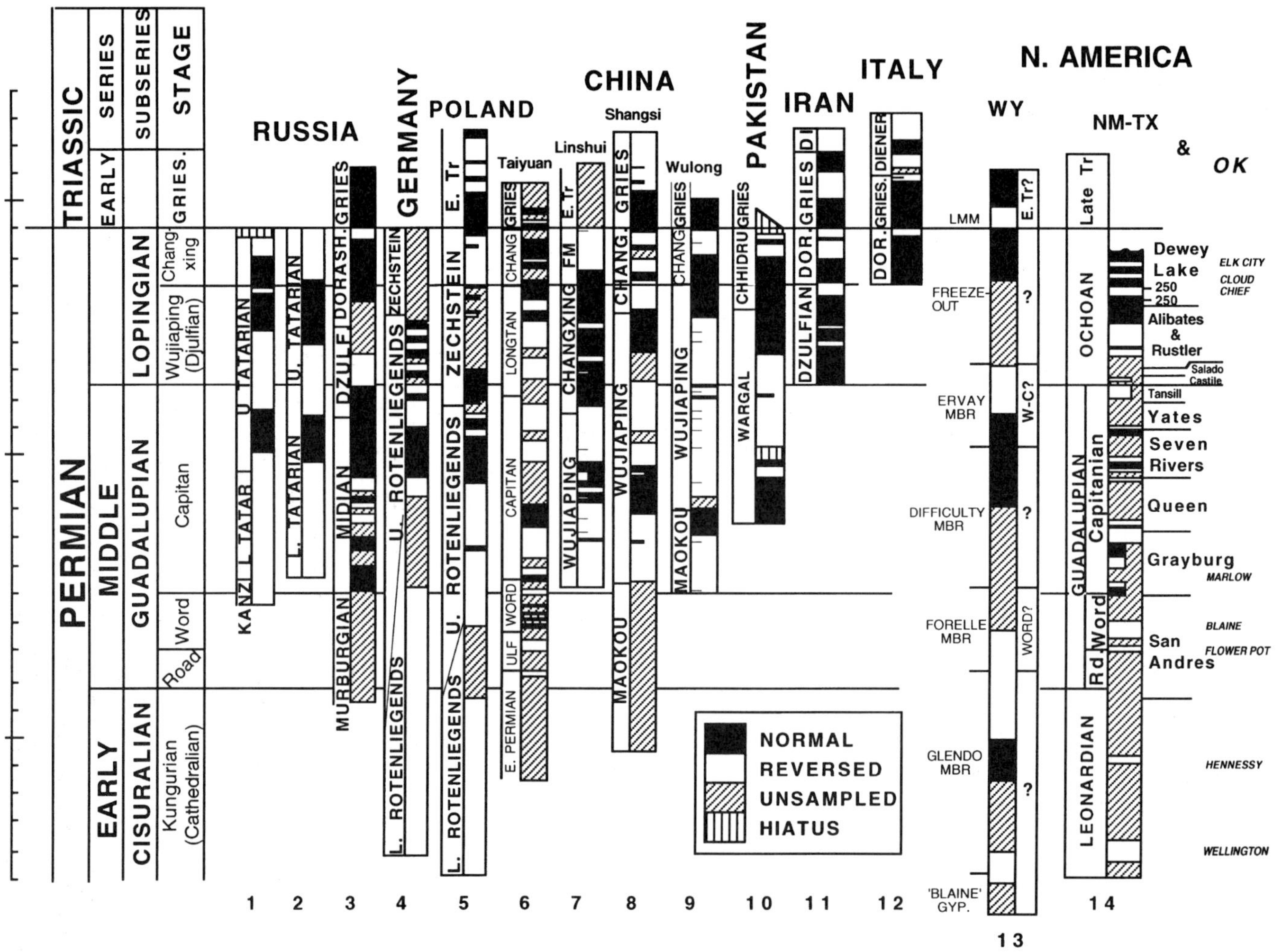

FIGURE 9a. The Dewey Lake, Alibates/Rustler, and Castile data are displayed with other Middle and Late Permian global magnetostratigraphic results. 1: Molostovsky (1983); 2: Gurevich and Stautsitays (1985); 3: Zakharov and Sokarev (1991) and Gialanella, et al. (1997); 4: Menning (1988); 5: Nawrocki (1999); 6: Embleton et al. (1996); 7: Heller et al. (1995); 8: Steiner et al. (1989); 9: Heller et al. (1995); 10: magnetostratigraphy of Haag and Heller (1991) biostratigraphic ages from Wardlaw and Pogue (1995); 11: Gallet et al. (2000); 12: Scholger et al. (2000); 13: Steiner, Hutna, and Lyford, in prep.; 14: This paper; Steiner, Christensen, and Parkin, in prep; and Peterson and Nairn, 1971.

2) Paleomagnetic poles and magnetostratigraphy demonstrate that the age of the Dewey Lake strata in northwestern Texas is probably early Late Permian, Wujiapingian Stage, through early late Late Permian, early Changsingian Stage.

3) A drastic change in lithology of the Dewey Lake strata at Caprock Canyons, northwest Texas, does not represent an unconformity or new formational boundary. As would be expected with the filling and desiccation of the Permian Basin, fluvial and aeolian deposits replaced tidal flat deposits in the northern reaches of the basin.

4) Magnetostratigraphy suggests that the Rustler Formation may correlate approximately to the Alibates Formation.

5) The Castile Formation is suitable for paleomagnetic and magnetostratigraphic study.

ACKNOWLEDGMENTS

Funding support for this work from NSF grants is gratefully acknowledged; initial phases of this study were supported by grant EAR-9614512 and later phases by grant EAR-9814637.

REFERENCES

Embleton, B. J.J., McElhinny, M.W., Ma, X., Zhang, Z., and Li, Z.X., 1996, Permo-Triassic magnetostratigraphy in China: the type section near Taiyuan, Shanxi Province, North China: Geophys. J. Int., v. 126, p. 382-388.

Gallet, Y., Krystyn, L., Besse, J., Saidi, A., and Ricou, L.-E., 2000, New constraints on the Upper Permian Lower Triassic geomagnetic polarity timescale from the Abadeh section (central Iran): J. Geophys. Res., v. 105, p. 2805-2815.

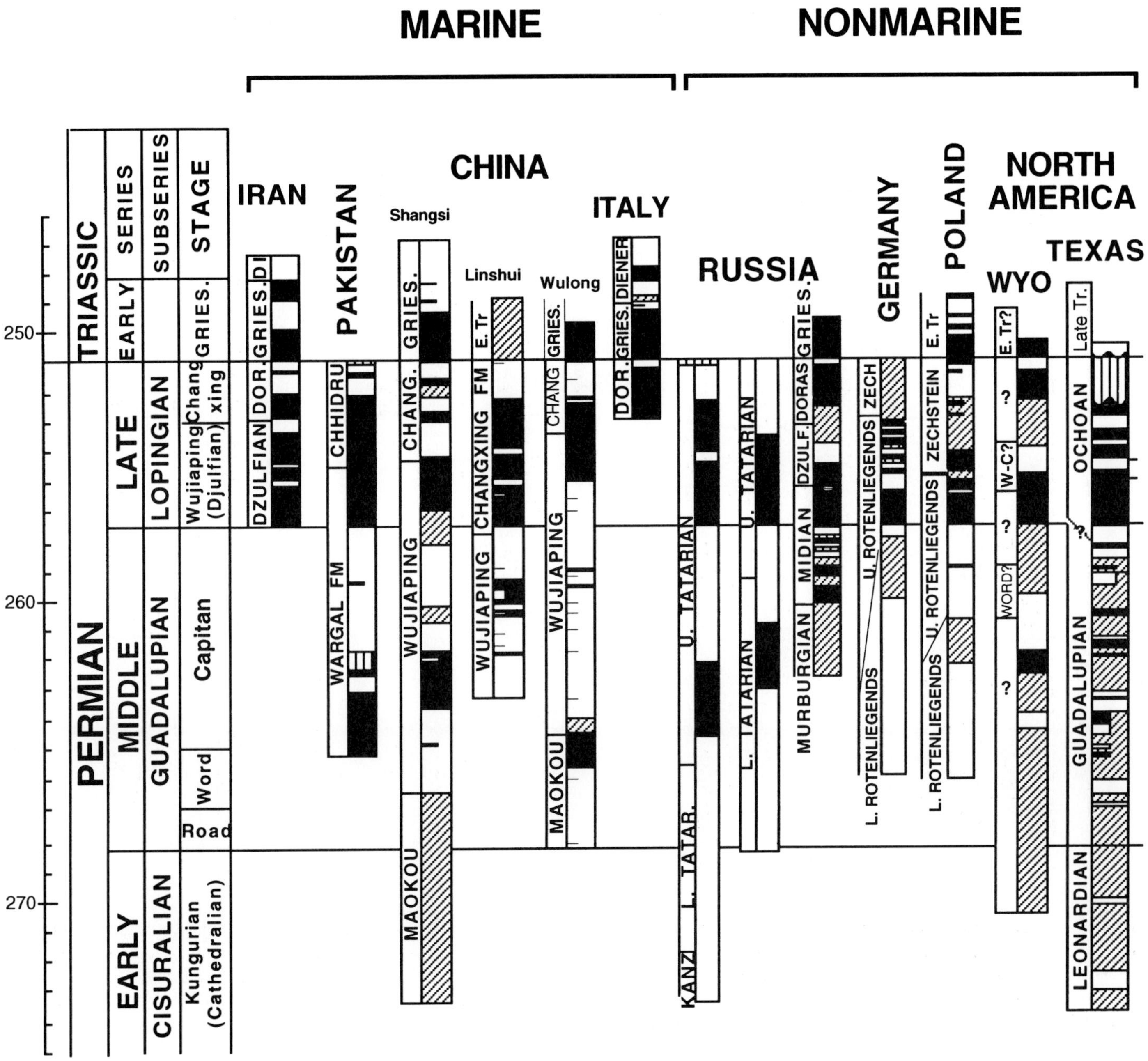

FIGURE 9b. All sequences have been adjusted in length such that their polarity sequences approximately match that of the Iranian sequence, considered to be the most representative (see text).

Gialanella, P.R., Heller, F., Haag, M., Nurgaliev, D., Borisov, A., Burov, B., Jasonov, P., Khasanov, D., Ibragimov, S., and Zharkov, I., 1997, Late Permian magnetostratigraphy on the eastern Russian platform: Geologie en Mijnbouw, v. 76, p. 145-154.

Gurevich, Ye..L., and Stautsitays, I.P., 1985, A paleomagnetic section in the Upper Permian and Triassic deposits of Novaya Zemlya: Int. Geol. Rev., 27, 168-177.

Haag, M., and Heller, F., 1991, Late Permian to Early Triassic magnetostratigraphy, Ear. Planet. Sci. Lett., v. 107, p. 42-54.

Heller, F., Chen, H., Dobson, J., Haag, M., 1995, Permian-Triassic magnetostratigraphy — new results from South China: Ear. Planet. Sci. Lett., v. 89, p. 281-295.

Lucas, S.G., and Anderson, O.J., Ochoan (Late Permian) stratigraphy and chronology, southeastern New Mexico and west Texas: NM Bur. Mines & Min. Res. Bull., v. 150, p. 29-36.

Menning, M., 1988, Magnetostratigraphic investigation of the Rotiegendes (200-252 Ma) of Central Europe: Z. Geol. Wiss., v. 16, p. 1045-1063.

Molina-Garza, R.S., Geissman, J.W., and Lucas, S.G., 2000, Paleomagnetism and magnetostratigraphy of uppermost Permian strata, southeast New Mexico, USA: correlation of the Permian-Triassic boundary in non-marine environments: Geophys. J. Int., v. 141, p. 778-786.

Molostovsky, E.A., 1983, Paleomagnetic Stratigraphy of the Eastern European Part of the USSR (in Russian): Saratov, Russia, Univ. of Saratov, 167 p.

Nawrocki, Jerzy, 1997, Permian to Triassic magnetostratigraphy from the Central European Basin in Poland: Implications on regional and worldwide correlations: Ear. Planet. Sci. Lett., v. 152, p. 37-58.

Peterson, D.N., and Nairn, A.E.M., 1971, Paleomagnetism of Permian red beds from the southwestern United States, Geophys. J. R.A.S., 23,191-206.

Renne, P.R., Steiner, M.B., Sharp, W.D., Ludwig, K.R., and Fanning, C.M., 1996, $^{40}Ar/^{39}Ar$ and U/Pb SHRIMP dating of the latest Permian tephras in the Midland Basin, Texas: EOS, Trans. Am. Geophys. U., v. 77, p. 794.

Roth, R.I., Newell, N.D., and Burma, B.H., 1941, Permian pelecypods in the lower Quartermaster, Texas: J. Paleo., v. 15, p. 312-317.

Schiel, K.M., 1987, Investigations of the Dewey Lake Formation, Delaware Basin, New Mexico: El Paso Geol. Soc. Guidebook, v. 18, p. 149-156.

Scholger, R., Mauritsch, H.J., and Brandner, R., 2000, Permian-Triassic magnetostratigraphy from the Southern Alps (Italy): Ear. Planet. Sci. Lett., v. 176, p. 495-508.

Steiner, M.B., Ogg, J.G., Zhang, Z., and Sun, S., 1989, The Late Permian/Early Triassic magnetic polarity time scale and plate motions of South China,): J. Geophys. Res., v. 94, p. 7343-7363.

Steiner, M.B., Christensen, D., and Parkin, D., in prep., Magnetostratigraphic dating of the Quartermaster (Dewey Lake) and the Alibates Formations, northwest Texas.

Steiner, M.B., Hutna, M., and Lyford, M., in prep., Preliminary magnetostratigraphic dating of the Goose Egg Formation, southeastern Wyoming.

Steiner, M.B., and Renne, P.R., in prep., Magnetostatigraphy of the lower Quartermaster (Dewey Lake) Formation, northwestern Texas.

Wardlaw, B., and Pogue, 1995, The Permian of Pakistan, *in* Scholle, P., Peryt, T.M., and Ulmer-Scholle, D.S., eds., The Permian of Northern Pangea, New York, Springer-Verlag, v. 2, p. 215-224.

Zakharov, Y.D., and Sokarev, A.N., 1991, Permian-Triassic paleomagnetism of Eurasia, *in* Kotaka, T., Dickens, J.M., McKenzie, K.G., Mori, K., Ogasawara, K., and Stanley, G.D. Jr., eds., Shallow Tethys 3., p. 313-323.

RATON-CLAYTON AND OCATE VOLCANIC FIELDS

J.C. AUBELE AND L.S. CRUMPLER
New Mexico Museum of Natural History and Science, 1801 Mountain Road NW, Albuquerque, New Mexico 87104; jaubele@nmmnh.state.nm.us

Abstract.—The Raton-Clayton (RCVF) and Ocate Volcanic Fields (OVF) constitute the most significant late Cenozoic volcanic activity east of the Rocky Mountains. A salient characteristic of volcanic fields is the relatively low volumetric rates of eruption compared with major volcanic centers. The RCVF consists of about 125 vents and, including outlying portions, covers nearly 20,000 km^2. Rocks range from about 9 Ma to less than 60,000 yrs, yielding a recurrence rate of approximately 1.3×10^{-5} volcanic events per year. The RCVF can be divided into three main phases of volcanic activity: (1) Raton, which includes basalt flows and rhyodacite domes; (2) Clayton, which includes voluminous basalt flows and the Sierra Grande shield volcano; and (3) Capulin, which includes cinder cones and flows centered on Capulin Volcano National Monument. Although widely cited, the previous age estimate of less than 10,000 yrs for Capulin Volcano was based on an interpreted correlation, not an absolute date. Two dating techniques, the cosmogenic helium technique and the Ar-Ar technique indicate a date for Capulin of about 58,000 yrs. Three major series of lava flow units were erupted from Capulin, with the second and third series erupted from the base of the cone at the "boca". It is likely that the eruption of the boca flow altered the profile of Capulin, subsequently its symmetry was restored by continued building of the cone. The Ocate Volcanic Field consists of at least 16 flows and about 50 associated vents ranging in age from 8.34±0.50 Ma to 0.81±0.14 Ma. Like the RCVF, three major topographic levels of lava flows relate to three age groups: (1) older than 5 Ma, which includes flows capping mesas with surfaces around 3000m in elevation and an apparent E-W fissure line of vents near Wagon Mound; (2) 5-4 Ma, which includes most of the flows around Agua Fria; and (3) younger than 4 Ma, which includes the most abundant flows in the OVF and such vents as Cerrito Pelon, Cerro Negro, Cerro del Oro and Maxon Crater. Flows from Maxon flowed 90 km eastward along the canyon cut by the Mora River where they occur 100m below the rim of the canyon and 125m above the present level of the river. The youngest flows in the field were erupted from Cerro del Oro, in the central part of Charette Mesa, northwest of Wagon Mound.

INTRODUCTION

The Raton-Clayton volcanic field (RCVF) is the eastern-most Cenozoic volcanic field in the United States. To the east, Cenozoic volcanic rocks are not encountered at this latitude again until the mid-Atlantic ridge. The RCVF lies at the northeastern end of the "Jemez lineament", an alignment of volcanic fields extending from the Pinacate field of Mexico through Arizona and into New Mexico. The Jemez Lineament has been cited by some as a structural anomaly, a hot spot trace, or a chance alignment of disparate fields arising from different melting anomalies. The Ocate volcanic field (OVF) lies immediately to the southwest along the eastern foothills of the Sangre de Cristo Mountains. Together these two fields constitute the most significant late Cenozoic volcanic activity east of the Rocky Mountains.

This paper is intended to be a review of work by many researchers in both of these volcanic fields. The paper also summarizes and coalesces present knowledge about the fields and all dates acquired on rocks from these fields. Stormer, publishing in the 1970's; Nielsen and Dungan, O'Neill, O'Neill and Mehnert, and Staatz in the 1980's; and Sayre, et al., Scott, et al., and Stroud in the 1990's have all significantly increased our knowledge and understanding of these volcanic fields. Sayre, et al. (1995) and Stroud (1996), have fundamentally changed the "common knowledge" of the date for Capulin Volcano. Geologist/NPS Park Ranger Allyson Mathis collected material from the researchers cited above for presentations to Monument visitors during her work in the 1990's at Capulin Volcano National Monument (A. Mathis, unpublished manuscript and personal communication, 1997).

Both fields contain volcanic rocks categorized by several researchers using a range of petrologic classification schemes, since new classifications are favored each decade. In general, the petrologic classifications used in this article are from Scott et al. (1990), Nielson and Dungan, (1985), Phelps et al. (1983), and Stormer (1972). The rock names used here do not, therefore, reflect the current IUGS chemical classification scheme.

CHARACTERISTICS OF VOLCANIC FIELDS

The volcanic fields of late Cenozoic age throughout the southwest are distinctive individually, in terms of the details of petrology, timing of eruptions, size, and morphology of vents; but overall, they are all characterized by many small centers of eruption of fundamentally basaltic, but ranging to more silicic, compositions. Much of what we know about volcanic fields is petrologic and stratigraphic, but studies of basaltic volcanism in general within the past three decades (Connor and Conway, 2000) have begun to explore the processes of volcanic eruption (well-recorded within volcanic fields), individual vents, the lava flows which make up the bulk of their erupted volumes, and the overall significance of fields in terms of their relationship to regional structures and patterns, and rates of activity.

Perhaps a salient characteristic of volcanic fields is the relatively low volumetric rates of eruption compared with major volcanic centers (Crumpler et al., 1992). Rates of 10^{-3} km^3 per year are so characteristic as to be practically diagnostic. Currently these low rates within volcanic fields are seen as evidence for relatively small amounts of melting in the underlying mantle over extended periods on the order of hundreds of thousands to millions of years. In contrast to major volcanic centers and edifices, such rates are so low that residence time in any likely shallow magma reservoirs exceeds the cooling time for the magma bodies. As a result, the individual batches of magma produced must find their way to the surface, ultimately resulting in a series of isolated eruptions that are unrelated to surrounding eruptions in terms of the details of their petrology. It is likely that volcanic fields represent melt production within many small areas

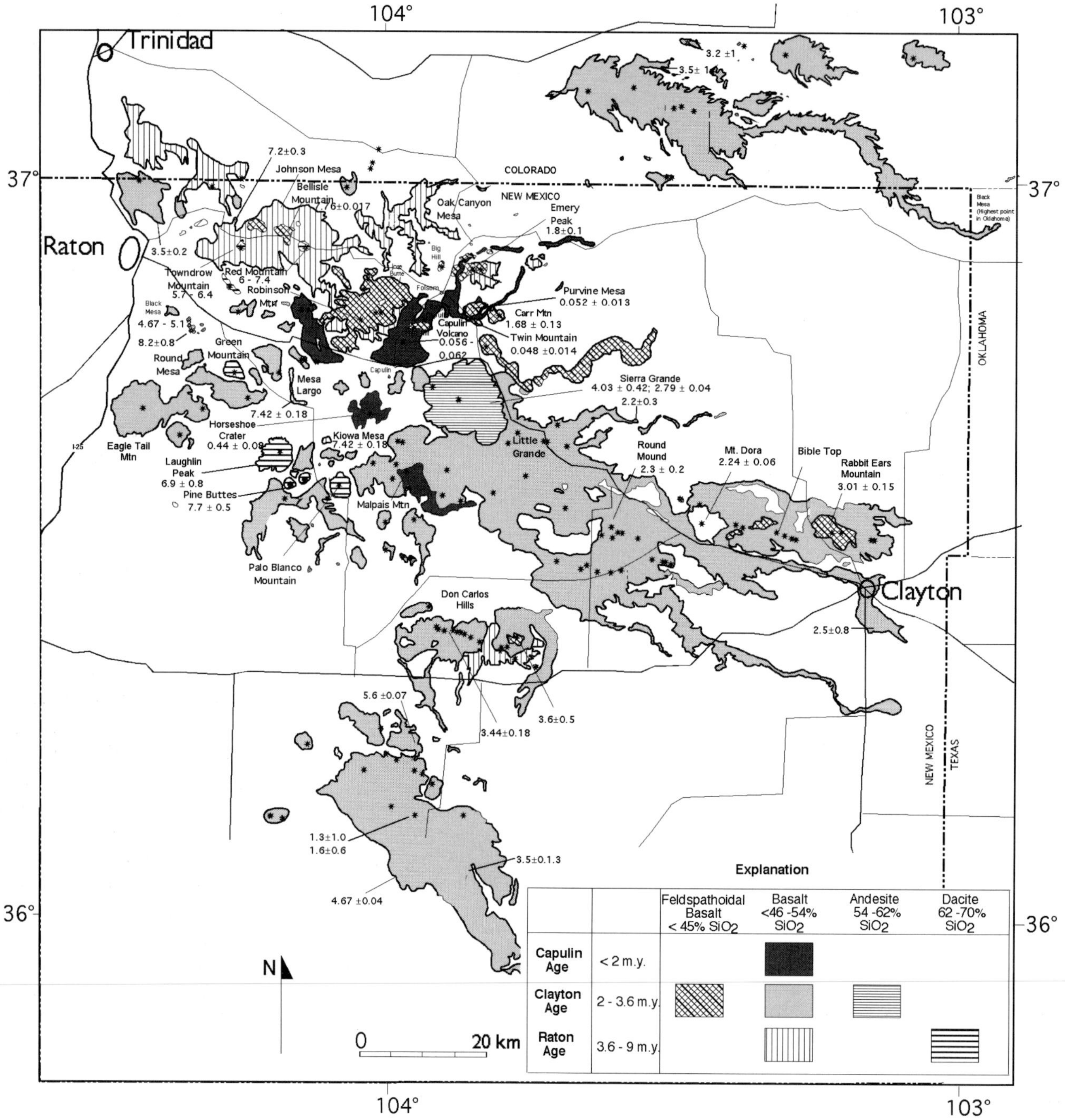

FIGURE 1. Geologic map of the Raton-Clayton volcanic field, showing the three principal phases of volcanism, the fundamental petrology and the currently known ages for volcanic units. Map redrafted with revisions after Luedke and Smith (1978).

of the regional mantle that are close to the solidus in response to regional thermal and local physiochemical anomalies.

RATON-CLAYTON VOLCANIC FIELD

The RCVF, delineated in Figure 1, consists of about 125 vents and extends from near Trinidad, Colorado, 140 km southeastward to Clayton, NM (Dungan, et al., 1989). Including outlying portions, the field covers nearly 20,000 km^2 (Stormer, 1987; Muehlberger et al., 1967). The oldest erupted rocks range from about 9 Ma and the youngest erupted less than 60,000 years ago (Sayre, et al., 1995; Stroud, 1996) (see Table 1), yielding a recurrence rate of approximately 1.3 x 10^{-5} volcanic events per year. Thus the eruption of Capulin over 60,000 years ago does not preclude future activity based on the overall recurrence intervals for the field as a whole.

One of the distinctive characteristics of the Raton-Clayton field is the presence of highly alkalic mafic lavas with SiO_2 contents as low as 36%, and ranging in composition from olivine melitite nephelinite to basanite (Gust, 1990; Phelps et al., 1983).

TABLE 1. Collected dates on rocks from the RCVF.

Name of Feature	Age (technique)	Reference
Raton Phase (9.0-3.6 Ma)		
Johnson Mesa	7.76 ± 0.017 Ma (Ar-Ar)	Stroud, 1996
Barrela Mesa	8.74 ± 0.04 Ma (Ar-Ar)	Stroud, 1996
Kiowa Mesa	7.42 ± 0.18 Ma (Ar-Ar)	Stroud, 1996
Mesa Larga	7.28 ±0.11 Ma (Ar-Ar)	Stroud, 1996
Mesa de Maya or Black Mesa	4.67-5.13 Ma (Ar-Ar)	Stroud, 1996
Red Mountain	6-7.4 Ma (K-Ar)	Scott, et al., 1990
	7.7-8.1 Ma (K-Ar)	Staatz, 1986, 1985
Dale Mountain	4.03 ± 0.12 Ma (Ar-Ar)	Stroud, 1996
Towndrow Peak	5.7-6.4 Ma (K-Ar)	Scott et al., 1990
Laughlin Peak	6.9 ± 0.80 Ma (K-Ar)	Scott et al., 1990
Pine Buttes	7.7 ± 0.5 Ma (K-Ar)	Scott et al., 1990
Kiowa Mesa	7.42 ± 0.18 Ma (Ar-Ar)	Stroud, 1996
Black Mesa	4.67 - 5.13 Ma (Ar-Ar)	Stroud, 1996
Mesa southeast of Black Mesa and north of Round Mesa	8.2 ± 0.8	Stormer, 1972
Clayton Phase (3.6 – 2.0Ma)		
Rabbit Ears Mountain	3.01 ± 0.15 Ma (Ar-Ar)	Stroud, 1996
Sierra Grande	4.0 – 2.6 Ma (Ar-Ar)	Stroud, 1996
Sierra Grande (base of NW flank)	1.9 ± 0.05 (Ar-Ar)	Stroud, 1996
Round Mound (Mt. Clayton)	2.3 ± 0.2 (Ar-Ar)	Stroud, 1996
Mt. Dora	2.24 ± 0.06 (Ar-Ar)	Stroud, 1996
Clayton basalt (distal end SE of Clayton)	2.5 ± 0.8 (Ar-Ar)	Stroud, 1996
Clayton basalt (east of Sierra Grande)	2.2 ± 0.3 (K-Ar)	Stormer, 1972
Don Carlos Hills	3.44 ± 0.18(K-Ar)	Stormer, 1972
	3.6 ± 0.5 (K-Ar)	
Southernmost vent	1.3 ± 1.0 (K-Ar)	Trauger, 1973
Roy flow	3.5 ± 1.3 (K-Ar)	Trauger, 1973
N. margin of Roy flow	5.62 ± 0.07	Stroud, 1996
S. margin of Roy flow	4.67± 0.04	Stroud, 1996
Clayton basalt in	3.5 ± 1.0 (K-Ar)	Hager, 1976
SE Colorado (extends into NM and OK)	3.2 ± 1.0 (K-Ar)	Hager, 1976
Capulin Phase (2.0-0.04 Ma)		
Horseshoe Crater	0.44 ± 0.08 Ma (Ar-Ar)	Stroud, 1996
Carr Mountain	1.68 ± 0.13 Ma (Ar-Ar)	Stroud, 1996
Emery Peak	1.8 Ma (Ar-Ar)	Stroud, 1996
Purvine Mesa	0.052 ± 0.13 Ma (Ar-Ar)	Stroud, 1996
Twin Mountain Mesa	0.048 ± 0.014 Ma (Ar-Ar)	Stroud, 1996
3rd flow unit, Capulin	0.225 ± 0.058 (Ar-Ar)	Stroud, 1996
5th flow unit, Capulin	0.056 ± 0.008 (Ar-Ar)	Stroud, 1996
Basalt in crater, Capulin	0.065 ± 0.049 (Ar-Ar)	Sayre et al., 1995
Boca, Capulin	0.050 ± 0.030 (Ar-Ar)	Stroud, 1996
Boca, Capulin	0.059 ± 0.006 Ma (3He:4He)	Sayre et al., 1995

Very few rocks of similar composition have been described from the Cenozoic volcanic fields of New Mexico. However, the field also contains centers, such as Sierra Grande, that erupted andesite, dacite or rhyodacite (up to nearly 70% SiO_2). Support for the probability of significant contamination in the petrogenesis of some of these rocks was suggested from the strontium isotope study by Jones et al. (1974).

Another distinctive characteristic of the field, and one it shares with the nearby Ocate Volcanic Field, is that the physiographic expressions of flows reflect their relative ages, with the oldest flows capping the highest mesas and the youngest flows following modern lowlands.

Phases of Volcanic Activity in the Raton-Clayton Volcanic Field

Traditionally, the RCVF has been divided into three main phases of volcanic activity (Baldwin and Muehlberger, 1959; Muehlberger et al., 1967; Calvin, 1987). These phases were initially determined by relative dating and compositional similarities. Although this fundamental sequence is still an appropriate model, the ages of these phases are now known to be older than previously believed based on new dates from Stroud (1996).

Raton Phase (9.0-3.6 Ma)

The Raton phase includes eruptions dated from 9.0 to 3.6 Ma (Stroud, 1996) and consists mostly of alkali olivine basalt lava flows and rhyodacite domes. The Raton age basalts are generally mesa-capping lava flows, which disconformably overlie the Paleocene Raton Formation and Cretaceous sandstones and shales, and reflect inverted topography. These include Johnson Mesa, Bartlett Mesa, Barrela Mesa, Horse Mesa, Kiowa Mesa, Mesa Larga, Mesa de Maya or Black Mesa and Kelleher Mesa.

Stroud (1996) investigated the erosion rates for the RCVF using the elevation difference between mesa tops and the local base area and estimating the rate of erosion with the assumption that erosion has been uniform since a flow's eruption. He found that erosion rates are greatest in the northwestern portion of the field (where many of the Raton age basalts are located) where erosion rates are as great as 115 m/Ma. In the east and central portions of the field, lava flows as old as 3.5 Ma have no significant erosional relief. This implies that, early in the RCVF history, uplift was occurring to the northwest.

The Raton-age Red Mountain rhyodacite, occurs as numerous volcanic domes in the western and central portion of the field (Scott and Pillmore, 1993); and includes volcanic centers such as Red Mountain, Towndrow Peak, Cunningham Butte, Green Mountain, Laughlin Peak, Raspberry Mountain, Pine Butte and Palo Blanco Mountain. Laughlin Peak is the only one of these eruptive centers that displays evidence for some pyroclastic activity. The vent edifice is primarily formed of rhyodacite vitrophyre flows and breccia, with a white rhyodacitic tuff found on the south flank (Staatz, 1985).

Between Laughlin Peak and Raspberry Mountain (porphyritic rhyodacite), in an area of approximately 12 km^2, occur many cross-cutting dikes and sills of different igneous composition and numerous thorium and rare-earth veins. The following description is from Staatz (1985). The veins lie along fracture zones and each consists of one large vein or a series of parallel smaller veins; the most common strike direction of the veins is parallel to that of the early faults in the region (between N30W and N80W). The suite of thorium and rare-earth minerals in these veins is unusual when compared with that of other thorium and rare-earth districts; in the Laughlin Peak area this suite consists of brockite, crandallite, and xenotime (all PO_4 minerals), and neither thorite nor monazite have been identified.

Clayton Phase (3.6-2.0 Ma)

The Clayton phase includes eruptions dated from 3.0 - 2.2 Ma (Stroud, 1996) and consists of three units: Clayton basalts (considered transitional olivine basalts by Dungan et al., 1989); basanites (such as Jose Butte and Robinson Peak); and the Sierra Grande andesite. The voluminous Clayton basalts encompass a number of eroded basaltic cinder cones or shield volcanoes that are located near Clayton, NM. These include Rabbit Ear Mountain, Bible Top Butte, and Sierra Clayton or Round Mound (all are cinder cones) and Mt Dora (or Cieneguilla del Burro Mountain) which is a shield volcano. The final eruption of the Clayton basalts was at Mt. Dora, dated at 2.24±0.06 Ma.

One of the greatest battles of Spanish New Mexico was fought near Rabbit Ear Mountain in 1717 between the Spanish and the Comanche; it was also the first "mountain" encountered by travelers on the Santa Fe trail as they approached the Rocky Mountains. Today, the basalt flows and vents of the extreme eastern margin of the field are still among the first visible sights that travelers on airline fights from Chicago to Albuquerque see as they enter New Mexico.

The easternmost edge of the RCVF is primarily dominated by Clayton age flows capping the sand and gravel of the Ogallalah Formation. For nearly the entire distance from Clayton to Des Moines, NM (named for the Iowa capitol), about 20 km, the highway is on Clayton age basalt flows for which this is the type area. The vent sources for these flows have not been positively identified, and may have come from vents in the central part of the field as well as from the eroded cinder cones in the eastern field.

Sierra Grande is an unusual volcano in that it most closely resembles a shield volcano in profile; and it consists almost entirely of 2-pyroxene (both clino- and orthopyroxene) andesite lava flows with compositions ranging from 60-62% SiO_2. Sierra Grande is the only eruptive center in the RCVF containing 2-pyroxene andesite. The volcano is about 10 km in diameter with a slope of about 8 degrees and a maximum elevation of 2650 m. Several small cinder cones are visible around its base. It is not a dome nor is it a true stratovolcano; the vent edifice can best be described as an andesite shield volcano. In this respect we believe that it is comparable to the large volcanoes of the Taos Plateau volcanic field, such as San Antonio Mountain and Ute Mountain, in which thick viscous lava flows dominate and very little pyroclastic material is evident. Stroud (1996) obtained 6 Ar-Ar dates from Sierra Grande; one returned an age of 4.03 ± 0.42 Ma, however the other 5 dates cluster between 2.88 and 2.41 Ma. An earlier K-Ar date of 1.9±0.05 was reported by Stormer (1972).

The outstanding example of a fissure eruption in this volcanic field are the Clayton-age Don Carlos Hills in southwestern Union County. As many as 16 individual vents have been identified along a line that extends about 22 km.

Capulin Phase (2.0-0.04 Ma)

The Capulin phase includes rocks dated from 1.7 - 0.05 Ma (Stroud, 1996) (although Luedke and Smith, 1978, using the originally interpreted Capulin date, use a cutoff of 0.01 for the beginning of the Capulin age). This phase primarily consists of olivine basalt to basaltic andesite erupted as cinder cones, late fissure flows, and at least one phreatomagmatic vent. The Capulin-age basalts are generally in the western and central portions of the field and occupy present day valley floors.

Carr Mountain (also known as Gaylord Mt.) and Emery Peak both represent Capulin-age feldspathoidal eruptions. Carr Mountain is classified as a nephelinite cinder cone with 35.9 % SiO_2. It is the most mafic/alkalic rock in the RCVF, with rocks that contain phenocrysts of hauyne, a rare mineral found only in highly ultramafic rocks. Capulin-age Purvine Mesa was formed by a fissure eruption; and Mud Hill is a phreatomagmatic tuff ring.

Capulin Volcano, Baby Capulin, Twin Mountain, Horseshoe Crater, The Crater, and Malpie Mountain are all basaltic andesite cinder cones. Relative ages based on field studies indicate that Mud Hill, Horseshoe Crater and Emery Peak pre-date Capulin, while Baby Capulin and Twin Peak post-date Capulin. Based on stratigraphic relationships of basalt flows, Baby Capulin may be the youngest vent in the field. Lava flows from Baby Capulin flowed about 10 km down the Dry Cimarron River.

Capulin Volcano

Capulin Volcano (*capulin* is Spanish for wild cherry or chokecherry— named for the bushes found in its sheltered crater) is the center of the most recent volcanic activity in the field. Recent work by Sayre et al. (1995), and Stroud (1996) has resulted in new dates indicating that Capulin erupted approximately 56,000 to 62,000 yrs B.P., somewhat older than the reported age of the famed Folsom paleosite (11,000 yrs B.P.) on which an earlier relative stratigraphic date was based.

Although widely cited, the previous age estimate of less than 10,000 years for the eruption of Capulin Volcano was obtained through an interpreted correlation of a Capulin lava flow unit with alluvium layers believed to be associated with the Folsom archaeological site. At the Folsom site, ^{14}C determinations were made on charcoal and bison bones within two layers of alluvium (Muehlberger, 1955; Anderson and Haynes, 1978; and Haynes et al., 1992). Nine miles from the Folsom site, a Capulin Volcano flow occurs between two alluvial deposits and Muehlberger (1955) interpreted these alluvium layers to be contiguous with the Folsom layers. Therefore, the Capulin eruption was assigned an age range between approximately 10,000 and 4,000 yrs B.P. (the average ^{14}C dates for the layers). Subsequent work (Anderson and Haynes, 1978) identified several distinct alluvial deposits and concluded that the basalt flow overlies an older alluvial sequence with a minimum age of 22,360 ± 1,160 yrs B.P.

Prior to 1995, no absolute age determinations existed for rocks from Capulin Volcano. In 1995 and 1996, two different techniques were used. Sayre et al. (1995) determined the age of a sample of Capulin basalt using the cosmogenic helium technique. This dating technique provides information on how long a rock sample has been within 1m of Earth's surface; and therefore is a useful technique for dating young volcanic flow units. Cosmic rays produce ^{3}He, normally a very scarce isotope, at a steady, known rate at the Earth's surface. The ^{3}He:^{4}He ratio is then used to determine the amount of time a rock surface has been exposed to the atmosphere, with a correction for altitude and latitude. Careful sampling is required to find a sample that has been at the surface in its present orientation since cooling. At Capulin, a sample was collected from the flow at the boca (a late-eruption site at the base of the cone). The results indicate that this late Capulin flow is 59,000 ± 6,000 yrs. Stroud (1996) used the Ar-Ar technique to date four samples from various locations on Capulin which yielded a weighted average of 58,000 ± 4,000 yrs. The Ar-Ar technique is analytically difficult for this flow due to its young age and abundant crustal and mantle xenoliths. It is encouraging that both techniques yielded comparable dates. The new dates are consistent with the apparent state of erosion of cones of similar age throughout the Southwest.

Capulin is a classic cinder cone with associated volumes of lava flows. The crater rim is approximately 1.7 km in circumference and stands 305 m high, with a crater depth of 125 m. Based on a recent detailed study (M. Ort, unpublished map and manuscript, 1997; W.O. Sayer and M. Ort, unpublished report for the National Park Service, 1999), a complicated series of lava flow units and cinder/spatter eruptions occurred during and after eruption of the cinder cone. Three major series of lava flow units erupted, the earliest flowed to the east-southeast of the cone, the second to the west and south, and the third to the west and north. The second and third series of flows were erupted from an area at the base of the cone called the boca ("mouth" in Spanish). During these eruptions several intermittent lava lakes formed at the lower flanks of the volcano and portions of the cone were "rafted" away on the surface of the flows.

Cinder Cone Formation

Cinder cones like Capulin form through explosive fragmentation of basaltic magma as it emerges at the surface. Both CO_2 and H_2O vapor are the most common gases in magmas, requiring less than 1% of the gas to cause dispersal of magma. The type of pyroclastic activity that results in cinder cone formation, known as "strombolian" from the type example Stromboli in Italy, is characterized by "fire fountains" of pyroclasts that may rise up to a thousand feet in the air. The shape of cinder cones reflects more than just a rubble pile, but is distinctive and defined by a well-organized interior bedding and structure. This structure results from the accumulation of ash and cinder in specific ways during strombolian eruptions. The thermal structure within the fire fountain (Head and Wilson, 1989) favors the deposition of fluid spatter, rootless flows, and welded cinder near the vent, and more rubble-like solid cinder and ash at greater distances. Both of these types of deposits may be seen at Capulin volcano. The erupted clasts follow ballistic, and characteristically parabolic, trajectories; and there is a spread in angle of eruption and therefore distance that the clasts travel. Thus, the accumulation of material tends to form a pile that is initially round. As the eruption continues, a number of things can happen to destroy this symmetry. For example, the low strength of the cone flanks insure that any significant increase in effusion within the shallow feeder conduit and summit crater may result in lateral collapse of the cone flanks resulting in a breached crater profile. For this reason, perfect cones like Capulin are rare. It is likely that the formation of the Boca flow significantly altered the profile of Capulin volcano during its eruption, resulting in displacement of the western cone flanks. Continued building of the cone, following this event, restored the cone symmetry more or less, although our detailed morphometric and structural analysis shows that the cone-like morphology is slightly skewed implying that the Strombolian vent migrated slightly to the west following the catastrophic breaching. An interesting feature of many of the cinder/spatter vents in the RCVF, including Capulin, is that the craters open to the southwest or west; suggesting a prevailing southwesterly to westerly wind direction during eruption and accumulation of the pyroclastic ash and cinder. This would also have the effect of making the southwest and west flanks (since they were lower) more susceptible to breaching and collapse.

Chico Sill Complex

Although the most prominent geomorphic features of the RCVF are the cinder/spatter cones, domes, and flows of the Raton, Clayton and Capulin age eruptions; the oldest dated igneous rocks in the region are from 37 to 20 Ma and occur in the Chico sill complex. The following description is from Scott et al. (1990). The Chico sill complex occupies a broad domical area, with an exposed area of about 363 km^2, in the southeastern part of the field between Raton and Eagle Tail Mesa. Cretaceous rocks have been arched upward nearly 300 m by the intrusion of many tabular sills of predominantly alkalic trachyte and phonolite. Several diatremes have also been found, one just east of Raton has been dated at 30 Ma. West of Cimarron a laccolithic intrusion, about 29 Ma, intrudes upper Cretaceous rocks. Mafic dikes are common from Raton southward and vary in age from 24 Ma to possibly as young as 5 Ma.

Turkey Mountains

Other areas of doming by probable igneous intrusions of comparable age to the Chico sill complex lie near and west of Cimarron, and form the Turkey Mountains 11 km west of Wagon Mound. Near Cimarron, numerous sills suggest the presence of a large subsurface intrusion that some workers have suggested is related to the Latir Peak volcanic field (Scott et al., 1990). The Turkey Mountains form a "bulls-eye" of uplifted sedimentary rock units and are clearly the result of the intrusion of a laccolith. Outcrops of syenite at the summit area of the Turkey Mountains have been dated at 29.0 ± 1/6 Ma using the zircon fission-track method (Scott at al., 1990).

OCATE VOLCANIC FIELD

Introduction

The Ocate Volcanic Field (OVF) (sometimes called Mora volcanic field) is located at the boundary between the Southern Rocky Mountains and the Great Plains physiographic provinces; and lies on the east flank of the Rio Grande rift (Fig. 2). The field consists of at least 16 basaltic to dacitic flows and about 50 associated vents ranging in age from late Miocene to Pleistocene (over an 8 million year period) (see Table 2). Regional uplift and subsequent erosion, as in the western RCVF, has resulted in inverted topography where the oldest lava flows occupy the highest local elevation and younger flows cap successively lower mesas. Three major levels of lava flow-capped mesas decrease in elevation from west to east. O'Neill and Mehnert (1988) therefore discuss the activity of the OVF in three age groups: older than 5 Ma; between 5 and 4 Ma; and younger than 4 Ma. The oldest flows are underlain by stream gravels and appear to define a single surface cut into the interior of the Sangre de Cristo Mountains during late Miocene time. This surface cuts across major Laramide structures. Subsequent episodic uplift and erosion marked at least three lower erosional surfaces capped by flows dated between 4.8-4.1Ma, 3.3-3.1Ma. and 2.2Ma. Of these three later surfaces, the oldest is truncated and warped, probably due to uplift of the Cimarron block with respect to the adjacent great Plains (O'Neill, 1988)

Based on geochemical and petrographic data, Nielsen and Dungan (1985) defined five major rock types for the OVF: alkali olivine basalt, transitional olivine-basalt, xenocrystic basaltic andesite, olivine andesite, and dacite. O'Neill and Mehnert (1988) grouped all of the volcanic rocks into two major groups: typical flood basalts from low-profile shield volcanoes and fissures; and alkali basalt, andesite, dacite and nepheline-normative basalt from shield volcanoes and composite cones.

The OVF extends from the southern part of the Cimarron Range of the Sangre de Cristo Mountains southeastward to the general vicinity of Wagon Mound, NM. Flows from the field extend as far south as, and partially surround, the Turkey Mountains. Flows from source vents, such as Maxon Crater (shown as Maxson Crater on topographic maps), southeast of the Turkey Mountains, flowed eastward following the modern drainage of the Mora and Canadian Rivers for some distance. The following description of the three age groups is from O'Neill and Mehnert (1988).

Volcanism older than 5 Ma

Volcanic rocks of the oldest age group cap the highest gravel-covered surfaces. The oldest dated flows were erupted 8.3 Ma in the northwestern part of the field. Remnants cap Sierra Montuosa Mesa and La Grulla Ridge (both are slightly over 3000m in elevation). Slightly younger flows (between 5 and 6 Ma) erupted from widely separated vents near Wagon Mound and along the west side of the field. The vents near Wagon Mound, on Jarosa and Santa Clara Mesas west of I-25 and Las Mesas del Conjelon,

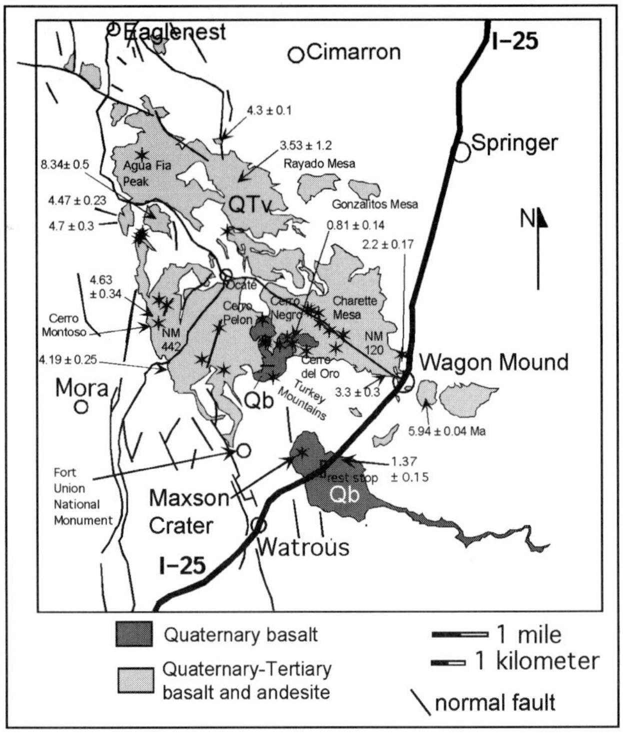

FIGURE 2. Geologic map of the Ocate volcanic field with currently known ages. Map redrafted and simplified from O'Neill and Mehnert (1988)

TABLE 2. Collected dates on rocks from the OVF.

Name of Feature	Age (technique)	Reference
Older Age Group (older than 5Ma)		
Sierra Montuoso	8.34 ± 0.50 (K-Ar)	O'Neill and Mehnert, 1998
Las Mesas del Conjelon (west mesa)	5.94 ± 0.40 (K-Ar)	O'Neill and Mehnert, 1998
Intermediate Age (5-4 Ma)		
Cerro Montoso	4.67 ± 0.32 (K-Ar)	O'Neill and Mehnert, 1998
Gonzalitos Mesa	4.52 ± 0. 34 (K-Ar)	O'Neill and Mehnert, 1998
Sawmill Ranch,	4.47 ± 0.23 (K-Ar)	O'Neill and Mehnert, 1998
S. of Agua Fria	4.7 ±0.3 (K-Ar)	Stormer, 1973
La Mesa	4.19 ± 0.25 (K-Ar)	O'Neill and Mehnert, 1998
El Cerro Colorado	4.12 ± 0.24 (K-Ar)	O'Neill and Mehnert, 1998
Urraca Mesa	4.3 ± 0.1 (K-Ar)	Hussey, 1971
Younger Age Group (younger than 4 Ma)		
White Peak on Ortega Mesa	3.53 ± 1.2 (K-Ar)	O'Neill and Mehnert, 1998
Santa Clara Mesa	3.3 ± 0.3 (K-Ar)	Stormer, 1973
Charette Mesa	3.07 ± 0.34 (K-Ar)	O'Neill and Mehnert, 1998
Vent NW of Wagon Mound	2.20 ± 0.17 (K-Ar)	O'Neill and Mehnert, 1998
Maxon Crater	1.37 ± 0.15 (K-Ar)	O'Neill and Mehnert, 1998
Cerro del Oro	0.81 ± 0.14 (K-Ar)	O'Neill and Mehnert, 1998

east of I-25, mark an apparent E-W fissure system that has been dated at 5.9 Ma. In general, these flows are aphanitic to porphyritic basalts containing both olivine and plagioclase phenocrysts. Chemically, they grade between calc-alkalic and tholeiitic.

Volcanism between 5 and 4 Ma

Volcanic rocks of the intermediate age group are abundant in the northern part of the OVF, including flows that cap Urraca and Fowler Mesas, in the NE, and Rayado and Gonzalitos Mesas, E-SE of the Cimarron Range. Most of the flows in and around Agua Fria Peak also appear to be of this age, as does the basalt cap of Gallina Mesa. In general, flows of this age group are generally olivine basalts that were erupted on an erosion surface that ranges in elevation from 3000 m south of the Cimarron Range to 2100 m on the adjacent Great Plains geographic province. Many of the vents associated with these flows are shield volcanoes: examples are Aspen Hill, Spruce Hill, Apache peak, and Agua Fria Peak. However, the intermediate age rocks are variable in chemistry, including andesite and dacite. Cerro Montoso, in the area of La Mesa, is a composite volcano. The cone is composed of scoria, volcanic breccia, and flow-banded andesite, with dacite flows emanating from its base. Geologic mapping of the basalt flows of El Cerro Colorado in Los Chupaderos Valley shows a high-angle normal fault that has offset flows of equivalent age by about 300 m.

Volcanism younger than 4 Ma

Basalts of the youngest age group are the most abundant flows in the OVF. This age group can, itself, be broken into three groups. The oldest consists of olivine basalt flows erupted onto gravel surfaces that are 50-75 m above the surrounding plains and are dated between 3.1 and 3.3 Ma. These include the flows that cap Charette Mesa and the upper flows of Ortega and Rivera Mesas. The next group consists of olivine basalt flows erupted onto gravel surfaces 15m above the surrounding plains. These include the flows from the vents named Cerrito Pelon and Cerro Negro, the flows of Le Febres area, and the flows from a small mesa north of Wagon Mound. One date of 2.2 Ma has been acquired for these flows. The youngest group of flows consists of olivine basalt with dates ranging from 0.8 to 1.4 Ma and erupted onto surfaces that are only slightly eroded and dissected. These include the flows of Cerro del Oro and its nearby vents, and Maxon Crater and its associated flows. The youngest flows in the field were erupted from the multiple Cerro del Oro scoria cone vents in the central part of Charette Mesa, northwest of Wagon Mound.

Maxon Crater (1.4 Ma) is a large basaltic shield volcano located about 12km south of Wagon Mound and west of I-25. Flows from Maxon flowed 90km eastward, through the canyon cut by the Mora River, to just beyond the confluence of the Mora and Canadian rivers, where the flows lie 100m below the rim of the canyon and 125m above the present level of the rivers.

REFERENCES

Anderson, A.D. and Haynes, C.V., Jr., 1978, How old is Capulin Mountain? correlation between Capulin Mountain volcanic flows and the Folsom type site, northeastern New Mexico *in* Shelton, N. ed., Proceedings of the First Conference on Scientific Research in the National Parks, p. 893-899.

Baldridge, W.S., 1990, Ocate, New Mexico, *in* Wood, C.A. and Kienle, J. Jr,. eds., Volcanoes of North America: Cambridge University Press, p. 293-294.

Baldwin, B. and Muehlberger, W.R., 1959. Geologic studies of Union County, NM.: NM Bur. Mines and Min. Res., Bull. 63, 171 p.

Calvin, E.M., 1987, A review of the volcanic history and stratigraphy of northeastern New Mexico, the Ocate and Raton-Clayton volcanic fields: NM Geological Society, 38th Field Conference, Guidebook, p. 83-85.

Connor, C. B., and Conway, F. M., 2000, Basaltic volcanic fields, *in* Sigurdsson, H., Houghton, B. F., McNutt, S. R., Rymer, H., and Stix, J., eds., Encyclopedia of Volcanoes: Academic Press, p. 331-344.

Crumpler, L. S., Aubele, J.C., and Condit, C.D., 1992, Volcanoes and neotectonic characteristics of the Springerville volcanic field, Arizona: New Mexico Geological Society, 45th Field Conference, Guidebook, p. 147-164.

Dungan, M.A., Thompson, R.A., and Stormer, J.S., 1989, Rio Grande rift volcanism: northeastern Jemez zone, New Mexico, *in* Field Excursions to Volcanic Terranes in the Western United States, vol. 1, Southern Rocky Mountain Region: NM Bureau Mines Min. Res., Mem. 46, p. 474-475.

Gust, D., 1990, Raton-Clayton, NM *in* C.A. Wood and J. Kienle, Jr. eds, Volcanoes of North America, Cambridge University Press p. 292-293.

Hager, M.W., 1976, Late Pliocene and Pleistocene history of the Donnelly Ranch vertebrate site, southeastern Colorado: Univ. Wyoming Contr. Geology, Spec. Paper, no. 2, 62p.

Haynes, C.V., Fr., Beukens, R.P., Jull, A.J.T., and Davis, O.K., 1992, New radiocarbon dates for some old Folsom sites, *in* Stanford, D.J. and Day, J.S., eds., Ice Age Hunters of the Rockies: Denver Museum of Natural History and University Press of Colorado, p. 83-100.

Head, J.W, III. and Wilson, L., 1989, Basaltic pyroclastic eruptions: influence of gas-release patterns and volume fluxes on fountain structure, and the formation of cinder cones, spatter cones, rootless flows, lava ponds and lava flows: Journal of Volcanology and Geothermal Research, v. 37, p. 261-271.

Hunt, A.P., Lucas, s.G., and Kues, B.S., 1987, Third Day Road Log, from Clayton to Raton, *in* NM Geological Society, 38th Field Conference, Guidebook, p.41-54.

Hussey, K.M., 1971, A K-Ar date on the Rocky Mountain pediment sequence, north-central NM: Isochron/West, no. 2, p. 45.

Jones, L.M., Walker, R.L. and Stormer, J.C., Jr., 1974, Isotope composition of strontium and origin of volcanic rocks of the Raton-Clayton district, northeastern NM: Geological Society of America Bulletin, v. 85, p. 33-36.

Kudo, A.M., 1976, A review of the volcanic history and stratigraphy of northeastern New Mexico: New Mexico Geological Society, 27th Field Conference, Guidebook, p. 109-110.

Luedke, R.G. and Smith, R.L., 1978, Map showing distribution, composition, and age of late-Cenozoic volcanic centers in Arizona and New Mexico: U.S. Geological Survey, Miscellaneous Geological Investigations Map I-1091, 2 sheets.

Muehlberger, W.R., 1955, Relative age of Folsom Man and the Capulin mountain eruption: Geological Society of America Bulletin, v. 66, p. 1600-1601.

Muehlberger, W.R., Baldwin, B., and Foster, R.W., 1967, High plains northeastern NM: Scenic Trips to the Geologic Past, no. 7: NM Bur. of Mines and Min. Res., 107p.

Nielsen, R.L. and Dungan, M.A., 1985, The petrology and geochemistry of the Ocate volcanic field, north-central NM: Geological Society of America Bulletin, v. 96, no. 3, p. 296-312.

O'Neill, J.M., 1988, Late Cenozoic physiographic evolution of the Ocate volcanic field: U.S.Geological Survey, Prof. Paper 1478-B.

O"Neill, J.M, 1989, Late Cenozoic volcanism, uplift, and erosion, Ocate volcanic field, north-central NM: summary: *in* Chapin, C.E., and Zidek, J., eds., Field Excursions to Volcanic Terranes in the Western United States, vol. 1: Southern Rocky Mountain Region: NM Bur. Mines and Min. Res., Mem. 46, p. 466-486.

O'Neill, J.M. and Mehnert, H.H., 1988, Petrology and physiographic evolution of the Ocate volcanic field, north-central New Mexico: U. S. Geological Survey Prof. Paper 1478-A.

Phelps, D.W., Gust, D.A., and Wooden, J.L., 1983, Petrogenesis of the mafic feldspathoidal lavas of the Raton-Clayton volcanic field, NM: Cont. Min. Petr. 84, p. 182-190.

Sayre, W.O., Ort, M.H., and Graham, D., 1995, Capulin Volcano: Park Science, v. 15, n. 2, p.10-11.

Scott, G.R., Wilcox, R.E., and Mehnert, H.H., 1990, Geology of volcanic and subvolcanic rocks of the Raton-Springer area, Colfax and Union Counties, NM: U.S. Geological Survey Prof. Paper 1507, 58p.

Staatz, M.H., 1985, Geology and description of the thorium and rare-Earth veins in the Laughlin Peak area, Colfax County, NM: U.S. Geological Survey Prof. Paper 1049E, p. E1-E32.

Staatz, M.H., 1986, Geologic map of the Pine Buttes quadrangle, Colfax County, NM: U.S. Geological Survey, Quadrangle Map GQ-1591, scale 1:24,000.

Stormer, J.C., 1972, The ages and nature of volcanic activity on the southern high plains, NM and Colorado: Geological Society of America, Bull. 83, p. 2443-2448.

Stormer, J.C., 1987, Capulin Mountain volcano and the RCVF, northeastern NM: Geological Society of America, Centennial Field Guide, vol. 2, Rocky Mountain Section, p. 421-424.

Stroud, J.R., 1996, The volcanic history and landscape evolution of the Raton-Clayton volcanic field [M.S. Thesis]: Socorro, NM Institute of Mining and Technology, 49p.

Trauger, F.D., 1973, K-Ar dates on a basalt flow and vent plug in northeast Harding County, northeastern NM: Isochron/West, no.7, p. 7-9.

NOTES ON UPPER PERMIAN AND LOWER TRIASSIC STRATA AND STRUCTURES IN EAST-CENTRAL AND CENTRAL NEW MEXICO

R. DOUGLAS MAIER
3762 Sierra Drive, Honolulu, HI 96816-3818

Abstract.—The Middle Triassic Anton Chico Member of the Moenkopi Formation accumulated in east-central and central New Mexico, for the most part, on a very low relief Permian surface that had not been previously deformed. This conclusion is indicated by a general lack of basal coarse-grained facies within the Anton Chico, shallow or no Triassic scouring below this horizon, and Middle Permian sandstone and stromatolite marker beds that can be traced laterally for over 150 km. Post-Triassic deformation appears to have followed preexisting trends, including a reactivated thrust belt. A regional oxidation-reduction front was discovered. Lineaments control the distribution and areas of sinks in east-central and central New Mexico, and one linea-ment zone can be correlated with those of the west-central United States. The degrees of thermal alteration of organic material in Middle Permian and Middle Triassic strata suggest they were once covered by thick deposits.

INTRODUCTION

Studies of Middle Permian and Middle Triassic strata in east-central and central New Mexico, as well as analyses of structures in these regions, have resulted in anumber of new findings and the compilation of an updated structure-contour map of the Permian-Triassic boundary in Guadalupe County (Fig. 1) and adjacent areas.

PREVIOUS WORK

Middle Permian stratigraphy in east-central and central New Mexico

Tait et al. (1962) correlated Middle Permian strata from the Permian Basin south-east of the study area in southeastern New Mexico to north-central New Mexico using well logs and outcrops. They showed that the relatively thin and clastic Bernal Formation gave way to predominantly evaporitic, thicker beds of the Artesia Group to the southeast. This was corroborated by Kelley (1972b) (Fig. 2). Both suggested that the Bernal Formation was equivalent in age to lower Artesia Group formations so that there was overstepping by overlying Triassic beds toward the northwest (Fig. 2). Lucas and Hayden (1991) proposed that "Artesia Formation" be used instead of "Bernal Formation" because of the relationships indicated by Tait et al. (1962). Stromatolite and sandstone marker beds were traced by Maier et al. (1986) in the upper Bernal (or Artesia) Formation and upper Artesia Group for several km (Fig. 2), suggesting that there was little or no uplift of Permian deposits prior to Triassic deposition. This will be discussed in more detail below.

Lower Triassic stratigraphy in east-central and central New Mexico

Gorman and Robeck (1946) separated the lower part of the Triassic section in north-central Guadalupe County into 4 units, a lower "purplish-red" sandstone, followed upward by a gray sandstone, then a shale unit, and above that by another gray sandstone bed (Fig. 3).

Stewart et al. (1972a, b) tentatively correlated the lowermost Triassic Moenkopi Formation across the Rio Grande from the Colorado Plateau to the west to the Sevilleta Grant, about 16 km northeast of Socorro. At this location they assigned the beds overlying the "Moenkopi (?)" to the Petrified Forest Member of the Chinle Formation, also extending the nomenclature from the west. Maier et al. (1986) stated that "Moenkopi-type" strata occur southeast of Socorro.

FIGURE 1. Location map.

Based on a Middle Triassic (Anisian) amphibian fossil found in the lower sandstone unit of Gorman and Robeck (1946), Lucas and Morales (1985) and Lucas et al. (1985) correlated this interval with the Moenkopi Formation. Later, Lucas and Hunt (1987) named this lower sandstone the Anton Chico Member of the Moenkopi Formation and Lucas (1991) proposed that the unit extends across west-central, south-central, and east-central New

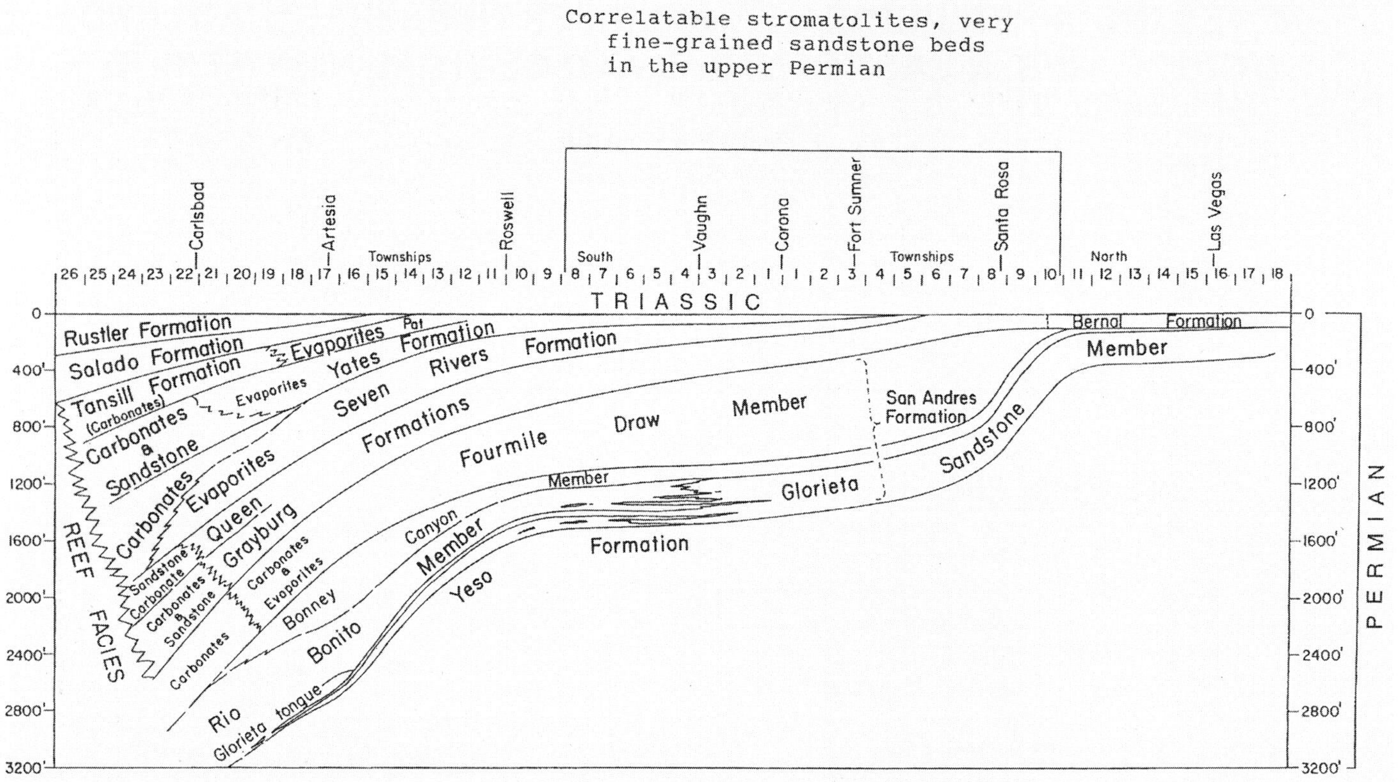

FIGURE 2. Cross section from Kelley (1972b).

Mexico (Fig. 3) because of lithology, stratigraphic position, and another Anisian amphibian fossil discovered approximately 25 km southeast of Socorro.

Permian-Triassic boundary surface

Foster et al. (1972) and Broadhead (1984) mapped the structure on the Permian-Triassic boundary surface in Guadalupe County. They used broader-based data than this study, which incorporates some of Broadhead's gamma-ray picks of the boundary, especially in eastern Guadalupe County.

Other regional structure data

Kelley (1972a) mapped surface structures in east-central New Mexico and concluded that they follow preexisting Pennsylvanian-Permian features, the north-south Pedernal high and the northeast-trending Sierra Grande arch (Fig. 4).

Maughan and Perry (1988) showed that within the west-central United States there are prominent northeast-southwest oriented lineaments, some extending to the north-central U.S. This study uncovered another long lineament in east-central and central New-Mexico that parallels those found by Maughan and Perry (1988).

In 1986, Maier et al. said that northeast-trending lineaments in east-central and central New Mexico control the areas of sinks and that Middle Permian and Middle Triassic beds were once covered by thick sedimentary deposits, as determined from thermal alteration values.

METHODS AND OBJECTIVES

Twenty-five stratigraphic sections of Middle Permian and Middle Triassic rocks were measured and described in the field in east-central and central New Mexico (Fig. 5). Cores from 8 drill holes in central northern Guadalupe County of the same interval were also documented. The purposes were to correlate Permian and Triassic strata in east-central and central New Mexico and to note any lateral differences in these beds.

The Permian-Triassic boundary was surveyed for 100 km in east-central New Mexico. This data set plus log (lithologic, radioactive, and electric) picks of this surface were used to construct a structure-contour map of the boundary in Guadalupe County and adjacent areas (Fig. 6). Seismic sections were studied from Guadalupe County for the purpose of integrating surface manifestations with subsurface structures.

Lineaments and circles were traced from black and white and color satellite images (black and white Landsat band 4 and Heat Capacity Mapping Mission (HCMM) thermal infrared images and Landsat false-color composite) of east-central and central New Mexico. The main objectives were to correlate sinks with lineaments and to show the spatial association of lineaments with those of the larger region (Western Interior U.S.).

Thermal alteration indices (TAI'S) were calculated from organic constituents of outcrops and core samples of Middle Permian and Middle Triassic strata in east-central and central New Mexico. The purpose was to estimate the former thickness ofrocks (that then captured heat) once covering these beds.

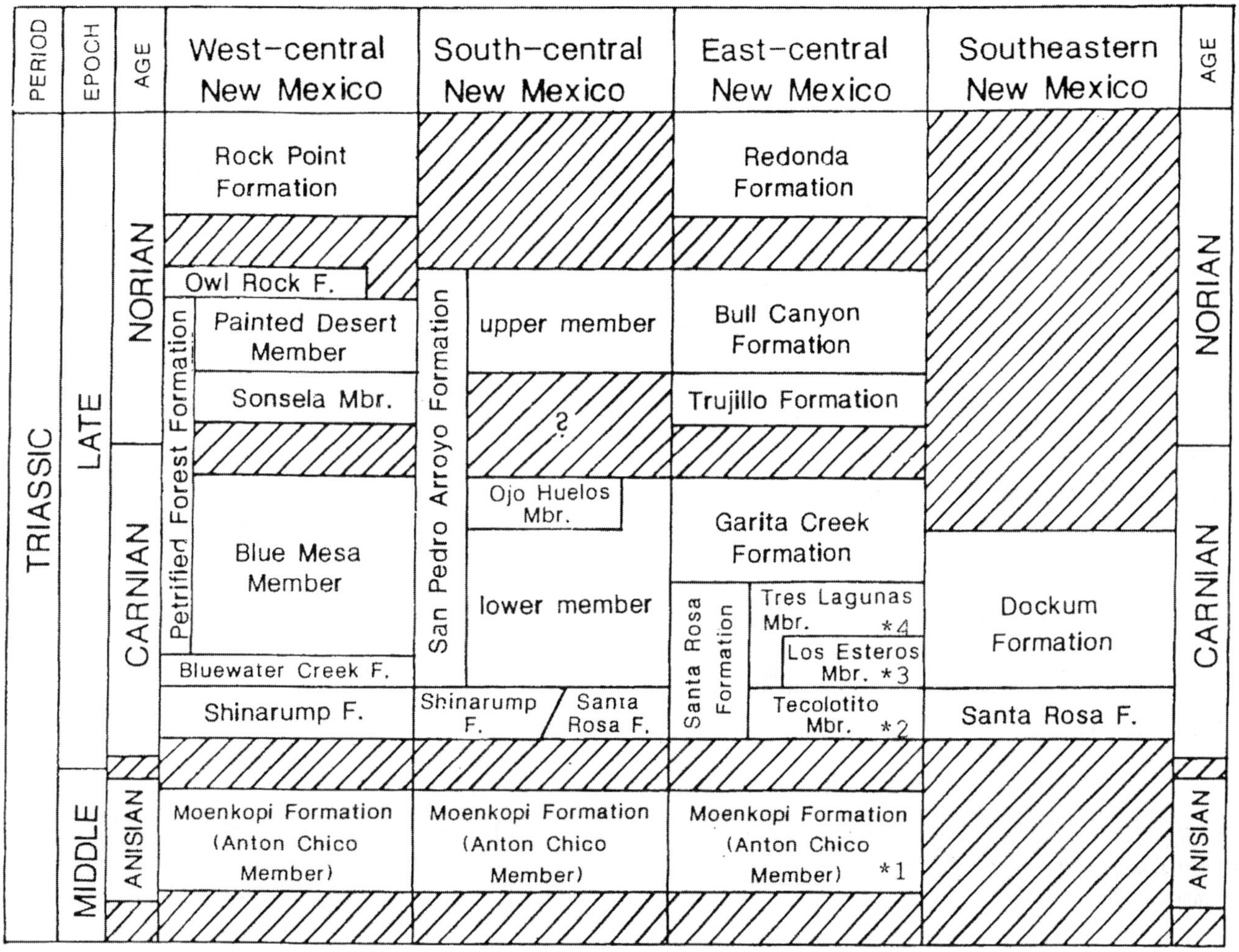

FIGURE 3. Triassic formation chart from Lucas (1991). * Unit numbers from Gorman and Robeck (1946).

RESULTS OF STUDY

Middle Permian stratigraphy

Stromatolites near the top of the Permian Bernal Formation-Artesia Group were found southward 160 km from the Aurora stratigraphic section location to the Fort Stanton outcrop. They extend eastward from the Barnham exposure about 100 km to near the Salado outcrop.

A continuous very fine-grained sandstone bed under the stromatolites can be traced along outcrops eastward from Aurora 40 km to the east of Colonias in the Pecos River canyon. The same sequence can be seen in the Pintada drainage, west of Santa Rosa, for 40 km, and at the West Salado measured section, 32 km southwest of Santa Rosa. The same sandstone unit was tentatively identified, although stromatolites were missing above it, in the Pecos River exposures southeast of Santa Rosa 10 km to the Puerta de Luna measured section. Other upper, very fine-grained sandstone beds were observed in local outcrops to the west and south but are less reliable markers.

Middle Triassic stratigraphy

As proposed by Lucas (1991), Unit 1 of Gorman and Robeck (1946) (his Anton Chico Member of the Moenkopi Formation) can be found in the entire study area. I also agree that the base of the Anton Chico is usually the first gray-red sandstone overlying the typically red-brown, finer-grained Bernal Formation, except where the latter is missing at Tejon.

However, I would place the upper limit of the Anton Chico below the first chert-bearing conglomerate, apparently occurring everywhere in the study area except at his Sevilleta Grant section and at the Derramadero outcrop of my study. It took some lateral tracing at the Sevilleta outcrop, located 8 km south of his section, to find chert clasts in what was the "usual" gray-red conglomerate of the Anton Chico. His unit 10 conglomerate of the Anton Chico occurs upsection from the Bernal at approximately the same level (within 6 m) as at my Sevilleta section and may correlate. This would reduce his measured thickness of the Anton Chico to 31 m, more typical of the member in central and east-central New Mexico. If the lowest chert conglomerate in the Bull Gap N sec-

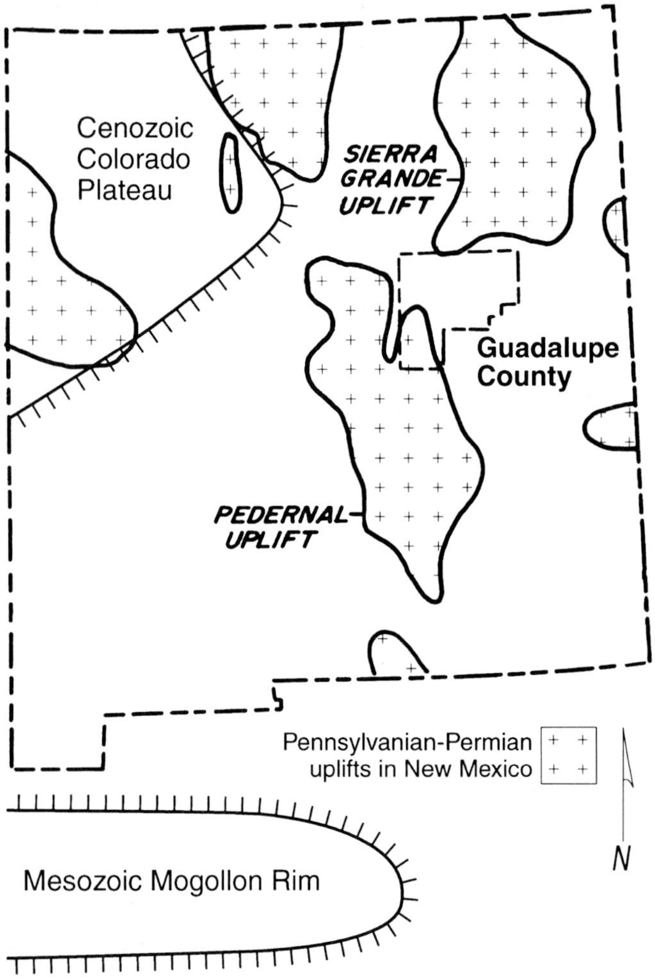

FIGURE 4. Selected tectonic features in and south of New Mexico.

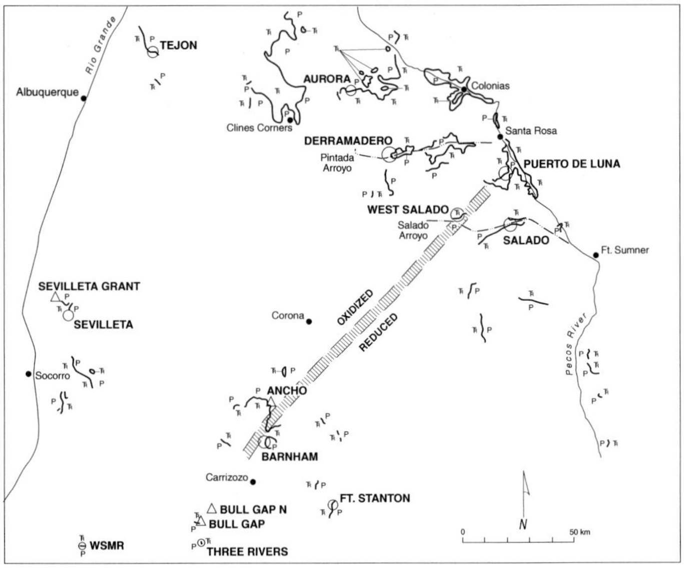

FIGURE 5. Permian-Triassic contact exposures, stratigraphic sections mentioned in this report. Triangles- Lucas (1991). Circles- this study.

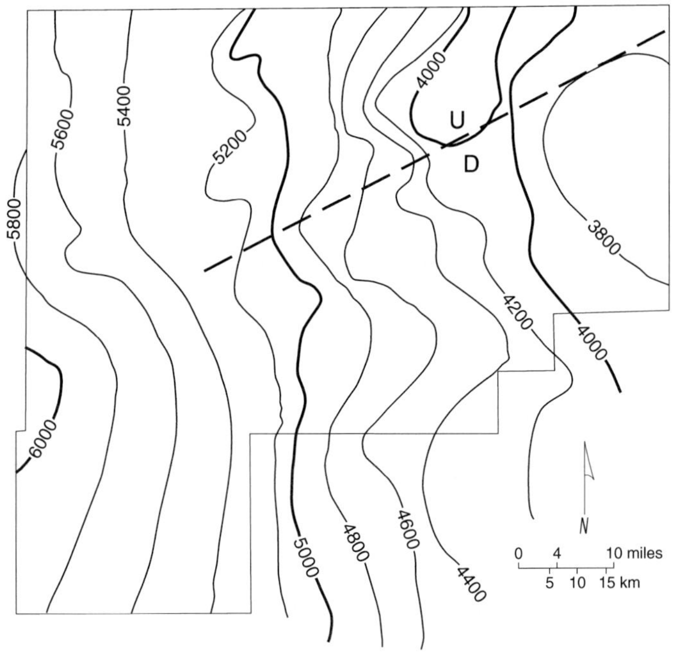

FIGURE 6. Structure-contour map of the Permian-Triassic contact in Guadalupe County and adjacent areas. Contour interval 200 feet (61 m).

tion is used as the same marker, it would make Lucas' Anton Chico 62 m thick, much more reasonable than the 101 m measured. The only other explanations would be a nearby active fault or sinkhole during Anton Chico deposition.

The chert conglomerate marks the base of Unit 2 of Gorman and Robeck (1946), which Lucas (1991) called the Tecolotito Member of the Santa Rosa Formation in east-central New Mexico and which he correlated with the Shinarump Formation to the west. I agree with this deignation and think that the term "Shinarump" should be used throughout the region for the lower chert conglomerate. My southernmost sections, WSMR, Three Rivers, and Fort Stanton, had by far the largest chert clast sizes, indicating a southerly source. I do not understand how 8 cm clasts got to Lucas' (1991) Ancho location.

A notable point is that the Anton Chico does not vary much in thickness throughout east-central and central New Mexico, roughly averaging 24 m. The maximum is at Lucas' (1991) Bull Gap N and Bull Gap (50 m at the latter) sections. The thinnest is at WSMR (3 m) and Three Rivers (6 m). This suggests that the Shinarump was deposited on a fairly level surface in the study area.

The Unit 3 shale of Gorman and Robeck (1946) (Los Esteros Member of the Santa Rosa Formation (Lucas 1991)) can be identified in all of east-central and central New Mexico where outcrop or drill hole data are available. The characteristic features are local mottling, calcareous nodules, bentonite, gypsum, and thin limestone interbeds. It is also varicolored and can contain a considerable amount of sandstone. Lucas' (1991) correlations of the Los Esteros and Tres Lagunas (Unit 4 of Gorman and Robeck (1946)) Members of the Santa Rosa Formation in east-central New Mexico westward with the lower San Pedro Formation then the lower Petrified Forest Formation are beyond the scope of this paper.

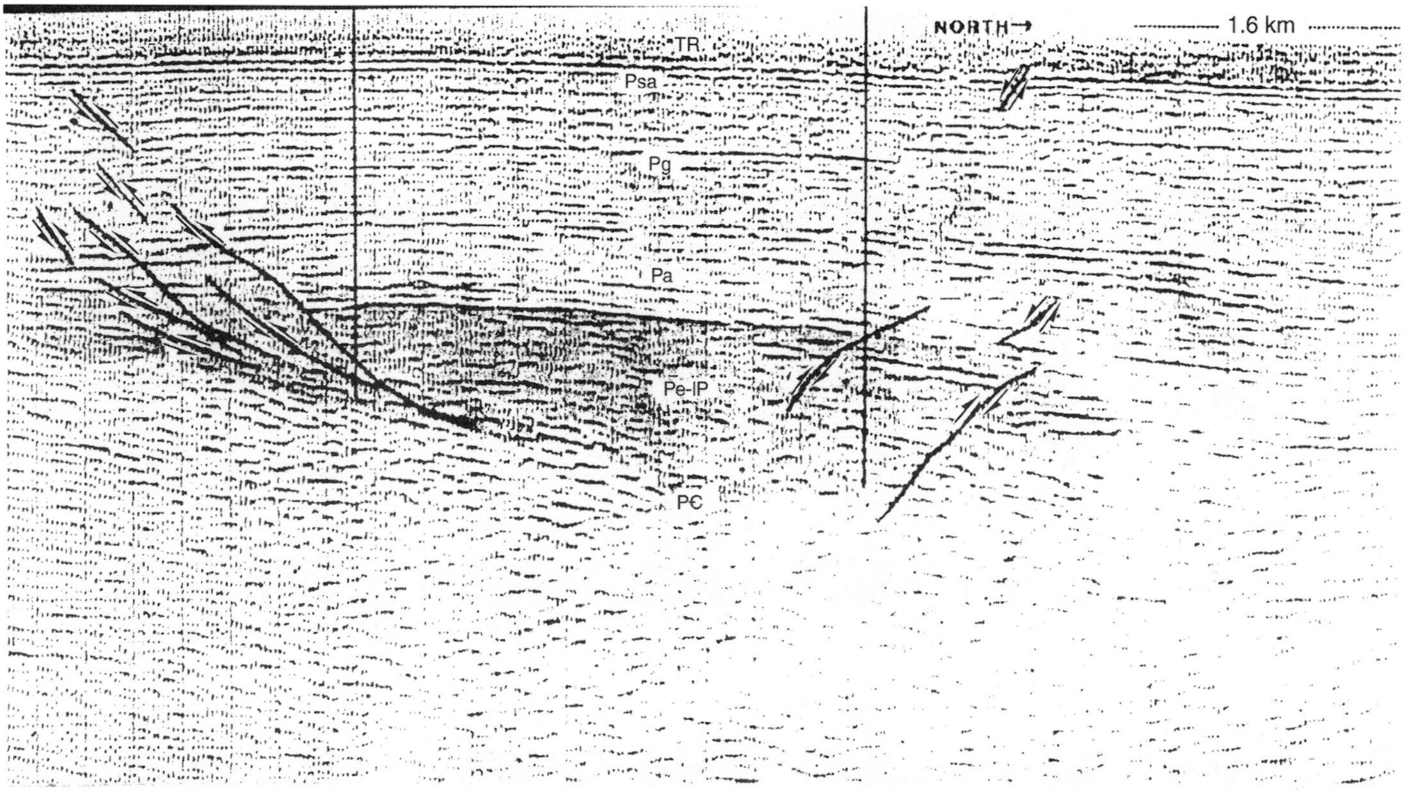

FIGURE 7. Seismic section of thrust faults in Guadalupe County. TR- Triassic; Psa- San Andres Formation; Pg- Glorieta Sandstone; Pa- Abo Formation; Pe-lP- Pennsylvanian-lower Permian. PC- Precambrian.

Oxidation-reduction front

Middle Permian and Middle Triassic strata are generally oxidized in east-central and central New Mexico. Exceptions occur at various southeasterly locations in the study area where the Bernal Formation and the Anton Chico are reduced. A northeast-southwest bar (Fig. 5) denotes where this front is projected. In Salado Arroyo just west of the West Salado section the lateral change from oxidized to the northwest to reduced to the southeast is exposed.

Permian-Triassic boundary surface

The main influences on the Permian-Triassic boundary in Guadalupe County (Fig. 6) are the resurgent north-south trending Pedernal arch to the west and northeast-oriented Sierra Grande arch to the north. A suspected fault was deduced from converging contours and follows the margin of the Sierra Grande arch.

Thrust faults

The fault mentioned just above is situated in the vicinity of a northeast-southwest belt of thrust faults that can be seen on seismic sections (Fig. 7) (Maier, 1987). They are southeast-vergent, and appear to be products of Laramide compression from the west that reactivated basement-cored faults.

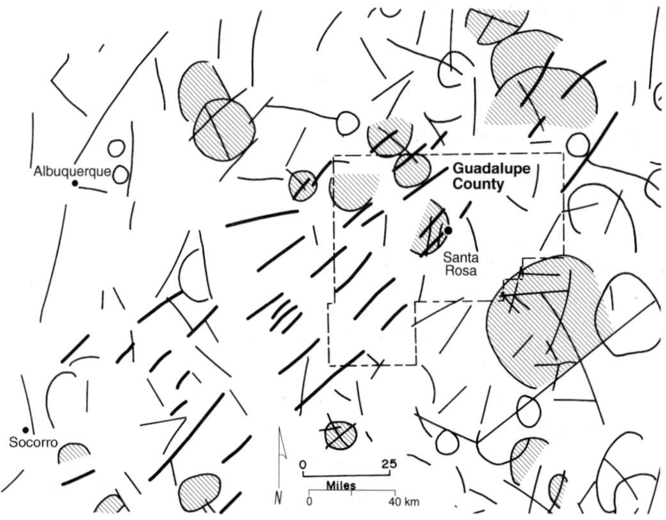

FIGURE 8. Lineaments and circular features from HCMM image. Darkened lines show prominent northeast-trending lineament zone. Hachures show sinks areally controlled by northeast-oriented lineaments.

Lineaments and circles

Recognizing regional-scale lineaments was best achieved by using relatively low resolution HCMM thermal infrared imagery. A major lineament extends northeast-southwest for 250 km across the study area (Fig. 8), paralleling other west-central U.S. long lineaments (Fig. 9). It is in an n=2 zone, described by the formula

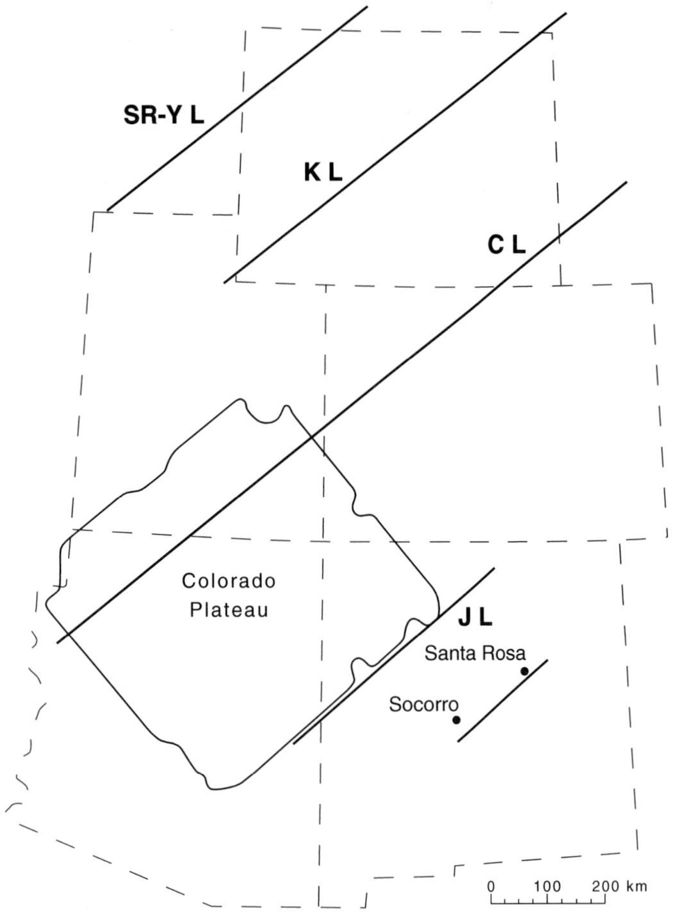

FIGURE 9. Mapped northeast-oriented lineaments in west-central U.S. from Maughan and Perry (1988) and Aldrich et al. (1986) (Jemez Lineament). Paralleling HCMM lineament zone in east-central and central New Mexico. SR-Y L- Snake River-Yellowstone Lineament; K L- Kaycee Lineament; C L- Colorado Lineament; J L- Jemez Lineament.

x/2 with x being 500 km for North America (Maier, 1993).

Sinks occur at the intersections of mostly northeast-trending lineaments (following generally the strike of the beds) and updip-trending lineaments to the north and northwest. Their areas are confined by the lengths of northeast-oriented lineaments and this phenomenon can be seen at variable scales, including that of aerial photography and Landsat and HCMM images (Fig. 8).

Thermal alteration indices

That thick deposits once overlaid Middle Permian and Middle Triassic strata in east-central and central New Mexico is indicated by elevated thermal alteration values that were found. The minimum thickness of these former beds can be estimated by application of formulas Pigott (1985) proposed:

$$T=164.4-19.39 \ln t$$
$$=164.4-19.39 \ln (250)$$
$$=57.54 \text{ C}$$
$$Doc=100 ((T-Ts)/dt/dZ)$$
$$=100 ((57.54-20)/3)$$
$$=4129 \text{ feet } (1228 \text{ m})$$

T is the threshold temperature at which organic matter will decompose to form hydrocarbons; t is the age of the bed in 10 years; Doc is the depth of the oil ceiling; Ts is the mean surface temperature; and dt/dZ is the thermal gradient.

Thermal alteration values of three or more were found in the palynological samples from east-central and central New Mexico that were processed (Dr. Cornell, personal communication) This suggests that the threshold temperatures had been reached (Waples, 1980). The Doc of 1230 m implies a major basin of deposition in this region. Presumably, this was during the Early Tertiary and/or Late Cretaceous.

ACKNOWLEDGMENTS

I would like to especially thank a few of the many people who contributed to this paper: Mr. Robert R. Norman, who introduced me to this area; Mrs. Dorothy B. Maier, my mother, a stalwart supporter; Dr. Frank E. Kottlowski, who gave encouragement and helped gain financial assistance from the New Mexico Bureau of Mines and Mineral Resources; and Dr. Spencer G. Lucas, who provided invaluable data and kind patience.

REFERENCES

Aldrich, M.J., Jr., Laughlin, A.W., Meade, J.S., and Peirce, H.W., The Jemez Lineament: structural boundaries and control on sedimentary facies, tectonism, and mineralization, *in* Aldrich, M.J. Jr., and Laughlin, A.W., eds., Proceedings of the Sixth International Conference on Basement Tectonics, p. 104-113.

Broadhead, R.F., 1984, Subsurface petroleum geology of the Santa Rosa Sandstone (Triassic), northeastern New Mexico: New Mexico Bureau of Mines and Mineral Resources, Circular 193, 23 p.

Foster, R.W., Frentress, R.M., and Riese, W.C., 1972, Subsurface geology of east-central New Mexico: New Mexico Geological Society, Special Publication 4, 22 p.

Gorman, J.M., and Robeck, R.C., 1946, Geology and asphalt deposits of north-central Guadalupe County, New Mexico: U.S. Geological Survey, Oil and Gas Investigations Preliminary Map 44.

Kelley, V.C., 1972a, Geology of the Fort Sumner sheet, New Mexico: New Mexico Bureau of Mines and Mineral Resources Bulletin 98, 55 p.

Kelley, V.C., 1972b, Outcropping Permian shelf formations of eastern New Mexico: New Mexico Geological Society, 23rd Field Conference, Guidebook, p. 72-78.

Lucas, S.G., 1991, Triassic stratigraphy, paleontology, and correlation, south-central New Mexico: New Mexico Geological Society, 42nd Field Conference, Guidebook, p. 243-259.

Lucas, S.G., and Hayden, S.N., 1991, Type section of the Permian Bernal Formation and the Permian-Triassic boundary in north-central New Mexico: New Mexico Geology, v.13, p. 9-15.

Lucas, S.G., and Hunt, A.P., 1987, Stratigraphy of the Anton Chico and Santa Rosa Formations, Triassic of east-central New Mexico: Journal of the Arizona-Nevada Academy of Science, v. 22, p. 21-33.

Lucas, S.G., Hunt, A.P., and Morales, M., 1985, Stratigraphic nomenclature and correlation of Triassic rocks of east-central New Mexico; a preliminary report: New Mexico Geological Society, 36th Field Conference, Guidebook, p. 171-184.

Lucas, S.G., and Morales, M., 1985, Middle Triassic amphibian from the basal Santa Rosa Formation, east-central New Mexico: New Mexico Geological Society, 36th Field Conference, p. 56-58.

Maier, R.D., 1987, Thrust faults in east-central New Mexico (abs.): American Association of Petroleum Geologists Bulletin, v. 71, no. 2, p. 241-242.

Maier, R.D., 1993, Orthogonal fractal structure (abs.): Eos, Transactions, American Geophysical Union, v. 74, no. 16, April 20 Supplement, p. 379-380.

Maier, R.D., Dyer, J.R., Pingitore, N.E., and Cornell, W.C., 1986, Structural history during and after the Late Permian in east-central and central New Mexico (abs.): New Mexico Geology, v. 8, no. 4, p. 97.

Maughan, E.K., and Perry, W.J., 1988, Lineaments and their tectonic implications, *in* the Rocky Mountains and adjacent plains region, in Peterson, J.A., ed., Paleotectonics and Sedimentation in the Rocky Mountains Region, American Association of Petroleum Geologists Memoir 41, p. 41-53.

Pigott, J.D., 1985, Assessing source rock maturity in frontier basins: importance of time, temperature, and tectonics: American Association of Petroleum Geologists Bulletin, v. 69, no. 8, p. 1269-1274.

Stewart, J.H., Poole, F.G., and Wilson, R.F., 1972a, Stratigraphy and origin of the Chinle Formation and related Upper Triassic strata in the Colorado Plateau region: U.S. Geological Survey Professional Paper 690, 336 p.

Stewart, J.H., Poole, F.G., and Wilson, R.F., 1972b, Stratigraphy and origin of the Triassic Moenkopi Formation and related strata in the Colorado Plateau region: U.S. Geological Survey Professional Paper 691, 195 p.

Tait, D.B., Ahlen, J.L., Gordon, A., Scott, G.L., Motts, W.S., and Spatler, M.E., 1962, Artesia Group of New Mexico and West Texas: American Association of Petroleum Geologists Bulletin, v. 46, no. 4, p. 504-517.

Waples, D.W., 1980, Time and temperature in petroleum formation: application of Lopatin's method to petroleum exploration: American Association of Petroleum Geologists Bulletin, v. 64, no. 6, p. 916-926.

At Revuelto Creek in Quay County, New Mexico, trough-crossbedded sandstones and conglomerates of the Trujillo Formation of the Chinle Group are well exposed. Paleocurrent azimuths measured in these strata indicate westerly and northwesterly flow of large, Late Triassic rivers that crossed West Texas and New Mexico toward a Pangean shoreline that lay near the present Utah-Nevada border.

TRIASSIC STRATIGRAPHY, BIOSTRATIGRAPHY AND CORRELATION IN EAST-CENTRAL NEW MEXICO

SPENCER G. LUCAS[1], ANDREW B. HECKERT[2] AND ADRIAN P. HUNT[3]
[1]New Mexico Museum of Natural History, 1801 Mountain Rd NW, Albuquerque, NM 87104;
[2]Department of Earth and Planetary Sciences, University of New Mexico, Albuquerque, NM 87131-1116;
[3]Mesalands Dinosaur Museum, Mesa Technical College, 911 South Tenth Street, Tucumcari, NM 88401

Abstract.—Triassic strata in east-central New Mexico are siliciclastic red beds of Middle and Late Triassic age. As much as ~500 m thick, the Triassic section is assigned to the Middle Triassic Moenkopi Formation (Anton Chico Member) and overlying Upper Triassic Chinle Group (Santa Rosa, Garita Creek, Trujillo, Bull Canyon and Redonda Formations). The Anton Chico Member of the Moenkopi Formation is as much as 46 m thick and is mostly trough crossbedded micaceous litharenite. It disconformably overlies Middle Permian red beds of the Artesia Group, and is disconformably overlain by the Upper Triassic Santa Rosa Formation. The Santa Rosa Formation consists of three members: (1) basal, Tecolotito Member, as much as 34 m thick and mostly grayish-orange, trough crossbedded and ripple laminated quartzarenite; (2) medial, Los Esteros Member, as much as 44 m thick and mostly red-bed mudstone; and (3) upper, Tres Lagunas Member, as much as 46 m thick and mostly orange and yellowish brown, crossbedded quartzarenite. The Garita Creek Formation conformably overlies the Tres Lagunas Member and is as much as 76 m thick and mostly red-bed mudstone. It is disconformably overlain by the Trujillo Formation, a medial sandstone complex of the Chinle Group as much as 68 m thick. The Bull Canyon Formation conformably overlies the Trujillo and is as much as 110 m thick and mostly red-bed mudstone. We name a new unit in the Bull Canyon Formation, the Saladito Point Bed, for a discrete, laterally persistent sandstone 12.2 m thick and approximately 30 m above the base of the formation. The Bull Canyon Formation is disconformably overlain by the Redonda Formation, which is up to 92 m thick and mostly laterally continuous, repetitive beds of red-bed sandstone and siltstone. The Middle Jurassic Entrada Sandstone disconformably overlies the Redonda Formation, although the Triassic section is also erosionally truncated locally by younger units throughout east-central New Mexico.

The Anton Chico Member of the Moenkopi Formation correlates with the uppermost Holbrook Member of the Moenkopi Formation in eastern Arizona. The occurrence of the capitosaurid amphibian *Eocyclotosaurus* in the Anton Chico Member indicates a Perovkan (Anisian) age. The erosional surface at the base of the Moenkopi Formation in east-central New Mexico is a compound unconformity corresponding to the Tr-0, Tr-1 and Tr-2 unconformities of earlier authors.

Chinle Group strata in east-central New Mexico correlate readily to the Chinle lithosome in West Texas, Oklahoma, Colorado and across the Colorado Plateau. These strata define three major sequences, in ascending order: the Shinarump-Blue Mesa (Santa Rosa-Garita Creek), Moss Back-Owl Rock (Trujillo-Bull Canyon) and Rock Point (Redonda) sequences. Unconformities at the base of the Santa Rosa, Trujillo and Redonda Formations correspond to the Tr-3, Tr-4, and Tr-5 unconformities, respectively. A lower-order sequence boundary may be present in the Chinle, expressed as a regionally persistent, planar-crossbedded litharenite exposed in east-central New Mexico (Saladito Point Bed), western New Mexico (Perea Bed) and in east-central Arizona at the Petrified Forest National Park (Flattops 2 and Painted Desert 3 sandstones).

Numerous lines of biostratigraphic evidence, principally tetrapod fossils, but also including megafossil plants, ostracodes and invertebrates, constrain the ages of Upper Triassic strata in east-central New Mexico and support the correlations advanced here. These indicate that the Tecolotito Member of the Santa Rosa Formation is of Otischalkian (early-late Carnian) age, and the Los Esteros and Tres Lagunas members and the Garita Creek Formation are of Adamanian (latest Carnian) age. The Tr-4 unconformity approximates the Carnian-Norian boundary, and the Trujillo and Bull Canyon Formations are of Revueltian (early-mid Norian) age. The Redonda Formation is of Apachean (late Norian-Rhaetian) age. The type Revueltian fauna is known from the Bull Canyon Formation in Revuelto Creek, and the type Apachean fauna is the vertebrate fauna of the Redonda Formation in Apache Canyon, both in Quay County.

INTRODUCTION

Exposures of Triassic strata in east-central New Mexico (San Miguel, Harding, Guadalupe, DeBaca, Roosevelt and Quay Counties) form the single largest outcrop belt of Triassic rocks in New Mexico (Fig. 1). These rocks are nonmarine siliciclastic red beds of Middle and Late Triassic age with a maximum thickness of ~500 m. Here, we review the lithostratigraphy, biostratigraphy and biochronology of these rocks to present a comprehensive correlation of the Triassic strata in east-central New Mexico.

PREVIOUS STUDIES

Marcou (1858) first identified Triassic strata in east-central New Mexico as the "New Red Sandstone" or "Keuper." He did so because these rocks underlie strata he considered Jurassic (Middle Jurassic Entrada through Lower Cretaceous Mesa Rica formations of current usage) and because the Triassic red beds resemble the nonmarine Triassic red beds of the Keuper in Germany and France.

Little new information on the Triassic of east-central New Mexico appeared until Darton (1922, 1928) used the name Santa Rosa Sandstone (it had actually first been used by Rich, 1921) to refer to the lower part of the Triassic section and "Chinle?" to the upper (Fig. 2). Regional geologic mapping by the U.S. Geological Survey in the 1940s (Gorman and Robeck, 1946; Dobrolvolny et al., 1946) substantiated Darton's broad divisions and further subdivided the Santa Rosa and Chinle Formations (Fig. 2). Griggs and Read (1959) reflected the Survey's conclusions, assigning the Triassic section in east-central New Mexico to the Santa Rosa, Chinle and Redonda Formations.

Kelley (1972a, b) mapped Triassic strata across much of Guadalupe and DeBaca counties and adjacent areas, proposing largely informal, but readily mappable, subdivisions of the Triassic rocks. At the same time, Gregory (1972) summarized the results of fossil collecting in Triassic strata in east-central New Mexico. This work dated back to the initial explorations of Case (1914) and demonstrated that substantial assemblages of Triassic fossil vertebrates were present in east-central New Mexico (also see Long and Murry, 1995 and Hunt, 1994, 1997a for reviews of the history

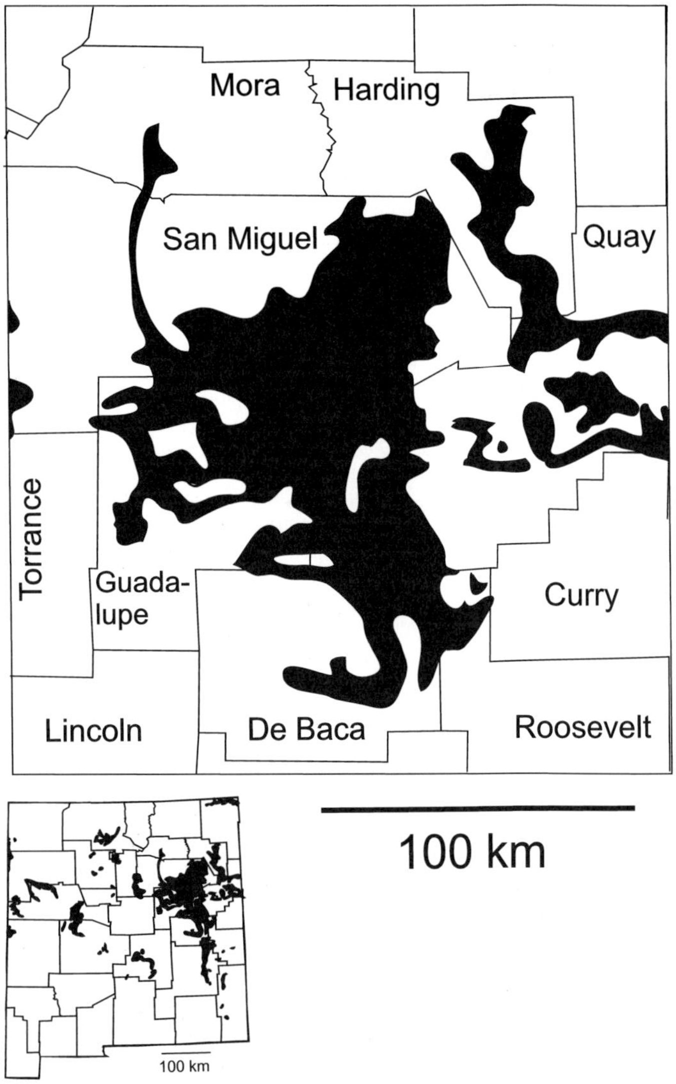

FIGURE 1. Index map showing the distribution of Triassic strata in east-central New Mexico.

of vertebrate fossil collecting in east-central New Mexico).

The 1980s-1990s saw an unprecedented explosion of knowledge of the Triassic of east-central New Mexico. These studies encompassed surface (Lucas et al., 1985a, 1994; Lucas and Hunt, 1987, 1989) and subsurface (Broadhead, 1984, 1985) stratigraphy, sedimentology (Granata, 1981; McGowan et al., 1983; Hester, 1988; Newell, 1993) and paleontological studies, including studies of invertebrate micro- and macrofossils, megafossil plants and vertebrate body and trace fossils (Table 1). This work has reshaped our understanding of Triassic stratigraphy, biostratigraphy and correlation in east-central New Mexico.

LITHOSTRATIGRAPHY

The Triassic section in east-central New Mexico (Fig. 3) has a maximum thickness of ~500 m. These strata are assigned to the Moenkopi Formation (Anton Chico Member) and overlying Chinle Group (Santa Rosa, Garita Creek, Trujillo, Bull Canyon and Redonda formations). The base of the Triassic section is a compound unconformity equivalent to the Tr-0 through Tr-2 unconformities of Pipiringos and O'Sullivan (1978). The erosional surface associated with the Tr-3 unconformity separates the Moenkopi from the overlying Chinle, and the Tr-4 and Tr-5 unconformities of Lucas (1993) are expressed as erosional surfaces beneath the Trujillo and Redonda formations, respectively. Where complete, the section is overlain by the Middle Jurassic Entrada Sandstone (a compound unconformity equivalent to the J-0 through J-2 unconformities of Pipiringos and O'Sullivan, 1978), although post-Jurassic erosion has locally truncated the section.

MOENKOPI FORMATION

The oldest Triassic strata in east-central New Mexico are assigned to the Anton Chico Member of the Moenkopi Formation of Lucas and Hunt (1987). As much as 46 m thick, the Anton Chico Member is mostly grayish-red and moderate reddish-brown, trough crossbedded micaceous litharenite; minor lithologies are red-bed mudstone, siltstone and intraformational conglomerate (Lucas and Hunt, 1987; Lucas and Hayden, 1991; Boy et al., 2001). The Anton Chico Member overlies red-beds of the Middle Permian (Guadalupian) Artesia Group at a substantial disconformity. Typically, the base of the Anton Chico Member is a conglomerate of Artesia Group rip-ups or a coarse sandstone that is color mottled and preserves root traces that are evidence of pedogenesis. This unconformity corresponds to the Tr-2 unconformity (and, by inference, the underlying Tr-1 and Tr-0 unconformities) of Pipiringos and O'Sullivan (1978). The base of the Tecolotito Member of the Santa Rosa Formation above the Anton Chico Member is also a substantial unconformity, the Tr-3 unconformity of Pipiringos and O'Sullivan (1978).

The Anton Chico Member is well exposed in the Sangre de Cristo front range and in the canyons of the Pecos River and its tributaries (Lucas et al., 1990; Lucas and Hayden, 1991; Boy et al., 2001). These outcrops extend as far southeast as Alamogordo Reservoir just north of Fort Sumner in DeBaca County (Lucas and Anderson, 1993). Figure 4 (also see the Appendix) illustrates a characteristic section of the Moenkopi Formation near Puerta de Luna in the Pecos River drainage in Guadalupe County.

The subsurface extent of the Anton Chico Member is more difficult to determine because it and the overlying, sandstone-dominated Tecolotito Member of the Santa Rosa Formation are not readily distinguished from each other on geophysical logs (Broadhead, 1984, 1985; Lucas and Hunt, 1987). In east-central New Mexico, the Anton Chico Member yields fossil plants and vertebrates (see below).

SANTA ROSA FORMATION

The lowermost part of the Chinle Group in east-central New Mexico is the Santa Rosa Formation, consisting of three members (in ascending order): Tecolotito, Los Esteros and Tres Lagunas. The Tecolotito Member disconformably overlies Moenkopi or older strata throughout most of east-central New Mexico, and the Tres Lagunas Member is conformably overlain by the Garita

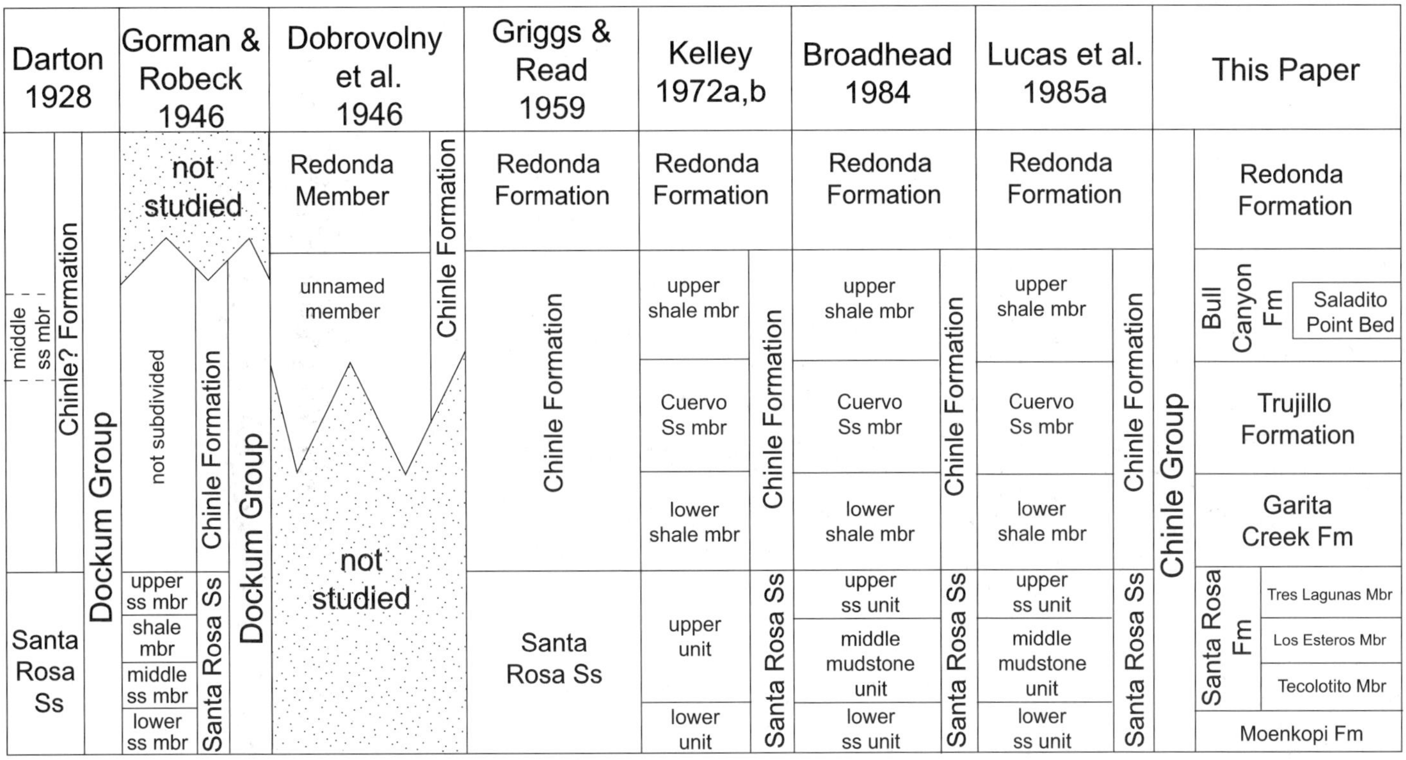

FIGURE 2. Nomenclature chart showing the evolution of Triassic stratigraphic nomenclature in east-central New Mexico.

Creek Formation in this area. The informal name "Canyon Sandstone" (Matthes, 1936) has been used for strata that clearly pertain to the Santa Rosa Formation (Fig. 5), and should be abandoned.

TECOLOTITO MEMBER

The basal interval of the Santa Rosa Formation is the Tecolotito Member of Lucas and Hunt (1987). It is as much as 34 m thick and consists mostly of grayish-orange to very pale orange, trough-crossbedded and ripple-laminated quartzarenite and lesser amounts of extrabasinal (mostly Paleozoic limestone, lithic and siliceous pebbles) conglomerate. The Tecolotito Member has a sharp, disconformable contact on the underlying Moenkopi Formation but grades upward into the overlying Los Esteros Member.

Outcrops of the Tecolotito Member in east-central New Mexico are confined to the canyon country of the Pecos River in Guadalupe and San Miguel counties and the hogbacks along the Sangre de Cristo front range in San Miguel County. The unit has a broad subsurface distribution; it is most of Broadhead's (1984) "lower sandstone unit" of the Santa Rosa Formation. However, the Tecolotito Member does not crop out in the Canadian River near Logan in Quay County (*contra* Finch et al., 1976; Finch and Wright, 1983; Lucas and Hunt, 1987). No fossils other than nondescript petrified wood have been identified in the Tecolotito Member.

TABLE 1. Paleontologic, biostratigraphic, and biochronologic studies of the Triassic in east-central New Mexico since 1985.

Plants:
Ash, 1988, 1989
Lucas et al., 1985d
Lucas, 2001

Calcareous microfossils:
Kietzke, 1987, 1989

Molluscs:
Good, 1989, 1993a,b, 1998
Kues, 1985

Fish:
Huber et al., 1993
Hunt, 1997b

Tetrapods (general):
Boy et al. (2001)
Carpenter and Parrish, 1985
Parrish and Carpenter, 1986
Lucas et al., 1985b,c
Hunt, 1991, 1994, 2001a, b
Hunt and Lucas, 1989, 1993a, b, 1997
Long and Murry, 1995
Murry, 1986, 1989
Hunt et al., 1989b
Lucas et al., 1998

Amphibians:
Hunt, 1989a, 1993, 1994
Hunt and Lucas, 1993b
Lucas and Morales, 1985

Phytosaurs:
Ballew, 1989
Hunt, 1989b,c 1994
Hunt and Lucas, 1993c
Heckert et al., 2001

Aetosaurs:
Heckert and Lucas, 1998
Heckert et al., 1996
Hunt and Lucas, 1991a
Hunt et al., 1993b

Dinosaurs:
Carpenter, 1997
Heckert et al., 2000
Hunt, 1989c
Hunt and Lucas, 1994
Hunt et al, 1998

Cynodonts:
Lucas and Oakes, 1988
Lucas et al., 1999b

Tetrapod biostratigraphy:
Hunt and Lucas, 1993a, 1997
Lucas and Hunt, 1993
Lucas, 1997, 1998
Murry, 1986, 1989
Parrish, 1989

Vertebrate Trace Fossils:
Cotton et al., 1997, 1998
Hunt et al., 1989a, 1993a, 2000
Hunt and Lucas, 2001
Lockley and Hunt, 1995, 199[illegible]
Lockley et al., 2000

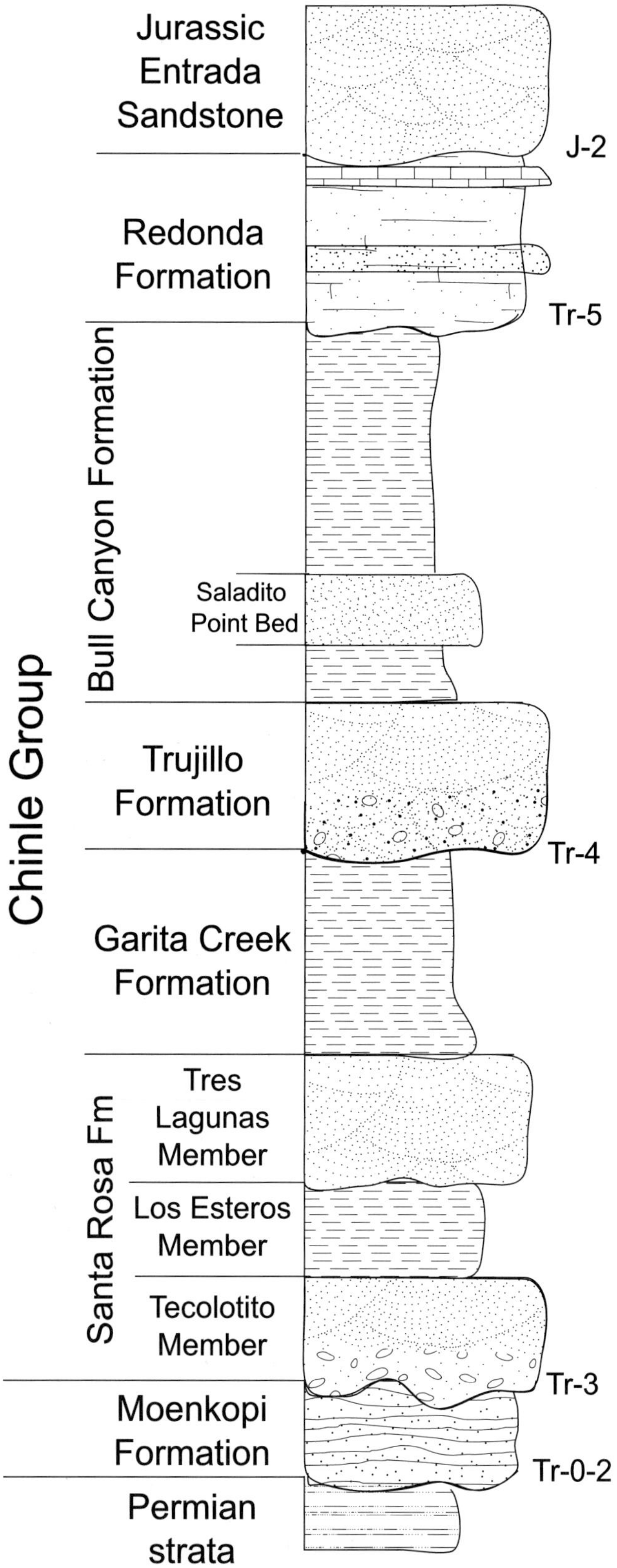

FIGURE 3. Schematic stratigraphic section showing generalized thickness and lithology of Triassic strata in east-central New Mexico.

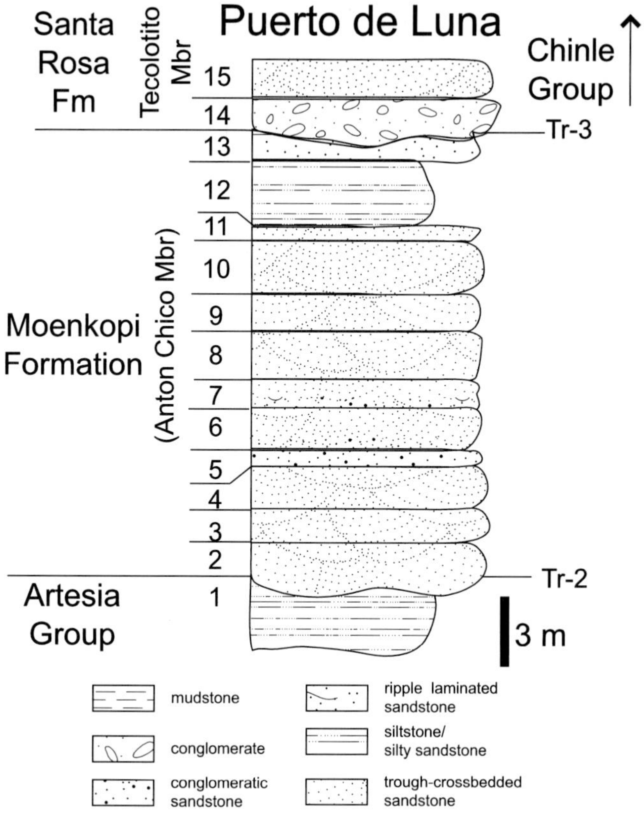

FIGURE 4. Measured stratigraphic section of the Moenkopi Formation near Puerto de Luna. See Appendix for precise location and description of numbered lithologic units.

LOS ESTEROS MEMBER

The medial member of the Santa Rosa Formation in east-central New Mexico is the Los Esteros Member of Lucas and Hunt (1987). The Los Esteros Member ranges in thickness from 0 to 44 m thick in this region. This unit is mostly mudstone and fine sandstone that forms a conspicuous slope between the ledge/cliff-forming Tecolotito (below) and Tres Lagunas (above) members. Mudstones of the Los Esteros Member are mostly reddish brown, but yellowish-brown and purplish-brown bands are also present. Sandstones are very fine- to fine-grained quartzarenites that are usually ripple-laminated or massive. The base of the Los Esteros Member is conformable, with typical Los Esteros mudstones thinly interbedded with sandstones of the upper part of the Tecolotito Member, or Tecolotito sandstones fining upward and/or grading upward into the Los Esteros Member.

In contrast, the contact of the Los Esteros Member with the overlying Tres Lagunas Member is an erosional unconformity. Thus, a thin limestone-pebble conglomerate is commonly present at the base of the Tres Lagunas Member, sharply overlying mudstone or very fine-grained sandstone of the Los Esteros Member. Local relief on this sharp surface is usually as much as 0.3 m (Lucas and Hunt, 1987, fig. 4). Indeed, at some locations (e.g., just north of Los Esteros Reservoir), the Los Esteros Member is absent, and the Tres Lagunas Member rests directly on the Teco-

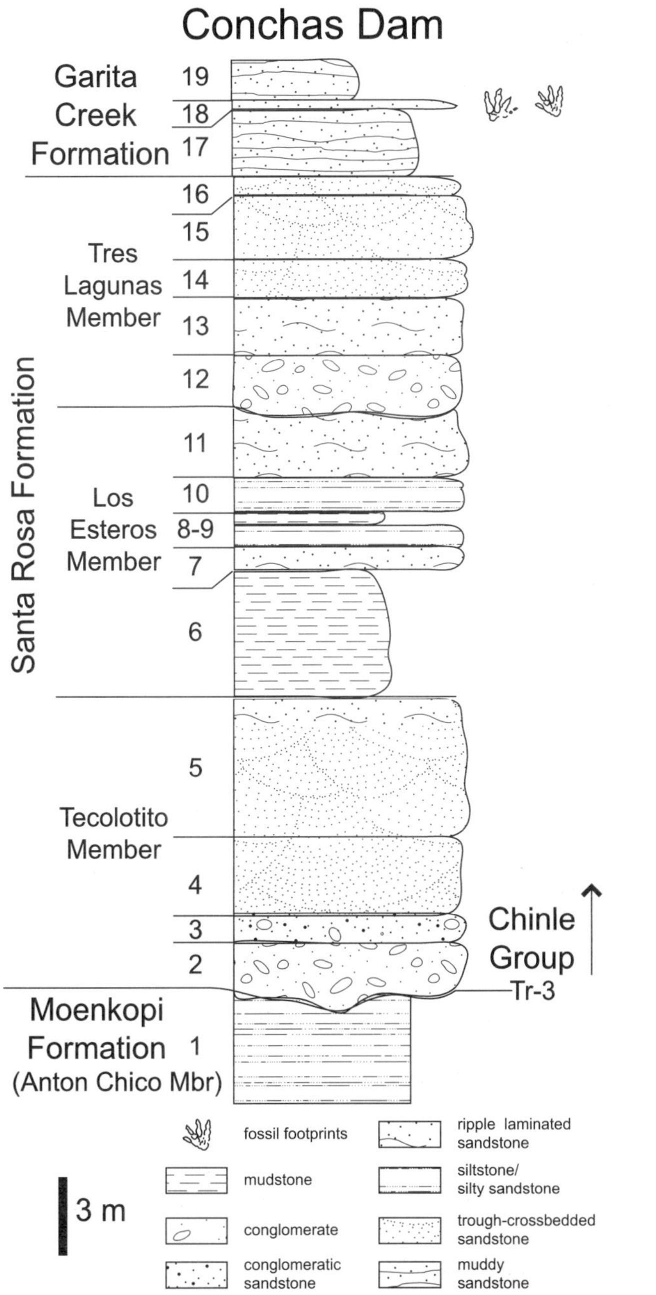

FIGURE 5. Measured stratigraphic section of the lower Chinle Group at Conchas Dam. See Appendix for precise location and description of numbered lithologic units.

lotito Member (Gorman and Robeck, 1946; Kelley, 1972b; Lucas and Hunt, 1987).

The outcrop belt of the Los Esteros Member parallels that of the underlying Tecolotito Member and is confined to the Pecos River valley and the Sangre de Cristo front range (Gorman and Robeck, 1946; Lucas and Hunt, 1987; Lucas et al., 1990). It has a much broader distribution in the subsurface, and corresponds to the informal "middle" member of the Santa Rosa Formation of Broadhead (1984). Fossil plants and vertebrates are known from the Los Esteros Member (see below).

TRES LAGUNAS MEMBER

The upper member of the Santa Rosa Formation in east-central New Mexico is the Tres Lagunas Member of Lucas and Hunt (1987). As much as 46 m thick, it is the most prominent and widely exposed member of the formation, and general perceptions of the Santa Rosa Formation have been based on the Tres Lagunas Member (e.g., Darton, 1928; Kelley, 1972a,b; Broadhead, 1984, 1985). The Tres Lagunas Member is mostly orange and yellowish-brown, medium-grained quartzarenite that displays planar and trough crossbeds and ripple laminae. It disconformably overlies the Los Esteros Member, and its upper contact is a transition from medium-grained quartzarenite to calcareous siltstone of the Garita Creek Formation.

The Tres Lagunas Member crops out in the Sangre de Cristo Front range, and throughout the Pecos River valley (Gorman and Robeck, 1946; Lucas and Hunt, 1987; Lucas et al., 1990). Its subsurface distribution extends eastward into Quay and Curry counties, but it is not present east of Clovis (Broadhead, 1984, fig. 9). No fossils are known from the Tres Lagunas Member other than petrified wood.

"CANYON SANDSTONE"

When Conchas Dam was built in the 1930s, the sandstone-dominated interval the dam was anchored in was referred to as the "Canyon Sandstone" (Matthes, 1936). Spiegel (1972) correlated this sandstone interval to the Trujillo Formation of West Texas, which is in the middle of the Chinle Group, but most workers (e.g., Griggs and Hendrickson, 1951; Wanek, 1962; Dane and Bachman, 1965; Kelley, 1972a; Lucas et al., 1985a) assigned it to the Santa Rosa Formation. Indeed, as Lucas et al. (2001b) demonstrate, all three members of the Santa Rosa Formation make up the "Canyon Sandstone" (Fig. 5). Surprisingly, Fritz (1991, p. 110) concluded that "it is not possible to determine whether or not the 'Canyon Sandstone' represents a member of the Santa Rosa Formation."

Nevertheless, that determination can be made in two ways: (1) a detailed measured section (Fig. 5, Appendix) reveals the characteristic lithologies, succession of lithologies and thicknesses of the three members of the Santa Rosa Formation at Conchas Dam; and (2) the succession of Triassic lithostratigraphic units above the "Canyon Sandstone," well exposed just east of the dam up to La Cinta Mesa (cf. Kelley, 1972a, fig. 1) is Garita Creek, Trujillo, Bull Canyon and Redonda formations, thus establishing that the "Canyon Sandstone" is the Santa Rosa Formation. The problem of correlation of the "Canyon Sandstone" created by Matthes (1936) and perpetuated by Spiegel (1972) and Fritz (1991) is thus readily solved by detailed stratigraphy. Obviously, the term "Canyon Sandstone" is superfluous and should be abandoned.

"LOGAN SANDSTONE"

Trauger et al. (1972) applied the name "Logan Sandstone" to the sandstone interval in which Ute Dam, near Logan in Quay County, is anchored. They did so because of uncertainty as to whether this sandstone represents the Santa Rosa Formation (Dane and Bachman, 1965; Finch et al., 1976; Finch and Wright, 1983) or the Trujillo Formation (Spiegel, 1972; Trauger et al., 1972). However, as Lucas et al. (1985a) noted, the "Logan Sandstone" is readily traced approximately 50 km eastward to Trujillo Camp, in West Texas, which is the type section of the Trujillo Formation of Gould (1907). "Logan Sandstone" thus is a synonym of Trujillo Formation.

GARITA CREEK FORMATION

Kelley (1972a,b) used the term "lower shale member of Chinle Formation" to refer to the mudstone-dominated interval of the Chinle Group immediately above the Tres Lagunas Member of the Santa Rosa Formation. Lucas and Hunt (1989) coined the term "Garita Creek Member" for this interval and designated a type section in the drainage of Garita Creek. As much as 76 m thick, the Garita Creek Formation is mostly grayish-red and moderate brown mudstone and lesser amounts of grayish-red, laminar sandstone and thin intraformational conglomerate. The unit typically forms a slope between the cliff/bench-forming Tres Lagunas Member of the Santa Rosa Formation (below) and the Trujillo Formation (above).

Garita Creek Formation outcrops are found throughout much of east-central New Mexico, especially along the Canadian Escarpment of San Miguel and Guadalupe Counties, and in the Pecos River drainage as far south as Fort Sumner. As stated above, the base of the Garita Creek Formation is conformable on the Tres Lagunas Member of the Santa Rosa Formation. The base of the Trujillo Formation overlying the Garita Creek is an unconformity (see below). Vertebrate fossils from the Garita Creek Formation are discussed below.

TRUJILLO FORMATION

As early as the 1920s, Darton (1928) recognized a persistent medial sandstone interval in Chinle Group strata in east-central New Mexico. Gould (1907) had named these strata the Trujillo Formation of the Dockum Group in West Texas, and Kelley (1972b) named them the Cuervo Sandstone Member of the Chinle Formation in east-central New Mexico. They are now assigned to the Trujillo Formation of the Chinle Group (Lucas and Hunt, 1989; Lucas, 1993, 1995). In east-central New Mexico, the Trujillo Formation is as much as 68 m thick and consists mostly of yellowish-gray, light olive gray and greenish-gray micaceous sandstone with lenses and beds of intrabasinal conglomerate and red-bed mudstone. Planar and trough crossbeds are the dominant bedforms. The clasts in Trujillo conglomerates are almost entirely rip-ups of pedogenic calcrete from underlying Chinle mudrocks.

The Trujillo Formation has a sharp, scoured and disconformable contact on the underlying Garita Creek Formation. Typically, Trujillo sandstone or conglomerate rests directly on Garita Creek mudstone at a sharp surface with up to 1 m of stratigraphic relief locally. Lucas (1993) identified the base of the Trujillo Formation in east-central New Mexico as the Tr-4 unconformity, a pervasive unconformity throughout the Chinle depositional basin. The Trujillo Formation is conformably overlain by the Bull Canyon Formation (see below).

The Trujillo Formation crops out extensively in east-central New Mexico, especially along the Canadian Escarpment in San Miguel and Guadalupe Counties and in the Canadian River drainage from Newkirk in Guadalupe County to Ute Dam in Quay County. The Trujillo Formation typically is a cliff-or bench-forming complex of sandstone/conglomerate bodies and intercalated mudrock that defends mesas and escarpments. Indeed, in east-central New Mexico, some of the flat geomorphic surfaces beneath badlands developed in the Bull Canyon Formation, such as the broad flats from Tucumcari to San Jon in Quay County, are underlain by the Trujillo Formation. In the subsurface, the Trujillo Formation extends eastward into West Texas (Broadhead, 1984, 1985).

A sparse, but age-diagnostic vertebrate fauna is known from the Trujillo Formation (Hunt, 2001a; also see below). Palynomorphs and megafossil plants have also been collected from the Trujillo Formation, principally in Texas (Ash, 1976; Dunay and Fisher, 1974, 1979; Litwin et al., 1991; Cornet, 1993).

BULL CANYON FORMATION

Kelley (1972a,b) used the term "upper shale member of Chinle Formation" to refer to the mudrock-dominated unit above the Trujillo Formation in east-central New Mexico. Lucas and Hunt (1989) named this unit the Bull Canyon Formation. In West Texas, the same unit had been included, in part, in the Trujillo Formation by Gould (1907) and referred to in its entirety as the Chinle Formation by Adams (1929), Adkins (1932) and Reeside et al. (1957). However, Chatterjee (1986) named this unit the Cooper Member of the Dockum Formation, failing to describe a type section and overlooking the fact that Cooper was a preoccupied name, having long been used as a stratigraphic term in South Carolina (Lucas et al., 1994).

Despite this, Lehman et al. (1992) renamed Chatterjee's Cooper Member the "Cooper Canyon Formation" and subsequently (Lehman, 1994) claimed that this name had priority over the Bull Canyon Formation of Lucas and Hunt (1989). The rule of priority, however, would require that Bull Canyon Formation has precedence as the lithostratigraphic name for the formation that immediately overlies the Trujillo Formation in east-central New Mexico and West Texas (Lucas et al., 1994).

In east-central New Mexico, the Bull Canyon Formation is as much as 110 m thick and is mostly grayish-red and moderate reddish-brown mudstone and lesser amounts of yellowish-gray to grayish-red, laminar to trough-crossbedded litharenite. Very minor lithologies are siltstone and intraformational conglomerate. The Bull Canyon Formation has a conformable, even intertonguing, relationship with the uppermost portion of the underlying Trujillo Formation (Granata, 1981; McGowan et al., 1983; Lucas and Hunt, 1989). The Redonda Formation overlies the Bull Canyon Formation throughout east-central New Mexico, and this contact has been described as conformable (Dobrovolny et al., 1946; Griggs and Read, 1959; Lucas et al., 1985a) or unconformable (Kelley, 1972a; Granata, 1981; Hester, 1988; Lucas, 1993; Hester and Lucas, 2001). We note the sharp lithologic con-

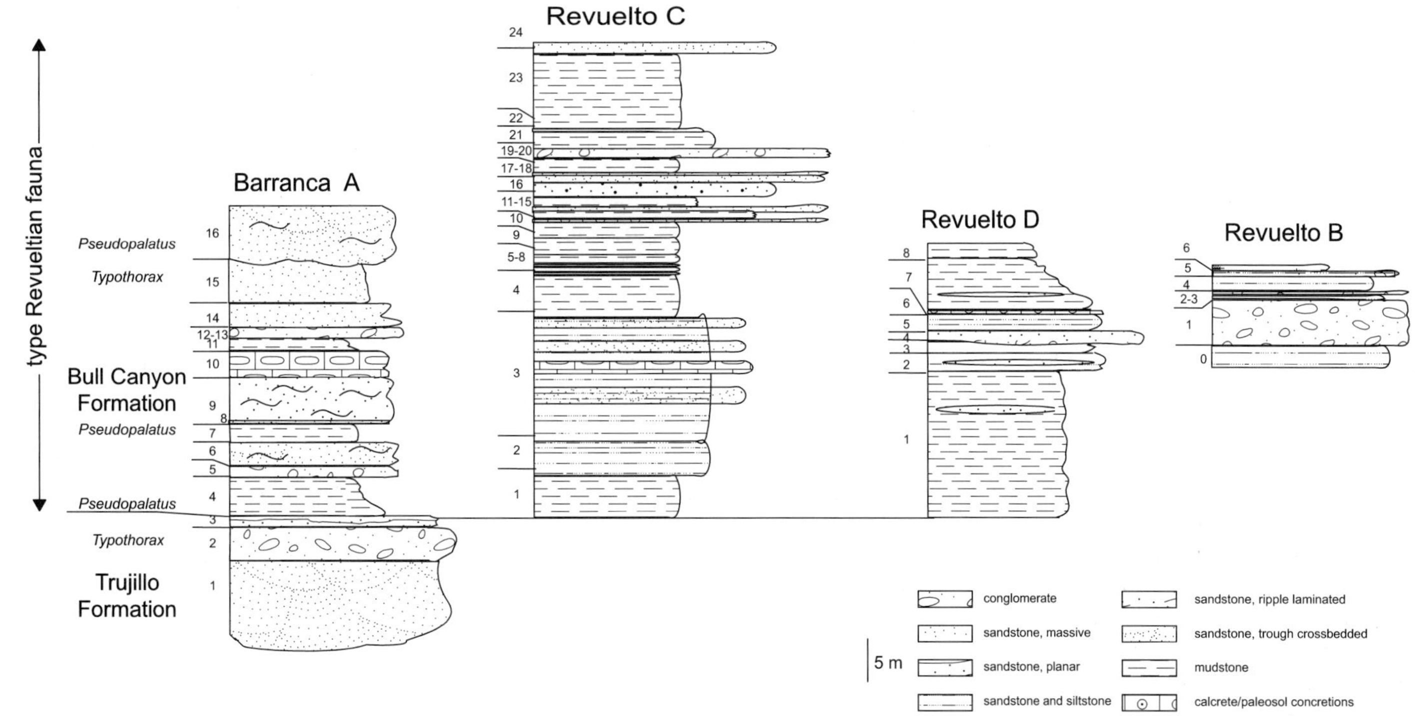

FIGURE 6. The stratigraphic range of the type Revueltian vertebrate assemblage is evident in measured stratigraphic sections of the Trujillo and Bull Canyon formations in the Barranca Creek (section A) and Revuelto Creek (sections B-D) drainages southeast of Tucumcari, Quay County. See Appendix for precise location and description of numbered lithologic units.

trast and change in depositional style that characterizes the Bull Canyon-Redonda contact, features well documented by Hester (1988), to conclude that the contact is disconformable. Indeed, for this reason, Lucas (1993) identified the base of the Redonda Formation as an unconformity, and based on the correlations of bracketing strata, identified it as the basinwide Tr-5 unconformity of the Chinle Group.

The Bull Canyon Formation crops out extensively in east-central New Mexico, especially along the Canadian Escarpment and Llano Estacado. Here, it forms mudstone-dominated slopes between the cliff/bench-forming Trujillo Formation (below) and the ledgy, repetitively-bedded Redonda Formation (above). Fossils of charophytes, land plants, ostracodes, unionids and vertebrates are present in the Bull Canyon Formation, including the type Revueltian fauna of Lucas and Hunt (1993) (Fig. 6), discussed in more detail below.

SALADITO POINT BED

We identify a persistent, bench-forming sandstone-dominated unit in the lower part of the Bull Canyon Formation as a new bed-level unit, the Saladito Point Bed (Figs. 7-8, Appendix). At its type section, the Saladito Point Bed is 12.5 m thick and is mostly pale reddish-brown, trough-crossbedded litharenite and grayish-red intraformational conglomerate. The bed is named for Saladito Point of Mesa Rica near the type section.

Kelley (1972a, p. 88) first drew attention to the Saladito Point Bed, referring to it as a "particularly well developed brown sandstone sequence 25-75 feet in thickness" and mapped its distribution in the vicinity of Newkirk and Cuervo in Guadalupe County at approximately 1:450,000 scale (Kelley, 1972a, fig. 1). The Saladito Point Bed crops out farther eastward, at least to Montoya in Quay County, and is present in the type section of the Bull Canyon Formation in Bull Canyon, Guadalupe County (Lucas and Hunt, 1989, fig. 3, units 9-10). This persistent, bench-forming unit low in the Bull Canyon Formation thus is of stratigraphic utility and merits formal recognition.

Interestingly, similar sandstone-dominated intervals are also present in correlative strata on the Colorado Plateau in west-central New Mexico and eastern Arizona. Cooley (1957) was the first to recognize this, and informally named a variety of sandstone units (the Chambers, Taaiylone, Zuni River and Perea beds) in the lower Painted Desert Member of the Petrified Forest Formation in west-central New Mexico and east-central Arizona in the drainages of the Little Colorado and Puerco Rivers. Billingsley (1985a,b) identified several persistent sandstone ledges in the Painted Desert Member of the Petrified Forest Formation as the Flattops sandstones (1-4 in ascending order, southern portion of the Petrified Forest National Park) and Painted Desert sandstones (1-2 of Billingsley, 1985a, 1-3 of Billingsley, 1985b) in the northern portion of the Petrified Forest National Park. We recognize Flattops Sandstone 2 and Painted Desert Sandstone 3 of Billingsley (1985b) as possible correlatives of the Saladito Point Bed. Lucas et al. (1997) designated a type section of the Perea Bed low in the Painted Desert Member of the Petrified Forest Formation near Fort Wingate, New Mexico. Although it is lithologically similar, the Correo

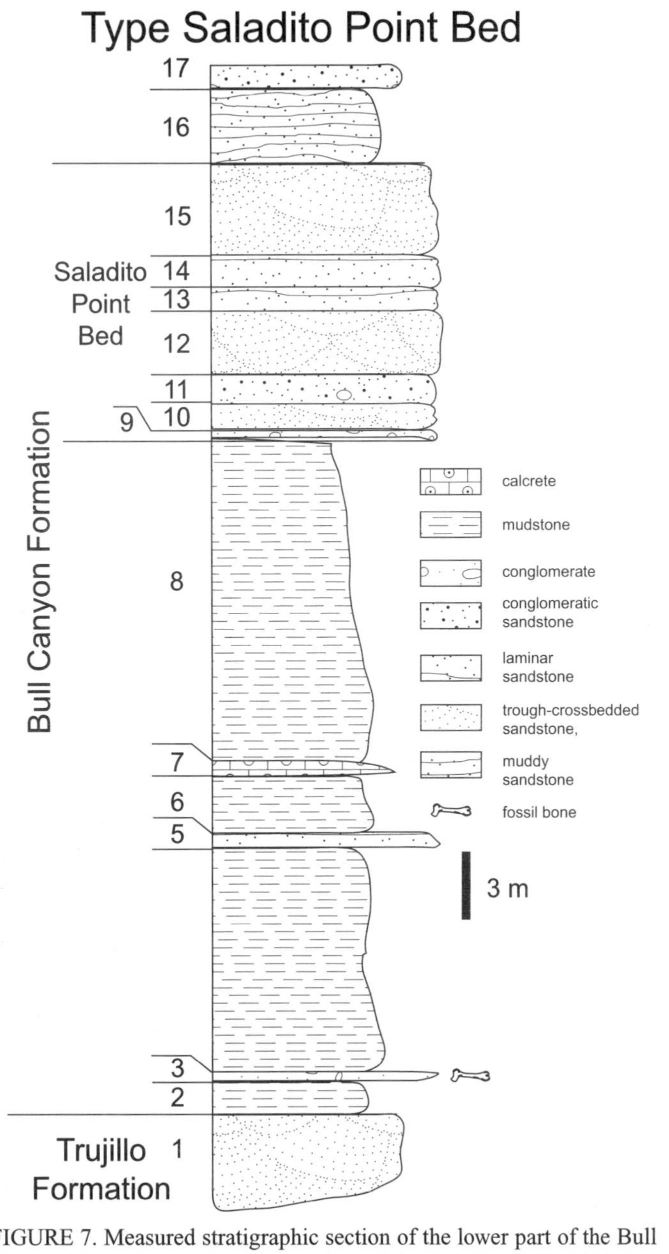

FIGURE 7. Measured stratigraphic section of the lower part of the Bull Canyon Formation at the type section of the Saladito Point Bed. See Appendix for precise location and description of numbered lithologic units.

FIGURE 8. Photograph of the type section of the Saladito Point Bed. T = Trujillo Formation, B = Bull Canyon Formation mudstones, S = Saladito Point Bed.

Bed (Lucas and Heckert, 1994) is much higher stratigraphically than these units and is not relevant to discussion here.

The Perea, Flattops 2, Painted Desert 3, Zuni River, Chambers and Taaiylone beds are similar to the Saladito Point bed lithologically and stratigraphically. All occur low in the mudrock-dominated interval of the second sequence of the Chinle Group (Lucas, 1993), either the Painted Desert Member on the Colorado Plateau or the Bull Canyon Formation on the High Plains. Typically, these sandstones are approximately 20-30 m above the highest sandstone of the Sonsela Member (Colorado Plateau) or Trujillo Formation (High Plains) and are 5-15 m thick. Lithologies characteristically are reddish-brown and pale green, planar to trough-crossbedded litharenites with minor intraformational conglomerate dominated by rip-up clasts of bentonitic mudstone and pedogenic calcrete nodules. Espegren (1985) interpreted Flattops Bed 2 as representing the deposits of a high sinuosity fluvial system with bed and suspended loads. Fossils are uncommon in these beds, but locally include abundant bivalves or fragmentary, badly weathered fossil bone. Each of these bed-level units can be traced laterally for km.

Beyond the lithostratigraphic similarity of these beds, note that this apparent outcrop trend, from east-central Arizona to east-central New Mexico, is a transect approximately 45° to 90° to the known depositional dip of the upper Chinle sediments based on numerous northerly paleocurrents (e.g., Stewart et al., 1972; Dubiel, 1989). This suggests that these regionally persistent sandstones may not simply be localized depositional features. Instead, they may represent an interval of lowered regional base level and subsequent recovery, when the rate of avulsion in Chinle streams was high relative to regional subsidence (accommodation) (e.g., Blakey and Gubitosa, 1984, fig. 5.1).

REDONDA FORMATION

The youngest Triassic strata in east-central New Mexico are assigned to the Redonda Formation of Dobrovolny et al. (1946). The formation is as much as 92 m thick and is laterally continuous, repetitively-bedded fine sandstones, siltstones and mudstones (Hester and Lucas, 2001). These are mostly moderate reddish-brown and form ribbed cliffs and steep slopes. As discussed above, the base of the Redonda Formation is a regional unconformity, and the top of the Redonda Formation is a regional unconformity overlain by Middle Jurassic strata; this is the J-2 unconformity of Pipiringos and O'Sullivan (1978).

The Redonda Formation crops out extensively in east-central New Mexico, particularly along the Canadian Escarpment and

Llano Estacado. Here it forms ledgy, repetitively-bedded slopes above the badlands of the Bull Canyon Formation (below) and cliffs of the Middle Jurassic Entrada Sandstone or, locally, younger strata (above). A diverse vertebrate fauna is known from the Redonda Formation (Gregory, 1972; Huber et al., 1993; Heckert et al., 2001; also see below).

BIOSTRATIGRAPHY AND BIOCHRONOLOGY

Triassic strata in east-central New Mexico yield megafossil plants, calcareous microfossils, unionid bivalves, vertebrate bones and tetrapod tracks (Table 1). Here, we review the stratigraphic distribution of these fossils and discuss their biochronological significance.

Megafossil Plants

No palynological studies have been undertaken in the Triassic strata of east-central New Mexico, but megafossil plants are present in various parts of the section. The stratigraphically lowest megafossil plants are from the Anton Chico Member of the Moenkopi Formation, a monotaxial assemblage of the bennettitalean *"Zamites" powellii* (Lucas, 2001). The only substantial Triassic paleofloral assemblage from east-central New Mexico is from the Los Esteros Member of the Santa Rosa Formation (Ash, 1972, 1988). This includes the fern *Cynepteris lasiophora*, the bennettitalean *Zamites powellii*, the conifer *Pelourdea poleoensis* and the *incertae sedis* taxa *Dinophyton spinosus* and *Samaropsis* sp. Stratigraphically higher plant records in the Triassic section in east-central New Mexico are sparse and isolated and include pith casts of *Neocalamites* in the Bull Canyon and Redonda formations and impressions of *Sanmiguelia* in the Bull Canyon Formation (Lucas et al., 1985a, d; Hunt, 1994).

Little is known of Moenkopi plants in the western United States (Ash and Morales, 1993), and the Moenkopi record of *"Zamites" powelli* is a substantial stratigraphic range extension of a characteristic Chinle Group plant taxon (Lucas, 2001). The flora from the Los Esteros Member is characteristic of the *Dinophyton* zone of Ash (1980), and this zone is present in Adamanian (upper Carnian) strata of the Chinle Group across a broad area (Lucas, 1997). *Sanmiguelia* from the Bull Canyon Formation is typical of Ash's (1980) *Sanmiguelia* zone, which is widely distributed in Revueltian-Apachean (Norian-Rhaetian) strata of the Chinle Group (Lucas, 1997).

Thus, the paleoflora of the Los Esteros Member indicates an Adamanian age, whereas the one record of *Sanmiguelia* from the Bull Canyon Formation is suggestive of a Revueltian-Apachean age. Much more will have to be known of Triassic fossil plants from east-central New Mexico before they can be of greater biostratigraphic utility.

Calcareous Microfossils

Kietzke (1987, 1989) documented charophytes (*Altochara* and *Stellatochara*), ostracods (*Darwinula* and *Lukevichinella*?) and "spirorbids" from the Bull Canyon and Redonda Formations in east-central New Mexico. These fossils are of limited value to biochronology, largely because their taxonomy is problematic. Nevertheless, Chinle Group records in Arizona of ostracods similar to those from the Bull Canyon and Redonda Formations are in Revueltian-Apachean strata (Lucas, 1997). This lends some weak support to regional correlations of both the Bull Canyon and Redonda formations based on tetrapods (see below).

Unionid Bivalves

Kues (1985) documented an extensive assemblage of unionids from the Bull Canyon Formation in Guadalupe County. Good (1993a,b, 1998) revised Chinle unionids and put them into a biostratigraphic context that demonstrates that they are of little or no biochronological value (also see Lucas, 1993, 1997). At best, the Bull Canyon Formation unionids are consistent with generally greater unionid abundances in the upper Chinle Group when compared to the lower Chinle. However, the taxa present (and note that we are very skeptical of the validity and utility of Good's [1998] Chinle unionid taxonomy) have long stratigraphic ranges in the Carnian-Norian Chinle Group and are not (at the species level) known elsewhere.

Vertebrate Fossils

Fossil vertebrates are found throughout the Triassic section in east-central New Mexico and provide the most robust biostratigraphy and biochronology of these strata. This record has been reviewed at length elsewhere (e.g., Hunt and Lucas, 1993a; Lucas, 1993, 1997, 1998, 2000; Hunt, 1994), so we provide only a brief summary:

1. The Anton Chico Member yields fossils of the temnospondyl *Eocyclotosaurus*, an index fossil of the Perovkan land-vertebrate faunachron (lvf) of early Anisian age (Lucas and Morales, 1985; Lucas and Hunt, 1987; Lucas, 1998, 2000; Boy et al., 2001). Other Moenkopi tetrapods from east-central New Mexico are fragmentary but include a *Stanocephalosaurus*-like temnospondyl (Boy et al., 2001) and a *Shansisuchus*-like erythrosuchid (Lucas et al., 1998), also consistent with a Perovkan age.

2. Santa Rosa Formation strata have not yielded tetrapods in east-central New Mexico other than fragmentary metoposaurid fossils indicative of a Late Triassic age. However, at Lamy in Santa Fe County, the Los Esteros Member yields the aetosaurs *Desmatosuchus haplocerus* and cf. *Stagonolepis wellesi*, an *Ischigualastia*-like dicynodont and the phytosaurs *Angistorhinus* and *Rutiodon*, taxa indicative of an early Adamanian age (Hunt and Lucas, 1995).

3. The Garita Creek Formation in east-central New Mexico yields a sparse tetrapod assemblage that includes *Desmatosuchus haplocerus*. This, and its stratigraphic position, suggest an Adamanian age (Hunt et al., 1989b). Near Lamy in Santa Fe County the Garita Creek Formation also contains the famous amphibian quarry of the temnospondyl *Buettneria perfecta* (Romer, 1939; Colbert and Imbrie, 1956; Hunt, 1993). An additional locality near Lamy yields numerous microvertebrate fossils, including teeth of the dinosaurs *Pekinosaurus olseni* and aff. *Galtonia* (Heckert

et al., 2000). Note also that previous reports of the aetosaur *Typothorax* from the Garita Creek Formation (Hunt et al, 1989b; Hunt and Lucas, 1993a, 1995) are of a new species distinct from the Revueltian (early-mid Norian) taxon *Typothorax coccinarum*. Thus, tetrapod fossils strongly support an Adamanian age for the Garita Creek Formation in east-central New Mexico.

4. Hunt (1991, 2001a) documents the Revueltian index taxa *Typothorax coccinarum* and *Pseudopalatus*-grade phytosaurs from the Trujillo Formation in eastern Quay County.

5. The vertebrate assemblage from the Bull Canyon Formation in east-central New Mexico is the type assemblage of the Revueltian lvf of Lucas and Hunt (1993) and is reviewed at length by Hunt (1994, 2001b).

6. The vertebrate assemblage of the Redonda Formation in east-central New Mexico is the basis of the Apachean lvf of Lucas and Hunt (1993). Huber et al. (1993), Hunt and Lucas (1993a) and Heckert et al. (1996, 2001) review this fauna.

FIGURE 9. Field photograph of the holotype slab of *Barrancapus cresapi.*

Tetrapod Footprints

Only a few tetrapod footprints are known from Triassic strata below the Redonda Formation in east-central New Mexico. These are a track of the ichnogenus *Brachychirotherium* from the Garita Creek Formation (Hunt and Lucas, 2001), and the holotype of the ichnotaxon *Barrancapus cresapi* from the Bull Canyon Formation (Hunt et al., 1993a; Fig. 9).

Barrancapus cresapi was named by Hunt et al. (1993a) for a small, quadrupedal trackway from the lower Bull Canyon Formation (Upper Triassic) of Quay County, New Mexico. The holotype slab (New Mexico Museum of Natural History [NMMNH] P-4782 is a plastotype) from NMMNH locality 55 preserves two parallel-trackways in convex hyporelief preserved in a lenticular, conglomeratic sandstone, which we interpret as representing a low-sinuosity channel with fluctuating discharge (Fig. 9).

The trackways display a range of extramorphological variation. The best preserved pedal impression has a length and width of 60 mm. It is pentadactyl and mesaxonic with broad, tapering digit impressions. The associated manus imprint is 40 mm long and 50 mm wide. This manus impression, as well as at least three others, are markedly entaxonic. The manus impression preserves ?five digit impressions with digit impressions I and V being the longest. *Barrancapus cresapi* is the only named track type from the Revueltian lvf (Norian) strata of western North America. Other tetrapod ichnotaxa from this time interval are restricted to swimming traces of phytosaurs(?) from near the type locality of *Barrancapus* and enigmatic tracks from the Painted Desert Member of the Petrified Forest Formation, Petrified Forest National Park, Arizona. The Painted Desert tracks have mean track (pedal track?) lengths and widths of 63 mm and 58 mm, respectively, which are very similar to values of *Barrancapus cresapi*.

The most notable feature of the morphology of *Barrancapus cresapi* is the large inferred size of digit I and its medial orientation. Among Triassic tetrapods, medial orientation of digit I is most closely matched in prosauropod dinosaurs. Two factors argue against *Barrancapus cresapi* representing a prosauropod: (1) the apparently large, inferred size of digit I; and (2) the hypothesis that the clawed pollex was held off the ground. However, at present it is most parsimonious to assign *Barrancapus cresapi* to the Prosauropoda, but we adopt a conservative approach and consider it as ?Prosauropoda. *Barrancapus cresapi* is the oldest track attributed to a prosauropod dinosaur. Other Revueltian prosauropod occurrences from western North America are limited to isolated specimens from the Bull Canyon Formation of Quay County, New Mexico and Garza County, Texas.

The Redonda Formation has an extensive tetrapod footprint assemblage of low diversity assigned to the ichnogenera *Rhynchosauroides* (rhnychocephalian), *Brachychirotherium* (aetosaur?), *Grallator* (theropod dinosaur) and *Pseudotetrasauropus* (prosauropod dinosaur) (Gregory, 1972; Hunt et al., 1989a, 1993a, 2000; Lockley et al., 2000; Lucas et al., 2001a). These tracks closely resemble other footprint assemblages from the Rock Point Formation of the Chinle Group on the Colorado Plateau, and thus support (but do not demonstrate) a Redonda-Rock Point correlation.

CORRELATION

Lithostratigraphic and biostratigraphic data allow a precise and detailed correlation of Triassic strata in east-central New Mexico to Triassic strata both to the west, in central and west-central New Mexico, and to the east, in West Texas (Fig. 10). Key points of this correlation are:

1. Moenkopi strata are continuous from west-central New Mexico to east-central New Mexico, and vary little in lithology and thickness across this transect. Moenkopi strata are not present in West Texas (Lucas and Anderson, 1992, 1993). Continuity of the Anton Chico Member of the Moenkopi Formation in New Mexico with the Holbrook Member in east-central Arizona can be demonstrated (Lucas and Hayden, 1989), and this further supports a Perovkan age assignment for the Anton Chico Member.

2. The base of the Chinle Group is a pervasive unconformity (Tr-3 unconformity) immediately overlain by an interval of quartzose sandstone and extrabasinal (siliceous) conglomerate referred to the Shinarump, Agua Zarca, Santa Rosa (Tecolotito

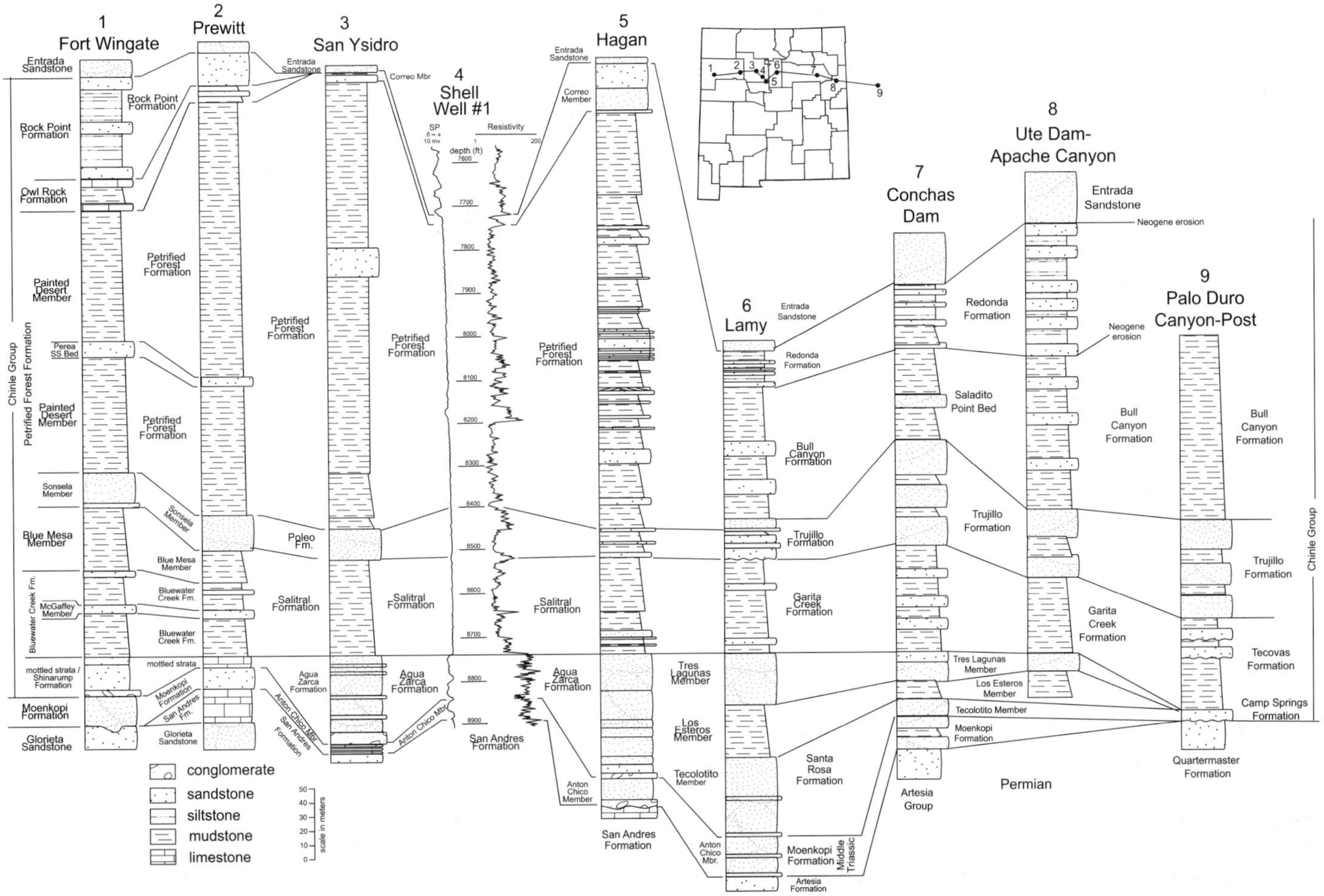

FIGURE 10. Correlation of Triassic strata from west-central New Mexico to West Texas. Sections from Fort Wingate to Lamy from Lucas et al. (1999). Section at Conchas Dam based on Wanek (1962), Lucas and Hunt (1987, 1989) and data presented here. Section at Ute Dam-Apache Canyon based on data in Dobrovolny et al. (1946), Finch et al. (1976) and Hunt (1994). Section at Palo Duro Canyon-Post based on data in Gould (1907), Hunt and Lucas (1991) and Lehman et al. (1992).

Member) and Camp Spring formations. These units are homotaxial and, within resolution, of the same age; note that along the transect depicted here, they are all overlain by units that yield Adamanian-age fossils. In West Texas, the Camp Springs Formation yields the phytosaur *Paleorhinus* and thus is of Otischalkian age (Hunt and Lucas, 1991b). Therefore, it is likely that the basal sandstone/conglomerate unit of the Chinle Group is of Otischalkian age from West Texas to west-central New Mexico.

3. Strata that lie below the Tr-4 unconformity along this transect yield Adamanian-age vertebrates (see below) that correlate them from west-central New Mexico to West Texas.

4. Significantly, the Adamanian section in east-central New Mexico is much thicker than the age equivalent section in West Texas (also see Lucas et al., 1994) or west-central New Mexico (Heckert and Lucas, 1996). The thickness variation between east-central New Mexico and West Texas is perhaps best explained as being controlled by the Frio uplift, a subdued Paleozoic positive that separates east-central New Mexico from West Texas; it is the basin divide between the late Paleozoic Tucumcari and Palo Duro basins (Broadhead, 1984). Thus, Moenkopi strata pinch out and Tecolotito member strata thin across the uplift, and Los Esteros and Tres Lagunas strata also pinch out (Broadhead, 1984, fig. 9). The Tecovas Formation in West Texas thus is apparently physically continuous with the Garita Creek Formation in east-central New Mexico.

In west-central New Mexico, the remnant Paleozoic Zuni uplift played a similar role in limiting the thickness of the Otischalkian-early Adamanian section. Here, the Shinarump Formation and associated "mottled strata" filled paleotopography developed in the Moenkopi Formation during development of the Tr-3 unconformity (Lucas and Hayden, 1989; Heckert and Lucas, 1996). Farther to the west, the Moenkopi Formation thickens again. Notably, Adamanian strata in west-central New Mexico also demonstrate that at least 50 m of strata were removed during the Tr-4 unconformity (Heckert and Lucas, 1996). Thus, structural features at the Chinle basin margin control the thickness of the Otischalkian-early Adamanian section here, and erosion associated with the development of the Tr-4 unconformity explains differences in the thickness of Adamanian strata from west-central to east-central New Mexico.

5. Across the transect (Fig. 10), a pervasive unconformity (Tr-4 of Lucas, 1991, 1993, 1997) underlies a medial Chinle Group

sandstone complex termed Sonsela, Poleo or Trujillo. Fossil vertebrates suggest that these are Revueltian (Norian)-age strata, but palynology in Texas suggests a Carnian age (Dunay and Fisher, 1979). The possibility that this sandstone complex is diachronous should be considered, as should the possibility that the late Carnian palynomorphs, which are from strata bearing intraformational clasts of late Carnian age, are reworked.

6. Thick sections of red-bed mudstones above the medial sandstone complex yield Revueltian-age fossils and are termed Painted Desert, Petrified Forest and Bull Canyon. Thickness variation across this transect (Fig. 10) probably reflects a combination of: (1) differential subsidence across the vast Chinle basin; and (2) differential erosion during development of the Tr-5 unconformity. Given the size of the basin, we suspect that differential subsidence is the more likely explanation.

7. Low in the Revueltian mudstone-dominated interval, persistent bed-level sandstone units correlative to the Saladito Point Bed include the Flattops 2, Painted Desert 3 and Chambers, Taaiylone, Chambers and Zuni River sandstones as well as the Perea Bed, all of east-central Arizona and west-central New Mexico. They probably represent a regional lowering of base-level during the early Revueltian, that lead to incision, followed by rising base level that lead to aggradation of high-sinuosity stream deposits.

8. Cyclically bedded strata (Rock Point and Redonda formations) are preserved at some localities but have been removed by post-Chinle erosion elsewhere. These youngest Chinle sediments yield Apachean-age fossils, including the type Apachean fauna from the Redonda Formation in Apache Canyon in east-central New Mexico.

Old ideas die hard, and one of the oldest incorrect ideas about Triassic deposition in the American Southwest is that two discrete depocenters existed, one in the Colorado Plateau region, the other in and around West Texas. To some extent this idea was an outgrowth of early, necessarily parochial stratigraphic studies, in which workers were confined to one or the other area by the logistical difficulties inherent to travel in the nineteenth and earlier part of the twentieth century. Notably, when earlier workers did cross from one study area to the other, some of the correlations advocated here were advanced (e.g., Case, 1928).

Unfortunately, the limited nature of earlier workers' studies was elaborated and perpetuated by parochial studies of local sedimentation patterns and an inability or unwillingness to recognize the lithologic similarity and continuity of Triassic strata from the Colorado Plateau to West Texas (e.g., McKee et al., 1959; Finch and Wright, 1983; McGowan et al., 1983; Johns and Granata, 1987; Dubiel, 1989; Lehman, 1994). Detailed lithostratigraphic and biostratigraphic studies demonstrate the interregional correlations (Lucas, 1993; Lucas et al., 1994), and recent sedimentological studies do not support the existence of a separate "Dockum basin" on the Southern High Plains (e.g., Riggs et al., 1996). The Triassic strata of the American Southwest are part of a single, vast depositional system, a Chinle basin, that extended from West Texas across New Mexico to Arizona, and northward into Nevada, Utah, Colorado and Wyoming.

ACKNOWLEDGMENTS

Numerous colleagues have worked with us on Triassic stratigraphic problems in east-central New Mexico. We thank all of them, especially O.J. Anderson and P. Huber. Numerous volunteers of the NMMNH, especially P. Bircheff, the Cotton family, N.V. "Dan" D'Andrea, R. Mongold, P. Sealey and the New Mexico Friends of Paleontology helped find and collect many of the fossils that made the biostratigraphy and biochronology advocated here possible. Orin Anderson and John Lorenz provided helpful reviews of the manuscript.

REFERENCES

Adams, J. E., 1929, Triassic of West Texas: American Association of Petroleum Geologists Bulletin, v. 12, p. 1045-1055.

Adkins, W. S., 1932, The Mesozoic systems in Texas: University of Texas. Bulletin 3232, p. 239-518.

Ash, S. R., 1972, Upper Triassic Dockum flora of eastern New Mexico and Texas: New Mexico Geological Society 23rd Field Conference, Guidebook, p. 124-128.

Ash, S. R., 1976, Occurrence of the controversial plant fossil *Sanmiguelia* in the Upper Triassic of Texas: Journal of Paleontology, v. 50, p. 799-804.

Ash, S. R., 1980, Upper Triassic floral zones of North America, *in* Dilcher, D. L., and Taylor, T. M., eds., Biostratigraphy of fossil plants: Stroudsburg, PA, Dowden, Hutchinson and Ross, p. 153-270.

Ash, S. R., 1988, Fossil plants from the mudstone member of the Santa Rosa Formation at the principal reference section: U.S. Geological Survey, Bulletin 1804C, p. 19-25.

Ash, S.R., 1989, A catalog of Upper Triassic plant megafossils of the western United States through 1988, *in* Lucas, S. G., and Hunt, A. P., eds., Dawn of the Age of Dinosaurs in the American Southwest: Albuquerque, New Mexico Museum of Natural History, p. 189-222.

Ash, S. R., and Morales, M., 1993, Anisian plants from Arizona: the oldest Mesozoic megaflora in North America: New Mexico Museum of Natural History and Science, Bulletin 3, p. 27-29.

Ballew, K. L., 1989, A phylogenetic analysis of Phytosauria from the Late Triassic of the western United States, *in* Lucas, S. G., and Hunt, A. P., eds., Dawn of the Age of Dinosaurs in the American Southwest: Albuquerque, New Mexico Museum of Natural History, p. 309-339.

Billingsley, G.H., 1985a, Geologic map of Petrified Forest National Park, Arizona: Unpublished Report to Petrified Forest Museum Association, Petrified Forest National Park, Arizona, 1 sheet.

Billingsley, G.H., 1985b, General stratigraphy of the Petrified Forest National Park, Arizona: Museum of Northern Arizona, Bulletin 54, p. 3-8.

Blakey, R. C., and Gubitosa, R., 1984, Controls of sandstone body geometry and architecture in the Chinle Formation (Upper Triassic), Colorado Plateau: Sedimentary Geology, v. 38, p. 51-86.

Boy, J. A., Schoch, R. R., and Lucas, S. G., 2001, The Moenkopi Formation in east-central New Mexico: Stratigraphy and vertebrate fauna: New Mexico Geological Society, 52nd Field Conference, Guidebook.

Broadhead, R. F., 1984, Subsurface petroleum geology of Santa Rosa Sandstone (Triassic), northeast New Mexico: New Mexico Bureau of Mines and Mineral Resources, Circular 193, 22 p.

Broadhead, R. F., 1985, Stratigraphy and petroleum geology of Dockum Group (Triassic), northeastern New Mexico: New Mexico Geological Society, 36th Field Conference, Guidebook, p. 307-314.

Carpenter, K., 1997, A giant coelophysoid (Ceratosauria) theropod from the Upper Triassic of New Mexico, U.S.A.: Neues Jahrbuch für Geologie und Paläontologie Abhandlungen, v. 205, p. 189-206.

Carpenter, K., and Parrish, J.M., 1985, Late Triassic vertebrates from Revuelto Creek, Quay County, New Mexico: New Mexico Geological Society, 36th Field Conference, Guidebook, p. 197-198.

Case, E. C., 1914, The red beds between Wichita Falls, Texas and Las Vegas, New Mexico in relation to their vertebrate fauna: Journal of Geology, v. 22, p.

243-259.

Case, E. C., 1928, Indications of a cotylosaur and of a new form of fish from the Triassic beds of Texas, with remarks on the Shinarump Conglomerate: University of Michigan, Contributions of the Museum of Paleontology, v. 3, no. 1, p. 1-14.

Chatterjee, S., 1986, The Late Triassic Dockum vertebrates: Their stratigraphic and paleobiogeographic significance, *in* Padian, K., ed., The beginning of the age of dinosaurs: Faunal change across the Triassic-Jurassic boundary: Cambridge, Cambridge University Press, p. 139-150.

Colbert, E. H., and Imbrie, J., 1956, Triassic metoposaurid amphibians: Bulletin of the American Museum of Natural History, v. 110, p. 403-452.

Cooley, M. E., 1957, Geology of the Chinle Formation in the upper Little Colorado drainage area, Arizona and New Mexico [M.S. thesis]: Tucson, University of Arizona, 317 p.

Cornet, B., 1993, Applications and limitations of palynology in age, climatic, and paleoenvironmental analyses of Triassic sequences in North America: New Mexico Museum of Natural History and Science, Bulletin 3, p. 75-93.

Cotton, W. D., Cotton, J. E., and Hunt, A. P., 1997, Social behavior of Cretaceous ornithopod dinosaurs of New Mexico: New Mexico Museum of Natural History and Science, Bulletin 11, p. 45-50.

Cotton, W. D., Cotton, J. E., and Hunt, A. P., 1998, Evidence for social behavior in ornithopod dinosaurs from the Dakota Group of northeastern New Mexico, U.S.A.: Ichnos, v. 6, p. 141-149.

Dane, C. H., and Bachman, G. O., 1965, Geologic map of New Mexico: Washington, D. C., U. S. Geological Survey, 2 sheets, scale 1:500,000.

Darton, N. H., 1922, Geologic structure of parts of New Mexico: U.S. Geological Survey, Bulletin 726-E, p. 275.

Darton , N. H., 1928, "Red beds" and associated formations in New Mexico, with an outline of the geology of the state: U. S. Geological Survey, Bulletin 794, 372 p.

Dobrovolny, E., Bates, R. L., and Summerson, C. H., 1946, Geology of northwestern Quay County, New Mexico: U. S. Geological Survey, Oil and Gas Investigations Map OM-62, 2 sheets.

Dubiel, R. F., 1989, Depositional and climatic setting of the Upper Triassic Chinle Formation, Colorado Plateau, *in* Lucas, S. G., and Hunt, A. P., eds., Dawn of the Age of Dinosaurs in the American Southwest: Albuquerque, New Mexico Museum of Natural History, p. 171-187.

Dunay, R. E., and Fisher, M. J., 1974, Late Triassic palynofloras of North America and their European correlatives: Review of Palaeobotany and Palynology, v. 17, p. 179-186.

Dunay, R. E., and Fisher, M J., 1979, Palynology of the Dockum Group (Upper Triassic), Texas, U.S.A.: Review of Palaeobotany and Palynology, v. 28, p. 61-92.

Espegren, W. A., 1985, Sedimentology and petrology of the upper Petrified Forest Member of the Chinle Formation, Petrified Forest National Park and vicinity, Arizona [M.S. Thesis]: Flagstaff, Northern Arizona University, 229 p.

Finch, W. I., and Wright, J. C., 1983, Measured stratigraphic sections of uranium-bearing Upper Triassic rocks in eastern New Mexico, west Texas, and the Oklahoma Panhandle with brief discussion of stratigraphic problems: U.S. Geological Survey, Open-file Report 83-701, 118 p.

Finch, W. I., Wright, J. C., and Davis, B. O., 1976, Unevaluated preliminary geologic cross section of uranium-bearing Upper Triassic rocks extending from Ute Reservoir, New Mexico, to Palo Duro Canyon, Texas U.S. Geological Survey, Open-file Report 76-205, 1 sheet.

Fritz, T., 1991, Depositional history of the mid-Late Triassic Santa Rosa Formation, eastern New Mexico [M.S. thesis]: Lubbock, Texas Tech University, 120 p.

Goddard, E. N., Trask, P. D., DeFord, R. K., Rove, O. N., Singewald, G. T. J., and Overbeck, R. M., 1984, Rock Color Chart: Boulder, Geological Society of America.

Good, S. C., 1989, Nonmarine Mollusca in the Upper Triassic Chinle Formation and related strata of the Western Interior, *in* Lucas, S. G., and Hunt, A. P., eds., Dawn of the Age of Dinosaurs in the American Southwest: Albuquerque, New Mexico Museum of Natural History, p. 233-248.

Good, S. C., 1993a, Molluscan paleobiology of the Upper Triassic Chinle Formation, Arizona and Utah [Ph.D. thesis]: Boulder, University of Colorado, 270 p.

Good, S. C., 1993b, Stratigraphic distribution of the mollusc fauna of the Chinle Formation and molluscan biostratigraphic zonation: New Mexico Museum of Natural History and Science, Bulletin 3, p. 155-159.

Good, S.C., 1998, Freshwater bivalve fauna of the Late Triassic (Carnian-Norian) Chinle, Dockum and Dolores formations of the southwest United States, *in* Johnston, P.O. and Haggart, J.W., eds., Bivalves: An Eon of Evolution—Paleobiological Studies Honoring Norman D. Newell: Calgary, University of Calgary Press, p. 223-249.

Gorman, J. M., and Robeck, R. C., 1946, Geology and asphalt deposits of north-central Guadalupe County, New Mexico: U. S. Geological Survey, Oil and Gas Investigations Preliminary Map OM-44.

Gould, C. N., 1907, The geology and water resources of the western portion of the Panhandle of Texas: U.S. Geological Survey, Water-supply Paper 191, 170 p.

Granata, G. E., 1981, Regional sedimentation of the Late Triassic Dockum Group, west Texas and eastern New Mexico [M.A. thesis]: Austin, University of Texas, 199 p.

Gregory, J. T., 1972, Vertebrate faunas of the Dockum Group, Triassic, eastern New Mexico and west Texas: New Mexico Geological Society, 23rd Field Conference, Guidebook, p. 120-123.

Griggs, R. L., and Hendrickson, G. E., 1951, Geology and ground-water resources of San Miguel County, New Mexico: New Mexico Bureau of Mines and Mineral Resources, Ground-water Report 2, 121 p.

Griggs, R. L., and Read, C. B., 1959, Revisions in stratigraphic nomenclature in Tucumcari-Sabinoso area, northeastern New Mexico: American Association of Petroleum Geologists Bulletin, v. 43, p. 2003-2007.

Heckert, A. B., and Lucas, S. G., 1996, Stratigraphic description of the Tr-4 unconformity, west-central New Mexico and eastern Arizona: New Mexico Geology, v. 18, no. 3, p. 61-70.

Heckert, A. B., and Lucas, S. G., 1998, First occurrence of *Aetosaurus* (Reptilia:Archosauria) in the Upper Triassic Chinle Group (USA) and its biochronological significance: Neues Jahrbuch für Geologie und Paläontologie Monatshefte, v. 1998, p. 604-612.

Heckert, A. B., Hunt, A. P., and Lucas, S. G., 1996, Redescription of *Redondasuchus reseri*, a Late Triassic aetosaur (Reptilia:Archosauria) from New Mexico (U.S.A) and the biochronology and phylogeny of aetosaurs: Geobios, v. 29, p. 619-632.

Heckert, A. B., Lucas, S. G., and Sullivan, R. M., 2000, Triassic dinosaurs in New Mexico: New Mexico Museum of Natural History and Science, Bulletin17, p. 17-26.

Heckert, A.B., Lucas, S.G., Hunt, A.P., and Harris, J.D., 2001, A giant phytosaur (Reptilia: Archosauria) skull from the Redonda Formation (Upper Triassic: Apachean) of east-central New Mexico: New Mexico Geological Society, 52nd Field Conference Guidebook.

Hester, P. M., 1988, Depositional environments of a Late Triassic lake, east-central New Mexico [M.S. thesis]: Albuquerque, University of New Mexico, 152 p.

Hester, P. M., and Lucas, S. G., 2001, Depositional environments of a Late Triassic lake, east-central New Mexico New Mexico: Geological Society, 52nd Field Conference Guidebook.

Huber, P., Lucas, S. G., and Hunt, A. P., 1993, Late Triassic fish assemblages of the North American Western Interior: Museum of Northern Arizona, Bulletin 59, p. 51-66.

Hunt, A. P., 1989a, Comments on the taxonomy of North American metoposaurs and a preliminary phylogenetic analysis of the family Metoposauridae, *in* Lucas, S. G., and Hunt, A. P., eds., Dawn of the Age of Dinosaurs in the American Southwest: Albuquerque, New Mexico Museum of Natural History, p. 293-300.

Hunt, A. P., 1989b, Cranial morphology and ecology among phytosaurs, *in* Lucas, S. G., and Hunt, A. P., eds., Dawn of the Age of Dinosaurs in the American Southwest: Albuquerque, New Mexico Museum of Natural History, p. 349-354.

Hunt, A. P., 1989c, A new ?ornithischian dinosaur from the Bull Canyon Formation (Upper Triassic) of east-central New Mexico, *in* Lucas, S. G., and Hunt, A. P., eds., Dawn of the age of dinosaurs in the American Southwest: Albuquerque, New Mexico Museum of Natural History, p. 355-358.

Hunt, A. P., 1991, The first tetrapod faunas from the Trujillo Formation (Late Triassic) of east-central New Mexico and their biochronological and paleoecological significance: New Mexico Geology, v. 13, p. 93.

Hunt, A. P., 1993, Revision of the Metoposauridae (Amphibia: Temnospondyli) and description of a new genus from western North America: Museum of

Northern Arizona, Bulletin 59, p. 67-97.

Hunt, A P., 1994, Vertebrate paleontology and biostratigraphy of the Bull Canyon Formation (Chinle Group: Norian), east-central New Mexico with revisions of the families Metoposauridae (Amphibia: Temnospondyli) and Parasuchidae (Reptilia: Archosauria) [Ph.D. dissertation]: Albuquerque, University of New Mexico, 403 p.

Hunt, A. P., 1997a, E.C. Case, J.T. Gregory and early explorations for fossil vertebrates in the Bull Canyon Formation (Upper Triassic) of eastern New Mexico: New Mexico Museum of Natural History and Science Bulletin 11, p. 15-24.

Hunt, A. P., 1997b, A new coelacanth (Osteichthyes: Actinistia) from the continental Upper Triassic of New Mexico: New Mexico Museum of Natural History and Science, Bulletin 11, p. 25-28.

Hunt, A.P., 2001a, Paleontology and age of the Upper Triassic Trujillo Formation, east-central New Mexico and west Texas: New Mexico Geological Society, 52nd Field Conference Guidebook.

Hunt, A. P., 2001b, The vertebrate fauna, biostratigraphy and biochronology of the type Revueltian land-vertebrate faunachron, Bull Canyon Formation (Upper Triassic), east-central New Mexico: New Mexico Geological Society, 52nd Field Conference Guidebook.

Hunt, A. P., and Lucas, S. G., 1989, Late Triassic vertebrate localities in New Mexico, *in* Lucas, S. G., and Hunt, A. P., eds., Dawn of the Age of Dinosaurs in the American Southwest: Albuquerque, New Mexico Museum of Natural History, p. 72-101.

Hunt, A. P., and Lucas, S. G., 1991a, A new aetosaur from the Upper Triassic of eastern New Mexico: Neues Jahrbuch für Geologie und Paläontologie Monatshefte, v. 1991, p. 728-736.

Hunt, A. P., and Lucas, S. G., 1991b, The *Paleorhinus* biochron and the correlation of the nonmarine Upper Triassic of Pangaea: Palaeontology, v. 34, p. 487-501

Hunt, A. P., and Lucas, S. G., 1993a, Triassic vertebrate paleontology and biochronology of New Mexico: New Mexico Museum of Natural History and Science, Bulletin 2, p. 49-60.

Hunt, A. P., and Lucas, S. G., 1993b, Late Triassic microvertebrate localities in New Mexico (USA): Implications for paleoecology: New Mexico Museum of Natural History and Science, Bulletin 3, p. 187-191.

Hunt, A. P., and Lucas, S. G., 1993c, A new phytosaur (Reptilia: Archosauria) genus from the uppermost Triassic of the western United States and its biochronological significance: New Mexico Museum of Natural History and Science, Bulletin 3, p. 193-196.

Hunt, A. P., and Lucas, S. G., 1994, Ornithischian dinosaurs from the Upper Triassic of the United States; *in* Fraser, N. C., and Sues, H.-D., eds., In the shadow of the dinosaurs: Early Mesozoic tetrapods: Cambridge, Cambridge University Press, p. 227-241.

Hunt, A. P., and Lucas, S. G., 1995, Vertebrate paleontology and biochronology of the Lower Chinle Group (Upper Triassic), Santa Fe County, north-central New Mexico: New Mexico Geological Society, 46th Field Conference, Guidebook, p. 243-246.

Hunt, A. P., and Lucas, S. G., 1997, Stratigraphy, paleontology and biochronology of the Upper Triassic Chinle Group in east-central New Mexico: Southwest Paleontological Symposium Proceedings, v. 1, p. 25-40.

Hunt, A. P., and Lucas, S. G., 2001, The first vertebrate track (*Brachychirotherium*) from the upper Carnian Garita Creek Formation, east-central New Mexico: New Mexico Geological Society, 52nd Field Conference, Guidebook.

Hunt, A. P., Lockley, M. G., and Lucas, S. G., 1993a, Vertebrate and invertebrate tracks and trackways from Upper Triassic strata of the Tucumcari basin, east-central New Mexico: New Mexico Museum of Natural History and Science, Bulletin 3, p. 199-201.

Hunt, A. P., Lucas, S. G., and Kietzke, K. K., 1989a, Dinosaur footprints from the Redonda Member of the Chinle Formation (Upper Triassic), east-central New Mexico; *in* Gillette, D. D. and Lockley, M. G., eds., Dinosaur tracks and traces: Cambridge, Cambridge University Press, p. 277-280.

Hunt, A. P., Lucas, S. G., and Reser, P. K., 1993b, A complete skeleton of the stagonolepidid *Typothorax coccinarum* from the Upper Triassic Bull Canyon Formation of east-central New Mexico, U.S.A.: New Mexico Museum of Natural History and Science, Bulletin 3, p. 209-212.

Hunt, A. P., Lucas, S. G., and Sealey, P. L., 1989b, Paleontology and vertebrate biochronology of the Upper Triassic Garita Creek Formation, east-central New Mexico: New Mexico Journal of Science, v. 29, p. 61-68.

Hunt, A. P., Lucas, S. G., Lockley, M. G. and Heckert, A. B., 2000, Occurrence of the dinosaurian ichnogenus *Grallator* in the Redonda Formation (Upper Triassic: Norian) of eastern New Mexico: New Mexico Museum of Natural History and Science Bulletin 17, p. 39-41.

Hunt, A. P., Lucas, S. G., Heckert, A. B., Sullivan, R. M., and Lockley, M. G., 1998, Late Triassic dinosaurs from the western United States: Geobios, v. 31, p. 511-531.

Johns, D. A., and Granata, G. E., 1987, Regional tectonic setting and depositional trends of the Dockum Group, West Texas and eastern New Mexico: Journal of the Arizona-Nevada Academy of Science, v. 22, p. 53-72.

Kelley, V. C., 1972a, Triassic rocks of the Santa Rosa area: New Mexico Geological Society, 23rd Field Conference, Guidebook, p. 84-90.

Kelley, V. C., 1972b, Geology of the Fort Sumner sheet, New Mexico: New Mexico Bureau of Mines and Mineral Resources, Bulletin 98, 55 p.

Kietzke, K.K., 1987, Calcareous microfossils from the Upper Triassic of northeastern New Mexico: New Mexico Geological Society, 38th Fall Field Conference Guidebook, p. 119-126.

Kietzke, K. K., 1989, Calcareous microfossils from the Triassic of the southwestern United States, *in* Lucas, S. G., and Hunt, A. P., eds., Dawn of the Age of Dinosaurs in the American Southwest: Albuquerque, New Mexico Museum of Natural History, p. 223-232.

Kues, B.S., 1985, Nonmarine molluscs from the Chinle Formation, Dockum Group (Upper Triassic), of Bull Canyon, Guadalupe County, New Mexico: New Mexico Geological Society, 36th Field Conference, Guidebook, p. 185-196

Lehman, T. M., 1994, The saga of the Dockum Group and the case of the Texas/ New Mexico boundary fault: New Mexico Bureau of Mines and Mineral Resources, Bulletin 150, p. 37-52.

Lehman, T., Chatterjee, S., and Schnable, J., 1992, The Cooper Canyon Formation (Late Triassic) of western Texas: Texas Journal of Science, v. 44, p. 349-355.

Litwin, R. J., Traverse, A., and Ash, S. R., 1991, Preliminary palynological zonation of the Chinle Formation, southwestern U.S.A., and its correlation to the Newark Supergroup: Review of Palaeobotany and Palynology, v. 68, p. 269-287.

Lockley, M. G., and Hunt, A. P., 1995, Dinosaur Tracks and Other Fossil Footprints of the Western United States: New York, Columbia University Press, 338 p.

Lockley, M. G., and Hunt, A. P., 1999, Dinosaur Tracks and Other Fossil Footprints of the Western United States. New York, Columbia University Press (revised paperback edition), 338 p.

Lockley, M. G., Lucas, S. G., and Hunt, A. P., 2000, Dinosaur tracksites in New Mexico: A review: New Mexico Museum of Natural History and Science, Bulletin 17, p. 9-16.

Long, R. A., and Murry, P. A., 1995, Late Triassic (Carnian and Norian) tetrapods from the southwestern United States: New Mexico Museum of Natural History and Science, Bulletin 4, p. 254 p.

Lucas, S. G., 1991, Correlation of Triassic strata of the Colorado Plateau and southern High Plains, New Mexico: New Mexico Bureau of Mines and Mineral Resources, Bulletin 137, p. 47-56.

Lucas, S. G., 1993, The Chinle Group: Revised stratigraphy and biochronology of Upper Triassic nonmarine strata in the western United States: Museum of Northern Arizona, Bulletin 59, p. 27-50.

Lucas, S. G., 1995, Triassic stratigraphy and chronology in New Mexico: New Mexico Geology, v. 17, p. 8-13, 17.

Lucas, S. G., 1997, The Upper Triassic Chinle Group, western United States, nonmarine standard for Late Triassic time, *in* Dickins, J. M., Yang, Z., Yin, H., Lucas, S. G., and Acharyya, S. K., eds., Permo-Triassic of the circum-Pacific: Cambridge, Cambridge University Press, p. 200-228.

Lucas, S. G., 1998, Global Triassic tetrapod biostratigraphy and biochronology: Palaeogeography, Palaeoclimatology, Palaeoecology, v. 143, p. 347-384.

Lucas, S. G., 2000, Tetrapod-based correlation of the nonmarine Triassic: Zentralblatt für Geologie und Paläontologie, v. 2000, no. 7-8, p. 497-521.

Lucas, S.G., 2001, The bennettitalean "*Zamites*" *powellii* from the Middle Triassic Moenkopi Formation, east-central New Mexico: New Mexico Geological Society, 52nd Field Conference Guidebook.

Lucas, S. G., and Anderson, O. J., 1992, Triassic stratigraphy and correlation, west Texas and eastern New Mexico, *in* Cromwell, D. W., Moussa, M. T., and Mazzullo, L. J., eds., Transactions Southwest Section AAPG, Midland, Texas, 1992: Midland, West Texas Geological Society, p. 201-207.

Lucas, S. G., and Anderson, O. J., 1993, Triassic stratigraphy in southeastern New

Mexico and southwestern Texas: New Mexico Geological Society, 44th Field Conference, Guidebook, p. 231-235.

Lucas, S. G., and Hayden, S. N., 1989, Triassic stratigraphy of west-central New Mexico: New Mexico Geological Society 40th Field Conference Guidebook, p. 191-211.

Lucas, S. G., and Hayden, S. N., 1991,Type section of the Permian Bernal Formation and the Permian-Triassic boundary in north-central New Mexico: New Mexico Geology, v. 13, p. 9–15.

Lucas, S. G., and Heckert, A. B., 1994, Triassic stratigraphy in the Lucero uplift, Cibola, Valencia and Socorro counties, New Mexico: New Mexico Geological Society 45th Field Conference, Guidebook, p. 241-254.

Lucas, S. G., and Hunt, A. P., 1987, Stratigraphy of the Anton Chico and Santa Rosa Formations, Triassic of east-central New Mexico: Journal of the Arizona-Nevada Academy of Science, v. 22, p. 21-33.

Lucas, S. G., and Hunt, A. P., 1989, Revised Triassic stratigraphy, Tucumcari basin, east-central New Mexico; *in* Lucas, S. G., and Hunt, A. P., eds., Dawn of the Age of Dinosaurs in the American Southwest: Albuquerque, New Mexico Museum of Natural History, p. 150-170.

Lucas, S. G., and Hunt, A. P., 1993, Tetrapod biochronology of the Chinle Group (Upper Triassic), western United States: New Mexico Museum of Natural History and Science, Bulletin 3, p. 327-329.

Lucas, S.G., and Morales, M., 1985, Middle Triassic amphibian from basal Santa Rosa Formation, east-central New Mexico: New Mexico Geological Society, 36th Field Conference, Guidebook, p. 56-58.

Lucas, S. G., and Oakes, W., 1988, A Late Triassic cynodont from the American South-West: Palaeontology, v. 31, p. 445-449.

Lucas, S. G., Anderson, O. J., and Hunt, A. P., 1994, Triassic stratigraphy and correlations, southern High Plains of New Mexico-Texas: New Mexico Bureau of Mines and Mineral Resources, Bulletin 150, p. 105-126.

Lucas, S. G., Estep, J. W., and Hunt, A. P., 1998, Giant erythrosuchid archosaur from the Middle Triassic Moenkopi Formation, east-central New Mexico: Geological Society of America Abstracts with Programs, v. 30, no. 6, p. 13-14.

Lucas, S. G., Heckert, A. B., and Anderson, O. J., 1997, Triassic stratigraphy and paleontology on the Fort Wingate quadrangle, west-central New Mexico: New Mexico Geology, v. 19, p. 33-42.

Lucas, S. G., Heckert, A. B., and Estep, J. W., 1999a, Correlation of Triassic strata across the Rio Grande rift, north-central New Mexico: New Mexico Geological Society, 50th Field Conference, Guidebook, p. 305-310.

Lucas, S.G., Hunt, A.P., and Huber, P., 1990, Triassic stratigraphy in the Sangre de Cristo Mountains, New Mexico: New Mexico Geological Society, 41st Field Conference, Guidebook, p. 305-318.

Lucas, S. G., Hunt, A. P., and Bennett, S. C., 1985b, Triassic vertebrates from east-central New Mexico in the Yale Peabody Museum: New Mexico Geological Society, 36th Field Conference, Guidebook, p. 199-203.

Lucas, S. G., Hunt, A. P., and Kietzke, K. K., 1985d, *Neocalamites* forest in the Upper Triassic of Bull Canyon, New Mexico: New Mexico Geological Society, 36th Field Conference, Guidebook, p. 8-9.

Lucas, S. G., Hunt, A. P. and Lockley, M. G., 2001a, Tetrapod footprint ichnofauna of the Upper Triassic Redonda Formation, Chinle Group, Quay County, New Mexico: New Mexico Geological Society, 52nd Field Conference, Guidebook.

Lucas, S. G., Hunt, A. P., and Morales, M., 1985a, Stratigraphic nomenclature and correlation of Triassic rocks of east-central New Mexico: A preliminary report: New Mexico Geological Society, 36th Field Conference, Guidebook, p. 171-184.

Lucas, S. G., Oakes, W., and Froehlich, J. W., 1985c, Triassic microvertebrate locality, Chinle Formation, east-central New Mexico: New Mexico Geological Society, 36th Field Conference, Guidebook, p. 205-212.

Lucas, S. G., Estep, J. W., Heckert, A. B., and Hunt, A. P., 1999b, Cynodont teeth from the Upper Triassic of New Mexico, U.S.A.: Neues Jahrbuch für Geologie und Paläontologie Monatsheft, v. 1999, p. 331-344.

Lucas, S. G., Hunt, A. P., Heckert, A. B., Kues, B. S., and McLemore, V. T., 2001b, Third-day road log, from Tucumcari to Conchas Dam, New Mexico: New Mexico Geological Society, 52nd Field Conference Guidebook.

Marcou, J., 1858, Geology of North America, With Reports on the Prairies of Arkansas and Texas, the Rocky Mountains of New Mexico, and the Sierra Nevada of California, Originally Made for the United States Government: Zurich, Zurcher and Furrer, 144 p.

Matthes, G. H., 1936, Problems at Conchas Dam: Civil Engineering, v. 6, p. 437-441.

McGowan, J. H., Granata, G. E., and Seni, S. J., 1983, Depositional setting of the Triassic Dockum Group, Texas Panhandle and eastern New Mexico, *in* Reynolds, M. W., and Dolly, E. D., eds., Mesozoic Paleogeography of the West-central United States: Denver, Society of Economic Paleontologists and Mineralogists, Rocky Mountain Section, p. 13-38.

McKee, E.D., Oriel, S.S., Ketner, K.B., MacLachlan, M.E.H., Goldsmith,J.W., MacLachlan, J.C., and Mudge, M.R., 1959, Paleotectonic maps of the Triassic System: U.S. Geological Survey, Miscellaneous Geological Investigations Map I-300, 33 p.

Murry, P. A., 1986, Vertebrate paleontology of the Dockum Group, western Texas and eastern New Mexico, *in* Padian, K., ed., The beginning of the age of dinosaurs: Faunal change across the Triassic-Jurassic boundary: Cambridge, Cambridge University Press, p. 109-137.

Murry, P. A., 1989, Geology and paleontology of the Dockum Formation (Upper Triassic), West Texas and eastern New Mexico, *in* Lucas, S. G., and Hunt, A. P., eds., Dawn of the Age of Dinosaurs in the American Southwest: Albuquerque, New Mexico Museum of Natural History, p. 102-148.

Newell, A. J., 1993, Depositional environment of the Late Triassic Bull Canyon Formation (New Mexico): implications for "Dockum Formation" paleogeography: New Mexico Museum of Natural History and Science, Bulletin 3, p. 359-368.

Parrish, J. M., 1989, Late Triassic tetrapods of the North American Southwest, *in* Lucas, S. G., and Hunt, A. P., eds., Dawn of the Age of Dinosaurs in the American Southwest: Albuquerque, New Mexico Museum of Natural History, p. 360-374.

Parrish, J. M., and Carpenter, K., 1986, A new vertebrate fauna from the Dockum Formation (Late Triassic) of eastern New Mexico, *in* Padian, K., ed., The beginning of the age of dinosaurs: Faunal change across the Triassic-Jurassic boundary: Cambridge, Cambridge University Press, p. 151-160.

Pipiringos, G. N., and O'Sullivan, G. S., 1978, Principal unconformities in Triassic and Jurassic rocks, western interior United States—a preliminary survey: U.S. Geological Survey, Professional Paper 1035-A, p. 1-29.

Reeside, J. B., Jr., Applin, P. L., Colbert, E. H., Hadley, H. D., Kummel, B., Lewis, P. J., Love, J. D., Maldanado-Koerdell, M., McKee, E. D., McLaughlin, D. B., Muller, S. W., Rodgers, J., Sanders, J., Silberling, N. J., and Waage, K., 1957, Correlation of the Triassic formations of North America exclusive of Canada: Geological Society of America Bulletin, v. 68, p. 1451-1514.

Rich, J. L., 1921, The stratigraphy of eastern New Mexico—a correction: American Journal of Science, series 5, v. 2, p. 295-298.

Riggs, N. R., Lehman, T. M., Gehrets, G. E., and Dickinson, W. R., 1996, Detrital zircon link between headwaters and terminus of the Upper Triassic Chinle-Dockum paleo-river system: Science, v. 273, p. 97-100.

Romer, A. S., 1939, An amphibian graveyard: Scientific Monthly, v. 49, p. 337-339.

Spiegel, Z., 1972, Cenozoic geology of the Canadian River valley, New Mexico: New Mexico Geological Society 23rd Field Conference, Guidebook, p. 118-119.

Stewart, J. H., Poole, F. G., and Wilson, R. F., 1972, Stratigraphy and origin of the Chinle Formation and related Upper Triassic strata in the Colorado Plateau region: U.S. Geological Survey, Professional Paper 690, 336 p.

Trauger, F. D., Mankin, C. J., and Brand, J. P., 1972, Road log of Tucumcari, Mosquero, and San Jon country: New Mexico Geological Society, 23rd Field Conference, Guidebook, p. 12-45.

Wanek, A. A., 1962, Reconnaissance geologic map of parts of Harding, San Miguel, and Mora counties, New Mexico: Washington, U. S. Geological Survey Oil and Gas Investigations Map OM 209, 1 sheet.

APPENDIX—DESCRIPTION OF MEASURED SECTIONS

PUERTO DE LUNA

Measured just east of NM Highway 31 between Santa Rosa and Puerto de Luna. Section base at UTM 53138E, 3863675N (zone 13, NAD 27) and top at 531480E, 3863734N. Strata are flat-lying.

unit lithology **thickness (m)**

Chinle Group: Santa Rosa Formation: Tecolotito Member:

15 Sandstone; very pale orange (10YR8/2) and pale yellowish brown (10YR6/2); very fine-to fine-grained; quartzarenite; calcareous; trough crossbedded. **not measured**

14 Conglomerate; moderate yellowish brown (10YR5/4) and pale yellowish brown (10YR6/2); clasts are up to 1.5 cm in diameter and are mostly limestone; matrix is a fine- to coarse-grained litharenite; massive. **1.3**

unconformity (Tr-3 unconformity of Pipiringos and O'Sullivan, 1978):

Moenkopi Formation: Anton Chico Member:

13 Sandstone; speckled very pale orange (10YR8/2) and pale yellowish brown (10YR6/2); fine-grained; hematitic quartzarenite; laminar with some low angle trough crossbeds. **1.4**

12 Sandy siltstone; grayish red (10R4/2); not calcareous; laminar (flaggy); forms a slope. **2.9**

11 Sandstone; same color and lithology as unit 7. **0.6**

10 Sandstone; same color and lithology as unit 9; scour base; paleoflow (based on crossbed dips) to N20°W. **2.3**

9 Sandstone; grayish red (10R4/2) to pale red (10R6/2); fine-grained; slightly micaceous litharenite; very calcareous; trough crossbedded. **1.6**

8 Sandstone; same color and lithology as unit 9 but with mud-chip conglomerate locally at base. **2.1**

7 Sandstone; same color and lithology as unit 9. **1.3**

6 Sandstone; pale red (10R6/2) and yellowish gray (5Y8/10; fine- to medium-grained; micaceous litharenite; very calcareous; trough crossbedded; lenses of mud-pellet conglomerate in lower 0.5 m; paleocurrent azimuths to N50°W. **1.8**

5 Conglomeratic sandstone; pale reddish brown (10R5/4); fine- to medium-grained; micaceous litharenite; very calcareous; mud pebbles up to 7 mm in diameter; trough crossbedded. **0.7**

4 Sandstone; grayish red (10R4/2); very fine-grained; micaceous litharenite; trough crossbedded. **2.0**

3 Sandstone; pale red (10R6/2); very fine-grained; micaceous litharenite; calcareous; trough crossbedded. **1.5**

2 Sandstone; very pale orange (10YR8/2) to yellowish gray (5Y8/1); fine-grained; micaceous litharenite; calcareous; trough crossbedded. **1.4**

unconformity (Tr-2 and older unconformities of Pipiringos and O'Sullivan, 1978)

Artesia Group:

1 Sandy siltstone; pale reddish brown (10R5/4); not calcareous; ripple laminated. **not measured**

CONCHAS DAM

Measured on the north side of the Canadian River near the first bend east of Conchas Dam. Section base at UTM 574796E, 3918639N (zone 13, NAD 27) and top at 574248E, 3918392N. Strata are flat-lying.

unit lithology **thickness (m)**

Chinle Group: Garita Creek Formation:

19 Sandy mudstone; same color and lithology as unit 17. **not measured**

18 Sandstone; grayish red (10R4/2); very fine-grained; litharenite; calcareous; ripple laminated bench; *Brachychirotherium* tracks. **0.4**

17 Sandy mudstone; grayish red (10R4/2); very calcareous. **2.7**

Santa Rosa Formation: Tres Lagunas Member:

16 Sandstone; same color and lithology as unit 14. **0.8**

15 Sandstone; grayish orange (10YR7/4); very fine grained; hematitic quartzarenite; calcareous; ripple laminated. **2.8**

14 Sandstone; grayish orange (10YR7/4); very fine-grained; hematitic quartzarenite; calcareous; trough crossbedded. **1.6**

13 Sandstone; pale yellowish brown (10YR6/2) and grayish orange (10YR7/4); very fine- to fine-grained; micaceous quartzarenite; platy to ripple laminated. **2.5**

12 Conglomerate; matrix supported; matrix is pale yellowish brown (10YR6/2); very fine to fine-grained; micaceous quartzarenite; clasts are light brownish gray (4YR6/1) limestone (calcrete) pebbles up to 5 cm in diameter; trough crossbedded; base of unit is a scour surface with as much as 1 m of relief locally. **2.3**

unconformity:

Los Esteros Member:

11 Silty sandstone; pale red (10R6/2) to pale yellowish brown (10YR6/2); very fine- to fine-grained; micaceous quartzarenite; calcareous; ripple laminated. **3.3**

10 Silty sandstone; pale reddish brown (10R5/4); very fine-grained; micaceous litharenite; calcareous; ripple laminated. **1.5**

9 Silty shale; grayish red (10R4/2); micaceous; not calcareous. **0.5**

8 Silty sandstone; pale reddish brown (10R5/4); very fine-grained; micaceous quartzarenite; ripple laminated with mudstone drapes. **0.9**

7 Sandstone; pale yellowish brown (10YR6/2) and grayish orange (10YR7/4); very fine-grained; micaceous quartzarenite; slightly calcareous; ripple laminated; forms a bench. **1.0**

6 Shale; grayish red (10R4/2) and yellowish gray (5Y7/2) mottling; not calcareous; some lenses of pebbly sandstone that are grayish orange (10YR7/4); fine-grained; clayey; micaceous litharenite. **5.5**

Tecolotito Member:

5 Sandstone; dark yellowish orange (10YR6/6) and pale yellowish brown (10YR6/2)fine to medium-grained; micaceous quartzarenite; calcareous; trough crossbedded; top ripple laminated. **6.1**

4 Sandstone; speckled very pale orange (10YR8/2) and pale yellowish brown (10YR6/2); fine- to medium-grained; quartzarenite; very calcareous; trough crossbedded. **3.5**

3 Conglomeratic sandstone; same colors as unit 4; medium to coarse grained; quartzarenite; calcareous. **1.3**

2 Conglomerate; pinkish gray (5YR8/1); matrix is fine- to coarse-grained quartzarenite; clasts are up to 15 cm in diameter and are extraformational lithics, fossiliferous Paleozoic limestones and quartzite. **2.0**

unconformity: (Tr-3 unconformity of Pipiringos and O'Sullivan, 1978)

Moenkopi Formation: Anton Chico Member:

1 Mudstone and siltstone; pale reddish brown (10R5/4); very calcareous; mostly covered. **5+**

SALADITO POINT

Measured just east of NM Highway 129 between Conchas Dam and Newkirk. Section base at UTM 566672E, 3885289N (zone 13, NAD 27) and top at 567o16E, 2885174N. Strata are flat-lying.

unit lithology **thickness (m)**

Chinle Group: Bull Canyon Formation:

17 Conglomeratic sandstone; grayish red (10R4/2); matrix is medium- to coarse-grained; clasts are limestone (calcrete); litharenite calcareous; trough crossbedded; paleocurrent azimuths (crossbed dips) are to 280°. **1.0**

16 Muddy sandstone; pale reddish brown (10R5/4); very fine-grained; litharenite; forms a slope. **3.3**

Saladito Point Bed (units 9-15 form type section)

15 Sandstone; pale reddish brown (10R5/4); very fine-grained; micaceous litharenite; calcareous; trough crossbedded. **4.0**

14 Sandstone; pale reddish brown (10R5/4); very fine-grained; micaceous litharenite; not calcareous; thinly laminated. **1.5**

13 Sandstone; pale reddish brown (10R5/4); fine-grained; micaceous litharenite; not calcareous; laminated. **1.1**

12 Sandstone; pale reddish brown (10R5/4); fine-grained; micaceous litharenite; trough crossbedded. **2.8**

11 Conglomeratic sandstone; grayish red (10R4/2); matrix is fine-grained micaceous litharenite; clasts are mudstone, siltstone and calcrete rip-ups up to 4 mm in diameter; calcareous; massive. **1.4**

10 Sandstone; pale reddish brown (10R5/4); very fine- to fine-grained; micaceous litharenite; calcareous; trough crossbedded. **1.1**

9 Conglomerate; same lithology as unit 11 but less matrix; trough crossbedded. **0.3**

Bull Canyon Formation:

8 Silty mudstone; pale reddish brown (10R5/4); calcareous; forms a long slope. **14.3**

7 Calcrete; nodular; pale reddish brown (10R5/4); nodules up to 4 cm diameter; coprolites. **0.7**

6 Silty mudstone; same color and lithology as unit 8. **2.5**

5 Sandstone; pale reddish brown (10R5/4) and light greenish gray (5GY8/1); very fine-grained; micaceous litharenite; very calcareous; thinly laminated and ripple laminated. **0.7**

4 Silty mudstone; same color and lithology as unit 8. **9.9**

3 Conglomerate; light brownish gray (5YR6/1); clasts are limestone and chert pebbles up to 1 cm in diameter; contains coprolites and bone fragments of metoposaurs and phytosaurs. **0.4**

2 Silty mudstone; pale red (10R6/2); calcareous. **1.4**

Trujillo Formation:

1 Sandstone; light olive gray (5Y6/1); fine-grained; subarkosic; not calcareous; trough crossbedded; floors the flats around base of the mesa. **Not measured**

BARRANCA CREEK, SECTION A

Section A was measured in the NE1/4 NE1/4 SW1/4 and NW1/4 NW1/4 SE1/4 Sec.

6 T11N, R33E, NW1/4 and SW1/4 SW1/4 SE1/4 Sec. 1, T10N, R32E and SE1/4 SE1/4 SW1/4 Sec. 1, T10N, 32E, Quay, County, New Mexico.

unit lithology **thickness (m)**

Chinle Group: Bull Canyon Formation:

16 Sandstone; moderate reddish brown (10R4/6); very fine grained; well sorted; rounded to subrounded; quartzose with mica, feldspar and lithic fragments; mildly calcareous; channelform with undulating erosional base; ripple laminar; trough and planar crossbeds. **6.0**

15 Sandstone and siltstone; sandstone is pale yellowish green (10GY7/2); very fine grained; well sorted; well rounded; quartzose; not calcareous; platy: siltstone is siltstone-silty mudstone; moderate reddish brown (10R4/6); very calcareous. **4.7**

14 Sandstone and siltstone-very fine sandstone; sandstone is pale green (5G7/2); very fine grained; moderately sorted; subrounded to subangular; quartzose; highly calcareous; siltstone-very fine sandstone is grayish red (10R4/2) and pale red purple (5RP6/2); poorly sorted; subrounded; quartzose; highly calcareous; several vertebrate localities including NMMNH localities 6 and 7. **3.0**

13 Conglomerate; light greenish gray (5G8/1); clasts are mainly limestone, few are siltstone (moderate reddish brown (10R4/6); well rounded up to 9 mm in diameter; matrix is sandstone; fine-coarse grained; poorly sorted; moderately rounded; quartzose with lithic fragments; highly calcareous; ledge-former, often forms top of small cliff. **0.5**

12 Conglomerate and sandstone; conglomerate as 13, but limestone clasts are light olive gray (5Y6/1); forms lower 0.4 m of unit; sandstone is moderate reddish brown (10R4/6); very fine grained; moderately sorted; subrounded; quartzose; highly calcareous; platy; 1 mm wide surface trails, sinuous and branching; phytosaur scrap. **0.6**

11 Mudstone (very slightly silty); moderate reddish brown (10 R 4/6); moderately calcareous. **1.5**

offset

10 Calcrete; moderate brown (5YR4/4); nodular; microcrystalline limestone; very laterally continuous. **3.0**

9 Siltstone and sandstone; siltstone is moderate reddish brown (10R4/6); structureless; highly calcareous; sandstone is 30-cm-thick layer 2 m from base of unit; sandstone is moderate reddish brown (10R4/6); very fine grained; moderately sorted; subrounded; ripple cross lamination; 8 mm diameter cylindrical, vertical burrows. **5.5**

8 Sandstone; moderate reddish brown (10R4/6); fine-very fine grained; poorly sorted; subrounded; quartzose; highly calcareous; structureless; pillowy. **0.5**

7 Silty mudstone; moderate reddish brown (10R4/6); highly calcareous; structureless. [interval of NMMNH localities 5, 36, 75] **2.0**

6 Sandstone; moderate reddish brown (10R4/6); very fine grained; poorly sorted; subrounded; quartzose; finely laminated; at 1.2 m from base of unit is 0.3 thick sandstone which is light greenish gray (5G8/1); very fine grained; poorly sorted; subrounded; quartzose; highly calcareous; platy, ripple laminar. **2.5**

offset [NMMNH locality 2 just above unit 5]

5 Sandstone and conglomerate; sandstone is moderate reddish brown (10 R 4/6)-light greenish gray (5GY8/1); fine-very fine grained; poorly sorted; subrounded; quartzose; platy; basal 10 cm is conglomerate; clasts are mainly light greenish gray (5GY8/1); limestone; matrix is like sandstone above but is coarse-very coarse grained; laterally at offset this unit is sandstone which is moderate reddish brown (10R4/6); fine-very fine grained; poorly sorted; subrounded; quartzose and lithic fragments; highly calcareous; bioturbated with 5 mm diameter cylindrical, vertical tubes. **1.1**

4 Silty mudstone and sandstone; silty mudstone is moderate reddish brown (10R4/6); moderately calcareous; sandstone is dark yellowish brown (10YR4/2); fine to very fine grained; poorly sorted; subrounded; quartzose; highly calcareous; ripple laminar; sandstone is top 30 cm of unit. **4.5**

3 Sandstone; moderate reddish brown (10R4/6); fine-medium grained; poorly sorted; subrounded; quartzose with lithic fragments; laminar and platy. **1.5**

2 Conglomerate; sandstone and siltstone; conglomerate is dark yellowish brown (10YR4/2); clasts are limestone up to 15 mm in diameter; matrix is sandstone; fine-coarse grained; poorly sorted; subrounded; quartzose; highly calcareous; sandstone is moderate brown (5YR3/4); fine-medium grained; poorly sorted; subrounded; quartzose with lithic fragments; laminated; siltstone is moderate reddish brown (10R4/6); highly calcareous; laminated; basal 1 m is conglomerate, overlying 2. 5 m is 20% sandstone 10% conglomerate 70% siltstone; conglomerates have many vertebrate fossils principally isolated phytosaur postcrania and *Typothorax* scute fragments. **3.5**

Trujillo Formation:

1 Sandstone; moderate brown (5YR4/4); medium grained well sorted; subrounded; quartzose with lithic fragments; highly calcareous; trough and planar crossbeds. **not measured**

REVUELTO CREEK, SECTION B

Section B was measured in NE1/4 NE1/4 SE1/4 Sec. 9, T10N, R33E.

unit lithology **thickness (m)**

Chinle Group: Bull Canyon Formation:

6 Soil. **0.6**

5 Siltstone and conglomerate; siltstone is pale red purple (5RP6/2) and small 1 mm diameter reduction spots of light greenish gray (5GY8/1); highly calcareous; conglomerate has pale brown (5YR5/2) clasts which are limestone and up to 6 mm in diameter; matrix is siltstone; pale red purple (5RP8/2); conglomerate is lens 0.15 m thick and contains bone pebbles. **0.6**

4 Siltstone; pale red purple (5RP6/2); highly calcareous; microbrecciated with stringers of grayish red purple (RP4/2). **1.5**

3 Conglomerate; clasts are pale red (5R6/2); limestone; matrix is sandstone pale green (5G7/2); fine-very fine grained; moderately sorted; subrounded; quartzose; highly calcareous. **0.3**

2 Siltstone; grayish red purple (5RP4/2); highly calcareous with floating pebbles (up to 2 mm in diameter) of very dusky red purple (5RP2/1); microbrecciated with small reduction spots of light greenish gray (5GY8/1); with 5 cm thick stringers of siltstone; dusky red (5R3/4); highly calcareous; main bone layer; unit is laterally discontinuous in scours. **0.3**

1 Conglomerate; clasts are light greenish gray (5G8/1); limestone; up to 18 mm in diameter; matrix is sandstone; moderate reddish brown (10R4/6); very fine grained; poorly sorted; well rounded; quartzose; conglomerate laterally continuous, thins laterally over 10 m to stacked lenses of conglomerate. **5.0**

0 Siltstone; moderate reddish brown (10R4/8) and locally grayish yellow green (5GY7/2); highly calcareous; structureless. **not measured**

REVUELTO CREEK, SECTION C

Section C was measured in NE1/4 NW1/4 SE1/4 Sec. 9, T10N, R33E, Quay County, New Mexico.

unit lithology **thickness (m)**

Chinle Group: Bull Canyon Formation:

24 Sandstone; pale yellowish green (10GY7/2); medium-fine grained; poorly sorted; well rounded; quartzose with lithic fragments; highly calcareous; ledge-former. **1.2**

23 Mudstone (slightly silty); moderate reddish brown (10R4/6); highly calcareous; lower 2.1 m has stringers of unit 22 lithology. **8.5**

22 Silty mudstone; pale yellowish green (10GY7/2); highly calcareous; erosional base. **0.15**

21 Mudstone (slightly silty); mottled light greenish gray (5G8/1) and grayish red purple (5RP4/2); highly calcareous. **1.8**

20 Conglomerate; clast are poorly sorted up to 18 mm in diameter; 70% limestone light olive gray (5Y6/1), 30% siltstone grayish red (10 R 4/1) and moderate reddish orange (10R4/6); matrix is sandstone; dark yellowish orange (10YR6/6); medium grained; poorly sorted; subrounded; quartzose; highly calcareous; ledge-former. **1.0**

19 Mudstone; pale red purple (5RP6/2); moderately calcareous; thins laterally to 0.3 m. **1.5**

18 Sandstone; greenish gray (5GY6/1); fine-very fine grained; poorly sorted; moderately rounded; quartzose; highly calcareous; local lenses of conglomerate (0.15 m thick) at base as 20. **0.30**

17 Sandstone; moderate brown (5YR3/4); fine to very fine grained; poorly sorted; subrounded; quartzose; highly calcareous. **0.8**

16 Conglomerate and sandstone; conglomerate has light olive gray (5Y6/1) clasts dominantly of limestone that are poorly sorted; matrix as 20; bone pebbles; sandstone is greenish gray (5GY6/1); fine grained; moderately sorted; well rounded; quartzose; highly calcareous; planar bedded; conglomerate grades up into sandstone. **1.05**

15 Mudstone; pale red purple (5RP6/2); very calcareous; scattered concretions of pale red purple (5RP6/2); highly calcareous; thins laterally over 10 m. **1.05**

14 Silty mudstone; pale red purple (5RP6/2); very calcareous. **0.6**

13 Silty mudstone; pale red purple (5RP6/2); very calcareous; calcareous concretions of pale red purple (5RP6/2) up to 0.3 x 1m; highly calcareous. **1.2**

12 Calcrete; moderate brown (5YR4/4); nodular; microcrystalline limestone; laterally continuous. **0.3**

11 Silty mudstone; pale red purple (5RP6/2); very calcareous; scattered concretions of pale red purple (5RP6/2); highly calcareous. **0.15**

10 Silty mudstone; reddish brown (10R4/6); very calcareous. **1.8**

9 Silty mudstone; pale red purple (5RP6/2); very calcareous; scattered concretions of pale red purple (5RP6/2); highly calcareous. **1.8**

8 Silty mudstone; reddish brown (10R4/6); very calcareous. **0.8**

7 Silty mudstone; pale red purple (5RP6/2); very calcareous; scattered concretions of pale red purple (5RP6/2); highly calcareous. **0.2**

6 Silty mudstone; reddish brown (10R4/6); very calcareous. **0.5**

5 Silty mudstone; pale red purple (5RP6/2); very calcareous; scattered concretions of pale red purple (5RP6/2); highly calcareous. **0.3**

4 Silty mudstone; moderate reddish brown (10R4/6); very calcareous. **3.0**

3 Siltstone and sandstone; siltstone is moderate reddish brown (10R4/6); highly

calcareous; sandstone from 3.0-4.5 from base of unit; sandstone is light greenish gray (5GY8/1); very fine grained; well sorted; moderately rounded; quartzose; highly calcareous; at 5.2 m calcareous cannonball concretions; at 6.7 m is 60 cm thick bed of siltstone; siltstone is moderate reddish brown (10R4/6) with small (1-2 mm diameter) reduction spots of light greenish gray (5GY8/1); at 7.9 m is 30 cm thick siltstone-very fine sandstone which is moderate reddish brown (10R4/6) with reduction spots (up to 3 mm diameter) of light greenish gray (5GY8/1); at 8.2-9.15 m is siltstone with concretions; siltstone is moderate reddish brown (10R4/6); concretions are light greenish gray (5G8/1) and up to 1 x 1.5 cm; both are highly calcareous. **9.15**

2 Siltstone; pale reddish brown (10R5/4) and pale red (10R6/2); highly calcareous; microbrecciated; wood locally and NMMNH locality. **2.5**

1 Mudstone and sandstone; moderate reddish brown (10R4/6). **not measured**

REVUELTO CREEK, SECTION D

Measured in NE1/4 NE1/4 SE1/4 sec. 16, T10N, R33E.

unit lithology **thickness (m)**

Chinle Group: Bull Canyon Formation:

9 Sandstone; pale red (5R6/2); very fine grained; moderately sorted; moderately rounded; quartzose; highly calcareous; contorted bedding and ripple lamination. **0.15**

8 Mudstone; pale red purple (5RP4/2); not calcareous; with siltstone stringers as unit 7 in basal 7.5 cm, 1.2 cm thick at 22 cm, 1.2 cm thick at 30 cm and 1 cm at 33 cm; laterally the basal siltstone thickens and coarsens to a conglomerate. **1.68**

7 Mudstone and siltstone; mudstone is slightly silty; pale yellowish green (10GY7/2) to pale red purple (5RP6/2); not calcareous; thin siltstone at 1.5 m from base; siltstone is light greenish gray (5GY8/1); highly calcareous; at 3 m from base are concretions like unit 6; many vertebrate fossils and coprolites (NMMNH locality 1). **5.8**

6 Concretions; moderate yellowish green (10GY6/4) unweathered, light olive gray (5Y5/2) weathered; highly calcareous; rounded concretions up to 15 cm x 10 cm. **0.15**

5 Siltstone; pale green (5G7/2) to pale red purple (5RP6/2); highly calcareous; in upper 7.5 cm are concretions which are muddy siltstone pale yellowish green (10GY7/2); highly calcareous. **2.0**

4 Sandstone and conglomerate; conglomerate clasts are pale green (5G7/2) and olive gray (5Y3/2); up to 10 mm in diameter; sandstone is mottled grayish red (10R4/2) and grayish red purple (5RP4/2); very fine grained; quartzose; highly calcareous; basal 3 cm is all conglomerate and above this is sandstone with scattered floating conglomerate clasts; conglomerate extends laterally up to 1.2 m. **1.2**

3 Siltstone; grayish red purple (5RP4/2); highly calcareous. **1.1**

2 Siltstone; grayish red purple (5RP8/1); moderately calcareous; stringers every 10-15 cm of sandstone; light greenish gray (5G8/1); very fine grained; well sorted; moderately rounded; quartzose; highly calcareous. **2.2**

1 Silty mudstone; moderate reddish brown (10R4/6); highly calcareous; 12.2-13.7 m from base are thin (0.5 cm thick) sandstone stringers; sandstone is light greenish gray (5GY8/1); very fine grained; well sorted; well rounded; quartzose; highly calcareous; ripple laminar. **16.75**

THE MOENKOPI FORMATION IN EAST-CENTRAL NEW MEXICO: STRATIGRAPHY AND VERTEBRATE FAUNA

JÜRGEN A. BOY[1], RAINER R. SCHOCH[2], AND SPENCER G. LUCAS[3]
[1]Institut für Geowissenschaften, Lehreinheit Paläontologie, Johannes-Gutenberg-Universität Mainz, 55913 Mainz, Germany;
[2]Humboldt Universität zu Berlin, Zentralinstitut Museum für Naturkunde, Invalidenstrasse 43, 10115 Berlin, Germany;
[3]New Mexico Museum of Natural History, 1801 Mountain Road NW, Albuquerque, NM 87104

Abstract.—The Anton Chico Member of the Moenkopi Formation in east-central New Mexico is 38-46 m thick and is divided into three main units that differ in bedforms, bed thickness and abundance of conglomerates. The following lithofacies can be distinguished: (1) floodplain siltstones, (2) fine- to medium-grained sandstones, probably deposited in broad channels, and (3) intraformational conglomerates with clasts of sandstone, siltstone and pedogenic concretions that mostly represent channel lags. Eight vertebrate-fossil-bearing levels are recognized; two of them are in sediments of ephemeral water bodies and yield articulated fossils of amphibians, whereas the other six levels represent intraformational, conglomeratic channel-fill deposits. One conglomeratic horizon is locally rich in isolated bones of amphibians and reptiles, and contains rare, well-preserved skulls. The Anton Chico fauna includes stereospondyl amphibians (*Eocyclotosaurus*, cf. *Stanocephalosaurus*), both known from the Holbrook Member of the Moenkopi Formation in Arizona, and archosauromorph reptiles (prolacertiforms, erythrosuchids, rauisuchians), which locally vary in abundance.

INTRODUCTION

In east-central New Mexico, the oldest Triassic strata were originally referred to the Santa Rosa Formation (Rich, 1921; Darton, 1928). Gorman and Robeck (1946) divided the Santa Rosa Formation into four members, and made it clear that the lowest of these units is lithologically distinct, being composed of purplish-red instead of greenish-gray sandstones, and containing intraformational conglomerates in contrast to mixed conglomerates typical of the three higher members (Finch and Wright, 1983; Lucas and Hunt, 1987).

Lucas and Morales (1985) realized that the lowest member of the Santa Rosa Formation of Gorman and Robeck (1946) closely resembles the Holbrook Member of the Moenkopi Formation in northern Arizona. Lucas and Hunt (1987) thus removed the lowest member from the Santa Rosa Formation and referred it to the Moenkopi Formation, coining the term Anton Chico Member (originally a formation) for the east-central New Mexican facies.

Correlation of the Anton Chico Member with the Holbrook Member was established by the occurrence of the temnospondyl amphibian *Eocyclotosaurus*, the most characteristic vertebrate in the Anton Chico and Holbrook members (Lucas and Morales, 1985; Schoch, 2000), which is also known from the Upper Buntsandstein of central Europe (e.g., Kamphausen, 1989). Both the European and Arizona deposits are of Perovkan (early Anisian) age, suggesting a similar age for the Anton Chico Member (Lucas, 1998).

The stratigraphy, sedimentology and fauna of the Anton Chico Member are still poorly known, as only small outcrop areas have been briefly studied (e.g., Lucas and Hayden, 1989, 1991). A field project focusing on stratigraphy and vertebrate paleontology of the Anton Chico Member started in 1999, and is now being supported by the Deutsche Forschungsgemeinschaft. This article summarizes the current state of knowledge by presenting: (1) a detailed stratigraphy of the Anton Chico Member, discussing the vertebrate-bearing deposits; and (2) reporting briefly the recognized taxa from a study area that extends along the Pecos River from about Tecolotito to Colonias (Fig. 1)

LITHOSTRATIGRAPHY AND VERTEBRATE-FOSSIL DEPOSITS

Stratigraphic Section

The lithostratigraphy and sedimentology of the Anton Chico Member have not been studied in detail. We measured three complete sections that are between 38 and 46 m in thickness. The sites are (from east to west) southeast of Colonias, south of Anton Chico and north of Tecolotito (Fig. 1). Four additional partial sections cover the lower part, two sections the upper part, and several additional sections cover different parts of the member. A comparison of these sections reveals that the Anton Chico Member can be divided into three intervals.

In the lower interval, the Anton Chico Member consists of siltstones and fine-grained sandstones that may be horizontally laminated and contain rare calcrete-bearing paleosols in their basal part. Their top portion usually contains siltstone and fine-grained sandstone deposited in temporary floodplain ponds or lakes. The pond deposits are lithologically difficult to distinguish from surrounding floodplain deposits that contain coprolites and isolated amphibian bones (Fig. 2, level 5). The lake sediments are fine-grained, well laminated and devoid of vertebrate fossils. They contain subaquatically formed invertebrate tracks of the *Isopodichnus* type and wave ripples at their base (see Smith, 1993b for a similar case).

2. Fine- to medium-grained sandstones and rare coarse-grained sandstones either have indistinct bedding or trough cross-beds. Small-scale ripples are absent. Presumably, these strata formed in wide, rather shallow channels in low sinuosity river systems. Thin beds of this lithofacies may also represent crevasse channels and crevasse-splay deposits.

3. Intraformational conglomerates have clasts of fine sandstone, coarse siltstone and pedogenic calcrete. These beds often have an erosional base. Laterally, they are intercalated with medium- to coarse-grained sandstones bearing siltstone clasts, into which they may grade. Locally, they form basal lag deposits of channel sandstones. More frequently, they are shallow channel deposits

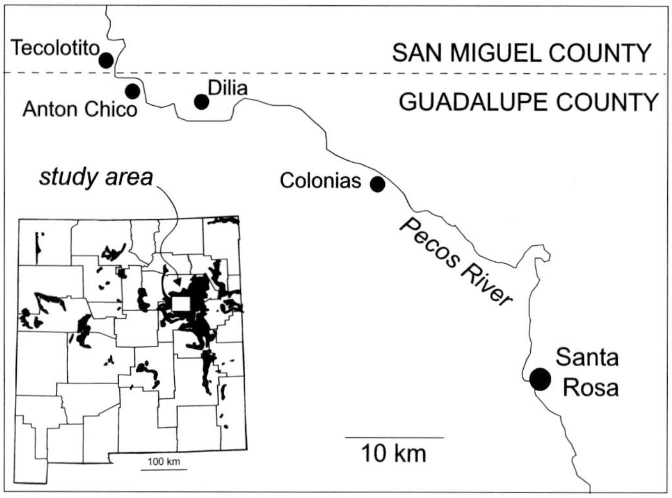

FIGURE 1. Distribution of Triassic strata in New Mexico (after Dane and Bachman, 1965), and detail of study area (from Tecolotito to Colonias) in San Miguel and Guadalupe Counties.

that may laterally intercalate with adjoining channel deposits, or they may overlie each other. The conglomerates usually, but not always, contain bones and vertebrate coprolites. In places, these components are accompanied by accumulations of black petrified (silicified) wood concentrated on single bedding planes.

Channelform sandstones dominate this facies. Overbank sediments represent only a small fraction of the section (around 10%), and conglomerates are confined to specific horizons. Many shallow channel deposits are filled with cross-bedded conglomerates or parallel-bedded overbank sediments. The general setting was that of a wide alluvial plain with branching, laterally shifting channels. Water levels changed according to long-term patterns, probably of seasonal origin, similar to megasequence 1 of the Pennsylvanian-Permian Cutler Formation in north-central New Mexico (Eberth and Miall, 1991).

The middle interval of the section is much more uniform than the lower. It consists of 1-4 m thick, fine- to coarse-grained sandstone layers. Bedforms are either indistinct, or large-scale cross-bedding that is partially trough-shaped and partially planar. In contrast to these wide-ranging channel-fills, silty overbank deposits are more localized, such as north of Tecolotito. Conglomerates are rare in this part of the section. At one site, conglomeratic lenses occur as lag deposits at the base of a channel sandstone. Only in the southern part of the study area are there abundant conglomeratic channel-fills. These may contain a few bones, but lack wood or coprolites (Fig. 2, level 6).

The upper part of the section is markedly thinner than the lower two parts and vertically more clearly subdivided than the middle part. There is a general trend from thin, cross-bedded fine-grained sandstones in the lower part to silty and fine-grained sandy overbank sediments in the upper part. Such fining-upward trends may have been produced by repeated cycles. Overbank deposits are horizontally laminated in their lower part, whereas in the upper part they lack clear stratification. Locally, especially in the southern part of the study area, the upper part of the Anton Chico Member may be rich in paleosols, as indicated by calcrete nodules and limonitic concretions. Conglomerates are laterally restricted, occurring in two different horizons. Bones were found in one place in interdigitating channel-fills (Fig. 2, level 7). In this part of the section, only one lacustrine deposit (near Tecolotito) was recognized, which is rich in vertebrates. Despite extensive prospecting, no other lacustrine horizons were found at this level. The predominance of overbank deposits suggests strong lateral restriction of channels between broad floodplains that formed interchannel flood basins. Sedimentation rates may have slowed down relative to accommodation as indicated by paleosols. The few available data match best with anastomosing and meandering river systems. These strata somewhat resemble fining-upward cycles (megasequence 2) of the Cutler Formation (Eberth and Miall, 1991), and similar cycles in the Molteno Formation of South Africa (Anderson et al., 1998).

OCCURRENCE OF FOSSIL VERTEBRATES

Fossil vertebrates are confined to discrete stratigraphic levels in the Anton Chico Member. In addition to the two formerly known levels, we were able to recognize six more, which are, however, less rich in bones (Fig. 2).

Level 1—in the southern part of the study area, small, thin conglomerate lenses contain bone fragments. In the northern part of the study area, larger lenses are locally rich in vertebrate bones and contain few coprolites and occasional wood accumulations.

Level 2—mainly in the northern part of the study area, thin conglomerate lenses locally contain bones, coprolites and wood.

Level 3—in the southern part of the study area, a reworked soil horizon, consisting of coarse silt and filled with pedogenic concretions and occasional silty/sandy clasts, is rich in bone fragments. In the northern part of the study area, a complex system of thin, interfingering channel fills persists over long distances and is the main source of bone in conglomerates. Bones are much more frequent than coprolites. Some large, irregularly dispersed, wood accumulations are found, and invertebrate trace fossils are also present.

Level 4—only present in the middle part of the study area, several shallow channel fills and conglomerate lenses are locally accompanied by black, fine- to coarse-grained, carbon-rich sandstones. Coprolites are rare, bones even more so, but if present they are well preserved and complete; some wood is present.

Level 5—broad outcrops of conglomerate are generally barren of fossils but sometimes have a top consisting of thin, gray to light brown sandstone and siltstone. The latter yields numerous coprolites and amphibian bones and, slightly higher in this unit, wood. This level is also represented by another, stratigraphically unclear locality, where an erythrosuchid was found (Lucas et al., 1998).

Level 6—only known from the southern part of the study area, it consists of single channel fills, a few meters wide and with a maximum thickness of 1.3 m. These fills include conglomerates that rarely contain bone fragments.

Level 7—in the southern and middle part of the area, conglomerate lenses normally do not yield fossils. At one site the lenses are relatively thick (2 m), formed by a local channel complex that yields isolated bones and small archosauromorph teeth.

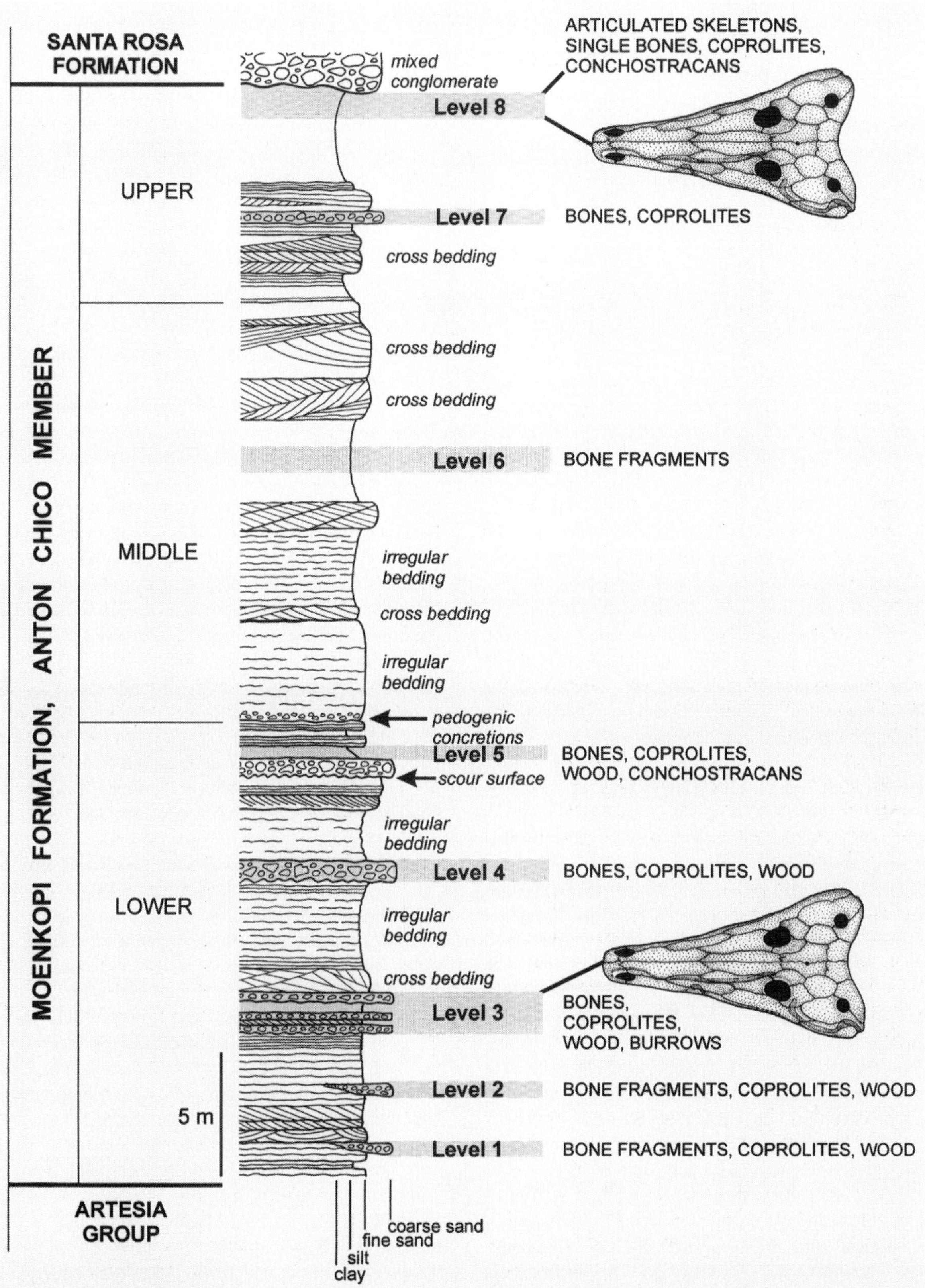

FIGURE 2. Stratigraphic section of the Anton Chico Member of the Moenkopi Formation in east-central New Mexico based on numerous field observations and a complete section measured on the Anton Chico Land Grant. The two skulls (*Eocyclotosaurus wellesi*) indicate the main vertebrate-bearing horizons.

Level 8—the top of the Anton Chico Member usually consists of gray to greenish siltstones and sandstones of varying thickness. These may contain thin conglomerate horizons, together with others rich in pedogenic pebbles. At one location, fine sandstones and coarse siltstones contain a rich deposit of coprolites, isolated bones and disarticulated skeletons.

VERTEBRATE FOSSIL DEPOSITS

The different fossil levels can be grouped into two types of deposit--ephemeral water bodies (levels 5 and 8) and conglomeratic channel fills (levels 1-4 and 6-7).

Ephemeral Water Bodies (Levels 5 and 8)

Short-lived ponds on floodplains were supplied with silty and fine sandy sediment during flooding phases. Compared with similar deposits in the Lower Permian (Sander 1987, 1989), the sediment in the Anton Chico Member is coarser, and evidence of a reducing environment (e.g., dark color, plant accumulation) is absent. This suggests only temporary water coverage and close proximity to a larger channel. Lithologically, there are no substantial differences from floodplain sediments. The frequency of aquatic tetrapods (*Eocyclotosaurus*), coprolites and conchostracans indicates quiet water conditions. Skeletons are likely to be disarticulated in warm climates and shallow water conditions (Wuttke, 1983; Elder and Smith, 1984). The sporadic occurrence of reptiles may be explained by reworking; this agrees with the much smaller average size of these elements compared to the amphibian remains, and with the predominance of teeth.

Conglomeratic Channel Fills (Levels 1-4 and 6-7)

Vertebrate-bearing, channel-fill deposits may be derived either from abandoned channels containing water, or crevasse-splay channels (Behrensmeyer, 1988). Incised, thin, lenticular channel-fills are amalgamated to form complex channel systems, reaching a thickness of more than 2 m. Yet, the precise distribution and preservation of vertebrate remains in channel fills has not been studied so far. Locally, bones and bone fragments are concentrated in the tops of uppermost conglomeratic units. However, the vertebrate fossil content varies in different conglomeratic horizons. The abundance of coprolites also varies considerably; most appear to have been produced by temnospondyl amphibians. Also, the relative frequency of amphibian and reptile remains varies from site to site.

Typically, intraformational conglomerates in the Anton Chico Member bear vertebrate bones (Lucas and Morales, 1985; Lucas and Hunt, 1987). In most cases, these are interpreted as channel-lag deposits, in which current orientation and mechanical wear preserved only the most resistant bones within a certain size range. Examples have been documented by Olson and Barghusen (1962) in the Lower Permian of Oklahoma and Turner (1981) in the Upper Permian of South Africa. In addition, Behrensmeyer (1988) lists thin intraformational conglomerates embedded in fine-grained channel-fills as containing vertebrate Lagerstaetten; their formation is explained by weathering and limited reworking in abandoned channels during sporadic flooding events. However, in channel-fill deposits of the Anton Chico Member, conglomerates predominate, whereas fine-grained clastics are absent. An alternative possibility is that these conglomerates are parts of crevasse-splay channels. This would match the first phase in the displacement of main channels, in which shallow, unstable channels form (Smith et al., 1989). The sediments of these unstable channels have been reported to contain coarsening-upward sequences (Sander, 1989), whereas deeper channels should be characterized by a fining-upward trend (Eberth and Miall, 1991). The channel fills of the Anton Chico Member show only a faint fining-upward trend, if any.

Behrensmeyer (1988) gave a generalized synopsis of the preservation of vertebrates in fluvial deposits. She arrived at the following results:

1. In wide-ranging sheet sandstones, formed by main channels, vertebrate remains are sparse.
2. In fine-grained overbank deposits, vertebrates are also rare. There, however, they may occur in local accumulations or clusters, which sometimes include articulated specimens.
3. Rich deposits of vertebrates are found in channel lags of active channels. Bones are usually sorted and mostly disarticulated, with corroded edges, and generally very fragmentary.
4. Vertebrate accumulations frequently occur in abandoned channels above channel-lags. There are two principal categories: (i) sorted and fragmented bones in nodule and mudclast conglomerates, and (ii) better preserved, partially articulated material in fine-grained units of channel-fills.

Our present findings confirm Behrensmeyer's (1988) interpretation of vertebrate deposits; her types (1) and (2) are common in the Anton Chico Member, while (3) is rare and (4, i) is absent. Type (4, ii) is the most abundant type in our study area. This is not surprising, because complex channel-fills are dominated by conglomerates in the Anton Chico Member.

MATERIAL AND VERTEBRATE FAUNA

The Moenkopi Formation was first recognized in New Mexico by S. G. Lucas in 1984 while collecting amphibian skull material of the genus *Eocyclotosaurus* (Lucas and Morales, 1985), known from the lower Anisian of central and western Europe (Welles and Cosgriff, 1965; Schoch and Milner, 2000). In Spring 1999, Lucas and Schoch collected material from various new localities that had been discovered by a private collector, A. Velasquez of Anton Chico. This material consisted of both isolated bones in conglomerates and articulated material in lake deposits. In Fall 2000, a field party of four people collected much more material in conglomerates, which is currently undergoing preparation. These finds form the basis of the present assessment of the vertebrate fauna of the Anton Chico Member. As the bulk of the vertebrate material is still to be prepared, we give only an incomplete, provisional list of taxa, which will probably be revised substantially.

Fishes

Fish skeletons and isolated bones have not been found in the Anton Chico Member. There are, however, traces of fishes in coprolites, documenting the presence of actinopterygians. Level 3 contains anisopolar, spiral-shaped coprolites (*sensu* McAllister, 1996). This coprolite type is known from the Permo-Carboniferous of central Europe, where it is assigned to xenacanthid sharks (Boy, 1998). The Upper Triassic Chinle Group has also produced this type of coprolite, and they were also referred to xenacanthids by Hunt (1992). Both curled and uncurled coprolites in the Anton Chico Member contain thick, unsculptured scales of paleoniscoid actinopterygians. Together with these, we identified conchostracans and ostracods in coprolites. In the better-studied Moenkopi Formation deposits of Arizona, fishes are relatively rare--they occur in sediments rather than coprolites, and include sharks, dipnoans and paleoniscoids (McKee, 1954). The lack of fish remains in Anton Chico sediments may be explained by the following reasons:

1. Collection of vertebrates focused on larger bones rather than small fragments. Small and unspectacular fish elements and scales may have been overlooked, which is the most likely explanation.
2. Conglomerates are likely to lack delicate elements such as fish bones.
3. Fishes may have had difficulties in reaching the higher parts of river systems, which are obviously documented in the Anton Chico Member.

Amphibians

Amphibians are the most abundant vertebrates in the Holbrook Member of the Moenkopi Formation in northern Arizona, where they reach a high morphological and taxonomic diversity (Welles, 1947; Welles and Cosgriff, 1965; Morales, 1987; Schoch, 2000). In the Anton Chico Member, they are abundant as specimens but taxonomically less diverse. At present, only capitosaur stereospondyls are known from the Anton Chico Member.

In conglomerates, amphibians are frequent, but not as dominant as in lake deposits. This may be explained by a longer transport of skeletons and single bones as documented by wear, or alternatively by greater distance from the actual habitats of stereospondyls. The presence and frequency of terrestrial forms such as archosaurs and prolacertiforms certainly demonstrates the proximity of a truly terrestrial ecosystem.

In both channel and lake deposits of the Anton Chico Member there is a wide range of amphibian sizes, from 100-400 mm skull length. In both types of deposits, a diversity of skeletal elements has been found. While the lake deposits contain semi-articulated specimens accumulated in great quantities, the conglomerates largely bear isolated bones, skull fragments or complete skulls in varying abundances. Sorting by transport certainly played a role, as the frequency of planar elements is clearly higher at some localities, whereas cylindrical vertebrae and limb fragments are more abundant at others.

Heylerosauridae

A new species of *Eocyclotosaurus* is the most common amphibian in the Anton Chico Member. This species is closely related to *E. wellesi,* recently described from the Holbrook Member of Arizona (Schoch, 2000). It will be described and discussed in a separate, more detailed article. *Eocyclotosaurus* occurs in both river channel and lake deposits, and in the conglomerates it varies locally in frequency. In lake deposits it is abundant, forming nearly 90% of the assemblage. *Eocyclotosaurus* is characterized by a long and slender snout, small orbits set relatively laterally and a broad posterior skull with closed otic fenestrae. The snout of the Anton Chico morph is much more slender than in the European species of *Eocyclotosaurus*. Articulated finds show that the mandible is unusually low with a large meckelian window, and the axial skeleton includes crescent-shaped intercentra typical of small, primitive capitosauroids.

Paracyclotosauridae

A second amphibian genus is known only from isolated and fragmentary skull and postcranial material. The distinctness of these finds is based on the morphology of the posterior skull table (especially the squamosal embayment) and the vertebral centra (intercentra). The cranial remains demonstrate this taxon to be an open-notched capitosauroid, similar to *Stanocephalosaurus* (Paracyclotosauridae) from the Holbrook Member of Arizona (Schoch and Milner, 2000). This is suggested by skull fragments of the otic region, the mandible and anterior palate, the latter being most diagnostic in capitosauroids. A second type of vertebral centrum is characterized by being more massive and a tendency to form a quasi-stereospondyl condition (formation of a closed disk), not by means of dorsal closure of the intercentral crescent, but by co-ossification of pleurocentra with each other and finally with the intercentrum. This is obviously a derived condition not found in other genera. Clearly, these intercentra are from a genus similar to *Stanocephalosaurus*, because they have the same relative position of the parapophysis (Schoch, 1999).

Unidentified large capitosauroids

Very large bones in the conglomerates suggest the presence of much larger stereospondyls. Among these finds are skull, mandible and girdle fragments suggesting a morphology different from that of the aforementioned genera. At this preliminary stage it is impossible to clarify the assignment of these large amphibians; they may either represent a third taxon, or merely large specimens of one or both other taxa. *Eocyclotosaurus wellesi* from Arizona is known to have grown much larger than its European relatives (Schoch, 2000).

Reptiles

The Anton Chico Member is unique among all Moenkopi members in having numerous reptile fossils. They clearly exceed the frequency of reptiles in the Holbrook Member, and at many sites (in intraformational conglomerates), reptile remains outnumber those of amphibians. The reptiles also are the largest vertebrate remains in the Anton Chico Member, reaching limb bone dimensions of more than half a meter.

The conglomerates have yielded numerous vertebral centra, limb bones and girdle elements of a range of reptiles differing in size and morphology. Skull fragments, teeth and neural arches are generally much rarer and more fragmentary. This largely isolated material is only diagnostic at the family level. The significance of this material is nevertheless remarkable: it is the only Lower/Middle Triassic archosaur material gathered in great quantity, and for the first time a North American terrestrial fauna may be compared to that of other continents.

Erythrosuchid

In 1997, a partially articulated skeleton of a very large archosaur was found in a lake deposit near Dilia, east of Anton Chico (Lucas et al., 1998). Revisting the site yielded more material, especially vertebrae and osteoderms, which reach the dimensions of the South African genus *Erythrosuchus*. The locality was on the slope of a shallow gully where bones had been weathering out of a light greenish-gray shale rich in plant debris, conchostracans (estherians) and vertebrate coprolites. We interpret this as an especially quiet-water phase of a small lake that must have persisted over a long time period. The erythrosuchid bones had obviously been under water, as the mandibular alveoli contain bivalves in life position.

Medium-sized primitive archosaur

The conglomerates are rich in vertebrae (centra) and appendicular elements of medium-sized archosaurs comparable to *Ticinosuchus ferox* in size and gross morphology. Humeri, femora, scapulae, ilia and fragmentary cranial elements suggest that they represent a somewhat more primitive archosaur than *Ticinosuchus*, but clearly more derived than *Erythrosuchus* and *Shansisuchus*. They form more than 70% of the reptile remains found in the conglomerates. The ilia, in particular, have a rather plesiomorphic morphology similar to the Russian genus *Vjushkovia*.

Advanced rauisuchians

A few isolated postcranial remains are distinct from the bulk of the archosaur material, and document the presence of a second taxon. The most definitive evidence is ischia that are remarkably similar to poposaurids and, especially, chatterjeeids (Long and Murry, 1995). The presence of such advanced archosaurs would be highly significant for the understanding of the radiation of early archosaurs, but at present the remains are too fragmentary to permit definite conclusions.

Prolacertiforms

Elongated neck vertebrae and thoracic vertebral centra suggest the presence of plesiomorphic, short-necked prolacertiforms. Only vertebrae have so far been identified, and they are from animals about the size of *Protorosaurus*. These finds are much rarer than the archosaur vertebrae and limb elements, but it is unclear which of the smallest bones, such as tail vertebrae and small limb elements, might belong to this taxon.

Reptile teeth

Teeth are extraordinarily rare, except in the floodplain pond deposits. In an amphibian-dominated pond deposit, several small (1-2 cm long) archosaur teeth were found. They are short and moderately wide, resembling phytosaur teeth. In most channel conglomerates, only jaw fragments have yielded teeth so far, and they are from rauisuchians. A small conglomerate locally rich in unusually small bones has also produced some archosaur teeth and one jaw fragment with a dentition resembling procolophonids that needs further investigation.

Reptile bones are abundant in the conglomerates, especially in Level 3. Locally they are more abundant than amphibians. The most common finds are vertebral centra and complete tail vertebrae of small archosaurs. Girdle elements are rarer and fragmentary, and limb bones are mostly present as smaller fragments of robust proximal and distal ends. Delicate skull fragments, neural arches and ribs are much more rare.

ACKNOWLEDGMENTS

We thank A. Velasquez and the owners of the Anton Chico Land Grant for their generosity, which made this work possible. Pete Reser provided valuable logistical support. Andrew Heckert, Adrian Hunt and Kate Zeigler reviewed the manuscript.

REFERENCES

Anderson, J. M., Anderson, H. M., and Cruickshank, A. R. J., 1998, Late Triassic ecosystems of the Molteno/Lower Elliot biome of Southern Africa: Palaeontology, v. 41, p. 387-421.

Behrensmeyer, A. K., 1988, Vertebrate preservation in fluvial channels: Palaeogeography, Palaeoclimatology, Palaeoecology, v. 63, p. 183-199.

Behrensmeyer, A. K., and Hook, R. W., 1992, Paleoenvironmental contexts of taphonomic models, *in* Behrensmeyer, A. K., Damuth, J. D., DiMichele, W. A., Potts, R., Sues, H.-D., and Wing, S. L., eds., Terrestrial Ecosystems through Time. Evolutionary Paleoecology of Terrestrial Plants and Animals: Chicago, University of Chicago Press, p. 14-136.

Boy, J. A., 1998, Möglichkeiten und Grenzen einer Ökosystem-Rekonstruktion am Beispiel des spätpaläozoischen lakustrinen Paläo-Ökosystems. 1. Theoretische und methodische Grundlagen: Paläontologische Zeitschrift, v. 72, p. 207-240.

Dane, C. H., and Bachman, G. O., 1965, Geologic map of New Mexico: Washington, D. C., U.S. Geological Survey, 2 sheets, scale 1:500,000.

Darton, N. H., 1928, "Red beds" and associated formations in New Mexico, with an outline of the geology of the state: U. S. Geological Survey, Bulletin 794, 372 p.

Eberth, D. A., and Miall, A. D., 1991, Stratigraphy, sedimentology and evolution of a vertebrate-bearing, braided to anastomosed fluvial system, Cutler Formation (Permian-Pennsylvanian), north-central New Mexico: Sedimentary Geology, v. 72, p. 225-252.

Elder, R. L., and Smith, G. R., 1984, Fish taphonomy and paleoecology: Geobios, Mémoire Spéciale 8, p. 287-291.

Finch, W. I., and Wright, J. C., 1983, Measured stratigraphic sections of uranium-bearing Upper Triassic rocks of the Dockum basin, eastern New Mexico, West Texas, and the Oklahoma Panhandle with brief discussion of stratigraphic problems: U. S. Geological Survey, Open-file Report 83-701, 118 p.

Gorman, J. M., and Robeck, R. C., 1946, Geology and asphalt deposits of north-central Guadalupe County, New Mexico: U. S. Geological Survey, Oil and Gas Investigations, Preliminary Map OM-44, scale 1:62,500.

Hunt, A. P., 1992, Late Pennsylvanian coprolites from the Kinney Brick Quarry, central New Mexico, with notes on the classification and utility of coprolites: New Mexico Bureau of Mines and Mineral Resources, Bulletin 138, p. 221-229.

Kamphausen, D., 1989, Die Schädel von *Eocyclotosaurus woschmidti* Ortlam (Amphibia, Stegocephalia) aus dem Oberen Buntsandstein (Trias) des Schwarzwaldes (SW-Deutschland): Stuttgarter Beiträge zur Naturkunde B, no. 149, 65 p.

Long, R. A., and Murry, P. A., 1995, Late Triassic (Carnian and Norian) tetrapods from the southwestern United States: New Mexico Museum of Natural History and Science Bulletin, v. 4, 254 p.

Lucas, S. G., 1998, Global Triassic tetrapod biostratigraphy and biochronology: Palaeogeography, Palaeoclimatology, Palaeoecology, v. 143, p. 347-384.

Lucas, S. G., and Hayden, S. N., 1989, Triassic stratigraphy of west-central New Mexico: New Mexico Geological Society 40th Field Conference Guidebook, p. 191-211.

Lucas, S. G., and Hayden, S. N., 1991,Type section of the Permian Bernal Formation and the Permian-Triassic boundary in north-central New Mexico: New Mexico Geology, v. 13, p. 9–15.

Lucas, S. G., and Hunt, A. P., 1987, Stratigraphy of the Anton Chico and Santa Rosa formations, Triassic of east-central New Mexico: Journal of the Arizona-Nevada Academy of Science, v. 22, p. 21-33.

Lucas, S. G. and Morales, M. 1985. Middle Triassic amphibian from the basal Santa Rosa Formation, east-central New Mexico. New Mexico Geological Society, 36th Field Conference, Guidebook, p. 171-184.

Lucas, S. G., Estep, J. W., and Hunt, A. P., 1998, Giant erythrosuchid archosaur from the Middle Triassic Moenkopi Formation, east-central New Mexico: Geological Society of America Abstracts with Programs, v. 30, no. 6, p. 13-14.

McAllister, J., 1996, Coprolitic remains from the Devonian Escuminac Formation, *in* Schultze, H.-P. and Cloutier, R., eds., Devonian fishes and plants of Miguasha, Quebec, Canada:. München, Pfeil, p. 328-347.

McKee, E. D., 1954, Stratigraphy and history of the Moenkopi Formation of Triassic age: Geological Society of America, Memoir 61, 133 p.

Morales, M., 1987, Terrestrial fauna and flora from the Triassic Moenkopi Formation of the southwestern United States: Journal of the Arizona-Nevada Academy of Science, v. 22, p. 1-19.

Olson, E. C., and Barghusen, H., 1962, Permian vertebrates from Oklahoma and Texas. Part I. Vertebrates from the Flowerpot Formation, Permian of Oklahoma: Oklahoma Geological Survey, Circular 59, 48 p.

Rich, J. L., 1921, The stratigraphy of eastern New Mexico – a correction: American Journal of Science, v. 2, p. 295-298.

Sander, P. M., 1987, Taphonomy of the Lower Permian Geraldine bonebed in Archer County, Texas: Palaeogeography, Palaeoclimatology, Palaeoecology, v. 61, p. 221-236.

Sander, P. M., 1989, Early Permian depositional environments and pond bonebeds in central Archer County, Texas: Palaeogeography, Palaeoclimatology, Palaeoecology, v. 69, p. 1-21.

Schoch, R. R., 1999, Comparative osteology of *Mastodonsaurus giganteus* (Jaeger, 1828) from the Middle Triassic (Lettenkeuper: Longobardian) of Germany (Baden-Württemberg, Bayern, Thüringen): Stuttgarter Beiträge zur Naturkunde B, no. 278, 175 p.

Schoch, R. R., 2000, The status and osteology of two new cyclotosaurid amphibians from the upper Moenkopi Formation of Arizona (Amphibia: Temnospondyli; Middle Triassic): Neues Jahrbuch für Geologie und Paläontologie Abhandlungen, v. 216, p. 387-411.

Schoch, R. R., and Milner, A. R., 2000, Stereospondyli: Handbuch der Paläoherpetologie, v. 3B, p. 1-203.

Smith, N. D., Cross, T. A., Dufficy, J. P., and Clough, S. R., 1989, Anatomy of an avulsion: Sedimentology, v. 36, p. 1-23.

Smith, R. M. H., 1993a, Vertebrate taphonomy of Late Permian floodplain deposits in the southwestern Karoo Basin of South Africa: Palaios, v. 8, p. 45-67.

Smith, R. M. H., 1993b, Sedimentology and ichnology of floodplain paleosurfaces in the Beaufort Group (Late Permian), Karoo Sequence, South Africa: Palaios, v. 8, p. 339-357.

Smith, R., and Kitching, J., 1997, Sedimentology and vertebrate taphonomy of the *Tritylodon* acme zone: A reworked palaeosol in the Lower Jurassic Elliot Formation, Karoo Supergroup, South Africa: Palaeogeography, Palaeoclimatology, Palaeoecology, v. 131, p. 29-50.

Turner, B. R., 1981, The occurrence, origin and stratigraphic significance of bone-bearing mudstone pellet conglomerates from the Beaufort Group in the Jansenville District, Cape Province, South Africa: Palaeontologica Africana, v. 24, p. 632-673.

Welles, S. P., 1947, Vertebrates from the upper Moenkopi Formation of northern Arizona: University of California Publications in Geological Sciences, v. 27, p. 241-289.

Welles, S. P., and Cosgriff, J., 1965, A revision of the labyrinthodont family Capitosauridae: University of California Publications in Geological Sciences, v. 54, p. 1-148.

Wuttke, M., 1983, Aktualistische Studien über den Zerfall von Wirbeltieren. Teil I. Anura: Senckenbergiana Lethaea, v. 64, p. 529-560.

At Sierrita de la Cruz, north of Amarillo, Texas, badlands are developed in the Tecovas (slopes) and Trujillo (sandstone cap) formations of the Chinle Group. These Upper Triassic strata are riverine floodplain and channel deposits that yield one of the world's most important fossil records of Late Triassic terrestrial life.

THE BENNETTITALEAN LEAF *"ZAMITES" POWELLII* FROM THE MIDDLE TRIASSIC MOENKOPI FORMATION, EAST-CENTRAL NEW MEXICO

SPENCER G. LUCAS
New Mexico Museum of Natural History, 1801 Mountain Rd. NW, Albuquerque, NM 87104

Abstract—Two localities in Middle Triassic (lower Anisian) strata of the Moenkopi Formation (Anton Chico Member) in east-central New Mexico yield numerous fossils of the bennettitalean *"Zamites" powellii* Fontaine. This record extends the temporal distribution of *"Z." powellii* from the late Carnian to the early Anisian, eliminating its use as a late Carnian index taxon. It also indicates that at least one Triassic bennettitalean taxon has a longer temporal range than previously known, and that the diversity of Middle Triassic bennettitaleans is not well documented.

INTRODUCTION

In the western United States, an extensive paleofora of Late Triassic age has long been known from nonmarine red beds of the Chinle Group (e.g., Ash, 1989; Lucas, 1995). However, the underlying Lower-Middle Triassic red beds of the Moenkopi Formation have to date yielded plants from only one locality in Arizona (Ash and Morales, 1993). This locality, in the lower Anisian Holbrook Member of the Moenkopi Formation, yielded the fungus *Polyporites*, the sphenophyte *Neocalamites*, the conifer *Araucarioxylon* and a new tree fern. With the exception of the new tree fern, these taxa are known from the Upper Triassic Chinle Group (Ash and Morales, 1993). Here, I describe a second Moenkopi Formation fossil plant occurrence that also yields a taxon well known from the Chinle Group, the bennettitalean leaf *"Zamites" powellii*. In this article, NMMNH = New Mexico Museum of Natural History, Albuquerque.

PROVENANCE

Alex Velásquez discovered numerous fossil leaves of *"Zamites" powellii* at two localities just northeast of Anton Chico in Guadalupe County, New Mexico (Fig. 1). Both localities are at the same stratigraphic level in the Anton Chico Member of the Moenkopi Formation of Lucas and Hunt (1987). These sites are NMMNH localities 3643 and 3644. The plant-bearing horizon is a 2-m thick, grayish red, ripple-laminated sandstone in approximately the middle of the Moenkopi section (Fig. 1). Nearby vertebrate-fossil localities that are stratigraphically higher in the Moenkopi Formation yield the temnospondyl *Eocyclotosaurus* (Lucas and Morales, 1985; Lucas and Hunt, 1987; Morales, 1987). This taxon is an index fossil of the Perovkan land-vertebrate faunachron of Lucas (1998), and this establishes the early Anisian age of the *Zamites powellii* localities. Indeed, the Anton Chico Member of the Moenkopi Formation in east-central New Mexico is a correlative of the early Anisian Holbrook Member of the Moenkopi Formation in eastern Arizona (Morales, 1987).

IDENTIFICATION

The NMMNH collection contains about 30 leaves from NMMNH localities 3643 and 3644 preserved as oxidized impressions. These fossils are catalogued as NMMNH P-20310, 33254 and 33279 through 33291. These leaves (Fig. 2) belong to a single taxon and are pinnate, ranging in length from 9 to 21 cm, and from 5 to 8 cm wide. The linear pinnae (leaflets) have essentially parallel margins and truncated apices (they are rectangular) and parallel veins that end at the pinna apices. The bases of at least some of the pinnae are contracted (e.g., Fig. 2F). In all features, these leaves closely resemble previously described specimens of *"Zamites" powellii* Fontaine (cf. Ash, 1975), to which they are assigned.

The name *Zamites* is placed here in quotation marks because of taxonomic problems that surround *Zamites powellii sensu* Ash (1975) that cannot be resolved here. Thus, *Zamites* has long been used as a broadly defined form genus, and not all species assigned to it, including *"Z." powellii* of Ash (1975), necessarily belong here. Indeed, not all taxa assigned to *Zamites* are necessarily benettitalean (J. Knaus , written commun., 2001). Recently, Weber and Zamudio-Varela (1995) proposed a new benettitalean genus, *Laurozamites*, for specimens from the Carnian of Sonora, Mexico. They reassigned *"Z." powellii sensu* Ash to *Laurozamites*. However, this reassignment may not be justified (B. Axsmith, written commun., 2000). Clearly, more systematic study is needed to resolve the taxonomic status of *"Z." powellii*. What is important here is that a plant morphotype previously thought characteristic of Chinle paleofloras is now known from the much older Moenkopi Formation.

DISCUSSION

"Zamites" powellii is one of the most common fossil plants found in the lower part of the Upper Triassic Chinle Group in the Western Unites States (Ash, 1975, 1989; Lucas, 1995). Ash (1980) indicated that this taxon is found in both his *Eoginkgoites* and *Dinophyton* Zones, which are of late Carnian age. *"Zamites" powellii* is also present in correlative strata of the Newark Supergroup—the Pekin Formation in North Carolina (Ash, 1975) and the Stockton Formation in Pennsylvania (Axsmith and Kroehler, 1989). The record of *"Z." powellii* in the Moenkopi Formation of east-central New Mexico thus represents a considerable extension of the temporal range of this taxon, back to the early Anisian. The use of *"Z." powellii* as an indicator of late Carnian time thus needs to be abandoned.

Ash (1975, 1980) noted that *"Zamites" powellii* is abundant in strata of the Chinle Group and apparently was one of the domi-

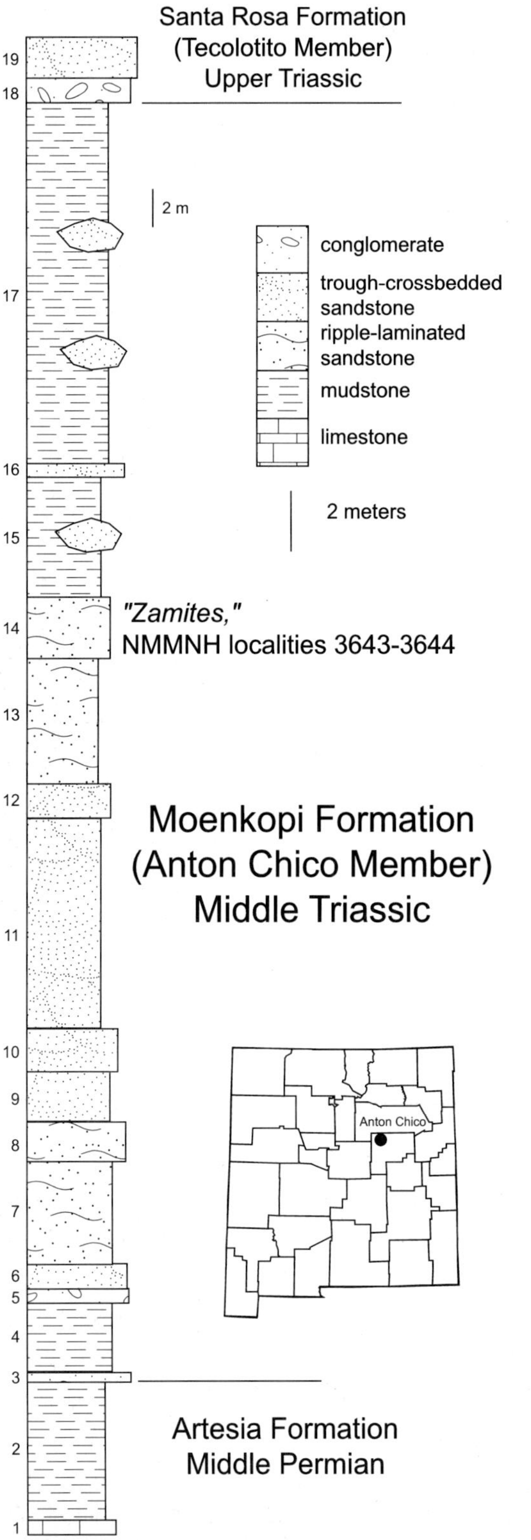

FIGURE 1. Stratigraphic section and location of the Moenkopi Formation "*Zamites*" locality near Anton Chico. See Appendix for description of measured section.

nants of the Chinle paleoflora. Its abundance near Anton Chico may also mean that it too was a dominant of the Moenkopi paleoflora. Indeed, the fossil plants from the Holbrook Member of the Moenkopi Formation in Arizona (Ash and Morales, 1993) and the correlative Anton Chico Member in New Mexico are mostly taxa known from the Chinle Group. This may suggest a long-lived (Anisian-Carnian) chronoflora probably was present near riverine environments in the western United States during the Middle-Late Triassic.

Crane (1986) and Axsmith et al. (1995) drew attention to the diversity and unusual morphology of some Late Triassic bennettitaleans, implying a much earlier origination of the group. The Middle Triassic record of "*Zamites*" indicates a much greater longevity of one Late Triassic bennettitatlean than previously reported. Indeed, the morphology of some Late Triassic bennettitaleans may not be so unusual if they are understood as members of a bennettitalean diversification whose roots are much older, in the Early or Middle Triassic.

ACKNOWLEDGMENTS

Alex Velásquez generously showed me the fossil plant localities he discovered, and he, Peter Reser and John Estep assisted in the field. Brian Axsmith, Lisa Boucher and Jane Knaus provided helpful comments on an earlier version of the manuscript.

REFERENCES

Ash, S.R., 1975, *Zamites poewlli* [sic] and its distribution in the Upper Triassic of North America: Palaeontographica B, v. 149, p. 139-152.

Ash, S.R., 1980, Upper Triassic floral zones of North America; *in* Dilcher, D.L. and Taylor, T. N., eds., Biostratigraphy of fossil plants: Stroudsburg, Dowden, Hutchinson & Ross, Inc., p. 153-170.

Ash, S.R., 1989, A catalog of Upper Triassic plant megafossils of the western United States through 1988; *in* Lucas, S.G. and Hunt, A.P., eds., Dawn of the age of dinosaurs in the American Southwest: Albuquerque, New Mexico Museum of Natural History, p. 189-222.

Ash, S. R., and Morales, M., 1993, Anisian plants from Arizona: The oldest Mesozoic megaflora in North America: New Mexico Museum of Natural History and Science Bulletin 3, p. 27-29.

Axsmith, B. J., and Kroehler, P. A., 1989, Upper Triassic *Dinophyton* zone plant fossils from the Stockton Formation in southeastern Pennsylvania: The Mosasaur, v. 4, p. 45-47.

Axsmith, B. J., Taylor, T. N., Delevoryas, T., and Hope, R. C., 1995, A new species of *Eoginkgoites* from the Upper Triassic of North Carolina: Review of Palaeobotany and Palynology, v. 85, p. 189-198.

Crane, P. R., 1986, The morphology and relationships of Bennettitales; *in* Spicer, R. A. and Thomas, B. A., eds., Systematic and taxonomic approaches to palaeobotany: The Systematics Association Special Volume 31, p. 163-175.

Lucas, S. G., 1995, Upper Triassic Chinle Group, western United States: A nonmarine standard for Late Triassic time; *in* Dickins, J. M., Yang, Z., Yin, H., Lucas, S. G. and Acharyya, S. K., eds., Late Palaeozoic and early Mesozoic circum-Pacific events and their global correlation: Cambridge, Cambridge University Press, p. 209-228.

Lucas, S.G., 1998, Global Triassic tetrapod biostratigraphy and biochronology: Palaeogeography, Palaeoclimatology, Palaeoecology, v. 143, p. 347-384.

Lucas, S.G., and Hunt, A.P., 1987, Stratigraphy of the Anton Chico and Santa Rosa formations, Triassic of east-central New Mexico: Journal of the Arizona-Nevada Academy of Science, v. 22, p. 21-33.

Lucas, S.G. and Morales, M., 1985, Middle Triassic amphibian from basal Santa Rosa Formation, east-central New Mexico: New Mexico Geological Society 36th Field Conference, Guidebook, p. 56-58.

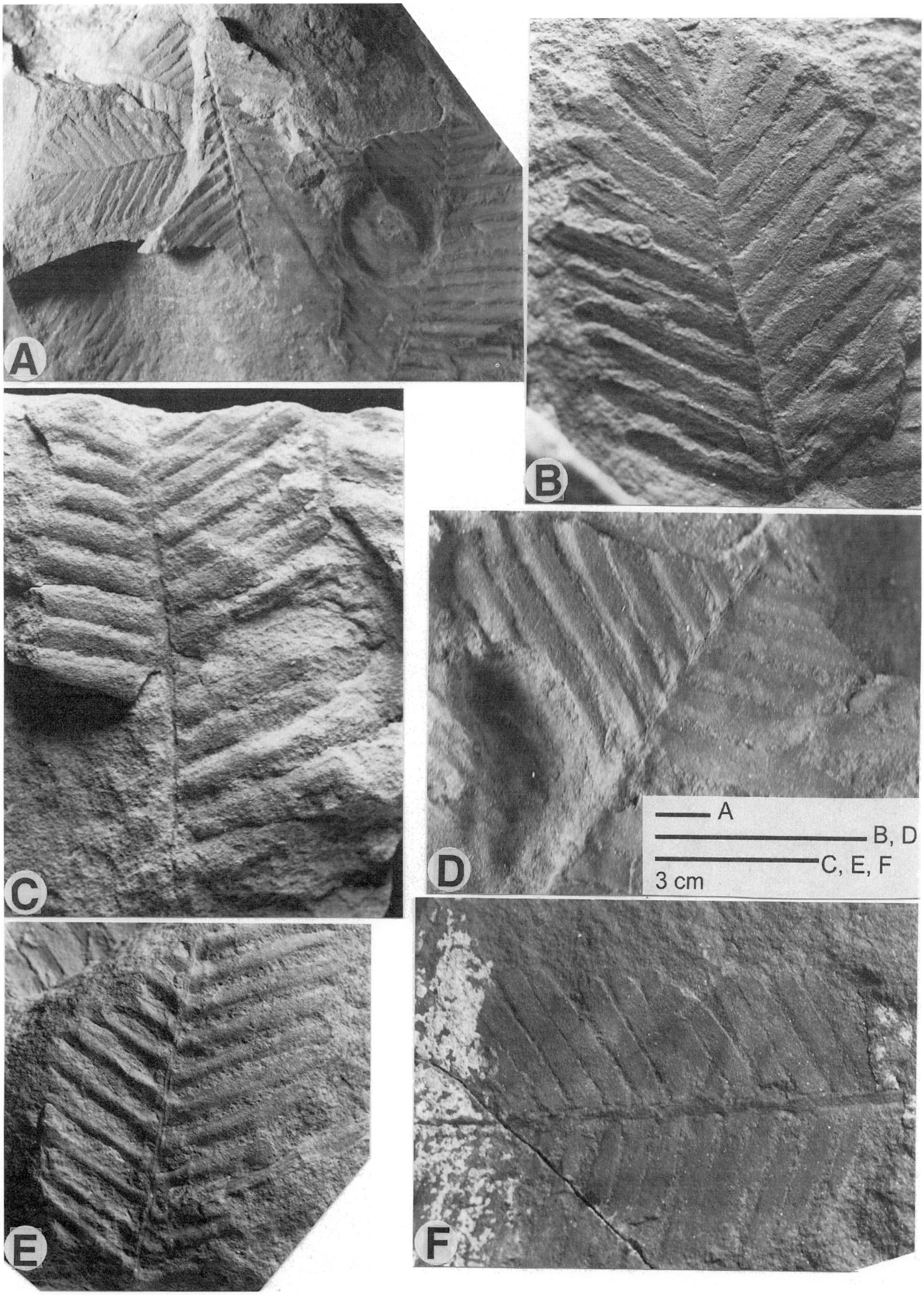

FIGURE 2. Selected specimens of leaves of "*Zamites*" *powellii* Fontaine from the Moenkopi Formation at NMMNH locality 3643. A, B, D, NMMNH P-33288. C, NMMNH P-33282. E, NMMNH P-33280. F, NMMNH P-33281.

Morales, M., 1987, Terrestrial fauna and flora from the Triassic Moenkopi Formation of the southwestern United States: Journal of the Arizona-Nevada Academy of Science, v. 22, p. 21-33.

Weber, R., and Zamudio-Varela, G., 1995, Laurozamites, a new genus and species of benettitalean leaves from the Late Triassic of North America: Revista Mexicana de Ciencias Geológicas, v. 12, p. 68-93.

APPENDIX-MEASURED STRATIGRAPHIC SECTION

Section measured at NMMNH locality 3643. Strata are essentially flat-lying, though some local slumping produces dips up to 14° to the N.

unit lithology **thickness (m)**

Santa Rosa Formation: Tecolotito Member:

19. Sandstone; yellowish gray (5Y7/2) with grayish orange (10YR7/4) mottles; medium to coarse grained; micaceous litharenite; some clay pebbles; calcareous; trough crossbedded. **2.0+**
18. Conglomerate; mottled pale yellowish brown (10YR6/2), pale red (10R6/2) and light greenish gray (5GY8/1); clasts are mostly rip-ups of Moenkopi siltstone and calcrete and Paleozoic limestone up to 15 cm in diameter; matrix is medium-to coarse-grained litharenite; calcareous; massive; scour base. **0.3-0.6**

Tr-3 unconformity

Moenkopi Formation: Anton Chico Member:

17. Mudstone and sandy mudstone; variegated pale reddish brown (10R5/4) and light olive gray (5Y6/1); contains a few lenses of grayish red (10R4/2) fine-grained, micaceous sandstone; some calcrete nodules; forms a slope. **18.2**
16. Sandstone; light greenish gray (5GY8/1); fine grained; micaceous litharenite; calcareous; low angle trough crossbeds; top bioturbated; forms a well indurated cuesta. **0.5**
15. Mudstone; pale reddish brown (10R5/4); slightly sandy; calcareous; contains a few lenses of grayish red (10R4/2) fine-grained, micaceous sandstone. **6.2**
14. Sandstone; grayish red (10R4/2); fine grained; micaceous litharenite; calcareous; ripple laminated; "*Zamites*" fossils (NMMNH localities 3643-3644). **3.0**
13. Sandstone; pinkish gray (5YR8/1); fine to medium grained; micaceous litharenite; calcareous; ripple laminated. **6.5**
12. Sandstone; pinkish gray (5YR8/1); fine grained; micaceous litharenite; calcareous; trough crossbedded; forms a bench. **1.6**
11. Sandstone; same color and lithotype as unit 9. **11.1**
10. Sandstone; pale reddish brown (10R5/4); very fine grained to fine grained; micaceous litharenite; trough crossbedded; forms a bench. **1.8**
9. Sandstone; pale red (10R6/2); fine grained; micaceous litharenite; calcareous; ripple laminated and small trough crossbedded; slope. **2.5**
8. Sandstone; yellowish gray (5Y8/1); fine to medium grained; micaceous litharenite; not calcareous; ripple laminated; forms an indurated bench. **1.8**
7. Sandstone; light greenish gray (5GY8/1); very fine grained; micaceous litharenite; calcareous; ripple laminated; forms a friable slope. **5.1**
6. Sandstone; yellowish gray (5Y8/1); fine grained; micaceous litharenite; calcareous; trough crossbedded. **1.3**
5. Conglomerate; mottled light brownish gray (5YR6/1); light olive gray (5Y6/1) and moderate yellowish brown (10YR5/4); clast supported; clasts are limestone pebbles up to 1 cm in diameter; massive. **0.4**
4. Siltstone; grayish red purple (5RP4/2); sandy; calcareous. **3.3**
3. Conglomerate; mottled grayish red (10R4/2), grayish orange (10YR7/4) and very pale orange (10YR8/2); clasts are siltstone chert and calcrete up to 6 mm in diameter; clast supported; calcareous. **0.2**

Tr-1 unconformity

Artesia Formation:

2. Shale; pale reddish brown (10R5/4); calcareous; calcrete nodules are pale reddish brown (10R5/4) to moderate reddish orange (10R6/6). **7.2**
1. Limestone; light bluish gray (5B7/1) ledge. **0.4**

NEW OCCURRENCE OF *TRILOPHOSAURUS* (REPTILIA: ARCHOSAUROMORPHA) FROM THE UPPER TRIASSIC OF WEST TEXAS AND ITS BIOCHRONOLOGICAL SIGNIFICANCE

ANDREW B. HECKERT[1], SPENCER G. LUCAS[2], ROBERT KAHLE[3] AND KATE ZEIGLER[1]
[1]Department of Earth and Planetary Sciences, University of New Mexico, Albuquerque, NM 87131-1116; [2]New Mexico Museum of Natural History, 1801 Mountain Rd NW, Albuquerque, 87104; [3]4305 Roosevelt Road, Midland, Texas 79703

Abstract.—We document a rich Upper Triassic bonebed in Borden County, Texas in either the uppermost Tecovas Formation or, more likely, the Trujillo Formation of the Chinle Group, that represents one of the highest stratigraphic occurrences of the aberrant archosauromorph reptile *Trilophosaurus*. The associated fauna includes indeterminate coprolites, unionid bivalves, osteichthyans, indeterminate reptiles, a possible ornithischian dinosaur, and the putative theropod dinosaur *Spinosuchus caseanus* Huene.

INTRODUCTION

Numerous workers have collected and described Upper Triassic fossils from various localities in West Texas, including Cope (1892), Case (1922, 1928, 1932), Gregory (1945), Sawin (1947), Murry (1982, 1986) and Chatterjee (1986) (Fig. 1). Among the fossils collected were the original specimens of the aberrant archosauromorph reptile *Trilophosaurus buettneri* Case, 1928, subsequently the subject of a monograph by Gregory (1945) and described further by Parks (1969) and Murry (1982). To date, all other verifiable *Trilophosaurus* records come from low in the Chinle Group in Texas (Gregory, 1945; Parks, 1969; Murry, 1982) or in Arizona and New Mexico (Long and Murry, 1995), with the exception of isolated teeth collected by Kirby (1989, 1991, 1993) in the Owl Rock Formation of north-central Arizona. Here, we outline the stratigraphy of a new locality yielding fossils of *Trilophosaurus*, describe the fauna of this locality, and comment on its biostratigraphic and biochronologic significance.

In this paper MNA = Museum of Northern Arizona, Flagstaff; and NMMNH = New Mexico Museum of Natural History and Science, Albuquerque.

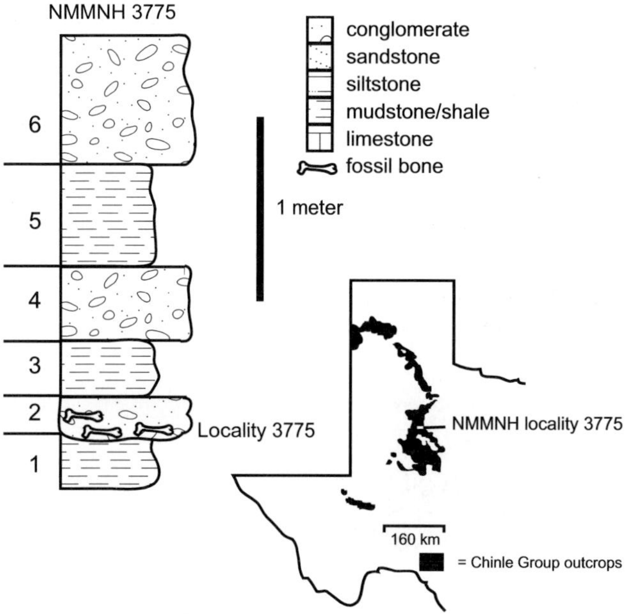

FIGURE 1. Index map and measured stratigraphic section of NMMNH locality 3775. For a description of stratigraphic units in the section see the Appendix. Detailed locality information is on file at NMMNH.

STRATIGRAPHY

One of us (RK) discovered New Mexico Museum of Natural History and Science (NMMNH) locality 3775 while working in West Texas in the early 1990s (Fig. 1). This locality yields numerous pinkish-gray disarticulated bones from a matrix of intraformational conglomerate. The principal fossiliferous horizon consists of a moderate brown, clast-supported conglomerate of rounded, flattened, intraformational mud pebbles up to 2 cm in diameter. The matrix is principally a rounded, moderately poorly sorted sublitharenite. This horizon is separated from the underlying moderate brown mudstone by an irregular contact that suggests slight scouring and erosion (Fig. 1). Most of the overlying strata are interbedded sandstone, conglomerate and mudstone. Unionid bivalves were collected from the uppermost conglomerate, which is a moderate yellowish-brown intraformational conglomerate dominated by clasts of calcrete and siltstone up to 1.5 cm in diameter.

The bedded intraformational conglomerates that dominate this outcrop strongly resemble the dominant lithologies of the Trujillo Formation of the Chinle Group (Lucas et al., 1994, 2001). However, the possibility exists that these strata instead represent an outcrop of the more areally limited coarse-grained channel bodies within the stratigraphically lower Tecovas Formation (Lucas et al., 1994 and sources cited therein). Because these are some of the only outcrops in an area characterized by broad flatlands, we remain unsure as to the exact stratigraphic position of these strata. However, we suspect that these deposits represent the base of the Trujillo Formation locally, and summarize arguments both for and against that assignment in the following paragraphs.

Trujillo Formation strata overlie a regional erosional surface termed the Tr-4 unconformity by Lucas (1993) that approximates the Carnian-Norian boundary. Although exposure in the vicinity of NMMNH locality 3775 is poor, strata to the north and west clearly pertain to the Bull Canyon Formation, which overlies the Trujillo throughout West Texas and east-central New Mexico. Similarly, the Trujillo Formation throughout this region is consid-

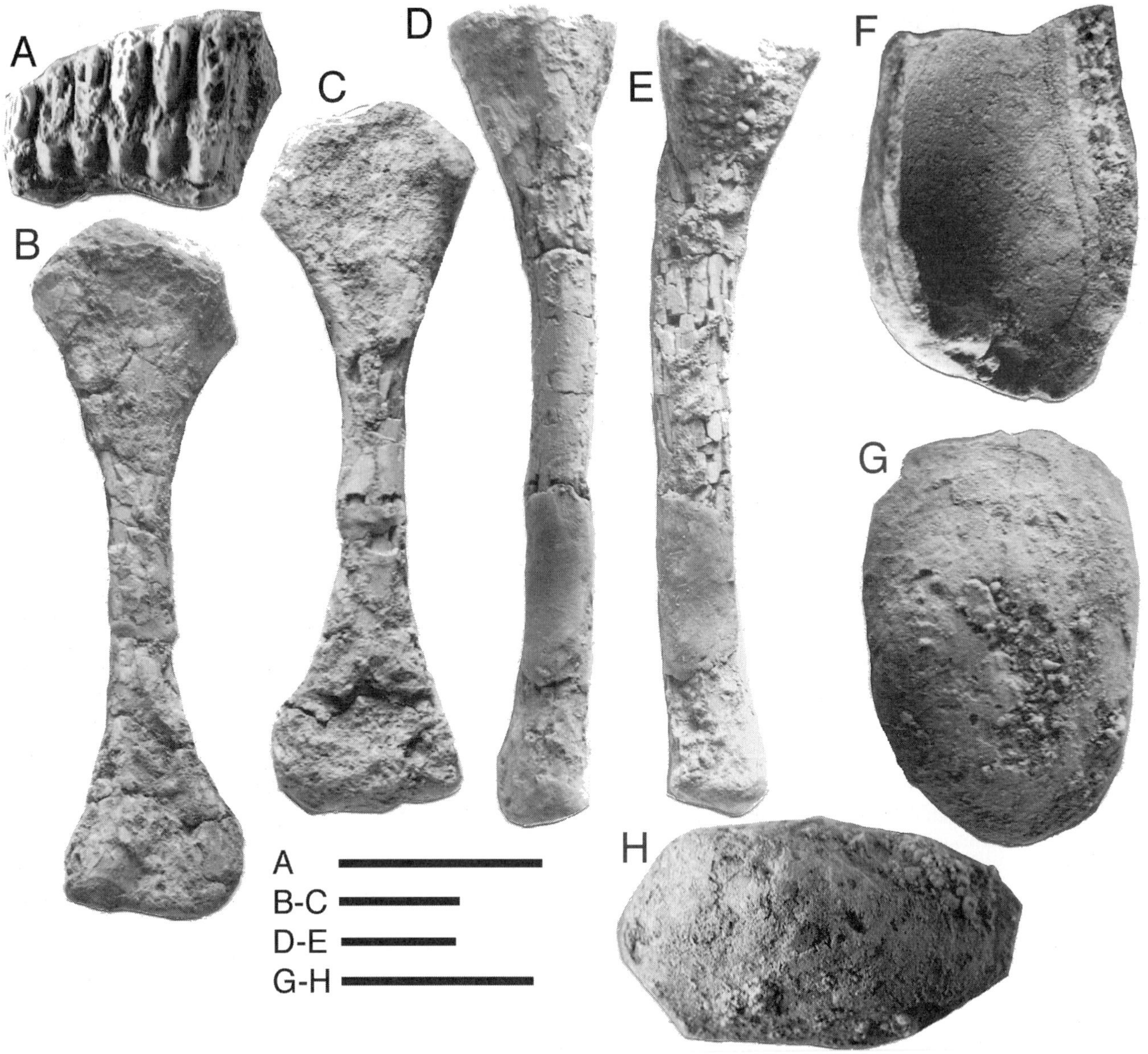

FIGURE 2. Photographs of representative fossils from NMMNH locality 3775. A-E. *Trilophosaurus* sp., A. NMMNH P-29959, maxillary? fragment in occlusal view; B-C. NMMNH P-34101 left humerus in (B) lateral and (C) medial views, D-E. NMMNH P-34102, right tibia in lateral (D) and medial (E) views; F-H. NMMNH P-29954, indeterminate unionid bivalves, incomplete valves in interior (F) and exterior (G-H) views. All scales = 2 cm.

erably more resistant than either the overlying Bull Canyon Formation or the underlying Tecovas Formation. Thus, the Trujillo Formation crops out as either a ledge-former above Tecovas Formation badlands, such as at Muchakooyoo Peak slightly to the southeast or, alternately, as a bench flooring broad plains beneath localized exposures and badlands of the Bull Canyon Formation. The strata at NMMNH locality L-3775 appear to represent the latter, as they are part of a broad upland surface that extends from Gail to the northeast southward to just northwest of Westbrook.

The only other possibility is that these strata represent a channel deposit in the Tecovas Formation. However, we note here that the broad flats of which this outcrop is a part suggest that it is not composed of isolated channel-fill lithologies, and instead represents more regional deposition of coarse-grained strata. Therefore, we tentatively assign NMMNH locality L-3775 to the Trujillo Formation.

PALEONTOLOGY

We have collected numerous macro- and microvertebrate fossils from NMMNH locality 3775. Although many of these fossils have yet to be prepared, it is evident that this locality yields a number of tetrapod taxa, dominated by the archosauromorph reptile *Trilophosaurus* (Fig. 2). The associated fauna includes indeterminate coprolites, unionid bivalves, osteichthyans, indeterminate reptiles, a possible ornithischian dinosaur, and probable theropods, including a possible occurrence of the putative theropod dinosaur *Spinosuchus caseanus* Huene.

Trilophosaurus

We have collected abundant material of *Trilophosaurus* from NMMNH locality 3775; these fossils include skull or lower jaw, dental, and postcranial material (Fig. 2A-E, 3A). The most diagnostic elements are an incomplete tooth-bearing element (probably either a left maxilla or right dentary) (Fig. 2A) and an associated, isolated tooth (Fig. 3A). Additional fossils of *Trilophosaurus*, all from a single field jacket, include a left humerus, a right tibia, a caudal vertebra, and a possible incomplete femur. We also collected as float or in screenwashing concentrate several vertebrae, a proximal? humerus fragment, a distal humerus, two distal ulnae, four proximal radii, a carpal I or V, a proximal and distal femur and an astragalus. For a complete list of specimens, see the table in the Appendix.

There are six teeth in the maxillary or dentary fragment (NMMNH P-29959) we assign to *Trilophosaurus* (Fig. 2A). The teeth are slightly damaged, but all are clearly transversely expanded and bear three cusps, one central and one each at the lingual and labial margins. The cusps are sharply pointed and aligned in a row connected by thin transverse ridges. The teeth are tightly packed and aligned medially, with narrower teeth anterior to broader teeth posteriorly, thus forming a laterally convex tooth margin. Comparison with illustrations of *Trilophosaurus buettneri* (Gregory, 1945, figs. 3, 6; Parks, 1969, fig. 2) suggests that this element probably represents the anterior portion of the tooth battery of a left maxilla. However, it is also possible that it is the anterior portion of the tooth battery of a right dentary. In general, the maxilla of *Trilophosaurus* demonstrates a greater increase in the width of teeth posteriorly than does the dentary, so this element is most likely a left maxilla.

One of the more striking features of NMMNH P-29959 is the close spacing of the teeth. Published illustrations and casts of comparative material of *Trilophosaurus buettneri* from the WPA localities indicate that the teeth are much closer together in NMMNH P-29959 than in the TMM specimens. Similarly, the teeth of NMMNH P029959 are also much closer together than in the holotype of *T. buettneri*. Given the large number of specimens from two size classes that Gregory (1945) examined (maxillae of at least four individuals and 15 lower jaws, Gregory, 1945, p. 291), we suspect that this difference has taxonomic significance, but withhold more definitive statements until associated material from locality 3775 is prepared.

A single tooth, NMMNH P-29960, associated with NMMNH P-29959, is similarly broad transversely, although the labial and medial cusps are broken off (Fig. 3A). Cingulae are present on both the anterior and posterior margins of this tooth, and evident on the teeth of NMMNH P-29959 as well. Clearly, this tooth is morphologically similar to those of NMMNH P-29959, with which it was associated. The tooth crown is relatively tall, and there is a substantial root that is nearly as tall as the crown. Where the central cusp is preserved, including NMMNH P-29960, it is particularly swollen mesio-distally, to the extent that the lower slopes of the central cusp divide the anterior and posterior cingulae into labial and lingual segments. This swollen base of the central cusp differs from that of both the type and WPA specimens of *Trilophosaurus* and appears to be unique to *Trilophosaurus* specimens from NMMNH locality 3775. As with the tooth spacing, this may have taxonomic significance, although we refrain from further comment until more material is prepared. Damage to teeth of NMMNH P-29959 and P-29960 prevents us from describing detailed features such as possible wear facets with confidence.

The humerus (NMMNH P-34101) is elongate, with slightly expanded proximal and distal ends (Fig. 2B-C). The deltopectoral crest forms a small, medially directed knob in the middle of the proximal expansion, and there is a thin but prominent process forming the posterior margin of the proximal expansion. The distal end is rotated slightly medially, so that the shaft appears somewhat twisted. The elongate ectepicondylar groove is poorly preserved but present. In distal view, the radial and ulnar condyles lie on the anterior half of the distal end, with the distal expansion almost entirely posterior to the ulnar condyle. This bone corresponds in all details to humeri of *Trilophosaurus buettneri* described by Gregory (1945, pl. 27[1-3]).

The tibia (NMMNH P-34102) is less well preserved (Fig. 2D-E). The proximal head is broadly expanded transversely, with prominent articular facets for the distal condyles of the femur on the posterior portion. The medial facet is slightly larger than the lateral facet. A weakly developed cnemial crest is present. The shaft is thin (although this may partly be an artifact of transverse crushing) and slightly convex anteriorly. The distal end is unexpanded, with a slight sulcus anterolaterally for reception of the ascending process of the astragalus. This bone, too, corresponds in all details to tibiae of *Trilophosaurus buettneri* described by Gregory (1945, pl. 30).

The associated femur? (NMMNH P-34104) from the same field jacket lacks both proximal and distal ends. The resulting shaft is considerably more stout than either of the other limb bones described, with a rounded center and flattened proximal and distal ends. The preserved bone is nearly as long as the associated tibia, and is thus similar in proportions to femora of *Trilophosaurus* (Gregory, 1945).

A nearly complete caudal centrum (NMMNH P-34103) from the same jacket is elongate, with a fused neural arch. The centrum is amphicoelous, with the anterior face slightly concave and the posterior face deeply so. In lateral view, the centrum appears slightly concave ventrally. The neural arch is long and low, and fused along approximately 80% of the centrum length. Most of the neural processes are broken, but the posterior zygopophyses slope abruptly dorsally at an angle approximately 45° to the dorsal surface of the centrum.

Ornithischia?

We recovered one possible ornithischian tooth from NMMNH locality 3775. This tooth, NMMNH P-34106, is triangular in outline (Fig. 3D), very slightly asymmetric in occlusal view (Fig. 3I-K), possesses denticles that are somewhat oblique to the tooth margin and has a basal constriction near the (broken) root (the posterior denticles are intact and not truncated). In these respects it conforms to descriptions of Upper Triassic ornithischian teeth given by Hunt and Lucas (1994). However, the tooth is also mod-

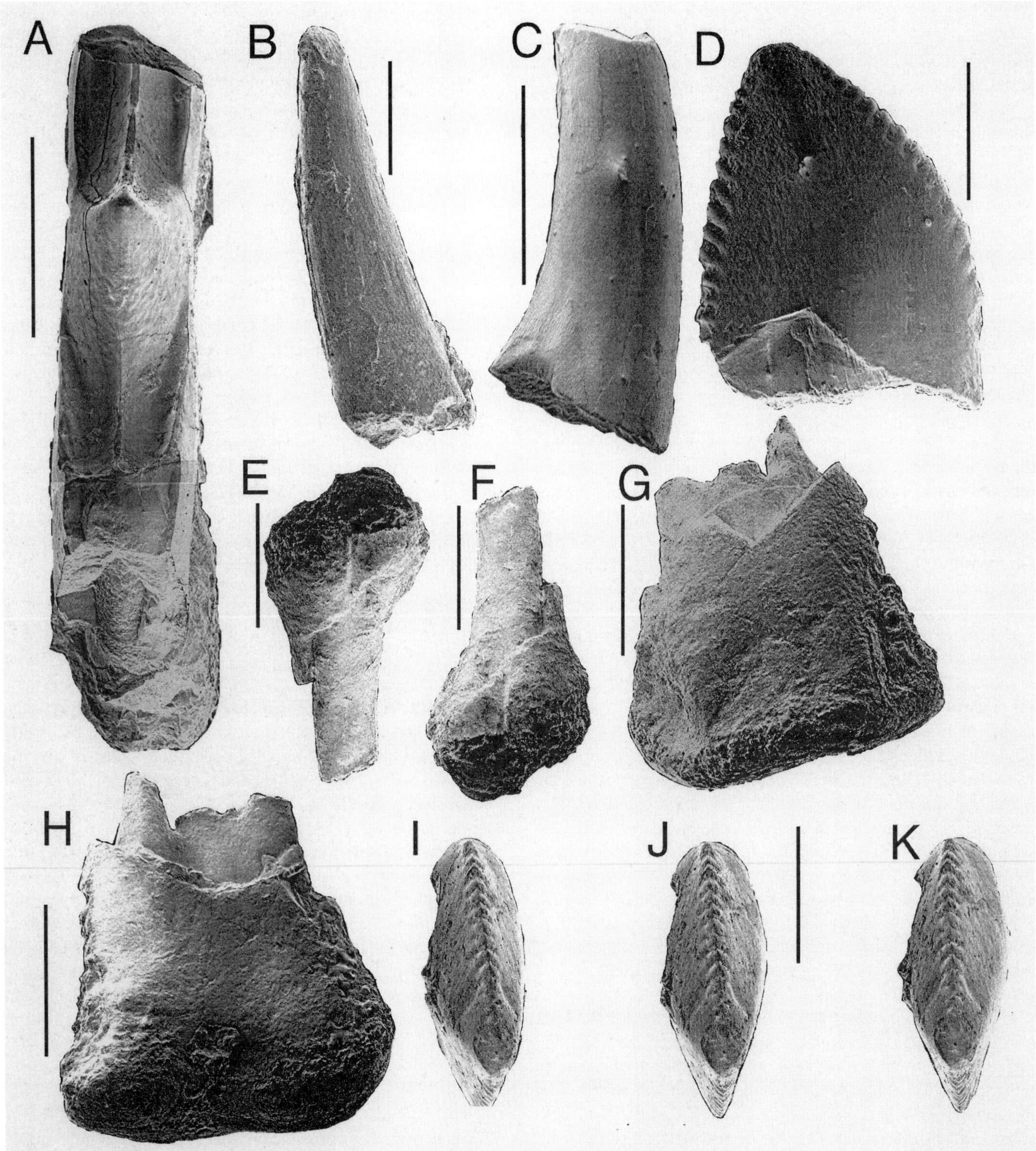

FIGURE 3. Scanning electron microphotographs of selected fossils from NMMNH locality 3775. A. NMMNH P-29960, tooth of *Trilophosaurus buettneri* photograph montage in occlusal view; B. NMMNH P-29956, possible reptile fang in labial? view; C. NMMNH P-34107, possible reptile tooth in lateral? view; D., I-K. NMMNH P-34106, possible ornithischian dinosaur tooth in lateral (D) and stereo occlusal (I-K) views; E-F. NMMNH P-34127, possible theropod metapodial or phalanx in dorsal (E) and ventral (F) views; G-H. NMMNH P-34125, possible theropod distal right humerus in lateral (G) and medial (H) views. Scale bars = 2 mm (A, E-F, and G-H) or 1 mm (B-D, I-K).

erately tall relative to its mesio-distal length and is only slightly asymmetric, and in these respects resembles teeth of basal sauropodomorphs, principally prosauropods. In either case, it clearly represents a herbivorous dinosaur.

If this tooth is an ornithischian, it most closely resembles *Revueltosaurus callenderi* Hunt, 1989 in its moderately tall, triangular crown, but is less asymmetric in occlusal view. It is less recurved than most *Revueltosaurus* presumed maxillary/dentary teeth but considerably more compressed than many *Revueltosaurus* premaxillary teeth. It is also significantly smaller than the holotype and referred specimens of that taxon, and appears to have more fine denticles. Notably, the holotype and paratype teeth of *Revueltosaurus* (Hunt, 1989, pl. 8e-h, pl. 9) have crowns 6-10 mm tall, while NMMNH P-34106 is no more than 3 mm tall (Fig. 3D), and is decidedly more labio-lingually compressed (Fig. 3I-K). NMMNH P-34106 is distinct from all other Triassic ornithischians, including *Galtonia*, *Tecovasaurus*, *Lucianosaurus*, and *Tecnhosaurus*. This tooth could represent a juvenile of *Revueltosaurus* or an unnamed taxon, so we tentatively refer it to Ornithischia? indet.

Prosauropod teeth are less variable than those of ornithischian teeth, but most are considerably taller than NMMNH P-34106. Teeth of the prosauropods *Plateosaurus* and *Sellosaurus* illustrated by Galton (1985) are generally taller than NMMNH P-34106, although posterior maxillary-dentary teeth of these taxa are grossly similar (Galton, 1985, figs. 3-4). The tooth is both less spatulate and has finer denticles than most Triassic prosauropods as well. If this tooth is indeed a prosauropod, it most closely resembles teeth of the Early Jurassic *Massospondylus* (Attridge et al., 1985) from the Elliot Formation in South Africa and the Kayenta Formation in north-central Arizona. Whether the tooth is prosauropod or ornithischian, it is a relatively rare record of a herbivorous dinosaur tooth in the Chinle.

Theropods

One of us (RK) has collected vertebrae possessing extremely tall neural spines from NMMNH locality 3775. These are similar to the vertebrae described by Case (1927) and later named *Spinosuchus caseanus* by Huene (1932). This is the first co-occurrence of *Spinosuchus* with *Trilophosaurus* and would appear to be evidence supporting Richards' (1999) recent hypothesis that *Spinosuchus* and *Trilophosaurus* are the same animal. However, we note that the holotype specimen of *Spinosuchus* photographically illustrated by Case (1927, pl. 1) differs greatly from illustrations of the vertebral column of *Trilophosaurus* (Gregory, 1945). *Trilophosaurus* does superficially resemble *Spinosuchus* in its possession of moderately tall neural spines in the caudal series. However, the dorsal vertebral series are entirely different, with *Trilophosaurus* bearing low neural spines over its comparatively shorter dorsal centra, as opposed to the tall neural spines on longer centra in *Spinosuchus*.

We also note here that the limb bones and vertebrae of *Trilophosaurus* are not hollow, and that we have recovered apparently hollow bones from this locality that appear to have been subsequently filled with matrix and/or calcite (e.g., NMMNH P-34123-34127: Fig. 3E-H). While the possibility remains that this is a preservational artifact, this mode of preservation is not evident on any of the limb bones referable to *Trilophosaurus*. In the nonmarine Upper Triassic, only pterosaurs and theropod dinosaurs possess hollow limb bones, and only theropods possess hollow vertebrae. Consequently, pending the preparation of additional material, we tentatively note the presence of a small theropod dinosaur, probably *Spinosuchus*, at NMMNH locality 3775. We also note here that, although some authors (e.g., Padian, 1986; Murry and Long, 1989; Long and Murry, 1995) do not consider *Spinosuchus* dinosaurian, Hunt et al. (1998) recently reviewed the material and concluded that, pending further preparation, the hollow centra and long bones of *Spinosuchus* are sufficient evidence to at least tentatively assign it to the Theropoda.

Reptilia indet.

Screenwashing matrix from the main bone-bearing horizon has yielded a diverse fauna of microvertebrate reptiles, most of which are indeterminate. Some of the more noteworthy fossils from this sample include at least two morphologies of elongate, recurved, fang-like reptilian teeth. One of these, NMMNH P-34107 (Fig. 3C), is very tall and thin, slightly recurved, and possesses very faint longitudinal striations. The second, represented by NMMNH P-29956 (Fig. 3B), is similarly tall and recurved but more robust and lacks longitudinal striations. Other specimens include a jaw fragment with three tooth sockets (NMMNH P-29957) and a small caudal? vertebra (NMMNH P-34117)

Osteichthyes indet.

Among the microvertebrates recovered thus far are a few scales and other fish elements. These include pitted scale fragments that may belong to semionotid fish, unornamented scales, and a very small, serrated scale still embedded in a small matrix block. More specimens will be necessary to make more precise determinations, but it appears that at least three different taxa of osteichthyan fish are represented at NMMNH locality 3775.

Coprolites

To date, we have recovered a few coprolites from NMMNH locality 3775. These fossils, two of which are catalogued as NMMNH P-29952, are small, cylindrical masses typical of coprolites found in the Chinle. We make no attempt to identify their perpetrator beyond the level of Vertebrata indet.

Invertebrates

Numerous mollusc shells occur above the bonebed in unit 6 of the measured section shown in Figure 1. These are largely fragmentary shells of unionid bivalves (Fig. 2F-H) catalogued as NMMNH P-29954.

DISCUSSION AND BIOCHRONOLOGICAL SIGNIFICANCE

Case (1928) first described fragmentary remains of *Trilophosaurus buettneri*, noting the particularly diagnostic, transversely expanded, lobed teeth with three prominent cusps. Following extensive work near Otis Chalk, Texas, Gregory (1945) published a monograph on the osteology of this unusual reptile, noting two size classes but preferring to maintain all of them in the species *T. buettneri*. The type specimens illustrated by Case (1928) are clearly of the smaller size class.

Parks (1969) wrote a thesis on the skull and dentition of *Trilophosaurus*, and Murry (1982) also characterized isolated remains. Murry (1987) named a second species of *Trilophosaurus*, *T. jacobsi*, for isolated tooth and jaw fragments collected from the base of the Chinle Group in east-central Arizona. Later, Sues and Olsen (1993) identified these teeth as belonging to a procolophonid. Long and Murry (1995) concurred, and re-assigned fossils of "*Trilophosaurus*" *jacobsi* to the procolophonid *Chinleogomphius jacobsi*. We concur with this assessment, noting that the asymmetric, pyramidal cusps of the posterior teeth of *C. jacobsi* strongly resemble those of procolophonids and are distinct from the conical cusps of *Trilophosaurus*.

Murry and Long (1989) identified a single occurrence of *T. buettneri* from high in the Blue Mesa Member of the Petrified Forest Formation, but did not illustrate or describe the specimen, nor is it included in Long and Murry's (1995) list of specimens referred to *Trilophosaurus*. Long and Murry (1995) documented other referred specimens of *Trilophosaurus* from Arizona, New Mexico, and Texas. All of the Arizona specimens are either from the *Placerias* quarry low in the Bluewater Creek Formation or are based on extremely fragmentary material higher in the section.

Omitted from Long and Murry's (1995) compendium were occurrences of *Trilophosaurus* documented by Kirby (1989, 1991, 1993). Kirby identified *Trilophosaurus* from the Owl Rock Formation near Wards Terrace in north-central Arizona. These records were based on an isolated tooth and several fragmentary skull bones. We concur that the incomplete tooth, MNA V7064, could represent *Trilophosaurus*, but it also closely resembles *Tricuspisaurus*, a possible procolophonid known from fissure fills of similar age in Britain (Robinson, 1957; Fraser, 1986). In particular, MNA V7064 and *Tricuspisaurus* are less transversely broad than referred specimens of *Trilophosaurus*, and thus more closely resemble teeth of procolophonids. Further, MNA V7064 is incomplete, and the presence of a third cusp, while probable, can only be inferred. If this specimen does pertain to *Trilophosaurus*, it represents the smaller of two body morphs, as it would be less than 2 mm wide were it complete (Gregory, 1945).

Regardless of whether the strata encompassing NMMNH locality 3775 are assigned to either the uppermost Tecovas Formation or the Trujillo Formation, they represent a range extension for the larger *Trilophosaurus* morph. The Tecovas Formation is of well-established Adamanian (latest Carnian) age, based on the occurrence of the Adamanian index taxa *Rutiodon* and *Stagonolepis* (Lucas and Hunt, 1993; Lucas et al., 1994). The Trujillo Formation is slightly younger. Tetrapod fossils, including the occurrence of *Typothorax coccinarum* (Hunt, 2001; Lucas et al., 2001), suggest that the Trujillo is of Revueltian (early-mid Norian) age, although palynological evidence indicates that the basal Trujillo could be as old as latest Carnian (Dunay and Fisher, 1974, 1979; Litwin et al., 1991; Cornet, 1993).

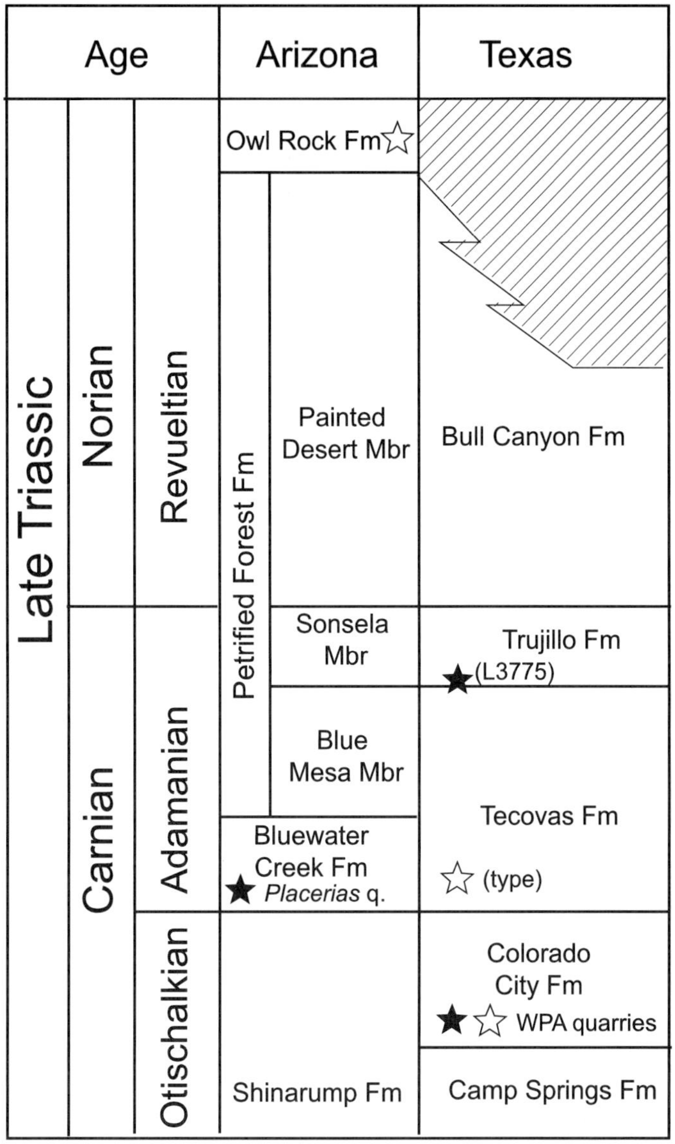

FIGURE 4. Generalized stratigraphic distribution of *Trilophosaurus* occurrences in the southwestern U.S. Solid stars = large morph, open-stars = small morph. See text for discussion.

All previous occurrences of *Trilophosaurus buettneri* documented by Long and Murry (1995) are from strata low in the Chinle Group of either Otischalkian (late early to early late Carnian) or early Adamanian (latest Carnian) age. The New Mexican occurrence is both fragmentary and stratigraphically low, occurring in the Los Esteros Member of the Santa Rosa Formation. The type locality, near Walker's Tank in Crosby County in the Tecovas Formation (Case, 1928), is stratigraphically low in the Tecovas Formation and of Adamanian age, based on the co-occurrence of the Adamanian index taxa *Rutiodon* (=*Leptosuchus*) and

Stagonolepis. The remaining Texas records referred to *Trilophosaurus* come from the Works Progress Administration (W.P.A.) quarries described by Gregory (1945), which are part of the type assemblage of the Otischalkian lvf (Lucas et al., 1993). Consequently, *Trilophosaurus* occurrences at NMMNH locality 3775 extend the range of *Trilophosaurus* to at least the latest Adamanian, and probably into the early Revueltian (Fig. 4). Although we do not rule out the possibility that the material here represents a distinct species of *Trilophosaurus*, it is the stratigraphically highest occurrence of the larger *Trilophosaurus* morph. If the material from NMMNH locality turns out to be a distinct species, we note that there may be some biochronological utility in having at least two stratigraphically superposed species of *Trilophosaurus*. However, *Trilophosaurus* is relatively rare in the Chinle and, so far as we can tell, the postcrania of older specimens and the NMMNH material are essentially identical, requiring relatively intact teeth to diagnose the two species.

ACKNOWLEDGMENTS

The Borden County Highway Department approved excavation of NMMNH locality 3775. J.D. Harris and H. Haubold assisted in the field. B. Cohen assisted with the microvertebrate washing and picking. S. Sucher prepared jacket 3 from this locality, including some of the material illustrated here. The NMMNH funded SEM work at UNM's Department of Earth & Planetary Sciences. D. Hill allowed us to borrow comparative material from the MNA. B. Edwards supplied comparative SEM photomicrographs of *Tricuspisaurus*. T. White and J.W. Merck Jr. were instrumental in acquiring a copy of Parks (1969). J.W. Merck, Jr. and A.P. Hunt provided helpful reviews of an earlier version of this manuscript.

REFERENCES

Attridge, J., Crompton, A. W., and Jenkins, F. A. Jr., 1985, The southern African Liassic prosauropod *Massospondylus* discovered in North America: Journal of Vertebrate Paleontology, v. 5, p. 128-132.

Benton, M. J., 1991, What really happened in the Late Triassic: Historical Biology, v. 5, p. 263-278.

Benton, M. J., 1994, Late Triassic to Middle Jurassic extinctions among continental tetrapods: testing the pattern; *in* Fraser, N. C., and Sues, H.-D., eds., In the Shadow of the Dinosaurs: Early Mesozoic Tetrapods: Cambridge, Cambridge University Press, p. 366-397.

Case, E. C., 1922, New reptiles and stegocephalians from the Upper Triassic of western Texas: Carnegie Institution Publication, v. 321, 84 p.

Ca se, E. C., 1927, The vertebral column of *Coelophysis* Cope: Contributions from the Museum of Paleontology, University of Michigan, v. 2, p. 209-222.

Case, E. C., 1928, Indications of a cotylosaur and of a new form of fish from the Triassic beds of Texas, with remarks on the Shinarump Conglomerate: University of Michigan Contributions of the Museum of Paleontology, v. 3, p. 1-14.

Case, E.C., 1932, A collection of stegocephalians from Scurry County, Texas: Contributions from the Museum of Paleontology, University of Michigan, v. 4, p. 1-56.

Chatterjee, S., 1986, The Late Triassic Dockum vertebrates: Their stratigraphic and paleobiogeographic significance, *in* Padian, K., ed., The Beginning of the Age of dinosaurs: Faunal Change Across the Triassic-Jurassic Boundary: Cambridge, Cambridge University Press, p. 139-150.

Cope, E. D., 1892, A contribution to the vertebrate paleontology of Texas: American Philosophical Society Proceedings, v. 30, p. 123-131.

Cornet, B., 1993, Applications and limitations of palynology in age, climatic, and paleoenvironmental analyses of Triassic sequences in North America: New Mexico Museum of Natural History and Science, Bulletin 3, p. 75-93.

Dunay, R. E., and Fisher, M. J., 1974, Late Triassic palynofloras of North America and their European correlatives: Review of Palaeobotany and Palynology, v. 17, p. 179-186.

Dunay, R. E., and Fisher, M. J., 1979, Palynology of the Dockum Group (Upper Triassic), Texas, U.S.A.: Review of Palaeobotany and Palynology, v. 28, p. 61-82.

Fraser, N. C., 1986, Terrestrial vertebrates at the Triassic-Jurassic boundary in south west Britain: Modern Geology, v. 10, p. 147-157.

Galton, P. M., 1985, Cranial anatomy of the prosauropod dinosaur *Sellosaurus gracilis* from the Middle Stubensandstein (Upper Triassic) of Nordwürttemberg, West Germany: Stuttgarter Beiträge zur Naturkunde B, v. 118, p. 1-39.

Goddard, E. N., Trask, P. D., DeFord, R. K., Rove, O. N., Singewald, G. T. J., and Overbeck, R. M., 1984, Rock Color Chart: Boulder, Geological Society of America.

Gregory, J. T., 1945, Osteology and relationships of *Trilophosaurus*: University of Texas Publication, v. 4401, p. 273-359.

Huene, F. v., 1932, Die fossile Reptil-Ordnung Saurischia, ihre Entwicklung und Geschichte: Monographien zur Geologie und Paläontologie, v. 1, p. 1-361

Hunt, A. P., 1989, A new ?ornithischian dinosaur from the Bull Canyon Formation (Upper Triassic) of east-central New Mexico, *in* Lucas, S. G., and Hunt, A. P., eds., Dawn of the Age of Dinosaurs in the American Southwest: Albuquerque, New Mexico Museum of Natural History, p. 355-358.

Hunt, A.P., 2001, Paleontology and age of the Upper Triassic Trujillo Formation, east-central New Mexico and West Texas: New Mexico Geological Society, 52nd Field Conference, Guidebook.

Hunt, A. P., and Lucas, S. G., 1994, Ornithischian dinosaurs from the Upper Triassic of the United States; *in* Fraser, N. C., and Sues, H.-D., eds., In the Shadow of the Dinosaurs: Early Mesozoic Tetrapods: Cambridge, Cambridge University Press, p. 227-241.

Hunt, A. P., Lucas, S. G., Heckert, A. B., Sullivan, R. M., and Lockley, M. G., 1998, Late Triassic dinosaurs from the western United States: Geobios, v. 31, p. 511-531.

Kirby, R. A., 1989, Late Triassic vertebrate localities of the Owl Rock Member (Chinle Formation) in the Wards Terrace area of northern Arizona, *in* Lucas, S. G., and Hunt, A. P., eds., Dawn of the Age of Dinosaurs in the American Southwest: Albuquerque, New Mexico Museum of Natural History, p. 12-28.

Kirby, R. A., 1991, A vertebrate fauna from the Upper Triassic Owl Rock Member of the Chinle Formation in northern Arizona [M.S. thesis]: Flagstaff, Northern Arizona University, 476 p.

Kirby, R. A., 1993, Relationships of Late Triassic basin evolution and faunal relationships in the southwestern United States: Perspectives from the upper part of the Chinle Formation in northern Arizona: New Mexico Museum of Natural History and Science, Bulletin 3, p. 232-242.

Litwin, R. J., Traverse, A., and Ash, S. R., 1991, Preliminary palynological zonation of the Chinle Formation, southwestern U.S.A., and its correlation to the Newark Supergroup: Review of Palaeobotany and Palynology, v. 68, p. 269-287.

Long, R. A., and Murry, P. A., 1995, Late Triassic (Carnian and Norian) tetrapods from the southwestern United States: New Mexico Museum of Natural History and Science, Bulletin 4, 254 p.

Lucas, S. G., 1993, The Chinle Group: Revised stratigraphy and biochronology of Upper Triassic strata in the western United States: Museum of Northern Arizona, Bulletin 59, p. 27-50.

Lucas, S. G., 1997, The Upper Triassic Chinle Group, western United States, a nonmarine standard for Late Triassic time, *in* Dickins, J. M., Yang, Z., Yin, H., Lucas, S. G., and Acharyya, S. K., eds., Permo-Triassic of the circum-Pacific: Cambridge, Cambridge University Press, p. 200-228.

Lucas, S. G., and Hunt, A. P., 1993, Tetrapod biochronology of the Chinle Group (Upper Triassic), western United States.: New Mexico Museum of Natural History and Science, Bulletin 3, p. 327-329.

Lucas, S. G., Anderson, O. J., and Hunt, A. P., 1994, Triassic stratigraphy and correlations, southern High Plains of New Mexico-Texas: New Mexico Bureau of Mines and Mineral Resources, Bulletin 150, p. 105-126.

Lucas, S. G., Hunt, A. P., and Kahle, R., 1993, Late Triassic vertebrates from the Dockum Formation near Otis Chalk, Howard County, Texas: New Mexico Geological Society, 44th Field Conference, Guidebook, p. 237-244.

Lucas, S. G., Heckert, A. B., and Hunt, A. P., 2001, Triassic stratigraphy, biostratigraphy and correlation in east-central New Mexico: New Mexico Geological Society, 52nd Field Conference, Guidebook.

Murry, P. A., 1982, Biostratigraphy and paleoecology of the Dockum Group, Triassic of Texas[Ph.D. dissertation]: Dallas, Southern Methodist University, 459 p.

Murry, P. A., 1986, Vertebrate paleontology of the Dockum Group, western Texas and eastern New Mexico, *in* Padian, K., ed., The Beginning of the Age of Dinosaurs: Faunal Change Across the Triassic-Jurassic Boundary: Cambridge, Cambridge University Press, p. 109-137.

Murry, P. A., 1987, New reptiles from the Upper Triassic Chinle Formation of Arizona: Journal of Paleontology, v. 61, p. 773-786.

Murry, P. A., and Long, R. A., 1989, Geology and paleontology of the Chinle Formation, Petrified Forest National Park and vicinity, Arizona and a discussion of vertebrate fossils of the southwestern upper Triassic, *in* Lucas, S. G., and Hunt, A. P., eds., Dawn of the Age of Dinosaurs in the American Southwest: Albuquerque, New Mexico Museum of Natural History, p. 29-64.

Padian, K., 1986, On the type material of *Coelophysis* Cope (Saurischia: Theropoda) and a new specimen from the Petrified Forest of Arizona (Late Triassic: Chinle Formation), *in* Padian, K., ed., The Beginning of the Age of Dinosaurs: Faunal Change Across the Triassic-Jurassic Boundary: Cambridge, Cambridge University Press, p. 45-60.

Parks, P., 1969, Cranial anatomy and mastication of the Triassic reptile *Trilophosaurus* [M.A. thesis]: Austin, University of Texas, 100p.

Richards, H. R., 1999, Is *Spinosuchus* a dinosaur?: Journal of Vertebrate Paleontology, v. 18, supplement, no. 3, p. 70A.

Robinson, P. L., 1957, An unusual sauropsid dentition: Zoological Journal of the Linnean Society, v. 43, p. 283-293.

Sawin, H. J., 1947, The pseudosuchian reptile *Typothorax meadei*: Journal of Paleontology, v. 21, p. 201-238.

Sues, H.-D., and Olsen, P. E., 1993, A new procolophonid and a new tetrapod of uncertain, possibly procolophonian affinities from the Upper Triassic of Virginia: Journal of Vertebrate Paleontology, v. 13, p. 282-286.

APPENDIX 1—
DESCRIPTION OF MEASURED SECTION L-3775

Section measured at UTM zone 14S 0252420E, 3625349N NA027. This is the first road cut on 1054S to Vealmoor south of U.S. 180, Borden County, Texas. Strata are flat-lying. Section measured by A.B. Heckert and S.G. Lucas, 22 April, 1999. Rock colors follow Goddard et al. (1984).

unit lithology **thickness (m)**

Chinle Group: Trujillo? Formation:

6 Conglomerate; moderate yellowish brown (10YR5/4) to grayish orange (10YR7/4); clasts are intraformational calcrete and siltstone pebbles up to 1.5 cm diameter; most clasts are less than 0.5 cm diameter, rounded; bed is well-indurated and caps outcrop; locally fossiliferous with unionid bivalves, highly calcareous. **0.7**

5 Mudstone; dark reddish brown (10R3/4) with some mottles of yellowish gray (5Y7/2); locally some conglomerate lenses which are also yellowish gray (5Y7/2) and contain siltstone rip-ups; mudstone is very slightly silty; mudstone is weakly calcareous; siltstone is calcareous. **0.55**

4 Conglomerate and sandstone; conglomerate is moderate brown (5YR4/4) with some sand-sized clasts; and is clast-supported; sand is rounded chert and quartzite; conglomerate clasts are claystone pebbles up to 0.5 cm diameter; weakly calcareous to calcareous; sand-sized fraction is coarse- to very coarse, rounded, moderately well-sorted sublitharenite. **0.4**

3 Mudstone; moderate brown (5YR4/4); very silty; slightly micaceous; calcareous. **0.3**

2 Conglomerate; moderate brown (5YR4/4); clasts are intraformational mud pebbles up to 2 cm diameter, rounded, flattened; clast-supported; matrix is rounded, moderately poorly sorted sublitharenite; slightly calcareous; this is the main bone-bearing interval; bone is pinkish gray (5YR8/1); some irregular scouring at base. **0.2**

1 Mudstone; moderate brown (5YR4/4) with some pale greenish yellow (10Y8/2) mottles; very silty; very calcareous, base of exposure. **not measured**

APPENDIX 2—
SPECIMENS REFERRED TO *TRILOPHOSAURUS* FROM NMMNH LOCALITY 3775

NMMNH P#	TAXON	DESCRIPTION
29959	*Trilophosaurus*	(1) Lower (?) jaw fragment
29960	*Trilophosaurus*	(1) Tooth fragment
29961	*Trilophosaurus*	(2) Complete caudal vertebrae
29962	*Trilophosaurus*	(2) One complete vertebra
29963	*Trilophosaurus*	(1) Proximal humerus (?) fragment
29964	*Trilophosaurus*	(1) Distal humerus
29965	*Trilophosaurus*	(2) Distal ulnas
29966	*Trilophosaurus*	(4) Proximal radii
29967	*Trilophosaurus*	(1) Carpal - I OR V
29968	*Trilophosaurus*	(1) Proximal femur
29969	*Trilophosaurus*	(1) Distal femur
29970	*Trilophosaurus*	(1) Astragalus
29972	*Trilophosaurus*?	(1) Carpal
29973	*Trilophosaurus*?	(1) Limb element (phalanx)
29974	*Trilophosaurus*?	(3) Podials
29975	*Trilophosaurus*?	(1) Metapodial end
34101	*Trilophosaurus*	(1) Left humerus
34102	*Trilophosaurus*	(1) Right tibia
34103	*Trilophosaurus*	(1) Caudal vertebra
34104	*Trilophosaurus*?	(1) Femur?
34105	*Trilophosaurus*	(1) Tooth fragment
34122	*Trilophosaurus*	(1) Caudal centrum

THE VERTEBRATE FAUNA, BIOSTRATIGRAPHY AND BIOCHRONOLOGY OF THE TYPE REVUELTIAN LAND-VERTEBRATE FAUNACHRON, BULL CANYON FORMATION (UPPER TRIASSIC), EAST-CENTRAL NEW MEXICO

ADRIAN P. HUNT
Mesalands Dinosaur Museum, Mesa Technical College, 911 South Tenth Street, Tucumcari, NM 88401

Abstract.— The Bull Canyon Formation of east-central New Mexico contains a large vertebrate fauna that is the type fauna of the Revueltian land-vertebrate faunachron. The Bull Canyon vertebrate fauna includes ?Colobodontidae indet., Redfieldiidae indet., *Semionotus* cf. *S. brauni*, ?*Tanaocrossus kallikoski*, Actinopterygii indet., ?Arganodontidae indet., *Quayia zideki*, *Apachesaurus gregorii*, Metoposauridae indet. (large), ?Squamata indet., Rhynchosauridae indet., *Vancleavia campi*, *Pseudopalatus pristinus*, *P. andersoni*, *P.* n. sp., *Typothorax coccinarum*, *Aetosaurus arcuatus*, Stagonolepididae ?gen. et sp. nov., *Paratypothorax* sp., *Desmatosuchus* n. sp., *Postosuchus kirkpatricki*, *Shuvosaurus inexpectatus*, Sphenosuchidae indet., Crocodylotarsi gen. et sp. nov., Herrerasauridae gen. et sp. nov. (2), *Gojirasaurus quayi*, Theropoda gen. et sp. nov., *Revueltosaurus callenderi*, *Lucianosaurus wildi* and *Pseudotriconodon chatterjeei*. The Revueltian can be divided into a lower R1 (Barrancan) defined by the FAD of *Pseudopalatus* and R2 (Lucianoan) defined by the FAD of *Lucianosaurus*. There is possibly an RO (Rainbowforestan) interval which has the co-occurrence of *Typothorax* and *Nicrosaurus* prior to the FAD of *Pseudopalatus*. The Revueltian represents most of Norian time.

INTRODUCTION

The Upper Triassic Bull Canyon Formation is widely exposed in east-central New Mexico. Vertebrate paleontologists have collected specimens from this formation for more than 80 years, but most of this collecting was sporadic, and most specimens were never described. However, prior to 1986, the few published accounts of the vertebrate fauna of the Bull Canyon Formation suggested that this unit is, at least locally, very fossiliferous and contained many new taxa (e.g., Case, 1922; Gregory, 1972; Lucas et al, 1985a,d; Parrish and Carpenter, 1986; Lucas and Oakes, 1988). These factors were the impetus for this study, which was part of a Ph.D. dissertation at the University of New Mexico (Hunt, 1994). From 1986 to 1991, I led field parties to collect from exposures of the Bull Canyon Formation in Quay County, New Mexico. This collecting effort was very successful and yielded about 2500 macro- and microvertebrate specimens. Subsequent to the development of this large fauna, Lucas and Hunt (1993) designated the fauna from the lower Bull Canyon Formation as the type fauna of the Revueltian land-vertebrate faunachron (lvf). Several publications have addressed aspects of the vertebrate paleontology of this extensive Bull Canyon fauna (Heckert and Lucas, 1998; Huber et al., 1991, 1993; Hunt, 1991c, 1992, 1993a-b, 1994, 1997a-b: Hunt and Lucas, 1989a-b, 1993b-d, 1997; Hunt et al., 1993a-b, 1988, 2001c; Lucas and Hunt, 1989a, 1993; Lucas and Oakes, 1998; Lucas et al., 1985a-d, 2000, 2001a). The purpose of this paper is to give an overview of the vertebrate fauna of the Bull Canyon Formation and to discuss its biostratigraphy and biochronology as it relates to the biochronology of the Revueltian lvf. Several new taxa from this fauna will be described elsewhere. Abbreviations are: AMNH, American Museum of Natural History, New York; CMNH, Carnegie Museum of Natural History, Pittsburgh; MDM, Mesalands Dinosaur Museum, Tucumcari; NMMNH, New Mexico Museum of Natural History and Science, Albuquerque; RHMP, Ruth Hall Museum of Paleontology, Ghost Ranch, Abiquiu; UCM, University of Colorado Museum, Boulder; UMMP, University of Michigan Museum of Paleontology, Ann Arbor; TTUP, Texas Tech University, Lubbock; YPM, Yale Peabody Museum, New Haven.

STUDY AREA

The majority of the specimens described here are from three areas in Quay and Guadalupe Counties, east-central New Mexico: (1) Bull Canyon; (2) Revuelto Creek area; and (3) Barranca Creek area (Fig. 1). For convenience, these areas will be referred to as Bull Canyon, the Barranca badlands and the Revuelto badlands (Fig. 2). The locations of some vertebrate localities have been published elsewhere, and these have been subsequently collected without the permission of land-managing agencies, so explicit locality information will not be presented here. The majority of fossils that form the basis of this study are reposited at NMMNH, and locality data are on file at that institution.

Bull Canyon is located near the border between Guadalupe and Quay Counties. Vertebrate fossil localities occur in secs. 10, 19-23, 27-29, 33, T9N, R26E, Guadalupe County. Bull Canyon is a 150-m-deep canyon that is a reentrant in the Llano Estacado. It exposes strata from the Trujillo Formation of the Chinle Group at its base through (in ascending order), the Bull Canyon and Redonda formations of the Upper Triassic Chinle Group, the Entrada Sandstone and Todilto Formation (Middle Jurassic), the Summerville and Morrison formations (Upper Jurassic) and

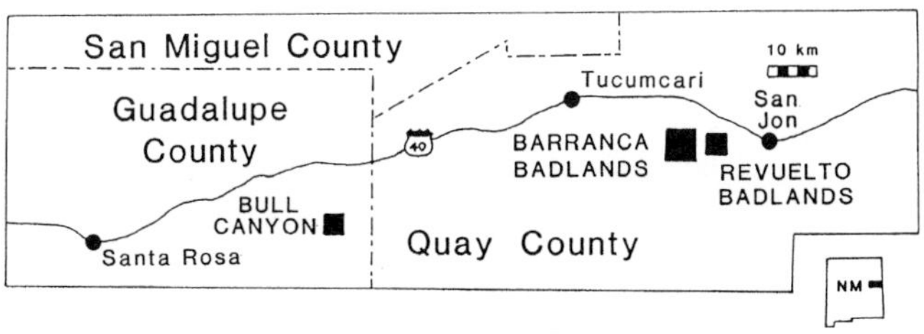

FIGURE 1. Principal collecting localities in the Bull Canyon Formation of east-central New Mexico.

FIGURE 2. Outcrops of the Bull Canyon Formation in east-central New Mexico. A. Revuelto Creek badlands. B. Barranca Creek badlands.

the Tucumcari Formation and Mesa Rica Sandstone (Lower Cretaceous)(Lucas et al., 2001b). There are extensive badlands of the Bull Canyon Formation at the base of the canyon, and the entire thickness of the formation is exposed, including its type section (Lucas and Hunt, 1989a). A significant number of vertebrate fossils from this area are derived from the upper half of the Bull Canyon Formation, whereas all the fossils from the Revuelto and Barranca badlands are from the lower half of the unit.

The Revuelto badlands are an area of low relief badlands in Quay County. They lie mainly to the east and partly to the west of Revuelto Creek, where it flows through exposures of the upper Trujillo Formation and lower Bull Canyon Formation. Vertebrate localities are in secs. 3-5, 8-11, 14-17, 20-22, T10N, R33E.

The Barranca badlands are an area of high-relief badlands in Quay County located just to the west of the Revuelto badlands. The Barranca badlands are to the southeast of Barranca Creek and expose the upper Trujillo Formation and the lower half of the Bull Canyon Formation. Vertebrate localities are in secs. 1-2, 11-12 T10N, R32E, sec. 36 T11N, R32E, secs. 6-7 T10N, R33E and secs. 30-31 R33E, T11N. The Barranca badlands are separated by a drainage divide from the Revuelto badlands.

GEOLOGIC SETTING

Tectonics

During the Late Triassic, the area that is currently New Mexico lay on the western equatorial edge of Pangaea at a paleolatitude of 5-15° N (Van der Voo et al., 1976). An evolving magmatic arch lay to the west and southwest of New Mexico, from western Nevada to southeastern Arizona (Coney, 1978; Dickinson, 1981). To the east and southeast there was arching and graben formation associated with the creation of the Gulf of Mexico (McGowan et al., 1983). New Mexico itself was tectonically stable in the Late Triassic, with no evidence of active faulting, folding or magmatic activity.

Paleogeography

Traditionally, the "Dockum" basin of eastern New Mexico and western Texas was perceived as being separated from a "Chinle" basin (Dubiel, 1989, fig. 2). There is no stratigraphic or sedimentologic evidence of such a drainage divide (Lucas, 1991). Most of New Mexico appears to have been an area of low relief in the Late Triassic, with a paleoslope to the north or northwest (e.g., Lucas and Anderson, 1993, fig. 4). The closest highlands would have been in north-central New Mexico (Chama) and southwestern Colorado (Colorado National Monument), where Chinle Group strata rest nonconformably on Precambrian basement.

Sedimentology

With the notable exception of Branson (1927, 1928; Branson and Mehl, 1929)(see reply by Reeside, 1928), all early workers considered the strata of the current Chinle Group to have been of broadly fluviolacustrine origin. With a few exceptions (e.g., Kiatta, 1960), detailed sedimentological studies have been conducted in the Chinle Group of eastern New Mexico and West Texas only in the past 12 years. The majority of these studies originated from two institutions, the University of Texas at Austin and Texas Tech University.

Several master's students of J. H. McGowan at Austin studied the sedimentology of the Chinle Group in eastern New Mexico (Granata, 1981) and West Texas (Seni, 1978; Boone, 1979) and subsequently co-authored papers on this topic (McGowan et al., 1979, 1983; Johns and Granata, 1987). The Austin sedimentologists subscribed to a single depositional model--the Chinle Group (their Dockum Group) of eastern New Mexico and West Texas was deposited in environments associated with a major, internally drained, lacustrine system (e.g., McGowan et al., 1983, figs. 25-27). Much of this model was developed from subsurface data with the explicit assumption that large thicknesses of mudrocks must be of lacustrine origin.

Another group of master's students of T. Lehman at Lubbock concentrated on deriving a depositional model based on outcrop studies. These students also studied the Chinle Group in both eastern New Mexico (Fritz, 1992) and West Texas (Frelier, 1987; May, 1988). They found no evidence of a major lacustrine system in the area and attributed virtually all Chinle Group strata to a fluvial (*sensu lato*) origin. Lucas and Anderson (1992, 1993) and Hunt et al. (2001c) have also challenged the lacustrine model, noting that paleontological and paleocurrent data are not consistent with this hypothesis.

Newell (1992, 1993) analyzed the sedimentology of the Bull Canyon Formation of eastern New Mexico and similarly reconstructed a broadly fluvial environment. In conclusion, the bulk of sedimentological and paleontological evidence indicates a fluvial environment for the Chinle Group of eastern New Mexico and West Texas. The only truly lacustrine strata in the Chinle Group in eastern New Mexico and west Texas are in the Redonda Formation (Hester, 1988; Hester and Lucas, 2001).

Only two of the above studies explicitly addressed the sedimentology of the Bull Canyon Formation in New Mexico (Granata, 1981; Newell, 1992), although Granata's conclusions were repeated by McGowan et al. (1979, 1983). Granata (1981, p. 56) dismissed the sedimentology of the Bull Canyon Formation in less than a page as "in general to be lacustrine mudstone." McGowan et al. (1983, fig. 9) explicitly presented Granata's views schematically and showed the Bull Canyon Formation as "lacustrine and prodelta mudstones."

Newell (1992) conducted a detailed sedimentological study of the lower Bull Canyon Formation in Quay, County, New Mexico, in conjunction with this study. Newell (1992, 1993) recognized channel, levee, and proximal and distal floodbasin deposits and demonstrated that the mudrocks do not represent a lacustrine environment. The Bull Canyon Formation formed in an aggrading alluvial floodplain. The channel deposits are the product of fixed, low-sinuosity, mixed-load streams characterized by fluctuating discharge (Newell, 1992, 1993). Floodbasin deposits represent overbank sheetfloods, crevasse channels and calcareous soils (Newell, 1992, 1993).

Taphonomy

Few early collectors made any taphonomic comments on the Bull Canyon Formation. Case (1922, p. 11) noted that rocks of the Chinle Group in West Texas and eastern and central New Mexico were formed in conditions unfavorable to the preservation of vertebrate fossils. He speculated that most of these rocks were formed in lacustrine offshore environments and that vertebrate fossils were preserved only locally in flood deposits. He interpreted very fossiliferous areas, such as Crosby County, Texas, as having formed on a rare "large river floodplain."

Newell (1992; Newell and Hunt, in prep.) recognized three taphofacies in the Bull Canyon Formation of the Barranca badlands: (1) channel-hosted assemblages; (2) floodplain (mudrock) assemblages; and (3) paleosol assemblages. Channel-hosted assemblages consist of abraded and/or fragmentary specimens principally of phytosaurs. Floodplain (mudrock) assemblages produce the bulk of the vertebrate specimens, including partial skeletons of large (dorsal centra > 2 cm in length) vertebrates (phytosaurs and aetosaurs). Paleosol assemblages are localized, but productive localities are dominated by fossils of terrestrial tetrapods, including partial skeletons of small animals (dorsal centra < 2 cm in length).

STRATIGRAPHY

Lithostratigraphy

Marcou (1855) first recognized the presence of Triassic rocks in east-central New Mexico, which included what is now known as the Bull Canyon Formation (Kues, 1985a, c). Marcou (1858) published a map that showed the Bull Canyon Formation (and other units of the Chinle Group) as New Red Sandstone (European stratigraphic name), which he identified as Triassic in age. In his text, Marcou (1858, p. 18) noted that the Redonda Formation and upper Bull Canyon Formation (e.g., at Pyramid Mountain) had "the same appearance as the upper part of the Keuper" [Carnian-Norian stratigraphic unit] in France. The Triassic age of these strata was the subject of heated debate for several years (Kues, 1985a, c).

Drake (1892) was the first to apply formal stratigraphic nomenclature to the Bull Canyon Formation in east-central New Mexico. He measured a series of stratigraphic sections in West Texas and eastern New Mexico and brought Cummins' (1889) name Dockum beds into Quay County. One of his measured sections is at the head of "east Fossil Creek." Marcou (1858) showed a "Fossil Creek" as a north-south flowing stream northeast of Mesa Redonda (Big Tucumcari Mountain of Marcou). The geographic location of Fossil Creek identifies it as Barranca Creek (compare Kues, 1985a, fig. 11 and Kues, 1985c, fig. 2), even though Marcou shows the location of the headwaters incorrectly. By inference, East Fossil Creek must be Revuelto Creek, which joins Barranca Creek near I-40. I also note that there are no other areas east of Fossil Creek (Barranca Creek) where Drake (1896, pl. 5) could have measured his thick section of "Dockum beds" (approximately 181 ft estimated from his plate) in Quay County. Drake (1896, plate 5) apparently measured 81 ft of lower beds, 25 ft of central beds and 75 ft of upper beds, but these thicknesses are much too thin to assign the strata to the Garita Creek, Trujillo and Bull Canyon formation. Indeed, the Garita Creek Formation does not crop out in this area, so the entire section was probably measured in the Bull Canyon and ?upper Trujillo formations.

The next major contribution to the stratigraphy of the Bull Canyon Formation was by Darton (1922, 1928), who recognized that the Dockum Group of east-central New Mexico consisted of a lower sandstone interval that he named the Santa Rosa Sandstone and an upper mud-rich interval. In 1928, he noted that the Dockum Group consisted of a lower Santa Rosa Sandstone, overlain by a tripartite sequence that he questionably assigned to the Chinle. The ?Chinle consisted of a basal shale interval, a medial sandstone-dominated interval and an upper shale interval. With the exception of the separation as the Redonda Formation of the upper part of this sequence by Dobrovolny et al. (1946), Darton's (1928) divisions have been used with minor adjustments until the present. Kelley (1972) named the medial sandstones the Cuervo Sandstone Member and referred to the units above and below it as the upper and lower shale members of the Chinle Formation. In 1989, Lucas and Hunt named the upper shale member the Bull Canyon Formation, the lower shale member the Garita Creek Formation and recognized that the Cuervo is Gould's (1907) Trujillo Formation. At its type section, the Bull Canyon Formation is 95 m of mudstone (80% of section), sandstone (16%) with minor amounts (2% each) of siltstone and conglomerate (Lucas and Hunt, 1989a). In West Texas, the Bull Canyon Formation is over 160 m thick (Lehman et al., 1992). Lucas (1993) raised the Chinle to Group status and included in it all the Upper Triassic formations in east-central New Mexico.

Chatterjee (1986) named a homotaxial equivalent of the Bull Canyon Formation in Garza County, Texas, the Cooper Member of the Dockum Formation. The "type section" of this unit is only 16 m thick and lacks an upper contact with Triassic strata. The name Cooper is preoccupied by several other stratigraphic units. Subsequently, Lehman et al. (1992) redefined the Cooper as the Cooper Canyon Formation with a new type section and suggested that this is the valid name for the upper mudrock-dominated part of the Dockum. Thus, Lehman et al. (1992; Lehman, 1994) suggested suppressing a validly defined stratigraphic name with a redefinition and renaming of an ill-defined and improperly named stratigraphic unit name. Clearly, as noted by Lucas and Anderson (1993, p. 60), "the Cooper Canyon Formation of Lehman et al. (1992) is simply an unneeded synonym of the Bull Canyon Formation of Lucas and Hunt (1989)"(Lucas et al., 1994).

It is worth noting that many of the vertebrate paleontologists who have worked in the Bull Canyon Formation were not aware of the local stratigraphy. Gregory (e.g., 1972) knew that he was collecting in the upper "Chinle" below the Redonda, but Carpenter and Parrish (1985; Parrish and Carpenter, 1986) and Murry (1982, 1986, 1987, 1989) wrote of the undivided Dockum Formation or Group.

Magnetostratigraphy

Reeve and Helsley (1972) measured a magnetostratigraphic section of the upper Bull Canyon Formation at Luciano Mesa. This section revealed mixed polarity zones, which they took to mean that this part of the Chinle Group was partially, or wholly, older than the Upper Triassic Newark Supergroup of the eastern United States, which had not yielded many reversed intervals prior to that time. At that time, the Newark Supergroup was considered Late Triassic in age, but it is now known to be Middle Triassic-Early Jurassic in age. Reeve and Helsley (1972) concluded that the Bull Canyon Formation was early Late Triassic in age. McIntosh et al. (1985) thought that this section was late Carnian in age based on the biostratigraphic correlations of Olsen et al. (1982). Witte et al. (1991) argued that the Bull Canyon Formation had been remagnetized and that the magnetostratigraphic polarity scheme was in error. Lucas et al. (1993b) noted that the Bull Canyon polarity sequence is compatible with Norian sequences of the Newark Supergroup and concluded that the Bull Canyon rocks had not been remagnetized.

Biostratigraphy/Biochronology (Previous work)

Marcou (1855, 1858) first indicated that the Bull Canyon Formation and super- and subjacent formations of the Chinle Group were Late Triassic in age. He noted what Ager (1993) later referred to as the persistence of facies, that is that certain facies and sequences of facies are correlatable over large areas of the Earth's surface. Thus, Marcou in eastern New Mexico saw redbeds (Chinle Group) overlain by yellow sandstone and shale (Entrada and Morrison Formations), in turn overlain by marine shale (Tucumcari Formation) that contain *Gryphaea*-like bivalves. This is reminiscent of the European section that consists of Upper Triassic redbeds (upper New Red Sandstone of older terminology) overlain by transitional marine beds (Rhaetic beds *sensu lato*) in turn overlain by marine Liassic strata containing *Gryphaea*. Thus, it is clear why Marcou (1855, 1858) referred the upper Chinle Group to the Upper Triassic by analogy with the European section. Even though the marine fossils from the Tucumcari Formation are Early Cretaceous in age, Marcou was still correct about the age of the Chinle Group. The sequence of facies is different between New Mexico and Europe, but it is still notable that the youngest redbeds in both areas are Late Triassic in age.

For 70 years it was generally accepted that Chinle Group strata in eastern New Mexico are Late Triassic in age. Huene (1922, 1927), however, suggested that part of the Chinle Group (his Dockum Formation) in West Texas (and by implication eastern New Mexico) is late Middle Triassic and part is early Late Triassic. In the United States, only Camp (1930), in his influential monograph on phytosaurs, followed this view. Camp (1930, table 1) considered the "Dockum" of eastern New Mexico to be equivalent to the upper Muschelkalk and Lettenkohle of Germany. He considered the ages of the German units to be late Middle Triassic and early Late Triassic, respectively, although they are now both considered to be of Middle Triassic age. The Huene/Camp correlation received no further support in this country.

The first major correlation of Triassic formations of North America by Reeside et al. (1957) included a section by Colbert and Gregory on vertebrate correlations. By this time, J. T. Gregory had collected a reasonably large vertebrate fauna from the Bull Canyon Formation to give him a more precise idea of its correlation. Colbert and Gregory (1957) noted that the fauna of the Bull Canyon Formation (their upper part of the clay member of the "Chinle" Formation below the Redonda) contained an advanced phytosaur comparable with those from Ghost Ranch and from the upper portion of the Petrified Forest Formation in northeastern Arizona. This taxon is now known as *Pseudopalatus* and does indeed occur at the Canjilon quarry at Ghost Ranch in the Petrified Forest Formation and in the Painted Desert Member of the Petrified Forest Formation in northeastern Arizona. Colbert and Gregory (1957) used the presence of this phytosaur and a perceived occurrence in eastern New Mexico above a Trujillo equivalent sandstone to correlate. They proposed that the Bull Canyon Formation correlated with (using current stratigraphic terminology) the Travesser Formation in northeastern New Mexico, the Petrified Forest Formation in north-central New Mexico and the Painted Desert Member of the Petrified Forest Formation and the Owl Rock Formation in northeastern Arizona. Their correlation with the Travesser was in error because they did not realize that an advanced phytosaur, now known as *Redondasaurus*, actually came from this formation rather than from the overlying Sloan Canyon Formation. Aside from this, the correlation of Colbert and Gregory is essentially the correlation advocated here.

Gregory (1957, 1969) essentially presented the same correlations, although he considered the Owl Rock Formation to be younger than the Bull Canyon Formation. This correlation scheme has been followed by most recent workers (e.g., Lucas et al., 1985c; Hunt and Lucas, 1989; Lucas and Hunt, 1989b; Murry, 1989; Lucas, 1993; Lucas and Hunt, 1993). Several workers have disputed these correlations. Long and Ballew (1985) misidentified two important specimens from the Bull Canyon Formation. They identified a paramedian osteoderm of *Paratypothorax* as *Stagonolepis* (=*Calyptosuchus*) and a skull of *Pseudopalatus* n. sp. as *Rutiodon* (their *Rutiodon* group A). Thus, they miscorrelated the Bull Canyon Formation with older strata (e.g., Tecovas Member of Dockum Formation) that do indeed contain *Rutiodon* and *Stagonolepis*. Carpenter and Parrish (1985) and Parrish and Carpenter (1986) misidentified another skull of *Pseudopalatus* n. sp. as *Rutiodon* and recognized the presence of *Desmatosuchus* in the Bull Canyon fauna. However, they did not realize that *Desmatosuchus* was not restricted to *Rutiodon*-bearing faunas, and so they followed the miscorrelation of Long and Ballew (1985). Chatterjee (1986, table 10.3) thought that the entire Chinle Group (his Dockum Formation) in eastern New Mexico was equivalent to the Bull Canyon Member of the Dockum Formation (his Cooper Member of the Dockum Formation), which contradicts all recent biostratigraphic and lithostratigraphic data (e.g., Lucas and Anderson, 1993). Following Long and Ballew (1985) and Chatterjee (1986), most recent authors have considered the Bull Canyon Formation and equivalent strata in the Chinle Group to be of Norian age either on the basis of invertebrate (Good, 1993) or vertebrate fossils (e.g., Lucas et al., 1985a; Ballew, 1989; Hunt and Lucas, 1989b; Murry, 1989; Lucas and Hunt, 1993).

HISTORY OF COLLECTING

Vertebrate fossils

Drake (1892, p. 244) was the first to mention vertebrate fossils ("reptilian bones") from the Bull Canyon Formation of New Mexico. He stated that they were "still [compared to west Texas] sparsely scattered through the conglomerates and sandstone." Lee (1907) reported finding Triassic vertebrates from the E. O. Davis ranch, 20 mi SE of Tucumcari, which were presumably from the Bull Canyon Formation.

E. C. Case was the first vertebrate paleontologist to investigate east-central New Mexico. He first collected the Bull Canyon Formation in 1912 (Case, 1914). That year he was traversing Permo-Triassic outcrops from north Texas to northeastern New Mexico to study their stratigraphic relationships. Case collected a number of fossils in this area. Specimen labels on vertebrate specimens at UMMP indicate collection dates of 1912 and 1915 (?misprint). Case (1922, p. 11) was not impressed by the Triassic paleontological resources of east-central New Mexico, stating that "it is evident that the great bulk of the Triassic beds revealed on the borders of the Staked Plains and in eastern and central New Mexico were deposited under conditions unfavorable to the preservation of vertebrate fossils." Case's (1914, 1922) publications, UMMP locality records and surviving field notes (Hunt, 1997) indicate that he principally collected in only one area in the Bull Canyon Formation in Quay County. This fossiliferous area is variously described as "the breaks of San Juan Arroya or Trujillo Creek" the "Badlands" (Case, 1914, p. 252) and lies 5 miles west of San Jon (Case, 1922, p. 11; UMMP specimen labels). Case (1914, figs. 4-7) illustrated several views of this area. Three lines of evidence pinpoint this area as approximately secs. 9-11 and 14-16, T10N, R33E, which is referred to in this report as part of the Revuelto badlands: (1) the prominent rock pillar illustrated by Case (1914, figs. 4-5) has now fallen but it was a prominent landmark in SW1/4 NE1/4 NE1/4 sec. 11, T10N, R33E (J. B. Cresap, oral commun., 1991); (2) J. T. Gregory, who later collected in this area, mentioned in his field notes for 1947 (Hunt, 1997b) that a local rancher named Cosner stated that the Michigan parties collected north of his house (house site is in NW1/4 SE1/4 NW1/4 SE1/4 sec. 15, T10N, R33E; J. B. Cresap, oral commun., 1986) and not to the west (i.e., not in Barranca badlands); and (3) UMMP 7441 is a collection of fragmentary specimens of seven taxa (six rare) of tetrapods, which only co-occur at NMMNH locality 1 (two taxa are unique to this site), which is in this area. Case's specimens from this area included UMMP 7274, 7313, 7438, 7441, 7442. Case evidently returned to this area later with Buettner and White, notably in July 1934 (Gregory, 1972; Hunt, 1997b).

An unknown collector obtained a phytosaur skull from near Santa Rosa, probably Bull Canyon (Gregory, 1972; J. T. Gregory, oral commun., 1986), which was initially at the New Mexico School of Mines (now New Mexico Institute of Mining and Technology) and was later transferred to the University of Chicago and ultimately to the Field Museum of Natural History. Mehl (1922) described this skull as a new species of *Machaeroprosopus*, *M. andersoni*. This skull is presumed to have come from the Bull Canyon Formation because this is the only Upper Triassic unit yielding vertebrate fossils near Santa Rosa and because it represents a grade of phytosaur that could only occur in that unit in that area.

J. T. Gregory of Yale University, and later of the University of California at Berkeley (from 1960 on), collected Triassic vertebrates from eastern New Mexico from 1947 until the early 1960's. However, most of his collections were from the Redonda Formation, and only in 1947 and 1948 did he collect much in the Bull Canyon Formation (Gregory, 1972; J. T. Gregory, oral commun., 1990; Hunt, 1997b). Although Gregory prospected several areas in Guadalupe and Quay Counties, the majority of YPM specimens from the Bull Canyon Formation come from three areas: (1) a few specimens from the base of Mesa Redonda; (2) a large collection from Bull Canyon; and (3) a large collection from the Revuelto badlands area (Cosner ranch).

Mesa Redonda specimens include an uncataloged phytosaur vertebra and distal humerus, and six *Apachesaurus* centra from YPM locality 6649. Note that YPM locality 6649 also includes specimens from the Bull Canyon Formation of Mesa Redonda, the Redonda Formation at "shark tooth hill" (e.g., uncataloged juvenile phytosaur dentary fragment) and many from the Redonda Formation of Apache Canyon (e.g., YPM 9897, phytosaur ilium).

All these specimens were apparently collected in 1958.

All the YPM specimens from Bull Canyon were collected in 1947 (Hunt, 1997b) and included a partial phytosaur skull and associated fragmentary skeleton (YPM 3300 from YPM locality 5831). J. T. Gregory went to the Revuelto badlands, then the Cosner ranch, in 1947 to collect fossil vertebrates and in 1948 to locate his localities on aerial photographs and to place them in stratigraphic context (J. T. Gregory, oral commun., 1986, 1990). In 1947, Gregory had plotted his localities on a road map of New Mexico, the best map at that time of Guadalupe and Quay Counties (J. T. Gregory, oral commun., 1986). Notable among Gregory's finds at Revuelto Creek were a large phytosaur skull with lower jaws (YPM 3293) and an aetosaur paramedian osteoderm with radial ornamentation (YPM 3695), both from YPM locality 5833. J. T. Gregory (oral commun., 1986, 1990) also collected two partial phytosaur skulls, one with a good palate and one with a good skull roof, which were "combined" and traded to the Cleveland Museum of Natural History (?specimens numbered 52 and 53 in Gregory's field notes: Hunt, 1997b). There is a much reconstructed, and possibly composite phytosaur skull on display at CMNH. However, CMNH has no record of a trade with YPM. CMNH 11162 is listed as a cast of *Phytosaurus* (*Machaeroprosopus*) *gregorii* (original at USNM), which was received from USNM in 1973. This cast could not be found, and the skull on display could have been identified as *P. gregorii* in older classifications (e.g., Gregory, 1962). Presumably the CMNH skull on display is from the Revuelto badlands.

Gregory did collect a couple of specimens from the Harmon ranch, just west of the Cosner ranch. The north end of the former Harmon ranch is the Barranca badlands of this report (J. Cresap, née Harmon, oral commun., 1986). However, I believe that Gregory prospected the less fossiliferous southern end of the ranch for the following reasons: (1) J. T. Gregory (oral commun., 1986, 1990) does not recall prospecting in a large area of badlands to the west of Revuelto; (2) Gregory evidently found few fossils on the Harmon ranch, whereas the Barranca badlands are very fossiliferous; and (3) access to the Barranca badlands was very difficult, whereas the southern end of the Harmon ranch was at the end of a county road. In 1983, P. A. Murry briefly prospected the Revuelto badlands (J. B. Cresap, oral commun., 1986). K. C. Carpenter visited the Revuelto badlands in 1982 to collect Triassic vertebrate fossils for the University of Colorado Museum. He collected a few specimens from UCM locality 82021, including a partial dinosaur skeleton (UCM 47221) and a partial phytosaur skull (UCM 48441)(Carpenter and Parrish, 1985; Parrish and Carpenter, 1986; Carpenter, 1997), and noted specimens nearby (UCM localities 82017-82020). Both Murry and Carpenter were revisiting an area whose location had been documented by Gregory (1972). From 1983 to 1986, field parties from the University of New Mexico under Drs. J. W. Froehlich, B. S. Kues and S. G. Lucas collected vertebrate fossils from the Bull Canyon area.

In 1986, I began working in the Revuelto badlands, initially with help from a UNM field party led by S. G. Lucas and J. W. Froehlich and later with members of the New Mexico Friends of Paleontology (FOP) and various volunteers. In 1987, we began prospecting the Barranca badlands, and this quickly became the main focus of collecting. The Revuelto area was depauperate in large specimens (e.g., phytosaur skulls), presumably because of previous collecting, but was rich in specimens of small terrestrial reptiles. A notable exception is NMMNH P-12964, a complete skeleton of *Typothorax coccinarum* (Hunt et al., 1993b) that was below the highly fossiliferous interval mentioned by Gregory (1972) and Carpenter and Parrish (1985). The Barranca badlands had never been collected and yielded many large specimens (e.g., 19 phytosaur skulls/lower jaws and partial phytosaur skeletons) as well as sites rich in terrestrial tetrapods. Field parties under my direction spent about 450-500 person-days collecting fossil vertebrates in the Revuelto and Barranca badlands from 1986 until 1991.

The reason why the highly fossiliferous and extensive Barranca badlands have not previously been collected is that access has always been difficult. The original highway US-66 followed the railroad for much of its route. However, it swung north from the Revuelto badlands and avoided the Barranca badlands, hence Case's failure to investigate this area as he followed the road along the railroad. The highway went north in order to avoid having to construct three bridges over the large drainages of Revuelto, Barranca and Plaza Larga Creeks, as did the railroad. Instead, the highway went north to where the three drainages coalesced, so only one bridge was necessary. The Barranca badlands were isolated from easy access by Barranca and Plaza Larga Creeks to the west and east and the railroad to the north. Now access is only feasible by driving across the now-abandoned railroad bridge across Plaza Larga Creek. This route was not possible until the abandonment of the railroad in the late 1960's. Thus, until recent years the difficulty of access hindered paleontological investigation of the Barranca badlands.

Invertebrate fossils

Invertebrate fossils are locally abundant in the Bull Canyon Formation, particularly in the lower half of the formation. Surprisingly, therefore, the first published reference to invertebrates from this formation is by Kues (1985b). Kues (1985b) described a large sample of unionid bivalves and gastropods from the lower Bull Canyon Formation in Bull Canyon. He recognized several taxa that included new forms designated by numbers. Kues (1985b) listed the unionid bivalves *Unio arizonensis*, *Unio* spp. 1-6, *Antediplodon dockumensis* and *Antediplodon* sp. and the hydrobiid gastropod *Triasamnicola pilsbyi* from the Bull Canyon Formation. Lucas et al. (1985c, fig. 4) illustrated the stratigraphic position of two of the localities that yielded these specimens. Carpenter and Parrish (1985) noted the presence of imbricated unionid shells in a channel in the lower Bull Canyon Formation. Carpenter (in Good, 1989, table 3) mentioned bivalves and a gastropod from the same area. Good (1993) considered invertebrate specimens from the Bull Canyon Formation as part of his *Antediplodon thomasi* biozone/paleobiogeographic unit.

The emphasis of this study was vertebrates, but numerous bivalve and gastropod specimens were collected from several localities in the lower half of the Bull Canyon Formation. Notable are juvenile unionids and many gastropods from NMMNH locality 1. All specimens are at NMMNH.

Microfossils

Kietzke (1987, 1989) described calcareous microfossils from two localities in the lower half of the Bull Canyon Formation. These microfossils include charophytes, ostracods and specimens originally identified as spirorbid worms, but which are now believed by some workers to be gastropods (Kietzke and Lucas, 1991). One locality (NMMNH locality 1) at Revuelto Creek yielded the charophytes *Altochara* sp. and *Stellatochara* sp. and the ostracodes *Darwinula* sp. A and *Gerdalia* sp. B. The other in Bull Canyon produced the ostracodes *Darwinula* sp. A, *Gerdalia* sp. B and ?*Lutkevichinella* sp. Both localities produced the spirorbid-like gastropods.

Trace fossils

Terrestrial trace fossils have largely been ignored until the last few years, and this is reflected in the sparse literature on the ichnology of the Bull Canyon Formation. Newell (1992, figs. 4.3c, 4.7a) illustrated simple and meniscate burrows from the Bull Canyon Formation, and Hunt et al. (1993) described surface trails and trackways. Hunt et al. (1993a) described abundant trails of the notostracan arthropod trackway *Acripes triassicus* and a new taxon of enigmatic vertebrate tracks (*Barrancapus cresapi*), which possibly pertains to a prosauropod (Hunt et al., 2001b). Root casts have been found in a few localities in the mudrock facies. However, it is clear that there is a need for further study of the ichnology of the Bull Canyon Formation.

Plant fossils

Drake (1892, p. 244) first mentioned plant fossils in the Bull Canyon Formation when he noted silicified wood in eastern Quay County. Case (1922, p. 8) noted wood from eastern New Mexico, that is probably from the Bull Canyon Formation. Since 1960, S. R. Ash has sporadically collected from the Chinle Group of eastern New Mexico (Ash, 1972). It is impossible to extract a floral list for the Bull Canyon Formation from Ash's work (e.g., Ash, 1972, 1987, 1989) because he did not distinguish between the "shale members/formations of the Chinle" (= Garita Creek and Bull Canyon Formations of current usage), and some specimens are reported only as being from the Chinle Group (Dockum Group of Ash). The only undoubted species to occur in the combined Garita Creek/Bull Canyon is the bennettitalean *Zamites powelli* (Ash, 1972, 1989). Lucas et al. (1985b, c) reported specimens of the horsetail *Neocalamites* sp. and the enigmatic ?angiosperm *Sanmiguelia* sp. from the Bull Canyon Formation at Bull Canyon. Newell (1992, fig. 4.3b) illustrated a fossil log from the Bull Canyon Formation and noted the presence of stumps and rootlets.

During the present study no attempt has been made to prospect systematically for plant fossils, but some specimens were noted or collected. Wood is fairly common in the lower part of the formation; it is often infested with heart-rot fungus (*Polyporites*). One 10 m log in the Barranca area (see below) is almost totally infested; isolated, spindle-shaped pieces of fungus-infected wood are common and are extremely abundant at NMMNH locality 1. One specimen worthy of note is a part of a "frond" of the enigmatic, putative angiosperm *Sanmiguelia* (NMMNH P-17597 from NMMNH locality 443). Representative plant specimens from the Bull Canyon Formation have been reposited at NMMNH.

SYSTEMATIC PALEONTOLOGY OF REVUELTIAN VERTEBRATE FOSSILS

Introduction

The Bull Canyon Formation has yielded a large, diverse vertebrate fauna. The following discussion includes all specimens from the Bull Canyon Formation. The type Revueltian fauna is from the lower Bull Canyon Formation in the Revuelto and Barranca badlands. However, as discussed below, the fauna of the upper Bull Canyon Formation is also considered to be of Revueltian age and thus is included in this discussion. Familial and higher level ranges are from the Fossil Record 2 (Benton, 1993; Gardiner, 1993; Milner, 1993; Schultze, 1993), and of lower taxonomic levels are from cited references.

Osteichthyes, Huxley, 1880
Actinopterygii, Klein, 1885
Perleidiformes Berg, 1937
?Colobodontidae Stensio, 1916
gen. et sp. indeterminate

Type Revueltian occurrence: NMMNH locality 1.

Referred specimens from type Revueltian: NMMNH locality 1; NMMNH P-4524, toothplates; NMMNH P-4850, toothplate (Fig. 3E).

First occurrence: Several genera with worldwide distribution (Triassic: Scythian-Anisian).

Last occurrence: Several species of *Colobodus* from Europe (Triassic: Carnian-Norian: Gardiner, 1993) and indeterminate toothplates from New Mexico (Triassic: late Norian-?Rhaetian: Huber et al., 1993).

Acme zone: Scythian-Anisian (Triassic)

Discussion: Several specimens from NMMNH locality 1 represent fragments of phyllodont toothplates. These specimens consist of multiple, superposed sets of replacement teeth. Such toothplates occur in eight teleost families (Johnson and Zidek, 1981), but Late Triassic specimens have usually been referred to the Perleididae (e.g., Murry, 1986, fig. 9.5A1-A2). However, I follow Gardiner (1993) in considering Perleididae to be a synonym of Colobodontidae.

Redfieldiformes Berg, 1940
Redfieldiidae Berg, 1940
gen. et sp. indeterminate

Type Revueltian occurrences: NMMNH localities 75 and 110.

Referred specimens from type Revueltian: NMMNH locality 75; NMMNH P-17250, body segments; NMMNH P-17255, head and body fragments: NMMNH locality 110; NMMNH P-4590, scales; NMMNH P-4640, skull fragment.

First occurrence: Several genera from Gondwana (Triassic: Scythian)

Last occurrence: *Redfieldius* from eastern North America (Jurassic: Hettangian).

Acme zone: Olenekian (Triassic)

Discussion: Scale squamation of a few specimens from the Bull Canyon Formation supports their assignment to the Redfieldiidae (Schaeffer and McDonald, 1978; Huber et al., 1993).

Halecostomi, Regan, 1923
Semionotiformes Arambourg and Bertin, 1958
Semionotidae Woodward, 1890
Semionotus* cf. *S. brauni

Type Revueltian occurrence: NMMNH locality 75.

Referred specimens from type Revueltian: NMMNH locality 75; NMMNH P-4185, body fragments; NMMNH P-17312, body fragments.

First occurrence: Genus from Anisian-Ladinian (Triassic) of Europe, North America and Africa, species from Hettangian (Jurassic) of eastern North America.

Last occurrence: Genus from eastern North America (Jurassic: Hettangian),

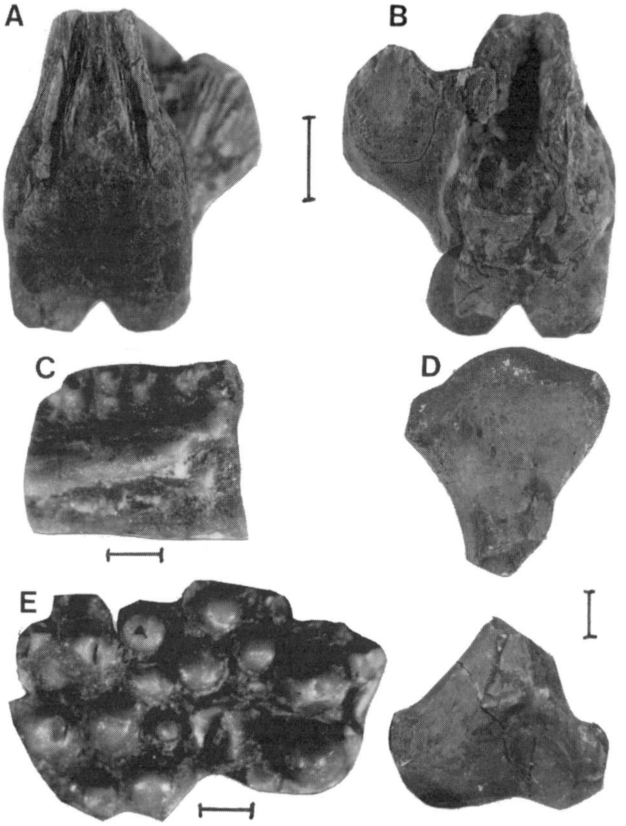

FIGURE 3. Fish and less derived reptiles from the Bull Canyon Formation. A-B. Holotype basisphenoid of *Quayia zideki* (NMMNH P-16888) in dorsal (A) and ventral (B) views. C. Dentary fragment of squamate (NMMNH P-4326) in medial view. D. Rhynchosaur humerus (UMMP 4771 [in part]) in posterior view. E. ?Colobodontid toothplate (NMMNH P-4850) in occlusal view. Scale bars are 1 cm (A-B, D) and 1 mm (C, E).

species from eastern North America (middle-late Norian).

Acme zone: Genus - Hettangian (Jurassic), species - Norian (Triassic).

Discussion: These two specimens represent a single species and resemble the *Semionotus brauni* group of Olsen et al. (1984) in having posteriorly directed spines on the dorsal midline scales (Huber et al., 1993).

Actinopterygii, Klein, 1885
incertae sedis
***?Tanaocrossus kalliokoskii* Schaeffer, 1967**

Type Revueltian occurrences: NMMNH localities 75 and 166.

Referred specimens from type Revueltian: NMMNH locality 75; NMMNH P-17306, skull; NMMNH P-17319, skeleton: NMMNH locality 166; NMMNH P-4860, skeleton.

First occurrence: Otischalkian (Triassic: late Carnian) of West Texas.

Last occurrence: Apachean (Triassic: late Norian-?Rhaetian) of southwestern Colorado.

Acme zone: ?Apachean (Triassic: late Norian-?Rhaetian).

Discussion: Huber et al. (1993) referred these specimens and AMNH 5662 from the Colorado City Formation of Howard County, Texas (Schaeffer, 1967, pl. 30, fig. 16) to *Tanaocrossus kalliokoskii* on the basis of similarities in the opercular, extrascapulars, dermopteric, elongate dorsal fin and elongate anal fin. Further preparation of the Bull Canyon specimens is required before this assignment can be fully evaluated.

Actinopterygii Klein, 1885
gen. et sp. indeterminate

Type Revueltian occurrences: NMMNH localities 1, 75, 81, 94, 97, 110, 369, 447, 501; UCM locality 82021; YPM locality 8199.

Referred specimens from type Revueltian: NMMNH locality 1; NMMNH P-4564, body segments and scales of large fish; NMMNH P-4707, scale; NMMNH 4711, scale; NMMNH P-4749, scale; NMMNH P-4759, scale; NMMNH P-4762, scale; NMMNH P-4765, scale; NMMNH P-4768, scale; NMMNH P-4748, scale; NMMNH P-4833, scale; NMMNH P-4834, scale; NMMNH P-4835, scale; NMMNH P-4840, scale; NMMNH P-4864, scale: NMMNH locality 75; NMMNH P-4184, body segments; NMMNH P-4186, body segments; NMMNH P-17199, head, body fragments; NMMNH P-17206, body segments; NMMNH P-17254, body segments; NMMNH P-17256, body fragment; NMMNH P-4885, partial body: NMMNH locality 81; NMMNH P-4177, headless three dimensional fish: NMMNH locality 94; NMMNH P-4116, partial body: NMMNH locality 97; NMMNH P-4124, body segment: NMMNH locality 110; NMMNH P-4964, jaw fragment: NMMNH locality 369; NMMNH P-4183, body segments: NMMNH locality 447; NMMNH P-16829, bone: NMMNH locality 501; NMMNH P-17297, vertebra: UCM locality 82021; UCM 52079, partial body and uncataloged isolated scales: YPM locality 5831; YPM 8199, fragment.

First occurrence: *Naxilepis* from Late Silurian of China.

Last occurrence: Numerous Recent species.

Acme zone: Not applicable.

Discussion: The Bull Canyon Formation ichthyofauna contains a number of indeterminate actinopterygian specimens. These include isolated scales and partial fish. Some specimens may become identifiable after further preparation, for example NMMNH P-4694, which is fragments of a large fish (scales about 10 mm in length). Some of the unornamented scales may pertain to semionotids.

Sarcopterygii Romer 1955
Actinistia Cope, 1871
Dipnoi Muller, 1844
Ceratodontoidei Vorobyeva and Obruchev, 1964
?Arganodontidae Martin, 1982

Type Revueltian occurrence: NMMNH locality 1.

Referred specimens from type Revueltian: NMMNH locality 1; NMMNH P-16590, skull fragments; NMMNH P-16889, skull fragments.

First occurrence: *Arganodus multicristatus* from the Lower Triassic of Malagasy.

Last occurrence: *Arganodus tiguidiensis* from the Late Jurassic-early Cretaceous of the Sahara.

Acme zone: Not applicable.

Discussion: A few skull fragments from NMMNH locality 1 are similar to specimens of lungfish (e.g., Schultze, 1981). Because of their age, these specimens are tentatively assigned to the Arganodontidae. Lungfish toothplates are notably absent from the Bull Canyon Formation, but they are locally common in correlative units (Long and Murry, 1989).

Coelacanthoidei Berg, 1937
Coelacanthidae Agassiz, 1843
***Quayia zideki*, Hunt, 1997**

Type Revueltian occurrences: NMMNH localities 1 and 462.

Referred specimens from type Revueltian: NMMNH locality 1; NMMNH P-4682, parasphenoid fragments; NMMNH P-4685, parasphenoid; NMMNH P-16588, scale; NMMNH P-16593, skull fragments; NMMNH P-16597, parasphenoid fragments; NMMNH P-16620, skull fragments; NMMNH P-16926, skull fragments; NMMNH P-16760, skull fragments; NMMNH P-16888 (Fig. 3A-B), basisphenoid lacking left antotic process (holotype); NMMNH P-16918, lower jaw; NMMNH P-16942, skull fragments; NMMNH uncataloged, three basisphenoids and fragments: NMMNH locality 462; NMMNH P-16873, skull fragments.

First occurrence: Revueltian (Triassic: early-middle Norian), New Mexico.

Last occurrence: Revueltian (Triassic: early-middle Norian), New Mexico.

Acme zone: Revueltian (Triassic: early-middle Norian), New Mexico.

Discussion: The basisphenoid of *Quayia* is about three quarters of the size of that of the living *Latimeria*. This indicates that it is the largest fish in the Bull Canyon ichthyofauna.

Quayia differs from all other coelacanths in the form of its basisphenoid, which is a generically diagnostic element (e.g., Schaeffer and Gregory, 1961). Notably, *Quayia* is distinct from *Chinlea* (Schaeffer, 1967, fig. 13A, pl. 27, 1-2), which is the genus into which all other Chinle Group coelacanth specimens have been rather arbitrarily placed. These differences are not ontogenetic, as the new genus is represented by a growth series with the posterior width varying from 20 mm to 36 mm. Diagnostic specimens of *Chinlea* are restricted to the Apachean (Late Norian-?Rhaetian) of western North America. The basisphenoid of *Quayia* differs from that element of *Chinlea* in the following features: (1) antotic processes are much narrower; (2) the portion of the pleurosphenoid that articulates with the antotic process is much smaller; (3) the sphenoid condyles do not flare posteriorly; (4) the lateral margins of the posterior part of the bone are parallel; (5) the basisphenoid extends considerably anterior to the antotic processes; (6) the basisphenoid narrows anteriorly; and (7) the dorsal depression between the antotic processes is much narrower. In overall morphology, the basisphenoid of *Quayia* is most similar to that element in *Spermatodus*, *Moenkopia*, *Latimeria* and *Chinlea* in having roughly rectangular antotic processes that have large overlap areas for the pleurosphenoid (Schaeffer and Gregory, 1961).

Family ?Coelacanthidae Agassiz, 1843

Type Revueltian occurrences: NMMNH localities 1 and 171.

Referred specimens from type Revueltian: NMMNH locality 1; NMMNH P-16586, skull fragment; NMMNH locality 171; NMMNH P-16706, skull fragments.

First occurrence: *Coelacanthus granulatus* from Ufimian (Permian) of Western Europe.

Last occurrence: *Heptanema* from Tithonian (Jurassic) of Germany.

Acme zone: Not applicable.

Discussion: A few skull fragments appear to have the characteristic ornamentation of coelacanth cranial elements, which consist of narrow, irregular channels (e.g., Elliot, 1987, fig. 1).

Amphibia Linnaeus, 1758
Temnospondyli Zittel, 1898
Metoposauridae Watson, 1919
***Apachesaurus gregorii* Hunt, 1993**

Type Revueltian occurrences: NMMNH localities 1, 110, 132, 160, 171, 467, 498, 499, 501, 518, 519: UMMP unnumbered.

Referred specimens from type Revueltian: NMMNH locality 1; NMMNH P-4670, intercentra; NMMNH P-16604, intercentra; NMMNH P-16901, skull: NMMNH locality 110; NMMNH P-4398, intercentrum: NMMNH locality 132; NMMNH P-4589, centra and skull fragments: NMMNH locality 160; NMMNH P-4571, intercentrum: NMMNH locality 171; NMMNH P-4855, intercentra: NMMNH locality 467; NMMNH P-16904, intercentrum: NMMNH locality 498; NMMNH P-17181, intercentrum: NMMNH locality 499; NMMNH P-17272, intercentrum: NMMNH locality 501; NMMNH P-17296, intercentrum: NMMNH locality 518; NMMNH P-17370, intercentra: NMMNH locality 519; NMMNH P-17178, intercentrum: NMMNH locality 527; NMMNH P-17491, intercentra: 5 miles west of San Jon in badlands; UMMP 7441 (in part), one intercentrum.

First occurrence: Adamanian (Triassic: late Carnian) of West Texas.

Last occurrence: Apachean (Triassic: late Norian-?Rhaetian) of New Mexico.

Acme zone: Revueltian-Apachean (Triassic: early Norian -?Rhaetian)

Discussion: *Apachesaurus* is the only identifiable amphibian in the Bull Canyon fauna. A nearly complete skull (NMMNH P-16901) and many diagnostic intercentra occur in the Bull Canyon Formation. This species of metoposaur is distinct in that it most commonly occurs in the paleosol-hosted taphofacies in association with terrestrial reptiles. The much larger *Metoposaurus* and *Buettneria* of older intervals of the Chinle are found exclusively in floodplain-hosted taphofacies in association with semi-aquatic phytosaurs. *Apachesaurus gregorii* has several morphologic features that suggest that it was more terrestrial in its habits than other metoposaurs, including a less-developed lateral line system and a more robust ilium (Hunt, 1993a; Hunt et al, 1993c).

***?Apachesaurus* sp.**

Type Revueltian occurrences: NMMNH localities 1, 122, 154, 171, 174, 182, 451, 452, 467, 486, 520, 523, 526, 527.

Referred specimens from type Revueltian: NMMNH locality 1; NMMNH P-4667, skull fragments; NMMNH P-4668, pectoral girdle fragments; NMMNH P-4669, mandible fragments; NMMNH P-16601, skull fragments; NMMNH P-16619, posterior mandible fragment; NMMNH P-16592, skull fragments; NMMNH P-16941, skull fragment; NMMNH P-16943, rib ends: NMMNH locality 122; NMMNH P-4439, partial clavicle: NMMNH locality 154; NMMNH P-16857, skull fragment: NMMNH locality 171; NMMNH P-4866, skull fragment; NMMNH P-4871, mandible fragment; NMMNH P-16727, interclavicle fragment; NMMNH P-16638, interclavicle fragment; NMMNH P-16893, posterior mandible fragment: NMMNH locality 174; NMMNH P-4872, mandible fragment; NMMNH P-16594, interclavicle fragment: NMMNH locality 182; NMMNH P-16595, interclavicle fragment: NMMNH locality 451; NMMNH P-16817, interclavicle fragment: NMMNH locality 452; NMMNH P-16857, skull fragment: NMMNH locality 467; NMMNH P-16906, mandible fragment; NMMNH locality 486; NMMNH P-17053, interclavicle fragment: NMMNH locality 520; NMMNH P-17192, skull fragment: NMMNH locality 523; NMMNH P-17453, mandible fragments: NMMNH locality 526; NMMNH P-17377, skull fragments; NMMNH P-17468, skull fragments: NMMNH locality 527, NMMNH P-17475, pectoral fragments.

First occurrence: Adamanian (Triassic: late Carnian) of West Texas.

Last occurrence: Apachean (Triassic: late Norian-?Rhaetian) of New Mexico.

Acme zone: Revueltian-Apachean (Triassic: early Norian -?Rhaetian)

Discussion: Many metoposaurid fragments occur in the Bull Canyon Formation. The size of these specimens is consistent with an assignment to *Apachesaurus*, the only diagnostic metoposaurid from this formation. However, these specimens can only be tentatively assigned to this genus because they lack its apomorphic characters.

Metoposauridae Watson, 1919
gen. et sp. indeterminate

Type Revueltian occurrences: NMMNH localities 175, 198, 465, 534.

Referred specimens from type Revueltian: NMMNH locality 175; NMMNH P-16567, waterworn skull fragments: NMMNH locality 198; NMMNH P-16558, ?skull fragments; NMMNH P-16583, skull fragments; NMMNH P-16585, skull fragment; NMMNH P-16602, skull fragment: NMMNH locality 465; NMMNH P-16882, skull fragment: NMMNH locality 534; NMMNH P-17388, skull fragments.

First occurrence: Several species from Otischalkian (Triassic: early late Carnian) and correlative strata worldwide. *Trigosternum* and the undescribed skull from Eschenau from the Middle Triassic Lettenkeuper of Germany do not pertain to the Metoposauridae (Hunt, 1993a; *contra* Milner, 1993).

Last occurrence: Apachean (Triassic: late Norian-?Rhaetian) of New Mexico.

Acme zone: Otischalkian-Adamanian (Triassic: early late Carnian to latest Carnian).

Discussion: Several specimens from the Bull Canyon Formation represent fragmentary skull fragments of individuals larger than known *Apachesaurus*. These specimens derive from intraformational conglomerates in the lower portion of the formation. They are probably reworked from late Carnian strata, but they could demonstrate the presence of a large metoposaur in the lowest portion of the Bull Canyon Formation.

Reptilia Laurenti, 1768
Lepidosauromorpha Benton, 1984
Lepidosauria Haeckel, 1866
?Squamata Oppel, 1811
Family indeterminate

Type Revueltian occurrences: NMMNH localities 1 and 110.

Referred specimens from type Revueltian: NMMNH locality 1; NMMNH P-4326, jaw fragment (Fig. 3C): NMMNH locality 110; NMMNH P-4965, jaw fragments.

First occurrence: Lepidosauria - *Saurosternon bainii* from Tatarian (Permian) of South Africa, Squamata - *Dorsetisaurus purbeckensis*, *Becklesius hoffstetteri* and *Saurillodon* spp. from the Oxfordian (Jurassic) of Portugal.

Last occurrence: Lepidosauria and Squamata - numerous species from Recent.

Acme zone: Not applicable.

Discussion: Jaw fragments from two localities in the Bull Canyon Formation exhibit homodont, pleurodont dentition and apparently represent squamates. Murry (1986) noted the presence of a similar jaw fragment from the Tecovas Formation of the Chinle Group in Texas. These specimens appear to indicate the presence of a Late Triassic lizard-like animal in the upper Chinle Group, but they might also pertain to the Kuehneosauridae, a non-squamate lepidosauromorph clade of Carnian-Rhaetian gliding reptiles.

Archosauromorpha Huene, 1946
Rhynchosauria Osborn, 1903
Rhynchosauridae Huxley, 1887
gen. et sp. indet.

Type Revueltian occurrences: NMMNH locality 1: UMMP locality 5 miles west of San Jon in Badlands.

Referred specimens from type Revueltian: NMMNH locality 1; NMMNH P-4686, right distal humerus; NMMNH P-4703, right distal humerus: 5 miles west of San Jon in badlands; UMMP 4771 (part), left distal and proximal humerus (Fig. 3D), small left distal humerus.

First occurrence: Two genera from *Cynognathus-Diademodon* Assemblage Zone (Triassic: Olenekian) of South Africa.

Last occurrence: Revueltian (Triassic: early-middle Norian), New Mexico.

Acme zone: Anisian-Ladinian (Triassic).

Discussion: The larger proximal and distal humeri of UMMP 4771 obviously represent the same taxon as NMMNH P-4703, being identical in size, preservation and morphology. A variety of evidence indicate that NMMNH locality 1 is the same as Case's locality 5 miles west of San Jon in Badlands (Hunt, 1994). These specimens pertain to a primitive reptile that has a humerus with greatly flared proximal and distal extremities. These specimens evidently represent a rhynchosaur very close in morphology to *Otischalkia* (Hunt and Lucas, 1991b, pl. 1, 4-5, fig. 2). The Bull Canyon Formation taxon differs from *Otischalkia* in the form of the ectepicondylar area--they apparently lack a supinator crest and may have had an enclosed ectepicondylar foramen.

UMMP 12972 is a complete humerus from the Tecovas Formation in West Texas that also represents a rhynchosaur. This specimen is Adamanian in age but is younger than the Otischalkian *Otischalkia*. These specimens indicate that a lineage of rhynchosaurs survived in North America through the late Carnian and into the Norian. The Bull Canyon specimens are the youngest known rhynchosaur specimens and demonstrate that rhynchosaurs persisted beyond the Carnian (*contra* Hunt and Lucas, 1991b).

Archosauromorpha Huene, 1946
incertae sedis
***Vancleavea campi* Long and Murry, 1995.**

Type Revueltian occurrence: NMMNH locality 522.

Referred specimens from type Revueltian: NMMNH locality 5; NMMNH P-20852, centra associated as stomach contents with skeleton of *Pseudopalatus andersoni*: NMMNH locality 522; NMMNH P-4984, partial articulated skeleton including limb, vertebra and osteoderms;

First occurrence: Late Carnian (Adamanian), Los Esteros Member of Santa Rosa Formation and Blue Mesa Member of Petrified Forest Formation.

Last occurrence: Late Norian-Rhaetian (Apachean) Redonda Formation and Rock Point Formation, New Mexico.

Acme zone: None

Discussion: Long and Murry (1995) named *Vancleavea campi* for a partial skeleton of an armored reptile from the Adamanian Blue Mesa Member of the Petrified Forest Formation in Petrified Forest National Park. This unusual reptile combines primitive-aspect postcrania with specialized osteoderms. The most typical osteoderms are rounded to diamond-shaped with a smooth dorsal surface. The most distinctive feature of the osteoderms is a median ridge that terminates in an arcuate anterior prong. Osteoderms of this morphology are found in New Mexico in the Los Esteros Member of the Santa Rosa Formation (Adamanian), Bull Canyon Formation (Revueltian) and Redonda and Rock Point formations (Apachean). The Redonda specimens had previously been assigned to a sphenosuchian (e.g., Hunt and Lucas, 1993d) on the basis of the anterior prong, which has been considered a crocodylomorph apomorphy (e.g., Clark et al., 2000). It is clear that one or more species of this clade of reptiles are present throughout the Adamanian-Apachean portions of the Chinle Group. Ongoing study indicates that some specimens assigned to Reptilia indet. from NMMNH locality 1 actually pertain to the same taxon as NMMNH P-4984.

Phytosauria Baur, 1898
Parasuchidae Lydekker, 1888
***Pseudopalatus pristinus* Mehl, 1928**

Type Revueltian occurrences: NMMNH localities 4, 13, 432, 442: YPM locality 5831.

Referred specimens from type Revueltian: NMMNH locality 4; NMMNH P-4654, skull: NMMNH locality 13; NMMNH P-4979, skull and partial skeleton including stomach contents (*Apachesaurus* intercentra): NMMNH locality 432; NMMNH P-4977, rostrum: NMMNH locality 442; NMMNH P-4980, half rostrum: YPM locality 5831; YPM 5831, skull and partial skeleton.

First and last occurrences: Numerous localities in Revueltian (early-middle Norian) strata in Arizona, New Mexico and Texas.

Acme zone: Revueltian (early-middle Norian).

Discussion: Diagnostic specimens of this dolichorostral (*sensu* Hunt, 1994) phytosaur are relatively uncommon in the Bull Canyon fauna. One skeleton (NMMNH P-4979) contains stomach contents consisting of *Apachesaurus* intercentra. These specimens are considered to be true gastric contents because: (1) they occur in the ribcage area; (2) there are only specimens of the phytosaur in this quarry (apart from the metoposaur intercentra); and (3) the intercentra are part of a single vertebral column.

***Pseudopalatus andersoni* (Mehl, 1922)**

Type Revueltian occurrences: NMMNH localities 5, 6, 12, 14, unknown FMNH locality.

Referred specimens from type Revueltian: NMMNH locality 5; NMMNH P-20852, skull and partial skeleton (20+ ribs, coracoid, 36+ osteoderms, 2 ilia, 3 cervical vertebrae, 9 dorsal vertebrae, 11 caudal vertebrae, pubis, 13+ neural spines, chevron, radius, ulna, 3 ?sacral vertebrae, 2 isolated teeth 5 cervical ribs, humerus, scapula, coracoid, tibia, astragalus, proximal femur) including stomach contents (one small proximal phytosaur pubis) and small centra of *Vancleavea campi*: NMMNH locality 6; NMMNH P-4973, skull; NMMNH locality 12; NMMNH P-4970, skull: NMMNH locality 14; NMMNH P-4972, skull in two pieces: UNM locality; UNM MV-650, skull (Fig. 4): Unknown FMNH locality that is probably Bull Canyon (Hunt, 1994); FMNH UC 396, skull (holotype).

First and last occurrences: Numerous localities in Revueltian (early-middle Norian) strata in Arizona, New Mexico and Texas.

Acme zone: Revueltian (early-middle Norian).

Discussion: *Pseudopalatus andersoni* is known from six skulls from the lower Bull Canyon Formation. One skull (UNM MV-650) (Fig. 4) was illustrated by Hunt and Lucas (1989b). NMMNH P-20852 is a skull and the majority of a skeleton that preserved stomach contents consisting of a proximal pubis of a smaller phytosaur and a series of vertebrae of *Vancleavea campi*.

Pseudopalatus n. sp.

Type Revueltian occurrences: NMMNH localities 2, 3 and 199: YPM locality 5833

Referred specimens from type Revueltian: NMMNH locality 2; NMMNH P-4256, skull and cervical series: NMMNH locality 3; NMMNH P-4239, anterior of skull and lower jaws: NMMNH locality 199; NMMNH P-16608, tip of rostrum: PM locality 5833; YPM 3293, skull and lower jaws.

First and last occurrences: Several localities in Revueltian (early-middle Norian) strata in Arizona and New Mexico.

Acme zone: Revueltian (early-middle Norian).

Discussion: *Pseudopalatus* n. sp. is a robust morph of the genus, analogous to *Rutiodon* (*Smilosuchus*) *gregorii*. Specimens of this species are more common in the Bull Canyon Formation than in any other units of the Chinle Group. Specimens from the Bull Canyon Formation have previously been assigned to the species "*Machaeroprosopus" gregorii* and have variously been placed in the genera *Machaeroprosopus* (Camp, 1930), *Phytosaurus* (Gregory, 1972), *Rutiodon* (Carpenter and Parrish, 1985; Parrish and Carpenter, 1986) or *Nicrosaurus* (Lucas et al., 1985a c). Three well preserved skulls of this species have been collected from the Revuelto and Barranca badlands.

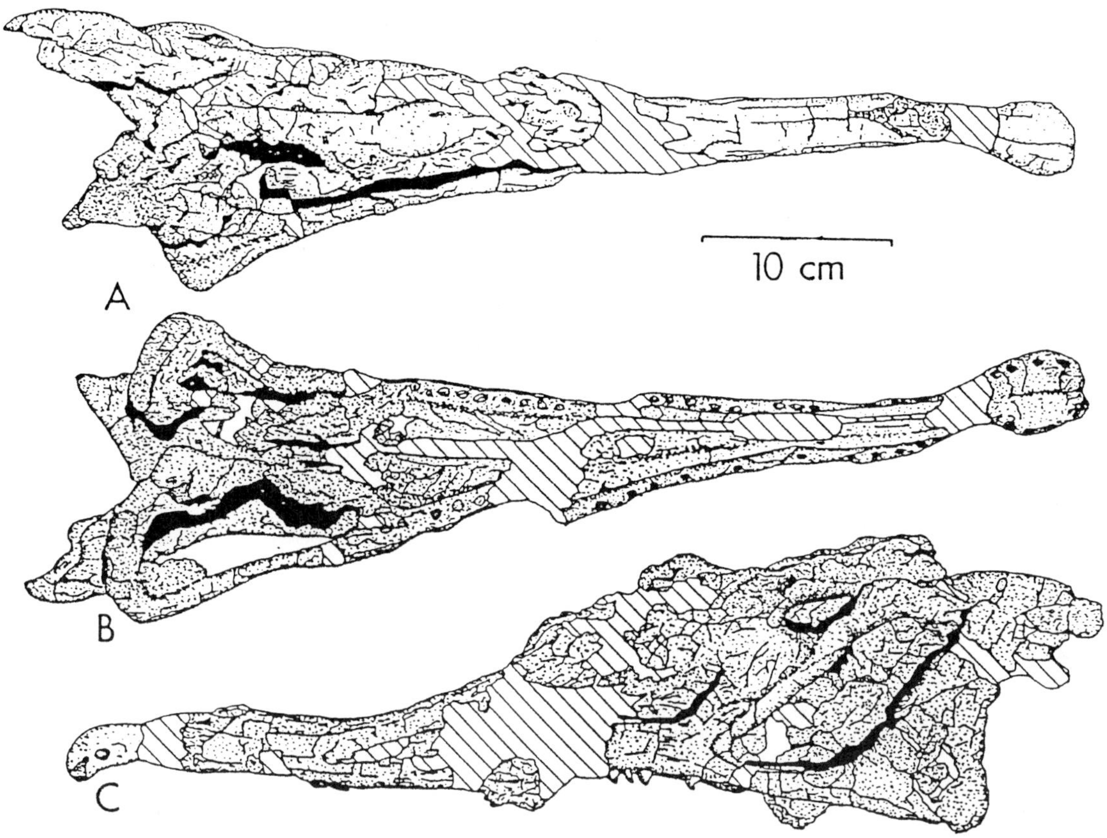

FIGURE 4. Skull of *Pseudopalatus andersoni* (UNM MV-650) from the Bull Canyon Formation at Bull Canyon in dorsal (A), ventral (B) and left lateral (C) views. Drawings by R. Pence.

?*Pseudopalatus* n. sp.

Type Revueltian occurrences: NMMNH locality 30: UCM locality 82021.

Referred specimens from type Revueltian: NMMNH locality 30; NMMNH P-4981, maxillary fragment: UCM locality 82021; UCM 48441, posterior skull.

First and last occurrences: Several localities in Revueltian (early-middle Norian) strata in Arizona and New Mexico.

Acme zone: Revueltian (early-middle Norian).

Discussion: Two skull fragments represent robust animals the size of *Pseudopalatus* n. sp. and are tentatively assigned to this taxon.

***Pseudopalatus* sp.**

Type Revueltian occurrences: NMMNH localities 1, 8, 9, 77, 89, 11, 122, 436, 542, 544-545.

Referred specimens from type Revueltian: NMMNH locality 1; NMMNH P-16507, juvenile right squamosal: NMMNH locality 8; NMMNH P-4974, lower jaw: NMMNH locality 9; NMMNH P-4971, lower jaws: NMMNH locality 77; NMMNH P-4194, tip of squamosal: NMMNH locality 89; NMMNH P-4978, lower jaw ramus: NMMNH locality 111; NMMNH P-4368, right squamosal: NMMNH locality 122; NMMNH P-4422, left squamosal: NMMNH locality 436; NMMNH P-4975, lower jaws: NMMNH locality 542; NMMNH P-17380, squamosal: NMMNH locality 544; NMMNH P-17323, lower jaws: NMMNH locality 545; NMMNH P-17317, left squamosal.

First and last occurrences: Numerous localities in Revueltian (early-middle Norian) strata in Arizona, New Mexico and Texas.

Acme zone: Revueltian (early-middle Norian).

Discussion: Several lower jaws and posterior skull fragments can be assigned to *Pseudopalatus*, but the species cannot be determined because of the lack of a rostrum.

Parasuchidae Lydekker, 1888
gen. et sp. indet.

Type Revueltian occurrences: NMMNH localities 1, 17, 19, 21-23, 25, 27, 30-31, 33-35, 37-43, 45-49, 51-54, 57, 59-67, 69-72, 74-81, 84, 86-89, 93-96, 100-104, 106, 110, 112-122, 124, 126, 128-129, 132-134, 136, 140-142, 146-148, 151-157, 160-166, 169-175, 177-179, 182, 185-186, 189, 191-192, 194-196, 198, 447-448, 450-452, 461-467, 497-500, 502, 506, 508, 510-512, 514, 520, 522-523, 525, 527, 529-530, 534-535-540, 542, 546, 548: YPM localities 5831, 5833, 5887, 6649.

Referred specimens from type Revueltian: NMMNH locality 1; NMMNH P-4697, osteoderm fragment; NMMNH P-4853, tooth (type B, sensu Hunt, 1994); NMMNH P-4863, teeth (type U, I); NMMNH P-4914, osteoderm fragment; NMMNH P-4942, rostral fragments; NMMNH P-4952, fibula fragments; NMMNH P-16505, large tooth (type B); NMMNH P-16508, tooth (type I); NMMNH P-16515, rib; NMMNH P-16517, osteoderm fragments; NMMNH P-16529, maxillary fragment; NMMNH P-16605, osteoderm fragment; NMMNH P-16613, tooth fragments (type U); NMMNH P-16614, teeth (type B, I); NMMNH P-16625, skull fragments; NMMNH P-16678, teeth (type I, U); NMMNH P-16688, teeth (type U): NMMNH locality 16; NMMNH P-4350, teeth (type U); NMMNH P-4352, osteoderm fragment: NMMNH locality 17; NMMNH P-4348, jaw fragment in conglomerate: NMMNH locality 18; NMMNH P-4648, osteoderms: NMMNH locality 19; NMMNH P-4343, skull fragments; NMMNH P-4649, tooth (type U): NMMNH locality 20; NMMNH P- 4340, small osteoderm: NMMNH locality 21; NMMNH P-4313, tip of snout: NMMNH locality 22; NMMNH P-4312, large humerus; NMMNH P-4344, osteoderm fragments: NMMNH locality 23; NMMNH P-17496, juvenile jaw; NMMNH P-4200, caudal central fragments; NMMNH P-4314, rib: : NMMNH locality 24; NMMNH P-4323, dorsal vertebrae fragments: NMMNH locality 25; NMMNH P-4327, 4 teeth (types B, U) in conglomerate; NMMNH P-16528, tooth (type I); NMMNH P-17313, partial centrum: NMMNH locality 27; NMMNH P-4281, osteoderm fragments; NMMNH P-4285, skull and teeth fragments; NMMNH P-4325, teeth (types B, U) and osteoderm fragments: NMMNH locality 30; NMMNH P-4334, paramedian osteoderm fragment: NMMNH locality 31; NMMNH P-4297, osteoderm fragment: NMMNH locality 33; NMMNH P-4462, tooth; NMMNH P-4562, cranial fragment: NMMNH locality 34; NMMNH P-4143, osteoderm fragments; NMMNH P-4295, skull fragment; NMMNH P-4317, osteoderm fragments; NMMNH P-4341, osteoderm fragments: NMMNH locality 35; NMMNH P-4212, tooth (type U); NMMNH P-4301, transverse process; NMMNH P-4303, dorsal centrum; NMMNH P-4306, osteoderm fragment; NMMNH P-4307, teeth (type U); NMMNH P-4335, rib: NMMNH locality 37; NMMNH P-4298, cervical centrum and neural arch fragments: NMMNH locality 38; NMMNH P-4290, cervical vertebra fragments: NMMNH locality 39; NMMNH P-4279, skull fragments (many): NMMNH locality 40; NMMNH P-4136, skull fragment; NMMNH P-4287, skull fragments: : NMMNH locality 41; NMMNH P-4651, humerus fragments with teeth marks: NMMNH locality 42; NMMNH P-4316, tooth (type B): NMMNH locality 43; NMMNH P-4267, dorsal centrum; NMMNH P-4268, skull fragment: NMMNH locality 45; NMMNH P-4291, ulna: NMMNH locality 46; NMMNH P-4330, rib: NMMNH locality 47; NMMNH P-4204, osteoderm fragments; NMMNH P-4321, osteoderm, neural spine fragments; NMMNH locality 48; NMMNH P-4206, rib and vertebral fragments: NMMNH locality 49; NMMNH P-4201, osteoderm fragments: NMMNH locality 51; NMMNH P-4276, dorsal centra, neural arch and rib fragments; NMMNH P-17248, proximal scapula (small); NMMNH P-17252, proximal pubis; NMMNH P-17261, distal femur; NMMNH P-17260, cervical vertebrae: NMMNH locality 52; NMMNH P-4264, interclavicle and clavicle fragments: NMMNH locality 53; NMMNH P-4134, tooth (type B); NMMNH P-4159, osteoderm fragments; NMMNH P-4158, rib fragment; NMMNH P-4167, tooth (type U) in sandstone; NMMNH P-4229, cervical vertebral fragments; NMMNH P-4320, osteoderm fragments; NMMNH P-4322, centrum: NMMNH locality 54; NMMNH P-4266, osteoderm fragments: NMMNH locality 57; NMMNH P-4220, metatarsal fragment; NMMNH P-4221, ilium fragment; NMMNH P-4222, osteoderm fragments; NMMNH P-17427, atlas; NMMNH P-17433, osteoderm fragments; NMMNH P-17434, rib fragments: NMMNH locality 59; NMMNH P-4154, skull fragments: NMMNH locality 60; NMMNH P-4236, proximal and distal tibia: NMMNH locality 61; NMMNH P-4133, osteoderm fragments: NMMNH locality 62; NMMNH P-4230, osteoderm fragments; NMMNH P-4646, tip of neural spine: NMMNH locality 63; NMMNH P-4244, ilium fragments: NMMNH locality 64; NMMNH P-4132, skull fragment NMMNH; P-4241, osteoderm fragment in conglomerate: NMMNH locality 65; NMMNH P-4247, dorsal vertebra: NMMNH locality 66; NMMNH P-4243, osteoderm impression: NMMNH locality 67; NMMNH P-4217: NMMNH locality 68; NMMNH P-4131, centrum fragments; NMMNH P-4215, osteoderm fragment: NMMNH locality 69; NMMNH P-4224, osteoderm fragment: NMMNH locality 70; NMMNH P-4137, proximal chevron fragment; NMMNH P-4148, vertebral, skull and tooth (type B) fragments; NMMNH P-4149, osteoderm fragment; NMMNH P-4153, top of neural spine; NMMNH P-4146, rostral crest fragment; NMMNH P-4192, skull fragments; NMMNH P-4213, osteoderm; NMMNH P-17311, rostrum without crest: NMMNH locality 71; NMMNH P-4225, cervical and dorsal vertebral fragments; NMMNH P-17351, cervical centrum: NMMNH locality 72; NMMNH P-4231, jaw fragment: NMMNH locality 74; NMMNH P-4152, tooth fragment (type U); NMMNH P-4156, osteoderm fragments: NMMNH locality 75; NMMNH P-4180, osteoderm fragment; NMMNH P-4187; teeth fragments (type B); NMMNH P-17343, vertebral fragments; NMMNH P-18187, calcaneum, metatarsals: NMMNH locality 76; NMMNH P-4182, juvenile rostrum: NMMNH locality 77; NMMNH P-4193, osteoderm fragments: NMMNH locality 78; NMMNH P-4189, osteoderm fragments: NMMNH locality 79; NMMNH P-4174, osteoderm fragments; NMMNH P-4172, tooth fragments: NMMNH locality 80; NMMNH P-4169, cervical vertebra; NMMNH P-4170, osteoderm fragment; NMMNH P-17397, proximal femur: NMMNH locality 81; NMMNH P-17331, basicranium: NMMNH locality 84; NMMNH P-4164, skull fragments; NMMNH P-4166, tooth fragment (type U): NMMNH locality 85; NMMNH P-4144, dorsal centrum and neural arch fragments: NMMNH locality 86; NMMNH P-4140, tooth (type U) in conglomerate; NMMNH P-16523, osteoderm fragments; NMMNH P-16524, teeth (type U, B); NMMNH P-16525, top of neural spine; NMMNH P-16526, skull fragments: NMMNH locality 87; NMMNH P-4139, osteoderm fragments: NMMNH locality 88; NMMNH P-4142, osteoderm fragments; NMMNH P-4647, ?dentary fragments: NMMNH locality 89; NMMNH P-4118, skull fragment; NMMNH P-17124, vertebral fragments; NMMNH P-17143, vertebral fragments; NMMNH P-17379, rib fragments: NMMNH locality 93; NMMNH P-4138, tooth fragment (type B): NMMNH locality 94; NMMNH P-4114, caudal centrum and chevron fragments: NMMNH locality 95; NMMNH P-4117, skull fragments: NMMNH locality 96, NMMNH P-4123, jaw fragments: NMMNH locality 100; NMMNH P-4128, rib fragments; NMMNH locality 101; NMMNH P-4109, paramedian osteoderm: NMMNH locality 102; NMMNH P-4108, vertebral fragments; NMMNH P-17333, osteoderm: NMMNH locality 103; NMMNH P-4107, vertebral fragments; NMMNH P-4338, 2 small osteoderms; NMMNH P-4650, teeth (type I): NMMNH locality 104; NMMNH P-4336, neural arch fragments; NMMNH P-4337, osteoderm fragments: NMMNH locality 106; NMMNH P-4645, tip of large rostrum: NMMNH locality 109; NMMNH P-4103, skull fragments: NMMNH locality 110; NMMNH P-4355, phalanx; NMMNH P-4356, skull fragments; NMMNH P-4357, dorsal centrum fragment; NMMNH P-4358, osteoderm fragments in conglomerate; NMMNH P-4485, distal femur; NMMNH P-4488, teeth fragments; NMMNH P-4529, osteoderm fragments; NMMNH P-4530, cervical rib fragments; NMMNH P-4531, centrum; NMMNH P-4533, ilium, osteoderm and limb fragments; NMMNH P-4535, teeth (type B, U); NMMNH P-4538, tooth (type B); NMMNH P-4540, juvenile dentary tip; NMMNH P-4567, ischium fragment; NMMNH P-4577, teeth (type U, I); NMMNH P-4578, teeth (type U); NMMNH P-4579, tooth (type U): NMMNH locality 112; NMMNH P-4390, 2 articulated dorsal centra: NMMNH locality 113; NMMNH P-4411, small osteoderm; NMMNH P-4516, teeth (type B); NMMNH P-4517, tooth (type B); NMMNH P-4518, proximal coracoid; NMMNH P-4519, jaw fragment: NMMNH locality 114; NMMNH P-4477, osteoderm fragments: NMMNH locality 115; NMMNH P-4363, tooth (type B); NMMNH P-4367, tooth (type I); NMMNH P-4397, tooth (type U); NMMNH P-4408, tooth (type U); NMMNH P-4410, teeth fragments; NMMNH P-4413, tooth (type U); NMMNH P-4631, osteoderm; NMMNH P-4632, teeth fragments (type B): NMMNH locality 116; NMMNH P-4395, jaw fragments; NMMNH P-4497, dorsal vertebral fragments; NMMNH P-4501, jaw fragment: NMMNH P-4503, osteoderm fragments: NMMNH locality 117; NMMNH P-4365, centrum; NMMNH P-4392, centrum; NMMNH P-4483, osteoderm fragment: NMMNH locality 118; NMMNH P-4361, osteoderm fragment in sandstone; NMMNH P-4467, tooth (type C): NMMNH locality 119; NMMNH P-4362, mandible fragment; NMMNH P-4432, tip of snout: NMMNH locality 120; NMMNH P-4360, skull fragment (nasal area); NMMNH P-4389; NMMNH P- 4389, dorsal centrum; NMMNH P-4441, teeth (types B, U); NMMNH P-4446, jaw fragments; NMMNH P-4528, many skull fragments and caudal centra; NMMNH P-4549, proximal humerus; NMMNH P-4550, distal coracoid; NMMNH P-4551, vertebral fragments: NMMNH locality 121; NMMNH P-4642, dentary fragments: NMMNH locality 122; NMMNH P-4354, lower jaws; NMMNH P-4400, small osteoderm; NMMNH P-4468, osteoderm; NMMNH P-4493, very large cervical vertebra; NMMNH P-4523, jaw fragments; NMMNH P-4524, osteoderm fragments; NMMNH P-4525, skull fragments; NMMNH P-4526, dorsal vertebra fragments; NMMNH P-4527, scapula fragments; NMMNH P-4608, many osteoderm fragments; NMMNH P-4609, osteoderm; NMMNH P-4610, caudal centrum; NMMNH P-4611, humerus and ?limb bone fragments; NMMNH P-4619, osteoderm fragments: NMMNH locality 124; NMMNH P-4623, teeth (type I, B); NMMNH P-4624, caudal centra; NMMNH P-4626, osteoderm: NMMNH locality 126; NMMNH P-4383, jaw fragments; NMMNH P-4620, teeth fragments (Types C, I, B, U); NMMNH P-4622, skull fragments; NMMNH P-4628, pelvic fragments; NMMNH P-4629, dorsal centra; NMMNH P-4633, centra; NMMNH P-4634, osteoderms; NMMNH P-4635, skull fragments: NMMNH locality 128; NMMNH P-4419, dorsal centrum; NMMNH P-4453, skull fragments; NMMNH P-4473, dorsal centra: NMMNH locality 129; NMMNH P-4455, dentary fragment; NMMNH P-4479, distal scapula; NMMNH P-4492, osteoderm fragments: NMMNH locality 132; NMMNH P-4424, osteoderm fragments; NMMNH P-4428, teeth (types B, I, U); NMMNH P-4574, osteoderm fragment: NMMNH locality 133; NMMNH P-4511, osteoderm fragments; NMMNH P-4514, jaw fragments; NMMNH P-4515, ilium fragments: NMMNH locality 134; NMMNH P-4464, tip of dentary; NMMNH P-4472, teeth (types B, U); NMMNH P-4505, tooth fragment (type ?C); NMMNH P-4506, neural arch fragments; NMMNH P-4507, osteoderm fragment; NMMNH P-4508, cranial fragment; NMMNH P-4612, osteoderm fragments; NMMNH P-4615, skull fragments; NMMNH P-4637, dentary fragment; NMMNH P-4638, neural arch fragments: NMMNH locality 134; NMMNH P-4656, tooth: NMMNH locality 136; NMMNH P-4403, tooth (type B): NMMNH locality 140; NMMNH P-4486, neural spine and caudal centrum; NMMNH P-4544, osteoderm fragment: NMMNH locality 141; NMMNH P-4421, cervical vertebra: NMMNH locality 142; NMMNH P-4409, tooth (type B): NMMNH locality 146; NMMNH P-4457, osteoderm fragments: NMMNH locality 147; NMMNH P-4451, osteoderm fragments; NMMNH P-4521, skull fragments; NMMNH P-4546, osteoderm fragments: NMMNH locality 148; NMMNH P-4450, tooth (type U); NMMNH

P-4613, neural arch; NMMNH P-4616, skull fragment; NMMNH P-4618, teeth (type U): NMMNH locality 151; NMMNH P-4480, teeth fragments (type B, U); NMMNH P-4502, centrum fragment: NMMNH locality 152; NMMNH P-4476, osteoderm fragment: NMMNH locality 153; NMMNH P-4563, radius fragments: NMMNH locality 154; NMMNH P-4564, osteoderm fragments and small osteoderm: NMMNH locality 155; NMMNH P-4548, osteoderm fragment: NMMNH locality 156; NMMNH P-4561, osteoderm fragments: NMMNH locality 157; NMMNH P-4554, dentary fragments: NMMNH locality 160; NMMNH P-4570, osteoderm fragment; NMMNH P-4572, skull fragments: NMMNH locality 161; NMMNH P-4640, teeth (type C, U): NMMNH locality 162; NMMNH P-16629, osteoderm fragment: NMMNH locality 163; NMMNH P-4976, jaw fragments: NMMNH locality 164; NMMNH P-4690, partial crested rostrum: NMMNH locality 165; NMMNH P-4883, tooth fragment (type U); NMMNH P-4906, teeth fragments (type U); NMMNH P-4907, osteoderm fragments; NMMNH P-16511, teeth (type I, B); NMMNH P-16536, osteoderm; NMMNH P-16537, ilium fragment; NMMNH P-16584, metacarpal fragment; NMMNH P-16587, osteoderm fragment; NMMNH P-16596, rib end: NMMNH locality 166; NMMNH P-16527, pubis fragments; NMMNH P-16624, teeth fragments (type U); NMMNH P-16626, dorsal vertebral and neural arch fragments: NMMNH locality 169; NMMNH P-4899, skull fragment; NMMNH P-4900, osteoderm fragment; NMMNH P-16609, tooth fragment (type U): NMMNH locality 170; NMMNH P-16532, osteoderm fragments; NMMNH P-16834, centra: NMMNH locality 171; NMMNH P-4881, teeth fragments (type U, I); NMMNH P-16531, tibia/fibula fragments; NMMNH P-16635, teeth fragments; NMMNH P-16636, osteoderm; NMMNH P-16639, osteoderm fragments; NMMNH P-16896, tooth fragments: NMMNH locality 172; NMMNH P-4880, teeth (type I): NMMNH locality 173; NMMNH P-4857, teeth (type U, I); NMMNH P-4868, jaw fragment; NMMNH locality 174; NMMNH P-4873, teeth (type I); NMMNH P-16610, tooth fragments: NMMNH locality 175; NMMNH P-4876, dorsal rib (large); NMMNH P-4923, skull fragment; NMMNH P-16522, tooth (type U); NMMNH P-16530, mandible fragment; NMMNH P-16680, jaw fragments: NMMNH locality 177; NMMNH P-16533, sacral vertebra: NMMNH locality 178; NMMNH P-4889, pelvic fragments; NMMNH P-4892, partial dorsal centrum: NMMNH locality 179; NMMNH P-16509, osteoderm fragment: NMMNH locality 182; NMMNH P-16501, tooth fragments (type U); NMMNH P-16535, neural arch fragments: NMMNH locality 185; NMMNH P-16503, teeth (type B); NMMNH P-16534, teeth fragment: NMMNH locality 186; NMMNH P-4921, tooth (type U): NMMNH locality 189; NMMNH P-4930, osteoderm fragment: NMMNH locality 191; NMMNH P-4926, tooth (type U): NMMNH locality 192; NMMNH P-4938, femoral fragments: NMMNH locality 194; NMMNH P-16506, dentary fragment; NMMNH P-16514, osteoderms: NMMNH locality 195; NMMNH P-16516, humerus fragments; NMMNH P-16520, teeth fragments (type U); NMMNH P-16521, osteoderm fragments: NMMNH locality 196; NMMNH P-16510, proximal chevron; NMMNH P-16513, proximal calcaneum: NMMNH locality 198; NMMNH P-16502, proximal scapula; NMMNH P-16518, teeth (type B, I); NMMNH P-16519, osteoderm fragments: NMMNH locality 434; NMMNH P-17133, radius; NMMNH P-17338, tooth fragments: NMMNH locality 447; NMMNH P-16816, skull fragments; NMMNH P-16825, osteoderm: NMMNH locality 448; NMMNH P-16832, proximal coracoid; NMMNH P-16830, osteoderm fragment: NMMNH locality 450; NMMNH P-16821, osteoderm fragment: NMMNH locality 451; NMMNH P-16819, centrum; NMMNH P-16827, cervical centrum: NMMNH locality 452; NMMNH P-16850, caudal vertebra; NMMNH P-16851, osteoderms; NMMNH P-16854, teeth fragments (type U, B); NMMNH P-16856, vertebral fragments; NMMNH P-16860, teeth (type U): NMMNH locality 455; NMMNH P-16848, quadrate: NMMNH locality 461; NMMNH P-16881, osteoderm fragments: NMMNH locality 462; NMMNH P-16874, metatarsal; NMMNH P-16875, teeth fragments (type C, I, B): NMMNH locality 463; NMMNH P-16845, dorsal centrum; NMMNH P-16878, rib fragments: NMMNH locality 464; NMMNH P-16837, skull fragments: NMMNH locality 465; NMMNH P-16884, osteoderm fragments: NMMNH locality 466; NMMNH P-16900, osteoderm fragments: NMMNH locality 467; NMMNH P-16903, tooth; NMMNH P-16907, osteoderm fragment: NMMNH locality 497; NMMNH P-17170, teeth fragments; NMMNH P-17176, osteoderm fragment; NMMNH P-17244, rib fragments: NMMNH locality 498; NMMNH P-17138, juvenile hind limb (femur, tibia, fibula fragments, astragalus); NMMNH P-17185, teeth fragments (type U); NMMNH P-17190, partial juvenile pelvis: NMMNH locality 499; NMMNH P-17140, caudal vertebral fragments; NMMNH P-17173, phalanx; NMMNH P-17267, occipital condyle; NMMNH P-17269, osteoderms; NMMNH P-17273, osteoderm fragments: NMMNH locality 500; NMMNH P-17197, teeth (type I): NMMNH locality 502; NMMNH P-17128, iliac blade: NMMNH locality 506; NMMNH P-17129, rib fragments; NMMNH P-17403, osteoderm fragment; NMMNH P-17404, centrum fragment; NMMNH P-17405, metapodial fragment; NMMNH P-17407, osteoderm fragment: NMMNH locality 508; NMMNH P-17149; NMMNH P-17196, osteoderm; NMMNH P-17123, osteoderm fragments; NMMNH P-17249, teeth (type U) in conglomerate: NMMNH locality 510; NMMNH P-17204, distal neural spine: NMMNH locality 511; NMMNH P-17125, jaw fragments; NMMNH locality 512; NMMNH P-17147, central fragments; NMMNH P-17155, distal fibula; NMMNH P-17156, limb fragments; NMMNH P-17171, caudal centrum; NMMNH P-17243, ?metatarsal fragment: NMMNH locality 514; NMMNH P-17121, osteoderm; NMMNH P-17142, vertebral fragments: NMMNH locality 518; NMMNH P-17329, iliac fragments; NMMNH P-17414, osteoderms and osteoderm fragments; NMMNH P-17420, vertebral fragments; NMMNH P-17426, tip of neural spine; NMMNH P-17324, fragmentary anterior skeleton (interclavicle, vertebrae, phalange fragments); NMMNH P-17257, fragmentary skeleton (vertebrae, pubis, scapula): NMMNH locality 520; NMMNH P-17188, rib fragments: NMMNH locality 522; NMMNH P-17315, impression of tooth (?type C) in sandstone; NMMNH P-17383, caudal vertebrae; NMMNH P-17409, dorsal centrum; NMMNH P-17425, centra; NMMNH P-17428, caudal centrum; NMMNH P-17445, tooth (type I): NMMNH locality 523; NMMNH P-17452, neural spine: NMMNH locality 525; NMMNH P-17441, osteoderm; NMMNH locality 527; NMMNH P-17341, skull fragment; NMMNH P-17473, osteoderm fragments; NMMNH P-17477, jaw fragment; NMMNH P-17481, tooth fragments; NMMNH P-17484, skull fragment; NMMNH P-17502, cervical centrum: NMMNH locality 529; NMMNH P-17308, bone fragment; NMMNH P-17309, proximal ulna (large); NMMNH P-17310, femoral fragment; NMMNH P-17363, rib fragment; NMMNH P-17366, pelvic fragment; NMMNH P-17367, pelvic fragment; NMMNH P-17369, osteoderms; NMMNH P-17372, jaw fragment; NMMNH P-17393, tooth (type B) in conglomerate; NMMNH P-17395, neural arch fragments: NMMNH locality 530; NMMNH P-17439, osteoderm; NMMNH P-17447, tooth (type C); NMMNH P-17448, caudal centrum: NMMNH locality 534; NMMNH P-17387, juvenile ilium; NMMNH P-17396, juvenile ilium: NMMNH locality 535; NMMNH P-17349, osteoderm; NMMNH P-17353, cervical fragment; NMMNH P-17357, osteoderm fragment; NMMNH P-17358, tooth fragments (type U, B): NMMNH locality 536; NMMNH P-17334, rib fragments: NMMNH locality 537; NMMNH P-17298, dorsal vertebral and neural arch fragments; NMMNH P-17300, osteoderms; NMMNH P-17394, skull fragments; NMMNH P-4839, distal neural spine; NMMNH P-17437, rib fragments; NMMNH P-17438, osteoderm fragments: NMMNH locality 538; NMMNH P-17335, tooth fragment; NMMNH P-17430, proximal scapula; NMMNH P-17431, centrum; NMMNH P-17432, proximal ischium; NMMNH P-17471, osteoderm fragments: NMMNH locality 539; NMMNH P-17401, dentary fragment: NMMNH locality 540; NMMNH P-17378, cervical centrum: NMMNH locality 542; NMMNH P-17410, tooth fragment (type B); NMMNH P-17411, osteoderms; NMMNH P-17412, vertebral fragments; NMMNH P-17417, iliac blade; NMMNH P-17421, proximal ulna: NMMNH locality 546; NMMNH P-17450, skull fragments: NMMNH locality 548; NMMNH P-17348, rostrum (small): San Juan (sic), NM: UMMP 7234, cervical centra: 5 miles west of San Jon in Badlands; UMMP 7313, interclavicle; UMMP 7441 (part), juvenile rostrum tip; UMMP 7442, interclavicle: YPM locality 5831; fragmentary skull: YPM locality 5833 (in part); unnumbered osteoderms, skull fragments, teeth, ribs, dorsal vertebra, proximal pubis, partial interclavicle: YPM locality 6649 (in part); unnumbered two caudal centra, end of humerus: YPM locality 5887, unnumbered sacral vertebra.

First occurrence: *Paleorhinus* from numerous localities worldwide from the late Carnian (Triassic).

Last occurrence: *Redondasaurus* spp. From Late Norian/?Rhaetian(Triassic) of New Mexico, Utah and ?Oklahoma, *Angistorhinopsis ruetimeyeri* from the late Norian/Rhaetian of Germany and Switzerland, indeterminate phytosaurs from the Cliftonian (Rhaetian) of New Jersey and Connecticut.

Acme zone: Late Carnian-late Norian/ ?Rhaetian (Triassic) of North America and late Carnian-middle Norian of Europe.

Discussion: Clearly, specimens of indeterminate phytosaurs are the most common vertebrate fossils in the Bull Canyon Formation, as they are in all formations of the Chinle Group. Presumably the majority of specimens represent *Pseudopalatus pristinus* or *P. andersoni* because they are not very robust, as would be expected for the postcrania of *Pseudopalatus* n. sp. Skeletons from the Canjilon quarry of north-central New Mexico as well as specimens collected in this study (notably NMMNH P-20852) are adequate to allow the study of the complete osteology of *Pseudopalatus*. Such an osteological study would allow the identification of some of the Bull Canyon specimens, but most are too fragmentary to ever be diagnostic.

The acronyms for tooth types follow Hunt (1994) who recognized that there are only four types of phytosaurian teeth (Hunt, 1994, fig. 2): (1) caniniform teeth (type C) are elongate, medio-laterally compressed blades with fully serrated anterior and posterior margins that occur in the anterior portion of the jaws in very robust phytosaurs. In cross-section, these teeth are slightly asymmetrical, and the ratios of height to maximum length to maximum width are approximately 3.4:2.0:1.2; (2) "unserrated" teeth (type U) that are round to oval in cross-section, elongate and sometimes slightly recurved and have longitudinal striations or flutings. They are present in the anterior portions of most phytosaurian jaws and typically have height-to-diameter ratios of approximately 3:1; (3) intermediate teeth with incomplete serrated margins (type I). These teeth are subrounded in cross-section, bicarinate or unicarinate, elongate and sometimes slightly recurved. They form a continuum with type U teeth and occur between them and the fourth type of teeth in toothrows. Type I teeth have length to diameter ratios from 3:1 to 2.2:1; (4) blade teeth (type B) occur in the posterior portions of the dentitions of most phytosaurs. These teeth do not form a continuum with type I teeth. Type B teeth are laterally compressed but markedly asymmetrical in cross-section (rounded D-shaped) and have fully serrated anterior and posterior margins. They are relatively low-crowned (height:length approximately 1.3:1 to 2:1) and are generally restricted to the maxilla and posterior lower jaw.

?Parasuchidae Lydekker, 1888
gen. et sp. indet.

Type Revueltian occurrences: NMMNH localities 1, 16, 33, 35, 41, 53-54, 58, 67, 92, 116, 122, 127, 129, 132, 135, 145, 147, 169, 157-158, 167, 448, 523.

Referred specimens from type Revueltian: NMMNH locality 1; NMMNH P-4789, tooth: NMMNH locality 16; NMMNH P-4353, proximal tibia: NMMNH locality 33; NMMNH P-4510, pelvis fragment with tooth marks: NMMNH locality 35; NMMNH P-4304, ?caudal vertebra: NMMNH locality 41; NMMNH P-4286, ?radius fragment: NMMNH locality 53; NMMNH P-4261, teeth: NMMNH P-4262, rib fragment: NMMNH locality 54; NMMNH P-4277, bone with invertebrate borings: NMMNH locality 58; NMMNH P-4250, tooth: NMMNH locality 67; NMMNH P-4218, jaw fragment with tooth; NMMNH P-4216, tooth: NMMNH locality 92; NMMNH P-4135, tooth: NMMNH locality 116; NMMNH P-4500, ?distal radius: NMMNH locality 122; NMMNH P-4543, ?humerus fragment; NMMNH P-4494, ?distal tibia; NMMNH P-4416, distal tibia; NMMNH P-4621, jaw fragment: NMMNH locality 127; NMMNH P-4593, distal limb: NMMNH locality 129; NMMNH P-4478, ?ilium fragment: NMMNH locality 132; NMMNH P-4587, zygapophysis: NMMNH locality 135; NMMNH P-4495, bone: NMMNH locality 145; NMMNH P-4454, ?pubic fragments: NMMNH locality 147; NMMNH P-4459, skull fragments: NMMNH locality 169; NMMNH P-16606, sacral vertebra: NMMNH locality 157; NMMNH P-4555, proximal rib: NMMNH locality 158; NMMNH P-4545, jaw fragment in sandstone: NMMNH locality 167; NMMNH P-16512, teeth: NMMNH locality 448; NMMNH P-16831, scapula fragment: NMMNH locality 523; NMMNH P-17454, osteoderm fragment.

First occurrence: *Paleorhinus* from numerous localities worldwide from the late Carnian (Triassic).

Last occurrence: *Redondasaurus* spp. From Late Norian/?Rhaetian(Triassic) of New Mexico, Utah and ?Oklahoma, *Angistorhinopsis ruetimeyeri* from the late Norian/Rhaetian of Germany and Switzerland, indeterminate phytosaurs from the Cliftonian (Rhaetian) of New Jersey and Connecticut.

Acme zone: Late Carnian-late Norian/ ?Rhaetian (Triassic) of North America and late Carnian-middle Norian of Europe.

Discussion: A large number of specimens from the Bull Canyon Formation show similarities in size and morphology to phytosaurs, but are too fragmentary to allow unequivocal assignment to the Parasuchidae.

Aetosauria Lydekker in Nicholson and Lydekker, 1889
Stagonolepididae Lydekker, 1887
Typothorax coccinarum Cope, 1875

Type Revueltian occurrences: NMMNH localities 1, 18, 22, 32, 33, 35, 38, 42-43, 46, 48, 49, 51, 53, 62, 64, 67, 70, 72, 75, 82, 86, 89, 94-95, 99, 107, 109, 110, 116-117, 125, 132, 134, 140, 146-148, 155, 157, 162, 165- 166, 169-171, 174, 177, 180, 182, 185, 194, 196-197, 401, 404, 420, 447, 449, 451-452, 454, 457, 465, 499-501, 505, 507, 513-515, 518-519, 528-530, 533-535, 537, 539, 541-542, 547, 2640: UCM locality 82021.

Referred specimens from type Revueltian: NMMNH locality 1; NMMNH P-4671, paramedian osteoderm fragments; NMMNH P-4672, lateral osteoderm fragment; NMMNH P-4674, paramedian osteoderm fragment; NMMNH P-4675, lateral osteoderm fragments; NMMNH P-4912, osteoderms; NMMNH P-4951, paramedian osteoderm fragment; NMMNH P-16556, lateral osteoderm fragments; NMMNH P-16557, lateral osteoderm fragments; NMMNH P-16559, paramedian osteoderm fragments; NMMNH P-16562, paramedian osteoderm fragments; NMMNH P-16570, paramedian osteoderm fragments: NMMNH P-16576, paramedian osteoderm fragment; NMMNH P-16578, paramedian osteoderm fragment; NMMNH P-16580, paramedian osteoderm fragments; NMMNH P-16581, lateral osteoderm fragments; NMMNH P-16695, paramedian osteoderm fragment; NMMNH P-16686, paramedian osteoderm fragment; NMMNH P-16913, lateral osteoderm fragment: NMMNH locality 18; NMMNH P-4346, paramedian osteoderm fragments: NMMNH locality 20; NMMNH P-4349: NMMNH locality 22; NMMNH P-4345, paramedian osteoderm fragment; NMMNH locality 25; NMMNH P-4982, paramedian osteoderm: NMMNH locality 28; NMMNH P-4333, paramedian osteoderm fragments: NMMNH locality 32; NMMNH P-4296, osteoderm fragments: NMMNH locality 33; NMMNH P-4513, paramedian osteoderm fragment: NMMNH locality 34; NMMNH P-4151, paramedian osteoderm fragments; NMMNH P-4162, paramedian osteoderm fragments; NMMNH P-4293, paramedian osteoderm fragments: NMMNH locality 35; NMMNH P-4210, paramedian osteoderm fragment: NMMNH locality 38; NMMNH P-4289, lateral osteoderm fragments; NMMNH P-4288, paramedian osteoderm fragment; NMMNH P-4311, paramedian osteoderm fragments: NMMNH locality 42; NMMNH P-4282, osteoderm fragments; NMMNH P-4284, paramedian osteoderm fragments: NMMNH locality 43; NMMNH P-4269, paramedian osteoderm fragments; NMMNH P-4283, paramedian osteoderm fragments: NMMNH locality 46; NMMNH P-4205, large paramedian osteoderm fragments: NMMNH locality 48; NMMNH P-4207, paramedian osteoderm fragments: NMMNH locality 49; NMMNH P-4208, lateral osteoderm fragment: NMMNH locality 51; NMMNH P-4275, paramedian osteoderm fragments: NMMNH locality 53; NMMNH P-4189, paramedian osteoderm fragments: NMMNH P-4161, paramedian osteoderm fragments; NMMNH P-4195, paramedian osteoderm fragments: NMMNH locality 62; NMMNH P-4954, osteoderm fragments: NMMNH locality 64; NMMNH P-4242, paramedian osteoderm fragment: NMMNH locality 67; NMMNH P-4176, paramedian osteoderm fragments in sandstone; NMMNH P-4214, paramedian osteoderm fragment: NMMNH locality 70; NMMNH P-4150, paramedian osteoderm fragments: NMMNH locality 72; NMMNH P-4232, paramedian osteoderm fragment: NMMNH locality 75; NMMNH P-17303, paramedian osteoderm fragments: NMMNH locality 82; NMMNH P-4175, paramedian osteoderm fragments: NMMNH locality 86; NMMNH P-4179, paramedian osteoderm fragments; NMMNH P-16632, paramedian osteoderm fragments: NMMNH locality 89; NMMNH P-17175, paramedian osteoderm fragment: NMMNH locality 94; NMMNH P-4115, paramedian and lateral osteoderm fragments: NMMNH locality 95; NMMNH P-4121, paramedian osteoderm fragment: NMMNH locality 99; NMMNH P-4125, paramedian osteoderm fragments: NMMNH locality 107; NMMNH P-4105, paramedian osteoderm fragment: NMMNH locality 109; NMMNH P-4102, paramedian osteoderm fragment: NMMNH locality 110; NMMNH P-4532, paramedian osteoderm fragment; NMMNH P-4539, paramedian osteoderm fragment; NMMNH P-4575, small paramedian osteoderm fragment; NMMNH P-4644, paramedian osteoderm fragments: NMMNH locality 116; NMMNH P-4499, paramedian osteoderm fragment: NMMNH locality 117; NMMNH P-4482, paramedian osteoderm fragment: NMMNH locality 125; NMMNH P-4382, paramedian osteoderm fragments: NMMNH locality 132; NMMNH P-4373, caudal vertebrae; NMMNH P-4378, centra; NMMNH P-4391, paramedian osteoderm fragments; NMMNH P-4594, centrum: NMMNH locality 134; NMMNH P-4401, paramedian osteoderm fragments; NMMNH P-4639, paramedian osteoderm fragment: NMMNH locality 140; NMMNH P-4541, paramedian osteoderm fragments: NMMNH locality 146; NMMNH P-4461, paramedian osteoderm fragments: NMMNH locality 147; NMMNH P-4458, paramedian osteoderm fragment: NMMNH locality 148; NMMNH P-4617, osteoderm fragment: NMMNH locality 155; NMMNH P-4565, paramedian osteoderm fragments: NMMNH locality 157; NMMNH P-4553, paramedian and lateral osteoderm fragments: NMMNH locality 162; NMMNH P-16628, lateral osteoderm fragments; NMMNH P-16815, paramedian osteoderm fragments: NMMNH locality 165; NMMNH P-4903, paramedian osteoderm fragment; NMMNH P-4904 lateral osteoderm fragments; NMMNH P-4953, many paramedian osteoderm fragments; NMMNH P-16569, associated paramedian osteoderms; NMMNH P-16572, paramedian osteoderm fragments; NMMNH P-16579, paramedian osteoderm fragment: NMMNH locality 166; NMMNH P-4865, cervical vertebra; NMMNH P-16565, paramedian osteoderm fragments: NMMNH locality 169; NMMNH P-4915, paramedian osteoderm fragments; NMMNH P-16618, paramedian osteoderm fragment;

NMMNH P-16575, paramedian osteoderm fragments; NMMNH P-16780, paramedian osteoderm fragments: NMMNH locality 170; NMMNH P-4897, paramedian osteoderm fragments; NMMNH P-16564, paramedian osteoderm fragments: NMMNH locality 171; NMMNH P-16634, lateral osteoderm fragments: NMMNH locality 174; NMMNH P-16566, paramedian osteoderm fragment: NMMNH locality 177; NMMNH P-16577, paramedian osteoderm fragment: NMMNH locality 180; NMMNH P-4911, paramedian osteoderm fragments: NMMNH locality 182; NMMNH P-4950, lateral osteoderm; NMMNH P-16582, paramedian osteoderm fragments: NMMNH locality 185; NMMNH P-16568, paramedian osteoderm fragment: NMMNH locality 194; NMMNH P-16561, paramedian osteoderm fragments: NMMNH locality 196; NMMNH P-16555, paramedian osteoderm fragment: NMMNH locality 197; NMMNH P-16545, paramedian osteoderm fragments; NMMNH P-16560, lateral osteoderm fragments: NMMNH locality 401; NMMNH P-17581, paramedian osteoderm fragments: NMMNH locality 404; NMMNH P-3679, paramedian osteoderm fragment: NMMNH locality 420; NMMNH P-17568, paramedian osteoderm fragments: NMMNH locality 447; NMMNH P-16820, paramedian osteoderm fragments: NMMNH locality 449; NMMNH P-16833, paramedian osteoderm fragments: NMMNH locality 451; NMMNH P-16822, paramedian osteoderm fragment: NMMNH locality 452; NMMNH P-16852, paramedian and lateral osteoderm fragments: NMMNH locality 454; NMMNH P-16868, paramedian and lateral osteoderms; NMMNH P-16869, paramedian osteoderm fragments: NMMNH locality 457; NMMNH P-16880, paramedian osteoderm fragments: NMMNH locality 465; NMMNH P-16885, paramedian osteoderm fragments: NMMNH locality 499; NMMNH P-17137, many paramedian osteoderm fragments; NMMNH P-17265, paramedian osteoderm fragments: NMMNH locality 500; NMMNH P-17263, osteoderm fragment: NMMNH locality 501; NMMNH P-17116, small paramedian osteoderm fragments; NMMNH P-17118, small paramedian osteoderm fragments; NMMNH P-17127, many small paramedian osteoderm fragments; NMMNH P-17214, small paramedian osteoderm: NMMNH locality 505; NMMNH P-17208, paramedian osteoderm fragments: NMMNH locality 507; NMMNH P-17210, paramedian osteoderm fragment: NMMNH locality 513; NMMNH P-17251, paramedian osteoderm fragments: NMMNH locality 514; NMMNH P-17120, paramedian osteoderm fragments: NMMNH locality 515; NMMNH P-17135, paramedian osteoderm fragments: NMMNH locality 518; NMMNH P-17136, paramedian osteoderm fragments; NMMNH P-17419, large paramedian osteoderm fragments; NMMNH P-17429, paramedian osteoderm fragments: NMMNH locality 519; NMMNH P-17180, paramedian osteoderm fragments: NMMNH locality 528; NMMNH P-17345, osteoderm fragment: NMMNH locality 529; NMMNH P-17365, lateral osteoderm fragment; NMMNH P-17368, paramedian osteoderm fragments; NMMNH P-17371, cervical vertebrae; NMMNH P-17391, paramedian osteoderm fragment: NMMNH locality 530; NMMNH P-17440, paramedian osteoderm fragment: NMMNH locality 533; NMMNH P-17307, paramedian osteoderm: NMMNH locality 534; NMMNH P-17385, paramedian osteoderm fragments: NMMNH locality 535; NMMNH P-17350, paramedian osteoderm fragments: NMMNH locality 537; NMMNH P-17299, paramedian osteoderm fragments: NMMNH locality 539; NMMNH P-17402, lateral osteoderm fragment; NMMNH P-17408, paramedian osteoderm fragments: NMMNH locality 541; NMMNH P-17340, lateral osteoderm fragment: NMMNH locality 542; NMMNH P-17413, paramedian osteoderm fragments: NMMNH locality 547; NMMNH P-17344, paramedian osteoderm fragments: NMMNH locality 2640; NMMNH P-12964, complete articulated skeleton: UCM locality 82021; UCM 47725, paramedian osteoderms.

First and last occurrences: Several localities in Revueltian (early-middle Norian) strata in Arizona and New Mexico.

Acme zone: Revueltian (early-middle Norian).

Discussion: Osteoderm fragments of *Typothorax coccinarum* are the most common generically identifiable specimens of aetosaurs from the Bull Canyon Formation. *Typothorax* dorsal paramedian osteoderms are readily identifiable, even from fragmentary specimens, because of two features. The dorsal ornamentation is a pattern of shallow, rounded pits that are irregularly arranged, and the paramedian osteoderms have a deep ventral keel so that the ventral margin of the osteoderm is strongly convex. Dorsal lateral osteoderms are characterized by equal-sized ventral and dorsal processes, by a rounded triangular dorsal/ventral outline and by a sharp angulation in anterior/posterior view. Virtually all aetosaur osteoderms of *Typothorax* from the Bull Canyon Formation are from the dorsal series.

Few specimens of *Typothorax* from the Bull Canyon Formation are not isolated osteoderms. A notable exception is NMMNH P-12964, which is a nearly complete articulated specimen of this species (Hunt et al., 1993b, Fig. 6). Preparation of this specimen is incomplete at this time. Therefore, only a few preliminary observations can be presented.

NMMNH P-12964 is an articulated skeleton lying on its right side. The body length is approximately 2.5 m, and the entire axial body is sheathed in armor plates. Paramedian osteoderms are dominantly flat and rectangular in shape. Lateral osteoderms are nearly flat in the caudal region to highly arched in the dorsal region. Most osteoderms have an ornamentation of round shallow pits. Some osteoderms have a raised boss among the pits. The body is very wide, and the widest paramedian osteoderms are almost as long as the femur. The tibia and fibula are almost half the length of the femur. The forelimb is shorter than the hind limb. The neck is extremely short, and each cervical vertebra is at most 20% of the length of the longest of the dorsal series. Dorsal ribs appear to be fused to the vertebrae. In *Typothorax*, the gradual transition between the dorsal and cervical armor shown by Long et al. (1989, fig. 3A) is erroneous as is the shortness of the tail. The tail actually accounts for about half the length of the animal, and the carapace narrows rapidly in the shoulder region. The dorsal carapace is strongly discoid. The skull is very small, with a length of about 20 cm, and the lower jaw is relatively shallow. Teeth are bulbous, with waisted crown bases and no serrations (Fig. 6A).

NMMNH P-12964 can be assigned to the stagonolepidid species *Typothorax coccinarum* because it possessed wide paramedian osteoderms with a deep ventral keel and a dorsal ornamentation of round pits. These characters are also present in *Redondasuchus reseri* (Hunt and Lucas, 1991a) and in an undescribed genus from the Los Esteros Member of the Santa Rosa Formation. However, *T. coccinarum*, as represented by NMMNH P-12964, possesses several autapomorphies including the broad, discoid carapace, short neck, short distal lower hind limbs and small skull.

The skull, and cervical and caudal vertebrae of NMMNH P-12964 are in articulation, but the thoracic cavity is breached and disorganized. Much of the dorsal series of paramedian osteoderms is visible in ventral view, and most of the dorsal vertebrae and ribs are either absent or are dislocated. One dorsal rib was found about 1 meter from the skeleton. Obviously, an extremely wide-bodied and dorsoventrally compressed animal such as *Typothorax* would be unstable if it lay on its side. The carapace would tend to fracture and come to equilibrium in a horizontal plane. This would explain the rotation of the thoracic region relative to the rest of the skeleton. However, the internal disruption of the body cavity is best explained by scavenger activity.

The skeleton is very complete and occurs in a fine-grained matrix (silty mudstone). These features suggest a low energy environment in which the skeleton was not moved. The combination of limited scavenger damage and fine preservation of much of the skeleton suggests a short period of subaerial exposure before rapid burial.

Some *Typothorax* specimens are small (= ?juvenile) (e.g., NMMNH P-17116, 17118, 17127, 17214)(Fig. 5C-H). These indicate that the ornamentation of stagonolepidid osteoderms did not change through ontogeny. Thus, small (= ?juvenile) specimens of *Typothorax* have the same ornament as large specimens.

?Typothorax sp.

Type Revueltian occurrences: NMMNH localities 110, 134, 499-500: UMMP locality San Juan (sic).

Referred specimens from type Revueltian: NMMNH locality 110; NMMNH P-4606, tooth: NMMNH locality 134; NMMNH P-4504, paramedian osteoderm fragment: NMMNH locality 499; NMMNH P-17282, many small lateral osteoderm fragments: NMMNH locality 500; NMMNH P-17241, astragalus: San Juan (sic), New Mexico; UMMP 7241 (in part), lateral osteoderm fragment.

First and last occurrences: Several localities in Revueltian (early-middle Norian) strata in Arizona and New Mexico.

Acme zone: Revueltian (early-middle Norian).

Discussion: Some osteoderm fragments from the Bull Canyon Formation have the general morphology of *Typothorax* osteoderms, but they are too fragmentary and/or weathered to allow exact identification.

Aetosaurus arcuatus (Marsh, 1896)

Type Revueltian occurrences: NMMNH localities 499-502, 519, 521.

Referred specimens from type Revueltian: NMMNH locality 499; NMMNH P-17279, chevrons: NMMNH locality 500; NMMNH P-17145, paramedian osteoderm fragment; NMMNH P-17193, lateral osteoderm fragments; NMMNH P-17233, vertebrae; NMMNH P-17234, paramedian osteoderm fragments; NMMNH P-17280, paramedian osteoderm fragments: NMMNH locality 501; NMMNH P-17117, many paramedian osteoderm fragments; NMMNH P-17126, paramedian and lateral osteoderm fragments; NMMNH P-17130, left femur; NMMNH P-17131, many osteoderm and limb fragments; NMMNH P-17163,

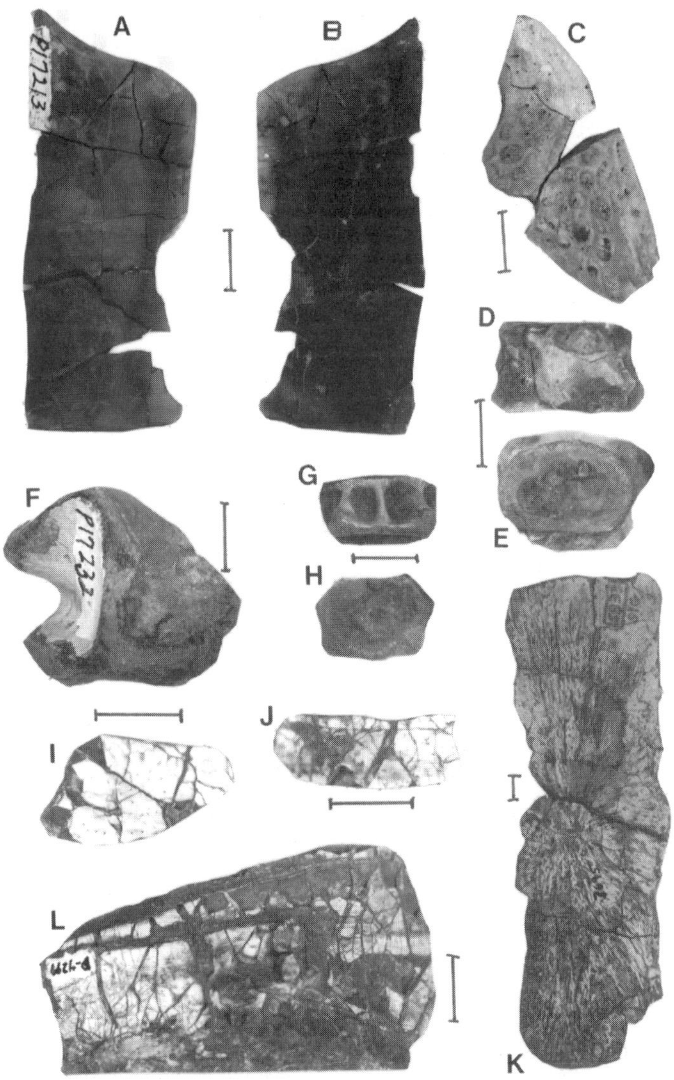

FIGURE 5. Aetosaurs from the Bull Canyon Formation. A-B. *Aetosaurus arcuatus*. A-B. Paramedian scute (NMMNH P-17213) in ventral (A) and dorsal (B) views. C-H. Juvenile *Typothorax coccinarum*, C. Lateral scute (NMMNH P-17161) in dorsal view. D-E, Cervical vertebra of NMMNH P-17231 in dorsal (D) and anterior (E) views. F. Astragalus(NMMNH P-17232) in dorsal view. G-H, Cervical vertebra (NMMNH P-17231) in ventral (G) and anterior (H) views. I. Dorsal paramedian scute of *Paratypothorax* sp. (YPM 3695) in dorsal view. J-L. Parts of skeleton of ?gen et sp. nov. (NMMNH P-4299). J. Lateral scute in dorsal view. K. ?Caudal paramedian scute in dorsal view. L. Articulated dorsal paramedian scutes. Scale bars are 1 cm.

FIGURE 6. Complete skeleton of *Typothorax coccinarum* (NMMNH P-12964) from the Bull Canyon Formation. A. Detail of skull region. B. Detail of foot.

paramedian osteoderm fragments; NMMNH P-17165, paramedian osteoderm fragment; NMMNH P-17167, paramedian osteoderm fragments; NMMNH P-17168, paramedian osteoderm fragment; NMMNH P-17174, 4 cervical vertebrae; NMMNH P-17212, paramedian osteoderm fragments; NMMNH P-17213, paramedian osteoderm (Fig. 5A-B); NMMNH P-17215, paramedian osteoderm fragment ; NMMNH P-17216, paramedian osteoderm; NMMNH P-17218, mold of lateral osteoderm; NMMNH P-17220, lateral osteoderm fragments; NMMNH P-17221, osteoderm fragments; NMMNH P-17228, paramedian osteoderms; NMMNH P-17230, proximal end of left femur; NMMNH P-17231, cervical centra; NMMNH P-17232, manus, pes and tarsus elements; NMMNH P-17286, paramedian osteoderm fragment; NMMNH P-17287, paramedian and lateral osteoderm fragments; NMMNH P-17236, pubis and ischium fragments: NMMNH P-17238, ulnae; NMMNH P-17239, metapodial; NMMNH P-17247, vertebrae, osteoderms: NMMNH locality 502; NMMNH P-17161, lateral and paramedian osteoderm fragments: NMMNH locality 519; NMMNH P-17162, lateral and paramedian osteoderm fragments: NMMNH locality 521; NMMNH P-17223, paramedian osteoderm; NMMNH P-17227, osteoderm fragments.

First and last occurrence: Early-mid Norian (Triassic) of Neshanician of eastern North America and Revueltian of western North America.

Acme zone: Early-mid Norian (Triassic)

Discussion: Hunt (1994) described a number of specimens of a small aetosaur that were subsequently identified by Heckert and Lucas (1998) as pertaining to *Aetosaurus arcuatus*. NMMNH P-17213, is a left dorsal, paramedian osteoderm that is 75 mm in length and 31 mm wide (Fig. 5A-B). The medial margin is thickened to 3 mm. The lateral margin is concave, and the width of the osteoderm increases anteriorly. A narrow anterior bar widens laterally. The dorsal ornamentation is a faint radial pattern. The osteoderm is slightly concave. Another dorsal paramedian osteoderm is NMMNH P-17165, which is 76 mm in length and 27 mm wide. This osteoderm has a small, triangular, anteriorly-directed prong on the anteromedial corner. Ventral osteoderms (e.g., NMMNH P-17167) are ovoid in shape with a maximum diameter of 10 mm.

NMMNH localities 499, 500, 501, 502, 519 and 521 are discrete areas along a 30-m-long outcrop. These localities yield a fauna of small terrestrial tetrapods. To facilitate discrimination of some individual skeletons, the outcrop was divided into several localities. However, it is clear that only two taxa of aetosaurs are present in this outcrop, one is a juvenile *Typothorax* and the other is *Aetosaurus arcuatus*. The majority of *Aetosaurus arcuatus* specimens are of similar size and may represent one scattered individual.

gen. et sp. nov.?

Type Revueltian occurrence: NMMNH locality 36.

Referred specimens from type Revueltian: NMMNH locality 36; NMMNH P-4299, one partial, articulated skeleton and parts of a smaller skeleton (Fig. 5J-L).

First and last occurrence: Revueltian (Triassic: early-middle Norian) of east-central New Mexico.

Acme zone: Revueltian (Triassic: early-middle Norian).

Discussion: All specimens of this ?new taxon were found in a calcareous concretionary layer that also includes NMMNH locality 75, which yields three-dimensional fish. The aetosaur specimens are partially articulated but are imbedded in concretionary material. Much preparation will be required before the osteology of this taxon can be adequately described. However, enough morphological features can be discerned in the osteoderms for some preliminary observations. Fused neural arches indicate that the specimens do not represent juveniles.

Paramedian osteoderms are very long and narrow. A mid-dorsal paramedian osteoderm is at least 70 mm in length and 20 mm in width. The dorsal ornamentation consists of very shallow, small round indentations and shallow elongate grooves which may be slightly radial in pattern. There is a raised anterior bar, and the lateral margin is slightly sinuous. Paramedian osteoderms are a rounded triangle in dorsal view. A mid-dorsal lateral osteoderm is 22 mm long and 16 mm wide. The ornamentation is shallow, small, round indentations. Small nearly square osteoderms (18 mm by 15 mm) of the caudal series or ventral cuirass have a shallow radial pattern. A maxilla fragment preserves a portion of a laterally-compressed, but unserrated tooth. The laterally compressed teeth are unusual for aetosaurs, and their morphology and size suggests an insectivorous diet. The majority of the other skeletal remains require preparation and include femora, limb and podial fragments, cervical and dorsal vertebrae (with fused neural arches), ilia, an articulated tarsus, and a partial, slightly disarticulated dorsal carapace.

This taxon appears to be distinct in possessing very wide and narrow dorsal paramedian osteoderms (length:width ratio at least 3.5:1) with a very shallow ornamentation and a slightly sinuous lateral margin. The ornamentation of the osteoderms suggests an affinity to *Aetosaurus arcuatus*.

***Paratypothorax* sp. Long and Ballew, 1985**

Type Revueltian occurrences: NMMNH locality 165: YPM locality 3695.

Referred specimens from type Revueltian: NMMNH locality 165; NMMNH P-4937, osteoderm fragments: YPM locality 5833; YPM 3695, complete paramedian osteoderm and fragment of the lateral margin of a second (Fig. 5K).

First occurrence: Adamanian (Triassic: late Carnian) from Arizona, New Mexico and Texas.

Last occurrence: Revueltian (Triassic: early-mid Norian) of Arizona, New Mexico, Texas and Colorado.

Acme zone: None.

Discussion: YPM 3695 is a thin, long paramedian osteoderm of the dorsal series of a stagonolepidid with a radial ornamentation (Fig. 5K). This specimen has previously been incorrectly identified as the stagonolepidid *Typothorax* cf. *coccinarum*, with "grave doubts" (Gregory, 1953, p. 12), the phytosaur *Phytosaurus* (Gregory, 1962) and the stagonolepidid *Stagonolepis* (=*Calyptosuchus*) *wellesi* (Long and Ballew, 1985). However, the radial ornamentation, thinness of the osteoderm and its length clearly demonstrate that this specimen pertains to *Paratypothorax*. This specimen cannot be assigned to a particular species of the genus in the absence of more complete material, notably lateral osteoderms.

?*Paratypothorax* sp.

Type Revueltian occurrences: NMMNH localities 113, 150, 169, 422.

Referred specimens from type Revueltian: NMMNH locality 113; NMMNH P-4387, paramedian osteoderm fragment: NMMNH locality 150; NMMNH P-4443, paramedian osteoderm fragment: NMMNH locality 169; NMMNH P-4898, paramedian osteoderm fragment: NMMNH locality 422; NMMNH P-3670, mid-cervical lateral spike.

First occurrence: Adamanian (Triassic: late Carnian) from Arizona, New Mexico and Texas.

Last occurrence: Revueltian (Triassic: early-mid Norian) of Arizona, New Mexico and Texas and Colorado.

Acme zone: None.

Discussion: Fragmentary paramedian osteoderms that apparently have a radial ornamentation are tentatively assigned to *Paratypothorax*. One lateral spike (NMMNH P-3670) is laterally compressed and may represent this genus.

***Desmatosuchus* n. sp.**

Type Revueltian occurrences: NMMNH localities 171, 420: UCM locality 82021.

Referred specimens from type Revueltian: NMMNH locality 171; NMMNH P-4894, lateral spike; NMMNH P-4895, lateral spike: NMMNH locality 420; NMMNH P-3668, small lateral spike: UCM locality 82021; UCM 47225, lateral osteoderm.

First and last occurrence: Revueltian (Triassic: early-middle Norian) of north-central and east-central New Mexico.

Acme zone: Revueltian (Triassic: early-middle Norian).

Discussion: The best specimen of *Desmatosuchus* in the Bull Canyon fauna is UCM 47225, a cervical lateral osteoderm. It differs from other specimens of this genus in having a recurved spike, but this could be a preservational artifact (it is not a preservational artifact). However, recently Heckert and Zeigler have collected new specimens from the Petrified Forest Formation of Rio Arriba County, New Mexico that demonstrate the presence of a new species of *Desmatosuchus* in the Revueltian to which UCM 47225 pertains. Parrish and Carpenter (1986) assigned paramedian osteoderm fragments to *Desmatosuchus*, but they actually pertain to *Typothorax* (see above). Fragmentary NMMNH specimens consist of broken lateral spikes. These are rounded in cross section, in contrast to the compressed spikes of an undescribed species of *Paratypothorax* from the Bull Canyon Formation of west Texas.

Stagonolepididae Lydekker, 1887
gen. et sp. indeterminate

Type Revueltian occurrences: NMMNH localities 57, 64, 75, 79, 90, 162-163, 165, 173, 181, 376, 447, 462, 497, 499: UMMP locality San Juan (sic).

Referred specimens from type Revueltian: NMMNH locality 57; NMMNH P-17436, paramedian osteoderm fragment: NMMNH locality 64; NMMNH P-4130, paramedian and lateral osteoderm fragments: NMMNH locality 75; NMMNH P-4188, rib fragment: NMMNH locality 79; NMMNH P-4171, osteoderm fragment; NMMNH P-4173, caudal centra: NMMNH locality 90; NMMNH P-17422, tibia: NMMNH locality 162; NMMNH P-16627, paramedian osteoderm fragment: NMMNH locality 163; NMMNH P-4339, osteoderm fragment: NMMNH locality 165; NMMNH P-4908, vertebral fragments; NMMNH P-4922, paramedian osteoderm fragments: NMMNH locality 173; NMMNH P-16546, paramedian osteoderm in conglomerate: NMMNH locality 181; NMMNH P-4940, ?proximal femur; NMMNH P-4941, proximal femur; NMMNH P-4943, limb fragment; NMMNH P-4944, articulated tail fragments: NMMNH locality 376; NMMNH P-14394, paramedian osteoderm fragment: NMMNH locality 447; NMMNH P-16826, paramedian osteoderm fragment: NMMNH locality 462; NMMNH P-16877, paramedian osteoderm fragment: NMMNH locality 497; NMMNH P-17172, paramedian osteoderm fragment: NMMNH locality 499; NMMNH P-17139, caudal centra: San Juan (sic), New Mexico; UMMP 7274 (in part), lateral osteoderm fragment.

First occurrence: Several genera from the Otischalkian (Triassic: late Carnian) of Texas and *Longosuchus* from Morocco.

Last occurrence: *Redondasuchus* and *Neoaetosauroides* from the Apachean (Triassic: late Norian-?Rhaetian) of New Mexico and *Neoaetosauroides* from the late Norian-?Rhaetian of Argentina.

Acme zone: None.

Discussion: A number of fragmentary and/or weathered osteoderms cannot be identified more closely than Stagonolepididae

?Stagonolepididae Lydekker, 1887

Type Revueltian occurrences: NMMNH localities 1, 134, 162.

Referred specimens from type Revueltian: NMMNH locality 1; NMMNH P-16891, dorsal centra: NMMNH locality 134; NMMNH P-4463, cervical centrum: NMMNH locality 162; NMMNH P-4861, caudal centra.

First occurrence: Several genera from the Otischalkian (Triassic: late Carnian) of Texas and *Longosuchus* from Morocco.

Last occurrence: *Redondasuchus* and *Neoaetosauroides* from the Apachean (Triassic: late Norian-?Rhaetian) of New Mexico and *Neoaetosauroides* from the late Norian-?Rhaetian of Argentina.

Acme zone: None.

Discussion: A few vertebral centra from the Bull Canyon Formation are similar to stagonolepidid centra but cannot be definitely assigned to the family.

Rauisuchidae Price, 1946
***Postosuchus kirkpatricki* Chatterjee, 1985**

Type Revueltian occurrence: NMMNH locality 51.

Referred specimens from type Revueltian: NMMNH locality 51; NMMNH P-4178, pubic and ilium fragments, dorsal and sacral vertebrae.

First occurrence: Adamanian (Triassic: late Carnian) of Texas and Arizona.

Last occurrence: Revueltian (Triassic: early-middle Norian) of New Mexico, Texas and Arizona. Note that I do not consider a CM specimen of a rauisuchian from the Whitaker quarry at Ghost ranch to pertain to *Postosuchus* (*contra* Long and Murry, 1995) and the RHMP ilium from the same locality that Long and Murry (1995) assign to *Postosuchus* is associated with sphenosuchian scutes and does not pertain to a rauisuchian (see below).

Acme zone: None

Discussion: NMMNH P-4178 is a partial pelvis and posterior dorsal vertebra of a large rauisuchian. The ilium has the characteristic large supracetabular shelf and vertical buttress of rauisuchians (*sensu lato*). The Rauisuchia is a group in a state of taxonomic confusion (e.g., Parrish, 1993; Long and Murry, 1995). The current usages of Rauisuchia, Rauisuchidae and Poposauridae are inconsistent. Herein, I refer to all rauisuchians (*sensu lato*) as rauisuchids.

Postosuchus kirkpatricki has a confusing and still unresolved taxonomy. Chatterjee (1985) named *Postosuchus* for numerous rauisuchian specimens from the Bull Canyon Formation of West Texas. It is clear that Chatterjee's (1985) reconstruction of the pelvis of *Postosuchus kirkpatricki* is based on three taxa from two localities of different ages (Long and Murry, 1995). Long and Murry (1995) recognized this confusion, but were not rigorous in their assignment of rauisuchian specimens to particular taxa. I note, for example, that the cervical series that they use as their basis for reconstructions of *Postosuchus* is TTUP 9235 and it pertains to a phytosaur (Long and Murry, 1995, fig. 128).

The most diagnostic element of NMMNH P-4178 is a partial ilium. Unfortunately, no complete ilium was found associated with the holotype of *Postosuchus* (smaller ilia represent *Shuvosaurus* =*Chatterjeea*) The only iliac fragment of *Postosuchus* from the Revueltian type locality is the paratype TTUP 9002 (Long and Murry, 1995, fig. 137B). Long and Murry (1995) assigned two rauisuchian ilia (UMMP 7266, 7333a), originally described by Case (1922), from the Adamanian of Texas, to *Postosuchus*.. The ilium of NMMNH P-4178 differs from UMMP 7266, the more complete of the two, in being larger, more robust and in having a ventrally deflected posterior process. However, the morphological differences between the ilia of NMMNH P-4178 and UMMP 7266 could be ontogenetic, so I refer NMMNH P-4178 to *Postosuchus kirkpatricki*.

Shuvosauridae Chatterjee, 1993
***Shuvosaurus inexpectatus* Chatterjee, 1993**

Type Revueltian occurrences: NMMNH localities 1, 110, 162, 529: UCMP locality 82081: UMMP locality 5 miles west of San Jon in badlands.

Referred specimens from type Revueltian: NMMNH locality 1; NMMNH P-4687, proximal femur; NMMNH P-4693, astragalus; NMMNH P-4695, femur; NMMNH P-4702, distal femur: NMMNH locality 110; NMMNH P-4601, conjoined ischia: NMMNH locality 162; NMMNH P-4859, conjoined ischia: NMMNH locality 529; NMMNH P-17373, ischia, tibia: UCM locality 82081; UCM 52080, right ilium, left pubic foot; UCM 52081, left premaxilla (Fig. 7D): 5 miles west of San Jon in Badlands; UMMP 7438, sacrum.

First occurrence: Adamanian (Triassic: late Carnian) of Arizona, New Mexico and Texas.

Last occurrence: Revueltian (Triassic: early-middle Norian) of Arizona, New Mexico and Texas.

Acme zone: None.

Discussion: Parrish and Carpenter (1986, fig. 11.8) described UCM 52081, an edentulous premaxilla. They compared UCM 52081 with various groups of reptiles that lack premaxillary teeth but could find no similarities. However, this specimen is identical to *Shuvosaurus* (Chatterjee, 1993). UCM 52081 shares the following characters with *Shuvosaurus*: (1) premaxilla is edentulous; (2) ventral margin of premaxilla is concave; and (3) narial opening is small and located at the posterior end of a large narial depression. Thus, it is clear that UCM 52081 represents *Shuvosaurus*.

Chatterjee (1993) named *Shuvosaurus inexpectatus* as an ornithomimosaur from the Revueltian Post quarry, the type locality of *Postosuchus*. Hunt et al. (1998) demonstrated that *Shuvosaurus* is not dinosaurian. Long and Murry (1995) named *Chatterjeea elegans* for the smaller specimens from the Post quarry that Chatterjee (1985) had included in *Postosuchus*. They noted that the holotype skull of *Shuvosaurus* was intermixed with probable *Chatterjeea* specimens that lacked cranial material, despite being relatively complete otherwise. They also noted that in size, proportions, and preservation, the holotype skull of *Shuvosaurus inexpectus* matches the holotype skeleton of *Chatterjeea elegans* (Long and Murry, 1995, p. 162). Yet, Long and Murry (1995) still named *Chatterjeea elegans*. The most powerful evidence that these two taxa actually both pertain to *Shuvosaurus inexpectatus* outside the Post Quarry is UCM locality 82081 in the Revuelto badlands. Here, an unmistakable premaxilla of *Shuvosaurus* (Parrish and Carpenter, 1986, fig. 11.8) is associated with the equally distinctive ilium of *Chatterjeea elegans* (Parrish and Carpenter, 1986, fig. 11.4). Thus, it seems that the most parsimonious conclusion is that *Chatterjeea elegans* Long and Murry, 1995 is a subjective junior synonym of *Shuvosaurus inexpectans* Chatterjee 1993.

Rauhut (1997) suggested that *Shuvosaurus* was indeed a theropod, but failed to demonstrate that it possessed any synapomorphies of this clade.

Family Rauisuchidae Price, 1946
gen. et sp. indeterminate

Type Revueltian occurrences: NMMNH localities 1, 47, 60, 110, 115, 120, 134, 153, 156, 161-163, 165, 171, 174, 176, 177, 182-183, 198, 425-426, 452, 454, 462, 498, 507, 510, 523, 526-527, 542, 582.

Referred specimens from type Revueltian: NMMNH locality 1; NMMNH P-4661, second sacral centrum: NMMNH P-4662, centra; NMMNH P-4718, tooth; NMMNH P-4786, tooth; NMMNH P-4787, tooth; NMMNH P-4788, tooth; NMMNH P-4893, dorsal vertebrae; NMMNH P-4913, cervical vertebrae; NMMNH P-16644, tooth; NMMNH P-16647, tooth; NMMNH P-16694, teeth; NMMNH P-16917, partial centrum: NMMNH locality 47; NMMNH P-4106, tooth: NMMNH locality 60; NMMNH P-4160, cervical vertebrae; NMMNH P-4238, tooth: NMMNH locality 110; NMMNH P-4418, centrum; NMMNH P-4537, teeth: NMMNH locality 115; NMMNH P-4852, tooth fragment: NMMNH locality 120; NMMNH P-4447, tooth: NMMNH locality 134; NMMNH P-4466, teeth; NMMNH P-4627, dorsal centrum: NMMNH locality 153; NMMNH P-4475, many teeth: NMMNH locality 156; NMMNH P-4558, tooth: NMMNH locality 161; NMMNH P-4641, tooth: NMMNH locality 162; NMMNH P-16655, sacral, iliac and vertebral fragments: NMMNH locality 163; NMMNH P-16672, vertebral, pelvic fragments: NMMNH locality 165; NMMNH P-4891, cervical vertebra; NMMNH P-4905, cervical vertebra; NMMNH P-4920, dorsal centrum; NMMNH P-16611, teeth: NMMNH locality 171; NMMNH P-16640, teeth; NMMNH P-16649, teeth; NMMNH P-16895, teeth: NMMNH locality 174; NMMNH P-16642, teeth: NMMNH locality 176; NMMNH P-16657, fragmentary skeleton: NMMNH locality 177; NMMNH P-16653, tooth: NMMNH locality 182; NMMNH P-16648, teeth: NMMNH locality 183; NMMNH P-4935, centra; NMMNH P-4946, large mandible: NMMNH locality 198; NMMNH P-16641, centrum: NMMNH locality 425; NMMNH P-4963, teeth: NMMNH locality 426; NMMNH P-17521, teeth: NMMNH locality 452; NMMNH P-4729, tooth fragment; NMMNH P-16853, tooth: NMMNH locality 454; NMMNH P-16864, centrum fragment; NMMNH P-16887, 2 cervical vertebrae: NMMNH P-16863, tooth: NMMNH locality 462; NMMNH P-16886, 3 dorsal vertebrae: NMMNH locality 498; NMMNH P-17262, vertebrae; NMMNH P-17182, caudal centra; NMMNH P-17183, centra: NMMNH locality 507; NMMNH P-17209, caudal centrum: NMMNH locality 510; NMMNH P-17198, caudal centrum: NMMNH locality 523; NMMNH P-17451, tooth: NMMNH locality 526; NMMNH P-17355, tooth; NMMNH P-17381, tooth; NMMNH P-17469, centrum fragments: NMMNH locality 527; NMMNH P-17493, teeth; NMMNH P-17356, proximal ischium: NMMNH locality 542; NMMNH P-17423, pelvic fragments, centra: NMMNH locality 582; NMMNH P-17787, tooth.

First occurrence: Several genera from the Anisian (Triassic) of Africa, Asia and Europe. Rauisuchidae as utilized here includes Poposauridae, Prestosuchidae, etc. of other workers, pending a thorough revision of these taxa.

Last occurrence: *Fasolasuchus tenax* from late Norian/?Rhaetian (Triassic) of Argentina and an undescribed taxon from the Apachean (late Norian/?Rhaetian) Whitaker quarry, New Mexico.

Acme zone: Anisian-Ladinian (Triassic).

Discussion: Indeterminate rauisuchian specimens include postcrania and teeth. I have adopted a conservative stance in identifying laterally compressed and serrated teeth. Parrish (1988) suggested that rauisuchian teeth could be distinguished from dinosaur teeth by the fact that the latter had concave posterior margins. This criterion results in the identification of all laterally compressed and serrated teeth from the Bull Canyon Formation as rauisuchian. Obviously, some of these teeth must pertain to theropods, but pending additional study, I cannot distinguish the dinosaur teeth, and I assign all such teeth to the Rauisuchia.

Family ?Rauisuchidae Price, 1946

Type Revueltian occurrence: NMMNH locality 16.

Referred specimens from type Revueltian: NMMNH locality 16; NMMNH P-4351, cervical vertebra.

First occurrence: Several genera from the Anisian (Triassic) of Africa, Asia and Europe. Rauisuchidae as utilized here includes Poposauridae, Prestosuchidae etc. of other workers pending a thorough revision of these taxa.

Last occurrence: *Fasolasuchus tenax* from late Norian/?Rhaetian (Triassic) of Argentina and undescribed taxon from the Apachean (late Norian/?Rhaetian) Whitaker quarry, New Mexico.

Acme zone: Anisian-Ladinian (Triassic).

Discussion: This specimen is similar to vertebrae of rauisuchians, but is not definitely assignable to this family.

Crocodylomorpha Hay, 1930
Sphenosuchia Bonaparte, 1971
Sphenosuchidae Huene, 1922
gen. et sp. indeterminate

Type Revueltian occurrences: NMMNH localities 1, 57, 124, 169, 459, 501: UMMP locality 5 miles west of San Jon in Badlands.

Referred specimens from type Revueltian: NMMNH locality 1; NMMNH P-4658, 2 cervical vertebrae; NMMNH P-4659, vertebrae; NMMNH P-4663, cervical, dorsal, caudal vertebrae; NMMNH P-4673, cervical vertebra; NMMNH P-16754, vertebrae; NMMNH P-16764, premaxilla: NMMNH locality 57; NMMNH P-4120, premaxilla, vertebral fragments; NMMNH P-17346, partial skeleton (cervical, dorsal, sacral and caudal vertebrae, partial pubes, ischia and ilia, premaxilla and maxilla fragments): NMMNH locality 124; NMMNH P-4573, sacral vertebrae: NMMNH locality 169; NMMNH P-4874, lower jaw: NMMNH locality 459; NMMNH P-16839, caudal centrum: NMMNH locality 501; NMMNH P-17229, osteoderm: 5 miles west of San Jon in badlands; UMMP 4771 (in part), three dorsal and one caudal centrum.

First occurrence: *Dyoplax arenaceus* from the late Carnian (Triassic) of Germany.

Last occurrence: *Kayentasuchus* sp. from the Sinemurian/Pliensbachian (Jurassic) of the western United States.

Acme zone: None.

Discussion: Several specimens in the Bull Canyon vertebrate fauna represent a sphenosuchian. The best specimens are NMMNH P-17346, which is a partial skeleton, and NMMNH P-4663, which is a partial vertebral column. The former specimen is the most diagnostic, although elements in common between the two specimens are nearly identical in size and morphology. NMMNH P-17346 contains several vertebrae that are indistinguishable from Chinle Group sphenosuchians that have been assigned to *Hesperosuchus agilis* (Colbert, 1952; Clark et al., 2000) and cf. *Sphenosuchus* sp. (Parrish, 1991). Unfortunately, the most potentially diagnostic elements in the Bull Canyon specimen are the fragmentary ilia, ischia and pubes. These elements are totally lacking in the holotypes of *Sphenosuchus* (Walker, 1990) and *Hesperosuchus* (Colbert, 1952). Parrish's (1991) specimen from Revueltian-age strata in Arizona includes portions of the pubes that are proximally, narrower anteroposteriorly. The ilium of the Bull Canyon Formation specimen indicates an imperforate, or nearly so, acetabulum, a less derived condition than seen in the European *Terrestrisuchus* (Crush, 1984), which is also of Norian age. There is a buttress above the acetabulum in NMMNH P-17346, which has previously not been recognized in the Sphenosuchidae. This buttress was the apparently the main reason behind Long and Murry (1995, fig. 146) assigning an ilium from the Whitaker quarry to *Postosuchus* despite the fact that the specimen was surrounded by osteoderms of a sphenosuchian. The Whitaker quarry specimen obviously pertains to the taxon described by Clark et al. (2000) from the same locality. In conclusion, a distinct sphenosuchid is present in the Bull Canyon Formation, but it is not represented by diagnostic specimens. Most Bull Canyon sphenosuchias specimens probably pertain to this taxon. However, NMMNH P-4673 is a cervical vertebra that has a much wider and less longitudinally arched centrum. This probably represents a second taxon.

Family ?Sphenosuchidae Huene, 1922

Type Revueltian occurrences: NMMNH localities 1, 156.

Referred specimens from type Revueltian: NMMNH locality 1; NMMNH P-4660, centrum: NMMNH locality 156; NMMNH P-4559, centrum.

First occurrence: *Dyoplax arenaceus* from the late Carnian (Triassic) of Germany.

Last occurrence: *Kayentasuchus* sp. from the Sinemurian/Pliensbachian (Jurassic) of the western United States.

Acme zone: None.

Discussion: These vertebrae are similar to specimens assigned above to the Sphenosuchidae but they are not diagnostic of that family.

Crocodylotarsi Benton and Clark, 1988
incertae sedis
gen. et sp. nov.

Type Revueltian occurrences: NMMNH localities 1, 162, 182, 467, 523, 527, 534-535: UMMP locality 5 miles west of San Jon in badlands.

Referred specimens from type Revueltian: NMMNH locality 1; NMMNH P-4691, left proximal femur; NMMNH P-4888, right proximal femur (large); NMMNH P-4902, partial dorsal vertebra; NMMNH P-4869, ?scapula blade; NMMNH P-16929, sacral vertebrae; NMMNH P-16920, proximal ischium; NMMNH P-16922, many osteoderm fragments; NMMNH P-16930, cervical vertebra; NMMNH P-16923, partial carapace; NMMNH P-16685, osteoderm: NMMNH locality 162; NMMNH P-4665, 11 vertebrae (cervical [1], dorsal [7], dorso-sacral [2], anterior caudal [1]): NMMNH locality 182; NMMNH P-16912, iliac blade: NMMNH locality 523; NMMNH P-17455, many osteoderm fragments; NMMNH P-17459, dorsal vertebra: NMMNH locality 467; NMMNH P-16932; small scavenged skeleton, portions of a large specimen and rauisuchian teeth: NMMNH locality 527; NMMNH P-17474, osteoderm fragments: NMMNH locality 534; NMMNH P-17384, osteoderm fragment: NMMNH locality 535; NMMNH P-17352, distal tibia: 5 miles west of San Jon in Badlands; UMMP 4771 (in part), proximal femur.

First and last occurrence: Revueltian (Triassic: early-middle Norian) of New Mexico and Arizona.

Acme zone: Revueltian (Triassic: early-middle Norian).

Discussion: This highly distinctive crocodylotarsan is distinguished by wide, rectangular paramedian osteoderms with an irregular pattern of deep pits and no lateral osteoderms, and a wide tarsus that has a small astragalar medial process and corresponding medial calcaneal concavity. It is known from several localities in the Revuelto and Barranca badlands and has recently been recognized from the lower Painted Desert Member of the Petrified Forest Formation at Petrified Forest National Park. This distinctive new taxon will be described in more detail elsewhere.

NMMNH P-16932 is the most complete specimen and consists of an articulated series of dorsal osteoderms with associated osteoderms and two ribs, many other elements of comparable size and a proximal femur and right astragalus and calcaneum of a larger individual. The articulated block consists of seven dorsal vertebrae. The anterior three vertebrae have associated paramedian osteoderms, one pair for each vertebra. The osteoderms are thick and have a deep ornamentation of rounded, irregularly placed pits. The posterior dorsal vertebra is heavily grooved, and a laterally compressed, serrated tooth (?rauisuchian) is preserved on this vertebra. Several skull fragments are present, one of which is a portion of a maxilla which preserved the tip of a replacement tooth. This tooth has large dorsally directed denticles. A fragment of a larger isolated tooth has a 5 mm wide root that flares to a 6 mm wide crown that has longitudinal striations.

Numerous partial osteoderms and one complete osteoderm are in the collection. The complete osteoderm is rectangular and is 64 mm long, 33 mm wide and 5 mm thick. The dorsal surface is covered by an irregular pattern of deep, rounded pits. The medial end of the osteoderm is thickened. The osteoderm thins at its lateral and anterior margins. A smooth strip runs along the anterior margin and broadens near the lateral margin.

This taxon represents a crocodylotarsan because it possesses: (1) a hemicylindrical calcaneal condyle for the fibula; (2) a flexed tibial facet on astragalus; (3) a single articulation between astragalus and calcaneum; and (4) single paramedian osteoderm per vertebra (Benton and Clark, 1988; Sereno, 1991). Further, it is assignable to a clade containing derived crurotarsans (= Suchia of Sereno, 1991) on the basis of an advanced "crocodile-normal" tarsus (Benton and Clark, 1988). However, the pit in the calcaneum to receive the astragalar process is shallower than in other members of this clade and thus is arguably less derived. The heavy dorsal armor is reminiscent of the aetosaurs, but there are no lateral osteoderms preserved in this taxon, and the morphology of the lateral margins of the paramedian osteoderms indicates that none were present. I note, however, that *Redondasuchus* has no lateral osteoderms. On the basis of the tarsal configuration, this taxon might represent a sister taxon of the Stagonolepididae.

The orientation of the large denticles on its teeth indicates that this taxon was herbivorous. This is consistent with the heavy dorsal armor, which is rare in carnivores that are not semiaquatic. The articulated specimen represents an animal about 1.5 m long. Other specimens indicate individuals about twice as large.

gen. et sp. nov.?

Type Revueltian occurrences: NMMNH localities 1, 171.

Referred specimens from type Revueltian: NMMNH locality 1; NMMNH P-4664, neural arches; NMMNH P-4680, distal femur; NMMNH P-4700, distal femur; NMMNH P-4955, caudal vertebra; NMMNH P-16687, osteoderm fragment; NMMNH P-16919, fragmentary skeleton; NMMNH P-16924, proximal ischium; NMMNH P-16925, part of carapace; NMMNH P-16928, proximal pubis; NMMNH P-16931, quadrate: NMMNH locality 171; NMMNH P-16730, neural spine.

First and last occurrence: Revueltian (Triassic: early-middle Norian) of New Mexico and Arizona.

Acme zone: Revueltian (Triassic: early-middle Norian).

Discussion: Several fragmentary specimens show similarities in morphology to the specimens of the new species discussed above.

Crocodylotarsi Benton and Clark, 1988
Family Indeterminate

Type Revueltian occurrences: NMMNH localities 1, 84, 124, 129, 162, 164-165, 171, 182, 518, 521, 523, 527, YPM locality 6649.

Referred specimens from type Revueltian: NMMNH locality 1; NMMNH P-4870, distal femur; NMMNH P-4676, distal femur; NMMNH P-4677, distal tibia; NMMNH P-4678, distal tibia; NMMNH P-4679, distal femur; NMMNH P-4681, distal femur; NMMNH P-4858, basicranium; NMMNH P-4955, 3 caudal vertebrae; NMMNH P-16670, distal tibia; NMMNH P-16691, neural arch fragment; NMMNH P-16692, neural arch fragment; NMMNH P-16726, central fragments; NMMNH P-16743, top of neural spine; NMMNH P-16898, fragmentary skeleton of ?new animal possibly intermixed with specimen of another genus; NMMNH P-16908, proximal tibia: NMMNH locality 84; NMMNH P-16661, centrum fragment: NMMNH locality 124; NMMNH P-4625, metapodial: NMMNH locality 129; NMMNH P-4491, vertebral fragment: NMMNH locality 162; NMMNH P-16630, vertebrae: NMMNH locality 164; NMMNH P-4692, proximal tibia: NMMNH locality 165; NMMNH P-4919, distal tibia: NMMNH locality 171; NMMNH P-16897, end of centrum: NMMNH locality 182; NMMNH P-4949, scapula fragment: NMMNH locality 518; NMMNH P-17416, metapodial: NMMNH locality 521; NMMNH P-17222, vertebral fragments: NMMNH locality 523; NMMNH P-17460, centrum fragment: NMMNH locality 527; NMMNH P-17361, partial astragalus; NMMNH P-17476, distal femur; NMMNH P-17501, distal tibia: YPM locality 6649; uncataloged caudal vertebra.

First occurrence: Several genera from the Anisian (Triassic) of Africa, Asia and Europe.

Last occurrence: Several Recent genera.

Acme zone: Late Carnian-late Norian (Triassic).

Discussion: A variety of specimens show similarities to crocodylotarsan taxa but cannot be identified further.

?Crocodylotarsi Benton and Clark, 1988

Type Revueltian occurrence: NMMNH locality 132.

Referred specimens from type: NMMNH locality 132; NMMNH P-4585, femoral shaft.

First occurrence: Several genera from the Anisian (Triassic) of Africa, Asia and Europe.

Last occurrence: Several Recent genera.

Acme zone: Late Carnian-late Norian (Triassic).

Discussion: NMMNH P-4585 is a femoral shaft that is reminiscent of crocodylotarsans.

Saurischia Seeley, 1888
Theropoda Marsh, 1881
Herrerasauridae Benedetto, 1973
gen et sp. nov. 1

Type Revueltian occurrences: NMMNH localities 163, 169, 176, 499, 503: UMMP locality San Juan (sic), New Mexico.

Referred specimens from type Revueltian: NMMNH locality 163; NMMNH P-4666, pubis fragments: NMMNH locality 169; NMMNH P-16607, tooth fragment: NMMNH locality 176; NMMNH P-16656, dorsal, caudal centra: NMMNH locality 499; NMMNH P-17258, fragmentary skeleton (vertebrae, scapula fragments): NMMNH locality 503; NMMNH P-17134, dorsal and caudal vertebral and pelvic fragments: San Juan (sic), New Mexico; UMMP 7274 (in part), two dorsal centra.

First and last occurrence: Revueltian (Triassic: early-middle Norian) of New Mexico.

Acme zone: Revueltian (Triassic: early-middle Norian).

Discussion: This is the most common dinosaur in the lower Bull Canyon Formation fauna. It is represented by three fragmentary skeletons (NMMNH P-16656, NMMNH P-17134, NMMNH P-17258) from the Revuelto and Barranca Creek badlands.

NMMNH P-17134 is a partial skeleton that includes both anterior and posterior dorsal centra. An anterior dorsal centrum is 47 mm long with a width of 34 mm whereas a posterior member of the series is 39 mm long and 39 mm wide. Thus, posterior dorsal centra are shorter and wider than anterior dorsal centra. NMMNH P-16656 also includes both anterior and posterior dorsal centra.

gen et sp. nov. 1?

Type Revueltian occurrences: NMMNH localities 1, 498.

Referred specimens from type Revueltian: NMMNH locality 1; NMMNH P-16946, centra: NMMNH locality 498; NMMNH P-17154, centra.

First and last occurrence: Revueltian (Triassic: early-middle Norian) of New Mexico.

Acme zone: Revueltian (Triassic: early-middle Norian).

Discussion: Two poorly preserved centra are tentatively assigned to this taxon.

gen et sp. nov. 2

Type Revueltian occurrences: NMMNH localities 110, 114, 124, 128

Referred specimens from type Revueltian: NMMNH locality 110; NMMNH P-4569, partial skeleton including dorsal centra, proximal left femur, partial astragalus, metatarsal fragments and phalanges (Fig. 7A-C): NMMNH locality 114; NMMNH P-4440, centrum fragment: NMMNH locality 124; NMMNH P-4380, vertebral fragment: NMMNH locality 128; NMMNH P-4375, vertebral fragment.

First and last occurrence: Revueltian (Triassic: early-middle Norian) of New Mexico.

Acme zone: Revueltian (Triassic: early-middle Norian).

Discussion: This taxon is restricted to the upper Bull Canyon Formation. Some specimens were previously attributed to *Coelophysis* (Lucas et al., 1985d, fig. 3D-O). It is a herrerasaurid because the ascending process of the astragalus does not form a wall that separates the anterior and dorsal surfaces (Novas, 1989).

?gen et sp. nov. 2

Type Revueltian occurrence: NMMNH locality 155.

Referred specimens from type Revueltian: NMMNH locality 155; NMMNH P-4547, phalanx.

First and last occurrence: Revueltian (Triassic: early-middle Norian) of New Mexico.

Acme zone: Revueltian (Triassic: early-middle Norian).

Discussion: NMMNH P-4547 is a phalanx that appears very similar to specimens referred to gen et sp. nov. 2 in the but it cannot be definitely be placed in this genus

Herrerasauridae Benedetto, 1973
indeterminate

Type Revueltian occurrences: NMMNH localities 1, 73, 110, 463.

Referred specimens from type Revueltian: NMMNH locality 1; NMMNH P-4927, posterior dorsal centrum: NMMNH locality 73; NMMNH P-17325, posterior dorsal centrum: NMMNH locality 110; NMMNH P-4766, posterior dorsal centrum: NMMNH locality 463; NMMNH P-16844, posterior dorsal vertebrae.

First occurrence: *Herrerasaurus* and *Staurikosaurus* from the Adamanian (late Carnian) of South America, *Caseosaurus* from the Adamanian of North America.

Last occurrence: *Chindesaurus* and two undescribed taxa from the Revueltian (early-middle Norian) of North America.

Acme zone: None.

Discussion: A number of isolated centra are identifiable as herrerasaurid posterior dorsal centra (Lucas et al., 1985d, fig. 3P-Q). These centra are relatively short and wide. The cross section through the middle of the centrum has a straight ventral margin with lateral constrictions. These centra are of a fairly uniform size and appear to represent a distinct herrerasaurid taxon of the size of *Chindesaurus bryansmalli* (Long and Murry, 1995) from the Revueltian of Petrified Forest National Park. These centra are typically about 20 mm long

Ceratosauria Gauthier, 1986
***Gojirasaurus quayi* Carpenter, 1997**

Type Revueltian occurrence: UCM locality 82021.

Referred specimens from type Revueltian: UCM locality 82821; UCM 47221, partial skeleton including right pubis, left tibia, four dorsal centra, one dorsal neural arch, four ribs, one metatarsal, one chevron and one tooth.

First and last occurrence: Revueltian (Triassic: early to middle Norian) of New Mexico.

Acme zone: Revueltian (Triassic: early to middle Norian).

Description: Parrish and Carpenter (1986, figs. 11.5B-D, 11.6, 11.7B, C, E, H and J) and Hunt (1994) described UCM 47221, and subsequently Carpenter (1997) named this partial skeleton *Gojirasaurus quayi*. This is one of the largest Late Triassic theropods.

Theropoda Marsh, 1881
incertae sedis
gen et sp. nov. 3

Type Revueltian occurrence: NMMNH locality 527

Referred specimens from type Revueltian: NMMNH locality 527; NMMNH P-17375, fragmentary skeleton including pubis, proximal femora, vertebrae (Fig. 7E-G).

First and last occurrence: Revueltian (Triassic: early to middle Norian) of New Mexico.

Acme zone: Revueltian (Triassic: early to middle Norian).

Diagnosis: This is a theropod that is distinguished from others by possession of pubes that are conjoined almost to their proximal ends and a pubic shaft that is straight, narrow and with a proximal extremity that flares both anteriorly and posteriorly.

Discussion: This taxon is only known from one partial skeleton from the lower Bull Canyon Formation in the Barranca badlands. NMMNH P-17375 consist of 10 dorsal central fragments, proximal femora and a pubis. Dorsal centra are extremely hollowed and are 33 mm in width. The anteroposterior width across the femoral heads is 33 mm. The femoral head is rectangular but is only slightly offset from the shaft. The proximal pubis is flared both anteriorly and posteriorly with an anteroposterior length of 41 mm. The pubic shaft is narrow and straight. The medial sheet of the pubis is broken but extends proximally beyond the onset of the proximal flaring of the element, e.g., the pubes are sutured together almost to their proximal ends. No obturator foramen is present.

Theropoda Marsh, 1881
indeterminate

Type Revueltian occurrences: 1, 55, 98, 108, 120, 122, 124, 134, 152, 166, 168, 171, 174, 176-177, 182, 187, 519, 526-527: UMMP localities: San Juan (sic), New Mexico; 5 miles west of San Jon in Badlands: YPM locality 5883.

Referred specimens from type Revueltian: NMMNH locality 1; NMMNH P-4783, cervical vertebra; NMMNH P-4657, centra; NMMNH P-4701, proximal tibia; NMMNH P-4704, tibia; NMMNH P-4688, centrum; NMMNH P-4928, centrum fragments; NMMNH P-16643, centrum fragment; NMMNH P-16645, centrum fragments; NMMNH P-16650, centrum fragment; NMMNH P-16652, centrum; NMMNH P-16666, proximal ischium; NMMNH P-16674, proximal pubis; NMMNH P-16713, cervical vertebra; NMMNH P-16728, centrum; NMMNH P-16787, centrum fragment; NMMNH P-16935, centrum: NMMNH locality 55; NMMNH P-4129, caudal centrum: NMMNH locality 98; NMMNH P-4126, proximal femur: NMMNH locality 108; NMMNH P-4652, ilium, ischium: NMMNH locality 120; NMMNH P-4596, phalanx: NMMNH locality 122; NMMNH P-4465, posterior dorsal centrum; NMMNH P-4588, metapodial; NMMNH P-4636, proximal ischium: NMMNH locality 124; NMMNH P-4427, phalanx: NMMNH locality 134; NMMNH P-4984, dorsal centrum; NMMNH P-4415, proximal femur; NMMNH P-4509, metapodial: NMMNH locality 152; NMMNH P-4474, dorsal centrum: NMMNH locality 166; NMMNH P-4948, centrum fragments; NMMNH locality 168; NMMNH P-4862, centrum fragments: NMMNH locality 171; NMMNH P-4856, vertebral fragments: NMMNH locality 174; NMMNH P-16658, vertebral, scapula fragments: NMMNH locality 176; NMMNH P-4884, phalanx: NMMNH locality 177; NMMNH P-4882, vertebrae: NMMNH locality 182; NMMNH P-16761, phalanx: NMMNH locality 187; NMMNH P-4932, centrum fragment: NMMNH locality 519; NMMNH P-17179, centrum fragments: NMMNH locality 526; NMMNH P-17470, centra: NMMNH locality 527; NMMNH P-17359, caudal centra: San Juan (sic), New Mexico; UMMP 7274 (in part), caudal vertebra: 5 miles west of San Jon in badlands; UMMP 7441 (in part), caudal centrum: YPM locality 5883; uncataloged dorsal centra (distorted).

First occurrence: Several taxa from the late Carnian (Triassic) of South America and North America.

Last occurrence: Many Recent taxa.

Acme zone: None.

Discussion: A large number of isolated elements, dominantly centra, cannot be identified more precisely than Theropoda. NMMNH P-4126 and NMMNH P-4415 are the proximal ends of two femora from the lower Bull Canyon Formation that do not pertain to any of the undescribed taxa.

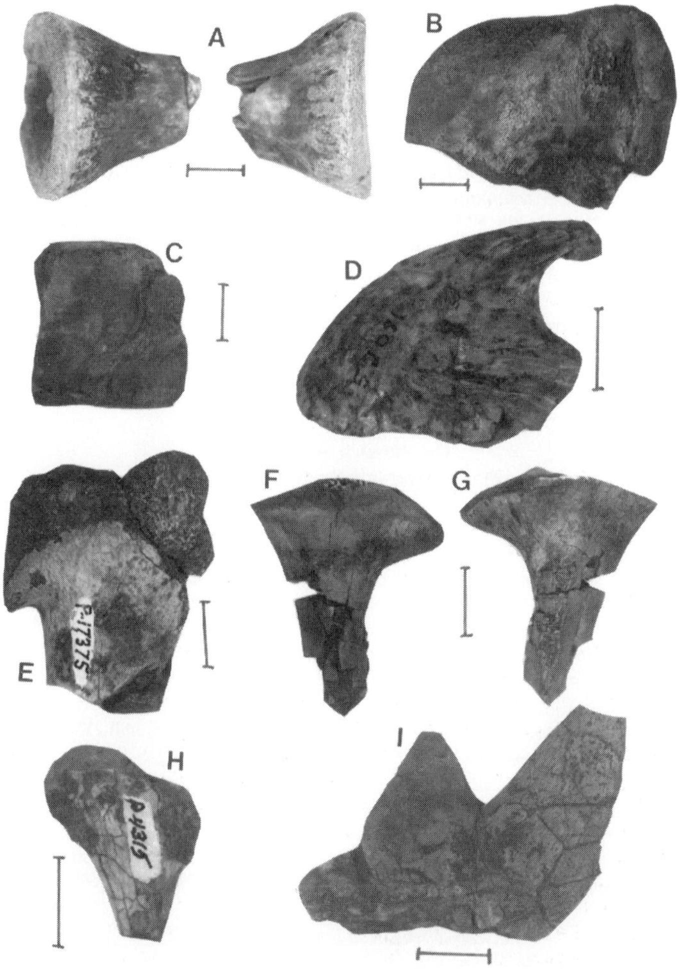

FIGURE 7. Theropods, *Shuvosaurus inexpectatus* and unidentified reptiles from the Bull Canyon Formation. A-C. Partial skeleton of Herrerasauridae gen. et sp. nov. 2 (NMMNH P-4569). A. Dorsal centra in lateral view. B. Posterior view of proximal end of left femur. C. Dorsal view of left astragalus. D. Lateral view of left premaxilla of *Shuvosaurus inexpectatus* (UCM 52081). E-G. Partial skeleton of Theropoda incertae sedis gen. et sp. nov.(NMMNH P-17375). E. Proximal left femur in anterior view. F-G. Proximal pubis in lateral (F) and medial (G) views. H. Proximal femur of unidentified reptile (NMMNH P-4315) in anterior view. I. Skull fragment of unidentified reptile, possibly turtle (NMMNH P-16697) in dorsal view. Scale bars are 1 cm.

?Theropoda Marsh, 1881
indeterminate

Type Revueltian occurrences: NMMNH localities 1, 16, 38, 73, 498-499, 507.

Referred specimens from type Revueltian: NMMNH locality 1; NMMNH P-4698, distal tibia; NMMNH P-16793, centrum end; NMMNH P-16940, vertebra: NMMNH locality 16; NMMNH P-4653, centrum: NMMNH locality 38; NMMNH P-17342, dorsal centrum; NMMNH locality 73; NMMNH P-4223, vertebrae: NMMNH locality 498; NMMNH P-17153, pelvic fragment: NMMNH locality 499; NMMNH P-17281, centrum: NMMNH locality 507; NMMNH P-17211, centrum.

First occurrence: Several taxa from the late Carnian (Triassic) of South America and North America.

Last occurrence: Many Recent taxa.

Acme zone: None.

Discussion: Most questionable theropod specimens are vertebral fragments that are waisted as in the Theropoda.

Ornithischia Seeley, 1888
Revueltosaurus callenderi Hunt, 1989

Type Revueltian occurrences: NMMNH localities 1, 132, 171, 498, 526.

Referred specimens from type Revueltian: NMMNH locality 1; NMMNH P-4957, incisiform tooth (holotype); NMMNH P-4958, dentary/maxillary tooth (paratype); NMMNH P-4959, premaxillary tooth (paratype); NMMNH P-4960, twenty eight premaxillary and dentary/maxillary teeth; NMMNH P-16573, teeth: NMMNH locality 132; NMMNH P-4426, tooth: NMMNH locality 171; NMMNH P-16637, tooth: NMMNH locality 498; NMMNH P-17187, tooth: NMMNH locality 526; NMMNH P-17362, teeth; NMMNH P-17382, tooth.

First and last occurrence: Revueltian (Triassic: early to middle Norian) of New Mexico and Arizona.

Acme zone: Revueltian (Triassic: early to middle Norian).

Discussion: The holotype and paratypes of *Revueltosaurus* are from NMMNH locality 1 in the Revuelto badlands, as are most of the remaining specimens. *Revueltosaurus* also occurs in the Barranca badlands and in the lower Bull Canyon Formation at Bull Canyon.

Lucianosaurus wildi Hunt and Lucas, 1994

Type Revueltian occurrence: NMMNH locality 110.

Referred specimens from type Revueltian: NMMNH locality 110; NMMNH P-18194, dentary/maxillary tooth (holotype); NMMNH P-18195, dentary/ maxillary tooth.

First and last occurrence: Revueltian (Triassic: early to middle Norian) of New Mexico and Arizona.

Acme zone: Revueltian (Triassic: early to middle Norian).

Discussion: Hunt and Lucas (1994) named *Lucianosaurus* for an isolated ornithischian tooth. This taxon only occurs in the upper Bull Canyon Formation (NMMNH locality 110).

Ornithischia Seeley, 1888
indeterminate

Type Revueltian occurrence: NMMNH locality 527.

Referred specimens from type Revueltian: NMMNH locality 527; NMMNH P-17483, proximal femur.

First occurrence: Various taxa from South America, North America and Africa from the late Carnian (Triassic).

Last occurrence: Various taxa worldwide from the Maastrichtian (Cretaceous)

Acme zone: None

Discussion: NMMNH P-17483 is a fragment of a femur with a prominent fourth trochanter that is similar to specimens of this element in early ornithischians (e.g., Santa Luca, 1984).

Therapsida Broom, 1905
Cynodontia Owen, 1861
Dromatheridae Gill, 1872
Pseudotriconodon chatterjeei Lucas and Oakes, 1988

Type Revueltian occurrence: NMMNH locality 110.

Referred specimens from type Revueltian: UNM MV-518.

First and last occurrence: Revueltian (Triassic: early to middle Norian) of New Mexico.

Acme zone: Revueltian (Triassic: early to middle Norian).

Discussion: *Pseudotriconodon chatterjeei* is known only from the holotype, which is from the upper Bull Canyon Formation (NMMNH locality 110) at Bull Canyon (Lucas et al., 1985d; Lucas and Oakes, 1988).

Reptilia Laurenti, 1768
indeterminate

Type Revueltian occurrences: NMMNH localities 1, 38, 53, 57, 70, 73, 78, 89, 95, 110, 122, 133, 135, 155, 165-167, 170-171, 174, 176-177, 179, 182, 189, 193, 196, 434, 451-456, 458, 462, 464, 498-503, 508, 518, 521-524, 526-527, 530, 534-535, 537, 539, UCM locality 52081, UMMP locality San Juan (sic), 5 miles west of San Jon in badlands.

Referred specimens from type Revueltian: NMMNH locality 1; NMMNH P-4791, tooth; NMMNH P-4792, tooth; NMMNH P-4793, tooth; NMMNH P-4794, tooth; NMMNH P-4795, tooth; NMMNH P-4796, tooth; NMMNH P-4797, tooth; NMMNH P-4798, tooth; NMMNH P-4799, tooth; NMMNH P-4800, tooth; NMMNH P-16806, cervical vertebra; NMMNH P-16696, zygapophysis; NMMNH P-4683, proximal femur; NMMNH P-4696, centra; NMMNH P-4699, cervical vertebrae; NMMNH P-4705, tooth; NMMNH P-4706, tooth; NMMNH P-4708, tooth; NMMNH P-4709, tooth; NMMNH P-4710, tooth; NMMNH P-4712, tooth; NMMNH P-4714, tooth; NMMNH P-4785, tooth; NMMNH P-4814, bone; NMMNH P-4815, tooth; NMMNH P-4818, tooth; NMMNH P-4819, phalanx; NMMNH P-4820, small limb end; NMMNH P-4821, small limb end; NMMNH P-4822, bone; NMMNH P-4823, small ?limb end; NMMNH P-4824, small vertebra; NMMNH P-4825, small ungual phalanx; NMMNH P-4826, tooth; NMMNH P-4827, tooth; NMMNH P-4828, tooth; NMMNH P-4829, tooth; NMMNH P-4830, tooth; NMMNH P-4831, tooth; NMMNH P-4837, tooth; NMMNH P-4838, tooth; NMMNH P-4841, tooth; NMMNH P-4842, tooth; NMMNH P-4843, tooth; NMMNH P-4844, tooth; NMMNH P-4845, tooth; NMMNH P-4848, tooth; NMMNH P-4849, tooth; NMMNH P-16603, rib; NMMNH P-16612, ?edentulous lower jaw; NMMNH P-16623, bone; NMMNH P-16660, limb end; NMMNH P-16663, distal humerus; NMMNH P-16664, proximal ischium; NMMNH P-16665, scapula fragment; NMMNH P-16667, metapodial; NMMNH P-16668, bone; NMMNH P-16669, ?femoral end; NMMNH P-16673, osteoderm fragment; NMMNH P-16677, jaw fragment; NMMNH P-16679, elongate caudal centrum; NMMNH P-16684, jaw fragment; NMMNH P-16690, pelvic fragment; NMMNH P-16693, tooth fragment; NMMNH P-16797, skull fragment; NMMNH P-16698, skull fragment; NMMNH P-16699; NMMNH P-16701, bone; NMMNH P-16702, centrum fragment; NMMNH P-16703, centrum fragment; NMMNH P-16704, centrum fragment; NMMNH P-16705, centrum fragment; NMMNH P-16707, osteoderm fragment; NMMNH P-16708, centrum fragment; NMMNH P-16709, centrum fragment; NMMNH P-16710, centrum fragment; NMMNH P-16711, centrum fragment; NMMNH P-16712, centrum fragment; NMMNH P-16714, vertebra; NMMNH P-16715, centrum fragment; NMMNII P-16716, elongate dorsal centrum; NMMNH P-16717, centrum fragment; NMMNH P-16718, centrum; NMMNH P-16719, centrum fragment; NMMNH P-16724, vertebra; NMMNH P-16725, osteoderms; NMMNH P-16729, distal scapula; NMMNH P-16731, skull fragments; NMMNH P-16732, vertebra; NMMNH P-16737, bone; NMMNH P-16738, bone; NMMNH P-16740, vertebra; NMMNH P-16741, caudal centrum; NMMNH P-16744, very elongate centrum; NMMNH P-16746, ?scapula fragment; NMMNH P-16747, vertebra; NMMNH P-16748, very elongate centrum; NMMNH P-16749, bone fragments; NMMNH P-16751, centrum; NMMNH P-16755, long bone with step fracture; NMMNH P-16756, cranial fragments; NMMNH P-16757, atlas intercentrum; NMMNH P-16758, bone; NMMNH P-16762, centrum fragment NMMNH P-16766, centrum fragment; NMMNH P-16767, centrum fragment: NMMNH P-16768, atlas; NMMNH P-16769, centrum fragment; NMMNH P-16770, centrum fragment; NMMNH P-16771, centrum; NMMNH P-16772, centrum fragment; NMMNH P-16773, centrum fragment; NMMNH P-16774, centrum fragment; NMMNH P-16775, centrum fragment; NMMNH P-16776, osteoderm; NMMNH P-16777, vertebra; NMMNH P-16778, vertebra; NMMNH P-16781, cervical centrum; NMMNH P-16782, skull fragment; NMMNH P-16783, skull fragment; NMMNH P-16784, caudal centrum; NMMNH P-16785, centrum; NMMNH P-16786, caudal vertebra; NMMNH P-16788, limb fragment; NMMNH P-16789, phalanges; NMMNH P-16790, cervical vertebra; NMMNH P-16791, cervical vertebra; NMMNH P-16792, caudal vertebra; NMMNH P-16794, tarsals; NMMNH P-16795, anterior dorsal centrum; NMMNH P-16796, centra; NMMNH P-16797, centrum fragment; NMMNH P-16798, proximal ischium; NMMNH P-16799, vertebra; NMMNH P-16801, centra; NMMNH P-16802, centrum; NMMNH P-16803, bone; NMMNH P-16804, bone; NMMNH P-16805, centrum; NMMNH P-16807, neural spine; NMMNH P-16808, bone; NMMNH P-16809, neural spine; NMMNH P-16810, centrum; NMMNH P-16811, centrum; NMMNH P-16890, vertebral column and osteoderms (part of UMMP 7441); NMMNH P-16909, limb fragments; NMMNH P-16910, scapula fragment; NMMNH P-16911, skull fragment; NMMNH P-16914, centrum fragment; NMMNH P-16915, vertebral fragment; NMMNH P-16916, vertebral fragment; NMMNH P-16921, osteoderm and limb fragments; NMMNH P-16927, ?fragmentary skeleton; NMMNH P-16934, neural process; NMMNH P-16938, centrum fragment; NMMNH P-16939, centrum fragment; NMMNH P-16944, centrum fragment; NMMNH P-16945, centrum fragments; NMMNH P-16947, centrum fragments: NMMNH locality 38; NMMNH P-4308, rib, vertebral fragments: NMMNH locality 53; NMMNH P-4315, proximal femur; NMMNH P-4157, ilium fragment: NMMNH locality 57; NMMNH P-4253, cervical vertebrae: NMMNH locality 70; NMMNH P-17374, centrum fragment: NMMNH locality 73; NMMNH P-17326, vertebral fragment; NMMNH P-17328, vertebral fragment: NMMNH locality 78; NMMNH P-4191, skull fragments: NMMNH locality 80; NMMNH P-4168, scapula fragments: NMMNH locality 89; NMMNH P-16671, distal humerus: NMMNH locality 95; NMMNH P-4112, bone; NMMNH locality 110; NMMNH P-4487, tooth; NMMNH P-4580, tooth; NMMNH P-4595, small bones; NMMNH P-4598, elongate caudals; NMMNH P-4603, partial quadrate; NMMNH P-4605, tooth; NMMNH P-4630, bone fragments: NMMNH locality 122; NMMNH P-4388, limb end; NMMNH P-4556, bone: NMMNH locality 133; NMMNH P-4420, quadrate: NMMNH locality 135; NMMNH P-4396, quadrate: NMMNH locality 155; NMMNH P-4566, vertebral fragments: NMMNH locality 165; NMMNH P-4909, bone; NMMNH P-4910, iliac fragment; NMMNH P-4934, skull fragment; NMMNH P-16589, centrum fragment; NMMNH P-16733, centrum fragment: NMMNH P-16750, centrum fragment: NMMNH locality 166; NMMNH P-16621, bone: NMMNH locality 167; NMMNH P-4924, vertebral fragment: NMMNH locality 170; NMMNH P-4896, neural arch: NMMNH locality 171; NMMNH P-16734, small centra; NMMNH P-16735, vertebral fragments; NMMNH P-16742, bone; NMMNH P-16752, centrum fragment; NMMNH P-16759, bone; NMMNH P-16936, phalanx: NMMNH locality 174; NMMNH P-16745, transverse process; NMMNH P-16763, osteoderm: NMMNH locality 176; NMMNH P-16654, cervical vertebra: NMMNH locality 177; NMMNH P-4886, teeth; NMMNH P-16892, elongate centrum: NMMNH locality 179; NMMNH P-4890, phalanx: NMMNH locality 182; NMMNH P-16662, metapodial fragment; NMMNH P-16736, small ?limb bone: NMMNH locality 189; NMMNH P-4931, skull fragment: NMMNH locality 193; NMMNH P-4929, bone: NMMNH locality 196; NMMNH P-16765, centrum: NMMNH locality 434; NMMNH P-17195, centrum: NMMNH locality 451; NMMNH P-16816, centrum; NMMNH P-16823, centrum: NMMNH locality 452; NMMNH P-16858, phalanx; NMMNH P-16859, vertebral fragments: NMMNH locality 453; NMMNH P-16843, distal femur: NMMNH locality 454; NMMNH P-16861, vertebral fragment; NMMNH P-16862, vertebral fragment; NMMNH P-16865, vertebra; NMMNH P-16867, rib fragment; NMMNH P-16870, bone: NMMNH locality 455; NMMNH P-16849, proximal femur: NMMNH P-locality 456; NMMNH P-16879, bone: NMMNH locality 458; NMMNH P-16846, phalanges; NMMNH P-16847, vertebral fragments: NMMNH locality 462; NMMNH P-16871, vertebral fragments; NMMNH P-16872, phalanges; NMMNH P-16876, ?skull fragment: NMMNH locality 464; NMMNH P-16838, centrum: NMMNH locality 498; NMMNH P-17150, vertebrae; NMMNH P-17184, dentary fragments; NMMNH P-17191, osteoderms: NMMNH locality 499; NMMNH P-17152, occipital condyle; NMMNH P-17160, elongate centra; NMMNH P-17264, ischium; NMMNH P-17266, caudal vertebra; NMMNH P-17268, neural spine; NMMNH P-17270, caudal centrum; NMMNH P-17271, centrum fragments; NMMNH P-17274, caudal centrum; NMMNH P-17275, centrum fragment; NMMNH P-17277, bone; NMMNH P-17278, proximal pubis; NMMNH P-17283, centrum fragments; NMMNH P-17284, chevron fragment; NMMNH P-17285, centrum fragment; NMMNH P-17289, elongate centrum; NMMNH P-17290, small limb bones; NMMNH P-17291, vertebra; NMMNH P-17292, vertebra: NMMNH locality 500; NMMNH P-17194, bone fragments: NMMNH locality 501; NMMNH P-17132; NMMNH P-17164, centra; NMMNH P-17242, elongate caudals; NMMNH P-17288, vertebral fragment; NMMNH P-17294, elongate vertebra; NMMNH P-17295, small limb end: NMMNH locality 502; NMMNH P-17158, centra (cervical, sacral): NMMNH locality 503; NMMNH P-17217, sacral vertebra: NMMNH locality 508; NMMNH P-17235, vertebral fragments: NMMNH locality 517; NMMNH P-17207, centrum: NMMNH locality 518; NMMNH P-17415, limb end: NMMNH locality 521; NMMNH P-17225, phalanx; NMMNH P-17226, small vertebrae: NMMNH locality 522; NMMNH P-4984, partially articulated skeleton; NMMNH P-17418, centrum; NMMNH P-17424, centrum fragment: NMMNH locality 523; NMMNH P-17456, centrum fragment; NMMNH P-17457, vertebral fragments; NMMNH P-17458, vertebral fragments; NMMNH P-17461, limb bone: NMMNH locality 524; NMMNH P-17398, maxilla fragment: NMMNH locality 526; NMMNH P-17376, tooth; NMMNH P-17462, centrum; NMMNH P-17463, pelvic fragments; NMMNH P-17464, centrum fragment; NMMNH P-17466, 3 small vertebrae; NMMNH P-17467, partial centrum: NMMNH locality 527; NMMNH P-17360, basicranium; NMMNH P-17480, sacral vertebra; NMMNH P-17482, radius; NMMNH P-17487, quadrate; NMMNH P-17492, cervical centrum; NMMNH P-17494, skull fragment; NMMNH P-17503, pelvis fragment: NMMNH locality 530; NMMNH P-17490, vertebral fragments: NMMNH locality 534; NMMNH P-17386, centrum; NMMNH P-17390, jaw fragment: NMMNH locality 535; NMMNH P-17354, small limb end: NMMNH locality 537; NMMNH P-17301, caudal vertebrae: NMMNH locality 539; NMMNH P-17400, centrum fragment: UCM locality 82081; UCM 52081, right premaxilla; uncataloged teeth, vertebrae: San Juan (sic), New Mexico; UMMP 7274 (in part), two different cervical vertebrae: 5 miles west of San Jon in Badlands; UMMP 4771 (in part); proximal and distal humerus, partial articulated caudal series with osteoderms, femoral fragment, anterior dorsal centrum.

First occurrence: ?*Westlothiaana lizziae* from the Visean (Carboniferous) of Scotland.

Last occurrence: Many Recent species.

Acme zone: None

Discussion: A large number of generally isolated bones represent indeterminate reptiles. Many specimens hint at a diversity of small, undescribed reptiles. For example, NMMNH P-4598, 16679, 16716, 16744, 16892, 17160, 17294 represent one or more taxa of small reptiles with very elongate centra, and NMMNH P-16660, 16663 and 16670 are limb ends of one or more small, primitive reptiles.

Three specimens deserve special comment. NMMNH P-16890 is vertebral centra and osteoderms of a small reptile. Part of this individual is present within UMMP 4771, which was collected 72 years before NMMNH P-16890. This is evidenced by morphology and preservation and provides further evidence that Case collected at NMMNH locality 1. The centra are moderately waisted and are reminiscent of sphenosuchid dorsal centra, but the armor is much more robust than is found in that family.

NMMNH P-16697 is a portion of the lateral margin of a reptile skull (Fig. 7I). It consists of a series of flattened spikes, each of which preserves several sutures. Longitudinal sutures run medially down the spikes and split into two near the tips so that they are Y-shaped in dorsal view. Proximally the central spike suture is crossed by another suture running parallel to the presumed skull margin. Superficially, this specimen resembles the skull margin of *Proganochelys* and some procolophonids. However, it differs significantly in the pattern of sutures. It may represent a turtle (Lucas et al., 2000).

NMMNH P-4315 is a proximal femur from NMMNH locality 53 (Fig. 7H). The femoral head is spherical and is distinctly offset from the shaft. It is reminiscent of the femur of the primitive turtle *Proganochelys* (Gaffney, 1990, fig. 163) but differs in having a more spherical head. NMMNH P-4315 differs in the same manner from a cynodont proximal femur (Jenkins, 1971, fig. 48) and also in lacking a large lesser trochanter. This femoral fragment obviously represents a distinct taxon, but its affinities are unclear.

BIOSTRATIGRAPHY OF THE BULL CANYON FORMATION

There are clearly two fossil assemblages present within the Bull Canyon Formation (Table 1). The lower assemblage occurs in the basal 50 m of the formation in the Revuelto and Barranca badlands and in the lower part of Bull Canyon. The upper assemblage is restricted to the upper part of the formation in Bull Canyon. Taxa in common between the faunas are Colobontidae, Redfieldiidae, *Apachesaurus*, Squamata, *Pseudopalatus*, *Typothorax*, *Shuvosaurus*, Rauisuchidae and Herrerasauridae (small).

Taxa restricted to the lower assemblage are *Semionotus*, ?*Tanaocrossus*, *Quayia*, Metoposauridae (large), *Pseudopalatus n. sp.*, Crocodylotarsi incertae sedis gen. et sp. nov., *Aetosaurus*, Stagonolepididae gen. et sp. nov., *Paratypothorax*, *Desmatosuchus*, *Postosuchus*, Sphenosuchidae, Herrerasauridae gen. et sp.

nov. 1 , Theropoda incertae sedis gen. et sp. nov. *3* and *Revueltosaurus*. Taxa restricted to the upper assemblage are, Herrerasauridae gen. et sp. nov. *2, Lucianosaurus* and *Pseudotriconodon*.

The lower assemblage is more diverse in all major elements (fish, amphibians and reptiles) and also includes a higher diversity of terrestrial animals. Part of this diversity is obviously related to topographic factors. The upper Bull Canyon Formation (and overlying Redonda Formation) underlie the Jurassic Entrada Sandstone, which is a cliff-forming unit in this area, which is usually topographically high (e.g., top of mesa, edge of Llano Estacado). Thus, the upper Bull Canyon Formation often forms the basal portion of a high relief slope. The next resistant sandstone unit below the Entrada is the Trujillo Formation, which underlies the Bull Canyon. The Trujillo is a topographically low unit in east-central New Mexico, just above the level of the Canadian River. Thus, tributaries of the Canadian draining the Llano Estacado and mesas in this area have to cut through the lower Bull Canyon Formation, and in several areas (Revuelto and Barranca badlands) large areas of badlands result. Obviously, the lower relief outcrops and badland exposures of the lower Bull Canyon Formation have been more explored than the higher relief outcrops of the upper part of the formation. However, similar distribution patterns also occurs within the Revueltian Painted Desert Member of the Petrified Forest Formation in northeastern Arizona, where not only are vertebrate fossils more common in the lower portion of the unit, but they are represented by a higher diversity of terrestrial animals (e.g., Hunt and Lucas, 1993c, 1994). Thus, the distribution pattern seen within the Bull Canyon Formation is pervasive in this stratigraphic interval of the Chinle Group. Hunt and Lucas (1993c) noted that this pattern is consistent with sequence stratigraphic models of the Chinle Group. All lower Revueltian faunas occur near the base of a transgressive systems tract near the bottom of the mud-dominated sequence above its basal sandstone complex (Sonsela Member in northeastern Arizona, Trujillo in east-central New Mexico). Upper Revueltian faunas occur near the top of a transgressive systems tract. It is probable that faunas preserved during the progression of a transgressive systems tract would show an increase in semiaquatic components relative to terrestrial components as the highstand was approached. Furthermore, relatively higher rates of sedimentation in the early transgressive systems tract would preserve more fossils and more complete fossils than the later transgressive systems tract or highstand tract. Thus, the sequence stratigraphic interpretation of the Chinle group may explain some aspects of tetrapod abundance in the Revueltian.

However, there are distinct faunal differences between the upper and lower Bull Canyon Formation. These include: (1) lower diversity of fish in the upper part of the formation; (2) absence of the large predators *Pseudopalatus* n. sp. and *Postosuchus* in the upper fauna; (3) a major change in the dinosaur fauna, with the theropods Herrerasauridae gen. et sp. nov. 1 , Theropoda incertae sedis gen. et sp. nov. 3 and the ornithischian *Revueltosaurus* of the lower assemblage being are replaced by the theropod Herrerasauridae gen. et sp. nov. 2 and the ornithischian *Lucianosaurus* in the upper assemblage; (4) the diversity of aetosaurs is reduced from five to one, with only *Typothorax* remaining in the upper part of the formation, and (5) restriction of several small reptiles to the lower assemblage (e.g., *Vancleavea*-like reptile and Crocodylotarsi incertae sedis gen. et sp. nov.)

BIOCHRONOLOGY OF THE REVUELTIAN LAND-VERTEBRATE FAUNACHRON

Lucas and Hunt (1993) named a series of land-vertebrate faunachrons (lvf) for the Late Triassic of western North America. The Revueltian lvf was based on the assemblage of the Bull Canyon Formation and Trujillo Formations in the Revuelto Creek area (Barranca and Revuelto badlands of this report). Lucas (1998) provided a formal definition of the Revueltian lvf and defined its type assemblage as that of the Bull Canyon Formation in east-central New Mexico (Quay and Guadalupe counties). Lucas and Hunt (1993) listed the index taxa of the Revueltian to be *Pseudopalatus* and *Typothorax,* and Lucas (1998) added *Aetosaurus* to

TABLE 1. Two faunas from the Bull Canyon Formation of east-central New Mexico.

VERTEBRATE FAUNAS OF BULL CANYON FORMATION	
Upper fauna	?Colobodontidae indet.
	Redfieldiidae indet.
	Apachesaurus gregorii
	?Squamata indet.
	Pseudopalatus pristinus
	P. andersoni
	Typothorax coccinarum
	Shuvosaurus inexpectatus
	Herrerasauridae gen. et sp. nov. 2
	Lucianosaurus wildi
	Pseudotriconodon chatterjeei
Lower fauna	?Colobodontidae indet.
	Redfieldiidae indet.
	Semionotus cf. *S. brauni*
	?*Tanaocrossus kallikoski*
	?Arganodontidae indet
	Quayia zideki
	Apachesaurus gregorii
	Metoposauridae indet. (large)
	?Squamata indet.
	Rhynchosauridae indet.
	Vancleavia campi
	Pseudopalatus pristinus
	P. andersoni
	P. n. sp.
	Aetosaurus arcuatus
	Typothorax coccinarum
	Desmatosuchus n. sp.
	Paratypothorax sp.
	Stagonolepididae ?gen. et sp. nov.
	Postosuchus kirkpatricki
	Shuvosaurus inexpectatus
	Crocodylotarsi gen. et sp. nov.
	Sphenosuchidae indet.
	Herrerasauridae gen. et sp. nov. 1
	Gojirasaurus quayi
	Theropoda gen. et sp. nov. 3
	Revueltosaurus callenderi

this list. Lucas (1998) defined the base of the Revueltian as the First Appearance Datum (FAD) of *Pseudopalatus*. Diverse vertebrate, pollen, magnetostratigraphic and sequence stratigraphic data indicate that the Revueltian is equivalent to most, if not all, of Norian time (e.g., Litwin, 1986; Ash, 1987; Litwin et al., 1991; Lucas, 1991, 1993,1998; Good, 1993; Hunt, 1994).

I propose to divide the Revueltian into two, and possibly three sub-lvfs based on the biostratigraphy presented above (Fig. 8). R1 (Barrancan) is based on the vertebrate fauna from Barranca Creek and it begins with the FAD of the phytosaur *Pseudopalatus* spp. Index fossils for R1 are *Quayia zideki*, *Pseudopalatus n. sp.*, Crocodylotarsi incertae sedis gen. et sp. nov., *Aetosaurus arcuatus*, Stagonolepididae gen. et sp. nov., Herrerasauridae gen. et sp. nov. 1, Theropoda incertae sedis gen. et sp. nov. 3 and *Revueltosaurus callenderi*. Other elements of the type vertebrate assemblage are Perleididae, Redfieldiidae, *Semionotus*, ?*Tanaocrossus, Apachesaurus*, Metoposauridae (large), Squamata, *Pseudopalatus andersoni* and *P. pristinus*, *Typothorax coccinarum*, *Shuvosaurus*, Rauisuchidae and Herrerasauridae (small), *Paratypothorax*, *Desmatosuchus*, *Postosuchus* and Sphenosuchidae. The principal correlatives are the Post quarry and associated faunas of the lower Bull Canyon Formation, Garza County, Texas and the lower fauna of the Painted Desert Member of the Petrified Forest Formation, Petrified Forest National Park and environs, Apache County, Arizona.

R2 (Lucianoan) is based on the fauna of the upper Bull Canyon Formation at Bull Canyon and Luciano Mesa, and it begins with the FAD of *Lucianosaurus*. Index taxa are Herrerasauridae gen. et sp. nov. *2*, and *Pseudotriconodon*. Other elements of the type fauna are Colobodontidae, Redfieldiidae, *Apachesaurus*, Squamata, *Pseudopalatus*, *Typothorax*, *Shuvosaurus*, Rauisuchidae and Herrerasauridae (small). The principal correlatives are the Canjilon quarry and other localities from the upper Petrified Forest Formation, Rio Arriba County, New Mexico and the Owl Rock Formation of Arizona (Kirby, 1989, 1991). This sub-lvf is easy to recognize on the basis of an aetosaur fauna that includes *Typothorax coccinarum*, but not *Aetosaurus arcuatus*, or *Paratypothorax* sp., but the FAD of *Lucianosaurus* is weak because of its restricted distribution and low abundance.

A third possible sub-lvf would be R0 (Rainbowforestan) based on the fauna of the Sonsela Member of the Petrified Forest in the Rainbow Forest area of Petrified Forest National Park (Hunt et al., 2001a). R0 would begin with the FAD of the phytosaur *Nicrosaurus* which co-occurs in the Sonsela with *Typothorax*, apparently in the absence of *Pseudopalatus* and so possibly prior to the FAD of *Pseudopalatus* (Hunt et al., 2001a). Possible corelatives would be the Trujillo Formation in the Plaza Larga-Barranca Creek area of Quay County, New Mexico (Hunt, 1991, 2001). If this is the case, then the Revueltian should be redefined on the basis of the FAD of *Typothorax coccinarum*. This would be preferable to the FAD of *Pseudopalatus* in any case because the stagonolepidid is represented by a far greater number of identifiable specimens in Revueltian faunas as demonstrated above.

ACKNOWLEDGMENTS

This project would not have been possible without the support and inspiration of Spencer Lucas. Joe Gregory very generously supplied me with field notes, locality information and access to specimens. I thank the members of my dissertatation committee Spencer Lucas, Barry Kues, Jiri Zidek, and Maya Elrick. Many individuals facilitated access to collections under their care including John Bolt, Gene Gaffney, Joe Gregory, Mike Greenwald, Greg Gunnell, John Ostrom, Kevin Padian, Michael Parrish, Peter Robinson, Mary Ann Turner and Michael Williams. Pete Reser, Ruby Williamson, Beryl Roach and Gary Morgan of NMMNH asssisted with curation and preparation. Numerous individuals, particularly from New Mexico Friends of Paleontology assisted with fieldwork including Dan D'Andrea, Phil Bircheff, Jack Clift, Bob Jackson, and Roger Mongold. New Mexico Museum of Natural History, New Mexico Friends of Paleontology and the Petrified Forest Museum Association supported this project. Many landowners provided generous access to their properties including John Bruce Cresap, O. B. Palmer, James Robbins, Jack Lewis and James Watson. Spencer Lucas and Andrew Heckert provided helpful reviews of this manuscript. A translation of Rauhut (1997) was made by O. W. M. Rauhut

	East-central NM	North-central NM	Northeastern AZ	West TX
Lucianoan (R2)	Upper Bull Canyon Formation	?Upper Petrified Forest Formation	Upper Painted Desert Member of Petrified Forest Formation, Owl Rock Formation	?Not present
Barrancan (R1)	Lower Bull Canyon Formation	Lower Petrified Forest Formation	Lower Painted Desert Member of Petrified Forest Formation	Lower Bull Canyon Formation
?Rainbowforestan (R0)	Trujillo Formation	Poleo Formation	Sonsela Member of Petrified Forest Formation	Trujillo Formation

FIGURE 8. Correlation of the Revueltian land-vertebrate faunachron in the Southwestern United States.

and obtained courtesy of the Polyglot Paleontologist website (www.uhmc.sunysb.edu/anatomicalsci/paleo).

BIBLIOGRAPHY

Ager, D. V., 1993, The nature of the stratigraphical record: New York, John Wiley, 151 p.

Ash, S. R., 1972, Upper Triassic Dockum flora of eastern New Mexico and Texas: New Mexico Geological Society, 23rd Field Conference, Guidebook, p. 124-128.

Ash, S. R., 1987, The Upper Triassic red bed flora of the Colorado Plateau, western United States: Journal of the Arizona-Nevada Academy of Science, v. 22, p. 95-105.

Ash, S. R., 1989, A catalog of Upper Triassic plant megafossils of the western United States through 1988, *in* Lucas, S. G. and Hunt, A. P., eds., The dawn of the age of dinosaurs in the American Southwest: Albuquerque, New Mexico Museum of Natural History and Science, p. 189-222.

Ballew, K. L., 1989, A phylogenetic analysis of Phytosauria from the Late Triassic of the western United States, *in* Lucas, S. G. and Hunt, A. P., eds., The dawn of the age of dinosaurs in the American Southwest: Albuquerque, New Mexico Museum of Natural History and Science, p. 309-339.

Benton, M. J., 1993a, Reptilia, *in* Benton, M. J., ed., The Fossil Record: London, Chapman and Hall, p. 681-715.

Benton, M. J., 1993b, ed., The Fossil Record: London, Chapman and Hall, 795 p.

Benton, M. J. and Clark, J. M., 1988, Archosaur phylogeny and the relationships of the Crocodylia: Systematics Association Special Volume, v. 35A, p. 295-338.

Boone, J. L., 1979, Lake margin depositional systems of the Dockum Group (Upper Triassic) in Tule Canyon, Texas Panhandle [M. S. thesis]: Austin, University of Texas, 147 p.

Branson, E. B., 1927, Triassic-Jurassic 'red beds' of the Rocky Mountain region: Journal of Geology, v. 35, p. 629-630.

Branson, E. B., 1928, "Triassic-Jurassic 'red beds' of the Rocky Mountain region": a reply: Journal of Geology, v. 36, p. 64-75.

Branson, E. B. and Mehl, M. G., 1929, Triassic amphibians from the Rocky Mountain region: University of Missouri Studies, v. 4, p. 155-239.

Camp, C. L., 1930, A study of the phytosaurs with description of new material from western North America: Memoirs of the University of California, v. 10, p. 1-175.

Carpenter, K., 1997, A giant coelophysoid (Ceratosauria) theropod from the Upper Triassic of New Mexico, U.S.A.: Neues Jahrbuch für Geologie und Paläontologie Abhandlungen, v. 205, p. 189-206.

Carpenter, K. and Parrish, J. M., 1985, Late Triassic vertebrates from Revuelto Creek, Quay County, New Mexico: New Mexico Geological Society, 36th Field Conference, Guidebook, p. 197-198.

Case, E. C., 1914, The red beds between Wichita Falls, Texas and Las Vegas, New Mexico in relation to their vertebrate fauna: Journal of Geology, v. 22, p. 243-259.

Case, E. C., 1920, Preliminary description of a new suborder of phytosaurian reptiles, with a description of a new species of *Phytosaurus*: Journal of Geology, v. 28, p. 524-535.

Case, E. C., 1922, New reptiles and stegocephalians from the Upper Triassic of western Texas: Carnegie Institution of Washington Publication, v. 321, p. 1-84.

Chatterjee, S., 1985, *Postosuchus*, a new thecodontian reptile from the Triassic of Texas, and the origin of tyrannosaurs: Philosophical Transactions of the Royal Society of London, Series B, v. 309, p. 395-460.

Chatterjee, S., 1986, The Late Triassic Dockum vertebrates: their stratigraphic and paleobiogeographic significance, *in* Padian, K., ed., The beginning of the age of dinosaurs: faunal change across the Triassic-Jurassic boundary: Cambridge, Cambridge University Press, p. 139-150.

Chatterjee, S., 1993, *Shuvosaurus*, a new theropod: National Geographic Research and Exploration, v. 9, p. 274-285.

Clark, J. M., Sues, H.-D. and Berman, D. S, 2000, A new specimen of *Hesperosuchus agilis* from the Upper Triassic of New Mexico and the interrelationships of basal crocdylomorph archosaurs: Journal of Vertebrate paleontology, v. 20, p. 683-704.

Colbert, E. H., 1952, A pseudosuchian reptile from northern Arizona: Bulletin of the American Museum of Natural History, v. 99, p. 563-592.

Colbert, E. H. and Gregory, J. T., 1957, Correlation of continental Triassic sediments by vertebrate fossils: Geological Society of America Bulletin, v. 68, p.1456-1467.

Colbert, E. H. and Imbrie, J.1956, Triassic metoposaurid amphibians: Bulletin of the American Museum of Natural History, v. 110, p. 403-452.

Coney, P. J., 1978, The plate tectonic setting of southeastern Arizona: New Mexico Geological Society, 29th Field Conference, Guidebook, p. 285-290.

Cummins, W. F., 1889, Report of geologist for northern Texas: Texas Geological Survey Progress Report, v. 1, p. 123-131.

Crush, P. J., 1984, A late Upper Triassic sphenosuchid crocodylian from Wales: Palaeontology, v. 27, p. 131-157.

Darton, N. H., 1922, Geologic structure of parts of New Mexico: U. S. Geological Survey Bulletin, v. 726F, 275 p.

Darton, N. H., 1928, "Red beds" and associated formations in New Mexico, with an outline of the geology of the state: U. S. Geological Survey Bulletin, v. 794, 372 p.

Dickinson, W. R., 1981, Plate tectonic evolution of the southern Cordillera: Arizona Geological Society Digest, v. 14, p. 113-135.

Dobrovolny, E. H., Summerson, C. H. and Bates, R. L., 1946, Geology of northwestern Quay County, New Mexico: U. S. Geological Survey, Oil and Gas Investigations Preliminary Map 62.

Drake, N. F., 1892, Stratigraphy of the Triassic of West Texas: Geological Survey of Texas Annual Report, v. 3, p. 227-247.

Dubiel, R. F., 1989, Depositional and climatic setting of the Upper Triassic Chinle Formation, Colorado Plateau, *in* Lucas, S. G. and Hunt, A. P., eds., The dawn of the age of dinosaurs in the American Southwest: Albuquerque, New Mexico Museum of Natural History and Science, p. 171-187.

Elliott, D. K., A new specimen of *Chinlea sorenseni* from the Chinle Formation: Journal of the Arizona-Nevada Academy of Sciences, v. 22, p. 85-94.

Frelier, A. P., 1987, Sedimentology, fluvial paleohydrology, and paleogeomorphology of the Dockum Formation (Triassic), west Texas [M. S. thesis]: Lubbock, Texas Tech University, 198 p.

Fritz, T. L., 1991, Depositional history of the mid-late Triassic Santa Rosa Formation, eastern New Mexico [M. S. thesis]: Lubbock, Texas Tech University, 120 p.

Gaffney, E. S., 1990, The comparative osteology of the Triassic turtle *Proganochelys*: Bulletin of the American Museum of Natural History, v. 194, 263 p.

Gardner, B. G., Osteichthyes: basal actinopterygians, *in* Benton, M. J., ed., The Fossil Record: London, Chapman and Hall, p. 611-619.

Good, S. C., 1989, Nonmarine Mollusca in the Upper Triassic Chinle Formation and related strata of the Western Interior. Systematics and distribution, *in* Lucas, S. G. and Hunt, A. P., eds., The dawn of the age of dinosaurs in the American Southwest: Albuquerque, New Mexico Museum of Natural History and Science, p. 233-248.

Good, S. C., 1993, Molluscan paleobiology of the Upper Triassic Chinle Formation, Arizona and Utah [Ph. D. dissertation]: Boulder, University of Colorado, 276 p.

Gould, C. N., 1907, The geology and water resources of the western portion of the panhandle of Texas: U. S. Geological Survey Water Supply Paper, v. 191, p. 1-70.

Granata, G. E., 1981, Regional sedimentation of the Late Triassic Dockum Group, West Texas and eastern New Mexico [M. S. thesis]: Austin, University of Texas, 199 p.

Gregory, J. T., 1953, *Typothorax* and *Desmatosuchus*: Postilla, v. 17, p.1-27.

Gregory, J. T., 1957, Significance of fossil vertebrates for correlation of Late Triassic continental deposits of North America: report of the 20th Session of the International Geological Congress 1956, Section II, p. 7-25.

Gregory, J. T., 1962, The genera of phytosaurs: American Journal of Science, v. 260, p. 652-690.

Gregory, J. T., 1969, Evolution und interkontinentale Beziehungen der Phytosauria: Palaontologische Zeitschrift, v. 43, p. 37-51.

Gregory, J. T., 1972, Vertebrate faunas of the Dockum group, Triassic, eastern New Mexico and West Texas: New Mexico Geological Society, 23rd Field Conference, Guidebook, p. 120-123.

Gregory, J. T. and Westphal, F., 1969, Remarks on the phytosaur genera of the Euopean Trias: Journal of Paleontology, v. 43, p. 1296-1298.

Heckert, A.B., and Lucas, S.G., 1998, First occurrence of *Aetosaurus* (Reptilia:Archosauria) in the Upper Triassic Chinle Group (USA) and its biochronological significance. Neues Jahrbuch für Geologie und Paläontologie Monatsheft, v. 1998, p. 604-612.

Heckert, A. B., Lucas, S. G. and Sullivan, R. M., 2000, Triassic dinosaurs in New Mexico: New Mexico Museum of Natural History Bulletin 17, p. 17-26.

Hester, P. M., 1988, Depositional environments in an Upper Triassic lake, east-central New Mexico [M. S. thesis]: Albuquerque, University of New Mexico, 154 p.

Hester, P. and Lucas, S. G., 2001, Depositional environments of a Late Triasic lake: New Mexico Geological Society, 52nd Field Conference, Guidebook, this volume.

Huber, P., Lucas, S. G. and Hunt, A. P., 1990, Late Triassic fish assemblages of the North American Western Interior and their biochronology: Symposium on Southwest Geology and Paleontology, p.10.

Huber, P., Hunt, A. P. and Lucas, S. G., 1991, Late Triassic fish from the Tucumcari basin, east-central New Mexico: New Mexico Geology, v. 12, p. 18.

Huber, P., Lucas, S. G. and Hunt, A. P., 1993, Late Triassic fish assemblages of the North American Western Interior and their biochronologic significance: Bulletin of the Museum of Northern Arizona 59, p. 51-66.

Huene, F. von, 1926, Notes on the age of the continental Triassic beds in North America with remarks on some fossil vertebrates: Proceedings of the United States National Museum, v. 69, p. 1-10.

Hunt, A. P., 1989a, A new ?ornithischian dinosaur from the Bull Canyon Formation (Upper Triassic) of east-central New Mexico, *in* Lucas, S. G. and Hunt, A. P., eds., The dawn of the age of dinosaurs in the American Southwest: Albuquerque, New Mexico Museum of Natural History and Science, p. 355-358.

Hunt, A. P., 1989b, Cranial morphology and ecology among phytosaurs, *in* Lucas, S. G. and Hunt, A. P., eds., The dawn of the age of dinosaurs in the American Southwest: Albuquerque, New Mexico Museum of Natural History and Science, p. 349-354.

Hunt, A. P., 1989c, The biochronological significance of Late Triassic metoposaurid amphibians: New Mexico Journal of Science, v. 29, p.117-118.

Hunt, A. P., 1990, Taxonomy and biochronological utility of phytosaurs in the Late Triassic of the western United States: Symposium on Southwestern Geology and Paleontology Abstracts, p. 11.

Hunt, A. P., 1991a, The early diversification pattern of dinosaurs in the Late Triassic: Modern Geology, v. 16, p. 43-60.

Hunt, A. P., 1991b, The first tetrapod faunas from the Trujillo Formation (Late Triassic) of east-central New Mexico and their biochronological and paleoecological significance: New Mexico Geology, v. 13, p. 93.

Hunt, A. P., 1991c, Two phytosaur (Reptilia: Archosauria) skeletons from the Bull Canyon Formation (Late Triassic) of east-central New Mexico with preserved stomach contents: New Mexico Geology, v. 13, p. 93.

Hunt, A. P., 1992, Aetosaurs from the Bull Canyon Formation (Late Triassic: Norian) of east-central New Mexico and their biochronological significance: New Mexico Geology, v. 14, p. 66.

Hunt, A. P., 1993a, A revision of the Metoposauridae (Amphibia: Temnospondyli) of the Late Triassic and description of a new genus from western North America: Bulletin of the Museum of Northern Arizona 59, p. 67-97.

Hunt, A. P., 1993b, Phytosaurs (Reptilia: Archosauria) from the Bull Canyon Formation of east-central New Mexico: taxonomy, biochronology and paleoecology: New Mexico Geology, in press.

Hunt, A. P., 1994, Vertebrate paleontology and biostratigraphy of the Bull Canyon Formation (Chinle Group: Norian), east-central New Mexico with revisions of the families Metoposauridae (Amphibia: Temnospondyli) and Parasuchidae (Reptilia: Archosauria) [Ph.D. dissertation]: Albuquerque, , University of New Mexico, 403 p.

Hunt, A. P., 1997a, A new coelacanth (Osteichthyes: Actinistia) from the continental Late Triassic of western North America. New Mexico Museum of Natural History and Science Bulletin.11, p.: 25-27.

Hunt, A. P., 1997b, E. C. Case, J. T. Gregory and early explorations for fossils vertebrates in the Bull Canyon Formation (Upper Triassic) of eastern New Mexico. New Mexico Museum of Natural History and Science Bulletin 11, p. 15-24.

Hunt, A. P., 2001, Paleontology and age of the upper Triassic Trujillo Formation, east-central New Mexico and West Texas: New Mexico Geological Society, 52nd Field Conference, Guidebook, this volume.

Hunt, A. P. and Lucas, S. G., 1989a, Large podokesaurid dinosaur from the Upper Triassic Bull Canyon Formation, east-central New Mexico: New Mexico Journal of Science, v. 29, p. 118.

Hunt, A. P. and Lucas, S. G., 1989b, Late Triassic vertebrate localities in New Mexico, *in* Lucas, S. G. and Hunt, A. P.,eds., The dawn of the age of dinosaurs in the American Southwest: Albuquerque, New Mexico Museum of Natural History and Science, p.72-101.

Hunt, A. P. and Lucas, S. G,. 1991a, A new aetosaur from the Redonda Formation (Late Triassic: middle Norian) of east-central New Mexico, USA: Neues Jahrbuch für Geologie und Palaontologie Monatshefte, v. 1991, p. 728-736.

Hunt, A. P. and Lucas, S. G. 1991b, A new rhynchosaur from the Upper Triassic of West Texas, and the biochronology of Late Triassic rhynchosaurs: Palaeontology, v. 34, p. 927-938.

Hunt, A. P. and Lucas, S. G., 1991c, The *Paleorhinus* biochron and the correlation of the nonmarine Upper Triassic of Pangaea: Palaeontology, v. 34, p. 487-501.

Hunt, A. P. and Lucas, S. G., 1993a, A new phytosaur (Reptilia: Archosauria) genus from the uppermost Triassic of the western United States and its biochronological significance. New Mexico Museum of Natural History and Science Bulletin, v. 3, p. 193-196.

Hunt, A. P. and Lucas, S. G., 1993b, Ornithischian dinosaurs from the Upper Triassic of the United States, *in* Fraser, N. C. and Sues, H.-D., eds., In the shadow of the dinosaurs: Cambridge, Cambridge University Press, p. 227-241.

Hunt, A. P. and Lucas, S. G., 1993c, Sequence stratigraphy and a tetrapod acme zone during the early Revueltian (Late Triassic: early Norian) of western North America. New Mexico Museum of Natural History and Science Bulletin, v. 3, p G46.

Hunt, A. P. and Lucas, S. G., 1993d, Triassic vertebrate paleontology and biochronology of New Mexico: New Mexico Museum of Natural History and Science Bulletin, v. 2, p. 49-60.

Hunt, A. P. and Lucas, S. G., 1997, Stratigraphy, paleontology and biochronology of the Upper Triassic Chinle Group in east-central New Mexico. Proceedings of the Southwest Paleontological Symposium, p. 25-40.

Hunt, A. P., Lockley, M. G. and Lucas, S. G., 1993a, Vertebrate and invertebrate tracks and trackways from Upper Triassic strata of the Tucumcari basin, east-central New Mexico, USA. New Mexico Museum of Natural History and Science Bulletin 3, p. 199-201.

Hunt, A. P., Lucas, S. G. and Reser, P. K., 1993b, A complete skeleton of the stagonolepidid *Typothorax coccinarum* from the Upper Triassic Bull Canyon Formation of east-central New Mexico. New Mexico Museum of Natural History and Science Bulletin 3: 209-212.

Hunt, A. P., Santucci, V. L. and Wall, W. P., 1993c, Paleoecology of Late Triassic metoposaurid amphibians: evidence from Petrified Forest National Park. Park Science, v. 13, p. 12.

Hunt, A. P., Lucas, S. G., Heckert, A. B., Sullivan, R. M. and Lockley, M. G. , 1998, Late Triassic dinosaurs from the western United States: Geobios, v. 31, p. 511-531.

Hunt, A. P., Lucas, S. G. and Heckert, A. B., 2001a, A Germanic-aspect phytosaur from the Upper Triassic Sonsela Sandstone (Chinle Group: Petrified Forest Formation), Petrified Forest, Arizona: Fossils of Arizona, in press.

Hunt, A. P., Lucas, S. G. and Heckert, A. B., 2001b, Does the enigmatic tetrapod ichnotaxon *Barrancapus cresapi* from the Upper Triasssic Bull Canyon Formation of east-central New Mexico represent the oldest prosauropod trackway?: New Mexico Geology, v. 23, p. 63.

Hunt, A. P., Lucas, S. G. and Heckert, A. B., 2001c, Paleocurrents in the Trujillo formation (Chinle Group; Upper Triassic), east-central New Mexico and the myth of the Dockum lake: New Mexico Geological Society, 52nd Field Conference, Guidebook, this volume.

Jenkins, F. A., Jr., 1971, The postcranial skeleton of African cynodonts: Peabody Museum of Natural History, Bulletin 36, 216 p.

Johns, D. A. and Granata, G. E., 1987, Regional tectonic setting and depositional trends of the Dockum Group, West Texas and eastern New Mexico: Journal of the Arizona-Nevada Academy of Science, v. 22, p. 53-72.

Johnson, G. D. and Zidek, J., 1981, Late Paleozoic phyllodont tooth plates: Journal of Paleontology, v. 55, p. 524-536.

Kelley, V. C., 1972, Geology of the Fort Sumner sheet, New Mexico: New Mexico Bureau of Mines and Mineral Resources Bulletin, v. 98, 55 p.

Kiatta, H. W., 1960, A provenance study of the Triassic deposits of northwestern Texas [M. A. thesis]: Lubbock, Texas Tech University, 63 p.

Kietzke, K. K., 1987, Calcareous microfossils from the Triassic of northeastern New Mexico: New Mexico Geological Society, 38th Field Conference, Guidebook, p. 119-126.

Kietzke, K. K., 1989, Calcareous microfossils from the Triassic of the southwestern United States, *in* Lucas, S. G. and Hunt, A. P., eds., The dawn of the age of dinosaurs in the American Southwest: Albuquerque, New Mexico Museum of Natural History and Science, p.223-233.

Kietzke, K. K. and Lucas, S. G., 1991, Triassic nonmarine "*Spirorbis*": gastropods not worms: New Mexico Geology, v. 13, p. 93.

Kirby, R. E., 1989, Late Triassic vertebrate localities of the Owl Rock Member (Chinle Formation) in the Ward Terrace area of northern Arizona, *in* Lucas, S. G. and Hunt, A. P., eds., The dawn of the age of dinosaurs in the American Southwest: Albuquerque, New Mexico Museum of Natural History and Science, p. 12-28.

Kirby, R. E., 1991, A new vertebrate fauna from the Upper Triassic Owl Rock Member of the Chinle Formation in northern Arizona [M.S.thesis]: Flagstaff, University of Northern Arizona, 476 p.

Kues, B. S., 1985a, Early geological explorations in northeastern and east-central New Mexico: New Mexico Geological Society, 36th Field Conference, Guidebook, p. 103-118.

Kues, B. S., 1985b, Nonmarine molluscs from the Chinle Formation, Dockum Group (Upper Triassic), of Bull Canyon, Guadalupe County, New Mexico: New Mexico Geological Society, 36th Field Conference, Guidebook, p. 185-196.

Kues, B. S., 1985c, Stratigraphy of the Tucumcari area: a historical account: New Mexico Geological Society, 36th Field Conference, Guidebook, p. 119-140.

Lee, W. T., 1907, Note on the red beds of the Rio Grande region in central New Mexico: Journal of Geology, v. 15, p. 52-58.

Lehman, T., 1994, The saga of the Dockum Group and the case of the Texas/New Mexico boundary fault: New Mexico Bureau of Mines and Mineral Resources Bulletin 150, p. 37-51.

Lehman, T., Chatterjee, S. and Schnable, J., 1992, The Cooper Canyon Formation (Late Triassic) of western Texas: Texas Journal of Science, v. 44, p. 349-355.

Litwin, R. J., 1986, The palynostratigraphy and age of the Chinle and Moenave Formations, southwestern U.S.A. [Ph. D. dissertation]: College Park, The Pennsylvania State University, 256 p.

Litwin, R. J., Traverse, A. and Ash, S. R., 1991, Preliminary palynological zonation of the Chinle Formation, southwestern U. S. A., and its correlation to the Newark Supergroup (eastern U. S. A.): Review of Palaeobotany and Palynology, v. 68, p. 269-287.

Long, R. A. and Ballew, K. L., 1985, Aetosaur dermal armor from the Late Triassic of southwestern North America, with special reference to material from the Chinle Formation of Petrified Forest National Park: Museum of Northern Arizona Bulletin, v. 54, p. 35-68.

Long, R. A. and Murry, P. A., 1995, Late Triassic (Carnian and Norian) tetrapods from the southwestern United States: New Mexico Museum of Natural History and Science Bulletin 4, 254 p.

Lucas, S. G., 1991, Correlation of Triassic strata of the Colorado Plateau and southern High Plains, New Mexico: New Mexico Bureau of Mines and Mineral Resources Bulletin, v. 137, p. 47-56.

Lucas, S. G., 1993, The Chinle Group: revised stratigraphy and chronology of Upper Triassic nonmarine strata in the western United States: Bulletin of the Museum of Northern Arizona 59, p. 27-50.

Lucas, S. G., 1998, Global Triassic tetrapod biostratigraphy and biochronology: Palaeogeography, Palaeoclimatology, Palaeoecology, v. 143, p. 347-384.

Lucas, S. G. and Anderson, O. J., 1992, Triassic stratigraphy and correlation, west Texas and eastern New Mexico, *in* Cromwell, D. W., Moussa, M. T. and Mazzullo, L. J., eds., Transactions southwest section AAPG, Midland, Texas, 1992: Midland, West Texas Geological Society, p. 201-207.

Lucas, S. G. and Anderson, O. J., 1993, Lithostratigraphy, sedimentation, and sequence stratigraphy of Upper Triassic Dockum Formation, West Texas, *in* Crick, R. E., ed., 1993 Southwest section geological convention, American Association of Petroleum geologists transactions and abstracts: Arlington, Southwest Section American Association of Petroleum Geologists, p. 55-65.

Lucas, S. G. and Hunt, A. P., 1989a, Revised Triassic stratigraphy in the Tucumcari Basin, east-central New Mexico, *in* Lucas, S. G. and Hunt, A. P., eds., The dawn of the age of dinosaurs in the American Southwest: Albuquerque, New Mexico Museum of Natural History and Science, p. 150-170.

Lucas, S. G. and Hunt, A. P., 1989b, Vertebrate biochronology of the Late Triassic: 28th International Geological Congress Abstracts, v. 2, p. 335-336.

Lucas, S. G. and Hunt, A. P., 1993, Tetrapod biochronology of the Chinle Group (Upper Triassic), western United States: New Mexico Museum of Natural History and Science Bulletin, v. 3, p. 327-329.

Lucas, S. G. and Oakes, W., 1988, A Late Triassic cynodont from the American South-West: Palaeontology, v. 31, p. 445-449.

Lucas, S. G., Hunt, A. P. and Bennett, S. C., 1985a, Triassic vertebrates from east-central New Mexico in the Yale Peabody Museum: New Mexico Geological Society Guidebook, 36th Field Conference, p. 199-203.

Lucas, S. G., Hunt, A. P. and Kietzke, K. K., 1985b, *Neocalamites* forest in the Upper Triassic of Bull Canyon, New Mexico: New Mexico Geological Society Guidebook, 36th Field Conference, p. 8-12.

Lucas, S.G., Hunt, A.P. and Morales, M., 1985c, Stratigraphic nomenclature and correlation of Triassic rocks of east-central New Mexico: New Mexico Geological Society Guidebook, 36th Field Conference, p. 171-184.

Lucas, S. G., Oakes, W. and Froehlich, J. W., 1985d, Triassic microvertebrate locality, Chinle Formation, east-central New Mexico: New Mexico Geological Society Guidebook, 36th Field Conference, p. 205-212

Lucas, S. G., Steiner, M. R., Huber, P. and Hunt, A. P., 1993, Magnetostratigraphy and paleomagnetic poles from Late Triassic-earliest Jurassic strata of the Newark basin: discussion: Bulletin of the Geological Society of America, v. 105, p. 1260-1261.

Lucas, S. G., Anderson, O. J. and Hunt, A. P., 1994, Triassic stratigraphy and correlations, southern High Plains of New Mexico-Texas: New Mexico Bureau of Mines and Mineral Resources Bulletin 150, p. 105-126.

Lucas, S. G., Heckert, A. B. and Hunt, A. P., 2000, A probable turtle from the Upper Triassic of east-central New Mexico. Neues Jahrbuch fur Palaontologie Monatshefte, v. 2000(5), p. 287-300.

Lucas, S. G., Heckert, A. B. and Hunt, A. P., 2001a, Triassic stratigraphy, biostratigraphy and correlation in east-central New Mexico: New Mexico Geological Society, 52nd Field Conference, Guidebook, this volume.

Lucas, S. G., Weadock, G., Kietzke, K. K., Hunt, A. P. and Kues, B. S., 2001b, Geology of the Ima NW quadrangle, east-central New Mexico: New Mexico Geological Society, 52nd Field Conference, Guidebook, this volume.

Marcou, J., 1855, Resume of a geological reconnaisance extending from Napoleon, at the junction of the Arkansas with the Mississippi, tothe Pueblo de Los Angeles, in California, *in* Whipple, A. W., Report of explorations for a railway route near the thirty-fifth parallel of latitude from the Mississippi River to the Pacific Ocean: 33rd Congress, 1st Session, House Executive Document 129, p. 40-48.

Marcou, J., 1858, Geology of North America, with two reports on the prairies of Arkansas and Texas, the Rocky Mountains of New Mexico, and the Sierra Nevada of California, originally made for the United States Government: Zurich, Zurcher and Furrer, 144 p.

May, B. S., 1988, Depositional envronments, sedimentology, and stratigraphy of the Dockum Group (Triassic) in the Texas Panhandle [M. S. thesis]: Lubbock, Texas Tech University, 180 p.

McGowan, J. H., Granata, G. E. and Seni, S. J., 1979, Depositional framework of the lower Dockum group (Triassic) Texas Panhandle: Texas Bureau of Economic Geology, Report of Investigations, v. 97, 60 p.

McGowan, J. H., Granata, G. E. and Seni, S. J., 1983, Depositional setting of the Triassic Dockum Group, Texas Panhandle and eastern New Mexico, *in* Reynolds, M. W. and Dolly, E. D., eds., Mesozoic paleogeography of the west-central United States: Denver, RMS-SEPM, p. 13-38.

McIntosh, W. C., Hargraves, R. B. and West, C. L., 1985, Paleomagnetism and oxide mineralogy of Upper Triassic to Lower Jurassic red beds and basalts in the Newark Basin: Bulletin of the Geological Society of America, v. 96, p. 463-480.

Mehl, M. G., 1922, A new phytosaur from the Trias of Arizona: Journal of Geology, v. 30, p. 144-157.

Mehl, M. G. 1928, *Pseudopalatus pristinus*, a new genus and species of phytosaur from Arizona: University of Missouri Studies, v. 3, p. 1-22.

Milner, A. R., 1993, Amphibian-grade Tetrapoda, *in* Benton, M. J., ed., The Fossil Record: London, Chapman and Hall, p. 665-679.

Murry, P. A., 1982, Biostratigraphy and paleoecology of the Dockum Group, Triassic of Texas [Ph. D. dissertation]: Dallas, Southern Methodist University, 459 p.

Murry, P. A., 1986, Vertebrate paleontology of the Dockum Group, western Texas and eastern New Mexico, *in* Padian, K., ed., The beginning of the age of dinosaurs: faunal change across the Triassic-Jurassic boundary: Cambridge, Cambridge University Press, p.109-137.

Murry, P. A., 1987, Notes on the stratigraphy and paleontology of the Upper Triassic Dockum Group: Journal of Arizona-Nevada Academy of Science, v. 22, p. 73-84.

Murry, P. A., 1989a, Geology and paleontology of the Dockum Formation (Upper Triassic), West Texas and eastern New Mexico, *in* Lucas, S. G. and Hunt, A. P., eds., The dawn of the age of dinosaurs in the American Southwest: Albuquerque, New Mexico Museum of Natural History and Science, p. 102-144.

Murry, P. A., 1989b, Paleoecology and vertebrate faunal relationships,of the Upper Triassic Dockum and Chinle Formations, southwestern United States, *in* Lucas, S. G. and Hunt, A. P., eds., The dawn of the age of dinosaurs in the American Southwest: Albuquerque, New Mexico Museum of Natural History and Science, p. 375-400.

Murry, P. A. and Long, R. A., 1989, Geology and paleontology of the Chinle Formation, Petrified Forest National Park and vicinity, Arizona and a discussion of vertebrate fossils of the southwestern Upper Triassic, *in* Lucas, S. G. and Hunt, A. P., eds., The dawn of the age of dinosaurs in the American Southwest: Albuquerque, New Mexico Museum of Natural History and Science, p. 29-64.

Newell, A. J., 1992, Sedimentological controls on vertebrate taphonomy in Triassic fluvial environments [Ph. D. Dissertation]: Belfast, Queen's University of Belfast, 350 p.

Newell, A. J., 1993, Depositional environment of the Late Triassic Bull Canyon Formation (New Mexico): implications for "Dockum Formation" paleogeography: New Mexico Museum of Natural History and Science Bulletin, v. 3, p. 359-368.

Novas, F. E, 1989, The tibia and tarsus in Herrerasauridae (Dinosauria, incertae sedis) and the origin and evolution of the dinosaurian tarsus: Journal of Paleontology, v. 63, p. 677-690.

Novas, F. E., 1992, Phylogenetic relationships of the basal dinosaurs, the Herrerasauridae: Palaeontology, v. 35, p. 51-62.

Olson, P. E., McCune, A. R. and Thomson, K. S., 1982, Correlation of the Early Mesozoic Newark Supergroup of eastern North America by vertebrates, principally fishes: American Journal of Science, v. 282, p. 1-44.

Parrish, J. M., 1989, Vertebrate paleoecology of the Chinle Formation (Late Triassic) of the southwestern United States: Palaeogeography, Palaeoclimatology, Palaeoecology, v. 72, p. 227-247.

Parrish, J. M., 1991, A new specimen of an early crocodylomorph (cf. *Sphenosuchus* sp.) from the Upper Triassic Chinle Formation of Petrified Forest National Park, Arizona: Journal of Vertebrate Paleontology, v. 11, p. 198-212.

Parrish, J. M., 1993, Phylogeny of the Crocodylotarsi, with reference to archosaurian and crurotarsan monophyly: Journal of Vertebrate Paleontology, v. 13, p. 287-308.

Parrish, J. M. and Carpenter, K., 1986, A new vertebrate fauna from the Dockum Formation (Late Triassic) of eastern New Mexico, *in* Padian, K., ed., The beginning of the age of dinosaurs: faunal change across the Triassic-Jurassic boundary: Cambridge, Cambridge University Press, p. 151-160.

Rauhut, O.W.M.,1997, Zur Schädelanatomie von *Shuvosaurus inexpectatus* (Dinosauria; Theropod), *in* Sachs, S., Rauhut, O.W.M. and Weigert, A., eds., 1. Treffen der desutchssprachigen Palaeoherpteologen,: Düsseldorf, Terra Nostra, p. 17-21.

Reeside, J. B. Jr., 1928, "Triassic-Jurassic 'red beds' of the Rocky Mountain region": a discussion: Journal of Geology, v. 36, p. 47-63.

Reeside, J. B., Jr., Applin, P. L., Colbert, E. H., Hadley, H. D., Kummel, B., Lewis, P. J., Love, J. D., Maldano-Koerdell, M., McKee, E. D., McLaughlin, D. B., Muller, S. W., Rodgers, J., Sanders, J., Silberling, N. J. and Waage, K., 1957, Correlation of the Triassic formations of North America exclusive of Canada: Geological Society of America Bulletin, v. 68, p. 1451-1514.

Reeve, S. G. and Helsley, C. E., 1972, Magnetic reversal sequence in the upper part of the Chinle Formation, Montoya, New Mexico: Geological Society of America Bulletin, v. 83, p. 3795-3812.

Santa Luca, A. P., 1984, Postcranial remains of Fabrosauridae (Reptila: Ornithischia) from the Stormberg of southern Africa: Palaeontologia Africana, v. 25, p. 151-180.

Schaeffer, B., 1967, Late Triassic fishes from the western United States: Bulletin of the American Museum of Natural History, v. 135, p. 285-342.

Schaeffer, B. and Gregory, J. T., 1961, Coelacanth fishes from the continetal Triassic of the western United States: American Museum Novitates, v. 2036, p. 1-36.

Schaeffer, B. and MacDonald, N. G., 1978, Redfieldiid fishes from the Triassic-Liassic Newark Supergroup of eastern North America:Bulletin of the American Museum of Natural History, v. 155, p.131-173.

Schultze, H.-P., 1993, Osteichthyes: Sarcopterygii, *in* Benton, M. J., ed., The Fossil Record: London, Chapman and Hall, p. 657-663.

Schultze, H.-P., 1981, das Schadeldach eines ceratontiden Lungfisches aus der Trias Suddeutschlands (Dipnoi, Pisces): Stuttgarter Beitrage zur Naturkunde Serie B (Geologie und Palaontologie), v. 70, p. 1-31.

Seni, S. J., 1978, Genetic stratigraphy of the Dockum Group (Triassic), Palo Duro Canyon, Panhandle Texas [M. S. thesis]: Austin, University of Texas, 157 p.

Sereno, P. C., 1991, Basal archosaurs: phylogentic relationships and functional implications: Society of Vertebrate Paleontology Memoir, v. 2, 53 p.

Sereno, P. C., Forster, C. A., Rogers, R. R. and Monetta, A. M., 1993, Primitive dinosaur skeleton from Argentina and the early evolution of Dinosauria: Nature, v. 361, p. 64-66.

Small, B. J., 1985, The Triassic thecodontian reptile *Desmatosuchus*: osteology and relationships [M. A. thesis]: Lubbock, Texas Tech University, 83 p.

Small, B. J., 1989, Aetosaurs from the Upper Triassic Dockum Formation, Post quarry, West Texas, *in* Lucas, S. G. and Hunt, A. P., eds., The dawn of the age of dinosaurs in the American Southwest: Albuquerque, New Mexico Museum of Natural History and Science, p. 301-308.

Van der Voo, R., Mauk, F. J. and French, R. B., 1976, Permian-Triassic continental configurations and the origin of the Gulf of Mexico: Geology, v. 4, p. 177-180.

Walker, A. D., 1961, Triassic reptiles from the Elgin area: *Stagonolepis*, *Dasygnathus* and their allies: Philosophical Transactions of the Royal Society of London, Series B, v. 244, p.103-204.

Walker, A. D., 1990, A revision of *Sphenosuchus acutus* Haughton, a crocodylomorph reptile from the Elliot Formation (Late Triassic or Early Jurassic) of South Africa: Philosophical Transactions of the Royal Society of London, Series B, v. 330, p. 1-120.

Wild, R., 1989, *Aetosaurus* (Reptilia: Thecodontia) from the Upper Triassic (Norian) of cene near Bergamo, Italy, with a revision of the genus: Revista del Museo Civico di Scienze Naturali "Enrico Caffi", v. 14, p. 1-24.

Witte, W. K., Kent, D. V. and Olsen, P. E., 1991, Magnetostratigraphy and paleomagnetic poles from the Late Triassic-earliest Jurassic strata of the Newark basin: Bulletin of the Geological Society of America, v. 103, p. 1648-1662.

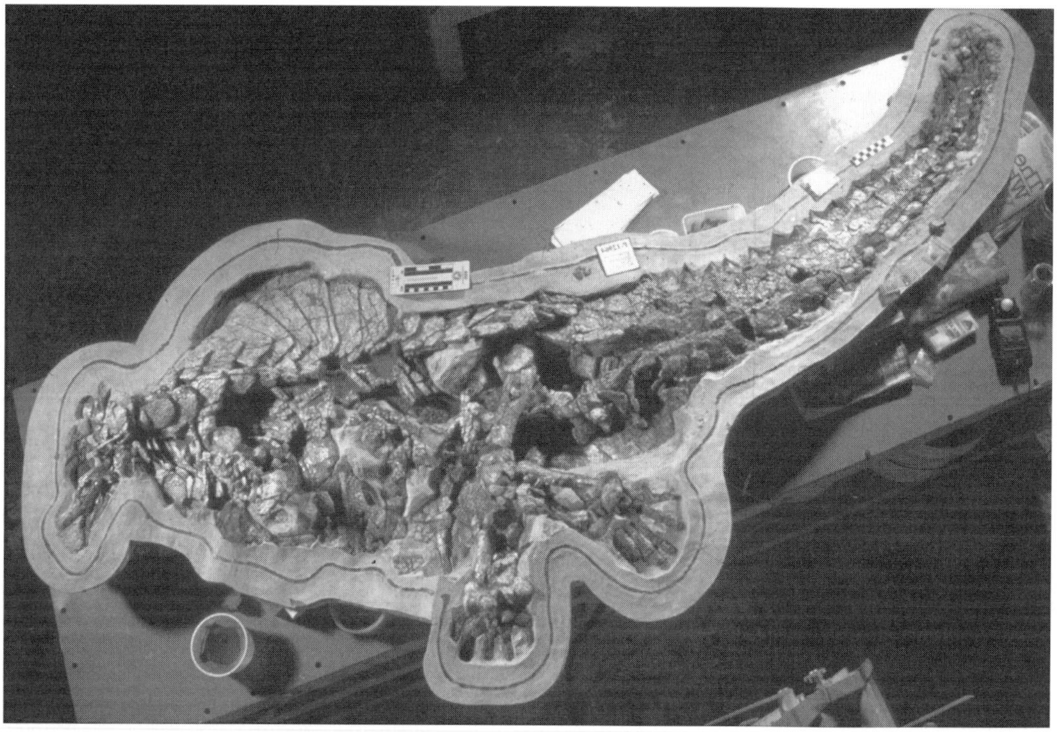

This complete skeleton of the armored reptile Typothorax coccinarum is being prepared at the New Mexico Museum of Natural History in Albuquerque. It was collected from the Upper Triassic Bull Canyon Formation near San Jon. Scale in cm.

LACUSTRINE DEPOSITIONAL ENVIRONMENTS OF THE UPPER TRIASSIC REDONDA FORMATION, EAST-CENTRAL NEW MEXICO

PATRICIA M. HESTER[1] AND SPENCER G. LUCAS[2]

[1]U.S. Bureau of Land Management, 435 Montaño NE, Albuquerque, NM 87107; [2]New Mexico Museum of Natural History, 1801 Mountain Road N.W., Albuquerque, NM 87104

Abstract.—The Redonda Formation of the Chinle Group represents deposition in lacustrine and lake margin depositional environments during the Late Triassic (Norian-Rhaetian?) in east-central New Mexico, but not in a single, large lake. Lake margin facies in the Redonda depositional basin correspond to basement highs that delineated the Paleozoic Tucumcari structural basin, indicating that older structures influenced deposition during the Late Triassic. Paleozoic structures were either topographically high, or a reactivation of the Tucumcari basin initiated changes in drainage systems. Coarse, terrigenous sediment was deposited in Gilbert deltas where fluvial systems flowed into the Redonda depositional system along its eastern edge. Delta progradation at the northern edge of the depositional system produced delta-front sheet sands that were deposited and reworked by waves or currents. Beach environments produced carbonate mud on shallow mudflats along some shorelines, whereas sand beaches prograded at other shorelines. In the deepest lake settings, deposition inevitably gave way to shallowing, and alternations of sandstone and mudstone recorded climatically induced changes in lake level and sediment influx throughout Redonda history. Evidence of Late Triassic climatically influenced lacustrine deposition at low paleolatitudes in the Southwest corresponds to other evidence for climate control of lacustrine deposits of similar age along the present East Coast. The frequency of the response in lakes of the Redonda depositional system is consistent with the interpretation that orbital forcing of climate influenced lacustrine sedimentation over a broad region of Pangea.

INTRODUCTION

Upper Triassic strata of the Chinle Group were deposited over a vast area in the western United States that extends from Nevada to Oklahoma, and from Texas to Wyoming (Lucas, 1993). Considerable sedimentological study of the Chinle Group has been undertaken on the Colorado Plateau (e.g., Stewart et al., 1972; Blakey and Gubitosa, 1983; Blodgett, 1984; Kraus and Middleton, 1987; Dubiel, 1989; Tanner, 2000) and in West Texas (McGowan et al., 1979, 1983; Granata, 1981). However, relatively little sedimentological study of the Upper Triassic strata in east-central New Mexico, between the Colorado Plateau and West Texas, has been attempted (a notable exception is Newell, 1993). Here, we present the results of a sedimentological study (Hester, 1988) of the Redonda Formation, which is the youngest Chinle Group stratigraphic unit in east-central New Mexico. This study concludes that Redonda deposition took place in an areally extensive lacustrine system of small lakes and lake margin facies, but not in a single, large lake, as had previously been concluded.

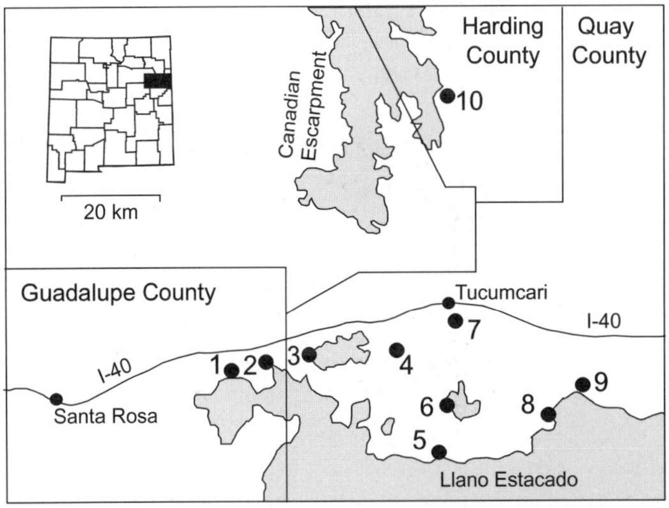

FIGURE 1. Map of part of east-central New Mexico showing location of sections of the Redonda Formation measured in this study. Sections are: 1 = Bull Canyon 1, 2 = Bull Canyon 2, 3 = Luciano Mesa, 4 = Pyramid Mountain, 5 = Ragland Hill, 6 = Mesa Redonda, 7 = Tucumcari Mountain, 8 = Apache Canyon, 9 = San Jon Hill, 10 = Gallegos.

PREVIOUS STUDIES

Dobrovolny et al. (1946) named, described and mapped the Redonda as a member of the Chinle Formation, documenting a variety of lithologies and thickness variations. From subsequent mapping (Wanek, 1962; Kelley, 1972), the Redonda is known to crop out over much of east-central New Mexico, including parts of San Miguel, Guadalupe, Quay and Harding Counties (Fig. 1). Fossils from the Redonda Formation include vertebrates, invertebrates and plants. Semionotid fishes and semiaquatic metoposaur and phytosaur fossils dominate the vertebrate fossil assemblages, and tracks represent rhynchocephalians, aetosaurs and dinosaurs (Gregory, 1972; Lucas et al., 1985; Hunt et al., 1989, 1993, 2000; Hunt, 1994; Lockley et al., 2000). Invertebrate fossils include abundant conchostracans and ostracods (Kietzke, 1987, 1989). Plant material is limited to pith casts of the calamitalean *Neocalamites* (Gregory, 1972; Lucas et al., 1985).

The tetrapod assemblage of the Redonda Formation is the basis of the Apachean land-vertebrate faunachron of Lucas and Hunt (1993), which is of late Norian and Rhaetian? age (Lucas, 1998). Magnetostratigraphy of the Redonda Formation is consistent with this age assignment, and indicates the Redonda Formation correlates to the Rock Point Formation of the Colorado Plateau (e.g., Steiner and Lucas, 2000).

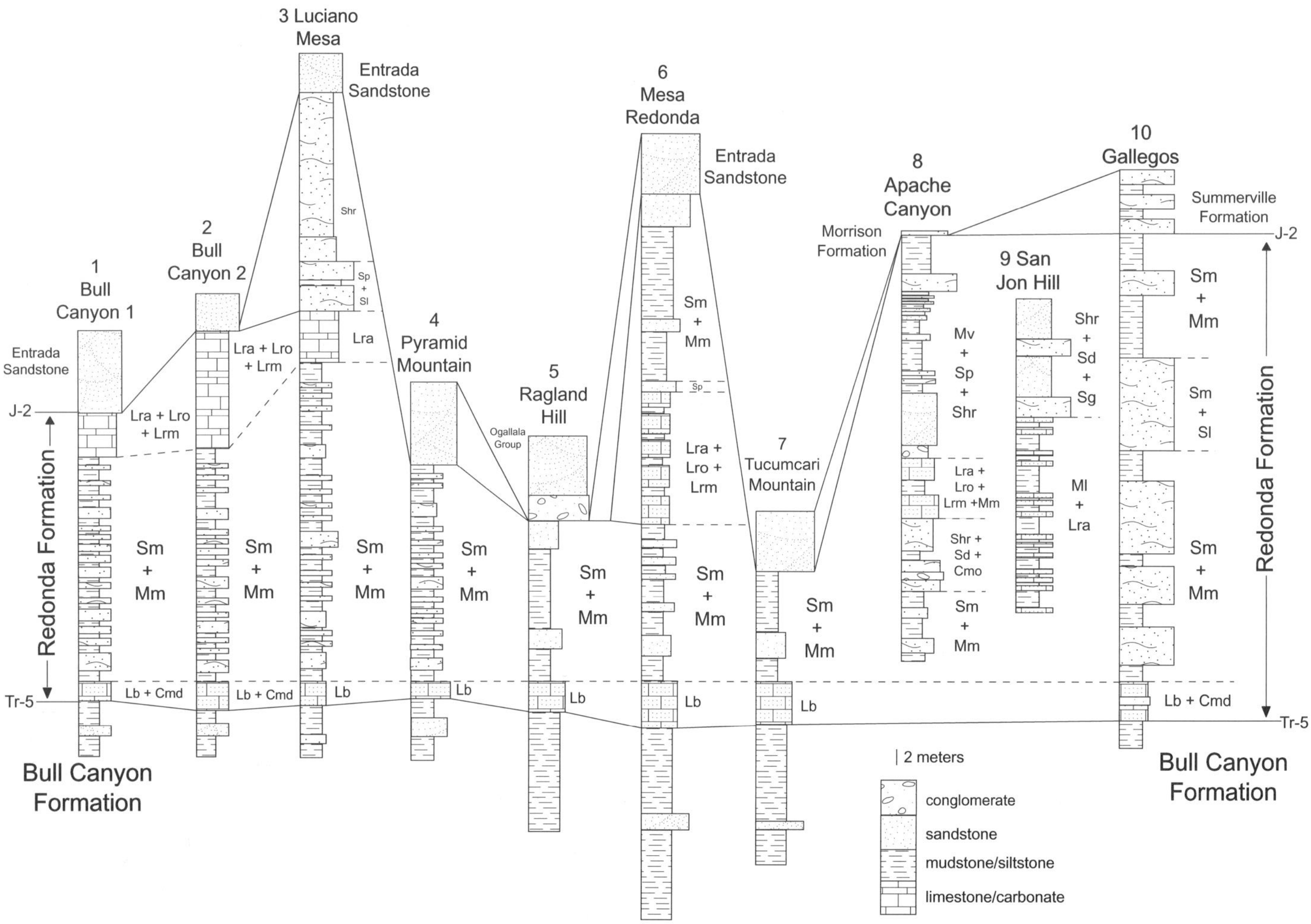

FIGURE 2. Measured stratigraphic sections and lithofacies of the Redonda Formation in east-central New Mexico. See Figure 1 for location of sections and text and Table 1 for lithofacies acronyms.

STRATIGRAPHY

The Redonda Formation is the youngest unit of the Chinle Group in east-central New Mexico, and overlies the Upper Triassic (Revueltian) Bull Canyon Formation (Lucas et al., 1985; Lucas and Hunt, 1989) (Fig. 2). In east-central New Mexico, the Bull Canyon Formation is as much as 110 m thick and is mostly grayish-red and moderate reddish-brown mudstone and lesser amounts of yellowish-gray to grayish-red, laminar to trough-crossbedded sandstone. Very minor lithologies are siltstone and intraformational conglomerate. Granata (1981) attributed both lacustrine and prodelta environments to the Bull Canyon Formation. However, the work of Newell (1993) and our own observations support a fluvial origin.

Dobrovolny et al. (1946) described the Redonda as ranging from 15 to 140 m in thickness and consisting of variegated shale, argillaceous limestone, siltstone and sandstone. Thickness of measured sections in this study ranges from 18 to 92 m (Fig. 2). Laterally continuous, interbedded fine sandstones and mudstones characterize the Redonda over most of the study area. This distinctive bedding makes the unit easily recognizable and prompted Griggs and Read (1959) to elevate the Redonda to formational rank (also see Lucas, 1993). The Redonda is unconformably overlain by Jurassic rocks (the Entrada, Summerville and Morrison formations) in nine of the ten measured sections. At one locality (Ragland Hill), the Redonda is unconformably overlain by the Neogene Ogallala Group.

The Bull Canyon/Redonda contact has been described as both unconformable (Granata, 1981; Kelley, 1972; Lucas, 1993) and conformable (Dobrovolny et al., 1946; Griggs and Read, 1959; Lucas et al., 1985). Granata (1981, p. 57) interpreted the contact as a paleosol, describing "variegated lavender and light green mudstone with possible root traces." Kelley (1972, p. 89) described fissures in the uppermost mudstone bed of the Bull Canyon Formation that are filled with Redonda sand in the upper meter of a "mottled greenish gray and lavender mudstone" and concluded that the contact is disconformable. Dobrovolny et al. (1946) defined the base of the Redonda as a variegated, purple calcareous siltstone. Lucas (1993) regarded the Redonda base as the Tr-5 unconformity correlative to the base of the Rock Point Formation on the Colorado Plateau.

We pick a distinctive pale red, laterally continuous, silty micrite as the Bull Canyon/Redonda contact in this study (Fig. 2). This unit is the variegated lavender and mottled green "mud-

FIGURE 3. Outcrops of the Redonda Formation (R) at Bull Canyon form ribbed slopes above Bull Canyon Formation (B) mudstones and sandstones that floor the canyon and below bold, light-colored cliffs of the Entrada Sandstone (E).

stone" described by earlier workers (Dobrovolny et al., 1946; Kelley, 1972; Granata, 1981). The bed clearly marks a profound change from discontinuous, lenticular sandstones and mudstones of the Bull Canyon Formation to the distinctive, laterally continuous beds of sandstone and mudstone characteristic of the Redonda Formation (Figs. 3, 4a). Therefore, we identify the base of the Redonda Formation as a disconformity.

REDONDA LITHOFACIES

We describe Redonda lithofacies (Table 1) according to their dominant lithology—conglomerate, sandstone, carbonate and mudstone. Lithofacies were established based on grain size, sedimentary structures and ichnofossils. Because some variation in lithofacies exists between localities, the descriptions clarify the variation.

Matrix-supported Conglomerate (Facies Cm)

Poorly-sorted conglomerates contain very coarse sand-to-gravel-size clasts supported in a mud matrix. Clasts include intraformational sedimentary rock fragments of micrite and siltstone, vertebrate teeth and bone fragments, fish scales, clay rip-up clasts and conchostracan hash. We use the organization of the coarse fraction and geometry of the beds to separate matrix-supported conglomerates into two subfacies, organized (Cmo) and disorganized (Cmd) matrix-supported conglomerates.

Concentration of clasts in the middle portion of individual beds and lateral discontinuity of beds define organized matrix-supported conglomerates (Cmo). Abundant vertebrate teeth and bone hash, micrite rock fragments and clay rip-up clasts comprise the coarse fraction. Clay rip-up clasts are aligned parallel to flow. This lithofacies occurs in thin (10-50 cm thick), localized beds that taper before pinching out into mudstone within 2-3 m of outcrop. Individual beds have a low angle of depositional dip, generally 3-5°. Lower contacts are sharp, and stretched clay rip-ups occur within the basal portion of individual beds. Upper contacts rapidly grade to mudstone.

Organized conglomerates are interpreted as localized, subaqueous debris flows, though some may be subaerial alluvial fan deposits. Indeed, subaqueous and subaerial debris flows may be difficult to distinguish, but the presence of mudstone interbeds is more common in subaqueous flows (Nemac and Steel, 1984). Subaqueous debris flows characteristically concentrate coarse material in the middle of the bed, responding to shear stress along upper and lower surfaces of the flow (Middleton and Hampton, 1973). Stretched clay rip-up clasts may represent this zone of shearing. Alignment of clay rip-up clasts suggests laminar flow at the time of deposition (Fisher, 1971). Localized, subaqueous debris flows have been described from Gilbert delta fronts (e.g., Postma, 1985) and from submarine fans (Hein, 1983). The alternative hypothesis, that these organized conglomerates are alluvial fan deposits, is thus less well supported.

Poorly-sorted clasts distributed in a mud-rich matrix, and lateral continuity of beds, characterize disorganized matrix-supported conglomerates (Cmd). From visual field estimates, matrix comprises 60% to 70% of the bed, increases upward within the bed and consists of calcareous mud, terrigenous silt and clay. A trend of decreasing clast size upward is seen. This lithofacies occurs in laterally extensive beds and is not confined to a particular locality. The coarse fraction ranges from very coarse sand (Bull Canyon) to gravel (Gallegos) (Fig. 2). Clasts include micrite rock fragments and clay rip-ups. Fossil debris is not present, in contrast to organized matrix-supported conglomerates. Post-depositional burrows are abundant in Cmd lithofacies. Thickness of disorganized conglomerate beds ranges from 40 cm to 1 m. The beds are laterally extensive for hundreds of meters along outcrop. Cmd occurs in single beds or amalgamated beds, both with irregular basal contacts. Upper contacts are sharp in single beds and irregular in amalgamated units.

Poor sorting and matrix-supported clasts suggest sediment gravity flow processes (e.g., Fisher, 1971; Middleton and Hampton, 1976; Lowe, 1982). Irregular basal contacts and clay rip-up clasts suggest scouring of underlying units. However, lateral extent of the lithofacies rules out a localized process and requires a mechanism that could transport material over some distance. This lithofacies thus represents subaqueous gravity flow or cohesive debris flows that were not localized. During their development, flow was turbulent. Scouring into the underlying unit exemplifies erosion and turbulent flow. Textural inversion of a mud matrix with disseminated clasts, upward increase in matrix and presence of post-depositional bioturbation indicate a subaqueous setting (Nemac and Steel, 1985).

The disorganized fabric suggests Cmd may represent proximal flows. However, the decrease in clast size upward indicates some sorting took place. In Walker's (1979) fan model, debris flows may be associated with feeder channels, but no channels were documented in the present study. Matrix-supported conglomerates may also be associated with sublacustrine fans, where debris flows mix with lake mud and are deposited in flat-lying lobes (Nemac et al., 1984).

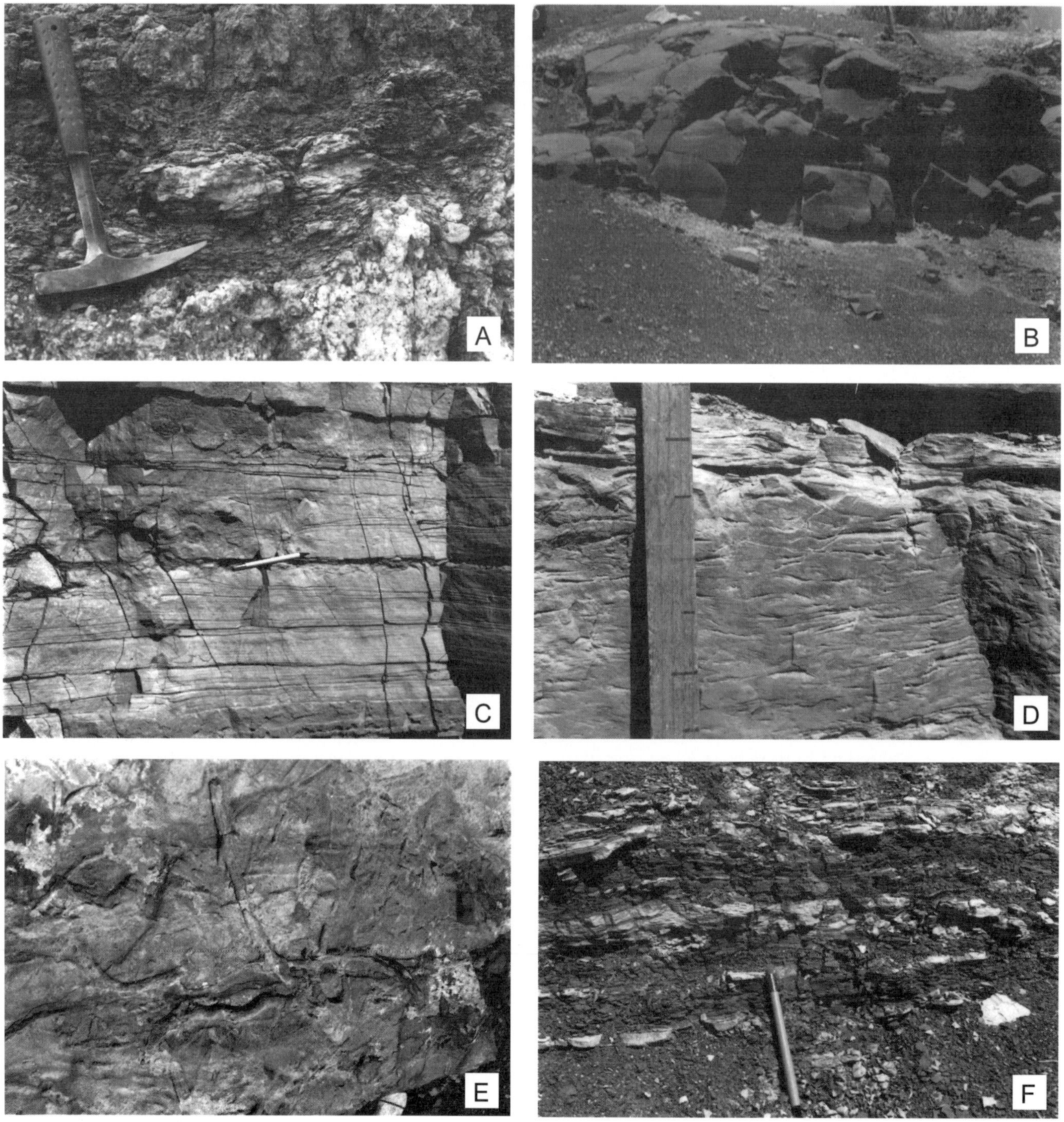

FIGURE 4. Selected examples of Redonda lithofacies. A. Basal marker bed of Redonda Formation at Mesa Redonda, lithofacies Lb. B. Lithofacies Sm at Bull Canyon. C. Lithofacies Sl at San Jon Hill. D. Lithofacies Shr at Apache Canyon. E. Lithofacies Lra at Bull Canyon. F. Lithofacies Ml at San Jon Hill.

Massive Sandstone (Facies Sm)

Massive sandstones are composed of very fine to medium, moderately sorted sand with minor amounts of coarse sand and occasional wood fragments. Vertical and sub-horizontal *Scoyenia* burrows occur throughout the otherwise structureless sandstones. Burrows contain meniscate backfill, accentuated by reduction halos along backfill and burrow walls. Intensity of bioturbation, thickness of individual sandstone beds and grain size vary by locality.

Sandstones occur as laterally continuous, 0.4-13 m thick, single bed units, although average thickness is 1.5 m. Thicker, medium-grained sandstones (Gallegos) contain fewer burrows. Some sandstone beds persist at the same stratigraphic level between localities for 24 km, attesting to their great lateral extent (Fig. 2). Indeed, persistent sandstone beds can be visually traced

TABLE 1. Redonda lithofacies.

Facies Code	Description	Geometry	Processes
Cmo	Organized, matrix-supported conglomerate; coarse fraction concentrated in center of bed; basal zone of shearing with "stretched" clay rip-ups.	Deposited in 10-50 cm thick, tapering wedges that pinch out into mudstone.	Sediment gravity flow, subaqueous debris flow.
Cmd	Disorganized, matrix-supported conglomerate; Clasts range from coarse sand to gravel; matrix increases upward in beds, comprises 60-70% of bed; post-depositional burrows.	Deposited in tabular, 0.4-1 m thick, single or amalgamated beds.	Sediment gravity flow, subaqueous, non-laminar debris flow.
Sm	Massive sandstone, very fine to medium; contains disseminated coarse sand and occasional wood; *Scoyenia* burrows; occasional sedimentary dikes. unconsolidated substrate.	Deposited in laterally continuous 0.4 to 13 m thick beds; beds continuous for km on strike.deposition on a weak,	Gravity settling of sand-sized grains ; bioturbation and sedimentation in equilibrium.
Sl	Laminated sandstone; fine, well-sorted sandstone; parallel and low-angle cross-stratification; minor ripple laminations	Laterally continuous beds from 20 to 60 cm thick.	Traction transport, high flow velocity, changing flow velocities.
Sp	Planar cross-stratified sandstone; fine to medium grained, moderately to well sorted sand; planar cross-stratification in sets from 10 to 30 cm thick. thickness 2-3.5 m.	Tabular, laterally continuous for hundreds of meters; thickness from 0.5 to 1.5 m; lenticular bodies pinch out into mudstone over 30 m;	Migration of two-dimensional mega-ripples under moderate flow velocity.
Sd	Disrupted sandstone; fine, moderately sorted sandstone; dish structures convolute bedding with parting lineations along bedding planes of exposed folds; ball-and-pillow structures rates.	Laterally discontinuous wedges; depositional dip 5-20 degrees; thickness 20-70 cm; wedges taper and pinch out into mudstone and sandstone lithofacies.	Destruction of primary bedforms by escaping pore fluid; deposition by high, unidirectional flow in beds that later slumped; deposition on weak substrate; high sedimentation
Sg	Normally graded sandstone; coarse to fine, poorly sorted sandstone; coarse tail grading; minor vertical and sub-horizontal burrows over 6 m of strike.	Laterally discontinuous wedges with depositional dip from 10 to 20 degrees; wedges range from 10 to 20 cm thick, taper and pinch out	Turbidity current deposition.
Shr	Horizontal and ripple-laminated sandstone; moderately well-sorted sandstone; asymmetrical ripples, horizontal laminations, parting lineations; climbing ripples	Laterally discontinuous wedges; thickness 20-70 cm; tabular, laterally continuous units up to 6 m thick composed of numerous, thin (5-10 cm) rippled beds; tabular, single beds are 0.1-1 m thick.	Traction transport; unidirectional high flow velocities; changing flow velocities; high sedimentation rates.
Lra	Asymmetrical, ripple-laminated micrite; calcareous mud with varying amounts of terrigenous silt and sand; wavy to lenticular bedding; ripple bedforms draped and interbedded with calcareous mud; or, ripple bedforms defined by terrigenous component of micrite; trace fossils (*Diplichnites*, *Cruziana*, *Pelecypodichnus*)	Laterally continuous for hundreds of meters along outcrop; thickness 0.1-3 m.	Precipitation of carbonate mud; occasional influx of terrigenous grains; unidirectional flow; diverse faunal activity.
Lro	Oscillation ripple-laminated micrite; calcareous mud with varying amounts of terrigenous silt and sand; symmetrical ripples; trace fossils as in Lra and tetrapod footprints.	Laterally continuous for hundreds of meters; thickness 0.1-3 m. grains;	Precipitation of carbonate mud; occasional influx of terrigenous bidirectional flow; diverse faunal activity; shallow to subaerial conditions.
Lrm	Mudcracked, ripple-laminated micrite; lithology as Lra and Lro; symmetrical and asymmetrical ripples; mudcracks; fauna as in Lro.	See Lra	Precipitation of carbonate mud with terrigenous influx; unidirectional and bidirectional flow; subaerial exposure.
Lb	Bioturbated, silty micrite; mottled to nodular; intensely bioturbated. of soupy, unconsolidated substrate.	Laterally continuous; thickness 1-1.5 m; traceable for km.	Precipitation of carbonate mud; settling of silt and clay; bioturbation
Mm	Massive mudstone; terrigenous silt and clay components; *Scoyenia* burrows.	Laterally continuous for hundreds of meters on strike; units are 1-9 m thick.	Suspension settling of silt and clay; quiet water; bioturbation.
Mv	Variegated mudstone; terrigenous silt and clay components; subtle color horizons; contains vertebrate bones, teeth, coprolites and fossil wood.	Laterally continuous units 0.1-3 m thick.	Settling from suspension in quiet water; diagenetic horizons; conditions for terrestrial flora and fauna.
Ml	Laminated mudstone; terrigenous silt and clay components; small, asymmetrical ripples.	Composite beds range from 0.1 to 1.5 m thick; composed of thin (1-4 cm) rippled beds; continuous for tens of meters.	Unidirectional flow.

along the Llano Estacado and Canadian escarpments between measured sections.

Lower contacts are irregular due to load casts, but spheroidal weathering accentuates the irregularity (Fig. 4B). Locally, small (2-3 cm wide, 10-100 cm long) sandstone structures taper downward and branch or are of consistent thickness as they project down into underlying mudstones. The longest of these structures appears slightly folded within a mudstone. Upper contacts are sharp. Burrows on the upper surface are filled with overlying mudstone.

Well-defined and preserved burrows, rather than a completely bioturbated fabric, imply continuous deposition with bioturbation and sedimentation at equilibrium (Ekdale et al., 1984). Fewer burrows in the thickest sandstones may indicate variability of sedimentation rates at the different localities. The lack of ichnofaunal diversity indicates some limiting conditions on organisms (Ekdale et al., 1984). *Scoyenia* burrows are attributed to soft-bodied animals whose oxygen requirement is less than that of hard-bodied organisms (Ekdale et al., 1984). Therefore, the dominance of bioturbation by soft-bodied organisms suggests that the oxygen content of the water could have been a limiting factor on the infauna during deposition of this lithofacies.

Onset of deposition of sand-sized material took place on an unconsolidated, weak substrate, creating conditions for load structures to form (Allen, 1982). The sandstone structures along basal contacts are interpreted as sedimentary dikes injected into muds. Sedimentary dikes may occur in overlying or underlying beds, and gentle folding of the longest dike indicates that mud continued to dewater after injection of the dike (Miall, 1984). These structures are interpreted as dikes because no pattern of occurrence was observed, and injection structures tend to have extreme length compared to width (Allen, 1982). Cessation of sand deposition was rapidly followed by settling from suspension of silt and clay, allowing mud to accumulate in burrows along the upper surfaces of sandstones (Ekdale et al., 1984).

Lack of traction structures within these sandstones could be attributed to: (1) destruction of primary bedforms by organisms, (2) homogeneity of sand or (3) deposition where currents or waves did not affect sediments. The well-preserved, discrete character of the burrows, rather than a completely bioturbated fabric, rules out the first possiblity. Although not abundant, the presence of some coarser material indicates transport energy sufficent to carry coarse sand. Wood fragments might float until they became waterlogged and sank. Once deposition took place, this coarser material was not reworked into traction structures, suggesting that once deposited, massive sandstones were not influenced by moving water.

Coarse sediments can be carried into lake waters by river plumes, turbidity or slumping (Sly, 1978). If episodic turbidity currents deposited these sands, some traction structures would be expected. Slumps would generate structures indicative of rapid deposition such as dewatering structures, but homogeneity of the sand and fine grain size might obscure such bedforms. Density flows from rivers transport fine sand and silt into a basin on a continuous basis (Pharo and Carmack, 1979). The Sm lithofacies contains evidence of continuous deposition (bioturbation), but lack of other structures makes the process that transported these sands difficult to identify.

Laminated Sandstone (Facies Sl)

This facies consists of fine-to-medium, well-sorted, calcite-cemented sandstone that contains parallel, low-angle cross-stratification (Fig. 4C) and minor amounts of ripple cross-stratification. Sets are 4-10 cm thick and contain parallel laminations. The lithofacies occurs in beds 20-60 cm thick that are laterally continuous for hundreds of meters along outcrop. Lower contacts are sharp or gradational from other sandstone lithofacies. Upper contacts are sharp.

Traction transport under high flow velocities operates to form parallel and low angle cross-stratification (Harms et al., 1982). Ripples formed in the same grain size imply changing flow velocities and periods of lower energy conditions.

Planar-stratified Sandstone (Facies Sp)

Fine-to-medium, moderately to well-sorted, calcite-cemented sandstones contain planar cross-stratification. Cross-stratification is in sets 10-30 cm thick. The lithofacies occurs as laterally continuous or discontinuous beds. Tabular sandstone is continuous along outcrop for hundreds of meters and 1.5-2 m thick. The lower contact of tabular sandstones (Luciano Mesa) is sharp above a carbonate lithofacies. The upper contact is sharp and overlain by facies Sl. Tabular sandstone lacks any internal geometry to suggest amalgamation of channels. Discontinuous sandstones (Apache Canyon and Mesa Redonda) occur in convex bodies that pinch out within 30 m of outcrop. Thickness is 2-3.5 m. Lower contacts are sharp and may contain coarse sand grains.

Disrupted Sandstone (Facies Sd)

Fine, moderately sorted, calcite-cemented sandstones contain convolute bedding, dish structures and ball-and-pillow structures. Sandstones were deposited in laterally discontinuous, wedge-shaped bodies that dip from 5° to 20° and pinch out laterally into other lithofacies within meters of outcrop. Thickness of wedges varies from 20 to 70 cm, and ball-and-pillow structures form irregular basal contacts with underlying mudstone interbeds. Upper contacts are sharp.

During deposition or shortly after burial, soft-sediment deformation occurred in water-laid sediments. High sedimentation rates promote loose packing of sand (Allen, 1982), and sediments lose strength as liquefaction causes grains to separate temporarily and disperse in pore fluid, promoting development of soft-sediment deformation. Primary structures in sandstones were distorted after deposition, forming dish structures and convoluted bedding. Dish structures form as escaping pore water from sediment loading disrupts laminated sands (Miall, 1984). Folded primary structures and dish structures suggest that these beds were initially deposited by unidirectional currents. High sedimentation rates created conditions for disruption after deposition. Ball-and-pillow structures form as sand is deposited on weaker,

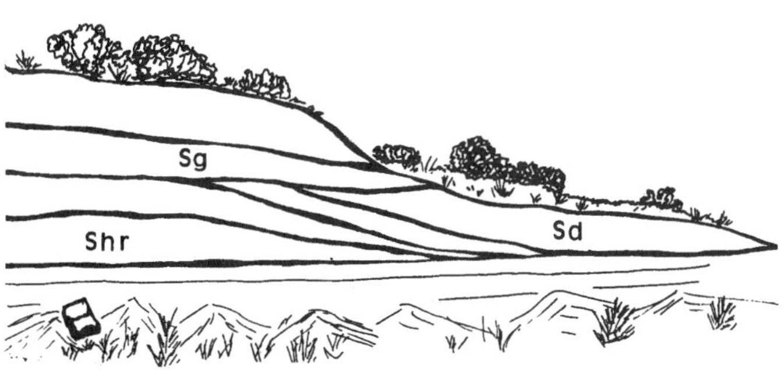

FIGURE 5. Drawing of the San Jon Hill locality, showing normally graded sandstone (Sg) in wedge-shaped bodies above disrupted (Sd) and horizontal and ripple-laminated sandstone (Shr). Sandstone interval is ~ 5 m thick.

unconsolidated mud (Miall, 1984). Unidirectional currents and high sedimentations rates operated in the formation of disrupted sandstones. Mudstone interbeds accumulated periodically as sand supply was cut off or flow energy reduced. When sand deposition resumed, muds were unconsolidated, allowing load structures to form. Environments compatible with development of these structures include deltas or river channels (Miall, 1984).

Normally Graded Sandstone (Facies Sg)

Coarse-to-fine, poorly sorted, calcite-cemented, normally graded sandstones contain fines throughout, with grading restricted to the coarse fraction (coarse-tail grading of Middleton, 1965). Minor vertical and sub-horizontal burrows are present. Grains include well-rounded, sedimentary rock fragments (micrite and siltstone), quartz, feldspar, oolites and fish scales.

This lithofacies occurs in laterally discontinuous beds that display wedge-shaped geometry along outcrop. Individual beds range in thickness from 10 to 30 cm and are generally separated by thin (1-2 cm) mudstone partings. Wedge-shaped bodies dip between 10° and 20°, taper and pinch out laterally within 6 m (Fig. 5).

Coarse-tail grading requires a range of available grain sizes and a sudden decrease in fluid competence, allowing grain settling in response to fluidal and gravitational forces (Middleton, 1965; Davis, 1983). Normally graded beds are associated with turbidite (underflow) deposition where denser river waters discharge into a lake (Hakanson and Jansson, 1983). The accumulation of packages of sandstone composed of normally graded beds implies sporadic discharge. Minor bioturbation indicates depositional processes dominated biogenic processes (Ekdale et al., 1984).

Horizontal and Ripple-laminated Sandstone (Facies Shr)

Fine, moderately well-sorted, calcite-cemented sandstones contain asymmetrical ripple cross-stratification and horizontal laminations. Surfaces of horizontally laminated beds contain parting lineations. Climbing ripples also occur with entire ripple form preserved (Fig. 4D). Lithofacies Shr occurs in both laterally continuous and discontinuous units. Discontinuous bodies are described in lithofacies Sg and Sd (Fig. 5). Climbing ripples are associated with these wedge-shaped beds.

Laterally continuous bodies (Luciano Mesa) may be as much as 6 m thick, but are composites of numerous, thin (5-10 cm) rippled beds. These laterally continuous units extend for hundreds of meters along outcrop. Small-scale scours (10 cm wide by 3 cm deep) are filled with thin (1-2 cm) rippled beds. Other laterally continuous units are thinner, ranging from 0.1 to 1 m thick and are single bed units (Apache Canyon).

Unidirectional flow, varying from high to low velocity, deposited these sandstones. In units having wedge geometry, direction of flow was at some angle to the dip of the beds. Climbing ripples associated with these units indicate periods of very rapid sediment supply from suspension combined with sufficient traction to produce and preserve entire ripple forms (Blatt et al., 1980). Horizontal stratification containing lineations indicates that in order to scour, flow must achieve a certain velocity for any given grain size (Blatt et al., 1980). Small scours in the tabular, composite interval suggest periods when flow velocity was increased to erode fine-grained sands. Increased flow velocity was sporadic, and once the scour developed, conditions of lower flow velocity returned.

Ripple-laminated Micrite (Facies Lr)

Ripple-laminated, calcareous mud containing varied amounts of terrigenous sand and silt comprises facies Lr. A diverse invertebrate ichnofauna occurs, including *Diplichnites*, *Cruziana* and *Pelecypodichnus* (Hester, 1988). Vertebrate tracks are abundant (Hunt et al., 1989, 1993, 2000; Lockley et al., 2000). Subfacies are defined by the presence of asymmetrical (Lra) and symmetrical (Lro) ripple laminations and mudcracks (Lrm). All three are in beds laterally continuous for hundreds of meters that range from 0.1 to 3 m thick.

Asymmetrical ripple-laminated micrites (Lra) may be composed of numerous, thin (1-2 cm) asymmetrical ripples with calcareous mudstone interbeds and drapes. Locally (Mesa Redonda), ripple-laminated micrites lack drapes or interbeds and contain the greatest amount of carbonate mud. The terrigenous component defines ripple cross-stratification that stands out in relief on outcrop. Upper contacts are sharp, but lower contacts may be sharp or gradational. The ichnofauna (Fig. 4E) includes numerous invertebrate trails and traces, similar to those described by Bromley and Asgard (1979) in Triassic lacustrine carbonates in Greenland.

Symmetrical, ripple-laminated silty/sandy micrite (Lro) contains symmetrical ripples, the amount of terrigenous material varies by locality (less abundant at Mesa Redonda), and reactivation surfaces can be seen. Mudcracked, ripple-laminated silty micrite (Lrm) gradationally overlies other micrite lithofacies. Tetrapod tracks of small rhynchocephalians and larger aetosaurs and dinosaurs are abundant in the micrites at Mesa Redonda. Lower contacts are gradational, whereas upper contacts are sharp.

Carbonate mud precipitation and terrigenous silt and sand deposited by unidirectional currents operated during the formation of lithofacies Lra. Units composed of mud drapes and ripples indicate that coarser, terrigenous material was not always available to form ripples. Precipitation of carbonate mud dominated during deposition of micrites that lack the wavy to lenticular bedding. When current energy was sufficient, terrigenous material was transported. A diverse and abundant ichnofauna indicates few limiting factors on organisms existed during deposition (Ekdale et al., 1984).

Symmetrical ripples (Lro) formed under oscillatory flow conditions produced by wave action (De Raff et al., 1977). Reactivation surfaces record a local and temporary shift of flow conditions (Allen, 1982). Wave-generated structures indicate that this lithofacies was deposited in shallow water. Local variation in the amount of terrigenous sediment suggests that Mesa Redonda was sediment starved relative to other locations during deposition of the micrite. Micrite and minor terrigenous silt and sand deposited by currents were subject to subaerial exposure when mudcracks formed (Lrm). Tetrapod tracks in this lithofacies further support identification of shallow to subaerial conditions.

Bioturbated Micrite (Facies Lb)

This lithofacies is only found at the base of the Redonda Formation. Massive, silty micrite displays an intensely bioturbated fabric that gives the facies a nodular or mottled appearance (Fig. 4A). Lithofacies Lb occurs as a laterally continuous unit, traceable for km. Thickness ranges from 1 to 1.5 m. The basal contact with underlying mudstone is gradational. Upper contacts are irregular when overlain by lithofacies Cmd or gradational where mudstones overlie the unit.

Precipitation of carbonate mud, bioturbation and deposition of terrigenous silt and clay are dominant processes contributing to this facies. Bioturbation was intense enough to destroy any primary structures. Mottled fabric can be produced by bioturbation of a very soft, highly liquified substrate (Ekdale et al., 1984) such as the bottom of a lake or pond. Pedogenic processes can also form calcareous nodules when K-soil horizons develop (Birkeland, 1974). Thus, facies Lb is interpreted here as primarily of lacustrine origin, though pedogenesis cannot be ruled out locally.

Massive Mudstone (Facies Mm)

Fine-grained rocks containing subequal parts of clay and silt are massive and contain silt-filled *Scoyenia* burrows. Lithofacies Mm occurs in laterally continuous beds from 1 to 9 m thick. Basal contacts with underlying units are sharp, and upper contacts appear irregular due to spheroidal weathering of overlying sandstone. Silt and clay settled from suspension in quiet water into an environment of faunal activity. Oxygen levels may have limited diversity, as in lithofacies Sm.

Variegated Mudstone (Facies Mv)

Fine-grained rocks that contain both clay and silt exhibit laterally continuous, subtle color banding (diagenetic?) that gives the lithofacies a stratified appearance. Vertebrate bones, teeth, coprolites and wood fragments occur scattered through the lithofacies. Thin, silty micrite lenses in the mudstones contain conchostracan molds.

Settling from suspension of clay and silt in quiet water formed this lithofacies. Carbonate precipitation in micrite beds containing fossils implies periods when standing bodies of water existed. Modern conchostracans live along the margins of large lakes and in small, temporary inland ponds ranging from fresh to brackish water (Webb, 1979). Color horizons in mudstones have been used as evidence of soil development (e.g., Kraus, 1987; Bown and Kraus, 1987). However, no root traces or other indications of pedogensis were identified.

Ripple-laminated Mudstone (Facies Ml)

Fine-grained rocks containing clay and silt display small, asymmetrical ripple laminations (Fig. 4F). Lithofacies Ml occurs as composite beds, composed of thin (1-4 cm thick) mudstone. Composite beds range in thickness from 0.1 to 1.5 m. Deposition of terrigenous clay and silt occurred under unidirectional flow. Fine grain size may indicate low flow velocities or reflect the size of sediment supply (Harms et al., 1982).

DEPOSITIONAL ENVIRONMENTS

Granata's (1981) interpretation of Redonda deposition as in a single, large lake cannot be confirmed. Instead, we interpret Redonda deposition as having taken place in a "lacustrine system" of small to moderate sized lakes (up to a few km? in diameter) and in lake margin and related environments. This is demonstrated by an analysis of the lateral occurrence and variation of lithofacies associations in the Redonda Formation, which identifies a large lacustrine depositional system with distal and proximal lake, shoreline, deltaic and fluvial environments.

Basal Marker Bed

The abrupt change from fluvial deposition of the Bull Canyon Formation to lacustrine deposition of the Redonda Formation occurs at the basal marker bed of the formation (Fig. 2). This marker bed is composed of silty, bioturbated micrite (Lb) and associated disorganized matrix-supported conglomerate (Lb + Cmd). Interpretation of this unit as a paleosol (Granata, 1981) that contains sandstone filled "fissures" (Kelley, 1972) supports a disconformable contact between the Bull Canyon and the Redonda formations.

Although detailed pedogenic analysis was outside the scope of the present study, an alternative interpretation of the basal marker bed is supported by field evidence. Lacustrine processes could account for the marker bed (Lb + Cmd and Lb). The change in depositional style from lenticular sandstones intercalated with mudstones of the Bull Canyon Formation into the laterally continuous, bioturbated, silty micrite (Lb) marks a profound change in stratigraphic architecture. Environments changed from variable to widespread consistency. One criterion used to delineate ancient lacustrine systems is lateral continuity of beds; intense bioturbation also is common (Picard and High, 1972).

Carbonate precipitation occurs in profundal and littoral lake environments (Dean, 1981; Picard and High, 1981), and terrigenous silt and clay, carried by rivers, can settle from suspension into deep lake environments. If these processes operated simultaneously, a mixture of silt, clay and carbonate mud would be deposited. Modern and ancient lakes show sediment distribution patterns that illustrate this combination (Picard and High, 1981, and references therein). In deep marine environments, this combination of sediments can result in a nodular texture (Jenkyns, 1978). Thus, the mottled and nodular appearance of Lb could represent a fabric developed by bioturbation of a soupy, unconsolidated substrate that existed on a lake bottom where carbonate mud and terrigenous silt and clay were accumulating.

The association of disorganized matrix-supported conglomerate with the silty micrite at Bull Canyon and Gallegos indicates that subaqueous sediment gravity flows influenced the environment during deposition of the marker bed. Matrix-supported conglomerates identified in Jurassic lake deposits in West Virginia occur in sublacustrine channels on lake slopes and on the non-channelized basin plain (Hentz, 1985). After deposition of the gravity flows, bioturbation continued, and burrows developed in the matrix-supported conglomerates. Matrix-supported conglomerates can be associated with sublacustrine fans (Nemac et al., 1984), where flows pick up mud from lake deposits. The lithologic similarity of Lb and the matrix of Cmd could be the result of such a process. Sediment gravity flows could be generated on unstable lake basin slopes or associated with oversteepening of delta slopes (Sly, 1978). Such flows can travel long distances on gentle slopes (Rodine, 1976). The occurrence of these conglomerates within the marker bed suggests that these processes were important early in the history of Redonda deposition.

The locality where Kelley (1972) described fissures was not located by us, but at Gallegos a structure interpreted as a sedimentary dike is associated with the marker bed. The structure originates from a massive sandstone, projects through a mudstone (where it appears contorted) and into the marker bed. If this structure represents a passively infilled fissure, then the fissure fill occurs between a mudstone and a massive sandstone and not within the marker bed. The slight folding of this structure suggests dewatering of the mudstone after emplacement of the sandstone, supporting the interpretation of a subaqueous, lake environment.

From evidence presented in this study, a lacustrine origin for the marker bed (Lb + Cmd) is reasonable. Field evidence is lacking for subaerial origin of the disorganized matrix-supported conglomerate. True root traces or bioturbation attributable to terrestrial fauna within this bed are absent, and this does not support a pedogenic origin.

Lacustrine Environments

Lacustrine environments recorded by parts of the Redonda Formation are represented by alternating massive mudstone and sandstone (Sm + Mm) (Fig. 2). A general trend in lacustrine systems is a decrease in grain size offshore (Picard and High, 1981), so we infer that Redonda mudstones represent distal environments, whereas sandstone deposition took place under more proximal conditions. The specific mechanisms that transported the sand and produced alternation of sand and mud are more difficult to identify.

The presence of disseminated coarse to medium grains within very fine to fine sandstone requires a physical mechanism capable of transporting these grain sizes. Physical processes that move coarse to medium sand into lakes include river plumes and turbidity currents. Rivers transport sediment into lakes, dispersing sediment as surface flows, interflows and underflows, depending on density contrasts (Hakanson and Jansson, 1983). Turbidity current underflows deposit a distinctive sequence (Strum and Matter, 1978). Pharo and Carmack (1979) distinguish turbidity processes from near bottom underflow by defining turbidity currents as episodic, and underflows as continuous. The massive sandstones (Sm) of the Redonda Formation contain no evidence of turbidite deposition such as graded bedding, so it is unlikely that turbidity currents carried sand into the environment. Continuous underflow could account for some of the features observed in the sandstones, because they can deposit chaotic, ungraded sand, silt and clay (Strum, 1979). The fabric of bioturbation developed in the massive sandstone (Sm) suggests an equilibrium between sedimentation and bioturbation (Ekdale et al., 1984), not episodic deposition. The area in a lake directly influenced by plumes is limited (Hakanson and Janson, 1983), and the wide exent of the sandstone is difficult to explain, even with multiple sources of sand (river plumes).

The widespread sandstones, however, probably represent transitional to shoreline lake environments rather than deeper water environments. Indeed, these sandstones are readily interpreted as the result of sheetflow in a subaerial and/or shallow, ephemeral lake setting. In modern lakes, sand-sized material occurs nearly exclusively in shallow shoreline or nearshore settings (Sly, 1978). Ancient lakes show similar sediment distribution patterns (Picard and High, 1981). Most sand-sized material is contributed by river inflows, and the remaining portion must be accounted for by shoreline processes, including current or wave activity (Sly, 1978).

A mixture of predominantly fine sand, silt and clay with minor medium sand can occur within transition zones of marine beaches where few inorganic primary structures are found (Reineck and Singh, 1973). Although lower energy conditions are found in lakes, shoreline processes differ little from marine counterparts (Collinson, 1978). Hence, the sandstones might represent widespread migration of a transitional facies. If so, the interbedded muds were deposited after lake expansion when coarser material was not being supplied to the middle of the basin, and suspension settling of only silt and clay occurred. Bioturbation is present in both sandstone and mudstone, suggesting that dissolved oxygen levels were sufficent to support burrowing organisms in both nearshore and profundal environments.

The interbedded pattern of sand and mud deposition suggests expansion and contraction of the lake. These changes occurred in a fairly regular pattern, and mud-filled burrows on the upper surfaces of sandstones suggest a rapid return to a quiet, mud-dominated environment.

Sheet sands have been interpreted as dune deposits in some other Late Triassic lakes (e.g., Stewart et al., 1972). Unlike the massive sandstone in the Redonda, such sandstones contain planar cross-stratification and developed during periods when the lake was dry. The fabric of the massive sandstone supports an interpretation of subaqueous depostion, not migrating dunes. Although eolian processes could transport the sand-sized material (Allen, 1982), deposition apparently took place by suspension settling, because no traction structures developed.

Laminated sandstone is associated with massive sandstone (Sm + Sl) at Gallegos (Fig. 2), indicating the influence of waves. A high sandstone/mudstone ratio (1.2), medium sand size and inorganic primary structures support a proximal shoreline position for Gallegos in contrast to other locations where Sm + Mm occurs.

No clear evidence of a delta was found at Gallegos, but the evidence listed above suggests that it was near a sediment source. Axelsson (1967) described large deltas in lakes that lack Gilbert-style morphology and resemble marine lobate deltas. River plumes feeding into a lacustrine basin initially deposit the coarsest load in a lobate body at the river mouth and spread out laterally into the basin, where finer grains are deposited (Hakanson and Jansson, 1983). Massive sandstones at Gallegos may have accumulated as river plumes deposited fine-to-medium sand near a river mouth. Waves or currents in the lake could then generate the bedforms visible in the thickest sandstone.

Shoreline Environments

Shallowing upward or a change in lake chemistry occurs as offshore mudstone and sandstone (Sm + Mm) grade into silty, ripple-laminated micrites (Lr) at Bull Canyon, Luciano Mesa and Mesa Redonda. Environments dominated by terrigenous material are overlain by silty, sandy micrites (Lr) at Apache Canyon. Lithofacies associations present at these locations indicate considerable variability in local shoreline environments.

Carbonate Shorelines

At Bull Canyon, Mesa Redonda and Luciano Mesa, there is an upward gradation from offshore mudstone and shallow water sandstone into ripple-laminated micrites (Fig. 2). Shallowing upward is indicated as lithofacies lacking traction structures are overlain by carbonate units containing traction-generated bedforms. Rippled beds that contain a greater diversity of ichnofauna relative to underlying offshore environments also support a shallowing-upward model. Similar ichnofossils occur in Triassic calcareous lake deposits of Greenland (Bromley and Asgard, 1979), where they represent the shallowest lacustrine conditions (Clemmensen, 1979). These impure micrites were deposited as a lake became shallower, allowing waves to affect bottom sediments (Howard and Reineck, 1972). Wavy to lenticular bedding at Bull Canyon and Luciano Mesa suggest periods of still water when carbonate mud accumlated and coarser material (silt) was not supplied. Changing wave energy would be likely in a lake because winds might change periodically in both pattern and velocity. Periods of subaerial exposure created conditions for mudcracks to develop in ripple-laminated micrites. At Apache Canyon, wave energy was dissipated enough for fragile conchostracans to be preserved in mudcracked units.

At Mesa Redonda, sequences of asymmetrical and oscillation, ripple-laminated micrites (Lra + Lro + Lrm) containing mud cracks and separated by mudstone breaks represent depth or wave energy changes. The presence of ostracods (*Darwinula*) in these fluctuating shoreline facies (Kietzke, 1987, 1989) provides some constraint on maximum depth. Modern *Darwinula* are most abundant at a depth of 6 m and absent in depths greater than 9 m (McGregor, 1969). Based on this, deposition of mudstone interbeds took place in water probably no deeper than 6 m and possibly much shallower. Sparse, terrigenous grains were occasionally supplied to an environment dominated by carbonate mud precipitation. Tetrapod trackways indicate either subaerial conditions or water shallow enough for the animals to walk on the substrate.

The uppermost micrite at Mesa Redonda is capped by stratified sandstone, indicating that a carbonate mudflat was channelized by streams. Massive sandstone and mudstone (Sm + Mm) overlie the stratified sandstone, indicating a return to terrigenous deposition and a long term change in base level.

Nearshore carbonate precipitation implies that the shallow waters were well aerated and sufficiently hard to precipitate carbonate. High water temperatures also promote the precipitation of carbonate (Hakanson and Jansson, 1983). Shallow-water micrites deposited in the Redonda suggest that clastic supply decreased (Picard and High, 1981), and lake conditions changed to promote carbonate precipation. A change to drier climatic conditions might also account for such a decrease. Fluvial discharge would decrease and reduce the amount of terrigenous material to the depositional system, although some terrigenous material would be supplied during sporadic rainfall events (Bloom, 1978).

High and Low-energy Beaches

Planar, cross-stratified sandstones capped by laminated sandstones (Sp + Sl) overlie the ripple-laminated micrites at Luciano Mesa, and document a change from carbonate to terrigenous environments. Four beach sequences can be seen (Fig. 2), but the first sequence is the best developed and suggests that, at times, fairly high energy levels existed. The gradation from megaripple bedforms into parallel laminations suggests an increase in hydrodynamic energy upward during deposition of the first sequence (Fig. 2).

Along marine shorelines, prograding beaches can deposit a specific vertical sequence of bedforms that develop in response to waves approaching and breaking along the shoreline (Clifton et al., 1971). The sequence on a marine beach would include (from bottom up) asymmetrical ripples, megaripples, planar beds and parallel laminations. In the Redonda Formation at Luciano Mesa, only megaripple bedforms and parallel laminations occur in the first beach unit. The comparison to marine environments is legitimate because the processes and products of clastic lake shoreline environments differ only by size of bedforms, sorting and maturity, which are reflections of lower energy levels along lacustrine beaches (Collinson, 1978). The significance of this beach deposit is the presence of bedforms that indicate high-energy shoreline processes. Because small fluctuations of lake level can rapidly shift shoreline environments (Picard and High, 1981), the complete prograding beach sequence described from marine environments is not present in the beach sequences at Luciano Mesa. The megarippled interval is absent in the upper two beach sequences, and ripple laminations are dominant. Grain size varies

FIGURE 6. Panoramic drawing of the Gilbert delta at San Jon Hill. Sandstone interval is ~ 5 m thick. Units 1-3 are individual delta fronts, and unit 4 is the topset interval.

little between individual sequences, suggesting an upward trend of decreasing hydrodynamic energy or increasing depth during deposition of these beach sequences.

Ripple-laminated sandstones with small scours overlie the uppermost beach sequence at Luciano Mesa, indicating a long term base-level change. Unlike the base-level change at Mesa Redonda, the environment at Luciano Mesa continued to be influenced by hydrodynamic processes. At times, vigorous current action generated higher transport and erosional energy conditions. Unusually large storms could have created scours; when flow velocities returned to normal, scours were infilled and capped with ripple-laminated beds.

Deltaic Environments

Deltaic processes controlled or influenced deposition of the Redonda Formation at San Jon Hill and Apache Canyon, the most eastern localities. Like shoreline environments, variability in these deltaic environments demonstrate complexity within the Redonda depositional system.

Gilbert Delta

A well-developed Gilbert delta was exposed at San Jon Hill (Figs. 5-7) but has been destroyed by recent highway construction. Its structure is tripartite, consisting of topsets, foresets (delta front) and bottomsets. The vertical changes expected by progradation of this delta type include coarsening upward of grain size and variation in bed geometry (Gilbert, 1885; Barrell, 1912; Stanley and Surdam, 1978; Postma, 1985). At San Jon Hill, the Gilbert style delta clearly prograded into a lacustrine basin, although it deviates from an ideal profile.

The lowest interval at San Jon Hill consists of ripple-laminated mudstone and asymmetrical, ripple-laminated micrites (Ml + Lra) and represents the bottomset portion of the delta, which has been modified by unidirectional flow. Ripple-laminated mudstones have been described from lake deposits associated with delta building in modern lakes (Strum and Matter, 1978), from shallow water environments in Green Lake, New York (Ludlam, 1974) and from profundal lake-slope deposits in Virginia (Hentz, 1985). Because this interval grades vertically and laterally (landward) into coarser sediments, a deltaic interpretation of the deposits at San Jon Hill is supported. Because a thick (13 m) interval is present, some of the lower deposits may represent lake environments that were indirectly influenced by the prograding delta.

A lenticular bed of sandy micrite occurs high in the interval of mudstones (Fig. 6) and may represent the onset of deposition dominated by deltaic processes. This bed is interpreted as a channelizing and scouring event that occurred within bottomset beds. As flow spilled out over the channel, waning flow of the event created the vertical sequence.

Three progradational packages of foreset beds overlie the bottomset beds (Fig. 6). Very finely to finely disrupted, and horizontal and ripple-laminated sandstone makes up the first prograding foreset. Medium-to-coarse, normally graded and horizontal and ripple-laminated sandstone was deposited as the delta prograded. Normally graded sandstone developed in response to changing fluvial discharge, possibly due to seasonally higher discharge. High rates of deposition, unidirectional flow and turbidites are consistent with delta building in lakes.

The geometry of the beds suggests prograding delta fronts building downflow into a lake basin. Slumping of a foreset bed occurred in unit 2 (Fig. 6), and horizontally laminated beds were folded. Similar traction structures on the delta front of lacustrine Gilbert deltas have been identified by other workers (e.g., Joplin and Walker, 1968; Stanley and Surdam, 1978). Accretion took place along inclined (10-20°) delta front surfaces. The geometry of these beds illustrates the expected inclined foreset surfaces of the ideal model, but dips from outcrop measurements do not represent the true angle, as exposure was at some angle to true dip.

The topset portion of the delta is unit 4 (Fig. 6). Lenticular bodies suggest stream channel influence characteristic of the ideal Gilbert delta model. No evidence of base-level changes can be seen. Thin mudstone interbeds within the graded portions may represent typical waning flow at the end of the depositional event. An estimate of water depth can be made from foreset thickness (Stanley and Surdam, 1978). At San Jon Hill, tracing one foreset bed to its bottomset suggests that water depth could have been as much as 6 m.

A number of Gilbert delta lobes prograded into the basin at San Jon Hill. A line drawing (Fig. 7) of a cliff exposure (just east of locality) illustrates a complex internal geometry with opposing dip directions. This geometry suggests that delta fronts prograded from various directions into the depositional system.

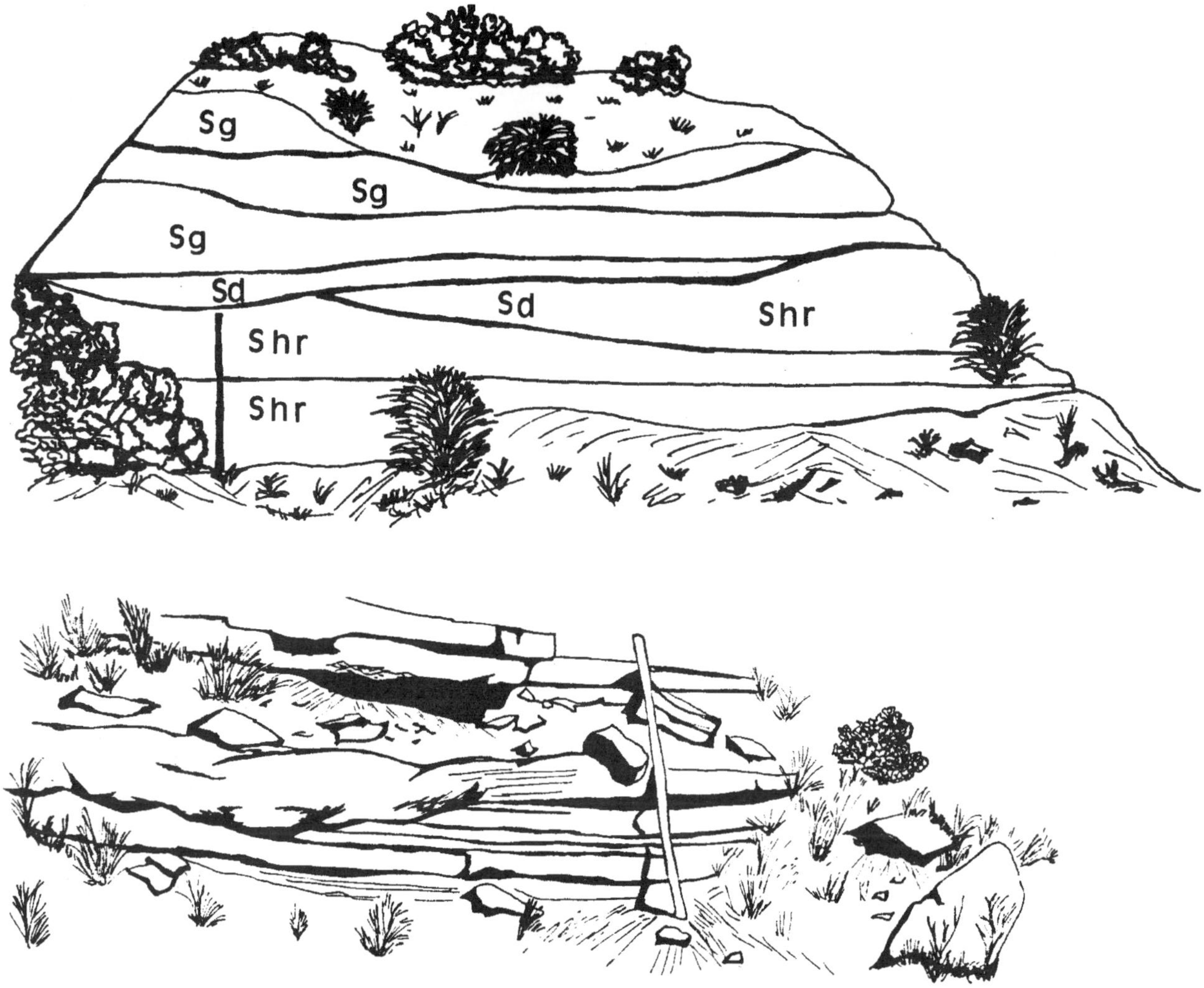

FIGURE 7. Top, the Gilbert delta front at San Jon Hill, illustrating the internal complexity of the prograding delta front. Bottom, drawing of the sublacustrine fan at Apache Canyon. Note small channel cutting into horizontally bedded units in sublacustrine fan system. Sticks are 1.5 m long.

Delta-front Fan

Sandstones and conglomerates interbedded with mudstones overlie the interval of massive sandstone and mudstone at Apache Canyon (Fig. 2). In contrast to San Jon Hill, deltaic processes are not clearly defined at Apache Canyon, but sedimentary processes are consistent with the environment of a delta-front fan. Alluvial fan deposition is also a possible source of these sediments, but delta-front fans can develop in lakes (Nelson, 1967; Strum, 1975), where sublacustrine fans are built predominantly by episodic high and low density turbidity currents (Rupke, 1978). Channel, levee and interchannel deposits of the fan system (Strum, 1975) produce interbedded sandstone and mudstone. No well-defined channels were identified within the interval at Apache Canyon. However, there are sandstones that truncate underlying interbedded sandstone and mudstone, suggesting channel deposits (Fig. 7). Localized debris flows carrying intraformational clasts deposited conglomerates on gently dipping (3-5°) fan surfaces. Deltaic processes ceased at the locality, and infilling of the basin continued with the deposition of silty and sandy micrites (Lr + Mm) in a carbonate-dominated shoreline. This change suggests that the supply of terrigenous sediment was cut off or at least slowed dramatically.

Fluvial Environments

Fluvial deposits overlie carbonate-dominated shoreline deposits at Apache Canyon (Fig. 2). Overbank fines and channel sands were deposited (Mv + Shr + Sp) after shorelines developed. Mud and fine sand accumulated in distal and proximal floodplain environments, and planar-stratified sandstone was deposited in channels by migrating bars. Carbonate mud accumulated, and conchostracans lived in floodplain lakes that were influenced by fluvial processes. Sand was occasionally deposited in small delta lobes within floodplain lakes as splays from channels developed during flood events. The sand component of these lobes and the channel sand are similar. Oolites deposited in mud were probably carried into the floodplain environment by eolian processes.

CONTROLS ON SEDIMENTATION

The offshore, shoreline, deltaic, fan and associated fluvial environments of the Redonda Formation comprise a complex, lacustrine depositional system (Fig. 8A). Interpretation of this system requires consideration of : (1) tectonic influences on basin formation and the locus of deposition, (2) lacustrine and fluvial processes responsible for the variations observed in the lithofacies and (3) climatic controls on sedimentation.

Tectonic Controls

The change from fluvial and floodplain deposition of the Bull Canyon Formation to the lacustrine and lacustrine margin deposition of the Redonda Formation suggests that either climatic or tectonic events, or both, were responsible. Climatic change alone cannot account for the change from through-flowing fluvial to essentially impounded lacustrine deposition. Without a closed basin, if only rainfall increased, rivers would continue to flow down gradient. Reconstruction of Redonda lacustrine environments suggests a depositional system of small and moderate sized lakes and lake margin environments over an area of at least 5000 km^2 (Fig. 8A).

Tectonic change that involved the alteration of drainage patterns is the most plausible explanation of the Redonda lacustrine system. Localities where shoreline environments existed during most of Redonda deposition include San Jon Hill, Apache Canyon and Gallegos, indicating that these locations were near the margins of the Redonda depositional basin. These locations show some correspondence to the basement highs that define the Tucumcari basin (Fig. 8B), suggesting that highlands defining the basin may have been inherited from the Paleozoic basin.

The tectonic events that may have reactivated the Paleozoic basinal topography (and possibly triggered alteration of drainage patterns) are speculative, but possible mechanisms include large-scale tectonic warping associated with rifting along the Gulf Coast, as suggested by McGowen et al. (1979, 1983), and development of the Chinle back-arc basin (Dickinson, 1981; Lawton, 1994). A more local contributing mechanism could have been salt dissolution of Permian evaporites in the subsurface.

Controls on Lacustrine and Fluvial Lithofacies

Once Redonda deposition commenced, physical controls determined by the waterbodies themsleves affected sedimentation. The variablity of environments at specific geographic locations in the lacustrine system can be discussed with emphasis on some physical lake processes.

Proximal Redonda deposition occurred near the margins of the Tucumcari basin, suggesting that the Redonda lacustrine system may have been at least 80 km wide. Evidence of a high energy beach at Luciano Mesa suggests that a lake there was capable of producing waves that could generate megaripples and a swash zone. Wave- or current-generated bedforms in predominantly massive sandstone at Gallegos indicate that this locality was also positioned where these processes could not produce sedimentary structures. In contrast to northern and western localities, the easternmost localities at San Jon Hill and Apache Canyon were sites where river-dominated deltas and delta-fans were deposited. The existence of shorelines dominated by waves or currents in one part of the basin and fluvially dominated shorelines in another, suggests complex patterns of sediment influx.

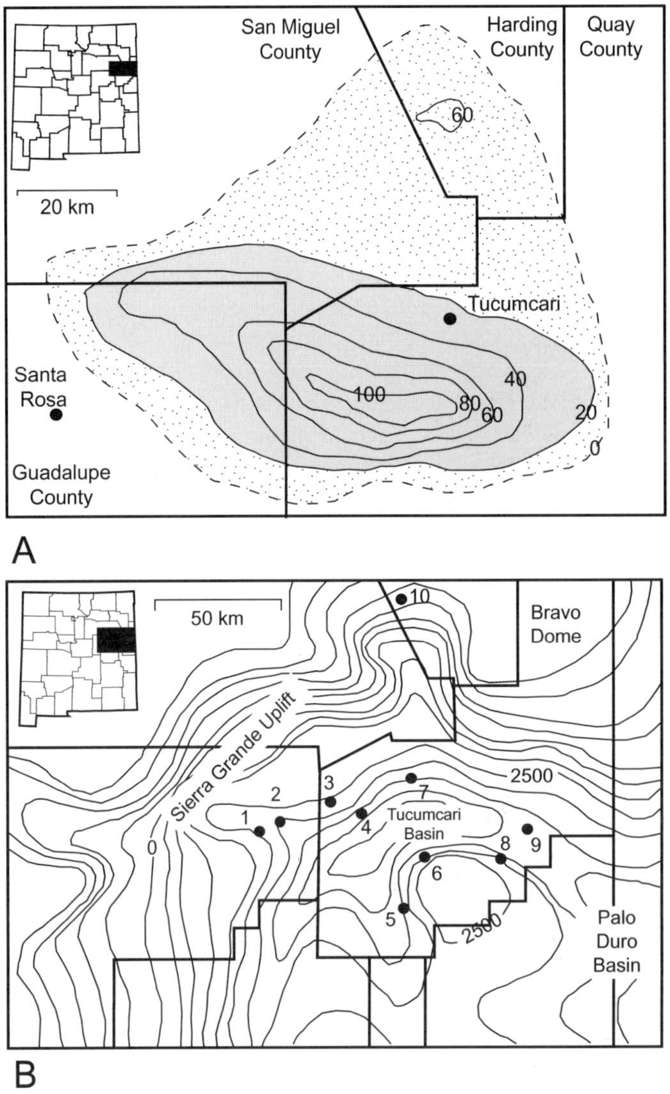

FIGURE 8. A. Isopach map of Redonda Formation (20 ft contour intervals) and outline of the minimum extent of the Redonda depositional system based largely on zero isopach line. B. Location of Redonda measured sections (Fig. 2) with respect to basement highs that define the late Paleozoic Tucumcari basin.

Orientation of the Redonda lacustrine system to prevailing wind patterns may have been an important factor in sedimentation (Sly, 1978; Fox and Davis, 1976), and Luciano Mesa may have been positioned along a shoreline that received much of the force of prevailing winds. The Redonda lake basin was situated between 9-11° north paleolatitude, which would place it in the zone of trade winds (Lamb, 1972). With prevailing winds blowing from the east-northeast, the eastern shorelines of lakes would have been sheltered.

Other shorelines would receive the effects of wind-generated waves within the lake. The sediments at Luciano Mesa preserve evidence of strong wave activity, but the location does not correspond to any basement high defining the Tucumcari basin. However, if basement highs controlled southern and western lake basin edges, it could represent an advancing southwestern or western shoreline taking the force of easterly winds. Northern shorelines (Gallegos) would have received intermediate wave energy. Because wind patterns would have been more complex than presented here, the model is a simplification. However, a low-energy, eastern-basin edge where deltas were river-dominated and only low-energy beaches developed is consistent with a northeasterly-easterly wind model.

The Gallegos locality may have been similarly located with respect to prevailing winds. If the massive sandstone was windblown, thickest accumulations would be along northern and eastern shorelines, consistent for the Gallegos locality but not for San Jon Hill, where no massive sandstone occurs. This does not rule out an eolian origin, because some massive sand accumulated at Apache Canyon. However, most sands are carried into lake basins by rivers (Sly, 1978), and there may have been a sediment source (river mouth?) near Gallegos. Rivers dominated delta building along the eastern basin edge by forming Gilbert deltas at San Jon Hill and sublacustrine fans at Apache Canyon. This contrasts with Gallegos, where river plumes may have carried coarse material into the basin. Fluvial systems that supplied sediment to the depositional system were carrying grains derived from sedimentary sources. Abundant delta sediments at San Jon Hill and fluvial systems to the north (Gallegos) suggest greater fluvial input from the northeastern and eastern sides of the depositional system. The San Jon Hill deltas were river dominated and built by underflows depositing normally graded beds. In contrast, the massive sandstone at Gallegos may have accumulated as sand from a discharging river plume settled by suspension. Fluvial response to lake water differed at these localities, because river water density differed, or lake waters changed through time.

Climate Controls

Climate variables can influence a lacustrine depositional system in many ways, including lake level, chemistry and stratification (e.g., Sly, 1978; Collinson, 1978; Eugster and Hardie, 1978). Changes in rainfall could produce lake expansion and contraction, changing base level and bringing deeper water environments into a particular geographic position within a lake. The massive mudstones of the Redonda Formation probably represent periods when lake levels were relatively high at any given locality. Decreasing rainfall would cause contraction of the lake basins and bring transitional environments into a lake location that had previously been farther from shore. Hence, changes in precipitation could indirectly produce a series of interbedded mudstones and sandstones at a given location within a lake.

Climate factors may also have promoted the precipitation of carbonate at different times during a lake's history. Shorelines are dominated by carbonate precipitation at Mesa Redonda. The bulk of primary lacustrine carbonates are inorganic chemical precipitates that are directly controlled by the physical factors of temperature and evaporation concentration (Kelts and Hsü, 1978). Although shallow water favors carbonate precipitation (Picard and High, 1981), the occurrence of carbonate-dominated shorelines overlain by clastic-dominated shorelines at Luciano Mesa indicates periods of lake history when carbonate precipitation was favored, regardless of a shallow-water setting. At Mesa Redonda, terrigenous intervals occur above and below a carbonate-dominated interval, suggesting clastic and chemical cycles of lake deposition.

Van Houten (1964) proposed that detrital and chemical lake cycles within the Lockatong Formation (Newark Supergroup, eastern United States) were a manifestation of Milankovitch cyclicity. Detrital cycles were deposited when the lakes were at their maximum level and had outlets. Chemical cycles record periods of decreased precipitation, contraction of lake levels and changes in lake chemistry through evaportion and concentration of salts. More recent work using Fourier analysis (e.g., Olsen and Kent, 1996) documents periodicities of Newark lake cycles that correspond to those predicted by orbital forcing. In the Newark basin, the Lockatong and Passaic formations span late Carnian to earliest Jurassic time, overlapping the late Norian-Rhaetian? when Redonda sediments were accumulating. If orbital forcing mechanisms controlled sedimentation of the Newark Supergroup lakes, some record could be expected in other lakes existing within similar climatic belts, such as the lakes of the Redonda depositional system.

In the Redonda Formation, evidence of base-level changes or stratification changes, possibly related to climate, are best observed in the cyclic alternation of offshore mud and transitional sand. Changes in heating insensity of northern continental masses caused directly by the precession of the equinoxes (20,000 year cycle) alter the intensity of low pressure cells that drive monsoonal winds and control rainfall at any given locality (Rossignol-Strick, 1985). Alternatively, the cyclic pattern of sandstone and mudstone deposition could also be a reflection of changes in river input and density stratification within a lake. If sediment plumes transported the fine sand now found in the massive sandstones, then climatic forcing would directly regulate this pattern. Chemical changes occur in a lake by evaporation, influence lake stratification and therefore fluvial response to a basin (Wright, 1977; Collinson, 1978; Hakanson and Jansson, 1983).

A detailed analysis of cycles like that done for the Lockatong and Passaic formations, is not possible in the Redonda Formation because the lack of control on time and sedimentation rates prohibits the application of Fourier analysis. The Redonda is very thin compared to the many km thick lake sediments of the Newark Supergroup, which record all frequencies of cycles predicted by orbital forcing. However, it would not be unreasonable for the effects of 20,000 and possibly 100,000 year cycles to be recorded in 92 m of sediment in the Redonda Formation. In fact, two different frequencies represented by thin sandstone beds deposited between four massive sandstones are found at Luciano Mesa (Fig. 2)—a pattern similar to that found in the Lockatong Formation and in Milankovitch bedding cycles.

CONCLUSIONS

A large array of relatively shallow lakes and lake margin environments covered east-central New Mexico during part of the latest Triassic. At times, sedimentary terranes contributed terrigenous detritus to this lacustrine depositional system, while at other times the precipitation of calcareous mud was the dominant sedimentary process. The northern and eastern margins of the depositional system were located near the Gallegos, Apache Canyon and San Jon Hill localities, which correspond to basement highs of the Paleozoic Tucumcari basin. This correspondence suggests that these structural features influenced sedimentation during Redonda deposition.

Variability of delta and shoreline lithofacies provides evidence for the distribution of wave energy and the effects of paleowinds. Higher wave energy existed along the northern and possibly southern or southwestern margins of the depositional system, where delta morphology and beach sediments were strongly influenced by lake processes. Deltas along the eastern edge of the basin were dominated by fluvial processes, suggesting lower wave energy. Terrigenous and carbonate cycles indicate periods in Redonda lake history when it was dominated by either physical or chemical processes under the strong influence of climatic change, especially chemical precipitation, but also changes in temperature and evaporation. Evidence for climatic control of deposition during the Late Triassic in New Mexico lends support to evidence of climatic forcing in the Newark basin to the east and suggests that orbital climatic forcing influenced Late Triassic lacustrine deposition over a broad geographic area.

ACKNOWLEDGMENTS

Comments and advice from Roger Anderson, Larry Smith and Robyn Wright Dunbar are appreciated. The Student Research Allocations Committee of the University of New Mexico supported part of this research. A Heckert, A. Hunt and L. Tanner provided helpful reviews of the manuscript.

REFERENCES

Allen, J. R. L., 1982, Sedimentary Structures, Their Character and Physical Basis, Vol. 1 and 2, Elsevier Developments in Sedimentology, v. 30A-30B.

Axelsson, V., 1967, The Laiture delta, a study of deltaic morphology and processes: Geografiska Annaler, v. 45A, p. 1-127.

Barrell, J., 1912, Criteria for the recognition of ancient delta deposits, Geological Society of American Bulletin, v. 23, p. 377-446.

Birkeland, P., 1974, Pedology, Weathering and Geomorphological Research. Oxford, Oxford University Press, 284 p.

Blakey, R. C., and Gubitosa, R., 1983, Late Triassic paleogeography and depositional history of the Chinle Formation, southern Utah and northern Arizona, *in* Reynolds, M. W., and Dolly, E. D., eds., Mesozoic paleogeography of the west-central United States: Denver, Rocky Mountain Section SEPM , p. 57-76.

Blatt, H., Middleton, G., and Murray, R., 1980, Origin of Sedimentary Rocks. Englewood Cliffs, Prentice-Hall, 782 p.

Blodgett, R. H., 1984, Nonmarine depositional environments and paleosol development in the Upper Triassic Dolores Formation, southwestern Colorado, in Brew, D. C., ed., Field Trip Guidebook, 37th Annual Meeting, Rocky Mountain Section, Geological Society of America, Durango, Colorado, p. 46-92.

Bloom, A., 1978, Geomorphology. Englewood Cliffs, Prentice-Hall, 510 p.

Bown, T., and Kraus, M., 1987, Integration of channel and floodplain suites. I. Developmental sequence and lateral relations of alluvial paleosols: Journal of Sedimentary Petrology, v. 57, p. 587-601.

Bromley, R., and Asgaard, U., 1979, Triassic freshwater ichnocoenoses from Carlsberg Fjord, east Greenland: Palaeogeography, Palaeoclimatology, Palaeoecology, v. 28, p. 39-80.

Clemmensen, L., 1979, Lacustrine facies and stromatolites from the Middle Triassic of east Greenland: Journal of Sedimentary Petrology, v. 48, p. 1111-1128.

Clifton, H. E., Hunter, R. E., and Phillips, R. L., 1971, Depositional structures and processes in the non-barred, high energy shoreline: Journal of Sedimentary Petrology, v. 41, p. 651-670.

Collinson, J., 1978, Lakes; in Reading, H. G., ed., Sedimentary Environments and Facies: Elsevier, p. 61.

Davis, R., 1983, Depositional Systems, a Genetic Approach to Sedimentary Geology. Englewood Cliffs, Prentice-Hall, 669 p.

Dean, W., 1981, Carbonate minerals and organic matter in sediments of modern north temperate hard-water lakes, in recent and ancient nonmarine depositional environments: Models for exploration: SEPM Special Publication 31, p. 349.

De Raaf, J., Boersma, J. and Van Gelder, A., 1977, Wave generated structures and sequences from a shallow marine succession, Lower Carboniferous, County Cork, Ireland: Sedimentology, v. 24, p. 451-483.

Dickinson, W., 1981, Plate tectonic evolution of the southern Cordillera: Arizona Geological Society Digest, v. 14, p. 113-135.

Dobrovolny, E., Summerson, G., and Bates, R., 1946, Geology of northwestern Quay County, New Mexico: U. S. Geological Survey, Oil and Gas Investigations Preliminary Map 62, sheet 1.

Dubiel, R. F., 1989, Depositional and climatic setting of the Upper Triassic Chinle Formation, Colorado Plateau, *in* Lucas, S. G., and Hunt, A. P., eds., Dawn of the Age of Dinosaurs in the American Southwest: Albuquerque, New Mexico Museum of Natural History, p. 171-187.

Ekdale, A., Bromley, R., and Pemberton, S., 1984, Ichnology, trace fossils in sedimentology and stratigraphy: SEPM Short Course 15, 317 p.

Eugster, H. R., and Hardie, L. A., 1978, Sedimentation in an ancient playa lake complex: The Wilkins Peak Member of the Green River Formation of Wyoming: Geological Society of America Bulletin, v. 86, p. 319-334.

Fisher, R., 1971, Features of coarse-grained, high-concentration fluids and their deposits: Journal of Sedimentary Petrology, v. 41, p. 916-927.

Fox, W., and Davis, R., 1976, Weather patterns and coastal processes, in beach and nearshore sedimentation: SEPM Special Publication 24, p. 1-23.

Gilbert, G. K., 1885, The topographic features of lake shores: U. S. Geological Survey, Annual Report 5, p. 69-123.

Granata, G. E., 1981, Regional sedimentation of the Late Triassic Dockum Group, West Texas and eastern New Mexico [M.A. thesis]: Austin, University of Texas, 199 p.

Gregory, J. T., 1972, Vertebrate faunas of the Dockum Group, Triassic, eastern New Mexico and West Texas: New Mexico Geological Society, 23rd Field Conference, Guidebook, p. 120-123.

Griggs, R. L., and Read, C. B., 1959, Revisions in stratigraphic nomenclature in Tucumcari-Sabinoso area, northeastern New Mexico: American Association of Petroleum Geologists, Bulletin, v. 43, p. 2003-2007.

Hakanson, L., and Jansson, M., 1983, Principles of Lake Sedimentology. Berlin, Springer-Verlag, 316 p.

Harms, J. C., Southard, J. B., and Walker, R. G., 1982, Structures and sequences in clastic rocks: SEPM Short Course 9, 112 p.

Hein, F. J., 1983, Depositional mechanisms of deep-sea clastic sediments, Cap Enrage Formation, Quebec: Canadian Journal of Earth Sciences, v. 19, p. 267-287.

Hentz, T., 1985, Early Jurassic sedimentation of a rift-valley lake: Culpepper basin, northern Virginia: Geological Society of America Bulletin, v. 96, p. 92-107.

Hester, P. M., 1988, Depositional environments of a Late Triassic lake, east-central New Mexico [M.S. thesis]: Albuquerque, University of New Mexico, 152 p.

Howard, J. D., and Reineck, H. E., 1972, Physical and biogenic sedimentary structures of the nearshore shelf: Senckenbergiana Maritim, v. 4, p. 81-124.

Hunt, A P., 1994, Vertebrate paleontology and biostratigraphy of the Bull Canyon Formation (Chinle Group: Norian), east-central New Mexico with revisions of the families Metoposauridae (Amphibia: Temnospondyli) and Parasuchidae (Reptilia: Archosauria) [Ph.D. dissertation]: Albuquerque, University of New Mexico, 403 p.

Hunt, A. P., Lockley, M. G., and Lucas, S. G., 1993, Vertebrate and invertebrate tracks and trackways from Upper Triassic strata of the Tucumcari basin, east-central New Mexico: New Mexico Museum of Natural History and Science, Bulletin 3, p. 199-201.

Hunt, A. P., Lucas, S. G., and Kietzke, K. K., 1989, Dinosaur footprints from the Redonda Member of the Chinle Formation (Upper Triassic), east-central New Mexico, *in* Gillette, D. D. and Lockley, M. G., eds., Dinosaur tracks and traces: Cambridge, Cambridge University Press, p. 277-280.

Hunt, A. P., Lucas, S. G., Lockley, M. G. and Heckert, A. B., 2000, Occurrence of the dinosaurian ichnogenus *Grallator* in the Redonda Formation (Upper Triassic: Norian) of eastern New Mexico: New Mexico Museum of Natural History and Science Bulletin 17, p. 39-41.

Jenkyns, H. C., 1978, Pelagic environments; *in* Reading, H. G., ed., Sedimentary Environments and Facies: Amsterdam, Elsevier, 557 p.

Johnson, T. C., 1984, Sedimentation in large lakes: Annual Review of Earth and Planetary Sciences, v. 12, p. 179-204.

Joplin, A. V., and Walker, R. G., 1968, Morphology and origin of ripple-drift cross-lamination, with examples from the Pleistocene of Massachusetts: Journal of Sedimentary Petrology, v. 38, p. 971-984.

Kelley, V. C., 1972, Triassic rocks of the Santa Rosa country: New Mexico Geological Society 23rd Field Conference, Guidebook, p. 84-90.

Kelts, K. and Hsü, K., 1978, Freshwater carbonate sedimentation; in Lerman, ed., Lakes, chemistry, geology and physics: Berlin, Springer-Verlag, p. 295-231.

Kietzke, K. K., 1987, Calcareous microfossils from the upper Triassic of northeastern New Mexico: New Mexico Geological Society, 38th Field Conference, Guidebook, p. 119-126.

Kietzke, K. K., 1989, Calcareous microfossils from the Triassic of the southwestern United States; *in* Lucas, S. G., and Hunt, A. P., eds., Dawn of the Age of Dinosaurs in the American Southwest: Albuquerque, New Mexico Museum of Natural History, p. 223-232.

Kraus, M., 1987, Integretation of channel and floodplain suites II. vertical relations of alluvial paleosols: Journal of Sedimentary Petrology, v. 57, p. 602-612.

Kraus, M., and Middleton, L., 1987, Dissected paleotopography and base level changes in a Triassic fluvial sequence: Geology, v. 15, P. 18-21.

Lamb, H. H., 1972, Climate: Present, Past and Future, v. 1, Fundamentals and Climate Now. London, Methuen & Co. Ltd., 610 p.

Lawton, T. F., 1994, Tectonic setting of Mesozoic sedimentary basins, Rocky Mountain region, United States, *in* Caputo, M. V., Peterson, J. A., and Franczyk, K. J., eds., Mesozoic Systems of the Rocky Mountain Region: Denver, RMS-SEPM, p. 1-25.

Lockley, M. G., Lucas, S. G., and Hunt, A. P., 2000, Dinosaur tracksites in New Mexico: A review: New Mexico Museum of Natural History and Science, Bulletin 17, p. 9-16.

Lowe, D. R., 1982; Sediment gravity flows, II, Depositional models with special reference to the Deposits of High-Density turbidity currents. Journal of Sedimentary Pretology, V. 52, pp. 279-297.

Lucas, S. G., 1993, The Chinle Group: Revised stratigraphy and biochronology of Upper Triassic nonmarine strata in the western United States: Museum of Northern Arizona, Bulletin 59, p. 27-50.

Lucas, S. G., 1998, Global Triassic tetrapod biostratigraphy and biochronology: Palaeogeography, Palaeoclimatology, Palaeoecology, v. 143, p. 347-384.

Lucas, S. G., and Hunt, A. P., 1989, Revised Triassic stratigraphy, Tucumcari basin, east-central New Mexico, *in* Lucas, S. G., and Hunt, A. P., eds., Dawn of the Age of Dinosaurs in the American Southwest: Albuquerque, New Mexico Museum of Natural History, p. 150-170.

Lucas, S. G., and Hunt, A. P., 1993, Tetrapod biochronology of the Chinle Group (Upper Triassic), western United States: New Mexico Museum of Natural History and Science, Bulletin 3, p. 327-329.

Lucas, S. G., Hunt, A. P., and Morales, M., 1985, Stratigraphic nomenclature and correlation of Triassic rocks of East-Central New Mexico: a preliminary report, in New Mexico Geological Society, Guidebook 36, pp. 171-184.

Ludlam, S. D., 1974, Fayetteville Green Lake, New York, the role of turbidity currents in lake sedimentation: Limnology and Oceanography, v. 19, p. 656-664.

McGowan, J., Granata, G., and Seni, S., 1979, Depositional framework of the lower Dockum Group (Triassic) Texas Panhandle: Texas Bureau of Economic Geology, Report of Investigations 97, 60 p.

McGowen, J., Granata, G., and Seni, S., 1983; Depositional setting of the Triassic Dockm Group, Texas Panhandle and eastern New Mexico; *in* Reynolds, M. W. and Dolly, eds., Mesozoic paleogeography of the west-central United States: Denver, Rocky Mountain Section SEPM, p. 13-38.

McGregor, D. L., 1969, The reproduction potential, life history and parasitism of the freshwater ostracod *Darwinula stevensoni* (Brady and Robertson), *in* Neale, J. W., ed., The taxonomy, morphology and ecology of recent ostracoda: Edinburgh, Oliver and Boyd, p. 195-221.

Miall, A. D., 1984, Principles of Sedimentary Basin Analysis. Berlin, Springer-Verlag, 490 p.

Middleton, G. V., ed., 1965, Primary Sedimentary Structures and their Hydrodynamic Interpretation—a Symposium: SEPM Special Publication 12, 265 p.

Middleton, G. V., and Hampton, M. A., 1973, Sediment gravity flows: Mechanics of flow and deposition: SEPM Pacific Section Short Course, 150 p.

Middleton, G. V., and Hampton, M. A., 1976, Subaqueous sediment transport and deposition by sediment gravity flows, *in* Stanley, D. J., and Swift, D. J. P., eds., Marine Sediment Transport and Environmental Management: New York, John Wiley, p. 197-218.

Nelson, B. W., 1967, Sediments of Crater Lake, Oregon: Geological Society of America Bulletin, v. 78, p. 833-848.

Nemac, W., and Steel, R. J., 1984, Alluvial and coastal conglomerates: Their significant features and some comments of gravelly mass-flow deposits, *in* Koster, E., ed., Sedimentology of Gravels: Canadian Society of Petroleum Geologists Memoir 10, p. 1-32.

Nemac, W., Steel, R. J., Porebski, S. J., and Sprinnangr, A., 1984, Domba conglomerate, Devonian, Norway: Process and lateral variability in a mass flow-dominated lacustrine fandelta, *in* Koster, E., ed., Sedimentology of Gravels: Canadian Society of Petroleum Geologists Memoir, p. 295-320.

Newell, A. J., 1993, Depositional environment of the Late Triassic Bull Canyon Formation (New Mexico): implications for "Dockum Formation" paleogeography: New Mexico Museum of Natural History and Science, Bulletin 3, p. 359-368.

Olsen, P., and Kent, D. V., 1996, Milankovitch climate forcing in the tropics of Pangaea during the Late Triassic: Palaeogeography, Palaeoclimatology, Palaeoecology, v. 122, p. 1-26.

Pharo, C. H., and Carmack, E. C., 1979, Sedimentation processes in a short residence time intermontane lake, Kamploos Lake, British Columbia: Sedimentology, v. 26, p. 523-541.

Picard, M. D., and High, L. R., 1972, Criteria for recognizing lacustrine rocks: SEPM Special Publication 16, p.

Picard, M. D., and High, L. R., 1981, Physical stratigraphy of ancient lacustrine deposits: SEPM Special Publication 31, p.

Postma, G., 1985, Slumps and their deposits in fan delta fronts: Geology, v. 12, p. 27-30.

Reineck, H. E., and Singh, I. B., 1973, Depositional Sedimentary Environments—With Reference to Terrigenous Clastics. Berlin, Springer-Verlag, 439 p.

Rodine, J. D., 1976, The ability of debris, heavily freighted with coarse clastic materials, to flow on gentle slopes: Sedimentology, v. 23, p. 213-234.

Rossignol-Strick, M., 1985, Mediterranean Quaternary sapropels, an immediate response of the African monsoon to variation of insolation: Paleogeography, Paleoclimatology, Paleoecology, v. 49, p. 237-257.

Rupke, N. A., 1978, Deep clastic seas; *in* Reading, H., ed., Sedimentary Environments and Facies. Amsterdam, Elsevier, p. 372-415.

Serruya, C., and Pollingher, U., 1983, Lakes of the warm belt. Cambridge, Cambridge University Press, 569 p.

Sly, P. G., 1978, Sedimentary processes in lakes, *in* Lerman, ed., Lakes, Chemistry, Geology, Physics. Berlin, Springer-Verlag, p. 65-85.

Stanley, K. O., and Surdam, R. C., 1978, Sedimentation on the front of Eocene Gilbert-type deltas, Washakie basin, Wyoming: Journal of Sedimentary Petrology, v. 48, p. 557-573.

Steiner, M., and Lucas, S. G., 2000, Paleomagnetism of the Late Triassic Petrified Forest Formation, Chinle Group, western United States: Further evidence of "large" rotation of the Colorado Plateau: Journal of Geophysical Research, v. 105, p. 25,791-25,808.

Stewart, J. H., Poole, F. G., and Wilson, R. F., 1972, Stratigraphy and origin of the Chinle Formation and related Upper Triassic strata in the Colorado Plateau region: U.S. Geological Survey, Professional Paper 690, 336 p.

Strum, M., 1975, Depositional and erosional sedimentary features in a turbidity current controlled basin (Lake Brienz): 9th International Congress of Sedimentologists, Nice, 1975, Theme 5 (2), p. 385-390.

Strum, M., 1979, Origin and composition of clastic varvcs, in Schluchter, M., ed., Moraines and Varves: Proceedings INQUA Symposium: Rotterdam, Balkema, p. 281-285.

Strum, M., and Matter, A., 1978, Turbidities and varves in Lake Brienz (Switzerland): Deposition of clastic detritus by density currents, *in* Matter, A., and Tucker, M. E., eds., Modern and Ancient Lake Sediments: International Association of Sedimentologists Special Publication 2, p. 147-168.

Tanner, L. H., 2000, Palustrine-lacustrine and alluvial facies of the (Norian) Owl Rock Formation (Chinle Group), Four Corners region, southwestern U.S.A.: Implications for Late Triassic paleoclimate: Journal of Sedimentary Research, v. 70, p. 1280-1289.

Van Houten, F. B., 1964, Cyclic lake sedimentation, Upper Triassic Lockatong Formation, central New Jersey and adjacent Pennsylvania, *in* Merriam, D. F., ed., Symposium on Cyclic Sedimentation: Kansas Geological Survey Bulletin 169, p. 496-531.

Walker, R. G., 1979,Facies models 8, turbidities and associated coarse clastic deposits, *in* Walker, R. G.,ed., Facies Models: Geological Society of Canada, p. 91-103.

Wanek, A. A., 1962, Reconnaissance geologic map of parts of Harding, San Miguel and Mora Counties, New Mexico: U.S.Geological Survey, Oil And Gas Investigations Map OM-208.

Webb, J. A., 1979, A reappraisal of the paleoecology of conchostracans: Neues Jahrbuch fur Geologie und Paläontologie Abhandlungen, v. 158, p. 259-275.

Wright, L. D., 1977, Sediment transport and deposition at river mouths: A synthesis: Geological Society of America Bulletin, v. 88, p. 857-868.

A GIANT PHYTOSAUR (REPTILIA: ARCHOSAURIA) SKULL FROM THE REDONDA FORMATION (UPPER TRIASSIC: APACHEAN) OF EAST-CENTRAL NEW MEXICO

ANDREW B. HECKERT[1], SPENCER G. LUCAS[2], ADRIAN P. HUNT[3] AND JERALD D. HARRIS[4]

[1]Department of Earth and Planetary Sciences, University of New Mexico, Albuquerque, NM 87131-1116; [2]New Mexico Museum of Natural History and Science, 1801 Mountain Rd NW, Albuquerque, 87104; [3]Mesalands Dinosaur Museum, Mesa Technical College, 911 South Tenth Street, Tucumcari, NM 88401; [4]Department of Earth and Environmental Sciences, University of Pennsylvania, Philadelphia, PA 19104

Abstract.—In the Summer of 1994, a field party of the New Mexico Museum of Natural History and Science (NMMNH) collected a giant, incomplete phytosaur skull from a bonebed discovered by Paul Sealey in east-central New Mexico. This bonebed lies in a narrow channel deposit of intraformational conglomerate in the Redonda Formation. Stratigraphically, this specimen comes from strata identical to the type Apachean land-vertebrate faunachron and thus of Apachean (latest Triassic: late Norian-Rhaetian) age. The skull lacks most of the snout but is otherwise complete and in excellent condition. As preserved, the skull measures 780 mm long, and was probably 1200 mm or longer in life, making it nearly as large as the holotype of *Rutiodon* (=*Machaeroprosopus*, =*Smilosuchus*) *gregorii*, and one of the largest published phytosaur skulls. The diagnostic features of *Redondasaurus* present in the skull include robust squamosal bars extending posteriorly well beyond the occiput and supratemporal fenestrae that are completely concealed in dorsal view.

The specimen was originally encased in a plaster jacket only marginally larger than the preserved skull. Still, the contents of the jacket reveal one of the densest accumulations of disarticulated bones in the Chinle Group, including a total of 275 other teeth, bones, and bone fragments, including a smaller phytosaur skull. The smaller skull is poorly ossified, distorted, and slightly disarticulated due to lack of fusion. We suspect that this specimen represents a subadult *Redondasaurus,* but it lacks the temporal region and is thus not identifiable at the genus level. Aetosaur scutes associated with the phytosaurs may represent the first record of *Neoaetosauroides* in North America and suggest correlation of the Apachean Redonda Formation with the Los Colorados Formation of Argentina.

INTRODUCTION

Phytosaurs were large, semiaquatic, carnivorous reptiles known from Upper Triassic strata in North America, Europe, Brazil, India, Thailand, North Africa, and Madagascar. The Chinle Group in eastern New Mexico yields numerous fossils, particularly skulls, of phytosaurs, including a skull from the Travesser Formation described by Stovall and Savage (1939) and a skull from the Redonda Formation described by Gregory (1957, 1972) named *Redondasaurus gregorii* by Hunt and Lucas (1993). Here, we describe a giant phytosaur skull and associated fossils collected from the Redonda Formation by parties of the New Mexico Museum of Natural History and Science in 1994 and briefly comment on its biostratigraphic and taphonomic importance. In this paper, NMMNH = New Mexico Museum of Natural History and Science, Albuquerque.

STRATIGRAPHY AND AGE

The fossils described here were collected from a single locality, NMMNH locality 4211, discovered by Paul Sealey stratigraphically high in the Redonda Formation in Apache Canyon, Quay County, New Mexico (Fig. 1). The fossils occur in a narrow (<3 m wide) channel deposit consisting of an intraformational conglomerate fining upward into a bentonitic mudstone. Clasts in the conglomerate are principally reworked clay pebbles and fossil reptile bones. This deposit is approximately 2.3 m below a prominent, ledge-forming sandstone, the "Redonda Bench," that serves as a marker bed locally. This is the Redonda Ledge marker bed of Gregory (1972).

Hester (1988) studied the Redonda Formation here and at the type locality (Mesa Redonda) and determined that it represents a series of lakes (fine-grained clastics and occasional carbonates) fed by numerous small streams and rivers (coarser-grained clastics). NMMNH locality 4211 appears to represent a relatively small-scale stream channel draining into one of the Redonda lakes.

Lucas and Hunt (1993; also see Lucas, 1998) established four land-vertebrate faunachrons (LVF) for the chronological intervals represented by successive faunas in the Chinle. The type fauna of the youngest of these intervals, the Apachean, was named for the fauna of the Redonda Formation in Apache Canyon. Index taxa that characterize the Apachean LVF are the phytosaur *Redondasaurus*, especially *Redondasaurus gregorii* Hunt and Lucas, 1993, and the aetosaur *Redondasuchus reseri* Hunt and Lucas, 1991. Other tetrapod taxa from the type fauna include the diminutive metoposaurid temnospondyl *Apachesaurus gregorii* Hunt (1993), a sphenodontian, a procolophonid, a rauisuchian, a large aetosaur, theropods, indeterminate cynodonts, and a giant, undescribed sphenosuchian (Hunt and Lucas, 1997; Lucas, 1998; Lucas et al., 1999). Occurrences of the phytosaur *Redondasaurus* correlate the Redonda Formation with the Travesser Formation in northeastern New Mexico, the Rock Point Formation in north-central New Mexico, including the famous Whitaker quarry (*Coelophysis*) bonebed at Ghost Ranch, and the basal Wingate Sandstone in southern Utah (Lucas et al., 1997a,b).

PALEONTOLOGY

Two jackets were collected from NMMNH locality 4211. The first was a small jacket containing a phytosaur ischium, although this jacket is apparently lost. The second jacket weighed several hundred pounds (200+ kg), and was known to contain a large, incomplete phytosaur skull and numerous other elements at the time of collection. Subsequent preparation of this jacket yielded the giant phytosaur skull described here (NMMNH P-31094: Fig. 2) as well as a "fauna" consisting of approximately 275 other ele-

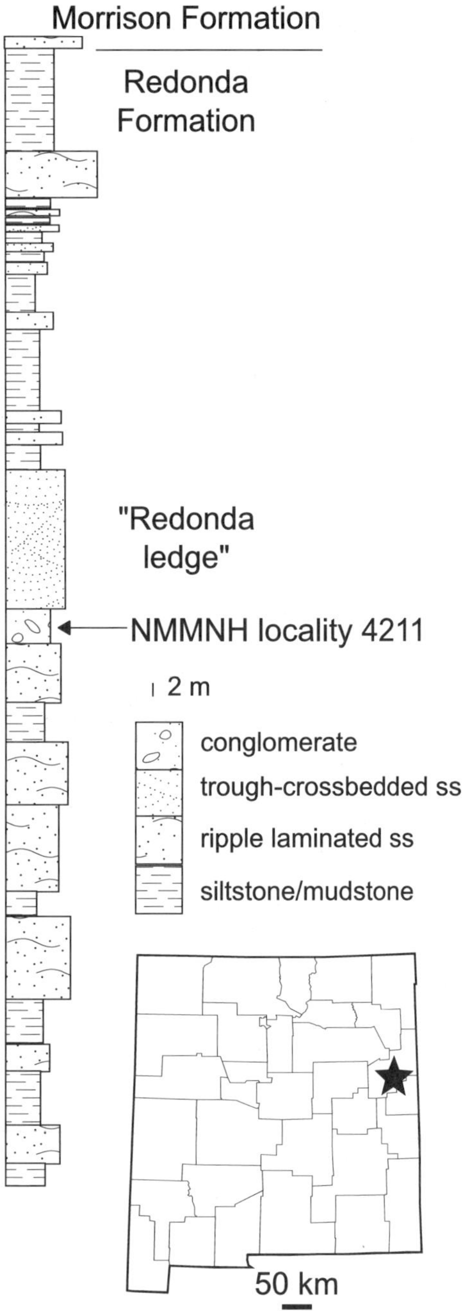

FIGURE 1. Index map and stratigraphic column showing the location of NMMNH locality 4211.

ments, principally teeth and bones of phytosaurs but also including bones of at least one aetosaur and scales of semionotid fish (Fig. 3). Perhaps the most surprising discovery was of a second, smaller, incomplete phytosaur skull (NMMNH P-34095) lying on the palate of the larger specimen (Figs. 2, 3C, 4). In this section we briefly describe the skulls (Figs. 2, 3C) and associated elements. The description of the larger skull, NMMNH P-31094 is the most detailed, yet is still preliminary pending more detailed comparisons with other phytosaurs.

DESCRIPTION OF THE GIANT SKULL

NMMNH P-31094 is a well-preserved phytosaur skull lacking only the anterior portion of the maxillae, much of the premaxillae, and any *in situ* teeth (Fig. 2; Table 1). The skull is very slightly crushed dorsally, with some shearing from left to right (Fig. 2D). The skull is robust, and measures nearly 560 mm across at the quadrates (89 mm interorbital width). The postnarial length of this specimen is approximately 440 mm (540 mm from the anterior margin of the naris). Using Gregory's (1962) calculation that most robust phytosaurs have a prenarial: postnarial length ratio of approximately 1.20-1.25:1, we very conservatively estimate that the complete skull was 1170 mm long. Although many older and more primitive phytosaurs, particularly numerous specimens of *Rutiodon* and *Angistorhinus* reached this length (see Gregory, 1962, fig. 4 for comparison), few phytosaur skulls are as broad and heavily constructed as NMMNH P-31094. Both the sutural arrangements and the rugose surface texture of the bones are clear. Here we focus on the exterior skull bones and their taxonomic significance.

The only remnants of the premaxillae are their elongate dorsal processes, which lie medial to the maxillae and anterior to the septomaxillae. They slope gently from the rim of the nares anteriorly and lack a well-developed, bulbous crest. However, in lateral profile they match the "crested" outline of *Redondasaurus bermani* and are more inclined than those of *R. gregorii*.

The preserved portions of the maxillae are broad in dorso-lateral view, but lack the pronounced lateral bulge seen in the holotypes of *Nicrosaurus kapffii* or *Rutiodon gregorii*, although this character appears to vary within these and other species (Hunt, 1994a; Hungerbühler and Hunt, 2000). The left preserves 14 alveoli and the right 17. The maxillae almost surround the relatively small antorbital fenestrae, which are medio-laterally narrow and antero-posteriorly elongate (Table 1).

The nasals are broad and surround the external nares except for the anterior border, which is comprised of the septomaxillae. At the anterior margin of the external nares, the nasals each form a slightly bulbous process, and the nares are slightly elevated by a narial rim. The nasals terminate shortly behind the posterior edge of the external nares and well anterior to the orbital region.

The elongate septomaxillae extend from their junction with the premaxillae and nasals posteriorly into the external nares. They are widest anterior to the projections of the nasals, and taper posteriorly.

The lacrimal forms the dorsal margin of the antorbital fenestra and extends posteriorly to the orbit. It contributes to the anterior margin of the orbit, but is not as extensively involved as those of smaller species of *Rutiodon* (=*Leptosuchus* of Long and Murry, 1995).

The prefrontals are broad anteriorly, and they taper posteriorly. Narrow lateral extensions of the prefrontals stretch posteriorly to the anterodorsal corner of the orbital rim.

The frontals are small and antero-posteriorly approximately the same length as the orbit. Like most of the skull bones, they are coarsely pitted, but they also posses finer pitting adjacent to the orbit.

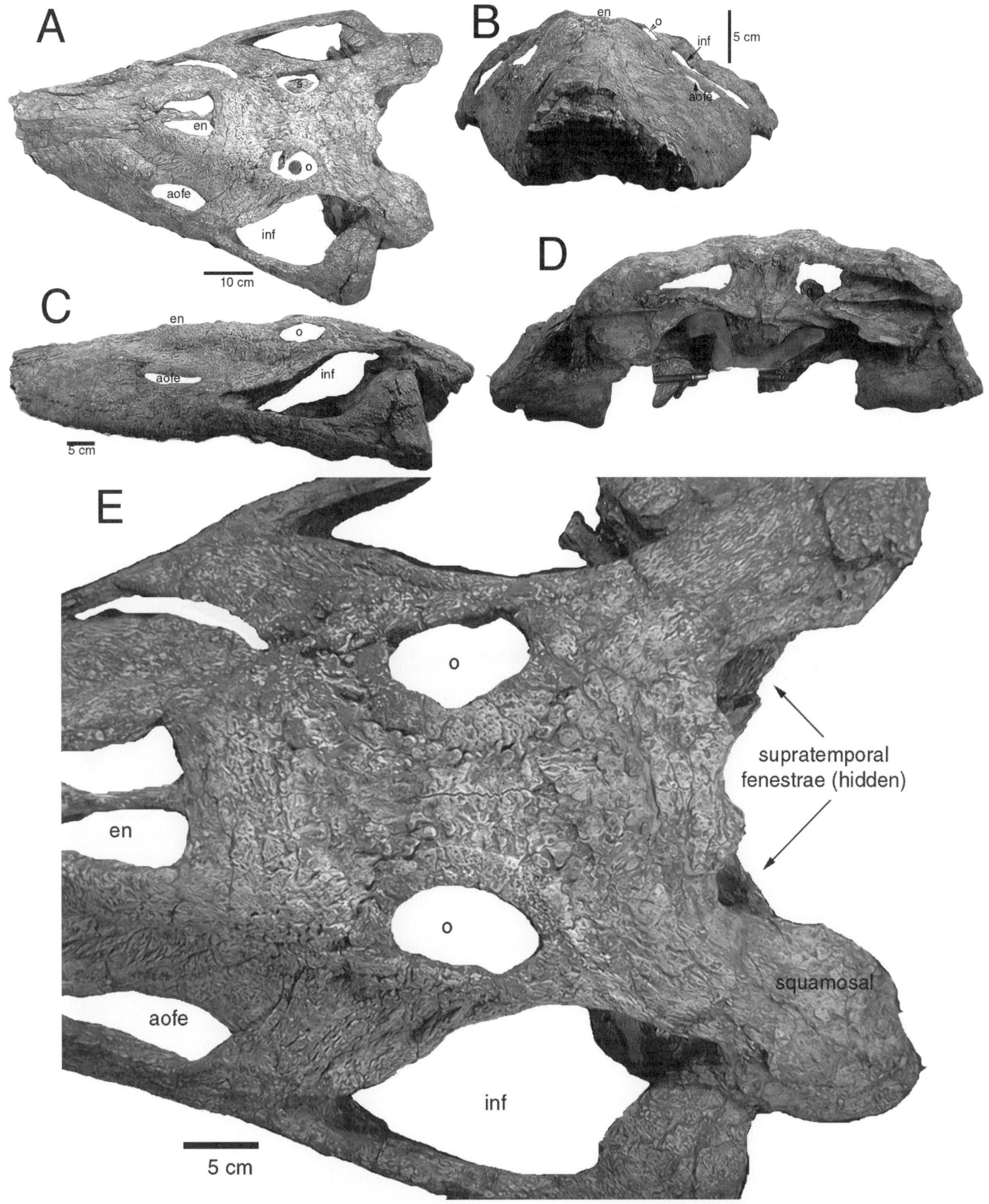

FIGURE 2. Photographs of the giant, incomplete phytosaur skull, NMMNH P-31094. A. Dorsal view showing elements remaining in orbits; B. Anterior view; C. Lateral view, D. Posterior view, metal is brackets of mount; E. Close-up of dorsal view with tooth and scute removed from orbits. Scale bars are 5 cm. Abbreviations: aofe = antorbital fenestra; en = external nares; inf = infratemporal fenestra; o = orbit; s = scute; t = tooth.

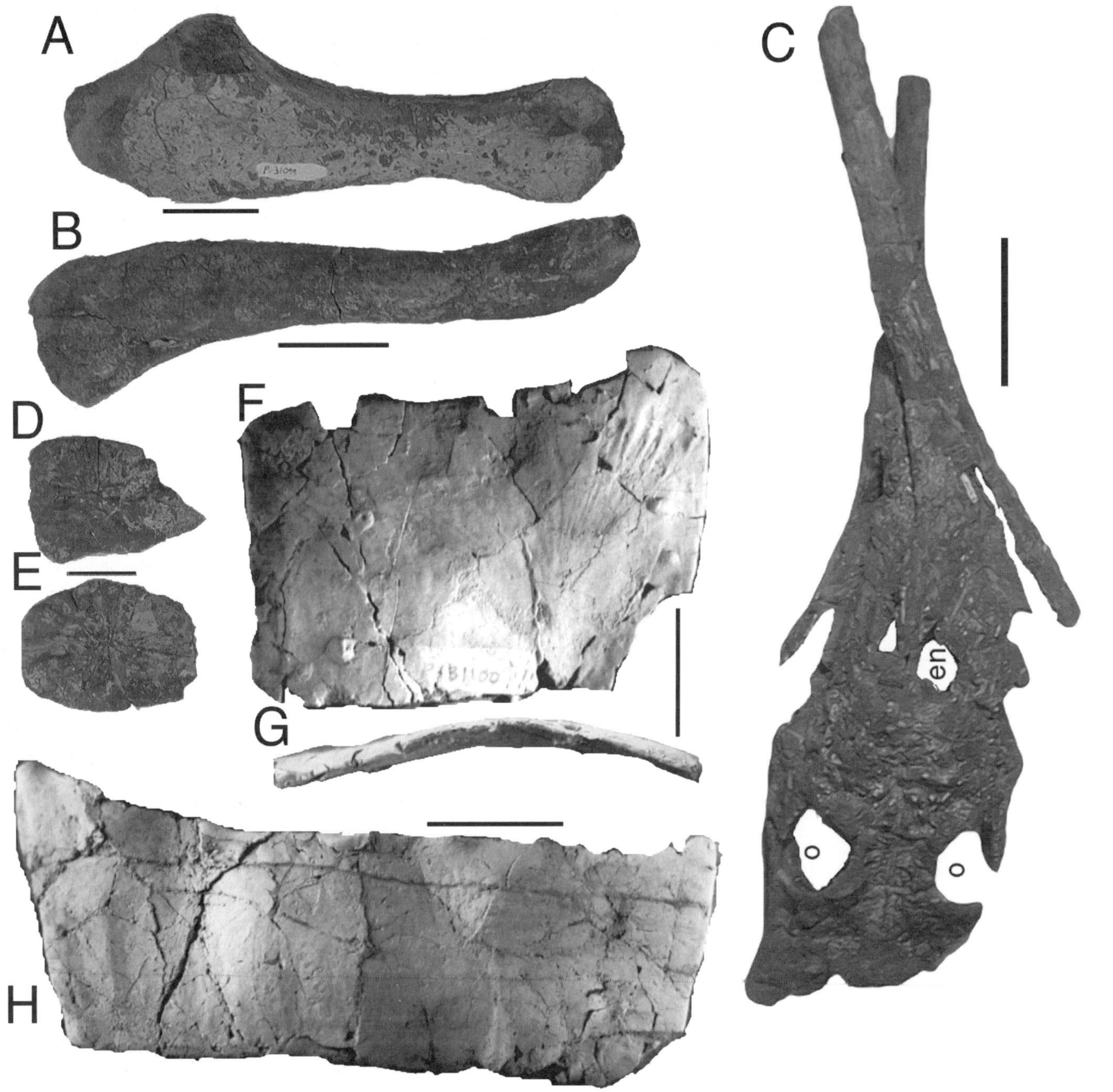

FIGURE 3. A small sample of the fossils found in the jacket with the giant phytosaur skull. A. NMMNH P-31099, large phytosaur right ulna in lateral view; B. NMMNH P-31098, small phytosaur right femur in posterior view; C. NMMNH P-31095, juvenile phytosaur skull in dorsal view; D-E. NMMNH P-31096, representative phytosaur scutes; F-G. NMMNH P-31100, small aetosaur paramedian? scute similar to *Neoaetosauroides* in dorsal (F) and anterior (G) views; H. NMMNH P-31100, larger aetosaur dorsal paramedian scute similar to *Neoaetosauroides* in dorsal view. Scale bars are 5 cm (A-B, D-E), 10 cm (C), and 2 cm (F-H).

Similarly, the relatively small postfrontals contribute to the posterior dorsal margin of the orbit. They also possess finer pitting adjacent to the orbit and coarser pitting posteriorly.

The parietals are relatively broad and flare posteriorly. Each has a slight posterior projection in the middle of its posterior margin. The parietals fully overlap the occiput and associated elements of the braincase, completely covering them in dorsal view.

The squamosals are broad and robust, extending posteriorly and laterally from the parietals. The posterior process of the squamosal is a thick, heavy flange of bone that does not taper posteriorly and is strongly downturned, extending ventrally below the upper extent of the quadratojugal, although some of this may be the result of postmortem crushing. The supratemporal fenestrae are depressed well below the level of the skull roof and are completely hidden in dorsal view by the squamosals and, to a lesser extent, the parietals.

The quadratojugals are broad in lateral view and taper slightly dorsally. They form the bulk of the posterior margin of the infratemporal fenestra and extend anteriorly across the posterior third of its ventral margin. The quadrates are massive and more than 400 mm apart. Each bears a broad articular surface marked by two condyles for the articulation with the lower jaw.

Anteriorly, the jugals are rectilinear and form the posterior margin of the antorbital fenestra and the ventral margin of the orbit. The dorsal process slants posteriorly and dorsally, forming a sharply acute angle around the anterior edge of the infratemporal fenestra. Posteriorly they are relatively slender across the anterior two-thirds of the ventral margin of the infratemporal fenestra.

SYSTEMATICS OF THE GIANT SKULL

Recent phytosaur classifications include Ballew (1989), Hunt (1994a,b), Long and Murry (1995), and Hungerbühler (1998). All are fraught with difficulties. Ballew (1989) is the only cladistic treatment, but she only examined phytosaurs from the American Southwest. Hunt's (1994a,b) treatment is the most comprehensive, but lacks a cladistic analysis and remains essentially unpublished. Long and Murry's (1995) taxonomy is not based on phylogenetic analysis and only superficially examines numerous issues of taxonomy and non-American specimens. Hungerbühler (1998) provides exhaustive descriptions of Norian phytosaur skulls from southwestern Germany, but is limited in its treatment of other forms and, like Hunt (1994a), remains unpublished. Consequently, it is difficult to assign generic, let alone specific, names with confidence to phytosaurs, in spite of more than 150 years of phytosaur collecting and perhaps 100 well-preserved skulls in Europe and North America alone. Here, we rely principally on Hunt's (1994a) description and comparison of phytosaur skulls, with reference to Ballew (1989) and older classifications, including Westphal (1976) and Gregory (1962). We rely on Long and Murry (1995) only for comparison to specimens they illustrate, as we find their approach to taxonomy at best problematic and, in cases, (e.g., the new genus *Arribasuchus buceros*), arbitrary.

Most workers since Gregory (1962) concur that primitive phytosaurs (*Paleorhinus* and *Angistorhinus* of most recent classifications) had broad supratemporal fenestrae at the level of the skull roof that are visible in dorsal view. The most derived phytosaurs have relatively smaller, depressed supratemporal fenestrae. These fenestra also tend to become partially to completely obscured by the squamosals and parietals in dorsal view. NMMNH P-31094 clearly fits into the derived phytosaurs based on its depressed supratemporal fenestrae that are completely concealed in dorsal view.

Named phytosaur taxa with depressed, concealed supratemporal fenestrae include *Coburgosuchus goeckeli* Heller, 1954,

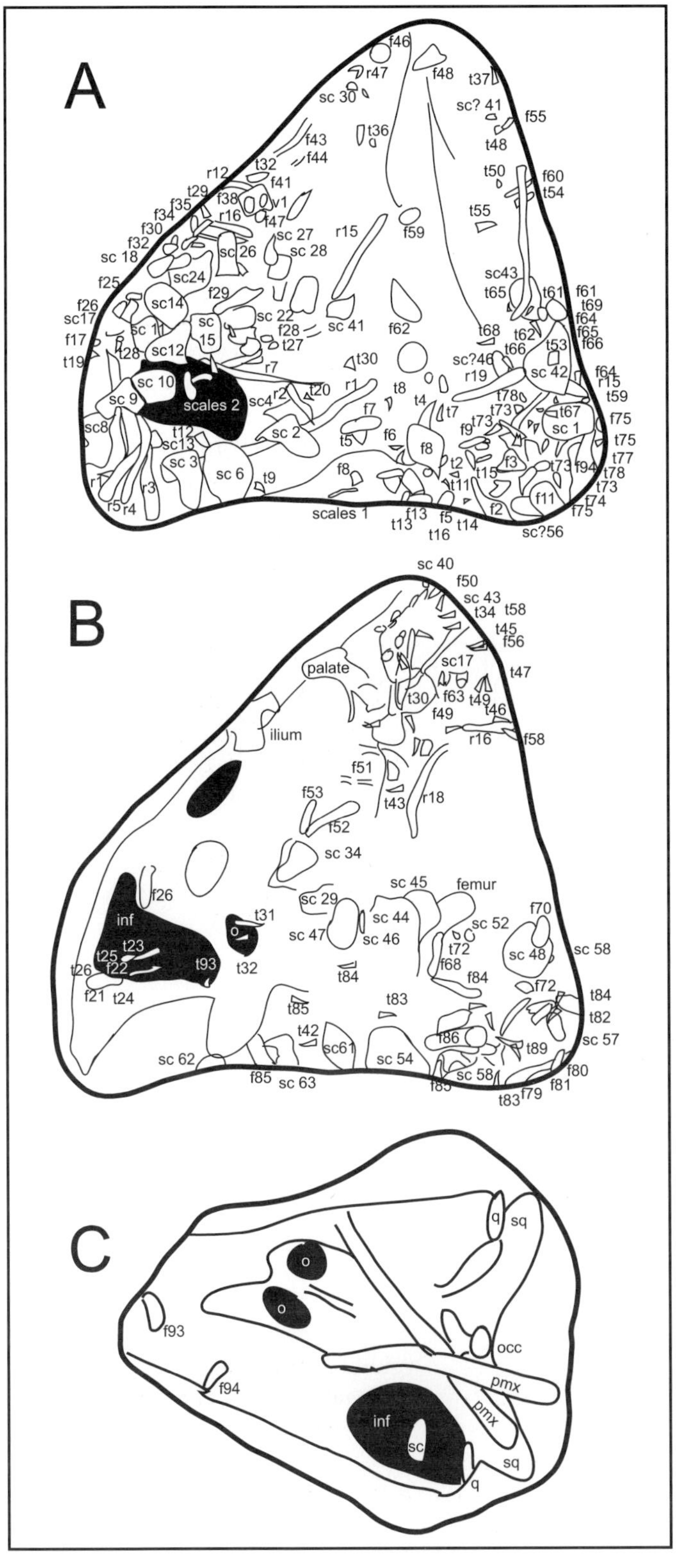

FIGURE 4: Three simplified, schematic sketches showing the distribution of bones found during preparation of the skull. A. Bones encountered dorsal to (stratigraphically below) the giant skull; B. Bones encountered adjacent to the skull on the dorsal surface; C. The position of the subadult phytosaur skull lying on the palate of the (upside-down) giant skull. Two patches of scales indicated. Shading indicates skull fenestra. Complete sketches on file at NMMNH. Abbreviations include: f = fragment, inf = infratemporal fenestra; o = orbit; occ = occipital condyle; pmx = premaxilla; q = quadrate; r = rib; sc = scute, sq = squamosal; t = tooth; v = vertebra. All elements numbered in the order of discovery (scute 1, 2, 3…etc).

Redondasaurus gregorii Hunt and Lucas, 1993, and *R. bermani* Hunt and Lucas, 1993. Additionally, Hunt (1994a) recognized another, robust morph from the Redonda Formation (NMMNH P-4983) with these characters. NMMNH P-31094 differs from all of these in that it possesses a relatively tiny antorbital fenestra. Indeed, no phytosaur illustrated by Gregory (1962), Westphal (1976) or Long and Murry (1995) has an antorbital fenestra as small relative to the nares in dorsal view as NMMNH P-31094. NMMNH P-31094 differs from *Coburgosuchus* in possessing squamosals that do not extend as far posteriorly and are proportionately broader, which we interpret as generally robust. NMMNH P-31094 is considerably more robust than the narrow-snouted (dolichorostral) *R. gregorii*, yet may not be as robust (brachyrostral) as Hunt's (1994a) robust morph exemplified by NMMNH P-4256 if the latter were scaled to the length of NMMNH P-31094. A true rostral crest is not as prominent as in many taxa, although the snout does taper from the nares anteriorly, as in *R. bermani*. Therefore, because we remain uncertain of its exact affinities, we refer this specimen to *Redondasaurus* sp. NMMNH P-31094 is most similar to *R. bermani* but possesses sufficient diagnostic features (autapomorphies), particularly regarding the reduction of the antorbital fenestra and breadth of the postorbital skull, to justify erection of a new specific name in most phytosaur classifications.

OTHER ELEMENTS FROM NMMNH LOCALITY 4211

NMMNH P-31094 was merely the largest element in an extraordinarily dense bonebed. Contents of the field jacket, aside from the giant skull, included another palatal skull element (P-31097), a large right ulna (P-31099; Fig. 3A), a small right femur (P-31098; Fig. 3B), an incomplete large right ilium (P-31101), a small incomplete interclavicle, (P-31102), two vertebrae, 27 ribs or rib fragments, 50 scutes, including three aetosaur scutes (P-31100; Fig. 3F-G) and a possible sphenosuchian scute, 93 phytosaur teeth, 94 other bones or bone fragments, and, most surprisingly, a second, smaller phytosaur skull (P-31095; Fig. 3C) wedged in the palate of the larger specimen.

The smaller skull (650 mm preserved length) is slightly disarticulated, and consists of most of the snout, including a short narial crest, and the orbital region of the skull roof (50 mm interorbital width). In gross morphology, the preserved portion of the skull resembles phytosaur skulls from the stratigraphically lower Canjilon and Snyder quarries in north-central New Mexico in possessing a prominent narial crest immediately anterior to the nares. We suspect that this specimen represents a subadult, or at least less mature, *Redondasaurus,* but it lacks the temporal region and is thus not identifiable at the generic level.

Most of the elements found in the jacket are phytosaurian, including the ulna, femur, ilium, interclavicle, vertebrae, and teeth. Others, including many of the ribs and fragments are not diagnostic below the level of Reptilia. Three of the 50 scutes are not phytosaurian, and instead represent a relatively rare occurrence of an aetosaur in the Redonda Formation. These scutes are wider than long, possess anterior bars, a very faint pattern of elongate pits and grooves, and very little if any dorsal boss (Fig. 3F-H). The longest scute (Fig. 3F-G) is particularly narrow, and has a width:length (W:L) ratio of less than 1.2:1. We interpret this scute as a right cervical? paramedian scute. A shorter scute (Fig. 3H) is approximately 2.5 times wider than long, with less distinct patterning and a very low dorsal boss near the posterior margin medial to the middle of the scute. We interpret this scute as a left dorsal paramedian scute. A third scute is smaller, but similar in most respects to the wider scute.

Aetosaurs characterized by very faint ornamentation include *Coahomasuchus kahleorum* Heckert and Lucas 1999, *Aetosaurus arcuatus* (Marsh, 1896), and *Neoaetosauroides engaeus* Bonaparte, 1967 (Heckert and Lucas, 1999, 2000). The scutes described here are more robust than those of *Coahomasuchus* and possess a more radial pattern of grooves and ridges. Scutes of *A.*

Table 1. Measurements of NMMNH P-31094

Feature	Measurement (in mm)
Length preserved skull (tip broken snout to posterior end squamosal)	782
Length preserved skull (tip broken snout to posterior edge occipital condyle)	649
Maximum width of skull (across quadratojugals)	560
Length post-snout (anterior border nares to poster edge squamosal)	539
Maximum length of naris	102
Maximum width of naris	66
Maximum length left orbit	83
Maximum width left orbit	59
Minimum interorbital width	89
Width cranial table between infratemporal fenestrae	238
Maximum length supratemporal fenestrae	55
Maximum width supratemporal fenestrae	39
Interfenestral width	70
Length ventral border left infratemporal fenestra	170
Length anterior border left infratemporal fenestra	180
Length posterior border left infratemporal fenestra	145
Length left antorbital fenestra	122
Maximum width left antorbital fenestra	41
Maximum width palate between tooth rows	290

arcuatus differ from these in being slightly smaller and lacking dorsal bosses. Additionally, many paramedian scutes of *Aetosaurus* have W:L ratios of 3.0 or higher. Scutes of *Neoaetosauroides* are not well-preserved in the holotype material but in general shape and morphology conform to those described here. Specifically, scutes of *Neoaetosauroides* have a faint radial pattern of pits and grooves and are relatively narrow (W:L < 3.0). Some scutes of *Neoaetosauroides* possess a low dorsal boss similar in shape and position to that illustrated here. Therefore, we tentatively assign these scutes to cf. *Neoaetosauroides* sp.

During preparation of the specimen we also recovered several clusters of fish scales. These scales are loosely articulated to associated. They are somewhat rhomboid and elongate, and in most respects conform to semionotid fish, which are relatively common in the Redonda Formation (Huber et al., 1993).

TAPHONOMY

This assemblage, with approximately 276 known teeth, bones, or bone fragments, all from an area of less than 1 m^2, represents one of the densest accumulations of bone in the Chinle Group. All elements are disarticulated, and they appear hydrodynamically sorted as the bulk of the elements represented are long bones, principally ribs and limb bones. However, even with extensive mapping of the bones as they were uncovered (Fig. 4), there is no obvious preferential orientation other than that the giant skull, the smaller skull, several ribs and the large ulna were all oriented roughly parallel to each other, although the ulna was resting against the giant snout.

There is absolutely no indication of articulation of any of the remains described here. Furthermore, the condition of the phytosaur skulls indicates that the animals were deceased and nearly completely desiccated and/or rotted prior to burial. Every aperture of the giant skull contained at least one allochthonous element, including but not limited to the femur in the left infratemporal fenestra, a scute in the right orbit, and several teeth (some removed during preparation) in the left orbit. Although many bones are broken, the assemblage was probably not transported very far before burial. There is no evidence of abrasion, and many of the broken or incomplete bones were damaged during collection, not deposition (note the large proportion of fragments adjacent to jacket walls in Figure 4). The break in the giant snout is fresh and probably was a result of Recent weathering, and the smaller snout was accidentally truncated while trenching around the jacket. Therefore, we suspect that the assemblage represents a short-term accumulation of dead and disarticulated individuals subsequently entrained and rapidly buried in a channel cut into floodplain mudstones. This accumulation thus well matches descriptions of channel lag deposits as characterized by Behrensmeyer et al. (1992).

The preserved elements indicate the presence of at least three individual phytosaurs, based on the giant skull, the subadult skull, and a tiny snout or jaw fragment too small to belong to either of the first two. In all probability, the minimum number of individual (MNI) phytosaurs was still higher, as even the largest limb and girdle elements appear too small for the giant skull, but are too large for the smaller one, and the unidentified palatal? element (NMMNH P-31097) likely represents another phytosaur. However, articulated phytosaur skeletons are rare and skull size:limb length ratios are essentially unknown.

Hunt et al. (1995) recognized six vertebrate taphofacies in the Redonda Formation. These included (1) nearshore clastic lacustrine; (2) carbonate lacustrine-margin; (3) beach conglomerate; (4) fluvial channels; (5) floodplain taphofacies; and (6) paleosol taphofacies. (Table 2). Of these, NMMNH locality 4211 clearly well-matches the fluvial channels, which Hunt et al. (1995, p. 32) describe as "intraformational conglomerates and lenticular sandstones, representing fluvial channel deposits, contain fragmentary and abraded bones; locally, small channels are full of well-preserved bones that are dominantly phytosaurian but also include ?poposaur and aetosaur specimens." These localities, while not areally extensive, are clearly an important source of fossil vertebrates.

Table 2. Taphofacies of the Redonda Formation

Taphofacies	Lithologic characteristics	Fossils
Nearshore lacustrine	Tabular sandstones and mudstones	Complete to disarticulated fish
Lacustrine margin	Carbonates, principally calcarenites	Vertebrate tracks, including *Brachychirotherium* *Pseudotetrasauropus* *Tetrasauropus* *Rhynchosauroides* *Grallator*
Beach	Tabular intraformational conglomerate	Fish scales, ichthyoliths, phytosaur teeth, rarely other tetrapods
Fluvial channel	Lenticular intraformational conglomerate and sandstone	Generally fragmentary and abraded bones, locally well-preserved material
Floodplain (proximal)	Mudrocks	Isolated phytosaur postcrania and skulls
Paleosol (distal floodplain)	Mottled mudrocks with calcareous nodules	Small tetrapods

CONCLUSIONS

The giant phytosaur here, which we refer to *Redondasuchus* sp., is another Apachean (latest Triassic) record of the genus from east-central New Mexico. Although phytosaur classification remains problematic, this skull is readily diagnosed as a highly derived phytosaur based on the presence of depressed supratemporal fenestrae that are completely concealed in dorsal view. Possible records of the aetosaur *Neoaetosauroides* from NMMNH locality 4211 are the first occurrence of that taxon outside the type area in the Los Colorados Formation (Ischigualasto basin) of Argentina and support cross-correlation of the type Apachean fauna with the Los Colorados, as suggested by Lucas (1998). Bonebeds such as NMMNH locality 4211 probably represent localized channel deposits near the margins of the Redonda Formation lakes.

ACKNOWLEDGMENTS

Paul Sealey discovered NMMNH locality 4211. The New Mexico Friends of Paleontology helped excavate the specimens described here. The Duke family graciously allowed access to their land and assisted in extricating the jacket. Mike Pierce, Matt Celeskey, and Daniel Weissmann of the NMMNH assisted with digital photography. A. Hungerbühler and R.M. Sullivan reviewed an earlier draft of this paper and provided helpful comments.

REFERENCES

Ballew, K. L., 1989, A phylogenetic analysis of Phytosauria from the Late Triassic of the western United States, *in* Lucas, S. G., and Hunt, A. P., eds., Dawn of the Age of Dinosaurs in the American Southwest: Albuquerque, New Mexico Museum of Natural History, p. 309-339.

Behrensmeyer, A. K., Damuth, J. D., DiMichele, W. A., Potts, R., Sues, H.-D., and Wing, S. L., 1992, Terrestrial ecosystems through time: Evolutionary paleoecology of terrestrial plants and animals: Chicago, The University of Chicago Press.

Bonaparte, J. F., 1967, Dos nuevas "faunas" de reptiles Triasicos de Argentina: Gondwana Symposium Proceedings and Papers, v. 1, p. 283-306.

Case, E. C., 1929, Description of the skull of a new form of phytosaur with notes on the characters of described North American phytosaurs: Memoirs of the University of Michigan Museums: Museum of Paleontology, v. 2, p. 56.

Gregory, J. T., 1957, Significance of fossil vertebrates for correlation of Late Triassic continental deposits of North America: Report of the 20th Session of the International Geological Congress 1956, Section II, p. 7-25.

Gregory, J. T., 1962, The genera of phytosaurs: American Journal of Science, v. 260, p. 652-690.

Gregory, J. T., 1972, Vertebrate faunas of the Dockum Group, Triassic, eastern New Mexico and West Texas: New Mexico Geological Society 23rd Fall Field Conference, Guidebook, p. 120-123.

Heckert, A. B., and Lucas, S. G., 1999, A new aetosaur (Reptilia: Archosauria) from the Upper Triassic of Texas and the phylogeny of aetosaurs: Journal of Vertebrate Paleontology, v. 19, no. 1, p. 50-68.

Heckert, A. B., and Lucas, S. G., 2000, Taxonomy, phylogeny, biostratigraphy, biochronology, paleobiogeography, and evolution of the Late Triassic Aetosauria (Archosauria:Crurotarsi): Zentralblatt für Geologie und Paläontologie Teil I 1998 Heft 11-12, p. 1539-1587.

Heller, F., 1954, Ein Parasuchier-Schädelrest aus dem Oberen Burgsandstein von Coburg: Geologisch Blätter für Nordost-Bayern, v. 3, p. 1-13.

Hester, P. M., 1988, Depositional environments in an Upper Triassic lake, east-central New Mexico [M.S. thesis]: University of New Mexico, 154 p.

Huber, P., Lucas, S. G., and Hunt, A. P., 1993, Late Triassic fish assemblages of the North American Western Interior: Museum of Northern Arizona Bulletin, v. 59, p. 51-66.

Hungerbühler, A., 1998, Cranial anatomy and diversityof the Norian phytosaurs of Southwestern Germany [unpublished Ph.D. dissertation]: Bristol, University of Bristol: 453 pp.

Hungerbühler, A., and Hunt, A. P., 2000, Two new phytosaur species (Archosauria, Crurotarsi) from the Upper Triassic of southwest Germany: Neues Jahrbuch für Geologie und Paläontologie Monatshefte, v. 2000, no. 8, p. 467-484.

Hunt, A. P., 1993, Revision of the Metoposauridae (Amphibia: Temnospondyli) and description of a new genus from western North America: Museum of Northern Arizona Bulletin, v. 59, p. 67-97.

Hunt, A. P., 1994a, Vertebrate paleontology and biostratigraphy of the Bull Canyon Formation (Chinle Group, Upper Triassic), east-central New Mexico with revisions of the families Metoposauridae (Amphibia: Temnospondyli) and Parasuchidae (Reptilia: Archosauria) [Ph.D. Dissertation thesis]: University of New Mexico, 404 p.

Hunt, A.P., 1994b, Phylogeny and biochronology of Phytosauria (Parasuchia): Journal of Vertebrate Paleontology, v. 14, p. 30A.

Hunt, A. P., and Lucas, S. G., 1993, A new phytosaur (Reptilia: Archosauria) genus from the Uppermost Triassic of the western United States and its biochronological significance: New Mexico Museum of Natural History and Science Bulletin, v. 3, p. 193-196.

Hunt, A. P., and Lucas, S. G., 1997a, Stratigraphy, paleontology and biochronology of the Upper Triassic Chinle Group in east-central New Mexico: Southwest Paleontological Symposium Proceedings, v. 1, p. 25-40.

Hunt, A. P., and Lucas, S. G., 1997b, Stratigraphy, paleontology and biochronology of the Upper Triassic Chinle Group in east-central New Mexico: Southwest Paleontological Society Proceedings, v. 1, p. 25-40.

Hunt, A. P., Heckert, A. B., Lucas, S. G., and Sealey, P. L., 1995, Vertebrate taphofacies of the Upper Triassic (Rhaetian) Redonda Formation, east-central New Mexico: New Mexico Geology, v. 17, no. 2, p. 32-33.

Long, R. A., and Murry, P. A., 1995, Late Triassic (Carnian and Norian) tetrapods from the southwestern United States: New Mexico Museum of Natural History and Science Bulletin, v. 4, p. 254 p.

Lucas, S. G., 1998, Global Triassic tetrapod biostratigraphy and biochronology: Palaeogeography, Palaeoclimatology, Palaeoecology, v. 143, p. 347-384.

Lucas, S. G., and Hunt, A. P., 1993, Tetrapod biochronology of the Chinle Group (Upper Triassic), western United States.: New Mexico Museum of Natural History and Science Bulletin, v. 3, p. 327-329.

Lucas, S. G., Estep, J. W., Heckert, A. B., and Hunt, A. P., 1999, Cynodont teeth from the Upper Triassic of New Mexico, U.S.A.: Neues Jahrbuch für Geologie und Paläontologie Monatsheft, v. 1999, no. 6, p. 331-344.

Lucas, S. G., Heckert, A. B., Anderson, O. J., and Estep, J. W., 1997a, Phytosaur from the Wingate Sandstone in southeastern Utah and the Triassic-Jurassic boundary on the Colorado Plateau: Southwest Paleontological Symposium Proceedings, v. 1, p. 49-59.

Lucas, S. G., Heckert, A. B., Estep, J. W., and Anderson, O. J., 1997b, Stratigraphy of the Upper Triassic Chinle Group, Four Corners Region: New Mexico Geological Society Guidebook, v. 48, p. 81-107.

Stovall, J. T., and Savage, D. E., 1939, A phytosaur in Union County, New Mexico, with notes on the stratigraphy: Journal of Geology, v. 47, p. 759-766.

Westphal, F., 1976, Phytosauria, *in* Kuhn, O., ed., Handbuch der Paläoherpetologie: Thecodontia: v. 13, Stuttgart, Gustav Fischer Verlag, p. 99-120.

TETRAPOD FOOTPRINT ICHNOFAUNA OF THE UPPER TRIASSIC REDONDA FORMATION, CHINLE GROUP, QUAY COUNTY, NEW MEXICO

SPENCER G. LUCAS[1], ADRIAN P. HUNT[2] AND MARTIN G. LOCKLEY[3]

[1]New Mexico Museum of Natural History, 1801 Mountain Road NW, Albuquerque, NM 87104; [2]Mesalands Dinosaur Museum, Tucumcari NM 88401; [3]Department of Geology, University of Colorado at Denver, PO Box 173364, Denver CO 80217

Abstract.—The tetrapod footprint ichnofauna from the Upper Triassic Redonda Formation in Quay County, east-central New Mexico, consists of four ichnogenera—*Rhynchosauroides* (rhynchocephalian), *Brachychirotherium* (aetosaur), *Grallator* (theropod) and *Pseudotetrasauropus* (prosauropod). A unique feature of the Redonda tetrapod ichnofauna is the absence of the sauropod track *Tetrasauropus* and the dominance of the prosauropod track *Pseudotetrasauropus*, which may indicate a local, mutual ecological separation of prosauropods and sauropods. This dominance of prosauropod tracks in the Redonda Formation is not mirrored by its body fossil record, which lacks prosauropods and is dominated by semiaquatic taxa. The Redonda tetrapod ichnofauna resembles other Chinle Group tetrapod ichnofaunas of the Apachean Rock Point sequence and well represents a global Late Triassic ichnofauna characterized by *Rhynchosauroides*, *Brachychirotherium*, *Grallator*, *Pseudotetrasauropus* and *Tetrasauropus*.

INTRODUCTION

In 1934, Robert Abercrombie discovered tetrapod footprints in Upper Triassic strata at Mesa Redonda, Quay County, New Mexico (Fig. 1). Gregory (1972) first mentioned this discovery, which was in strata now termed Redonda Formation of the Chinle Group. Gregory (1972) also noted similar track occurrences in the Redonda Formation at Apache Canyon, Quay County (Fig. 1).

Collecting during the 1980s and 1990s revealed both localities in Quay County to be rich sources of tetrapod footprints (Hunt et al., 1989, 1993, 2000; Lockley and Hunt, 1995; Cotton et al., 1996, 1997, 1998; Lockley et al., 2000). Here, we review the tetrapod ichnofauna of the Redonda Formation and discuss its significance. MDM = Mesalands Dinosaur Museum, Tucumcari; NMMNH = New Mexico Museum of Natural History, Albuquerque; UM = University of Michigan, Ann Arbor.

GEOLOGICAL CONTEXT

Redonda tetrapod tracks come from two collecting areas, Mesa Redonda and Apache Canyon in Quay County (Fig. 1). At Mesa Redonda, tracks are found around the entire perimeter of the mesa, though most collections come from limestones and siltstones stratigraphically high in the Redonda Formation in sec. 27, T9N, R31E (Fig. 1).

The Apache Canyon localities are also stratigraphically high in the Redonda Formation, occurring in the sandstone interval Gregory (1972) referred to as the "Redonda ledge" (Fig. 1). The localities are widely distributed along the Llano Estacado edge, and include sites in sec. 9, T8N, R33E (NMMNH localities 445, 446) and in sec. 28, T9N, R33E (NMMNH locality 1471). Redonda Formation deposition took place in and around a large lake (Hester, 1988), and the tetrapod tracks occur in lake-margin facies (Hester, 1988).

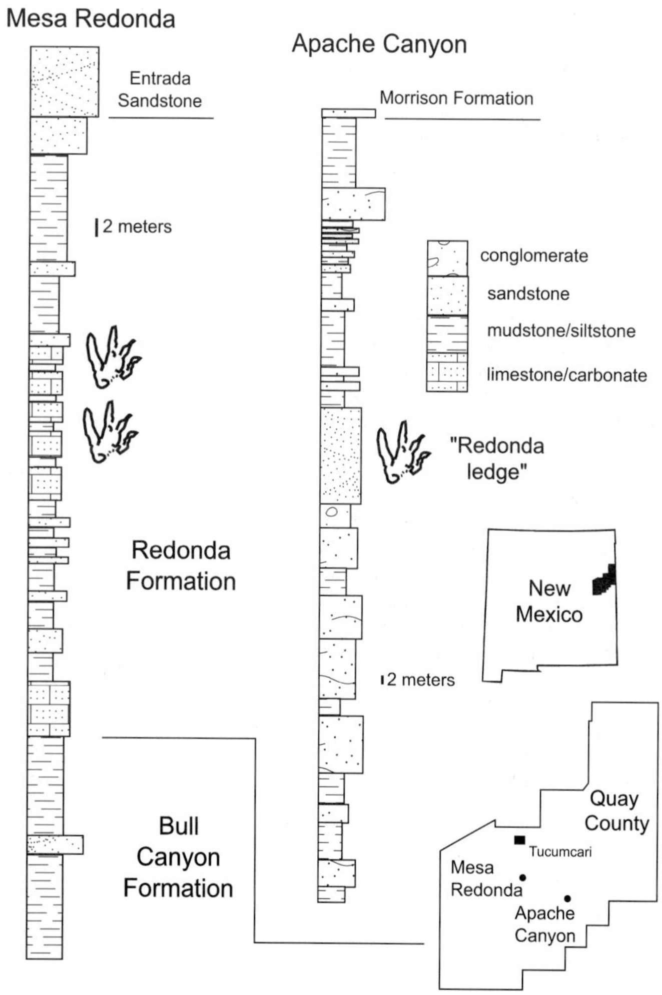

FIGURE 1. Location map and stratigraphic sections of Redonda Formation tetrapod footprint localities in Quay County, New Mexico.

TETRAPOD ICHNOTAXA

Redonda tetrapod footprints represent only four ichnogenera (Fig. 2). Here, we briefly review these ichnogenera and the Redonda specimens.

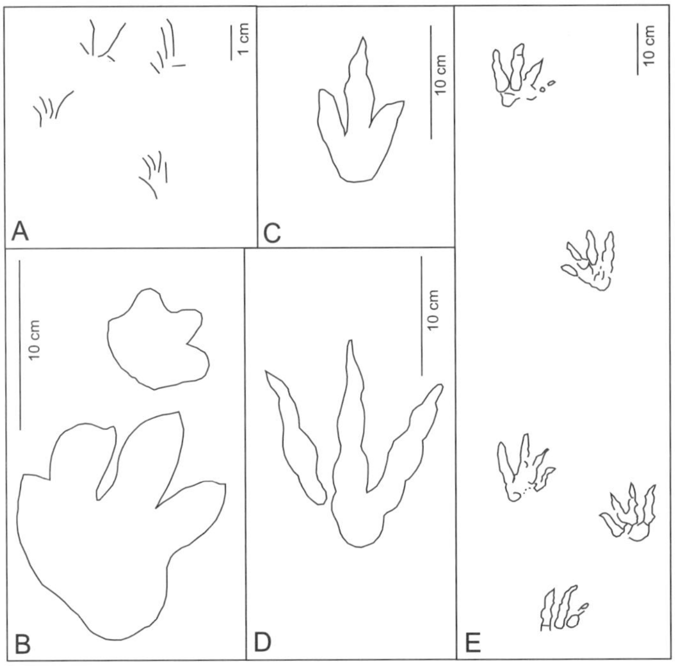

FIGURE 2. Outline drawings of the four Redonda tetrapod ichnotaxa. A, *Rhynchosauroides*. B, *Brachychirotherium*. C, *Grallator*. D, *Pseudotetrasauropus* (tridactyl extramorphological variant) E., trackway of *Pseudotetrasauropus*.

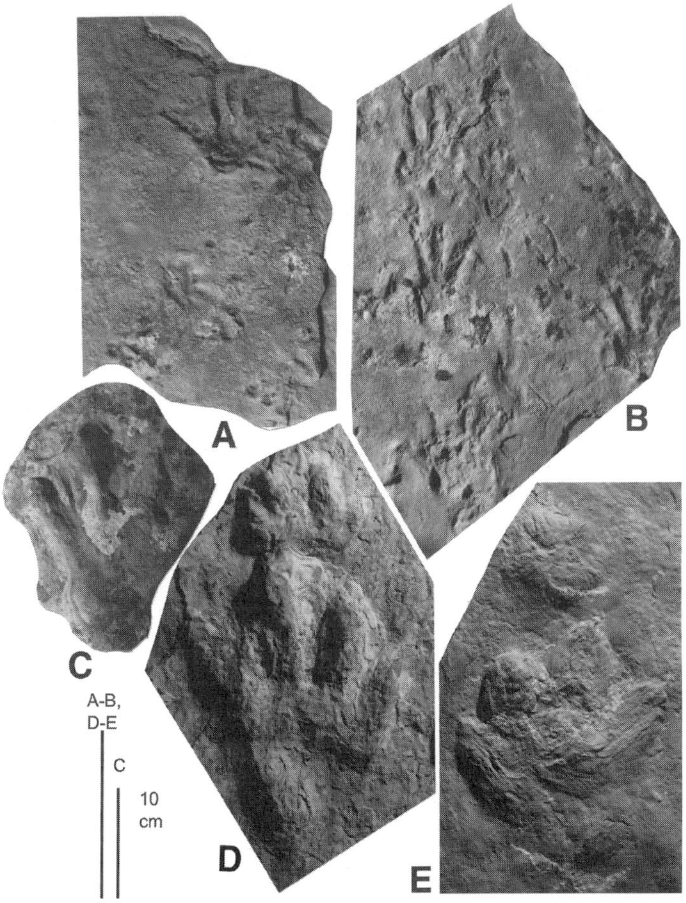

FIGURE 3. Selected tetrapod footprints from the Redonda Formation. A-B, *Rhynchosauroides*, MDM 208 from Apache Canyon 1. C, *Brachychirotherium*, NMMNH P-22426 from NMMNH locality 1477. D, *Brachychirotherium*, MDM 211 from Apache Canyon 2. E, *Brachychirotherium*, UMMP 16160 from Mesa Redonda.

Rhynchosauroides

Most *Rhynchosauroides* tracks from the Redonda Formation are relatively small (pes up to 2 cm long) tracks of a quadruped in which the pentadactyl manus is much smaller than the pentadactyl pes. Both manus and pes are longer than wide, and have long, thin digits that taper to clawed tips. The digits are straight or slightly curved, and length increases from digit I to digit IV, with digit V being as small or smaller than digit I. The animal usually walked in a digitigrade posture as heel (palm) impressions are rare. A large number of ichnospecies of *Rhynchosauroides* have been proposed (Baird, 1964; Haubold, 1971, p. 46), but we make no attempt to assign the Redonda specimens to an ichnospecies.

Rhynchosauroides tracks are well represented in the Redonda Formation, especially at Apache Canyon, and most are extramorphological variants that preserve four or fewer digits (Fig. 2A). Others, however, are much larger (pes length ~5 cm) that have distinct heel impressions (Fig. 3A-B). Probably, the Redonda *Rhynchosauroides* specimens encompass two distinct ichnospecies. *Rhynchosauroides* tracks are generally lizard-like and best match the foot structure of Triassic rhynchocephalians (Baird, 1964), which are also well known from body fossils (especially sphenodont teeth) in the Chinle Group.

Brachychirotherium

Brachychirotherium tracks from the Redonda Formation are relatively large tracks (pes ~18 cm long) of a quadruped with a pentadactyl pes much larger than the pentadactyl manus. The digits are thick, with rounded tips, and the "heel" of the pes is broad and rounded. Actually, this "heel" impression is that of a metatarsal pad, so that there are four digits anterior to it, and occasionally digit V is impressed posteriorly. A large number of ichnospecies of *Brachychirotherium* have been named (Haubold, 1971, p. 56-57), but we make no attempt to assign an ichnospecific name to the Redonda specimens.

These specimens (Fig. 3C-E) well preserve the thick digits I-IV of the pes, and some specimens (Fig. 3D-E) show a poorly differentiated manus impression immediately anterior to the pes. *Brachychirotherium* tracks are those of a crurotarsan archosaur, and those from the Chinle Group are most reasonably assigned to an aetosaur (Lockley and Hunt, 1995). Aetosaurs are among the most common body fossils of tetrapods from the Chinle Group.

Grallator

Grallator is the name we apply to tridactyl tracks of small theropod dinosaurs found in the Redonda Formation (Fig. 4D-E). These relatively small tracks (pes length ~8-12 cm) have a very long middle toe and long, thin digits with clawed tips, and distinct, well-differentiated digital pads in most examples. Lockley

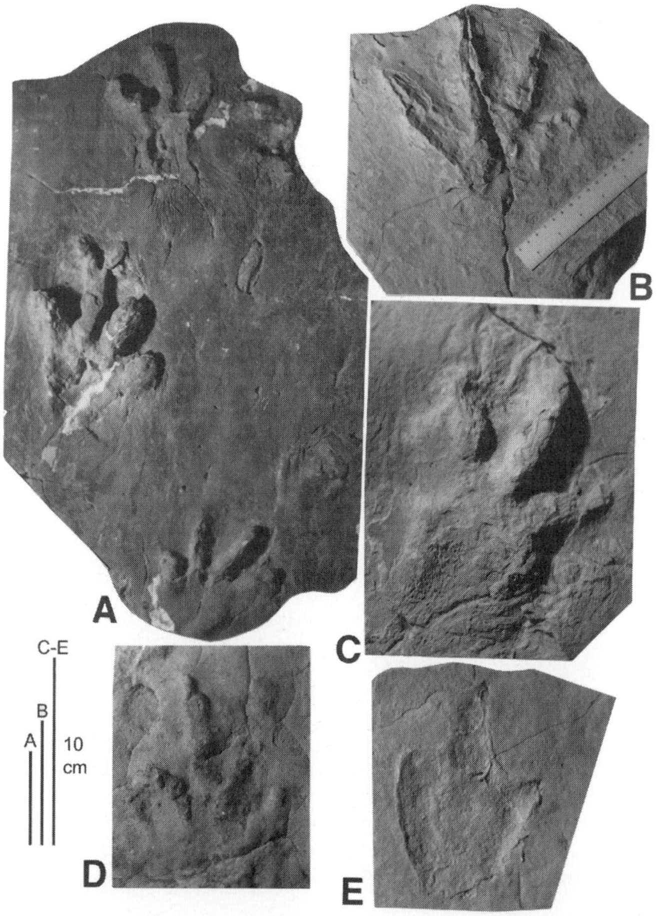

FIGURE 4. Selected tetrapod footprints from the Redonda Formation. A, *Pseudotetrasauropus*, NMMNH P-22416 from NMMNH locality 446. B, *Pseudotetrasauropus*, NMMNH P-22415 from NMMNH locality 445. C, *Pseudotetrasauropus*, NMMNH P-14151 from NMMNH locality 1477. D, *Grallator*, MDM 209 from Apache Canyon 2. E, *Grallator*, UMMP 16161 from Mesa Redonda.

et al. (2000) and Hunt et al. (2000) recently reviewed *Grallator* records from the Redonda Formation, noting that some specimens earlier assigned to this ichnogenus by Hunt and Lucas (1989), Hunt et al. (1989) and Lucas (1993) are actually tridactyl extramorphological variants of *Pseudotetrasauropus* (Fig. 2D, Fig. 4C) Thus, contrary to earlier statements (e.g., Hunt et al., 1989), tracks of *Grallator* are not the most common tracks in the Redonda Formation; instead, tracks of *Pseudotetrasauropus* are most abundant.

PSEUDOTETRASAUROPUS

Relatively large tracks (pes length ~ 12-16 cm) of a biped from the Redonda Formation are assigned by us to *Pseudotetrasauropus* (Fig. 4A-C). These tracks are tetradactyl, with widely diverging, long slender digits that usually taper to sharp, pointed claws. On one Redonda specimen there is an unusual extramorphological variant — a bifurcation of a digit tip (Fig. 4A), which may simply represent the pentration and extraction phases of the claw (cf. Peabody, 1959). Cotton et al. (1997, 1998) mapped an area at Apache Canyon of ~ 10 m^2 that shows evidence of preferred orientation of seven *Pseudotetrasauropus* trackways, probable evidence of gregarious behavior.

Pseudotetrasauropus is the most common track known from the Redonda Formation (Lockley et al., 2000). It resembles somewhat larger prosauropod tracks termed *Otozoum* (Gand et al., 2000), and well matches the foot structure of a prosauropod dinosaur (Lockley and Hunt, 1995).

DISCUSSION

The significance of the Redonda Formation tetrapod ichnofauna is threefold:

1. A unique feature of the Redonda tetrapod ichnofauna is the absence of the sauropod track *Tetrasauropus*, which is the dominant track at Peacock Canyon in the Upper Triassic Sloan Canyon Formation of northeastern New Mexico and in the Upper Triassic Sheep Pen Sandstone at Furnish Canyon in southeastern Colorado (Lockley et al., this guidebook). This may indicate some sort of mutual ecological separation between prosauropods and sauropods in the Redonda depositional basin, though the two ichnogenera do co-occur elsewhere.

2. The dominance of prosauropod tracks in the Redonda Formation is not mirrored by its body fossil record, which lacks prosauropods and is dominated by semiaquatic phytosaurs and metoposaurs.

3. The Redonda tetrapod ichnofauna resembles other Chinle Group tetrapod ichnofaunas of the Apachean Rock Point sequence. These ichnofaunas are characterized by *Rhynchosauroides*, *Brachychirotherium*, *Grallator* and *Pseudotetrasauropus* or *Tetrasauropus* (e.g., Lockley and Hunt, 1995). Indeed, the Redonda tetrapod ichnofauna well represents a global Late Triassic ichnofauna characterized by *Rhynchosauroides*, *Brachychirotherium*, *Grallator*, *Pseudotetrasauropus* and *Tetrasauropus*. Such ichnofaunas are known from the lower Stormberg Group of southern Africa (Ellenberger, 1970; Olsen and Galton, 1984; Raath et al., 1990; Raath, 1996), the upper Keuper and equivalent strata in western Europe (e.g., Haubold, 1971, 1984; Lockley and Meyer, 2000; Gand et al., 2000) and the Upper Triassic portion of the Newark Supergroup in eastern North America (e.g., Silvestri and Szajna, 1993; Olsen et al., 1998) as well as the Chinle Group in the American Southwest.

ACKNOWLEDGMENTS

We are grateful to the Abercrombie and Duke families for access to Redonda Formation tracksites on their land. Andrew Heckert and John Foster reviewed the manuscript.

REFERENCES

Baird, D., 1964, Dockum (Late Triassic) reptile footprints from New Mexico: Journal of Paleontology, v. 38, p. 118-125.

Cotton, W. D., Cotton, J. E., and Hunt, A. P., 1997, Social behavior of Cretaceous ornithopod dinosaurs of New Mexico: New Mexico Museum of Natural History and Science, Bulletin 11, p. 45-50.

Cotton, W. D., Cotton, J. E., and Hunt, A. P., 1998, Evidence for social behavior in ornithopod dinosaurs from the Dakota Group of northeastern New Mexico, USA: Ichnos, v. 6, p. 141-149.

Cotton, W. D., Hunt, A. P., Cotton, J. E., and Lockley, M. G., 1996, An addition to the vertebrate ichnofauna of the Redonda Formation (Upper Triassic), east-central New Mexico: New Mexico Geology, v. 18, p. 56.

Ellenberger, P., 1970, Les niveaux paléontologiques de première apparition des mammifères primordiaux en Afrique du Sud et leur ichnologie: Establissement de zones stratigraphiques detaillées dans le Stormberg du Lesotho, (Afrique du Sud) (Triassique supérieur à Jurassique; *in* Haughton, S. H., ed., International Union of Geological Sciences, Second Symposium on Gondwana Stratigraphy and Paleontology. Pretoria, Council for Scientific and Industrial Research, p. 347-370.

Gand, G., Vianey-Liaud, M., Demathieu, G., and Garric, J., 2000, Deux nouvelles traces de pas de dinosaures du Trias Supérieur de la Bordure Cévenole (La Grand-Combe, sud-est de la France): Geobios, v. 33, p. 599-624.

Gregory, J. T., 1972, Vertebrate faunas of the Dockum Group, Triassic, eastern New Mexico and west Texas: New Mexico Geological Society, 23rd Field Conference, Guidebook 23, p. 120-123.

Haubold, H., 1971, Ichnia amphibiorum et reptilorum fossilum: Encyclopedia of Paleoherpetology, v. 18, 124 p.

Haubold, H., 1984, Saurierfährten. Wittenberg Lutherstadt, A. Ziemsen Verlag, 231 p.

Hester, P. M., 1988, Depositional environments in a Late Triassic lake, east-central New Mexico [M. S. thesis]: Albuquerque, University of New Mexico, 153 p.

Hunt, A.P. and Lucas, S.G., 1989, Late Triassic vertebrate localities in New Mexico; *in* Lucas, S. G., and Hunt, A. P., eds., Dawn of the Age of Dinosaurs in the American Southwest. Albuquerque, New Mexico Museum of Natural History, p. 72-101.

Hunt, A. P., Lucas, S. G., Lockley, M. G., and Heckert, A. B., 2000, Occurrence of the dinosaurian ichnogenus *Grallator* in the Redonda Formation (Upper Triassic: Norian) of eastern New Mexico: New Mexico Museum of Natural History and Science, Bulletin 17, p. 39-41.

Hunt, A. P., Lucas, S. G., and Kietzke, K. K., 1989, Dinosaur footprints from the Redonda Member of the Chinle Formation (Upper Triassic), east-central New Mexico; *in* Gillette, D. D., and Lockley, M. G., eds., Dinosaur tracks and traces: Cambridge, Cambridge University Press, p. 277-280.

Hunt, A. P., Lockley, M. G., and Lucas, S. G., 1993, Vertebrate and invertebrate tracks and trackways from Upper Triassic strata of the Tucumcari basin, east-central New Mexico: New Mexico Museum of Natural History and Science, Bulletin 3, p. 199-201.

Lockley, M. G., and Hunt, A. P., 1995, Dinosaur tracks and other fossil footprints of the western United States. New York, Columbia University Press, 338 p.

Lockley, M. G., and Meyer, C., 2000, Dinosaur tracks and other fossil footprints of Europe. New York, Columbia University Press, 323 p.

Lockley, M. G., Lucas, S. G., and Hunt, A. P., 2000, Dinosaur tracksites in New Mexico: A review: New Mexico Museum of Natural History and Science, Bulletin 17, p. 9-16.

Lucas, S. G., 1993, Dinosaurs of New Mexico. Albuquerque, New Mexico Academy of Science, 130 p.

Olsen, P. E., and Galton, P. M., 1984, A review of the reptile and amphibian assemblages from the Stormberg of South Africa, with special emphasis on the footprints and the age of the Stormberg: Palaeontologia Africana, v. 25, p. 87-110.

Olsen, P. E., Smith, J. B., and McDonald, N. G., 1998, Type material of the species of the classic theropod footprint genera *Eubrontes, Anchisauripus,* and *Grallator* (Early Jurassic, Hartford and Deerfield basins, Connecticut and Massachusetts, U.S.A.): Journal of Vertebrate Paleontology, v. 18, p. 586-601.

Peabody, F., 1959, Trackways of living and fossil salamanders: University of California, Publications in Zoology, v. 63, no. 1, 72 p.

Raath, M. A., 1996, Earliest evidence of dinosaurs from central Godwana: Memoirs of the Queensland Museum, v. 39, p. 703-709.

Raath, M. A., Kitching, J. W., Shone, R. W., and Rossow, G. J., 1990, Dinosaur tracks in Triassic Molteno sediments: The earliest evidence of dinosaurs in South Africa: Palaeontologia Africana, v. 27, p. 89-95.

Silvestri, S. M. and Szajna, M. J., 1993, Biostratigraphy of vertebrate footprints in the Late Triassic section of the Newark Basin, Pennsylvania: Reassessment of stratigraphic ranges: New Mexico Museum of Natural History and Science, Bulletin 3, p. 439-445.

THE LATE TRIASSIC SAUROPOD TRACK RECORD COMES INTO FOCUS: OLD LEGACIES AND NEW PARADIGMS

MARTIN G. LOCKLEY[1], JOANNA L. WRIGHT[1], ADRIAN P. HUNT[2] AND SPENCER G. LUCAS[3]
[1]Geology Department, University of Colorado at Denver, PO Box 173364, Denver, Colorado, 80217;
[2] Mesalands Dinosaur Museum, Mesa Technical College, 911 Tenth Street, Tucumcari, New Mexico 87401,
[3]New Mexico Museum of Natural History, 1801 Mountain Road NW, Albuquerque NM 87104

Abstract.—Vertebrate tracks assigned to the ichnogenera *Tetrasauropus* and *Pseudotetrasauropus* (cf. *Otozoum*) have traditionally been attributed to robust and gracile prosauropods, respectively. Based on track morphology, we propose that the former ichnogenus (*Tetrasauropus*), as represented in the uppermost Triassic (Norian-Rhaetian) of western North America, and other regions, can be attributed to the Sauropoda. The suggestion that sauropod tracks exist in the Upper Triassic is no longer at variance with the body fossil record, as it was until recently. Previous failure to attribute *Tetrasauropus* to the Sauropoda provides a good example of the reluctance of ichnologists to make interpretations that do not concur with the skeletal record. While such caution is advisable in the still-specialized and immature science of vertebrate ichnology, it may also lead us to misinterpret important, even abundant, ichnological evidence that is easily accessible. Hindsight has revealed many other instances of the predictive potential of ichnology, and the reluctance of mainstream paleontology to accept predictions that have not been absolutely confirmed by osteological evidence.

Parallel evolutionary trends towards larger size in theropod tracks (e.g., *Grallator*), prosauropod tracks (*Pseudotetrasauropus* and *Otozoum*) and presumed sauropod tracks (*Tetrasauropus* and *Brontopdus*, or *Brontopodus*-like ichnotaxa) suggest that the interpretations proposed here are the most parsimonious and internally consistent. They indicate a convergent unfolding of evolution among all saurischian clades during the Late Triassic into the Jurassic.

INTRODUCTION

It has long been known that the tracks of dinosaurs and other vertebrates may occur in strata that lack body fossils. In fact, skeletal remains of a particular trackmaker, known only from some corresponding ichnospecies (or higher ichnotaxon) designation, may be completely unknown in the body fossil record. This has been referred to as the "Cinderella syndrome" (Lockley, 1998). Similarly, because tracks are often abundant, particular types may occur in strata that predate the earliest skeletal remains that have been reliably attributed to corresponding skeletal forms.

This seems to have happened in the case of various sauropodomorph track types from South Africa and North America, and we reinterpret a number of tracks assigned to "prosauropod ichnotaxa" as sauropodan. Our conclusions are based on a number of lines of evidence, including the discovery of body fossils of true sauropods in the Triassic (Buffetaut et al., 2000). However, equally important from the ichnological perspective is that, despite a significant literature that assumes a prosauropod affinity for the tracks, based on age and size, the track types in question (ichnogenus *Tetrasauropus sensu lato*) show morphological features of true sauropod tracks. Therefore, they cannot be considered prosauropodan (or "pre-sauropodan") without taking the unwarranted step of proposing a hypothetical sauropod-like, non-sauropodan trackmaker that coexisted with the earliest true sauropods. It is true that large prosauropods like *Riojasaurus* may have been the trackmakers, but this assumption simply postulates more sauropod-like (large) and less sauropod-like (small) forms (VanHeerden, 1997), still leaving open the question of whether there is also a record of true sauropod tracks at this time.

Other lines of evidence that support the interpretation of at least the North American *Tetrasauropus* trackmakers as sauropods is their small size, in comparison to post-Triassic sauropod tracks, and their co-occurrence with other small saurischians (theropods and true prosauropods). However, all three saurischian groups show consistent and convergent trends in increasing maximum size during the latest Triassic and earliest Jurassic.

The implications of these conclusions are twofold. First, in a strictly paleontological sense, we learn an object lesson about our reluctance to accept track evidence, no matter how abundant or compelling, unless it is backed up by the corresponding skeletal evidence of matching trackmakers, in strata of the same or similar age (Lockley, 1998, 1999a). Indeed, we may now conclude that the track evidence of Late Triassic sauropods has been staring us in the face for some time, and that despite mentioning or hinting at the possibility of such intepretations (Ellenberger, 1972; Conrad et al., 1987), the main reason for our reluctance to advance such conclusions has been the lack of corresponding skeletal remains.

Second, in a more general sense we must recognize that such instances of "reinterpretation" point to various inherent biases in science. Mainstream paleontology generally regards the hard "material" evidence of skeletal remains as reliable and views "soft," ichnological evidence as ambiguous and unreliable, despite repeated examples of track discoveries (e.g., *Chirotherium*) that are retrospectively acknowledged as prophetic and highly significant. Further examples are given in the discussion below. For these reasons we tend to approach the objects of our study with excessive skepticism. Although the inherent nature of track evidence requires a certain caution, too often such restraints create self-fulfilling prophecies and we fail to appreciate fully what the track evidence has to offer. This paper aims to rectify one such oversight.

HISTORY OF RESEARCH

Although the possibility of Upper Triassic sauropods had been speculated upon, based on cladistic extrapolations (e.g., Wilson and Sereno, 1998; Upchurch, 1993), it has not been proven with actual fossils until comparatively recently (Buffetaut et al., 2000). The existence of Upper Triassic prosauropods, and their tracks, however, has been known for some time. Even the famous *Chirotherium* ichnites, discovered in Europe during the 1820s and 1830s, were interpreted as prosauropodan (by Baron von Nopsca) despite being incompatible morphologically and, as we now know, too old: i.e., Early to Middle Triassic in age (see Lockley and Meyer, 2000 for recent summary). This interpretation left us with the legacy of perceiving a relationship between prosauropods and chirothere tracks (Olsen and Galton, 1984; Conrad et al., 1987; VanHeerden, 1997). Such perceptions are understandable given that various chirothere and chirothere-like tracks are found throughout the Triassic, and overlap the time range of well-known Late Triassic prosauropods such as *Plateosaurus*.

The first serious study of Upper Triassic track successions in which abundant sauropodomorph tracks occur was conducted by Ellenberger (1972, 1974) in southern Africa. Due to the remote location of most of his sites, the publication of his main work (in French), in an obscure monograph series, and the unproven assumption that his description of these ichnofaunas was "over-split" (Olsen and Galton, 1984), many of his conclusions have been ignored or not taken as seriously as they deserve (Lockley and Meyer, 2000). Subsequent work, particularly in North America (Lockley and Hunt, 1995; Lockley et al., 2000), but also in Europe (Lockley et al., 1996; Lockley and Meyer, 2000), confirms the validity of much of Ellenberger's work. More important than the possibility of different interpretations is the need to take the actual track evidence seriously and study it as objectively as possible. This is what we have attempted to do in the sections that follow.

Among the most significant of Ellenberger's conclusions were those that pointed to the widespread distribution of sauropodomorph trackmakers in the Late Triassic and Early Jurassic. He described *Sauropodopus* (Ellenberger, 1972, p. 73), which, as the name implies, is a relatively large track (pes length 26 cm) of possible sauropodan affinity, though this was questioned by Olsen and Galton (1984) and Lockley and Meyer (2000, p. 87-89). Ellenberger (1972, p. 88-91) also described *Deuterosauropodupus*, a larger trackmaker (pes length 40 cm) that he considered better adapted to quadrupedal progression than *Sauropodopus*. Thulborn (1990) included *Deuterosauropodupus* (but not *Sauropodopus*) in his catalog of miscellaneous tracks attributed to sauropods.

Of much greater interest are the two related ichnogenera, *Tetrasauropus* and *Pseudotetrasauropus* (Ellenberger, 1972), both of which have been widely accepted as probable examples of prosauropod/sauropodomorph tracks and recognized in both North America and Europe (Thulborn, 1990; Lockley and Hunt, 1995; Lockley et al., 1996, 2000; Hunt et al., 2000; Gand et al., 2000; Lockley and Meyer, 2000). In general, *Tetrasauropus* was made by a quadrupedal animal with large fleshy feet (pes length 44 cm), in which only the distal ends of the digits (claws) are separate from the plantar surface of the foot. In contrast, *Pseudotetrasauropus* represents a predominantly bipedal animal that had all pes digits widely separated and characterized by distinctive digital pads (Fig. 1). Ellenberger (1972, p. 59) noted that *Pseudotetrasauropus* is very similar to the probable prosauropod ichnogenus *Otozoum* from the Lower Jurassic (see Olsen and Galton, 1980; Thulborn, 1990; and Lockley and Meyer, 2000 for further re-evaluation of this material). A smaller track of this same general type was assigned by Ellenberger (1972) to the ichnogenus *Kalosauropus* (Lockley and Meyer, 2000, p. 91-93).

Although the ichnological community has limited access to the original South African specimens, some replicas are available in the University of Montpellier collections, and it is now clear that *Pseudotetrasauropus* (which is similar to *Otozoum*) is a good match for the prosauropod foot. Digit I in this ichnogenus is shorter, and usually much less deeply-impressed than digits II-IV. In contrast, in *Tetrasauropus* digits I, II and III are almost equally large and pronounced, as is typically seen in the tracks of later (Jurassic and Cretaceous) sauropods. In an attempt at standardization, the two ichnogenera *Tetrasauropus* and *Pseudotetrasauropus* have been applied to purported sauropodomorph tracks from North America and Europe (Lockley and Hunt, 1995; Lockley et al., 1996, 2000). However, due to the lack of skeletal evidence, it was assumed that they represent robust and gracile prosauropods, respectively. As indicated below, further examination of the abundant material now available suggest that the robust, fleshy forms are better assigned to true sauropods.

Baird (1980) described the trackway of a small quadruped from the Navajo Sandstone (Lower Jurassic) of Arizona as an example of a prosauropod trackway and named it *Navahopus*. He made much of the purported impression of a large pollex claw characteristic of prosauropods. Some authors (e.g., Olsen and Galton, 1984; Thulborn, 1990) have accepted the purported pollex claw impression as evidence of prosauropod affinity. Others, however, including ourselves, have not. The claw impression is by no means clearly a morphological feature, and as pointed out by Lockley and Hunt (1995), the trackway is typical of mammal-like reptiles, whose footprints are the dominant type in the Navajo dune facies. As noted below, the introduction of this name, and its subsequent use by Olsen and Galton (1984), has lead to various ichnotaxonomic complications.

Prior to our present claim that Upper Triassic *Tetrasauropus* tracks from western North America are sauropodan in origin, the oldest tracks attributed to sauropods from North America were those reported by Foster et al. (2000) from a poorly-dated (?late Middle or ?early Late Jurassic) part of the Entrada Formation, near Escalante, Utah. Elsewhere in the world the oldest tracks attributed to sauropods include examples from the Lower Jurassic of Morocco (Jenny and Josen, 1982; Ishigaki, 1988) and from the Lower Jurassic of Italy (Lanzinger and Leonardi, 1991; Leonardi and Avanzini, 1994).

ICHNOTAXONOMIC USAGE

When first studied in North America, the tracks (up to about 30 cm long) that we now assign to *Tetrasauropus* and which "at first sight....resemble those of small sauropods." (Fig. 2; Conrad et al., 1987, p. 132) were tentatively assigned to the ichnogenus *Chirotherium*. This error was based more on interpretation of the manus than on pes morphology. As pointed out by H. Haubold (personal communication, 1997), *Chirotherium* pes digits are straight and not curved outward as in *Tetrasauropus*. Conrad et al. (1987) also suggested that the tracks might have been made

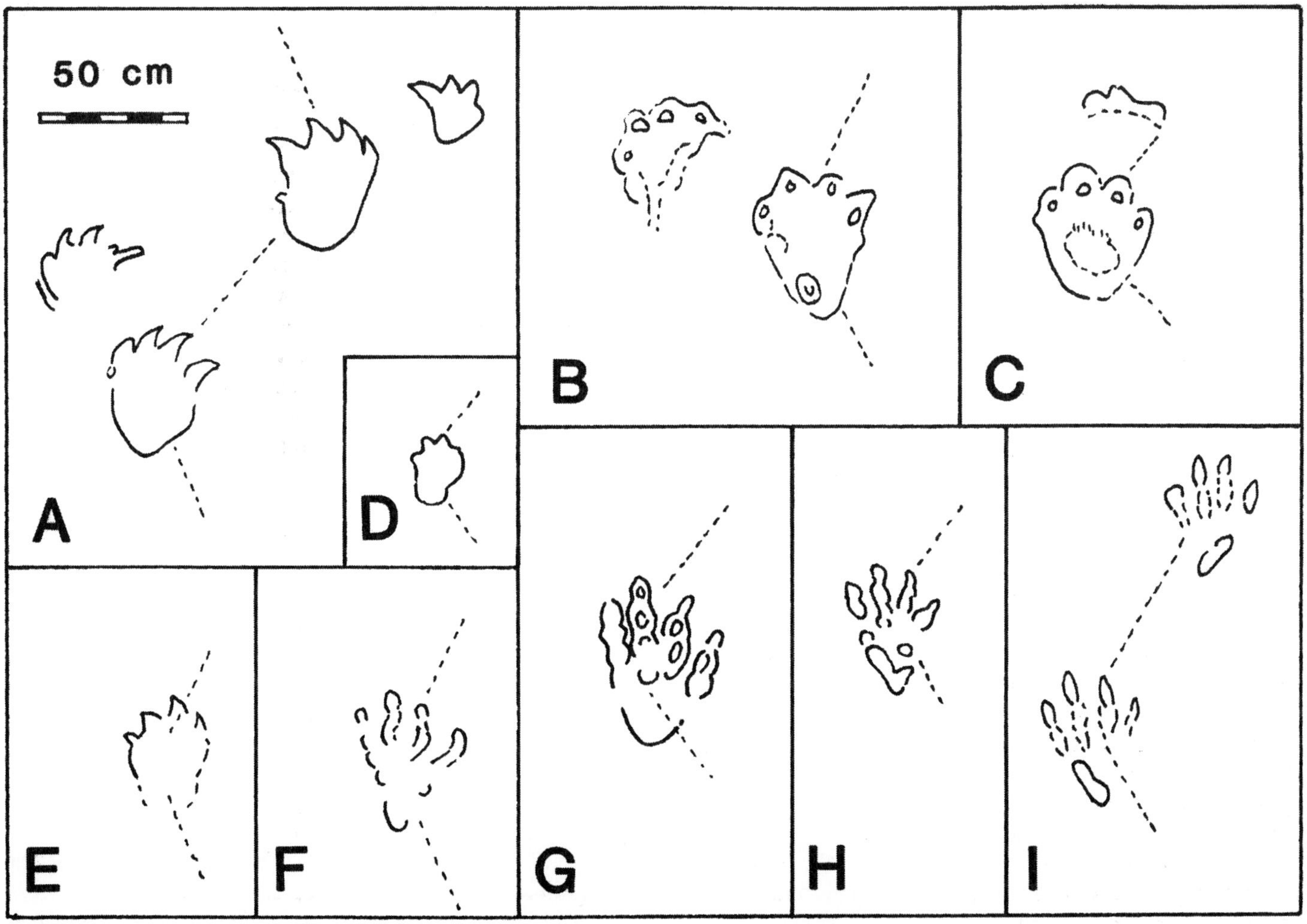

FIGURE 1. *Tetrasauropus* and *Pseudotetrasauropus* tracks, based on type specimens from southern Africa after Ellenberger (1972). Note that the latter ichnogenus resembles *Otozoum* (redrawn after Lockley and Meyer, 2000). A. *Tetrasauropus unguiferous*, B. *Pseudotetrasauropus jaquesi*, C. *P. elegans*, D. *P. dulcis*, E. *P. acutungis*, F. *P. angustus*, G. *P. mekalingensis*, H. P. *bidedoidea*, I. *Otozoum*.

by phytosaurs, because they were abundant on an exposure of exhumed fluvial channel deposits.

As more of these tracks came to light, and their similarity to *Tetrasauropus* became apparent, Lockley and Hunt (1995, p. 91) explicitly asked "What do prosauropod tracks look like?" and concluded that the label *Chirotherium* was questionable and best replaced by *Tetrasauropus* – implying sauropodomorph rather than phytosaur affinity. At the same time it became clear that the ichnogenus *Pseudotetrasauropus* was also common in North America and that it too was of probable prosauropod affinity, and similar to younger (Early Jurassic) *Otozoum*. One of the most significant differences between the South African and North American assemblages, besides the greater ichnotaxomic diversity generated by Ellenberger's liberal naming procedures, is that both *Tetrasauropus* and *Pseudotetrasauropus* are generally represented by smaller forms in North America.

A further level of complexity was added to the prosauropod tracks debate when Baird (1980) described *Navahopus* as a prosauropod track. This caused Olsen and Galton (1984) to consider "*Tetrasauropus* a synonym of *Navahopus* and although the former has historical priority (they) erected the ichnofamily Navahopidae to accommodate both ichnogenera" (Lockley et al., 1996, p. 36). We consider this conclusion untenable because *Navahopus* is evidently the track of a mammal-like reptile similar to *Brasilichnium* (Lockley and Hunt, 1995). It is not morphologically similar to *Tetrasauropus*, nor does it originate from deposits of the same age.

One of the most significant implications of regarding *Tetrasauropus* as a probable sauropod track is that we are encouraged to compare it with well-known sauropod tracks such as *Brontopodus* (Fig. 3; Farlow et al., 1989). This raises the possibility that *Brontopodus* could be a junior synonym of *Tetrasauropus* (Ellenberger, 1972). However, as discussed below, although the similarities are marked, especially in pes morphology, there are noteworthy differences in the shape of the manus and the overall trackway configuration.

It is worth noting that the type specimen of *Tetrasauropus* (Fig. 2) from South Africa shows claws that turn inward, not outward, as in most North American examples. This may indicate some subtle, or perhaps quite significant, differences in the morphology of the trackmakers, as discussed below, but may simply represent variable substrate conditions. Variation in the angle of claw impressions is found among Jurassic and Cretaceous sauro-

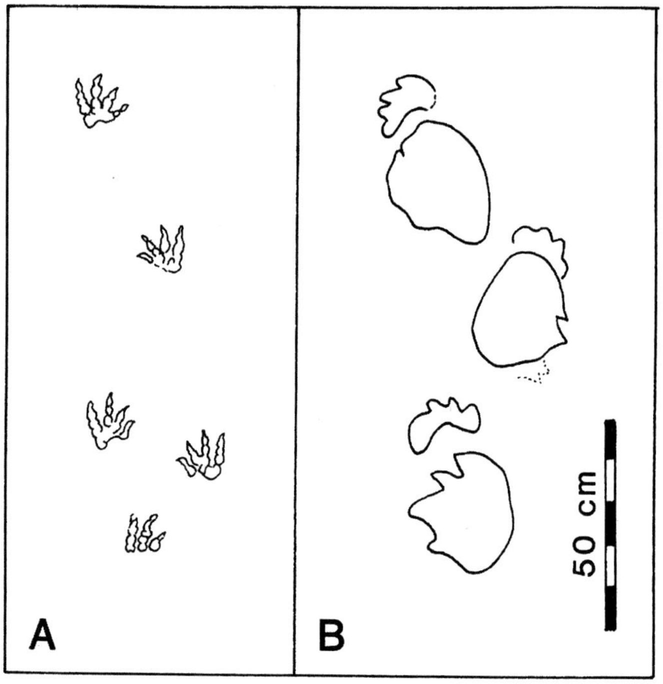

FIGURE 2. *Pseudotetrasauropus* (A) and *Tetrasauropus* (B) tracks based on North American sample (modified after Lockley and Hunt, 1995)

pod tracks, though most point outward or forward, not medially.

DATABASE

Upper Triassic track evidence from the western United States, compiled by the present authors and their colleagues, provides a coherent body of data which makes it possible to base ichnological conclusions, censuses and interpretations on a large data set and a significant body of literature. Almost all of the tracksites are found in the upper part of the Chinle Group (Rock Point sequence and Apachean land-vertebrate faunachron, *sensu* Lucas 1991,1993, 1998). In recent years, several additional tracksites have been reported in the Rock Point sequence, including the important Furnish Canyon site described below.

Upper Triassic rock successions in the western United States that are replete with diverse vertebrate track assemblages have been recorded in three large regions within the Chinle Group outcrop belt (i.e., the southern Colorado Plateau region, the northern Colorado Plateau region, and the eastern region, *sensu* Lockley and Hunt, 1995). *Tetrasauropus* and *Pseudotetrasauropus* tracks occur in both the northern and eastern regions, notably around Dinosaur National Monument (Lockley et al., 1992) and Gateway, Colorado. In eastern New Mexico, track-bearing outcrops have been reported from east of Folsom, New Mexico, downstream in the Dry Cimarron Valley to the vicinity of Kenton, Oklahoma, as well as in southeastern Colorado (Conrad et al., 1987; Lockley and Hunt, 1993, 2000). For example, a hitherto undescribed site in Furnish Canyon, Colorado northeast of Kenton, is characterized by large bedding plane exposures imprinted with long, sub-parallel trackways attributed to the ichnogenus *Tetrasauropus* (Fig. 4). Late Triassic track assemblages similar to those found in the Dry Cimarron Valley also occur elsewhere in east-central New Mexico, notably at Redonda Mesa and Apache Canyon (Hunt et al., 2000; Lockley et al., 2000).

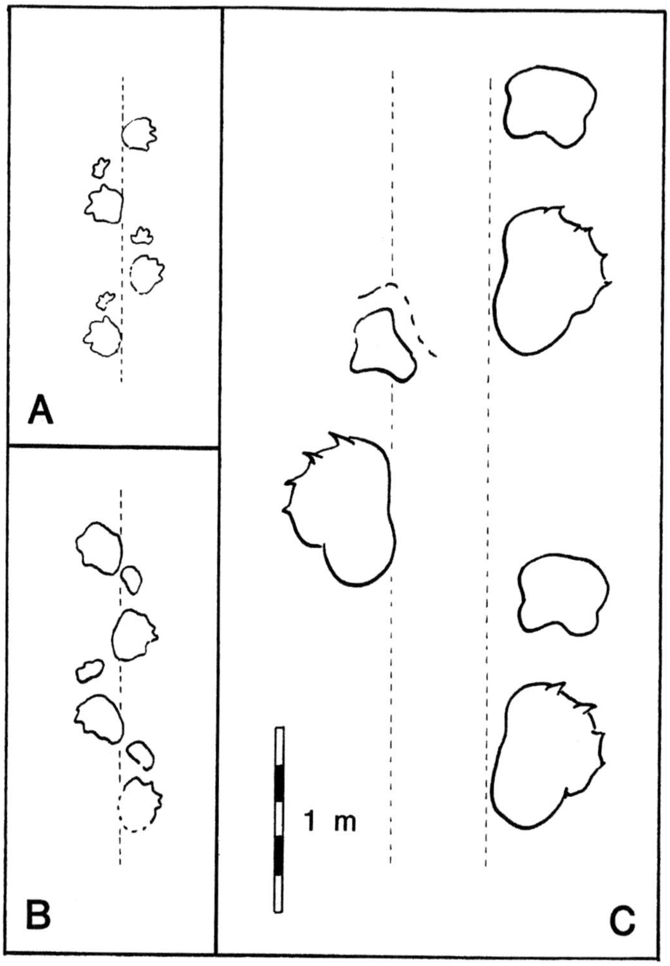

FIGURE 3. Comparison between North American *Tetrasauropus* (A and B from Cub Creek and Peacock Canyon, respectively) and *Brontopodus birdi* (C) from the Cretaceous of Texas (modified after Lockley and Hunt, 1995, and Farlow et al., 1989).

TETRASAUROPUS MORPHOLOGY

Tetrasauropus unguiferous, the type ichnospecies of the ichnogenus *Tetrasauropus,* has a large pes (44 cm long) with a smaller manus situated antero-lateral to the pes. The digits on both the manus and the pes appear to curve inward toward the trackway midline. The step is short (79-83 cm); i.e., less than twice foot length. Pace angulation is about 100-110° (Ellenberger, 1972, fig. 36; Lockley and Meyer, 2000, fig 4.7). It is not possible to determine if the trackway is narrow- or wide-gauge. However, if the partial type trackway, shown in Figure 1A, were extrapolated, using the pace distances and angulation angles indicated by Ellenberger's diagram, the trackway would appear to be narrow-gauge or very slightly wide-gauge (i.e., with little or no space between the inner margins of the pes prints). The space between the manus prints, however, is quite considerable. A second ichnospecies (*T. gigas*) originally assigned to ichnogenus *Tetrasauro-*

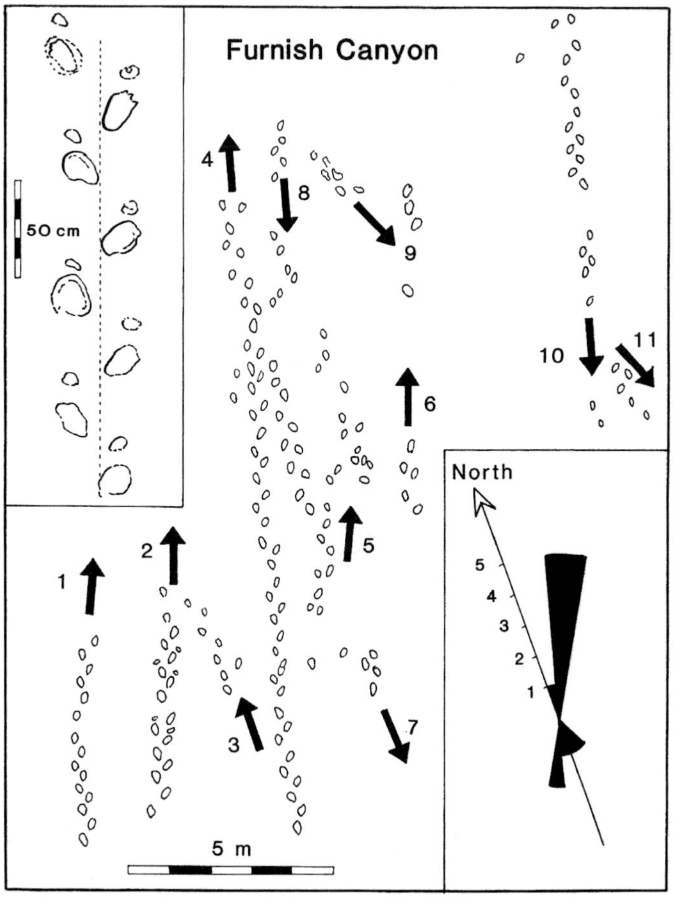

FIGURE 4. Map of Furnish Canyon *Tetrasauropus* tracksite, southeastern Colorado. Note preferred NNE and SSW trackway trends. Inset show details of trackway 2 (based on CU Denver tracing number 508).

FIGURE 5. Photograph of *Tetrasauropus* trackway from Peacock Canyon. Compare with Lockley (1986, pl. 1, top left).

pus, is here considered a synonym of *Pentasauropus*, a suggestion endorsed by Ellenberger (Lockley and Meyer, 2000, p. 88).

Tracks now assigned to *Tetrasauropus* (Fig. 5) from North America were first illustrated by Lockley (1986, pl. 1, fig. 16) and Conrad et al. (1987, figs. 4-5), and assigned tentatively to the ichnotaxon *Chirotherium* sp. In an attempt to relate abundant North American ichnofaunas to comparable material described from other regions, many additional ichnites were subsequently illustrated (cf. Lockley and Hunt, 1995, fig 3.13 and Lockley et al., 2000, fig 4) and assigned to the ichnogenus *Tetrasauropus*. We note, however, that the North American ichnites have pes digit impressions that consistently curve outwards. Whether the North American ichnites really belong in this ichnogenus (*Tetrasauropus*) depends on the degree to which the material is considered similar to material from southern Africa. As pointed out by Lockley and Meyer (2000), no one is really familiar with large samples of material from both regions. It is nonetheless our opinion that the North American material is not attributable to *Chirotherium*, and that its similarity to South African *Tetrasauropus* is noteworthy.

In the North American sample assigned to *Tetrasauropus*, the size of the tracks ranges from about 20 –30 cm (pes length) with quite a large differential between the size of the manus (small) and the pes (large). Pes pace angulation varies (120-140°) as does manus pace angulation (90-140°). It is also noticeable that the long axis of the pes is rotated outwards much more than in the type trackway from southern Africa (compare Fig. 1A with 2B and 3A,B). The North American trackways are also consistently narrow-gauge. Differences between North American and South African *Tetrasauropus*, especially in size, digit curvature directions and, to a lesser extent, probable pes axis alignment and trackway gauge, may argue for placing the North American ichnites in a different ichnotaxon, at least at the ichnospecies, if not at the ichnogenus level. However, in the interests of caution, no formal ichnotaxonomic revision is attempted here

In contrast to *Tetrasauropus, Pseudotetrasauropus* almost invariably appears to have been made by a bipedal trackmaker (Ellenberger, 1972; Lockley and Hunt, 1995; Lockley and Meyer, 2000), causing Olsen and Galton (1984) to suggest, erroneously in our opinion, that the tracks were made by bipedal chirotheres of the ichnogenus *Brachychirotherium* (VanHeerden, 1997, p. 253) The *Tetrasauropus* pes track is morphologically very similar to later sauropod tracks such as *Brontopodus* and related forms. However, the manus is different in that the individual digits, though relatively blunt, are more clearly defined. Viewed from the morphodynamic perspective proposed by Lockley (1999a,b), there is a coherent relationship between saurischian feet and

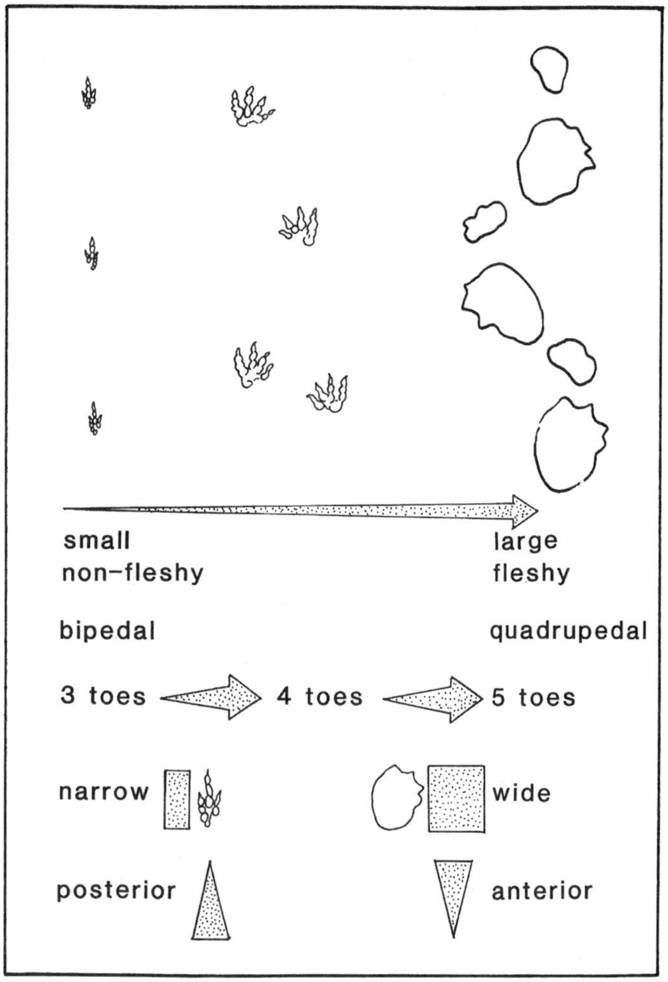

FIGURE 6. Morphodynamic relationships among saurischian (theropod, prosauropod and sauropod) tracks based on Triassic ichnofaunas of western North America. Note increase in size, fleshiness, number of digits, and trend from bipedal narrow trackways (posterior emphasis) to wide quadrupedal configuration (anterior emphasis).

tracks attributed to theropods, prosauropods and sauropods, as follows. They widen from three to four to five functional toes and become progressively wider and more fleshy, as the animals in question attain progressively larger maximum sizes. The fleshiness is also added progressively from the posterior as the morphological center of gravity of the whole animal moves forward. This anteriorization trend--also observed in other dinosaurian clades—is also very evident in the shift from obligatory bipedal to facultative biped to obligatory quadruped, and is accompanied by a general increase in overall body size, a lengthening of limbs and, perhaps counter-intuitively, a corresponding shortening of the step. This can be seen by arranging typical Late Triassic theropod, prosauropd and sauropod tracks (*Grallator*, *Pseudotetrasauropus* and *Tetrasauropus,* respectively) along this morphodynamic gradient (Fig. 6).

Most notable among these trends is the shift to quadrupedal locomotion in *Tetrasauropus*. Again, using the morphodynamic perspective, the earliest quadrupeds in many archosaurian and dinosaurian clades appear to have belonged to the small manus type (cf. Peabody, 1948; Lockley, 1999a). For example, there is a clear pattern of increasing manus size among sauropods during the Jurassic and Cretaceous, with the former small manus varieties being attributed to "primitive" diplodocids, which generally had narrow gauge trackways, and the latter, large manus varieties being attributed to "derived" brachiosaurids/titanosaurids, which generally had wider gauge trackways (Wilson and Carrano, 1999). North American *Tetrasauropus* ichnites fit this predictable pattern. They are smaller than most post-Triassic sauropod tracks, have a narrow-gauge trackway (primitive condition), and the manus is small and incompletely enclosed in a fleshy pad in comparison with later sauropods (also a primitive condition). The femoral morphology of the Triassic sauropod *Isanosaurus* (Buffetaut et al., 2001) suggests a diplodocid-like configuration of the distal condyles in which the femora are angled slightly inward towards the trackway midline, thus retaining the narrow gait (and gauge) characteristic of the primitive condition in saurischians.

It is also noteworthy that many *Tetrasauropus* trackways show partial or complete pes overlap of the manus. This, in addition to the strong outward rotation of the hind feet, is very characteristic of sauropod trackways. However, as noted above, overlap configurations, variabilities and other trackway parameters in *T. unguiferous*—the type *Tetrasauropus* trackway—are not known with certainty.

PALEOECOLOGICAL CENSUS

As noted above, tracks are abundant at many Chinle Group sites, and data are relatively easy to collect quickly. Even though there is doubt about the affinity of trackmakers, distinctive track types can be recorded and assigned to ichnotaxonomic categories. We also advocate the making of site maps so that other researchers can assess the significance of the track data. This has been done at many sites. For example, the map of the Cub Creek site near Dinosaur National Monument (Lockley and Hunt, 1995, fig 3.23) shows the distribution of 35 trackways attributed to dinosaurs (*Grallator* and *Pseudotetrasauropus*) and other archosaurs (*Brachychirotherium*). Two of the 22 known sites in this area reveal the presence of *Tetrasauropus*, and a census based on 111 trackways from all tracksites in the region reveals a total of 16 trackways of this type (see Lockley and Hunt, 1995, fig 3.24). We suggest one revision to the conclusions presented by Lockley et al. (1992a, Fig. 5) in which three trackways, from site 91-6, are attributed to phytosaurs (Fig. 7; Conrad et al., 1987). These trackways are probably best assigned to *Tetrasauropus*. Here, the relationship between trackways with manus-pes separation (1 and 2) and overlap (3) is clearly seen as a function of stride length.

In the region of Gateway, Colorado, similar trackways have been reported, both in the upper part of the Chinle Group (Rock Point sequence) and in the basal Wingate, which, based on the presence of *Brachychirotherium,* are also Late Triassic in age, as is the basal Glen Canyon Group (or Navajo-Nugget sequence) at Dinosaur National Monument (Lockley et al., 1992a). Many of these trackways are deep, and details of foot morphology are poorly preserved. Thus, few have been described in detail. Nevertheless, at least one visually distinctive trackway has been illus-

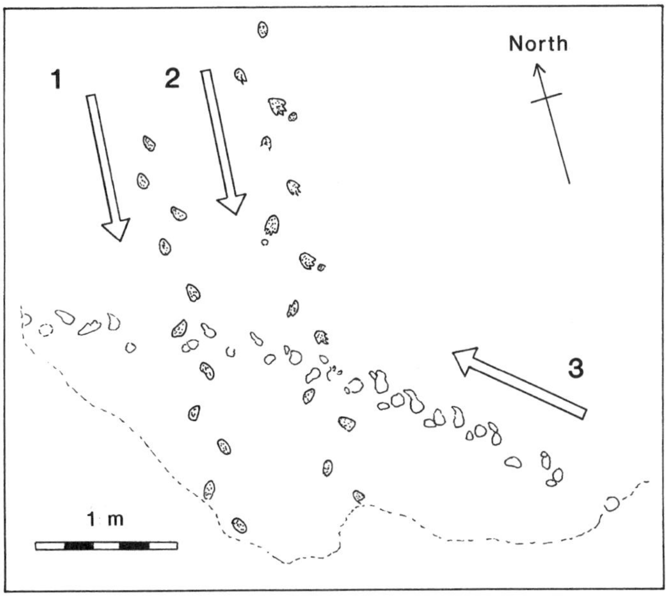

FIGURE 7. Dinosaur National Monument site 91-6 shows three trackways that we assign to the ichnogenus *Tetrasauropus* (after Lockley et al., 1992). Note the two parallel trackways (1 and 2), which are shallow, show a long step, and have no pes on manus overlap, and the third (3), which is deep, and shows a short step with partial pes on manus overlap.

trated (Lockley and Hunt, 1995, fig 4.4). Although some of these trackways appear to be have been made by bipeds, there are many sites, including Dinosaur National Monument sites (Cub Creek and 91-6), Peacock Canyon, New Mexico and Furnish Canyon, Colorado where it is easy to distinguish both manus and pes tracks (Figs. 2-7). Thus, it appears that the *Tetrasauropus* morphology is relatively common in the Late Triassic. This conclusion is supported by an examination of sites in northeastern New Mexico and southeastern Colorado.

In a recent review of dinosaur tracksites in New Mexico (Lockley et al., 2000, Table 1), we reported on the number of trackways attributed to dinosaurs in the following four categories: *Grallator* and cf. *Grallator*, *Pseudotetrasauropus* and *Tetrasauropus*, attributed respectively to small and medium-sized theropods and gracile and robust prosauropods. We present here a revised version of this tabulation (Table 1) to include the data from five sites: Redonda Mesa (RM), Apache Canyon (AC), Peacock Canyon (PC), Sloan Canyon (SC) and Furnish Canyon (FC). We revisited the Peacock Canyon site, where some of the best *Tetrasauropus* tracks occur, and also identified at least two good examples of *Pseudotetrasauropus*. Other changes to the table were made based on re-examination of available data.

Table 1. Distribution of four purported dinosaur track types from five sites in northeastern New Mexico and southeastern Colorado. (Updated after Lockley et al., 2000, table 1). RM= Redonda Mesa; AC = Apache Canyon; PC = Peacock Canyon; SC = Slaon Canyon; FC = Furnish Canyon.

Ichnotaxon and inferred trackmaker	Number of trackways reported from:					
	RM	AC	PC	SC	FC	TOTAL
Grallator small theropod	(1)	1	2	1	-	15
Grallator larger theropod	-	-	-	11	-	11
Pseudotetrasauropus prosauropod	1	7	2	1	-	11
Tetrasauropus sauropod	-	-	40	1	12	53
Subtotals	2	8	44	24	12	90

From the tabulated data, it is easy to determine that *Tetrasauropus* trackways are quite common, and that they are among the dominant elements at both the Peacock Canyon and Furnish Canyon sites, though non-dinosaurian trackmakers such as *Rhynchosauroides* (a lacertilian) are common at the former site. A total of 53 *Tetrasauropus* trackways have been recorded to date. When combined with at least 16 from the Dinosaur National Monument area and at least five from Gateway, Colorado, we estimate a sample in excess of 70 trackways. In all areas where *Tetrasauropus* tracks occur they are associated with abundant *Grallator* tracks and several *Pseudotetrasauropus* trackways. Thus we infer that the *Grallator-Pseudotetrasauropus-Tetrasauropus* ichnological assemblage is a widespread and significant indicator of the archosaurian/dinosaurian ichnofauna of the entire region in which upper Chinle Group (Rock Point sequence) strata crop out.

If we rank the four track types listed in Table 1 by relative abundance, we find that the sample (N= 90) is dominated by *Tetrasauropus* (59%), with theropod tracks (two ichnospecies of *Grallator sensu lato*) comprising 29% and *Pseudotetrasauropus* comprising 12%. To this census for northeastern New Mexico and Furnish Canyon, Colorado, we should add at least 20 trackways known from the Sheep Pen Sandstone of the tristate area, near Kenton, Oklahoma (Conrad et al., 1987; Lockley and Hunt, 1995; Lockley et al., 2000). This increases the census to 110 trackways, of which about 47% are *Tetrasauropus*, 43% theropod and 10% attributed to prosauropods (i.e., *Pseudotetrasauropus*). This census is probably a better reflection of the high proportion of *Grallator* tracks found at most Late Triassic sites, and, moreover, once presumed non-dinosaurian ichnotaxa are removed, the census size and percentages are very similar to estimates reported from the Dinosaur National Monument area (Fig. 8).

Our reinterpretation of *Tetrasauropus* as a probable sauropod track remains consistent with inferences made by Lockley et al (2000, p. 12), who stated that "small dinosaurian trackmakers were more widely distributed geographically than large trackmakers." By this we mean to stress that small animals are more common than large ones (Bonner, 1988). This is still essentially true because small *Grallator* and *Pseudotetrasauropus* tracks occur at four of the five sites tabulated above, whereas *Tetrasauropus* occurs at only three. This difference is even greater in the case of the data cited for the Dinosaur National Monument area, where *Tetrasauropus* occurs at only 2 of 22 sites. Because sauropod tracks are statistically most abundant at low latitude sites (Lockley et al., 1994), it may be worth plotting the paleogeographic distribution (and abundance) of *Tetrasauropus* tracksites as more data become available. Present evidence suggest that *Tetrasauropus* tracks are somewhat larger and more abundant in the Dry Cimarron Valley than in the Gateway and Dinosaur National Monument areas.

In addition to the South African sites, we know of at least four other regions where dinosaur tracks of the *Tetrasauropus* type have been reported. In South Wales (Lockley et al., 1996) a census based on 88 trackways of presumed dinosauran origin

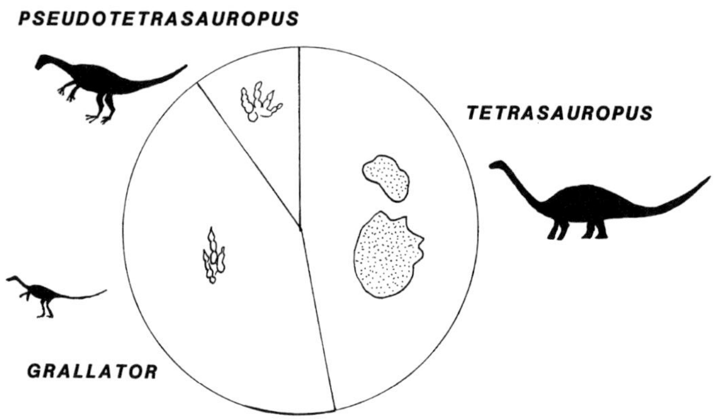

FIGURE 8. Proportions of inferred dinosaurian ichnotaxa (n =110) from northeastern New Mexico and southeastern Colorado. The data compare favorably with similar census data from strata of the same age from the Dinosaur National Monument area. See text for details.

reveals 66 theropod trackways (75%), 18 attributed to *Pseudotetrasauropus* and *Otozoum* (i.e., prosauropods) and 4 attributed to *Tetrasauropus*. Furrer (1984, 1993) reported a large *Tetrasauropus*-like trackway in the Upper Triassic of the Swiss Alps, Jenkins et al. (1994) reported a similar trackway from the Upper Triassic of Greenland (also illustrated by Lockley and Meyer, 1995, fig. 4.14), and trackways that are remarkably similar to both *Tetrasauropus* and *Pseudotetrasauropus* have been reported from the Italian Alps (Mietto, 1991; Lockley and Meyer, 2000). Some of the rounded tracks, which resemble their North American counterparts (cf. *Tetrasauropus*), have been attributed to sauropodomorphs (prosauropods).

DISCUSSION

While there is always doubt about the affinity of tracks, the progress made in vertebrate ichnology in recent years has allowed for the definition of distinctive track assemblages in well-defined stratigraphic intervals. The Upper Triassic of the western United States, in particular the Rock Point sequence of the Chinle Group (equivalent to the Apachean faunachron), is one such interval. There seems little doubt that abundant small *Grallator* tracks (many in the range of 6-10 cm foot length) represent the early radiation of small theropods, and the presence of a few larger tracks of this type in the latest Triassic indicates a phylogenetic size increase in a few groups. Similarly, it is now well established that tracks that are morphologically similar to *Pseudotetrasauropus* (Ellenberger, 1972) or *Otozoum* (Hitchcock, 1858) are highly reminiscent of prosauropod foot morphology. Again, these tracks are mostly small in Upper Triassic sequences and become progressively larger as one traces them into the Lower Jurassic. Similarly, the rounded (oval) fleshy foot tracks assigned to the ichnogenus *Tetrasauropus*, although large in the South African and Swiss samples (foot length up to 40-50 cm), are generally much smaller in the North American, Welsh, Greenland and Italian samples (15-30 cm). Given that post-Early Jurassic sauropod tracks reach much larger maximum sizes (80-100 cm), we again find evidence of a systematic size increase, which follows the trends observed for theropod and prosauropod tracks. Given that we know of no published objections to the interpretation of such *Tetrasauropus*-like track morphologies as sauropodan, when found in post-Triassic deposits, it is most parsimonious to assume that those found in the Late Triassic belong to sauropodan clades, as now supported by body fossil evidence. The previously-mentioned alternative interpretation--that they represent a hitherto-unknown robust prosauropod with fleshy feet--is not supported by much evidence of true prosauropods with sauropod-like feet. Nonetheless, we cannot entirely rule out the possibility of candidates such as *Riojasaurus* and *Blikanasaurus* in the group that Van Heerden (1997, p. 259) calls "heavy-bodied prosauropods."

It is not unusual to find that large ichnofaunas provide new evidence of the spatial and temporal distribution of important vertebrate groups that extends the knowledge derived from the skeletal record. Such data should be welcomed by paleontologists of an osteological bent. Too often such evidence is seen as contradictory to the skeletal record, when in fact it merely supplements it and extends temporal and spatial range boundaries, *without* seriously invalidating the general ranges established using body fossils. Nevertheless, when ichnological data extend the range of important groups closer to their origins, i.e., before the earliest known skeletal occurrence, such data usually prove controversial.

For example, abundant ichnological data now exist to indicate that tracks of shorebirds exist from the earliest Cretaceous (Berriasian) onwards, becoming very abundant, especially in Asia, by the late Neocomian (Lockley et al., 1992b; Lockley and Rainforth, in press). Skeletal evidence for these shorebird, or shorebird-like forms is not available, or at best very scare and ambiguous, even though there is a steady increase in the discovery of other bird types proving high avian diversity early in the Cretaceous. Many other examples could be cited, including trackway evidence, which is the oldest evidence of Devonian tetrapods, and australopithecine tracks that, at the time of discovery, were older than known skeletal remains (Lockley, 1999a).

The interpretation of *Tetrasauropus* as a sauropodan track has several significant implications. First, as noted, *Tetrasauropus* can now be regarded as the oldest sauropod track type reported from North America. In fact, assuming an age of about 210 ma, these tracks are about 50 million years older than the un-named sauropod trackway reported from the late Middle to early Late Jurassic (165-160 Ma) of Utah (Foster et al., 2000). Of equal significance is the implication that this interval defines a North American sauropod hiatus (*sensu* Lucas and Hunt, 1989). The original concept of a sauropod hiatus was applied to the chronostratigraphic interval between the late Albian and the Campanian-Maastrictian boundary, an interval of about 25 million years (100-75 Ma). It now appears that this could be considered the "second sauropod hiatus," the "first sauropod hiatus" being the longer interval between the late Norian-Rhaetian and the Callovian-Oxfordian occurrences.

Given that on a global scale sauropods were an extant group from the latest Triassic to the latest Cretaceous, these hiatuses evidently result from sauropod facies preferences and other large scale paleobiogeographical and geological influences on sauropod distribution. It is already known that sauropods preferred low latitude habitats, where carbonate precipitation prevailed in "semi-arid" climatic regimes as evaporation often exceeded precipitation (Lockley et al., 1994). Retallack (1997, p 355) has confirmed that "sauropod soils" (indicating the substrates preferred by prosauropds as well as sauropods) are typically "red, weakly clayey and strongly calcareous paloeosols (aridisols, calcic alfisols)." In many details upper Chinle Group sedimentary facies are similar to those of the Upper Jurassic Morrison Formation, which represents a typical sauropod habitat. Calcareous paleosols are present, and there is evidence of periodic aridity. Indeed, the uppermost Chinle Group facies underlie the massive Glen Canyon Group sequence, which represents sand-dominated deserts. Thus, it appears that after appearing in relatively large numbers, in late Chinle (Apachean) time, sauropods then disappeared from North America as excessively arid conditions developed. As part of any future study of the facies preferences of sauropods in North America, it may be fruitful to consider the relationship of stratal assemblages to paleolatitude (cf. Anderson and Lucas, 1996, fig. 3).

The possibility of the discovery of Upper Triassic brontosaur footprints is perhaps not surprising, given that we now know of coeval skeletal remains. From the viewpoint of the formulation of new scientific paradigms, however, it is significant that ichnological evidence, which suggests the existence of more than 70 early sauropod trackways, found in Upper Triassic strata of three states (New Mexico, Utah and Colorado), has until now failed to generate the most parsimonious morphological interpretation, i.e., that they are most probably sauropodan in origin. Failure to consider seriously this possibility and present it, as done here, reflects several influences on how we do science. First, it reflects a lack of experience and boldness on the part of ichnologists, ourselves included, in presenting interpretations that previously would have appeared controversial. Second, it reflects the need for the acquisition of a large database before the "big picture" emerges. And third, as noted previously, it reflects the convention of using osteological evidence as the ultimate arbitor in deciding at least the temporal, if not the spatial, range of major vertebrate groups in the fossil record.

ACKNOWLEDGMENTS

We thank James Farlow, Indiana University (Purdue-Fort Wayne), John Foster, University of Wyoming and Andrew Heckert, University of New Mexico for their helpful reviews and suggestions.

REFERENCES

Anderson, O. J. and Lucas, S. G., 1996, The base of the Morrison Formation (Upper Jurassic) of northwestern New Mexico and adjacent areas: Museum of Northern Arizona, Bulletin 60, p. 443-456.

Baird, D., 1980, A prosauropod dinosaur trackway from the Navajo Sandstone (Lower Jurassic) of Arizona; in Jacobs, L.L., ed., Aspects of Vertebrate History: Essays in Honor of Edwin Harris Colbert: Flagstaff, Museum of Northern Arizona Press, p. 219-230.

Bonner, J. T., 1988, The Evolution of Complexity. Princeton, Princeton University Press. 260 p.

Buffetaut, E., Suteethorn, V., Cuny, G., Tong, H., Le Loeuff, J., Khansubha, S. and Jongautcharlyaku, S., 2000, The earliest known sauropod dinosaur: Nature, v. 407, p. 72-74.

Conrad, K., Lockley, M. G., and Prince, N. K., 1987, Triassic and Jurassic Vertebrate-dominated trace fossil assemblages of the Cimarron Valley region: Implications for paleoecology and biostratigraphy: New Mexico Geological Society, 38th Field Conference, Guidebook, p. 127-138.

Ellenberger, P., 1972, Contribution a la classification des piste de vertebres du Trias: les types du Stormberg d'Afrique du Sud (1): Palaeovertebrata, Memoire Extraordinaire, 152 p

Ellenberger, P., 1974, Contribution a la classification des pistes de vertebres du Trias; les types du Stormberg d'Afrique du Sud, 2e partie: Palaeovertebrata, Memoire Extraordinaire, 170 p.

Farlow, J. O., Pittman, J. G., and Hawthorne, M. J., 1989, *Brontopdus birdi*, Lower Cretaceous sauropod footprints from the Gulf Coastal Plain; *in* Gillette, D. D. and Lockley, M. G., eds., Dinosaur Tracks and Traces: Cambridge, Cambridge University Press, p. 371-394.

Foster, J., Hamblin, A., and Lockley, M. G., 2000, The oldest evidence of a sauropod dinosaur in the western United States and other important vertebrate trackways from Grand Staircase-Escalante National Monument, Utah: Ichnos, v. 7, p. 169-181.

Furrer, H., 1984, Fundstelle von Dinosaurierfahrten am Piz dal Diavel: Bunfner Zeit, v. 25, p. 1-9.

Furrer, H., 1993, Entdecking und Untersuchung der Dinosaurierfahrten in National Park Zernz, Switzerland: Cratschla, Edizuns Special 1, 12 p.

Gand, G. Vianey-Liaud, M. Demathieu. G and Garric, G., 2000, Deux nouvelles traces des pas de dinosaures du Trias Supérieur de la Bordure Cévenole (La Grande-Combe, Sud-Est de la France): Geobios, v. 33, p. 599-624.

Hitchcock, E., 1858, Ichnology of New England. A report on the sandstone of the Connecticut Valley especially its fossil footmarks. Boston: W. White (Reprinted by Arno Press in the Natural Sciences in America Series), 192 p.

Hunt, A. P. Lucas, S. G. Lockley, M. G. and Heckert, A. B., 2000 Occurrence of the dinosaurian ichnogenus *Grallator*, in the Redonda Formation (Upper Triassic: Norian) of Eastern New Mexico: New Mexico Museum of Natural History and Science, Bulletin 17, p. 39-41.

Ishigaki, S., 1988, Les empreintes de Dinosaures du Jurassique Inférieur. Du haunt Atlas central marocain: Notes Service Géologique Maroc, v. 44, p. 79-86.

Jenkins, F. A. Shubin, N. H., Amaral, W. W., Gatesy, S. M., Schaff, C. R., Clemensen, L. B. Downs, W. R., Davidson, A. R., Bonde, N., and Osbaeck, F., 1994, Late Triassic continental vertebrates and depositional environments of the Fleming Fjord Formation, Jameson Land, east Greenland: Meddelelser om Grønland Geoscience, v. 32, p. 3-25.

Jenny, J. and Jossen, J-A., 1982, Découverte d'empreintes de pas de Dinosauriens dand le Jurassique Inférieur (Pleinsbachien) du haute- Atlas central (Maroc): Compte Rendu Academie Scienecs Paris, v. 294, p. 223-226.

Lanzinger, M. and Leonardi, G., 1991, Piste di dinosauri del Giurassico Inferiore ai Lavini di Marco (Trento); *in* Muscio, G., ed, Dinosaurs: Il Mondo dei Dinosauri: Le Scienze Quaderno, Milan, p. 89-94.

Leonardi, G., and Avanzini, M., 1994, Dinosauri in Italia; *in* Ligabue, G., ed., Il Tempo del Dinosauri: Le Scienze Quaderno, Milan, p. 69-81.

Lockley, M. G., 1986., Dinosaur tracksites: A field guide published in conjunction with the First International Symposium on Dinosaur Tracks and Traces: University of Colorado at Denver, Geology Department Magazine Special Issue no. 1, 56 p.

Lockley, M. G., 1998, The vertebrate track record: Nature, v. 96, p. 429-432.

Lockley, M. G., 1999a, The Eternal Trail: A Tracker Looks At Evolution. New York, Perseus Books, 334 p.

Lockley, M. G., 1999b, Einblicke in die Gestaltebiologie der Dinosaurier anhand ihrer Fahrtenspuren: Tycho de Brahe-Jahrbuch Fur Gotheanismus, p. 134-166.

Lockley, M. G., and Hunt, A. P., 1993, A new Late Triassic Tracksite from the Sloan Canyon Formation (type section) Cimarron Valley, New Mexico: New Mexico Museum of Natural History and Science, Bulletin 3, p. 279-283.

Lockley, M. G., and Hunt, A. P., 1995, Dinosaur Tracks and Other Fossil Footprints of the Western United States. New York, Columbia University Press, 338 p.

Lockley, M. G., and Meyer, C. A., 2000, Dinosaur Tracks and Other Fossil Footprints of Europe. New York, Columbia University Press, 323 p.

Lockley, M. G., and Rainforth, E., in press, The tracks record of Mesozoic birds and pterosaurs: An ichnological and paleoecological perspective; *in* Chiappe, L. and Witmer, L. M., eds., Mesozoic Birds Above the Heads of Dinosaurs: Berkeley, University of California Press.

Lockley, M. G., Lucas, S. G., and Hunt, A. P., 2000, Dinosaur tracksites in New Mexico: A review: New Mexico Museum of Natural History and Science, Bulletin 17, p. 9- 16.

Lockley, M. G., Santos, V. F., and Hunt, A. P., 1993, A new Late Triassic tracksite in the Sheep Pen Sandstone, Sloan Canyon, New Mexico: New Mexico Museum of Natural History and Science, Bulletin, 3, p. 285-288.

Lockley, M. G., Conrad, K., Paquette, M., and Hamblin, A., 1992a, Late Triassic vertebrate tracks in the Dinosaur National Monument area: Utah Geological Survey, Miscellaneous Publications 92-3, p. 383-391.

Lockley, M. G., King, M. J. Howe, S., and Sharp T., 1996, Dinosaur tracks and other archosaur footprints from the Triassic of South Wales: Ichnos, v. 5, p. 23-41

Lockley, M. G., Meyer, C., Hunt, A. P., and Lucas, S. G., 1994, The distribution of sauropod tracks and trackmakers: Gaia: Revista de Geociencias, Museu Nacional de Historia Natural, Lisbon, Portugal, v. 10, p. 233-248

Lockley, M. G., Yang, S-Y., Matsukawa, M., Fleming, F. and Lim, S-K., 1992b, The track record of Mesozoic birds: Evidence and implications: Philosophical Transactions of the Royal Society of London, v. 336, p. 113-134

Lucas, S. G., 1991, Correlation of the Triassic strata of the Colorado Plateau and southern High Plains, New Mexico: New Mexico Bureau of Mines and Mineral Resources Bulletin 137, p. 47-56

Lucas, S. G., 1993, The Chinle Group: Revised stratigraphy and biochronology of upper Triassic nonmarine strata in the western United States: Museum of Northern Arizona, Bulletin 59, p. 27-50.

Lucas, S. G., 1998, Global Triassic tetrapod biostratigraphy and biochronology: Paleogeography, Paleoclimatology, Palaeoecology, v. 143, p. 347-384.

Lucas, S. G., and Hunt, A. P., 1989, *Alamosaurus* and the sauropod hiatus in the Cretaceous of the North American Western Interior: Geological Society of America, Special Paper 238, p. 75-85.

Mietto, P., 1991, Impronte de dinosauri nel Triassico superiore delle Dolomiti, *in* Dinosaurs: Il mondo dei dinosauri: Published jointly by the museums of Trento, Torino, Udine and the Faculty of Mathematical Sciences, Rome, p. 83-106

Olsen, P. E. and Galton. P., 1984, A review of the reptile and amphibian assemblages from the Stormberg of Southern Africa., with special emphasis on the footprints and the age of the Stormberg: Paleontologia Africana, v. 25, p. 87-110.

Peabody, F., 1948, Reptile and amphibian trackways from the Lower Triassic Moenkopi Formation of Arizona and Utah: University of California, Department of Geological Sciences, Bulletin, 27, p. 295-468.

Retallack, G. J., 1997, Dinosaurs and dirt; *in* Wolberg, D. L., Stump, E. and Rosenburg, G. D., eds., Dinofest International: Proceedings of a Symposium Held at Arizona State University: Tempe, ASU, p. 345-359.

Thulborn, R. A., 1990, Dinosaur Tracks. London, Chapman & Hall, 410 p.

Upchurch, P., 1995, Evolutionary history of sauropod dinosaurs: Philosophical Transactions of the Royal Society of London, v. 349B, p. 365-390.

Van Heerden, J., 1997, Prosauropods; *in* Farlow J. O. and Brett-Surman, M. K., eds., The Complete Dinosaur: Purdue, Indiana University Press. p. 242-263.

Wilson, J. A., and Carrano, M. T., 1999, Titanosaurs and the origin of wide gauge trackways: A biomechanical and systematic perspective on sauropod locomotion: Paleobiology, v. 25, p. 252-267.

Wison, J. A., and Sereno, P. C., 1998, Early evolution and higher-level phylogeny of sauropod dinosaurs: Society of Vertebrate Paleontology. Memoir 5, 68 p.

GEOLOGY OF THE IMA NW QUADRANGLE, EAST-CENTRAL NEW MEXICO

SPENCER G. LUCAS[1], GARY WEADOCK[1], KENNETH K. KIETZKE[1], ADRIAN P. HUNT[2] AND BARRY S. KUES[3]
[1]New Mexico Museum of Natural History, 1801 Mountain Road NW, Albuquerque, NM 87104; [2]Mesa Technical College, Tucumcari, NM 88401;
[3]Department of Earth and Planetary Sciences, University of New Mexico, Albuquerque, NM 87131

Abstract.—The Ima NW quadrangle, located along part of the eastern border of Guadalupe County with Quay County, straddles the edge of the Southern High Plains (Llano Estacado) and the drainage divide between the Canadian and Pecos Rivers. No substantial geologic structures are exposed on the quadrangle, which is underlain by an essentially flat-lying section of Upper Triassic, Middle-Upper Jurassic, Lower Cretaceous and Neogene sedimentary rocks ~330 m thick. About half of that section (~170 m) is Upper Triassic nonmarine red beds of the Trujillo, Bull Canyon and Redonda formations of the Chinle Group. A much thinner (~63 m thick) section of overlying Jurassic strata belongs to the Entrada (Slick Rock and Exeter members), Todilto (Luciano Mesa Member), Summerville and Morrison formations. The Lower Cretaceous section is also thin (~30 m) and assigned to the upper Albian Tucumcari and Mesa Rica formations. Fossils from the Bull Canyon, Todilto, Tucumcari and Mesa Rica formations on the Ima NW quadrangle are abundant and of biostratigraphic significance. Neogene strata on the quadrangle belong to the Ogallala Formation, which underlies the Llano Estacado, and alluvial, eolian and colluvial units of Pleistocene age that are mostly exposed along Bull Canyon and Alamogordo Creeks.

INTRODUCTION

The Ima NW quadrangle is a U. S. Geological Survey 7.5-minute topographic map located along part of the border of Guadalupe and Quay Counties about 37 km east of Santa Rosa (Fig. 1). In 1983, we began studying the stratigraphy and paleontology of the Mesozoic rocks exposed on the quadrangle. This resulted in geologic mapping of the quadrangle (Lucas et al., 1994). Here, we present a review of the geology of the Ima NW quadrangle. NMMNH = New Mexico Museum of Natural History, Albuquerque and UNM = University of New Mexico, Albuquerque.

PREVIOUS STUDIES

The majority of previous work in the Ima NW area has been related either to the paleontology of the Upper Triassic Bull Canyon Formation, the Middle Jurassic Todilto Formation or to Mesozoic stratigraphy in general. Mehl (1922) first described a Triassic fossil from Guadalupe County. This phytosaur skull had imprecise published provenience, but all recent workers agree that it derives from Bull Canyon (e.g., Hunt and Lucas, 1989). In 1947, J. T. Gregory of Yale University collected fossils from Triassic strata in Bull Canyon and conducted limited stratigraphic studies (Gregory, 1972; Hunt, 1997). In 1983, another phytosaur skull was discovered, and a field party from the University of New Mexico subsequently collected this specimen. This ushered in a period of extensive Triassic fossil collecting in Bull Canyon by field crews from Albuquerque in the early-mid 1980s (e.g., Kues, 1985; Lucas et al., 1984,1985a, b, c, d; Hunt, 1994).

Koerner (1930) was the first to collect Jurassic fish from the Todilto Formation on the Ima NW quadrangle. The rich fossil fish localities near Bull Canyon were later referred to as Bull Canyon, Merrill Ranch and Hornsby Ranch. Fossil fish were later collected by Sternberg, Dunkle, Pierce and Gregory in 1947, Henderson (date unknown), Ash in 1958 and Schaeffer in 1964 (Dunkle, 1942; Schaeffer and Patterson, 1984; Lucas et al., 1985c; Lucas and Kietzke, 1986). In 1983 and 1984, the University of New Mexico made extensive collections of Todilto fish (Lucas et al., 1985c). Subsequently, some fish were collected on the Ima NW quadrangle for NMMNH and by commercial collectors. The intensive paleontological studies of the University of New Mexico in the early-mid 1980s included detailed stratigraphic studies and mapping (e.g., Lucas et al., 1985b, c, 1994) that provide much of the basis of this article.

The geologic map of Dobrovolny et al. (1946) at a scale of 1:62,500 encompassed the eastern edge of the Ima NW quadrangle. Lucas et al. (1994) mapped the geology of the quadrangle at a 1:24,000 scale. That map, somewhat simplified, is reproduced here (Fig. 2).

OVERVIEW

The Ima NW quadrangle is located at a significant physiographic junction--it straddles the boundary between the Southern High Plains (Llano Estacado) and the drainage divide between the Canadian and Pecos Rivers (Fig. 2). Thus, most of the northern third of the quadrangle is Bull Canyon, a 128-152 m deep canyon developed in Mesozoic bedrock by Bull Canyon Creek, a tributary of the Canadian River. The northeastern part of the quadrangle is Luciano Mesa, the westernmost edge of the Llano Estacado, whereas the western third of the quadrangle is a Quaternary upland surface.

The southern two-thirds of the Ima NW quadrangle are dominated by the broad, shallow drainage of Alamogordo Creek, a tributary of the Pecos River. Triassic and Jurassic strata are locally exposed here, particularly along the western escarpment of the Alamogordo Valley. However, most of the rocks exposed in the southern two-thirds of the quadrangle are Quaternary alluvial units.

STRUCTURE

The Ima NW quadrangle mostly exposes flat-lying Mesozoic and Neogene sedimentary rocks. The only geologic structure recognized on the surface of the quadrangle is a down-to-the-northwest normal fault in the Upper Triassic strata near the head of

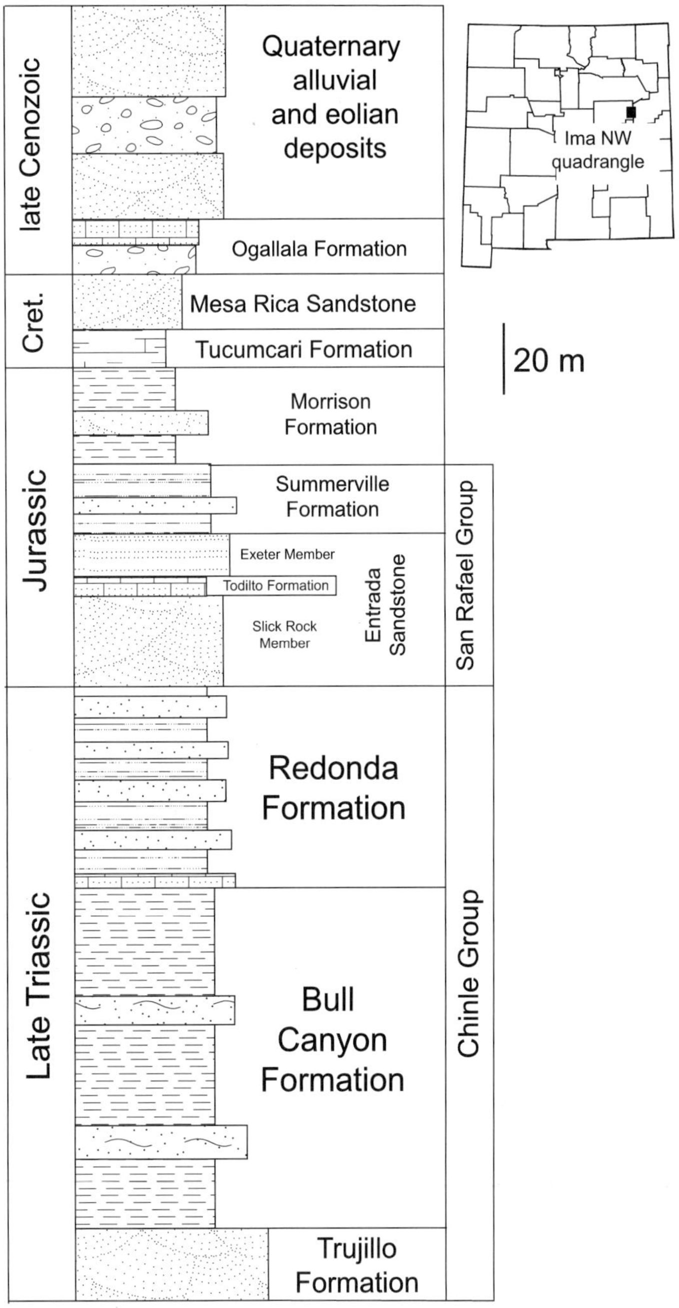

FIGURE 1. Location of Ima NW quadrangle and generalized stratigraphic section of strata exposed on the quadrangle.

Bull Canyon (Fig. 2). We interpret this as a Laramide (Late Cretaceous-Paleogene) structure because it is parallel to the SW-NE trending Laramide structures found elsewhere in east-central New Mexico (Hunt and Lucas, 2001).

MESOZOIC STRATIGRAPHY

Triassic Strata

The oldest rocks exposed on the Ima NW quadrangle are Upper Triassic strata of the Trujillo, Bull Canyon and Redonda formations of the Chinle Group.

Trujillo Formation

The Trujillo Formation of Gould (1907) (=Cuervo Sandstone Member of Chinle Formation of Kelley, 1972b) is only exposed in the southwestern portion of the Ima NW quadrangle along Alamogordo Creek and its tributaries. Here, approximately 6-20 m of the upper part of the formation crop out and are yellowish-gray, trough-crossbedded micaceous litharenites.

In east-central New Mexico, the Trujillo Formation is the medial sandstone complex of the Chinle Group, and is as much as 68 m thick (Kelley, 1972a, b; Broadhead, 1984; Lucas and Hunt, 1989; Lucas et al., 2001). Sandstones like those exposed on the Ima NW quadrangle are the dominant lithotype, although thin reddish brown mudstones and intrabasinal (mostly calcrete-pebble) conglomerates are common. Tetrapod fossils from the Trujillo Formation in eastern Quay County indicate it is of Revueltian (early-middle Norian) age (Lucas and Hunt, 1993; Hunt and Lucas, 1993; Hunt, 1994, 2001), but we collected no fossils from the Trujillo Formation on the Ima NW quadrangle.

Bull Canyon Formation

The Bull Canyon Formation of Lucas and Hunt (1989) (= upper shale member of the Chinle Formation of Kelley, 1972a, b) is the most extensively exposed Upper Triassic unit on the Ima NW quadrangle. Badlands developed in the Bull Canyon Formation form the floor of Bull Canyon in the northern part of the quadrangle. Most of the Bull Canyon Formation crops out here; on the Ima NW quadrangle, the unit has a maximum thickness of 82 m without its base exposed. At the type section of the Bull Canyon Formation, a few km north of the Ima NW quadrangle (sec. 4, T12N, R24E), the entire formation, between the Trujillo and Redonda Formations, is 95 m thick (Lucas and Hunt, 1989).

On the Ima NW quadrangle, as elsewhere in east-central New Mexico, the Bull Canyon Formation is mostly grayish-red and moderate reddish-brown mudstone, minor sandstones that are typically yellowish-gray to grayish-red laminar and crossbedded litharenites, and a few beds of intraformational conglomerates (Fig. 3A). The Trujillo-Bull Canyon contact is picked at the base of the first thick, mudstone interval above the highest Trujillo Formation sandstone. A persistent, bench-forming sandstone as much as 10 m thick and about 25 m below the top of the formation supports an areally extensive Quaternary floodplain terrace in upper Bull Canyon.

This sandstone bench and strata just below it are extremely fossiliferous, and the NMMNH records list more than 50 discrete fossil localities from these strata (Lucas et al., 1985d). Fossils are of plants, especially pith casts of the calamitalean *Neocalamites* (Fig. 3B; Lucas et al., 1985a), invertebrates, especially unionid bivalves and gastropods (Fig. 4; Kues, 1985), and of vertebrates, especially phytosaurs and other tetrapods (e.g., Lucas et al., 1985d; Hunt and Lucas, 1989, 1997; Hunt, 1994, 2001). The tetrapods indicate a Revueltian (early-middle Norian) age for the Bull Canyon Formation on the Ima NW quadrangle.

Redonda Formation

The youngest Triassic strata on the Ima NW quadrangle are assigned to the Redonda Formation of Dobrovolny et al. (1946).

About 55 m of Redonda strata are present, mostly interbedded moderate orange, fine-grained sandstones and moderate reddish-brown mudstones with lesser beds of pale red calcareous siltstone, pebbly mudstone and light greenish gray silty micrite. Redonda beds are laterally continuous and cyclically bedded (Fig. 3C-D), reflecting their deposition in an extensive (minimum area 5000 km^2) lacustrine system (Hester, 1988; Hester and Lucas, 2001). The basal bed of the Redonda is a distinctive bluish gray to purple silty limestone above reddish brown mudstone at the top of the Bull Canyon Formation.

The Redonda Formation is well exposed on the Ima NW quadrangle, especially along the flanks of Bull Canyon (Figs. 2, 3C-D). We collected no biostratigraphically significant fossils from the Redonda Formation on the Ima NW quadrangle. However, in Quay County it yields the type vertebrate-fossil assemblage of the Apachean (late Norian-Rhaetian?) land-vertebrate faunachron (Lucas and Hunt, 1993; Hunt, 1994).

Jurassic Strata

Jurassic rocks exposed on the Ima NW quadrangle are assigned to the Entrada, Todilto, Summerville and Morrison Formations. The Entrada, Todilto and Summerville Formations comprise the San Rafael Group in east-central New Mexico (Lucas and Anderson, 1998). They are best exposed around Bull Canyon and on the flanks of Luciano Mesa in the northern part of the quadrangle.

Entrada Sandstone

The Entrada Sandstone forms the bold, light-colored cliffs that ring Bull Canyon (Fig. 3C-D) and also defends part of the escarpment west of Alamogordo Creek. The Slick Rock and Exeter members of the Entrada sandstone are present.

The bulk of the Entrada Sandstone is the Slick Rock Member, as much as 26 m of yellowish-orange and pale orange subarkosic sandstone that typically displays large-scale trough crossbedding and laminar bedding (Fig. 3E). The Slick Rock Member rests with profound disconformity on the Upper Triassic Redonda Formation. This is the J-2 unconformity of Pipiringos and O'Sullivan (1978) that encompasses a hiatus of approximately 40 million years from the latest Triassic (age of the Redonda Formation) to the Callovian (Middle Jurassic), which is the age of the Slick Rock Member.

The Todilto Formation or the Exeter Member of the Entrada Formation disconformably overlies the Slick Rock Member. The Exeter Member is up to 10 m thick and is lithologically similar to the Slick Rock Member (Fig. 3G). Like Lucas et al. (1985c, 1987) and Lucas and Anderson (1998), we regard the Exeter as an upper tongue of the Entrada Sandstone that post-dates deposition of the Todilto Formation. No fossils are known from the Entrada Sandstone on the Ima NW quadrangle, so its age is based on regional stratigraphic relationships (Lucas et al., 1985c).

Todilto Formation

The Todilto Formation is well exposed along the walls of Bull Canyon, where it is up to 3 m of dark gray, kerogenic limestone with minor interbeds of sandstone and shale at its base (Fig. 3E-F). Lucas et al. (1985c) and Lucas and Kietzke (1986) provided a detailed description of the Todilto Formation on the Ima NW quadrangle. Large numbers of fossil fishes come from the Todilto Formation here, especially along the southwestern edge of Bull Canyon (Koerner, 1930; Dunkle, 1942; Schaeffer and Patterson, 1984; Lucas et al., 1985c). Furthermore, pinchouts of the Todilto Formation into the Entrada eolianites are evident along the southern edge of Bull Canyon (Figs. 3G, 5).

Across northern New Mexico, the Todilto Formation consists of two members, the Luciano Mesa (lower limestone) and Tonque Arroyo (upper gypsum) members (Lucas et al., 1995; Lucas and Anderson, 1998). The type section of the Luciano Mesa Member is on the Ima NW quadrangle (Lucas et al., 1995).

Summerville and Morrison Formations

Jurassic strata above the Exeter Member of the Entrada Sandstone are poorly exposed along the flanks of Luciano Mesa, where they are mostly covered by soil and colluvium (Figs. 2, 3H). Because of their poor exposure, we have mapped the Summerville and Morrison formations together as a single unit (Fig. 2).

However, the two units are easily discriminated in measured sections (Fig. 6). Thus, Summerville strata are red-bed mudstones, siltstones and fine (gypsiferous?) sandstones that directly overlie sandstone of the Exeter Member. A few thin, prominent beds of red chalcedony ("agate beds") are present (Fig. 6). Maximum Summerville thickness is approximately 20 m.

The Morrison Formation on the Ima NW quadrangle is as much as 30 m thick and consists of green smectitic claystone and trough-crossbedded subarkosic sandstone. The Morrison-Summerville contact is placed at the base of the first bed of green smectitic mudstone above Summerville red beds. Regional correlations indicate the Morrison Formation outcrops at the Ima NW quadrangle are part of the Brushy Basin Member (Lucas and Anderson, 1998). Fragmentary bones of sauropods and *Stegosaurus* are present in Jurassic strata on the Ima NW quadrangle (Fig. 6; Lucas et al., 1985c, fig. 25).

Cretaceous Strata

Two Lower Cretaceous units are exposed on the Ima NW quadrangle along the flanks of Luciano Mesa, the Tucumcari and Mesa Rica Formations.

Tucumcari Formation

The Tucumcari Formation is pale green, yellow, purple and gray, slope-forming shale up to 13 m thick that rests directly on trough-crossbedded sandstone or green smectitic mudstone of the Morrison Formation. It represents shoreface to offshore marine environments and has a diverse molluscan fauna dominated by large valves of the gryphaeid oyster *Texigryphaea pitcheri* (Fig. 7). At a locality in the SW1/4 NE1/4 SW1/4 sec. 23, T9N, R6E, the Tucumcari Formation yields hundreds of valves of *T. pitcheri* from a biostrome in the upper part of the formation (Fig. 5). Ammonites and bivalves from the Tucumcari Formation in Quay County indicate it is of late Albian age (Kues et al., 1985; Kues, 1997; Kues and Lucas, 2001).

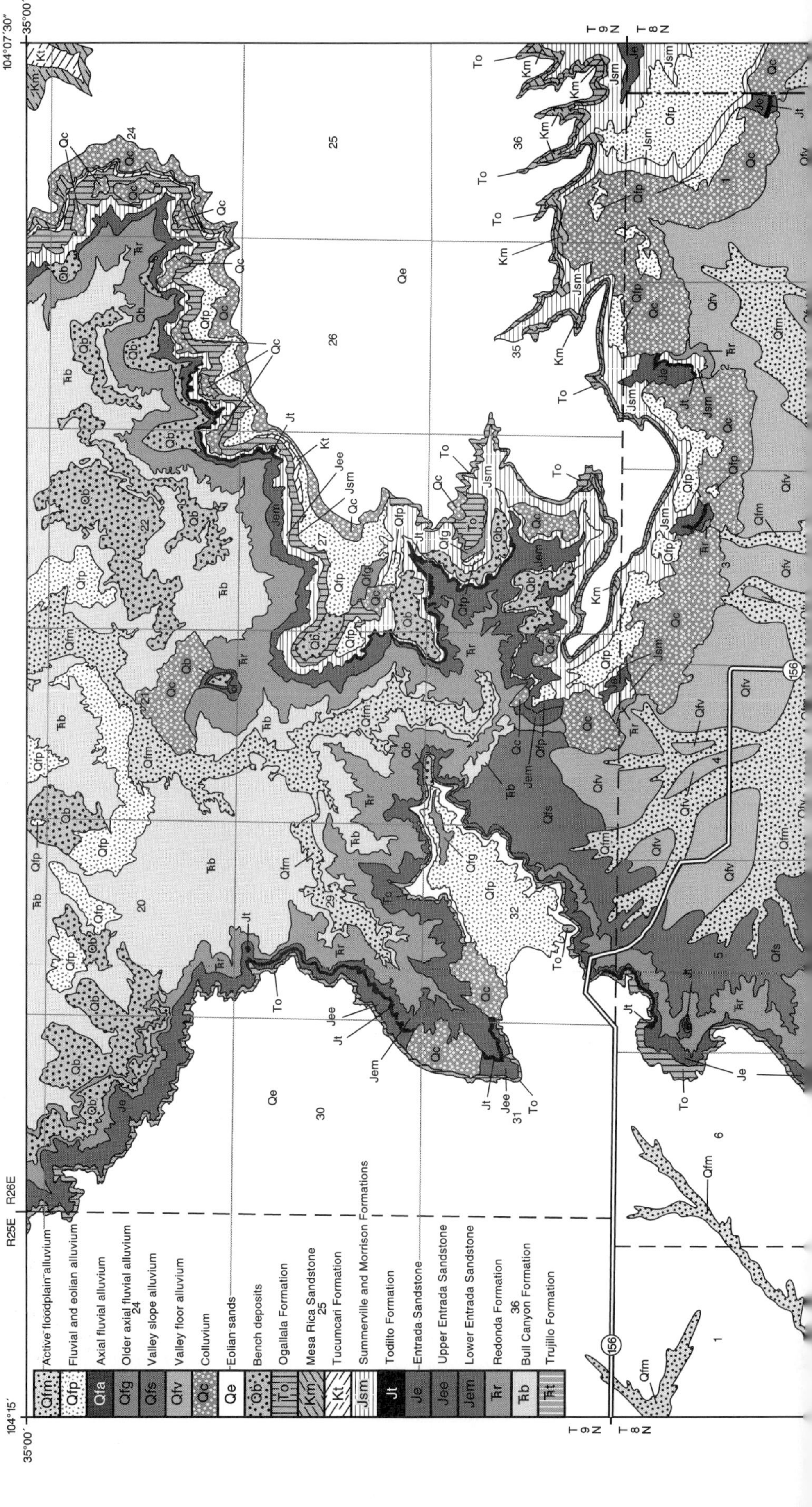
104°07'30"
35°00'
104°15'
35°00'
R25E
R26E
T9N
T8N
Qfm Active floodplain alluvium
Qfp Fluvial and eolian alluvium
Qfa Axial fluvial alluvium
Qfg Older axial fluvial alluvium
Qfs Valley slope alluvium
Qfv Valley floor alluvium
Qc Colluvium
Qe Eolian sands
Qb Bench deposits
To Ogallala Formation
Km Mesa Rica Sandstone
Kt Tucumcari Formation
Jsm Summerville and Morrison Formations
Jt Todilto Formation
Je Entrada Sandstone
Jee Upper Entrada Sandstone
Jem Lower Entrada Sandstone
TRr Redonda Formation
TRb Bull Canyon Formation
TRt Trujillo Formation

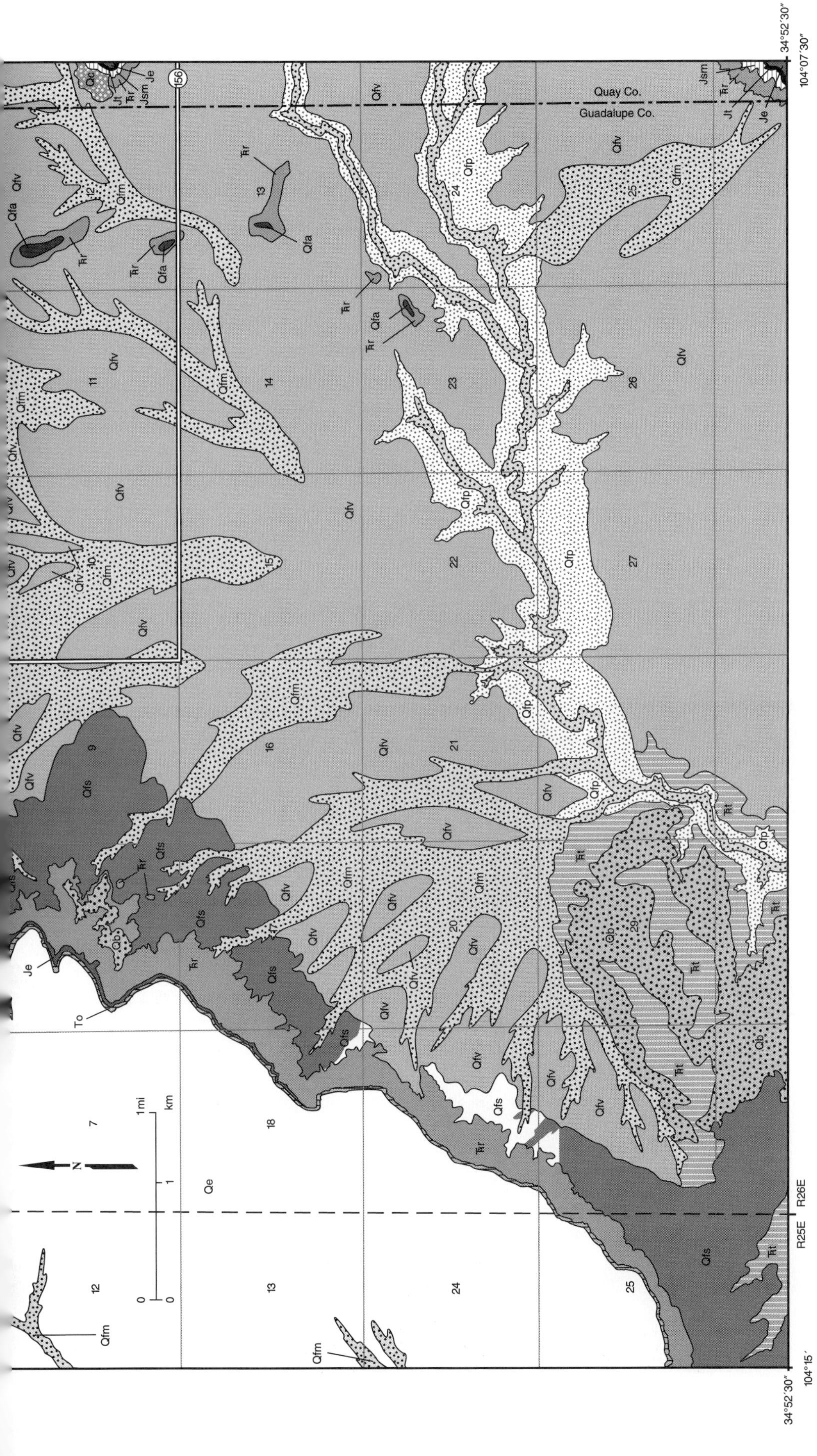

FIGURE 2. Geologic map of the Ima NW quadrangle. Simplified from Lucas et al. (1994).

A
B
E
R
C
E
R
D
T
S
E
F
E
S
G
H

FIGURE 4. Selected nonmarine bivalves (A-H) and gastropods (I-K) from the upper part of the Bull Canyon Formation on the Ima NW quadrangle. All figures x 1 unless otherwise indicated. A-B. "*Unio*" *arizonensis* Henderson, left valve, UNM 8403. C. "*Unio*" sp. 1, left valve, UNM 8450. D. "*Unio*" sp. 5, left valve, showing possible phytosaur tooth mark, UNM 8393. E. "*Unio*" sp. 3, left valve, UNM 8371. F-G. *Antediplodon dockumensis* (Simpson), left valve with incomplete posterior margin, UNM 8356 (F), and dorsal view of right valve, showing beak lirae, UNM 8352 (G). H. "*Unio*" sp. 6, left valve, UNM 8402. I-K. *Triasamnicola pilsbryi* Yen and Reeside (all x 2), apertural view, UNM 8469 (I), adapertural view, UNM 8465 (J), and oblique apical view, UNM 8468 (K).

Mesa Rica Sandstone

The Mesa Rica Sandstone is up to 17 m thick and consists chiefly of yellowish-brown to pale orange, trough-crossbedded sandstone that rests sharply on shale of the Tucumcari Formation. This is a minimum thickness, as the top of the unit is eroded off on the quadrangle. The Mesa Rica Sandstone is the basal unit of the Dakota Group in east-central New Mexico, and typically is approximately 30-40 m thick (Lucas and Kisucky, 1988). At Luciano Mesa, a specimen of the late Albian index ammonite *Mortoniceras equidistans* (Kues et al., 1985, fig. 9I) was collected from the basal Mesa Rica Sandstone just above the *Texigryphaea* biostrome mentioned above (Fig. 6).

FIGURE 3. Photographs of selected outcrops on the Ima NW quadrangle. A. Typical mudstone-dominated strata of the Upper Triassic Bull Canyon Formation. B. Upright, *in situ* pith casts of the calamitalean *Neocalamites* in the Bull Canyon Formation (see Lucas et al., 1985a). C. Cyclically-bedded strata of the Redonda Formation (R) overlain by the Entrada Sandstone (E). D. Cyclically-bedded strata of the Redonda Formation (R) overlain by the Entrada Sandstone (E). E. Slick Rock Member of Entrada Sandstone (S) overlain by Todilto Formation (T). F. Todilto Formation. G. Pinchout surface of Todilto Formation between Slick Rock (S) and Exeter members (E) of Entrada Sandstone. H. Slope-forming Morrison Formation.

CENOZOIC STRATA AND GEOMORPHOLOGY

Cenozoic strata on the Ima NW quadrangle are of Neogene (including Quaternary) age. The Ima NW quadrangle is located on the northwestern border of the Llano Estacado. Quaternary evolution of the Bull Canyon and Alamogordo Valley drainage basins has been characterized by several episodes of entrenchment, partial backfilling and valley floor stability. Evidence of Quaternary processes is recorded in a stepped sequence of graded valley-margin geomorphic surfaces of both constructional and erosional origin that are inset against the Ogallala Formation and older strata. In Bull Canyon, geomorphic surfaces take the form of large bedrock benches overlain by thin alluvium. Conversely, Alamogordo Valley has few bedrock benches, but numerous alluvial deposits and transport-limited slopes. Neogene strata recognized in the area include five Pleistocene valley-fill units, two eolian units, a colluvial unit and the Miocene Ogallala Formation. Thus, there are two principal kinds of Neogene deposits on the Ima NW quadrangle: (1) remnants of the formerly extensive Ogallala Formation; and (2) a suite of younger valley border and valley floor alluvial and stream deposits and associated geomorphic surfaces.

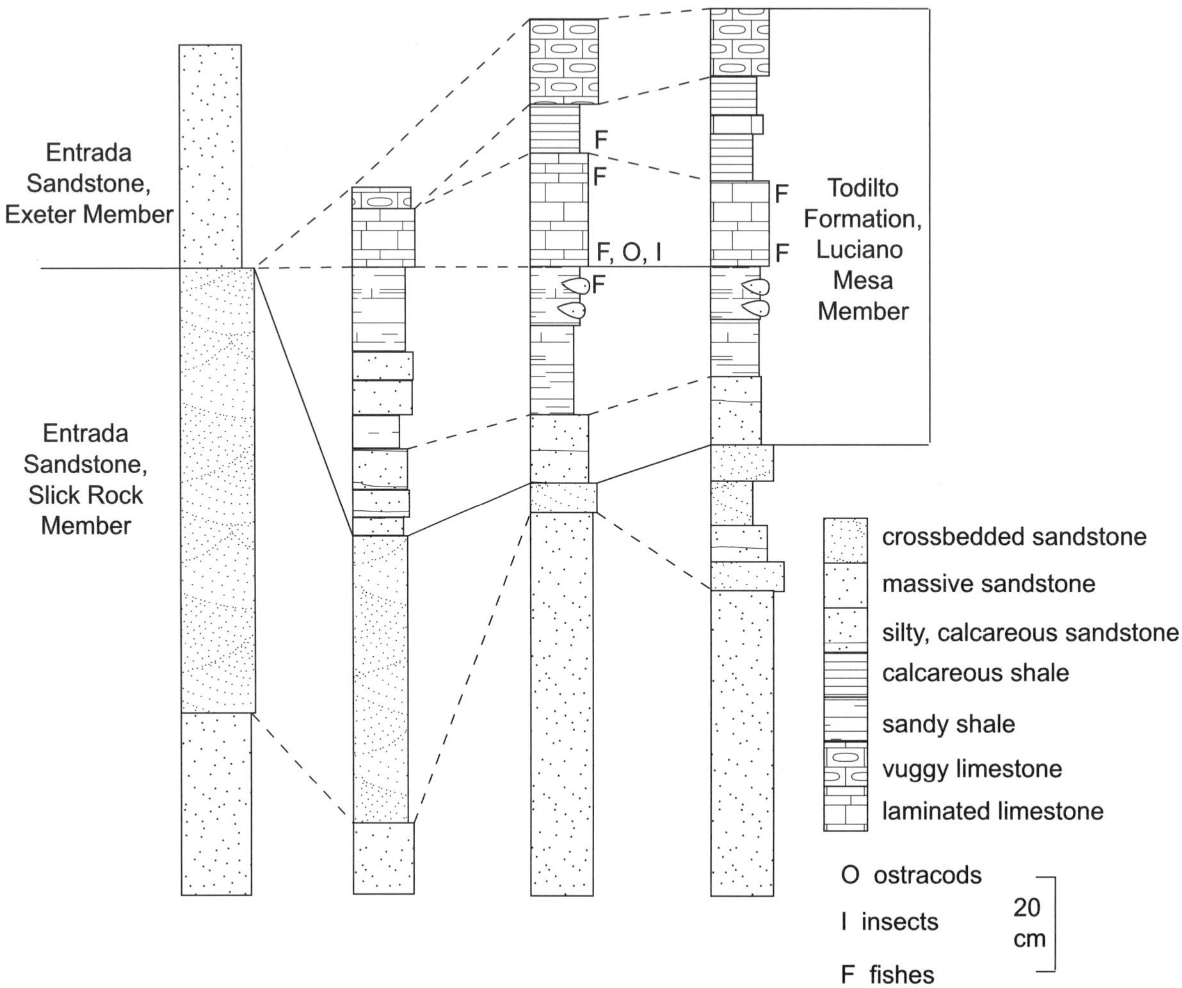

FIGURE 5. Jurassic stratigraphic sections showing the pinchout of the Todilto Formation in SW ¼ SW ¼ sec. 28, T9N, R26E.

Ogallala Formation

On the Ima NW quadrangle, the escarpment of Luciano Mesa and the upland surface to the west of Bull Canyon are underlain by the Miocene Ogallala Formation (cf. Hawley, 1984). The Ogallala here is 9-14 m thick and consists of a lower interval of conglomerate and an upper interval of white to tan calcrete, the "Caprock caliche."

The basal conglomerate of the Ogallala Formation is massive, moderately sorted and dominated by pebbles and cobbles of metamorphic (quartzite, gneiss) and sedimentary (sandstone) rocks. It has a carbonate matrix and is weakly cemented. The overlying calcrete is light colored (white to tan), thick (up to 10 m), highly indurated and has lenses of sand and gravel. The top of the Ogallala Formation is a plain covered by a thin veneer of eolian sand.

Older Alluvium (Qfg, Qfp)

Older alluvium includes deposits associated with at least three Pleistocene episodes of valley incision and partial backfilling, prior to the formation of the present landscape. Ancestral floodplain positions that formed local base levels ranged from 20-100 m above the present active floodplain. Older valley-fill alluvium represents aggradational events or cycles culminated by long periods of base level stability and geomorphic surface formation.

Unit Qfg contains stream deposits laid down during dissection of the Ogallala plain. This sediment is composed of large, rounded gravel clasts (1-15 cm in diameter) in a matrix of medium-grained sand and calcium carbonate. The matrix is pale tan to white, whereas the gravels are of various colors and lithologies. Clasts include granite, gneiss, Jurassic and Cretaceous sandstone as well as clasts derived from the Ogallala Formation. On the margin of Luciano Mesa, this unit forms elongated terraces that overlie Morrison Formation sandstone and ranges from 2 to 25 m in thickness.

At its most complete outcrops, Qfg has three distinct units. The lower unit is clast-supported and imbricated gravel (1-10 cm diameter clasts) in a calcium carbonate-cemented sand matrix. The middle coarse sand has abundant calcium carbonate in the matrix, is not imbricated, but does have inclined bedding dipping to the northwest. The upper section is poorly sorted gravel (clasts 5-15 cm in diameter), sand and silt containing only small amounts of calcium carbonate. Generally, these three cycles of sedimenta-

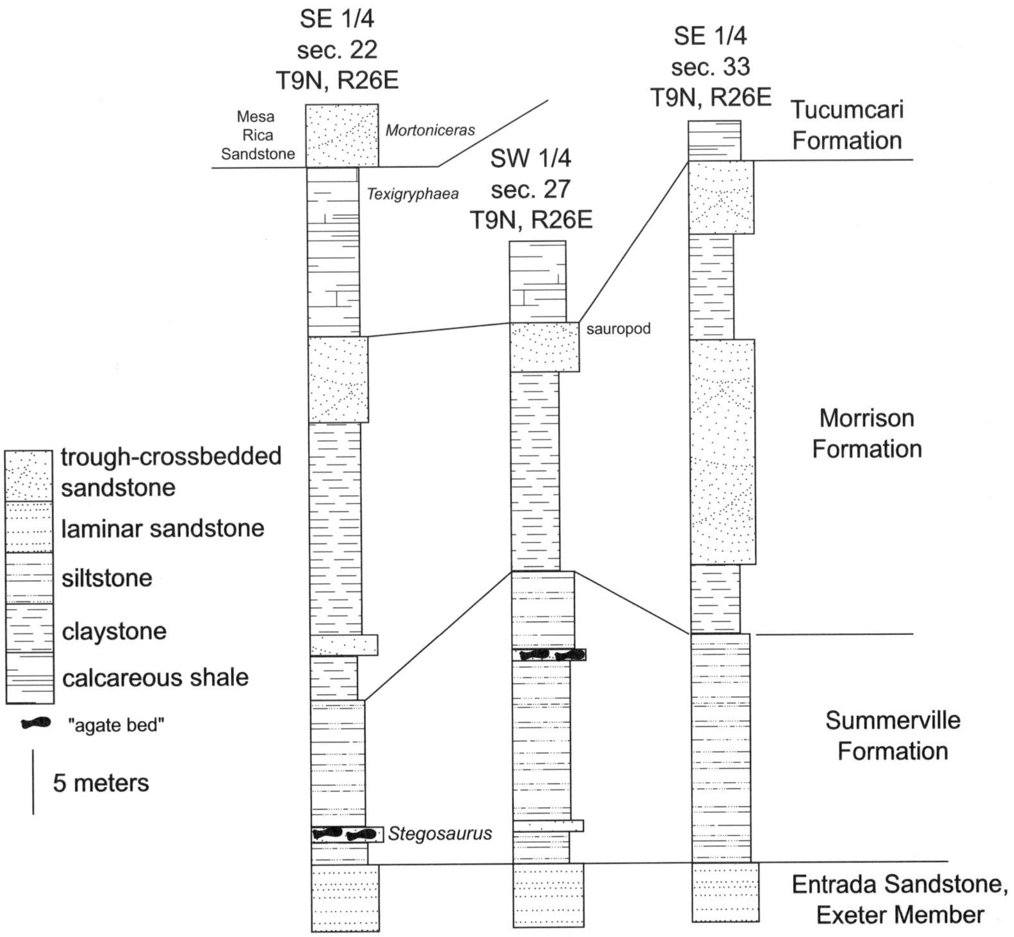

FIGURE 6. Sections of the Jurassic-Lower Cretaceous along the western edge of Luciano Mesa.

tion fine upward, although the uppermost unit is so poorly sorted that discrimintaing fluvial from eolian sands is difficult.

At the base of Qfg, recent gullying has exposed scour marks in the Morrison sandstone. These large scours (1 m deep) suggest a high-energy fluvial depositional environment for Qfg gravels. The orientation of scour marks as well as clast imbrication indicate a southerly flow for this early to middle Pleistocene river system.

Unit Qfp is primarily sandy fluvial and eolian deposits associated with graded surfaces formed during previous periods of Bull Canyon dissection and partial backfilling, with intermediate episodes of landscape stabilization and soil development. Many layers contain buried soils composed of fine-grained sediments, and, typically large amounts of pedogenic carbonate. Most of the buried soils are well developed and have stage II to stage III carbonate morphology (terminology of Gile et al., 1966). Unit Qfp contains fine- to coarse-grained river and floodplain sediments plus silty eolian material. Sediments contained in Qfp are interpreted to have been derived from local bedrock sources, including Chinle through Morrison strata in addition to some reworked Qe, Qrg and Ogallala sediments. Thickness of Qfp is variable and ranges from 2 to12 m.

Due to the discontinuous nature of these deposits, it is difficult to tell exactly how far local base level has dropped since the Pleistocene, although in central Bull Canyon 20 m of relief between Pleistocene and modern base levels are common. In central Bull Canyon, regional slope of the upper section of Qfp suggests a reversal in stream flow direction, from south flowing to north flowing.

Younger Alluvium (Qfs, Qfv, Qfm)

Younger valley-fill deposits are interpreted to be associated with episodes of valley incision and partial backfilling during late Pleistocene and Holocene time. Ancestral floodplain positions appear to have been 5 to 10 m above the modern channels. Soils are weakly developed. Younger valley-fill deposits represent aggradational events with few periods of base-level stability and/or soil formation.

Valley-slope alluvium (Qfs) is associated with alluvial aprons and graded surfaces formed during previous periods of aggradation of Alamogordo Valley. Sediments are generally coarse-grained sand and gravelly sand to silty clay.

On the western edge of Alamogordo Valley, Qfs is typically fine-grained sediments of arroyos and alluvial fans having a regional slope to the east, into central Alamogordo Valley. Sediments are locally derived from bedrock sources including Chinle through Morrison strata as well as reworked Ogallala Formation. Generally, Qfs sediments packages become finer upward locally,

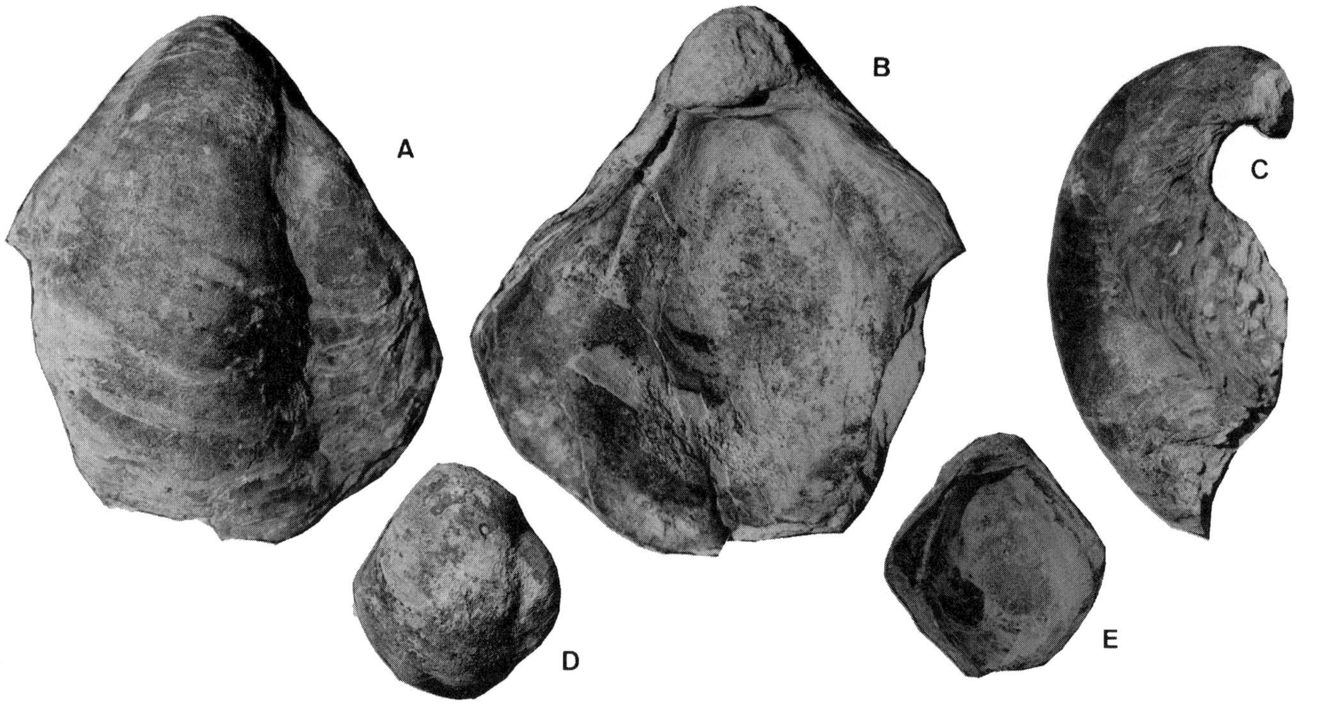

FIGURE 7. *Texigryphaea pitcheri* (Morton) from the Tucumcari Formation at Luciano Mesa, NE1/4 SW1/4 sec. 23, T9N, R26E. All figures x 1. A-C, external, internal and side views of large left valve, UNM 13137. D-E, external and internal views of small left valve, UNM 13138.

and they can contain laterally continuous buried soils with abundant disseminated calcium carbonate indicative of weak soil development. Ancestral local base level positions on the western slopes of Alamogordo Valley are approximately 5 to 10 m above the present floodplain. Thickness of these deposits at this location ranges from 3 to 10 m.

In contrast to Qfs on the western border of Alamogordo Valley, sediments preserved in the northeastern section are very coarse-grained sand and gravel. The sandy sediment was derived from local bedrock sources, including Chinle through Morrison strata; coarser gravels are primarily reworked Ogallala pebbles and cobbles. Thickness of the gravel deposits is approximately 3-4 m. As with Qfs on the western margin of the Alamogordo Valley, the base level position of Qfs in the east-central section of the Ima NW quadrangle is 15 m above the modern active channel and floodplain. Remnants of Qfs in the east-central part of Alamogordo Valley presently defend slopes and hilltops of less resistant material. These gravel-topped hills form an arcuate outcrop pattern near the eastern border of Alamogordo Valley.

Unit Qfv is confined to the floor of Alamogordo Valley and contains eolian and fluvial-arroyo sediments having surfaces graded to a Holocene, but not modern base level. Most of Qfv contains fine-grained sands and silts with discontinuous lenses of small gravel, and minor amounts of calcium carbonate. Material composing Qfv was derived from Chinle and reworked older Pleistocene sediments. Generally, Qfv sediment packages fine upward, although the poorly sorted alluvial fan and arroyo-distributary deposits may have an eolian component. In most sections of Alamogordo Valley, Qfv intertongues with Qfm, and overlaps basal sections of Qfs, making precise determination of contacts uncertain. Soils are typically immature and have only minor accumulations of calcium carbonate.

Modern floodplain alluvium, associated with surfaces graded to the present local base level, constitute Qfm. These sediments include fine-grained fluvial and floodplain deposits and coarse-grained alluvial fan/distributary channel sediments containing minor amounts of calcium carbonate. In both Alamogordo Valley and Bull Canyon, the composition of Qfm is largely reworked Triassic through Quaternary sediments. Generally, these sediments fine upwards from small gravels to medium sands. The thickness of unit Qfm is typically 0.5- 3 m.

In Alamogordo Valley, Qfm takes the form of partially channelized fans and distributary arroyos that are not integrated with the trunk stream of Alamogordo Creek. Instead, these distributaries terminate as large lobes of coarse-grained material and higher, abandoned floodplain surfaces. On the southwestern edge of Alamogordo Valley, the distance from the Llano Estacado escarpment to the axial drainage is short, and ephemeral streams connect to Alamogordo Creek. Throughout most of Alamogordo Valley, Qfm intertongues and overlaps with Qfv and locally overlaps and dissects lower sections of Qfs.

Unit Qfm in Bull Canyon is limited to small ephemeral streams (e.g., Bull Canyon Creek) and to arroyos originating on weathering limited bedrock slopes along valley margins. Compared to Alamogordo Valley, Bull Canyon has much less surface area covered by active channelized drainage. This is probably due to several factors, including greater canyon relief, well-developed Holocene channel patterns, and a lower local base level. These deposits overlap and dissect older Quaternary alluvium as Bull Canyon stream entrenchment migrates upstream.

ACKNOWLEDGMENTS

We are grateful to J. T. Gregory, J. Hawley, N. Mateer and B. Schaeffer for helpful discussions and information. Numerous UNM students assisted in the field, and several landowners, especially P. Houlihan, generously granted access to their property. S. Connell and A. Heckert provided helpful reviews of the manuscript

REFERENCES

Broadhead, R. F., 1984, Subsurface petroleum geology of Santa Rosa Sandstone (Triassic), northeast New Mexico: New Mexico Bureau of Mines and Mineral Resources, Circular 193, 22 p.

Dobrovolny, E, Bates, R. L., and Summerson, C. H., 1946, Geology of northwestern Quay County, New Mexico: U. S. Geological Survey Oil and Gas Investigations, Map OM-62, 2 sheets.

Dunkle, D. H., 1942, A new fossil fish of the family Leptolepidae: Cleveland Museum of Natural History Science Publications, v. 8, p. 61-64.

Gile, L. H., Peterson, F. F., and Grossman, R. B., 1966, Morphological and genetic sequences of carbonate accumulation in desert soils: Soil Science, v. 101, p. 347-360.

Gould, C. N., 1907, The geology and water resources of the western portion of the Panhandle of Texas: U.S. Geological Survey, Water-supply Paper 191, 170 p.

Gregory, J. T., 1972, Vertebrate faunas of the Dockum Group, Triassic, eastern New Mexico and west Texas: New Mexico Geological Society, 23rd Field Conference, Guidebook, p. 120-123.

Hawley, J. W., 1984, The Ogallala Formation in eastern New Mexico, *in* Wheatland, G. A., ed., Proceedings of the Ogallala Aquifer Symposium II: Lubbock, Texas Tech University Water Resources Center, p. 157-176.

Hester, P. M., 1988, Depositional environments in an Upper Triassic lake, east-central New Mexico [M. S. thesis]: Albuquerque, University of New Mexico, 154 p.

Hester, P. M., and Lucas, S. G., 2001, Depositional environments of a Late Triassic lake, east-central New Mexico: New Mexico Geological Society, 52nd Field Conference Guidebook.

Hunt, A P., 1994, Vertebrate paleontology and biostratigraphy of the Bull Canyon Formation (Chinle Group: Norian), east-central New Mexico with revisions of the families Metoposauridae (Amphibia: Temnospondyli) and Parasuchidae (Reptilia: Archosauria) [Ph.D. dissertation]: Albuquerque, University of New Mexico, 404 p.

Hunt, A. P., 1997, E. C. Case, J. T. Gregory and early explorations for fossils vertebrates in the Bull Canyon Formation (Upper Triassic) of eastern New Mexico: New Mexico Museum of Natural History and Science, Bulletin 11, p. 15-24.

Hunt, A. P., 2001, Paleontology and age of the Upper Triassic Trujillo Formation, east-central New Mexico and West Texas: New Mexico Geological Society, 52nd Field Conference Guidebook.

Hunt, A. P., and Lucas, S. G., 1989, Late Triassic vertebrate localities in New Mexico; *in* Lucas, S. G., and Hunt, A. P., eds., Dawn of the Age of Dinosaurs in the American Southwest: Albuquerque, New Mexico Museum of Natural History, p. 72-101.

Hunt, A. P. and Lucas, S. G., 1993, Triassic vertebrate paleontology and biochronology of New Mexico: New Mexico Museum of Natural History and Science, Bulletin 2, p. 49-60.

Hunt, A. P. and Lucas, S. G., 1997, Stratigraphy, paleontology and biochronology of the Upper Triassic Chinle Group in east-central New Mexico: Proceedings of the Southwest Paleontological Symposium, p. 25-40.

Hunt, A. P. and Lucas, S. G., 2001, The Laramide Tucumcari structural zone, east-central New Mexico: New Mexico Geological Society, 52nd Field Conference, Guidebook.

Kelley, V. C., 1972a, Triassic rocks of the Santa Rosa area: New Mexico Geological Society, 23rd Field Conference, Guidebook, p. 84-90.

Kelley, V. C., 1972b, Geology of the Fort Sumner sheet, New Mexico: New Mexico Bureau of Mines and Mineral Resources, Bulletin 98, 55 p.

Koerner, H. E., 1930, Jurassic fishes from New Mexico: American Journal of Science, v. 19, p. 463.

Kues B. S., 1985, Nonmarine molluscs from the Chinle Formation, Dockum Group (Upper Triassic), of Bull Canyon, Guadalupe County, New Mexico: New Mexico Geological Society, 36th Field Conference, Guidebook, p. 185-196.

Kues, B.S., 1997, New bivalve taxa from the Tucumcari Formation (Cretaceous, Albian) New Mexico, and the biostratigraphic significance of the basal Tucumcari fauna: Journal of Paleontology, v. 71, p. 820-839.

Kues, B. S., and Lucas, S. G., 2001, Fauna of nearshore sandy facies of the Tucumcari Formation (Lower Cretaceous, Albian), Quay County, New Mexico: New Mexico Geological Society, 52nd Field Conference, Guidebook.

Kues, B. S., Lucas, S. G., Kietzke, K. K., and Mateer, N. J., 1985, Synopsis of Tucumcari Shale, Mesa Rica Sandstone and Pajarito Shale paleontology, Cretaceous of east-central New Mexico: New Mexico Geological Society, 36th Field Conference, Guidebook, p. 261-281.

Lucas, S. G., and Anderson, O. J., 1998, Jurassic stratigraphy and correlation in New Mexico: New Mexico Geology, v. 20, p. 97-104.

Lucas, S.G., and Hunt, A.P., 1989, Revised Triassic stratigraphy, Tucumcari basin, east-central New Mexico; *in* Lucas, S. G., and Hunt, A. P., eds., Dawn of the Age of Dinosaurs in the American Southwest: Albuquerque, New Mexico Museum of Natural History, p. 150-170.

Lucas, S. G., and Hunt, A. P., 1993, Tetrapod biochronology of the Chinle Group (Upper Triassic), western United States: New Mexico Museum of Natural History and Science, Bulletin 3, p. 327-329.

Lucas, S. G., and Kietzke, K. K., 1986, Stratigraphy and petroleum potential of the Jurassic Todilto Formation in northeastern New Mexico; *in* Ahlen, J. L., Hanson, M. E. and Zidek, J., eds., Southwest Section of AAPG Transactions and Guidebook of 1986 Convention, Ruidoso, New Mexico: Socorro, New Mexico Bureau of Mines and Mineral Resources, p. 121-127.

Lucas, S. G., and Kisucky, M. J., 1988, Type and reference sections of the Tucumcari, Mesa Rica and Pajarito formations, Cretaceous of east-central New Mexico: New Mexico Geology, v. 10, p. 82-89.

Lucas, S. G., Anderson, O. J., and Pigman, C., 1995, Jurassic stratigraphy in the Hagan basin, north-central New Mexico: New Mexico Geological Society, 46th Field Conference, Guidebook , p. 247-255.

Lucas, S. G., Heckert, A. B., and Hunt, A. P., 2001, Triassic stratigraphy, biostratigraphy and correlation in east-central New Mexico: New Mexico Geological Society, 52nd Field Conference, Guidebook.

Lucas, S. G., Hunt, A. P., and Hayden, J. N., 1987, Type section of Exeter Member of Entrada Sandstone, Jurassic of northeastern New Mexico: New Mexico Geological Society, 38th Field Conference Guidebook 38, p. 17-18.

Lucas, S. G., Hunt, A.P., and Kietzke, K.K., 1985d, *Neocalamites* forest in the Upper Triassic of bull Canyon, New Mexico: New Mexico Geological Society, 36th Field Conference, Guidebook, p. 8-9.

Lucas, S. G., Hunt, A. P., and Morales, M., 1985b, Stratigraphic nomenclature and correlation of Triassic rocks of east-central New Mexico: A preliminary report: New Mexico Geological Society, 36th Field Conference, Guidebook, p. 171-184.

Lucas, S. G., Kietzke, K. K., and Hunt, A. P., 1985c, The Jurassic system in east-central New Mexico: New Mexico Geological Society, 36th Field Conference, Guidebook, p. 213-242.

Lucas, S.G., Kietzke, K.K., Sobus, J., Weadock, G., Mateer, N. J., and Hunt, A.P., 1984, Upper Triassic-Upper Jurassic stratigraphy, fossil vertebrates and depositional environments, Bull Canyon, Guadalupe County, East-Central New Mexico: Geological Society of America, Abstracts with Programs, v. 16, p. 245.

Lucas, S. G., Oakes, W., and Froehlich, J. W., 1985d, Triassic microvertebrate locality, Chinle Formation, east-central New Mexico: New Mexico Geological Society, 36th Field Conference, Guidebook, p. 205-212.

Lucas, S. G., Weadock, G. L., Kietzke, K. K., and Hunt, A. P. 1994, Geologic map of the Ima NW quadrangle, Guadalupe and Quay Counties, New Mexico: New Mexico Bureau of Mines and Mineral Resources Open-File Report NMBM-0399, 3 p., scale 1:24,000.

Mehl, M. G., 1922, A new phytosaur from the Trias of Arizona: Journal of Geology, v. 30, p. 144-157.

Pipiringos, G. N., and O'Sullivan, R. B., 1978, Principal unconformities in Triassic and Jurassic rocks, Western Interior, U.S.: A preliminary survey: U.S. Geological Survey, Professional Paper 1035-A, 29 p.

Schaeffer, B. and Patterson, C., 1984, Jurassic fishes from the western United States with comments on Jurassic fish distribution: American Museum Novitates, v. 2796, 86 p.

At Bull Canyon in eastern Guadalupe County, the Middle Jurassic Todilto Formation is a thin (up to 2 meters) bed of dark, kerogenic limestone between lower and upper eolian sandstone tongues of the Entrada Sandstone. Indeed, it is possible here to walk the Todilto to its pinchout into the eolianites, and thus examine the margin of the vast Todilto salina.

JURASSIC STRATA IN EAST-CENTRAL NEW MEXICO AND THEIR REGIONAL SIGNIFICANCE

SPENCER G. LUCAS[1] AND LEE A. WOODWARD[2]
[1]New Mexico Museum of Natural History, 1801 Mountain Rd. NW, Albuquerque, NM 87104; [2]Department of Earth and Planetary Sciences, University of New Mexico, Albuquerque, NM 87131

Abstract.—The Jurassic section in east-central New Mexico is as much as 330 m thick and consists of Middle and Upper Jurassic strata of the San Rafael Group (Entrada, Todilto and Summerville formations) and overlying Morrison Formation. The Entrada Sandstone is as much as 53 m of eolianite sandstone assigned to the lower, Slick Rock Member and the upper, Exeter Member, which is an upper tongue of the Entrada that postdates Todilto deposition. The Todilto Formation is 2-7 m of organic-rich limestone of the Luciano Mesa Member that disconformably (J-3 unconformity?) overlies the Slick Rock Member of the Entrada. Summerville Formation strata are up to 40 m of repetitively-bedded sandstone and siltstone that disconformably overlie the Todilto and grade into/intertongue with the Exeter Member. The Morrison Formation disconformably (J-5 unconformity) overlies the Summerville and consists of a basal Salt Wash Member (up to 50 m of arkosic sandstone/conglomerate) and an upper, Brushy Basin Member (up to 150 m of smectitic claystone); locally, a 10-15-m-thick, trough-crossbedded sandstone at the top of the Morrison may be an equivalent of the Jackpile Member to the west.

The absence of the basal, Dewey Bridge Member of the Entrada Sandstone indicates that east-central New Mexico was landward of marine influence of the Carmel seaway during the Bajocian-early Callovian, and that the J-2 unconformity at the Entrada base represents a longer hiatus in east-central New Mexico than it does to the west. Todilto deposition took place in a paralic salina with little or no direct connection to the coeval Curtis-Sundance seaway. Summerville tidal flat facies extended across northern New Mexico, much of Colorado and Utah, northeastern Arizona and adjacent areas during the Callovian-Oxfordian transition, and graded southward and southeastward into eolian dunes. The J-5 unconformity at the base of the Salt Wash Member of the Morrison Formation represents a significant tectonic reorganization of the Jurassic depositional systems. Deposition of the Brushy Basin Member of the Morrison Formation did not take place in a single, large lake (Lake T'oo'dichi'), but instead on a vast floodplain dotted with smaller lakes.

Regional eastward-trending piercing lines between the Colorado Plateau and east-central New Mexico can be defined by stratigraphic truncations and depositional pinchouts in Jurassic strata. They only allow 5-20 km of right slip between the eastern margin of the Colorado Plateau and the craton in post-Jurassic time. Arguments that 60-170 km of right slip occurred here during the Laramide orogeny (Late Cretaceous-Paleogene) or Neogene thus are untenable.

INTRODUCTION

Jurassic strata in east-central New Mexico (Fig. 1) are mostly siliciclastic nonmarine sediments of Middle and Late Jurassic age. As much as 320 m thick, these strata have been much less studied than the economically-significant Jurassic section on the Colorado Plateau. Nevertheless, Jurassic strata in east-central New Mexico were deposited by the same depositional systems as rocks on the Colorado Plateau. Therefore, the Jurassic section in east-central New Mexico is critical to a broad understanding of regional Jurassic deposition in the Western Interior. It is also of significance to interpreting the post-Jurassic regional tectonic history between the Colorado Plateau and the Southern High Plains. Here, we summarize Jurassic stratigraphy in New Mexico and discuss its significance for regional depositional and tectonic history

STRATIGRAPHY

Marcou (1858) first identified Jurassic strata in east-central New Mexico, and their stratigraphic relationships and nomenclature have been studied since Lee (1902) (Fig. 2). Here, we summarize Jurassic stratigraphy in east-central New Mexico to further justify the nomenclature and correlations advocated by Lucas and Anderson (1998). Jurassic strata in east-central New Mexico are assigned to the San Rafael Group (Entrada, Todilto and Summerville formations) and overlying Morrison Formation.

SAN RAFAEL GROUP

Entrada Sandstone

The oldest Jurassic strata in east-central New Mexico belong to the Middle Jurassic (Callovian) Entrada Sandstone. As much as 53 m thick, the Entrada consists mostly of yellowish-gray, white and pale reddish-brown, fine-grained subarkosic sandstone (Mankin, 1958). Prevalent bedforms are large-scale trough crossbeds, although laminar bedding is locally common.

The Entrada Sandstone disconformably overlies Upper Triassic strata of the Chinle Group (Fig. 3). This unconformity encompasses a hiatus of about 40 million years and is the J-2 unconformity of Pipiringos and O'Sullivan (1978). In east-central New Mexico, the Entrada consists of two members, the lower, Slick Rock Member, which is mostly trough-crossbedded eolianites, and the upper, Exeter Member, which is an upper tongue of the Entrada that postdates deposition of the Todilto Formation (Lucas et al., 1985, 1987a, b) (Fig. 4). The Todilto Formation disconformably rests on the Slick Rock Member as far east as the Todilto pinchout near the Guadalupe-Quay County border (Lucas et al., 1985; Lucas and Kietzke, 1986). The Exeter Member intertongues with or grades into the overlying Summerville Formation elsewhere.

Todilto Formation

In east-central New Mexico, the Todilto Formation consists only of its lower, limestone member, the Luciano Mesa Member of Lucas

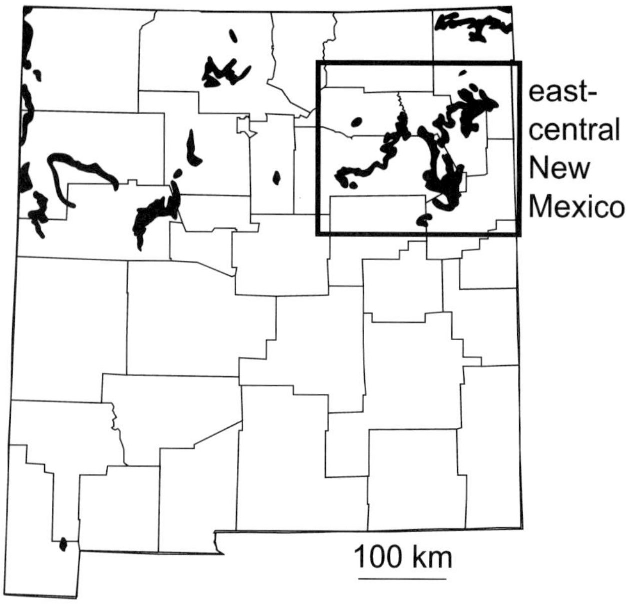

FIGURE 1. Distribution of Jurassic strata in New Mexico (after Dane and Bachman, 1965) and location of east-central New Mexico.

et al. (1995). The upper, gypsum member (Tonque Arroyo Member of Lucas et al., 1995) apparently was not deposited in east-central New Mexico. The Todilto Formation crops out in San Miguel, Guadalupe and western Quay County, where it is 2-7 m thick and is mostly organic-rich, varved limestone with minor intercalations of gritty, sapropelic shale and medium-grained sandstone (Figs. 3-4). The contact with the underlying Slick Rock Member of the Entrada Sandstone is a sharp surface of lithologic contrast with local reworking of Entrada sand into the basal Todilto Formation (Lucas et al., 1985; Lucas and Kietzke, 1986). As Lucas and Anderson (1998) suggested, the base of the Todilto is probably a regional unconformity correlative to the J-3 unconformity of Pipiringos and O'Sullivan (1978). Similarly, the Todilto Formation is sharply overlain by either the Summerville Formation or the Exeter Member of the Entrada Sandstone, and this too may be a regional unconformity.

Summerville Formation

Probably the most confusing Jurassic stratigraphic unit in east-central New Mexico is the interval here referred to the Summerville Formation (Figs. 2-3). These rocks have formerly been included in the Morrison Formation or termed Bell Ranch Formation, a parochial term introduced by Griggs and Read (1959). Furthermore, some workers have applied the terms Ralston Creek Formation or Wanakah Formation to these strata.

Summerville strata are as much as ~ 40 m thick and typically consist of repetitively bedded sandstone and siltstone. Individual beds are laterally continuous and vary in thickness from 0.1 to ~ 5 m. Colors range from very pale orange to light brown, and gypsum beds or nodules are common. At some localities, the Summerville can be divided into two units, a lower sandstone-dominated interval and an upper, red-bed siltstone and mudstone interval. The contact with the underlying Entrada or Todilto formations is sharp (disconformity?), and the Summerville intertongues/grades laterally into the Exeter Member of the Entrada Sandstone locally. The base of the overlying Morrison Formation is picked at either the base of a trough-crossbedded pebbly subarkosic sandstone or at the base of a green smectitic claystone.

Anderson and Lucas (1992) first justified use of the term Summerville Formation in east-central New Mexico by demonstrating the continuity and consistent lithologic character of Summerville strata from their type area in southeastern Utah across the Colorado Plateau, Rio Grande rift and onto the Southern High Plains (also see Anderson and Lucas, 1994, 1995, 1996, 1997, 2000; Lucas and Anderson, 1997; Lucas et al., 1995, 1999). The terms Bell Ranch, Ralston Creek and Wanakah are unnecessary synonyms of Summerville.

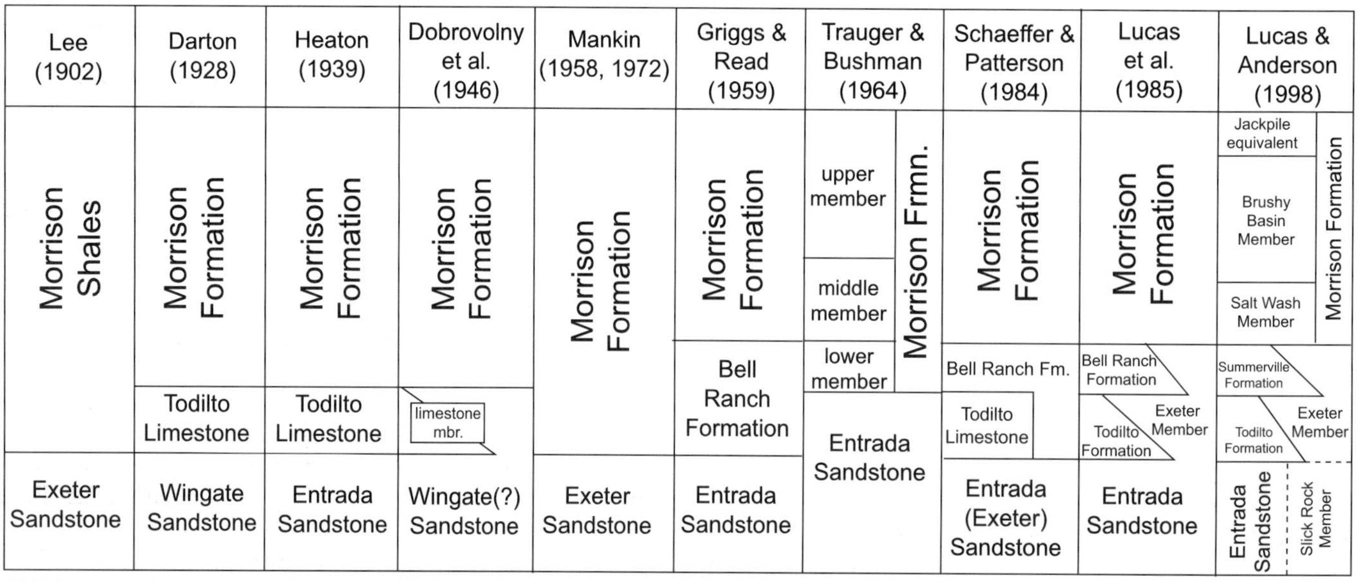

FIGURE 2. Development of Jurassic stratigraphic nomenclature in east-central New Mexico. The nomenclature advocated here is that of Lucas and Anderson (1998).

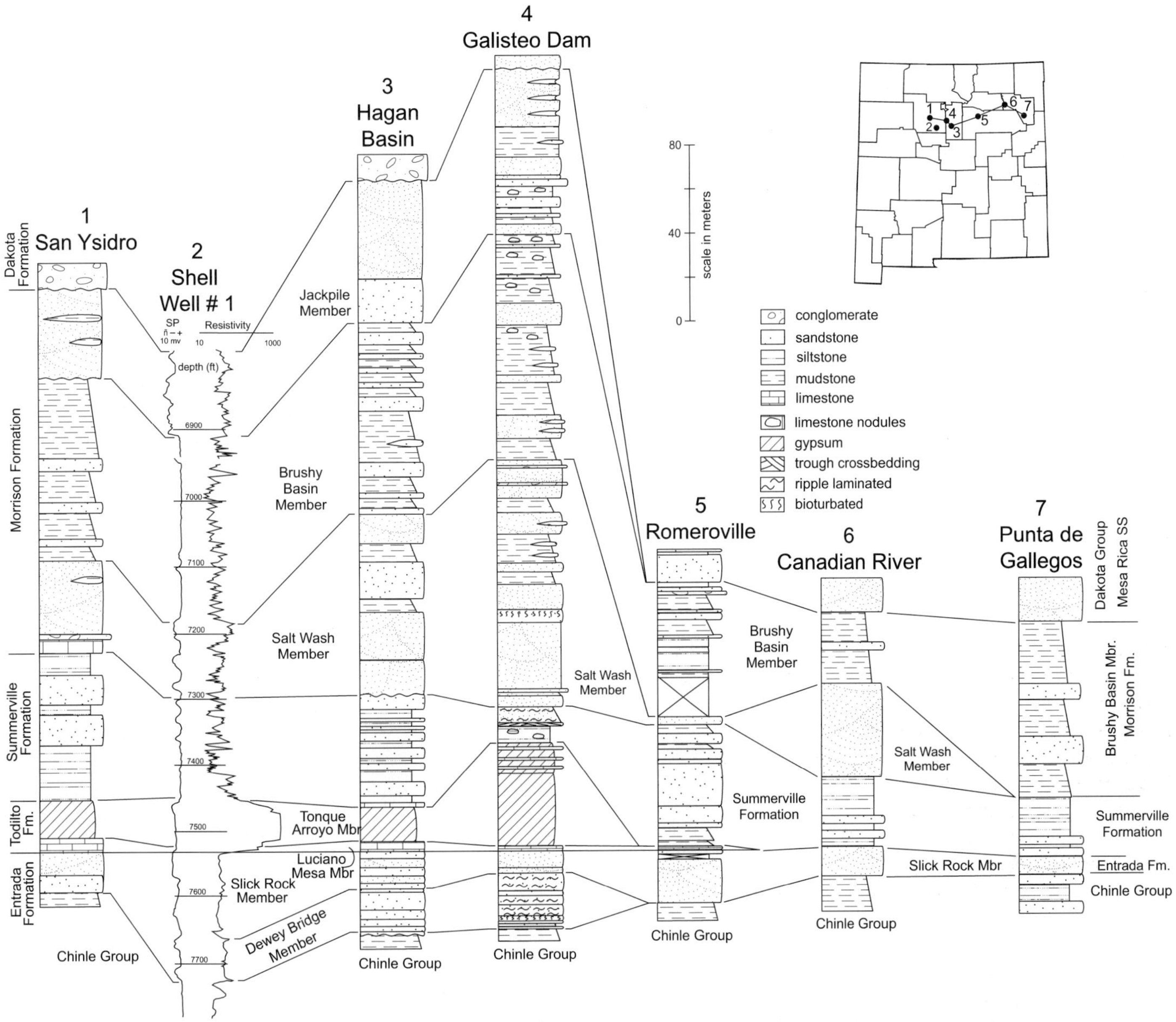

FIGURE 3. Correlation of Jurassic stratigraphic sections from the Colorado Plateau (San Ysidro) across the Rio Grande rift (Shell Well # 1, Hagan basin, Galisteo Dam) to east-central New Mexico (Romeroville Gap, Canadian River, Punta de Gallegos). Sections are from Lucas et al. (1999) except for sections at Canadian River (from Wanek, 1962) and Punta de Gallegos (from Lucas et al., 1985).

Morrison Formation

The Morrison Formation disconformably overlies the Summerville Formation in east-central New Mexico (Figs. 2-3). This is the J-5 unconformity of Pipiringos and O'Sullivan (1978). Lower Cretaceous strata (Tucumcari or Mesa Rica formations) disconformably overlie the Morrison Formation throughout this region; this is the K-0 unconformity of Pipiringos and O'Sulliuvan (1978).

The lower part of the Morrison Formation is as much as 50 m of subarkosic sandstone and conglomerate, the Salt Wash Member. The bulk of the formation is the overlying Brushy Basin Member, as much as 150 m of green and variegated, smectitic claystone and minor sandstone. Locally, the top of the Morrison Formation is a white, trough-crossbedded sandstone up to 15 m thick that may be a correlative of the Jackpile Member to the west (Holbrook et al., 1987).

REGIONAL SEDIMENTOLOGIC AND PALEOGEOGRAPHIC IMPLICATIONS

Jurassic strata in east-central New Mexico were deposited near or at the southeastern edge of the vast depositional basins that covered much of the Western Interior during the Middle and Late Jurassic. The older of these basins, here termed the San Rafael basin, is filled by the San Rafael Group, a westward thickening prism of sediments of mostly shallow marine, eolian and tidal flat origin. The San Rafael basin was a retroarc foreland basin that extended from Oklahoma-Nebraska on the east to Idaho-Nevada

FIGURE 4. Dark, organic-rich limestone of the Luciano Mesa Member of the Todilto Formation above Slick Rock Member of Entrada Sandstone and beneath Exeter Member of Entrada; at Bull Canyon, T9N, R26E, Guadalupe County.

on the west (e.g., Kocurek and Dott, 1983; Lawton, 1994). The younger, Late Jurassic basin, the Morrison basin, had a similar geographic extent but was the site of fluvial deposition of the Morrison Formation by rivers derived principally from the uplifted and volcanically active arc terrane to the west.

Although most workers acknowledge that the Jurassic strata of east-central New Mexico were deposited as part of these vast depositional systems, the full implications of this have not always been appreciated. Here, we briefly review these implications with regard to the: (1) landward extent of the Middle Jurassic Carmel transgression; (2) Todilto salina deposition; (3) full extent of Summerville and related eolian facies; (4) tectonic reorganization of the depositional system at the beginning of Morrison time; and (5) deposition of the Brushy Basin Member of the Morrison Formation.

Carmel Transgression

During Bajocian time, a shallow sea transgressed into the San Rafael basin, and its southern tongue in Utah deposited the Carmel Formation (Fig. 5A). The Dewey Bridge Member of the Entrada Sandstone is laterally equivalent to the Carmel Formation and represents broad supratidal (sabkha) and coastal plain deposits of the Carmel seaway (e.g., Kocurek and Dott, 1983). The southeasternmost extent of the Dewey Bridge Member is in the Rio Grande valley of north-central New Mexico at Placitas, in the Hagan basin and at Galisteo Dam, about 450 km southeast of the Carmel shoreline near Moab, Utah.

The absence of the Dewey Bridge Member of the Entrada Sandstone in east-central New Mexico thus means that this region was landward of the marine influence (coastal plain environments) of the Carmel seaway. In east-central New Mexico, the base of the Entrada is the base of the Slick Rock Member, deposits of the vast early Callovian sand sea (erg) that covered much of the American Southwest after the retreat of the Carmel seaway (Blakey et al., 1988). This means that the J-2 unconfomity at the

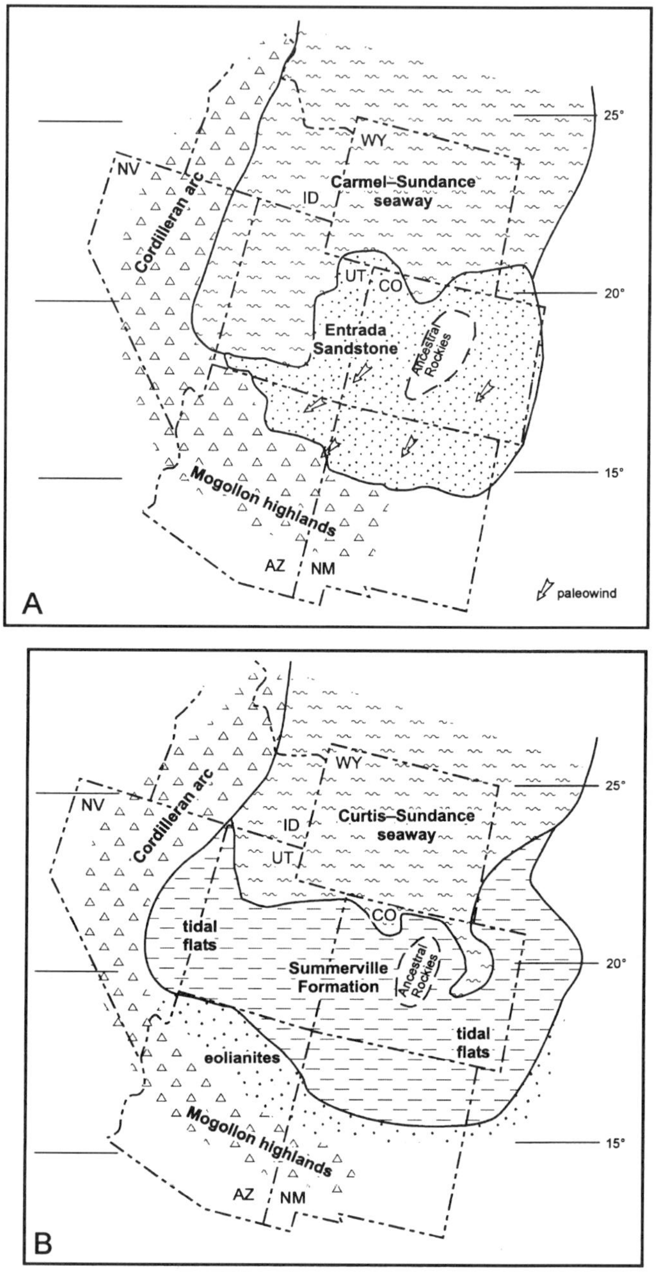

FIGURE 5. Simplified paleogeographic maps of part of San Rafael Group deposition (after Kocurek and Dott, 1983). A. Carmel-Entrada time (Bajocian-early Callovian). B. Summerville time (late Callovian-Oxfordian).

base of the Entrada represents a longer hiatus in east-central New Mexico than to the west in the Rio Grande rift and on the Colorado Plateau.

Todilto Deposition

Anderson and Kirkland (1960) early recognized the continuity of deposition of the Todilto Formation from the Four Corners region to as far southeast as eastern Guadalupe County (Fig. 6). However, until the 1980s most workers regarded the Todilto as deposits of a marine embayment of the Curtis seaway (e.g.,

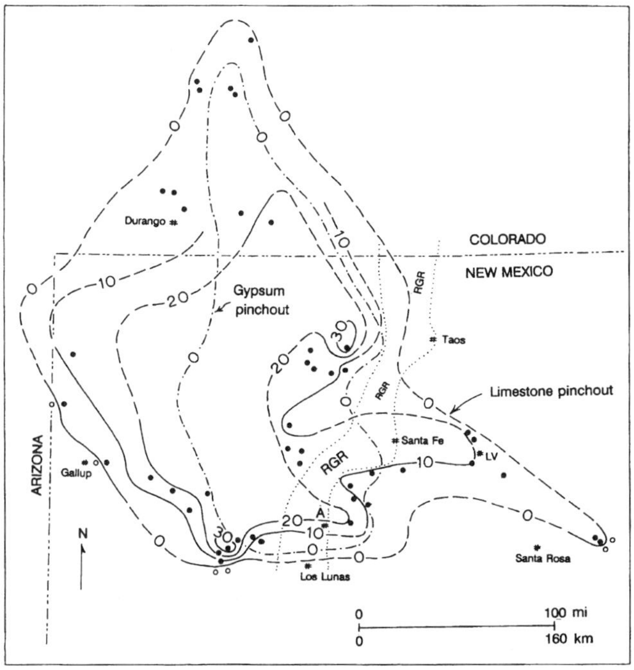

FIGURE 6. Approximate depositional limits of Jurassic Todilto limestone member (Luciano Mesa Member) and overlying Todilto gypsum member (modified from Ash, 1958; Kirkland et al., 1995). Dotted outline is Rio Grande rift (RGR). A = Albuquerque, LV = Las Vegas. Structure contours for Todilto limestone member are in feet.

Harshbarger et al., 1957; Anderson and Kirkland, 1960; Ridgley and Goldhaber, 1983). Nevertheless, several lines of evidence suggest that the Todilto depositional basin had little connection to the Jurassic seaway for most of its history and instead was a vast, paralic salina:

1. No direct stratigraphic continuity of Todilto strata and marine Jurassic strata exists, either on outcrop or in the subsurface. The Todilto evidently pinches out around its basin periphery into eolianites (Lucas et al., 1985; Anderson and Lucas, 1992).

2. No clearly marine fauna or flora are known from the Todilto Formation. Instead, a very low diversity fish fauna, as is characteristic of saline lakes, is known from the Todilto (Barbour and Brown, 1974; Lucas et al., 1985). Indeed, the low diversity fish and invertebrate fauna of the Todilto are strikingly similar to that of Quaternary salinas in Australia (e.g., Warren, 1982; Warren and Kendall, 1985). Armstrong (1995) claimed that the dasyclad algae found in the Todilto near Grants, New Mexico, indicate a marine environment, but today these blue green algae tolerate a wide range of salinity from fresh to hypersaline waters.

3. Carbon and sulphur isotope ratios calculated for Todilto limestones have a wide range of values compatible with a nonmarine, marine or mixed waterbody (Kirkland et al., 1995). However, strontium isotope ratios for the Todilto do not match those of sediments deposited from normal marine Callovian seawater (Kirkland et al., 1995).

Reconstructing the Todilto paleoenvironment as a salina is consistent with all data. Todilto deposition was initiated by marine flooding of an extensive area of low relief as the Curtis-Sundance transgression began (Fig. 5B). This supports the contention that the Todilto base, as far east as Guadalupe County, is the J-3 unconformity, which is the transgressive unconformity at the base of the Curtis Formation (Pipiringos and O'Sullivan, 1978). Following the flooding event, the Todilto basin was isolated from the seaway by coastal ergs. Freshwater stream runoff, influx of seawater by seepage through the erg, and possible short-lived overtopping of the erg replenished and maintained the Todilto salina. Increased aridity promoted evaporation, which eventually produced a smaller, evaporitic basin in which gypsum precipitated.

Summerville Deposition

During the Callovian-Oxfordian transgression and regression of the Curtis seaway, restricted marine and tidal flat deposits accumulated across much of Utah, Colorado and adjacent areas, including east-central New Mexico (Fig. 5B). The resulting broadly deposited lithosome merits one name, Summerville Formation, to express its continuity and lithologic integrity across this broad area. It has received numerous other names, including Wanakah, Beclabito, Bell Ranch and Ralston Creek, all of which should be abandoned.

Along the southern margin of Summerville deposition, and also immediately after Summerville deposition ended, eolian sediments accumulated. These eolianites were landward of the Summerville tidal flats, and in some areas encroached on them after Summerville deposition ceased. In east-central New Mexico, they are the Exeter Member of the Entrada Sandstone, strata homotaxial to the Bluff Sandstone and Acoma Tongue of the Zuni Sandstone in west-central New Mexico. The general similarity of depositional patrterns along the southern and southeastern edge of the Summerville tidal flat system is striking, though many details remain to be deciphered.

J-5 Unconformity

The onset of Morrison Formation deposition represents a significant tectonic reorganization of the Jurassic depositional system in the Western Interior. Prior to Morrison deposition, sedimentation in the San Rafael basin was by eolian, shallow marine and tidal flat systems in a relatively quiescent tectonic setting with source areas along the basin's western, southern and eastern perimeters. However, in Morrison time, deposition was primarily by easterly flowing rivers derived from the uplifted arc terrane to the west culminated by broad floodplains with a significant influx of altered volcanic ashes from the west (Fig. 7).

The J-5 unconformity at the base of the Salt Wash Member of the Morrison Formation thus is a tectonosequence boundary. It can be traced from the Salt Wash type section near Moab, Utah, across the Colorado Plateau and the Rio Grande rift onto the Southern High Plains of east-central New Mexico.

In east-central New Mexico, the thickness of the Salt Wash Member is highly variable, and at some localities it is absent, so the Brushy Basin Member rests directly on the Summerville Formation. We interpret this as a consequence of the hiatus associated with the J-5 unconformity. Thus, Salt Wash fluvial depos-

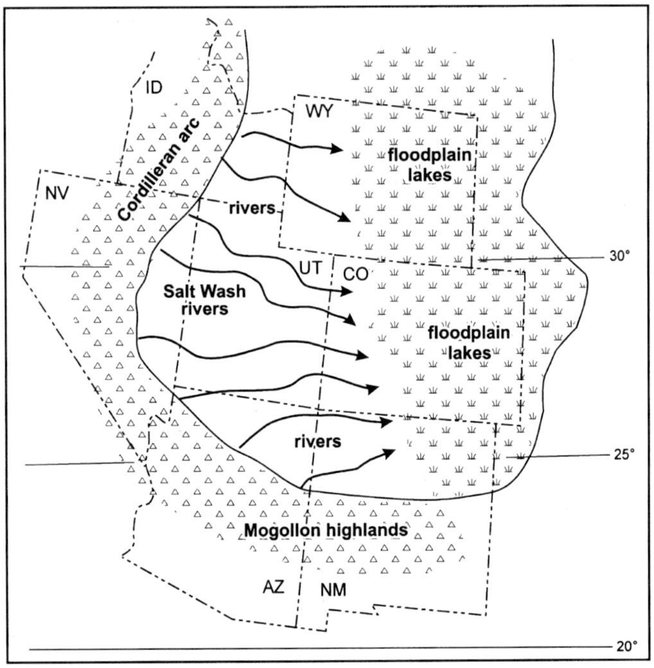

FIGURE 7. Simplified paleogeographic map of Late Jurassic Morrison Formation deposition.

its filled in topographic lows eroded into underlying San Rafael Group deposits during the J-5 hiatus. On the topographic highs, Morrison deposits did not accumulate until later, when Brushy Basin Member claystones blanketed the region.

Brushy Basin Member Deposition

Brushy Basin Member strata are the bulk of the Morrison Formation in east-central New Mexico. Indeed, the Brushy Basin Member is the most widely distributed interval of the Morrison, being found essentially across its entire depositional basin, from Utah to Nebraska and from New Mexico to Alberta (Peterson, 1972).

Significantly, Turner and Fishman (1991) argued that Brushy Basin Member deposition took place in a large lake in the Four Corners region, as much as 500 km in maximum dimension (NW-SE). This lake, termed Lake T'oo'dichi' (T'oo'dichi' is Navajo for "bitter water") was thought to be shallow, perhaps periodically drying up, with widely varying water chemistry, including high pH. However, most of the Brushy Basin outcrop belt lies outside of the margins of hypothesized Lake T'oo'dichi', and, indeed, the evidence presented to support the presence of the lake in the Four Corners region was seriously flawed (Anderson and Lucas, 1997).

Lacustrine deposits are present in the Brushy Basin Member throughout its outcrop belt, and these represent small, shallow, alkaline playa lakes surrounded by broad mudflats that laterally passed into adjacent alluvial environments (e.g., Dunagan, 2000). Brushy Basin Member facies in east-central New Mexico indicate this type of depositional setting (Lucas et al., 1985). The model of Lake T'oo'dichi' is not valid for Brushy Basin deposition in the Four Corners, nor does it explain Brushy Basin Member deposition across its vast outcrop area.

REGIONAL TECTONIC IMPLICATIONS

Jurassic stratigraphic data from east-central New Mexico have important implications for regional tectonic analyses concerning the timing and amount of movement of the Colorado Plateau during post-Jurassic time. It has been accepted since the 1950s (Kelley, 1955) that during the Laramide orogeny of Late Cretaceous and Paleogene age, the Colorado Plateau moved northeast with respect to the Rocky Mountain foreland to the north and the craton to the east (Fig. 8). The amount of resultant right slip on the eastern side of the plateau has been a topic of debate, with estimates ranging from 5 to 170 km.

Chapin and Cather (1981) estimated that the Colorado Plateau moved 60 to 120 km to the north-northeast relative to the crustal block to the east. They based their estimate largely on the presumed amount of crustal shortening across thrust and reverse faults to the north of the plateau in the Wyoming province (Fig. 8); the shortcoming of this approach was that it did not consider continuity of Mesozoic strata eastward from the plateau onto the craton (e.g., McKee et al., 1956; Hook et al., 1983). Chapin and Cather (1981) proposed that the decoupling occurred within a north-trending zone as wide as 100 km, marked by the Nacimiento fault system on the west, with most of the motion on

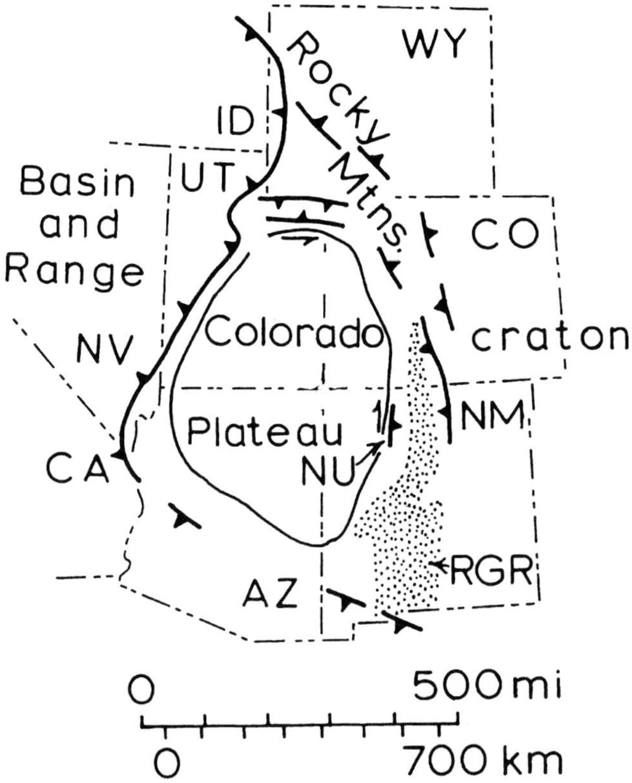

FIGURE 8. Location of Colorado Plateau and adjacent tectonic provinces. NU = Nacimiento uplift and RGR = Rio Grande rift (stippled).

faults that are now buried by the sedimentary fill of the Neogene Rio Grande rift. However, determining the amount of Laramide strike slip along the eastern margin of the Colorado Plateau and the Rio Grande rift region is complicated by the likelihood that strike-slip displacement may have occurred on some of the faults during the Precambrian, during the late Paleozoic, and again during the Neogene.

Following Chapin and Cather's (1981) model, Karlstrom and Daniel (1993) proposed that Proterozoic rocks demonstrated 100 to 170 km of Laramide right slip along several north-striking faults that they suggested were precursors of the Rio Grande rift. They proposed that the intersection of a regionally subhorizontal isobaric metamorphic surface with subvertical stratigraphic contacts and preexisting structures created piercing lines that could be used to indicate net displacement since about 1.4 Ga. Displacements of about 15 km across the Tusas-Picuris fault (Karlstrom and Daniel, 1993) and 37 km across the Pecos-Picuris fault (Miller et al., 1963) appear to be reasonable estimates, but these amounts of offset were constrained only as post-1.4 Ga. Miller et al. (1963) inferred that the right slip on the Pecos-Picuris fault probably occurred during the Precambrian.

Woodward et al. (1997) presented data from Jurassic isopach maps by McKee et al. (1956) that defined regional, east-trending piercing lines that permit 5 to 20 km of right slip between the eastern margin of the Colorado Plateau and the craton to the east (Fig. 9). However, Cather (1999) re-interpreted Jurassic and Cretaceous stratigraphic data that he felt were permissive or supportive of a minimum right slip of ~85 km on the eastern side of the Colorado Plateau. He included 13 km of well demonstrated right shift on the east side of the Defiance uplift in his estimate. The Defiance uplift, however, is within the Colorado Plateau, so it is irrelevant to the displacement on the eastern side of the Plateau.

Cather's (1999) paper was discussed by Ingersoll (2000), Lucas et al. (2000) and Woodward (2000). The thrust of Ingersoll's (2000) criticism was: (1) Cather's stratigraphic data were completely compatible with other models for Laramide deformation in this area, models not cited by Cather; (2) Cather's compilation does not provide a definitive choice among the various models; (3) interpretations of deflections of stratigraphic piercing lines were influenced by Cather's favored model; and (4) no lower limit on right slip was set by the stratigraphic data presented by Cather.

Lucas et al. (2000) noted that in Cather's (1999) paper the differences in depiction of data and analysis of previous workers stem primarily from: (1) selective use of part of, not the entire, database; (2) drawing trend lines on the basis of one data point; and (3) mistaken interpretation of data points. They presented precise examples to illustrate the problems in Cather's (1999) paper.

Woodward (2000) pointed out two major errors that caused Cather's (1999) estimate of a minimum of ~85 km of right slip to be highly questionable. A major problem concerns his failure to recognize that vertical surfaces as well as linear features can be used to determine lateral separation along major strike-slip faults. Also, his assumption that Phanerozoic pre-Laramide strike-slip motion on these faults was necessarily sinistral (Cather, 1999, p. 849) is not warranted (Ye et al., 1996; Woodward et al., 1999). In a paper dealing with phases of Laramide deformation in north-central New Mexico, Erslev (2001) concluded that the complexity of sedimentary patterns noted by Cather (1999) and the lack of consistent fault separations interpreted by Cather allow the possibility of no right lateral slip.

STRATIGRAPHIC DATA AND TECTONIC ANALYSIS

Lines that can provide piercing points where they meet fault surfaces include stratigraphic pinchouts and marker beds truncated beneath an angular unconformity (Crowell, 1959). In northern New Mexico, Jurassic strata are truncated southward by a regional, low-angle unconformity beneath overlying Cretaceous strata (e.g., Dobrovolny et al., 1946; Silver, 1948).

Paleotectonic maps of the Jurassic System by McKee et al. (1956) show zero isopach lines for three stratigraphic sequences (intervals B, C, and D of McKee et al., 1956) where the Entrada Sandstone, Todilto and Summerville formations, and the Morrison Formation are truncated by overlying Cretaceous strata. Although the precise placement of the zero isopach lines for the Entrada Sandstone (Fig. 9A) and the Morrison Formation (Fig. 9B) can be questioned, the overall patterns (Lucas et al., 1985) clearly indicate that 5 to 20 km of right slip can be accommodated along the Rio Grande rift, but certainly not 60 to 170 km. In a structural sense, isopach lines on maps represent surfaces of specific vertical dimensions that will show lateral separation if offset by strike-slip faulting. Isopach lines shown for Jurassic strata (Fig. 9) thus confirm a maximum of 5 to 20 km of right slip between the Colorado Plateau and the craton to the east.

Near Carthage, New Mexico, about 88 km south of the regional wedge-edge of the Morrison Formation (Fig. 9B), ~1 to 3.3 m of possible Morrison strata lie above Triassic Chinle Group strata and below the Cretaceous Dakota Formation (Hunt and Lucas, 1987). There is no indication of any Morrison Formation strata between the regional wedge edge, shown in Figure 9B, and Carthage or south of Carthage. Hunt and Lucas (1987) suggested that the Morrison Formation was preserved from erosion either by being deposited in a topographic low or by a local downwarp before deposition of the overlying Dakota Formation. These limited, thin outcrops do not define a regional piercing line and therefore cannot be used to determine strike-slip movement. Cather (1999) mistakenly assumed that this outcrop defined the regional southern wedge-edge of the Morrison and used it to estimate an unwarranted large amount of right slip across the Rio Grande rift.

One of the most impressive piercing lines involving Jurassic strata is the wedge edge of the lower, limestone-dominated Luciano Mesa Member of the Todilto Formation (Fig. 6). This thin (mostly <6 m), highly distinctive unit represents deposition in a huge saline lake basin during the Callovian (see above). Isopachs of the Luciano Mesa Member reveal no offset of it or its southward depositional pinchout (zero isopach) across the Rio Grande rift. Restoration of 60 to 170 km of right slip within the Rio Grande rift creates a paleodepositional geometry for the Todilto basin that is very unlikely (Woodward et al., 1997, 1999).

In summary, regional eastward-trending piercing lines defined by stratigraphic truncations and depositional patterns in Jurassic

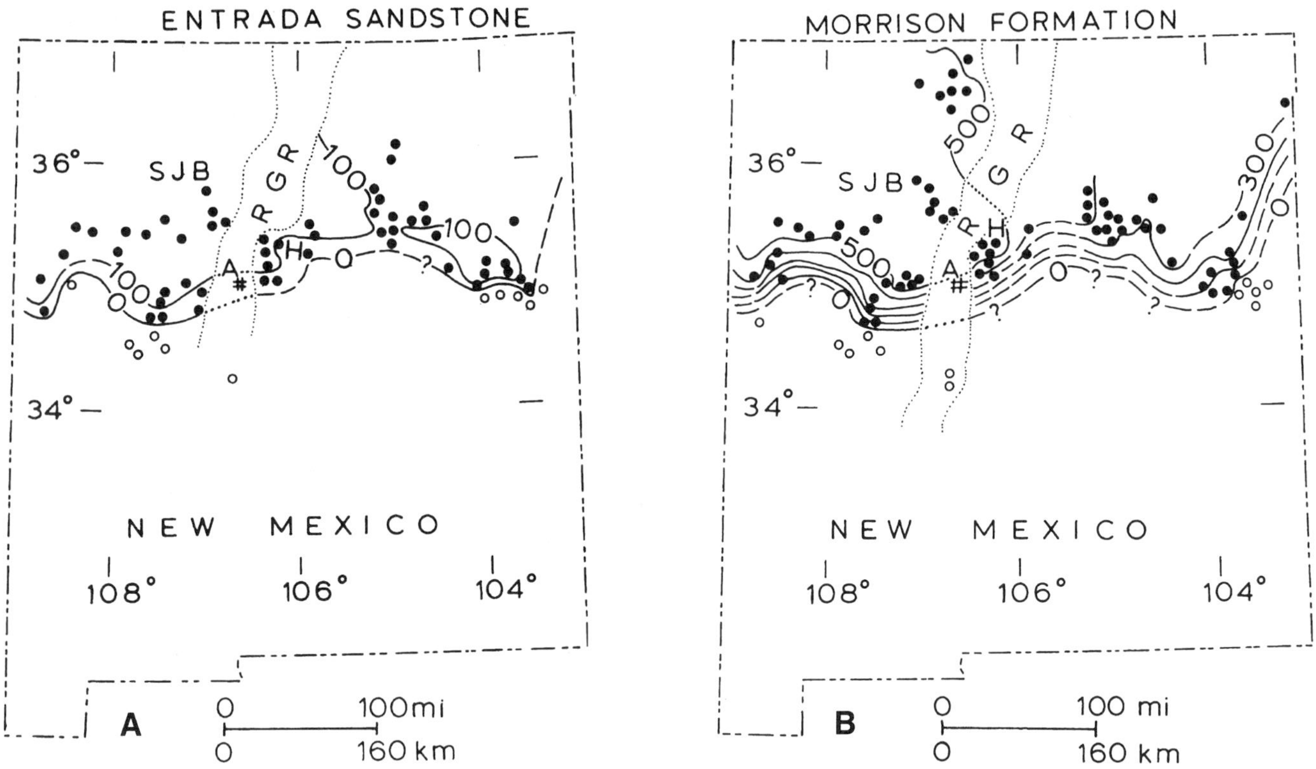

FIGURE 9. A. Thickness of Entrada Sandstone (interval B of McKee et al., 1956) showing wedge edge (zero isopach line). Contour interval is 100 ft (30 m). Solid circles show where Entrada Sandstone is present; open circles show where Entrada is absent, and older rocks are overlain by younger Jurassic or Cretaceous strata. B. Thickness of Morrison Formation and correlative strata (interval D of McKee et al., 1956) showing wedge edge (zero isopach line). Contour interval 100 ft (30 m). Solid circles show where Morrison is present; open circles show where Morrison is absent and older rocks are overlain by Cretaceous strata. A = Albuquerque, H = Hagan basin, RGR = Rio Grande rift (dotted outline), and SJB = San Juan basin. Modified from McKee et al. (1956).

strata in north-central New Mexico allow 5 to 20 km of right slip between the eastern margin of the Colorado Plateau and the craton to the east in Laramide (Late Cretaceous-Paleogene) and younger time. Estimates of 60-170 km of Laramide right slip by previous workers are not supported by Jurassic stratigraphic data. Restoration of 60-170 km of right slip creates lateral offset of the north-south extent of the Jurassic Todilto Formation, producing an untenable paleodepositional basin geometry.

ACKNOWLEDGMENTS

Orin Anderson, Adrian Hunt and Ken Kietzke were instrumental in the development of the ideas presented here. Reviews of the manuscript by Orin Anderson and Barry Kues are appreciated.

REFERENCES

Anderson, O. J., and Lucas, S. G., 1992, The Middle Jurassic Summerville Formation, northern New Mexico: New Mexico Geology, v. 14, p. 79-92.

Anderson, O. J., and Lucas, S. G., 1994, Middle Jurassic stratigraphy, sedimentation and paleogeography in the southern Colorado Plateau and southern High Plains; *in* Caputo, M. V., Peterson, J. A., and Franczyk, K. J., eds., Mesozoic systems of the Rocky Mountain region: Denver, RMS-SEPM, p. 299-314.

Anderson, O. J., and Lucas, S. G., 1995, Base of the Morrison Formation, Jurassic, of northwestern New Mexico and adjacent areas: New Mexico Geology, v. 17, p. 44-53.

Anderson, O. J., and Lucas, S. G., 1996, Stratigraphy and depositional environments of Middle and Upper Jurassic rocks, southeastern San Juan Basin, New Mexico: New Mexico Geological Society, 47th Field Conference, Guidebook, p. 205-210.

Anderson, O. J., and Lucas, S. G., 1997, Lake T'oo'dichi' and the Brushy Basin Member of the Morrison Formation: New Mexico Geological Society, 48th Field Conference, Guidebook, p. 2

Anderson, O. J., and Lucas, S. G., 2000, Jurassic stratigraphy at Placitas, New Mexico and its regional significance: Geological Society of America, Abstracts with Programs, v. 32, no. 7, p. A-458.

Anderson, R. Y., and Kirkland, D. W., 1960, Origin, varves and cycles of Jurassic Todilto Formation: American Association of Petroleum Geologists Bulletin, v. 44, p. 37-52.

Armstrong, A. K., 1995, Facies, diagenesis and mineralogy of the Jurassic Todilto Limestone Member, Grants uranium district, New Mexico: New Mexico Bureau of Mines and Mineral Resources, Bulletin 153, 41 p.

Ash, H. O., 1958, The Jurassic Todilto Formation of New Mexico [M.S. thesis]: Albuquerque, University of New Mexico, 63 p.

Barbour, C. D., and Brown, J. H., 1974, Fish species diversity in lakes: The American Midland Naturalist, v. 108, p. 423-489.

Blakey, R. C., Petrson, F., and Kocurek, G., 1988, Synthesis of late Paleozoic and Mesozoic eolian deposits of the Western Interior of the United States: Sedimentary Geology, v. 56, p. 3-125.

Cather, S.M., 1999, Implications of Jurassic, Cretaceous, and Proterozoic piercing lines for Laramide oblique-slip faulting in New Mexico and rotation of the Colorado Plateau: Geological Society of America Bulletin, v. 111, p. 849-868.

Chapin, C.E., and Cather, S.M., 1981. Eocene tectonics and sedimentation in the Colorado Plateau-Rocky Mountain area; *in* Dickinson, W.R., and Payne, M. D., eds., Relations to tectonics to ore deposits in the southern Cordillera: Arizona Geological Society Digest, v. 14, p. 173-198.

Crowell, J.C., 1959, Problems of fault nomenclature: American Association of Petroleum Geologists Bulletin, v. 43, p. 2653-2674.

Dane, C. H., and Bachman, G. O., 1965, Geologic map of New Mexico: Washington, D. C., U. S. Geological Survey, scale 1:500,000.

Darton, N. H., 1928, "Red beds" and associated formations in New Mexico, with an outline of the geology of the state: U. S. Geological Survey, Bulletin 794, 356 p.

Dobrovolny, E., Bates, R.L., and Summerson, C.H., 1946, Geology of northwestern Quay County, New Mexico: U.S. Geological Survey, Oil and Gas Investigations Map OM-62, scale 1:63,360.

Dunagan, S. P., 2000, Lacustrine carbonates of the Morrison Formation (Upper Jurassic, Western Interior), east-central Colorado, U.S.A.; *in* Gierlowski-Kordesch, E. H., and Kelts, K. R., eds., Lake basins through Space and Time: AAPG Studies in Geology 46, p. 181-188.

Erslev, E. E., 2001, Multistage, multidirectional Tertiary shortening and compression in north-central New Mexico: Geological Society of America Bulletin, v. 113, p. 63-74.

Griggs, R. L., and Read, C. B., 1959, Revisions in stratigraphic nomenclature in Tucumcari-Sabinoso area, northeastern New Mexico: American Association of Petroleum Geologists Bulletin, v. 43, p. 2003-2007.

Harshbarger, J. W., Repenning, C. A., and Irwin, J. H., 1957, Stratigraphy of the uppermost Triassic and Jurassic rocks of the Navajo country: U. S. Geological Survey, Professional Paper 291, p. 1-74.

Heaton, R. L., 1939, Contribution to Jurassic stratigraphy of Rocky Mountain region: American Association of Petroleum Geologists Bulletin, v. 23, p. 1153-1177.

Holbrook, J. M., Wright, R., and Kietzke, K. K., 1987, Stratigraphic relationships at the Jurassic-Cretaceous boundary in east-central New Mexico: New Mexico Geological Society, 38th Field Conference, Guidebook, p. 161-165.

Hook, S.C., Molenaar, C.M., and Cobban, W.A., 1983, Stratigraphy and revision of nomenclature of upper Cenomanian to Turonian rocks of west-central New Mexico; *in* Hook, S.C., ed., Contributions to Mid-Cretaceous paleontology and stratigraphy of New Mexico – Part II: New Mexico Bureau of Mines and Mineral Resources, Circular 185, p. 7-28.

Hunt, A.P., and Lucas, S.G., 1987, Southernmost outcrops of the Morrison Formation in the Carthage area, Socorro County, New Mexico: New Mexico Geology, v. 9, p. 58-62.

Ingersoll, R.V., 2000, Implications of Jurassic, Cretaceous, and Proterozoic piercing lines for Laramide oblique-slip faulting in New Mexico and rotation of the Colorado Plateau: Discussion: Geological Society of America Bulletin, v. 112, p. 796-797.

Karlstrom, K. E., and Daniel, C. G., 1993, Restoration of Laramide right-lateral slip in northern New Mexico by using Proterozoic piercing points: Tectonic implications from the Proterozoic to the Cenozoic: Geology, v. 21, p. 1139-1142.

Kelley, V.C., 1955, Regional tectonics of the Colorado Plateau and relationship to the origin and distribution of uranium: University of New Mexico, Publications in Geology, no. 5, 120 p.

Kirkland, D.W., Denison, R.E., and Evans, R., 1995, Middle Jurassic Todilto Formation of northern New Mexico and southwestern Colorado: Marine or nonmarine?: New Mexico Bureau of Mines and Mineral Resources, Bulletin 147, 37 p.

Kocurek, G., and Dott, R. H., Jr., 1983, Jurassic paleogeography and paleoclimate of the central and southern Rocky Mountain region; *in* Reynolds, M. W., and Dolly, E. D., eds., Mesozoic Paleogeography of the West-central United States, Denver, Rocky Mountain Section SEPM, p. 101-118.

Lawton, T. F., 1994, Tectonic setting of Mesozoic sedimentary basins, Rocky Mountain region, United States; *in* Caputo, M. V., Peterson, J. A., and Franczyk, K. J., eds., Mesozoic Systems of the Rocky Mountain Region: Denver, RMS-SEPM, p. 1-25.

Lee, W. T., 1902, The Morrison shales of southern Colorado and northern New Mexico: Journal of Geology, v. 10, p. 36-58.

Lucas, S. G., and Anderson, O. J., 1997, The Jurassic San Rafael Group, Four Corners region: New Mexico Geological Society, 48th Field Conference Guidebook, p. 115-132.

Lucas, S. G., and Anderson, O. J., 1998, Jurassic stratigraphy and correlation in New Mexico: New Mexico Geology, v. 20, p. 97-104.

Lucas, S. G., and Kietzke, K. K., 1986, Stratigraphy and petroleum potential of the Jurassic Todilto Formation in northeastern New Mexico; *in* Ahlen, J. L., Hanson, M. E. and Zidek, J., eds., Southwest Section of AAPG Transactions and Guidebook of 1986 Convention, Ruidoso, New Mexico: Socorro, New Mexico Bureau of Mines and Mineral Resources, p. 121-127.

Lucas, S.G., Anderson, O.J., and Black, B. A., 2000, Implications of Jurassic, Cretaceous, and Proterozoic piercing lines for Laramide oblique-slip faulting in New Mexico and rotation of the Colorado Plateau: Discussion: Geological Society of America Bulletin, v. 112, p. 789-790.

Lucas, S. G., Anderson, O. J., and Pigman, C., 1995, Jurassic stratigraphy in the Hagan basin, north-central New Mexico: New Mexico Geological Society, 46th Field Conference, Guidebook , p. 247-255.

Lucas, S. G., Estep, J. W. and Anderson, O. J., 1999, Correlation of Jurassic strata from the Colorado Plateau to the High Plains, across the Rio Grande rift, north-central New Mexico: New Mexico Geological Society, 50th Field Conference, Guidebook, p. 317-326.

Lucas, S. G., Hunt, A. P., and Hayden, J. N., 1987a, Type section of Exeter Member of Entrada Sandstone, Jurassic of northeastern New Mexico: New Mexico Geological Society, 38th Field Conference Guidebook 38, p. 17-18.

Lucas, S. G., Hunt, A. P. and Hayden, S. N., 1987b, The Triassic System in the Dry Cimarron Valley, New Mexico, Colorado and Oklahoma: New Mexico Geological Society, 38th Field Conference Guidebook, p. 97-117.

Lucas, S. G., Kietzke, K. K., and Hunt, A. P., 1985, The Jurassic System in east-central New Mexico: New Mexico Geological Society, 36th Field Conference, Guidebook, p. 213-243.

Mankin, C. J., 1958, Stratigraphy and sedimentary petrology of Jurassic and pre-Graneros Cretaceous rocks, northeastern New Mexico [Ph.D. dissertation]: Austin, University of Texas, 208 p.

Mankin, C. J., 1972, Jurassic strata in northeastern New Mexico: New Mexico Geological Society, 23rd Field Conference, Guidebook, p. 91-97.

Marcou, J., 1858, Geology of North America, with two reports on the prairies of Arkansas and Texas, the Rocky Mountains of New Mexico, and the Sierra Nevada of California, originally made for the United States Government: Zurich, Zurcher and Furrer, 144 p.

McKee, E.D., Oriel, S.S., Swanson, V.E., MacLachlan, M.E., MacLachlan, J.C., Ketner, K.B., Goldsmith, J.W., Bell, R.V., Jameson, D.J., and Imlay, R.W., 1956, Paleotectonic maps of the Jurassic System: U.S. Geological Survey, Miscellaneous Investigations Map I-175, scale 1:5,000,000.

Miller J.P., Montgomery, A., and Sutherland, P.K., 1963, Geology of part of the Sangre de Cristo Mountains, New Mexico: New Mexico Bureau of Mines and Mineral Resources, Memoir 11, 106 p.

Peterson, J. A., 1972, Jurassic System; *in* Mallory, W. W., ed., Geologic atlas of the Rocky Mountain region: Denver, Rocky Mountain Association of Geologists, p. 177-189.

Pipiringos, G. N., and O'Sullivan, R. B., 1978, Principal unconformities in Triassic and Jurassic rocks, Western Interior, U.S.: A preliminary survey: U.S. Geological Survey, Professional Paper 1035-A, 29 p.

Ridgley, J. L., and Goldhaber, M., 1983, Isotopic evidence for a marine origin of the Todilto Limestone, north-central New Mexico: Geological Society of America, Abstract with Programs, v. 15, p. 414.

Schaeffer, B., and Patterson, C., 1984, Jurassic fishes from the western United States, with comments on Jurassic fish distribution: American Museum Novitates, no. 2796, 86 p.

Silver, C., 1948, Jurassic overlap in western New Mexico: American Association of Petroleum Geologists Bulletin, v. 32, p. 68-81.

Trauger, F. D., and Bushman, F. X., 1964, Geology and groundwater in the vicinity of Tucumcari, Quay County, New Mexico: New Mexico State Engineer, Technical Report 30, 178 p.

Turner, C. E., and Fishman, N. S., 1991, Jurassic Lake T'oo'dichi': A large alkaline, saline lake, Morrison Formation, eastern Colorado Plateau: Geological Society of America Bulletin, v. 103, p. 538-558.

Wanek, A. A., 1962, Reconnaissance geologic map of parts of Harding, San Miguel and Mora Counties, New Mexico: U.S. Geological Survey, Oil and Gas Investigations Map OM-208.

Warren, J. K., 1982, The hydrological setting, occurrence and significance of gypsum in late Quaternary salt lakes in south Australia: Sedimentology, v. 29, p. 609-637.

Warren, J. K., and Kendall, C. S. G., 1985, Comparison of sequences formed in marine sabkha (subaerial) and saline (subaqueous) settings—modern and ancient: American Association of Petroleum Geologists Bulletin, v. 69, p. 1013-1023.

Woodward, L.A., 2000, Implications of Jurassic, Cretaceous, and Proterozoic piercing lines for Laramide oblique-slip faulting in New Mexico and rotation of the Colorado Plateau: Discussion: Geological Society of America Bulletin, v. 112, p. 783-785.

Woodward, L.A., Anderson, O.J., and Lucas, S.G., 1997, Mesozoic stratigraphic constraints on Laramide right slip on the east side of the Colorado Plateau: Geology, v. 25, p. 843-846.

Woodward, L.A., Anderson, O.J., and Lucas, S.G., 1999, Late Paleozoic right–slip faults in the ancestral Rocky Mountains: New Mexico Geological Society, 50th Field Conference Guidebook, p. 149-153.

Ye, H., Royden, L., Burchfiel, C., and Schuepbach, M., 1996, Late Paleozoic deformation of interior North America: The greater ancestral Rocky Mountains: American Association of Petroleum Geologists Bulletin, v. 80, p. 1397-1432.

A BRACHIOSAURID DINOSAUR (SAURISCHIA: SAUROPODA) FROM THE UPPER JURASSIC OF EAST-CENTRAL NEW MEXICO

ADRIAN P. HUNT

Mesalands Dinosaur Museum, Mesa Technical College, 911 South Tenth Street, Tucumcari, NM 88401

Abstract.—The Upper Jurassic Morrison Formation of northeastern New Mexico has yielded several dinosaur bones during the past 100 years. A newly collected specimen from the Brushy Basin Member of Quay County is a partial humerus of a brachiosaurid dinosaur. This occurrence of *Brachiosaurus* sp .is the first for the state of New Mexico.

INTRODUCTION

Upper Jurassic strata are widely exposed in northeastern New Mexico and include the Summerville and Morrison formations (Lucas et al., 1985; Lucas and Woodward, 2001). Stanton (1905) first reported a tetrapod fossil from the Brushy Basin Member of the Morrison Formation in Quay County. Since that time, a number of dinosaur bones and tracks have been reported from this part of the state (Table 1). Here, I report on a newly collected specimen that represents an addition to the dinosaur fauna of the area and indeed is the first report of the sauropod family Brachiosauridae from New Mexico.

GEOLOGIC CONTEXT

This specimen derives from the Brushy Basin Member of the Morrison Formation in Quay County. The Brushy Basin Member in this area consists mainly of green mudstone. The Brushy Basin is the most extensive and fossiliferous member of the Morrison Formation in the western United States.

TABLE 1. Jurassic tetrapods of New Mexico (Lucas and Hunt, 1985; Hunt and Lucas, 1993; Lucas et al., 1996; Lockley et al., 2000; Lucas and Heckert, 2000; this paper). ** indicates occurrence in northeastern New Mexico.

JURASSIC TETRAPODS OF NEW MEXICO

Formation	Member	Tetrapod taxa
Morrison Formation	Jackpile Member	Indeterminate dinosaurs
	Brushy Basin Member	*Allosaurus* sp.
		Large allosaurid
		Apatosaurus sp.
		Camarasaurus supremus
		Diplodocus carnegie
		Diplodocus (= *Seismosaurus*) *hallorum*
		Brachiosaurus sp.
		Stegosaurus sp.**
		Camptosauridae indet.**
		Theropod eggshell
	Salt Wash Member	Indeterminate dinosaurs
		dinosaur skeleton observed in field
Summerville Formation		?*Camarasaurus* sp.
		?*Megalosauripus* sp.
		aff. *Therangospodus* sp.**

DESCRIPTION

This specimen is in a private collection and is currently on loan to the Mesalands Dinosaur Museum. It consists of four pieces of a large limb bone. Several other pieces were discarded after collection. The largest piece consists of a shaft with broken ends that gently flairs proximally (Fig. 1A-B, D). This fragment is 870 mm long and has maximum and minimum diameters of 380 mm and 280 mm, respectively. The broken end of the shaft has the cross-section of a rounded rectangle with dimensions of 210 and

FIGURE 1. Left humerus of *Brachiosaurus* sp. from the Brushy Basin Member of the Morrison Formation, Quay County, New Mexico. A. Proximal fragment in cranial view. B. Proximal fragment in distal view. C. Distal fragment in cranial view. D. Proximal fragment in medial view. Scale bar is 10 cm.

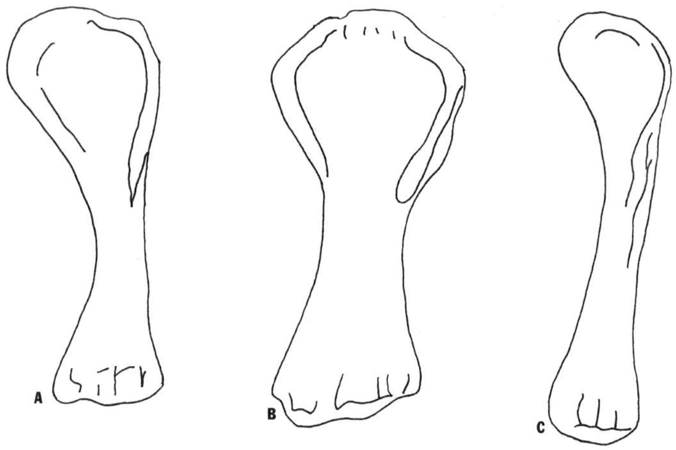

FIGURE 2. Late Jurassic sauropod left humeri in cranial view (after McIntosh, 1990). Not to scale. A. *Camarasaurus*. B. *Apatosaurus*. C. *Brachiosaurus*.

320 mm. Proximally, the bone widens, one side becoming more convex and the other more concave, so that the cross-section is like a rounded crescent. A vertical ridge about 380 mm long is developed on the edge of one side, and in cross-section the lateral side of the bone is vertical and less rounded. The vertical ridge is about 100 mm high and is composed of a higher segment about 280 mm long and a lower segment about 100 mm long that is connected to a rugose area on the lateral wall of the bone.

The second large fragment is a roughly fan-shaped piece, 400 mm long, 430 mm in maximum diameter and 205 mm thick (Fig. 1C). The top and bottom surfaces of this bone are slightly concave and convex, respectively and only one of the sides preserves a margin. The two smaller pieces are a flattened shaft fragment 240 mm long, 220 mm wide and 140 mm thick and a fragment of the end of a rounded condyle 220 mm long, 150 mm wide and 110 mm thick.

DISCUSSION

Based on size, occurrence and age this bone obviously pertains to a sauropod dinosaur. Furthermore, it is a limb bone and a left humerus based on the flairing ends and prominent cranio-lateral crest (deltopectoral crest). Humeri are known for all the principal taxa of sauropods from the Morrison Formation. Camarasaurids, diplodocids and *Haplocanthosaurus* have hour-glass-shaped humeri with a short, broad shaft separating two widened ends (Fig. 2). The Quay County specimen, however, exhibits a relatively long, parallel-sided shaft. Among the sauropods this morphology is found only in the Brachiosauridae (Fig. 2; Janensch, 1961, fig. 1a). Simililarities between the Quay County humerus and *Brachiosaurus* include prominent groove on the cranial surface medial to the deltopectoral crest, axis of deltopectoral crest in cranial view is medial to the humeral margin and shaft is narrowest at deltopectoral crest rather than more distally. The deltopectoral crest is not as prominent as in other specimens of *Brachiosaurus* (e.g., Janensch, 1961, fig. 1c), but this could be an ontogenetic feature. Based on the ratio of minimum humeral shaft diameter: humeral length compared to other brachiosaurid humeri, the Quay County humerus would have been about 2650 mm long, which would be as long or longer than the holotype of *Brachiosaurus brancai* (Paul, 1988, table 1). However, so little is known about variation in large sauropods that I conservatively assign this specimen to *Brachiosaurus* sp.

Brachiosaurids are rare components of Late Jurassic dinosaur faunas in North America. Currently, *Brachiosaurus* is known from Colorado and Utah (McIntosh, 1990). This new specimen is a first occurrence for New Mexico and represents one of the largest dinosaurs known from the state. This specimen suggests that the Morrison Formation of northeastern New Mexico has the potential to yield significant specimens.

ACKNOWLEDGMENTS

Jim Haller kindly let me study this important specimen and Spencer Lucas provided a helpful review.

REFERENCES

Hunt, A. P. and Lucas, S. G., 1993, Jurassic vertebrates of New Mexico: Bulletin of New Mexico Museum of Natural History and Science, v. 2, p. 71-75.

Janensch, W., 1961, Die gliedmaszen und gliedmaszengurtel der sauropoden der Tendagauru-schichten: Palaeontographica Supplement 7, p. 177-235.

Lockley, M. G., Lucas, S. G. and Hunt, A. P., 2000, Dinosaur tracksites in New Mexico: a review: Bulletin of New Mexico Museum of Natural History and Science, v. 17, p. 9-16.

Lucas, S.G., and Hunt, A.P. ,1985, Dinosaurs from the Upper Jurassic Morrison Formation in New Mexico: New Mexico Journal of Science, v. 25, p. 1-12.

Lucas, S. G. and Heckert, A. B., 2000, Jurassic dinosaurs in New Mexico: Bulletin of New Mexico Museum of Natural History and Science, v. 17, p. 43-45.

Lucas, S. G. and Woodward, 2001, Jurassic strata in east-central New Mexico and their regional significance: New Mexico Geological Society, 52nd Field Conference Guidebook.

Lucas, S. G., Williamson, T. E., Estep, J. W., Hunt, A. P. and Anderson, O. J., 1996, Jurassic vertebrates from New Mexico: Museum of Northern Arizona Bulletin 60, p. 235-241.

McIntosh, J. S., 1990, Sauropoda, *in* Weishampel, D. B., Dodson, P. and Osmolska, H., eds., The Dinosauria: Berkeley, University of California Press, p. 345-401.

Paul, G. S., 1988, The brachiosaur giants of the Morrison and Tendaguru with description of a new subgenus, *Giraffatitan*, and a comparison of the world's largest dinosaurs: Hunteria, v. 2, n. 3, p. 1-14.

Stanton, T. W., 1905, The Morrison Formation and its relations with the Comanche Series and the Dakota Formation: Journal of Geology, v. 13, p. 657-669.

CRETACEOUS STRATIGRAPHY AND BIOSTRATIGRAPHY, EAST-CENTRAL NEW MEXICO

SPENCER G. LUCAS[1], BARRY S. KUES[2] AND ADRIAN P. HUNT[3]
[1]New Mexico Museum of Natural History, 1801 Mountain Rd. NW, Albuquerque, NM 87104; [2]Department of Earth and Planetary Sciences, University of New Mexico, Albuquerque, NM 87131; [3]Mesalands Dinosaur Museum, Tucumcari, NM 88401

Abstract.—Cretaceous strata exposed in east-central New Mexico belong to the (in ascending order) Tucumcari, Mesa Rica, Pajarito, Romeroville, Graneros and Greenhorn formations. The Mesa Rica, Pajarito and Romeroville formations are the Dakota Group in this region. The Tucumcari Formation is as much as 25 m thick and is mostly marine shale and minor bioturbated sandstone and limestone. The Mesa Rica Sandstone is mostly cross bedded and ripple laminated quartzarenite as much as 43 m thick. The Pajarito Formation is as much as 30 m thick and is mostly silty mudstone and bioturbated quartzarenite. The Romeroville Sandstone is <10 m of bioturbated and cross bedded quartzarenite. The Graneros Shale is as much as 67 m of shale, and the Greenhorn Formation consists of the Lincoln (up to 34 m of shale and calcarenite), Hartland (up to 11 m of shale and bentonite) and Bridge Creek (up to 9 m of limestone and shale) members. We document a specimen of the late Albian ammonite *Mortoniceras leonense* from the lower Mesa Rica Sandstone. The Tucumcari, Mesa Rica and Pajarito formations are of late Albian age, whereas the Romeroville, Graneros and (most of) the Greenhorn formations are of Cenomanian age. The youngest Cretaceous strata preserved in east-central New Mexico are early Turonian beds of the upper part of the Bridge Creek Member of the Greenhorn Formation.

INTRODUCTION

Cretaceous strata exposed in east-central New Mexico (Fig.1) are mostly rocks of Early Cretaceous (late Albian) age and a few outliers of early Late Cretaceous age. First studied in 1853 by Marcou, these strata have been more intensively studied since the 1970s. Here, we review their lithostratigraphy and biostratigraphy.

LITHOSTRATIGRAPHY

Cretaceous strata in east-central New Mexico belong to the following formations (in ascending order): Tucumcari, Mesa Rica, Pajarito, Romeroville, Graneros and Greenhorn (Fig. 2). The Mesa Rica, Pajarito and Romeroville formations are the Dakota Group in east-central New Mexico. Here, we summarize the lithostratigraphy of these units.

Tucumcari Formation

Marcou (1889) first used the term "Tucumcari Beds" to refer to the strata at Pyramid Mountain southwest of Tucumcari (sec. 19, T9N, R29W) where he collected the type specimens of "*Gryphaea dilatata* var. *tucumcarii*" (= *Texigryphaea pitcheri*) (Marcou, 1855, 1858). Lucas and Kisucky (1988) designated this the type section of the Tucumcari Formation (Shale). At the type section, the Tucumcari Formation is ~9 m thick and is mostly olive-gray to yellowish-brown silty mudstone and shale with minor beds of yellowish-brown and yellowish-orange, muddy bioturbated sandstone (Lucas and Kisucky, 1988; Kues and Lucas, 2001). Another minor lithotype of the Tucumcari Formation is dark yellowish-brown, fossiliferous limestone.

These lithotypes characterize the Tucumcari Formation throughout its outcrop belt in east-central New Mexico. Maximum thickness is ~25 m, and the Tucumcari Formation everywhere rests disconformably on the Upper Jurassic Morrison Formation. Locally, a 1-2 m thick bed of sandstone at the base of the Tucumcari Formation is the Campana Sandstone Bed of Holbrook et al. (1987). The Mesa Rica Sandstone conformably to disconformably overlies the Tucumcari Formation throughout east-central New Mexico.

Outcrops of the Tucumcari Formation form a short, often covered, slope beneath cliffs of the Mesa Rica Sandstone. The Tucumcari Formation is present throughout much of Quay County and extends into adjacent parts of Guadalupe County.

Mesa Rica Sandstone

The most conspicuous Cretaceous unit in east-central New Mexico is the Mesa Rica Sandstone, which forms bold cliffs that cap the edges of the Llano Estacado and its northerly erosional

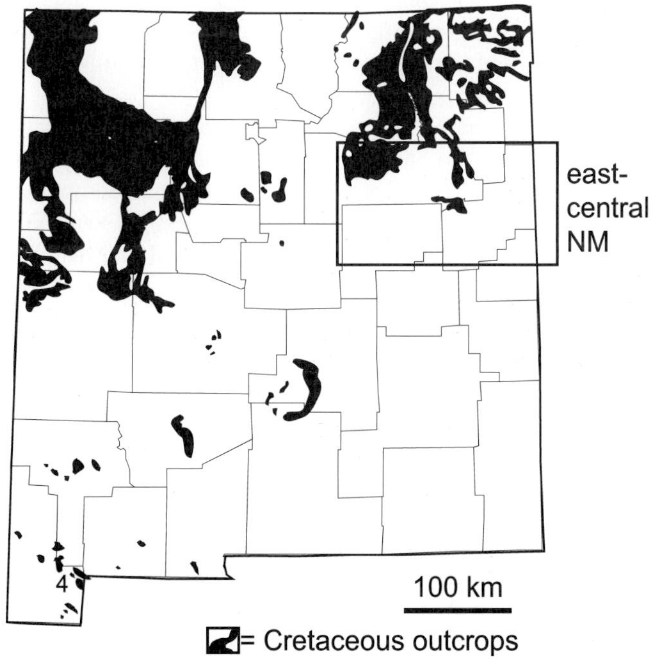

FIGURE 1. Distribution of Cretaceous strata in New Mexico (after Dane and Bachman, 1965) and location of east-central New Mexico.

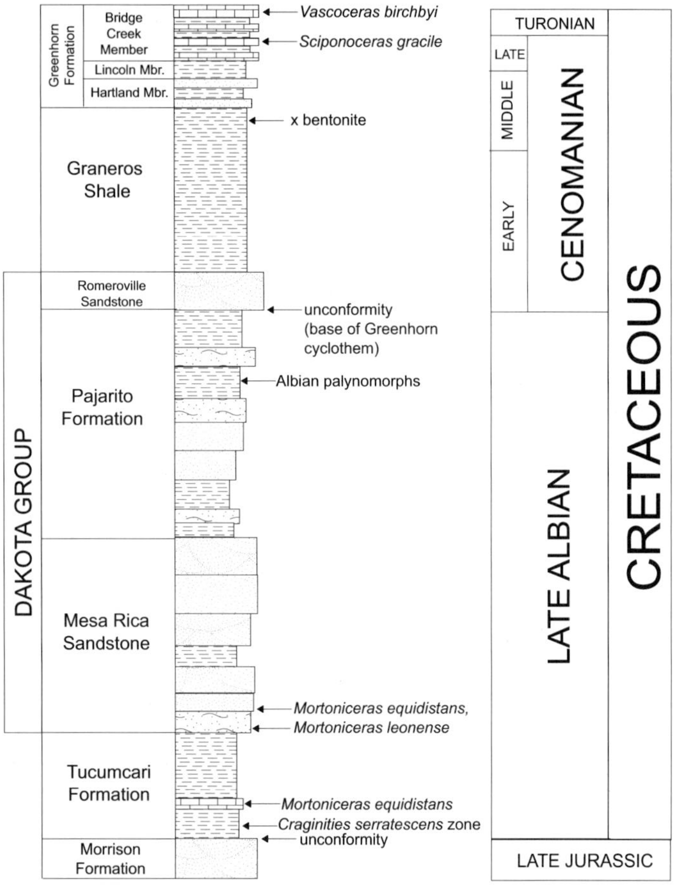

FIGURE 2. Generalized Cretaceous stratigraphic section in east-central New Mexico, key biostratigraphic indicators and age assignments.

outliers in the Canadian River drainage, such as Mesa Redonda, Tucumcari Mountain and Liberty Mesa in Quay County and Mesa Rica in Quay and Guadalupe Counties. It also forms bold cliffs along the edge of the Canadian escarpment from San Miguel to Harding Counties..

Dobrovolny and Summerson (in Dobrovolny et al., 1946) named the Mesa Rica Sandstone, and Lucas and Kisucky (1988) designated and described a type section on the southwestern face of Mesa Rica (sec. 5, T11N, R26E) in eastern Guadalupe County. Several workers earlier referred to the Mesa Rica Sandstone informally as the lower sandstone member (unit) or lower braided alluvial interval of the Dakota Group (Sandstone) (e.g., Bejnar and Lessard, 1976; Gilbert and Asquith, 1976; Stone, 1984).

At its type section, the Mesa Rica Sandstone is 24.5 m thick and consists of two sandstone intervals separated by light-gray and purple, variegated mudstone. The sandstones are yellowish-brown and yellowish-orange, very fine- to fine-grained quartzarenites that are trough cross bedded, ripple-laminated and laminated with some minor bioturbation. Maximum thickness of the Mesa Rica Sandstone is 43 m, and throughout east-central New Mexico its lithotypes are similar to those of the type section.

The Mesa Rica Sandstone grades upward into the Pajarito Formation. However, the placement and nature of its basal contact has been the subject of some disagreement. Normally, the contact between the Mesa Rica and underlying Tucumcari Formation is readily placed at the sharp contact of the Mesa Rica basal sandstone bed on the uppermost shale bed of the Tucumcari Formation. However, at San Jon hill, south of San Jon in the roadcuts of NM Highway 39 (sec. 19, T9N, R31E and vicinity), a 5.6-m thick, massive-to-laminar sandstone full of shell debris is at the contact of the Tucumcari and Mesa Rica formations. Largely because the fossils from this sandstone are both abundant (normally, the Mesa Rica Sandstone is sparsely fossiliferous) and mostly represent the same taxa present in underlying strata of the Tucumcari Formation (see below), several workers have assigned this sandstone to the Tucumcari Formation (Griggs and Read, 1959; Scott, 1970, 1974; Mateer, 1985, 1987). However, formations are cartographic units having contacts mapped at points of significant lithologic contrast. Therefore, as advocated by Lucas and Kisucky (1988), we include this fossiliferous sandstone at San Jon Hill in the Mesa Rica Sandstone.

Pajarito Formation

Dobrovolny and Summerson (in Dobrovolny et al., 1946) introduced the name Pajarito Shale. Lucas and Kiscuky (1988) designated and described a type section just south of Mesa Rica in the Pajarito Creek drainage of western Quay County (sec. 6, T11N, R29E) and also described a reference section that well exposes the top of the Pajarito Formation along a tributary of Burro Creek in Harding County (sec. 8, T18N, R27E).

At the type section, the Pajarito is at least 29 m thick and consists of interbedded yellowish-orange and brown, quartzose, generally bioturbated sandstone and light-gray silty mudstone. The formation is as much as 30 m thick in east-central New Mexico and forms a slope, or ribbed cliff, between the Mesa Rica Sandstone and the overlying Romeroville Sandstone. The Pajarito Formation is the middle shale member (unit) or meanderbelt interval of the Dakota Sandstone (Group) of some previous workers (Bejnar and Lessard, 1976; Gilbert and Asquith, 1976; Stone, 1984). The contact of the Romeroville Sandstone above the Pajarito Formation is a disconformity.

Romeroville Sandstone

Kues and Lucas (1987) and Lucas and Kisucky (1988) used the term Romeroville Sandstone to refer to the upper unit of the Dakota Group in northeastern New Mexico, and Lucas (1990) formally defined the unit. Dobrovolny et al. (1946) referred to this unit as the Dakota Sandstone, and it was the upper sandstone member (unit) or marine sand interval of the Dakota Sandstone (Group) of some earlier workers (Bejnar and Lessard, 1976; Gilbert and Asquith, 1976; Stone, 1984).

At its type section at Romeroville Gap near Las Vegas, San Miguel County, the Romeroville Sandstone is 3.5 m thick and consists of pale orange and dark yellowish-orange, trough cross bedded and bioturbated (esp. *Ophiomorpha*), quartzarenite sandstone. A thickness of <10 m is characteristic of the Romeroville Sandstone throughout east-central New Mexico. The Romeroville Sandstone has a sharp, scoured base, so it overlies the Pajarito Formation with evident disconformity. Indeed, locally, the Romeroville Sandstone is absent, and the Graneros Shale rests directly on the Pajarito Formation (Lucas et al., 2000).

Graneros Shale

Outside of the Raton basin, outcrops of post-Romeroville Upper Cretaceous strata are rare in east-central New Mexico. This is because Neogene erosion associated with deposition of the Ogallala Group has removed most of the post-Romeroville Cretaceous section. The exceptions to this are Upper Cretaceous outliers near Roy in Harding County (T20N, R25-26E) and in the Bonita fault zone southeast of Tucumcari in Quay County (T8N, R32E, and T9N, R33E) (Dobrovolny et al., 1946; Wanek, 1962; Lucas et al., 2000; Lucas and Heckert, 2001).

At the sections near Roy, the Graneros Shale is poorly exposed, and has an estimated thickness of 20 m. The Graneros Shale is much better exposed in Quay County, where it is ~67 m thick and consists mostly of dark gray, light brownish-gray and grayish-orange shale. Subordinate rock types are a few thin (<0.3 m thick) beds of calcarenite, siltstone and air fall bentonitic ash. The Graneros Shale forms slopes and low, rolling hills in east-central New Mexico. It conformably overlies the Romeroville Sandstone or disconformably overlies the Pajarito Formation where the Romeroville is absent. The Greenhorn Formation conformably overlies the Graneros Shale.

Greenhorn Formation

The youngest Cretaceous strata preserved in east-central New Mexico are assigned to the Upper Cretaceous Greenhorn Formation. Greenhorn strata overlie the Graneros outcrops near Roy in Harding County and in the Bonita fault zone of Quay County (Dobrovolny et al., 1946; Wanek, 1962; Lucas et al., 2000; Lucas and Heckert, 2001).

Three Greenhorn members are present in east-central New Mexico: (1) the basal, Lincoln Member is a few persistent ledges of calcarenite and slopes of calcareous shale, 34 m thick in the Bonita fault zone, but only 8.3 m thick near Roy; (2) the medial Hartland Member is bentonitic shale with three distinctive bentonite beds, 9 m thick in the Bonita fault zone and 11 m thick near Roy; and (3) the upper, Bridge Creek Member is interbeds of ledge-forming limestone and medium-gray calcareous shale, at least 4.1 m thick near Roy and at least 9 m thick in the Bonita fault zone. The Ogallala Group rests directly on the Bridge Creek Member at both locations and the uppermost beds of the Bridge Creek Member have been eroded away, so its thickness is a minimum.

BIOSTRATIGRAPHY

Tucumcari Formation

The Tucumcari Formation, in its main outcrop belt in Quay and eastern Guadalupe Counties, contains two, biostratigraphically distinct marine assemblages (Kues, 1997; Kues and Lucas, 2001). The basal assemblage from San Jon hill includes a late occurrence of the oyster *Ceratostreon texanum* (Roemer), and is succeeded slightly higher by the lowest occurrence of *Peilinia levicostata* Kues, an interval equivalent to the upper Albian *Craginites serratescens* zone of Young (1967). Brand and Mattox (1972) reported a fragmentary specimen of *C.* cf. *C. serratescens* (Cragin) from the basal Tucumcari at San Jon Hill, but we have not collected specimens of this ammonite. Older Albian strata, some containing *Texigryphaea navia* (Hall) and ammonites of the *Adkinsites bravoensis* zone, are present in Lower Cretaceous outliers in Roosevelt County, New Mexico (Kues, 1986), and in some of the playa lake exposures of west-central Texas discussed by Brand (1953).

The stratigraphically higher assemblage that occurs in the Tucumcari Formation, and above the basal assemblage at San Jon hill, is probably entirely within the *Mortoniceras equidistans* zone. Young (1967) and Cobban (1985) noted that in many areas of Texas, *M. equidistans* (Cragin) occurs with *Eopachydiscus marcianus* (Shumard), but extends higher in the section than that species. However, Young (1967, table 4) identified only the *M. equidistans* zone above the *C. serratescens* zone within the Tucumcari Formation, and this appears to us to be correct. Although *E. marcianus* (and its synonym *E. laevicaniculatum* [Roemer in Lasswitz]) have been reported from the Tucumcari Formation (Kues et al., 1985), separate zones for *E. marcianus* and *M. equidistans* cannot be distinguished. *M. equidistans* is present near the base of the Tucumcari Formation at Pyramid Mountain (Kues and Lucas, 2001), and co-occurs with *E. marcianus* as high as the basal unit of the overlying Mesa Rica Sandstone (Cobban, 1985, pl. 1). The Tucumcari Formation therefore is of early late to middle late Albian age (Fig. 2).

Mesa Rica Sandstone

Most of the Mesa Rica Sandstone represents a south-southeasterly prograding fluvio-deltaic complex (Holbrook and Dunbar, 1992). Only the lower part of the Mesa Rica Sandstone yields marine fossils that are of biostratigraphic significance, and these are indicative of the upper Albian *Mortoniceros equidistans* zone. *Mortoniceras equidistans* is found in the basal Mesa Rica Sandstone at several localities (Kues et al., 1985; Cobban, 1985; Lucas and Hunt, 2000), and a bivalve-dominated assemblage that shares most taxa with the underlying Tucumcari Formation is present in the lowermost sandstone bed of the Mesa Rica Sandstone at San Jon hill (Kues, 2001).

Here, we document a specimen of *Mortoniceras leonense*, MDM 212, from the basal Mesa Rica Sandstone just south of Tucumcari in T10N, R29E (Fig. 3). This specimen is moderately evolute, has a squarish whorl section and sparse ribs and a diameter of 250 mm. The 17 widely spaced ribs are rectiradiate and have strong nodate ventrolateral tubercles and somewhat weaker umbilical tubercles. The ventral keel becomes weak on the outer whorl. The specimen well matches but is much larger than specimens Cobban (1987) referred to *M.* aff. *M. leonense* from the Mojado Formation in southwestern New Mexico.

This fossil is an important addition to knowledge of the morphology of *M. leonense*. Conrad (1857, p. 160, pl. 16, figs. 2a-b) introduced the taxon as *Ammonites leonense*, from Leon Springs, just west of the present town of Fort Stockton, Pecos County, Texas. Adkins (1927, p. 43, pl. 5, fig. 4) provided the only published photograph of the holotype and established its stratigraphic provenance as upper Albian, Duck Creek-equivalent beds. Early authors assigned the species to *Pervinquieria* (a junior synonym of *Mortoniceras*: Wright et al., 1996) or to *Schloenbachia* (valid species of which are not known from North America: Wright et al., 1996) and reported it from the Duck Creek and overlying Fort Worth formations of Texas and Oklahoma. Adkins (1928, p. 231) pointed out that many of the Fort Worth records (e.g., Adkins and Winton, 1919; Bullard, 1925, 1926; Winton, 1925) actually are of *Pervinquieria maxima* Lasswitz (now *Mortoniceras* (*Angolaites*) *maximus*: Kennedy et al., 1998, p. 29). Slocki (1967, p. 195) and McGill (1967, p. 228) reported *Mortoniceras leonense* as a common species in the Fort Worth Formation in northeastern Texas, but it is not certain if their specimens are true *M. leonense* or *M.* (*A.*) *maximus*. Cobban (1987) referred *Ammonites leonense* to *Mortoniceras* (it belongs in the subgenus *Mortoniceras*: Cobban, 1987; Wright et al., 1996).

The holotype of *M. leonense* illustrated by Conrad (1857) and redescribed by Cobban (1987, p. 8) was said by Adkins (1928, p. 230) to be a weathered "cast" (steinkern) of a relatively small specimen, probably a juvenile. The umbilical area is covered by matrix, and no sutures are preserved. The specimen has a diameter of ~ 83 mm, an umbilical ratio of 0.43, sparse, rectiradiate ribs, prominent ventrolateral and umbilical tubercles and a squarish cross section. The inner whorl of MDM 212 is very similar to the holotype, with a diameter of 92 mm, and an umbilical ratio of 0.44. The suture line of the Mesa Rica *M. leonense* is characteristic of *Mortoniceras* (e.g., Young, 1957; Cobban, 1987; Wright et al., 1996; Kennedy et al., 1998) in having deeply incised lobes and saddles with a broad E/L saddle, a narrow and deep L and a massive U2 (Fig. 3). However, the saddles are not bifid as in most other *Mortoniceras*, though L is irregularly trifid.

Mortoniceras leonense has been reported in faunal lists of the Tucumcari Formation by several authors (see Kues et al., 1985), but it has not been adequately documented by descriptions or illustrations. Its co-occurrence with *M. equidistans* in the basal Mesa Rica Sandstone is not surprising in view of the fact that Adkins (1927, p. 44) reported that it occurs with *M. equidistans* and two other *Mortoniceras* species (*M. whitei*, *M. kiliani*) now considered synonyms *of M. equidistans* (Cobban, 1985) at its type locality in West Texas. Although the basal Mesa Rica Sandstone is clearly in the *M. equidistans* range zone, its upper part lacks marine fossils, so its precise age relative to late Albian ammonite zones cannot be determined. However, it is likely that the entire Mesa Rica Sandstone encompasses no more that the upper part of the *M. equidistans* zone and succeeding *M.* (*Angolaites*) *lasswitzi* zone, given the apparent latest Albian age of the overlying Pajarito Formation (Fig. 2).

Pajarito Formation

The most common fossils in the Pajarito Formation are ornithopod dinosaur tracks (see Hunt and Lucas, 1998, for a review), but they do not provide a basis for a precise age assignment. Baltz (1990) reported Albian palynomorphs from the Pajarito Formation, and the presence of the marine bivalve *Peilinia levicostata* (reported as *Lopha quadriplicata* by Dobrovolny et al., 1946) in the Pajarito is also consistent with a late Albian age (Kues, 1997). Therefore, we assign the entire Pajarito Formation a late Albian age (Fig. 2).

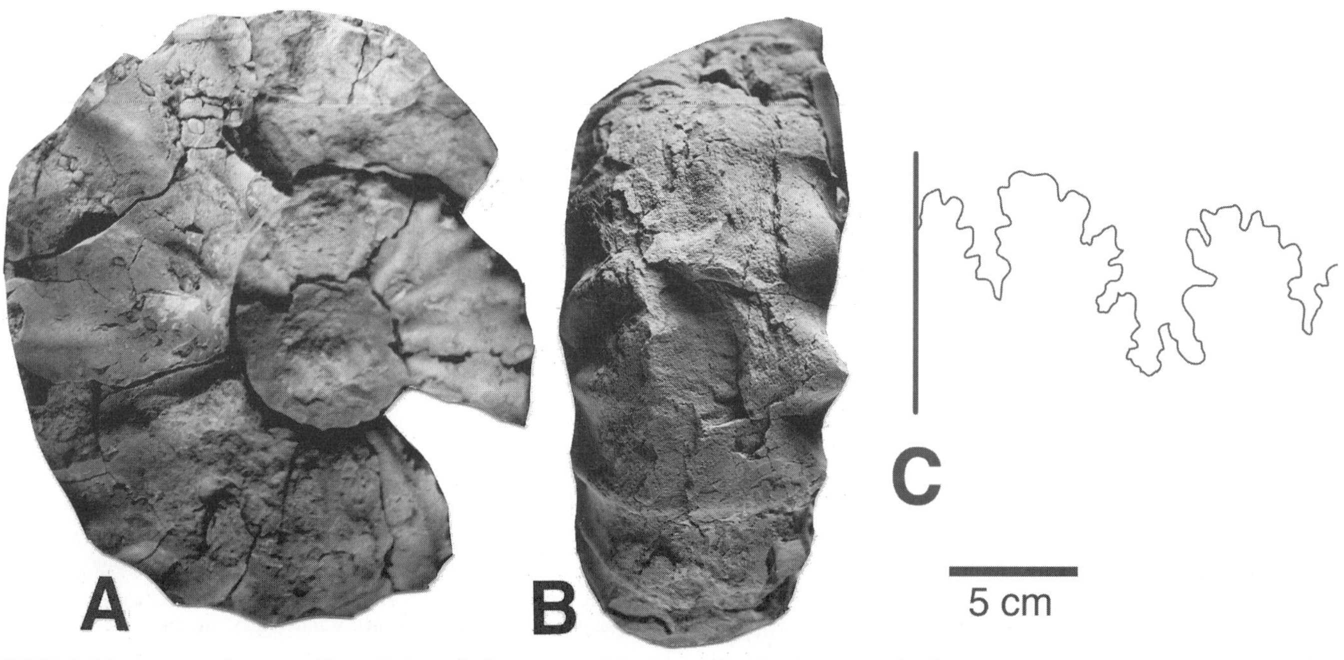

FIGURE 3. *Mortoniceras leonense* (Conrad) from the lower part of the Mesa Rica Sandstone south of Tucumcari, Quay County, MDM 212, lateral (A) and venter (B) views and suture line.

Romeroville Sandstone

No fossils of biostratigraphic significance are known from the Romeroville Sandstone. It disconformably overlies the upper Albian Pajarito Formation and is conformably overlain by the early Cenomanian Graneros Shale. Furthermore, the Romeroville Sandstone is the basal transgressive unit of the Greenhorn cyclothem of marine flooding of the Western Interior basin (Lucas, 1990; Lucas et al., 1998). Thus, it is most likely that the Romeroville Sandstone is of earliest Cenomanian age (Fig. 2).

Graneros Shale

No fossils of biostratigraphic value have been collected from the Graneros Shale in east-central New Mexico. Therefore, its age must be based on data from nearby Graneros sections in the Raton basin of northeastern New Mexico (Kauffman et al., 1969). These data indicate the Graneros Shale encompasses much of the lower Cenomanian-to lower middle Cenomanian interval of the "*Inoceramus" bellvuensis* through *Acanthoceras bellense* zones (also see Kauffman et al., 1993). Therefore, we assign an early to early middle Cenomanian age to the Graneros Shale in east-central New Mexico (Fig. 2).

Greenhorn Formation

Fossils from the Lincoln Member of the Greenhorn Formation in east-central New Mexico are shark's teeth and the characteristic small oyster *Ostrea beloiti* Logan (Lucas et al., 2000; Lucas and Heckert, 2001). No fossils are known from the Hartland Member, and ammonites and inoceramids are documented from the Bridge Creck Member (Lucas et al., 2000; Lucas and Heckert, 2001). These fossils are consistent with regional data that indicate the Lincoln, Hartland and lower Bridge Creek members are late Cenomanian in age, whereas the upper Bridge Creek member is early Turonian (e.g., Kennedy et al., 2000). At Roy, the youngest Bridge Creek bed is in the early Turonian zone of *Vascoceras birchbyi*, and in the Bonita fault zone, it is apparently the same age.

ACKNOWLEDGMENTS

Orin Anderson and William A. Cobban provided helpful comments on the manuscript.

REFERENCES

Adkins, W. S., 1927, The geology and mineral resources of the Fort Stockton quadrangle: University of Texas, Bulletin 2738, 166 p.

Adkins, W. S., 1928, Handbook of Texas Cretaceous fossils: University of Texas, Bulletin 2838, 385 p.

Adkins, W. S., and Winton, W. M., 1919, Paleontological correlation of the Fredericksburg and Washita formations in north Texas: University of Texas, Bulletin 1945, 128 p.

Baltz, E. H., 1990, Palynomorphic evidence of the age of Dakota Sandstone at Montezuma, New Mexico: New Mexico Geological Society, 41st Field Conference, Guidebook, p. 96.

Bejnar, C.R., and Lessard, R.H., 1976, Paleocurrents and depositional environments of the Dakota Group, San Miguel and Mora Counties, New Mexico: New Mexico Geological Society, 27th Field Conference, Guidebook, p. 157-163.

Brand, J. P., 1953, Cretaceous of Llano Estacado of Texas: Bureau of Economic Geology, Report of Investigations 20, 59 p.

Brand, J. P., and Mattox, R. B., 1972, Pre-Dakota Cretaceous formations in northwestern Texas and northeastern New Mexico: New Mexico Geological Society, 23rd Field Conference, Guidebook, p. 98-104.

Bullard, F. M., 1925, Geology of Love County, Oklahoma: Oklahoma Geological Survey, Bulletin 33, 77 p.

Bullard, F. M., 1926, Geology of Marshall County, Oklahoma: Oklahoma Geological Survey, Bulletin 39, 101 p.

Cobban, W. A., 1985, The Lower Cretaceous ammonite *Schloenbachia leonensis* Conrad var. *equidistans* Cragin: U.S. Geological Survey, Bulletin 1641-A, 4 p.

Cobban, W. A., 1987, Ammonite faunas of the Sarten Sandstone (Cretaceous), Luna County, New Mexico: U. S. Geological Survey, Bulletin 1641-B, 17 p.

Conrad, T. A., 1857, Descriptions of Cretaceous and Tertiary fossils; *in* Emory, T A., Report on the United States and Mexican boundary survey, made under the direction of the Secretary of the Interior: U. S. 34th Congress 1st Session, Senate Executive Document 108, p. 141-174.

Dane, C. H., and Bachman, G. O., 1965, Geologic map of New Mexico: U.S. Geological Survey, scale 1:500,000.

Dobrovolny, E., Summerson, C. H., and Bates, R. L., 1946, Geology of northwestern Quay County, New Mexico: U.S. Geological Survey, Oil and Gas Investigations Preliminary Map 62, scale 1:62,500.

Gilbert, J. L., and Asquith, G. B., 1976, Sedimentology of braided alluvial inter val of Dakota Sandstone, northeastern New Mexico: New Mexico Bureau of Mines and Mineral Resources, Circular 150, 16 p.

Griggs, R. L., and Read, C. B., 1959, Revisions in stratigraphic nomenclature in Tucumcari-Sabinoso area, northeastern New Mexico: American Association of Petroleum Geologists Bulletin, v. 43, p. 2003-2007.

Holbrook, J. M., and Dunbar, R. W., 1992, Depositional history of Lower Cretaceous strata in northeastern New Mexico: Implications for regional tectonics and depositional sequences: Geological Society of America Bulletin, v. 104, p. 802-813.

Holbrook, J. M., Wright, R., and Kietzke, K. K., 1987, Stratigraphic relationships at the Jurassic-Cretaceous boundary in east-central New Mexico: New Mexico Geological Society, 38th Field Conference, Guidebook, p. 161-165.

Hunt, A. P., and Lucas, S. G., 1998, Tetrapod ichnofaunas from the Lower Cretaceous of northeastern New Mexico, USA: New Mexico Museum of Natural History and Science, Bulletin 14, p. 163-167.

Kauffman, E.g., Powell, J. D., and Hattin, D. E., 1969, Cenomanian-Turonian facies across the Raton basin: The Mountain Geologist, v. 6, p. 93-118.

Kauffman, E.g., Sageman, B. B., Kirkland, J. I., Elder, W. P., Harries, P. J., and Villamil, T., 1993, Molluscan biostratigraphy of the Cretaceous Western Interior basin, North America: Geological Association of Canada, Special Paper 39, p. 397-434.

Kennedy, W. J., Walaszczyk, I., and Cobban, W. A., 2000, Pueblo, Colorado, USA, candidate global boundary stratotype section and point for the base of the Turonian Stage of the Cretaceous, and for the Middle Turonian substage, with a revision of the Inoceramidae (Bivalvia): Acta Geologica Polonica, v 50, p. 295-334.

Kennedy, W. J., Cobban, W. A., Gale, A. S., Hancock, J. M., and Landman, N. H., 1998, Ammonites from the Weno Limestone (Albian) in northeast Texas: American Museum Novitates, no. 3236, 46 p.

Kues, B. S., 1986, Paleontology and correlation of a Lower Cretaceous (Albian) outlier in Roosevelt County, New Mexico: New Mexico Geology, v. 8, p. 88-94.

Kues, B.S., 1997, New bivalve taxa from the Tucumcari Formation (Cretaceous, Albian) New Mexico, and the biostratigraphic significance of the basal Tucumcari fauna: Journal of Paleontology, v. 71, p. 820-839.

Kues, B. S., 2001, Late Albian marine invertebrate fauna from the basal Mesa Rica Sandstone at San Jon Hill, New Mexico: New Mexico Geological Society, 52nd Field Conference, Guidebook.

Kues, B. S., and Lucas, S. G., 1987, Cretaceous stratigraphy and paleontology in the Dry Cimarron Valley, New Mexico, Colorado, and Oklahoma: New Mexico Geological Society, 38th Field Conference, Guidebook, p. 167-198.

Kues, B. S., and Lucas, S. G., 2001, Fauna of nearshore sandy facies of the Tucumcari Formation (Lower Cretaceous, Albian), Quay County, New Mexico: New Mexico Geological Society, 52nd Field Conference, Guidebook.

Kues, B. S., Lucas, S. G., Kietzke, K. K., and Mateer, N. J., 1985, Synopsis of Tucumcari Shale, Mesa Rica Sandstone and Pajarito Shale paleontology, Cretaceous of east-central New Mexico: New Mexico Geological Society 36th Field Conference, Guidebook, p. 261-281.

Lucas, S. G., 1990, Type and reference sections of the Romeroville Sandstone (Dakota Group), Cretaceous of northeastern New Mexico: New Mexico Geological Society, 41st Field Conference, Guidebook, p. 323-326.

Lucas, S. G., and Heckert, A. B., 2001, Stratigraphy and biostratigraphy of an Upper Cretaceous outlier near Roy, Harding County, New Mexico: New Mexico Geological Society, 52nd Field Conference, Guidebook.

Lucas, S.G., and Hunt, A.P., 2000, The Lower Cretaceous ammonite *Mortoniceras equidistans* (Cragin) from the Mesa Rica Sandstone, Quay County, New Mexico: New Mexico Museum of Natural History and Science, Bulletin 16, p. 97-100.

Lucas, S. G., and Kisucky, M. J., 1988, Type and reference sections of the Tucumcari, Mesa Rica and Pajarito formations, Cretaceous of east-central New Mexico: New Mexico Geology, v. 10, p. 82-89.

Lucas, S. G., Anderson, O. J., and Estep, J. W., 1998, Stratigraphy and correlation of Middle Cretaceous rocks (Albian-Cenomanian) from the Colorado Plateau to the Southern High Plains, north-central New Mexico: New Mexico Museum of Natural History and Science, Bulletin 14, p. 57-66.

Lucas, S. G., Estep, J. W., Pence, R., and Hunt, A. P., 2000, Upper Cretaceous outlier in Quay County, New Mexico: Stratigraphy and biostratigraphy: New Mexico Museum of Natural History and Science, Bulletin 16, p. 121-127.

Marcou, J., 1855, Résumé éxplicatif d'une carte geologique des Etats-Unis et des provinces ánglaisés de l'Amerique du Nord, avec un profil géologique allant de la Vallée du Mississippi aux côtes du Pacifique et une planche des fossiles: Société Géologique de France Bulletin, Series 2, v. 12, p. 813-936.

Marcou, J., 1858, Geology of North America, with two reports on the prairies of Arkansas and Texas, the Rocky Mountains of New Mexico, and the Sierra Nevada of California, originally made for the United States Government. Zurich, Zürcher and Furrer, 144 p.

Marcou, J., 1889, American Neocomian and Chalk of Arkansas: American Geologist, v. 4, p. 357-367.

Mateer, N. J., 1985, Pre-Graneros Cretaceous stratigraphy of northeastern New Mexico: New Mexico Geological Society, 36th Field Conference, Guidebook, p. 243-246.

Mateer, N. J., 1987, The Dakota Group of northeastern New Mexico and southern Colorado: New Mexico Geological Society, 38th Field Conference, Guidebook, p. 223-236.

McGill, D. W., 1967, Washita Formation, north Texas, correlated to Georgetown Limestone, central Texas; *in* Hendricks, L., ed., Comanchean (Lower Cretaceous) stratigraphy and paleontology of Texas: Permian Basin Section SEPM, Publication 67-8, p. 219-239.

Scott, R. W., 1970, Stratigraphy and sedimentary environments of Lower Cretaceous rocks, southern Western Interior: American Association of Petroleum Geologists Bulletin, v. 54, p. 1225-1244.

Scott, R. W., 1974, Bay and shoreface communities in the Lower Cretaceous: Lethaia, v. 7, p. 315-330.

Slocki, S. F., 1967, Physical stratigraphy of the Georgetown Limestone equivalents in Tarrant, Denton, and Cooke Counties, Texas; *in* Hendricks, L., ed., Comanchean (Lower Cretaceous) stratigraphy and paleontology of Texas: Permian Basin Section SEPM, Publication 67-8, p. 183-216.

Stone, W. J., 1984, Dakota aquifer system in New Mexico; *in* Jorgenson, D. G., and Signor, D. C.,eds., Geohydrology of the Dakota aquifer: National Water Well Association, Proceedings of the C.V. Theis Conference, Oct. 5-6, 1982, p. 62-68.

Wanek, A. E., 1962, Reconnaissance geological map of parts of Harding, San Miguel, and Mora Counties, New Mexico: U.S. Geological Survey, Oil and Gas Investigations Map OM-208.

Winton, W. M., 1925, The geology of Denton County: University of Texas, Bulletin 2544, 86 p.

Wright, C. W., Calloman, J. H. and Howarth, M. K., 1996, Treatise on Invertebrate Paleontology Part L Mollusca 4: Boulder and Lawrence, Geological Society of America and University of Kansas, 302 p.

Young, K., 1957, Upper Albian (Cretaceous) Ammonoidea from Texas: Bureau of Economic Geology, Report of Investigations 28, 33 p.

Young, K., 1967, Ammonite zonations, Texas Comanchean (Lower Cretaceous); *in* Hendricks, L., ed., Comanchean (Lower Cretaceous) stratigraphy and paleontology of Texas: Permian Basin Section SEPM, Publication 67-8, p. 65-70.

ALBIAN-CENOMANIAN DEPOSITIONAL CYCLES TRANSGRESSED FROM CHIHUAHUA TROUGH TO WESTERN INTERIOR

R.W. SCOTT[1], J.M. HOLBROOK[2], M.J. EVETTS[3] AND F.E. OBOH-IKUENOBE[4]

[1]RR3 Box 103-3, Cleveland, OK 74020; [2]Southeast Missouri State University, Cape Girardeau, MO 63701; [3]1227 Venice Street, Longmont, CO 80501; [4]University of Missouri, Rolla, MO 65409

Abstract.—The Upper Albian to Lower Cenomanian stratigraphic record in the Chihuahua Trough at El Paso, Texas is more than 250 m thick and records six depositional cycles averaging 1.6 million years in duration. Five of these cycles are identified in the much thinner Albian-Lower Cenomanian stratigraphic sections of northeastern New Mexico, Colorado, Kansas, and Wyoming, and the cycle of the Mesilla Valley Formation correlates into the regional unconformity between the Glencairn and Mesa Rica Formations.

Three transgressive-progradational cycles were deposited continuously from Texas north to Wyoming and Montana and the associated seaways provided opportunities for south to north biotic exchange. The first early Late Albian connection is represented by correlation of the two cycles of the Del Norte and Smeltertown formations at El Paso with the Tucumcari/Purgatoire/Kiowa/Skull Creek formations in New Mexico, Colorado, and Kansas, and the Thermopolis Formation in Wyoming and Montana. Part of the third cycle represented by the Muleros Formation may be partly represented locally in Kansas-Colorado, but its record is absent in most places.

The second potential connection is represented by the Anapra Formation at El Paso, the Pajarito Formation in New Mexico and the Van Bibber Member of the Muddy Formation in eastern Colorado. The Anapra correlates with the Pawpaw and Main Street formations in north Texas, which includes the range of the uppermost Albian oyster in the Pajarito, *Peilinia levicostata* Kues [formerly *Rastellum quadriplicata* (Shumard)]. But the Van Bibber contains a sparse non-marine biota indicating that little northward exchange occurred across the Transcontinental Arch with deposition of the Mowry Formation in Wyoming and Montana. The third connection in the Early Cenomanian is indicated by the Del Rio/Buda formations in El Paso and the lower part of the Graneros Shale northward. However, during this time little biotic exchange occurred, and the foraminiferal faunas are quite different. Full and open exchange was resumed during the Middle Cenomanian when the Thatcher Member of the Graneros was deposited.

The geographic pathways of the ancestors of the endemic ammonites and inoceramids of the Mowry Formation in Wyoming and Montana are as yet unclear. It is possible that these species are relics of the earlier Late Albian connection between the Caribbean and North Temperate Provinces. The geographic range of one of the ammonites is expanded by a new occurrence in the basal Cenomanian section in northwestern France (Cobban and Amedro, personal commun., 2001). Most of the benthic foraminifers in the Mowry are from the northern area, but a sparse planktic foraminiferal assemblage indicates a filtered connection with the Tethyan Realm. This connection was not necessarily southward with the Gulf Coast Caribbean Province.

INTRODUCTION

Traditional sequence models of the mid-Cretaceous North American Western Interior Seaway recognize that the first north-south connection between the Boreal and Caribbean Provinces was made during the early part of the Late Albian (about 103 to 99 Ma) (Kauffman, 1984; Scott et al., 1998). During the latest Albian these biotic provinces became separated. This seaway was opened again in the Middle Cenomanian about 94 Ma. These two continental-wide transgressions correspond to sequences Al 8 and Ce 3 of Hardenbol et al. (1998), respectively; the timing of maximum flooding of these sequences is 101.06 Ma and 95.23 Ma. New stratigraphic field evidence supports the hypothesis that an additional major (3rd-order) transgressive/regressive episode between these two floodings spanned the Early/Late Cretaceous boundary in the Western Interior Basin. The evidence further suggests that transgression was sufficient to reconnect the Tethyan and Boreal seas for at least a brief period. The regional scale, timing, and mechanics of this proposed seaway connection are very poorly constrained. This cycle has not been incorporated into the existing stratigraphic framework of major transgressive/regressive Western Interior cycles.

The suggestion of an intermediate connection during the Albian/Cenomanian interval was made by Cobban and Kennedy (1989) to account for the dispersal of Caribbean engonocerid ammonites into the Boreal Mowry seaway. They also suggested that the correlative Albian/Cenomanian boundary defined in Europe was low in the Mowry Formation below the well-dated Clay Spur bentonite and close to the Arrow Creek bentonite dated at 98.52+/-0.41 Ma (Obradovich, 1993). The result of this suggestion was to increase the age of the Albian/ Cenomanian boundary nearly 2 my. However, this suggested correlation has yet to be tested utilizing other fossil groups such as dinoflagellates, foraminifers and radiolaria.

Here, we propose tentative stratigraphic correlations of Albian/ Cenomanian strata from the Chihuahua Trough at the Texas-Chihuahua border to New Mexico, Kansas and Colorado based on preliminary data. We are in the process of testing the hypothesis that an additional extensive transgressive/regressive cycle near the Early/Late Cretaceous boundary connected the southern and northern basins.

An age model of the fossils and depositional cycles of middle Cretaceous strata has been developed based on more than 40 outcrop and core data from around the world and calibrated to the time scale of Harland et al. (1990) (Scott et al., 2000). Ranges of more than 1100 bioevents, planktic and benthic foraminifers, nannofossils, dinoflagellates, ammonites, bivalves, echinoids, calcareous algae, magnetochrons, geochemical events, and sequence stratigraphic markers have been integrated by the technique of graphic correlation. Albian depositional cycles were defined in the Washita Group of north Texas and labeled WA1-6 (Scott et al., 1994; Scott et al. 2000). This data set includes the core of Cretaceous strata in the Amoco Bounds well at the Kansas-Colorado state line (Scott et al., 1994, 1998). Foraminiferal data from the

Mowry and Graneros formations in Kansas and Colorado have been graphed to this data set so that the ages of Western Interior marker fossils and depositional cycles can constrain regional correlations. This new data set and age model is the MIDK4 composite standard, which will be used to make preliminary estimates of the ages of mid-Cretaceous depositional sequences from the U.S.-Mexico border to northern Colorado.

STRATIGRAPHIC DATA

Biostratigraphy

Fossils that have biostratigraphic utility in these strata in Colorado, New Mexico, and Kansas are uncommon and not diverse. Benthic foraminiferal zones were used by Eicher (1962, 1965, 1967) to correlate the Mowry and Belle Fourche formations from Colorado to Montana. These zones in the Canadian Plains region are conceptually bodies "...of strata distinguished by a particular foraminiferal fauna that empirically is distinctive enough to have at least local biostratigraphic value" (Caldwell et al., 1978, p. 501). These assemblage zones are defined by the composite group rather than by first or last occurrences. The *Miliammina manitobensis* Zone consists of at least 14 principal taxa and extends from the Interior Plains of Canada to Colorado (McNeil and Caldwell, 1981). It characterizes strata above the Viking Sandstone and up to the Fish Scale Beds in Canada and the Shell Creek and Mowry Formations in the U.S. This zone occurs with the ammonites *Neogastroplites haasi* Reeside and Cobban, *N. cornutus* (Whiteaves), and *N. muelleri* Reeside and Cobban (Caldwell et al., 1978), which were originally correlated with Upper Albian ammonites in Europe. But Cobban and Kennedy (1989) reinterpreted these zones as Lower Cenomanian because the metengonocerid species are similar to Cenomanian species. However, additional data are needed to evaluate and clarify this new correlation. Our new data from the Mowry and basal Graneros formations in the sections at the I-70 roadcut west of Denver and at Turkey Creek near Morrison establish the presence of the *M. manitobensis* Zone across the Mowry-Graneros contact supporting correlation of this contact with the top of the Mowry in Wyoming.

The Graneros Shale in Colorado and southern Wyoming is divided into two foraminiferal zones characterized by low diversity of Boreal agglutinated foraminifera: the lower *Trochammina mellariolum* Zone and the upper *Evolutinella apricarius* Zone (Eicher, 1962, 1965, 1967; Caldwell et al., 1993). Both zones are used mainly in the U.S. part of the Western Interior, where the *T. mellariolum* Zone encompasses the Lower Cenomanian part of the Graneros and the *E. apricarius* zone spans from below the Middle Cenomanian Thatcher bed to the top of the Graneros (Caldwell et al., 1993). Both zones contain sparse numbers of cosmopolitan, shallow-dwelling planktic foraminifers; however the Thatcher is evidence that the seaway opened to Caribbean waters introducing Tethyan biota, including nannofossils into the Western Interior (Cobban and Scott, 1972; Scott et al., 1994, 1998).

Ammonite zones are not well developed in most of the section, however the Graneros Shale has four zones in ascending order: *Conlinoceras tarrantense* (Adkins), *Acanthoceras granerosense* Cobban and Scott, *Acanthoceras muldoonense* Cobban and Scott, and *Acanthoceras amphibolum* Morrow (Cobban and Scott, 1972; Cobban, 1990). In the Thatcher Member at Pueblo the *C. tarrantense* Zone is represented by *Conlinoceras gilberti* Cobban and Scott with *Borissiakoceras compressum* Cobban and *Inoceramus eulessanus* Stephenson. This zone is also present in the uppermost Woodbine Formation in north Texas and the basal Boquillas Formation at Cerro de Cristo Rey, and is correlated with the Middle Cenomanian. The *A. granerosense* and *A. amphibolum* zones span from just above the Thatcher to the top of the Graneros and are correlated with the Middle Cenomanian. The age span of these three zones in the MIDK4 composite standard is from 93.43 to 92.62 Ma.

Three Gulf Coast Upper Albian ammonite zones are known in this region below the Graneros: *Craginites serratescens* (Cragin), *Eopachydiscus marcianus* (Shumard), and *Mortoniceras equidistans* (Cragin). These three species define successively younger range zones in the lower Upper Albian Duck Creek Formation in Texas (Young, 1986; Kauffman et al., 1993), where they occur just above the top of *Hysteroceras* sp. cf. *varicosum* (Sowerby), which defines a zone in the lower Upper Albian in Europe, so the *E. marcianus* Zone is correlated with the *Callihoplites auritus* Zone in Europe (Hancock et al., 1993; Kennedy et al., 1999). Both *E. marcianus* and *M. equidistans* occur in the upper parts of the Tucumcari Formation and basal part of the Mesa Rica Sandstone, and *E. marcianus* is in the Glencairn Formation in northeastern New Mexico and southeastern Colorado with *Inoceramus comancheanus* Cragin and/or *Inoceramus bellvuensis* Reeside (Cobban, 1985, 1987a; Kues et al., 1985; Kues and Lucas, 1987). Both inoceramids occur in the Glencairn as far north as Bellvue, Colorado near the Northern Front Range section (Reeside, 1923), so the *M. equidistans* Zone can be correlated to this section. The *M. equidistans* Zone also occurs in the Smeltertown Formation at Cerro de Cristo Rey (Cobban, 1985), where it characterizes depositional cycle Washita (WA) 2 of the middle-upper Duck Creek Formation. Consequently, we correlate cycle WA2 in the upper part of the Smeltertown with the Glencairn in New Mexico and Colorado. In the MIDK4 composite standard the age of *E. marcianus* is 101.39-101.32 Ma, *M. equidistans* is 101.01-100.55 Ma, *I. comancheanus* is 101.48-101.39 Ma, and *I. bellvuensis* is 101.47-101.39 Ma. Note, however, that Cobban (1985) reports that ranges of *E. marcianus* and *M. equidistans* overlap and *M. equidistans* ranges higher.

Two of the Western Interior Cretaceous palynomorph biozones (Nichols et al., 1982) can be recognized by our preliminary data. The dinocyst, *Ovoidinium verrucosus* (Cookson and Hughes) of the *Spinidinium vestitum* Assemblage Zone occurs in the basal Graneros Shale in the Tollgate Canyon section in the Dry Cimarron Canyon. This zone spans the Bear River Formation and Aspen Shale in western Wyoming and it is found in the Thermopolis Shale, Skull Creek Shale, and Mowry Formation (Nichols et al., 1982). Several spore and pollen taxa suggest the presence of the nonmarine *Appendicisporites unicus* Interval Zone (Nichols et al., 1982) in the Van Bibber Shale Member in the Interstate 70 roadcut and in sections of the Pajarito and Dry Creek Canyon members in the Dry Cimarron Valley. Albian pollen were found in the Pajarito Formation at Romeroville Gap, New Mexico (Lucas, 1990).

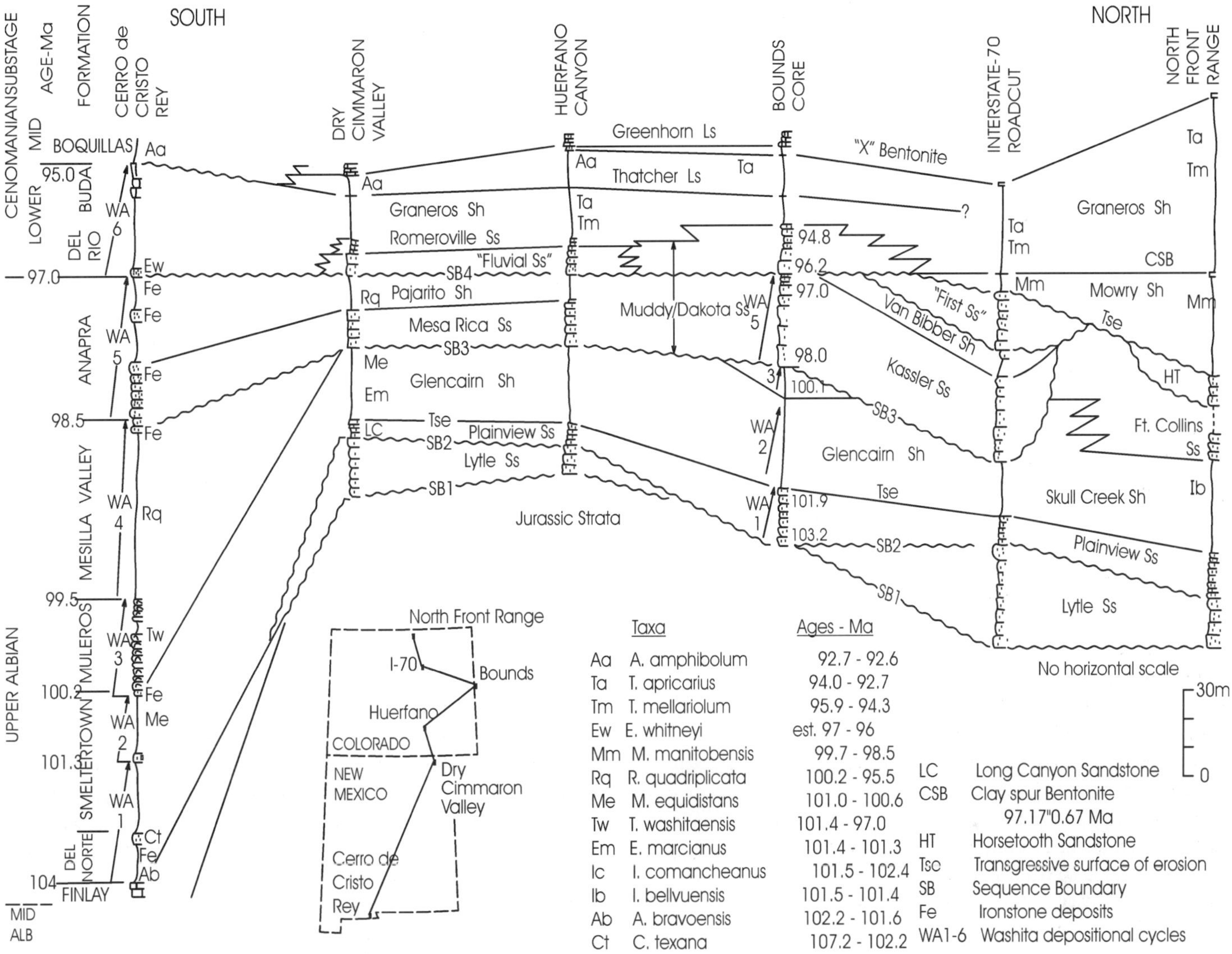

FIGURE 1. Regional north-south stratigraphic cross section on the datum of the Albian-Cenomanian boundary placed at SB4 and the Clay Spur Bentonite bed. Long arrows indicate generalized shoaling up depositional cycles. No horizontal scale.

Cerro de Cristo Rey, New Mexico Section

The Upper Albian-Lower Cenomanian reference section of the Chihuahua Trough is documented at Cerro de Cristo Rey uplift on the border between Doña Ana County, New Mexico and Ciudad Juarez, due west of El Paso, Texas (Lovejoy, 1976). This section is approximately 265 m thick and composed of seven lithostratigraphic units that yield age-diagnostic ammonites and bivalves that enable correlation of the section with the Trinity River Valley section near Fort Worth (Scott et al., 1994, 1996, 2000) and with the eastern New Mexico sections (Cobban, 1985, 1987a, b; Kues, 1986, 1997; Kues and Lucas, 1993; Lucas, 1991; Lucas and Estep, 1998). Six depositional cycles are distinguished by shale to ferruginous sandstone successions from about 20 to nearly 100 m thick (Fig.1). The ferruginous sandstones tend to contain taphonomically mixed assemblages suggesting transportation and depositional bypassing. These cycles appear to represent deepening to shoaling base levels. Diagnostic ammonites and oysters indicate biostratigraphic correlations with depositional cycles of shale to limestone in north Texas (Scott et al., 2000) and in northern New Mexico (Kues and Lucas, 1987; Cobban, 1985). We correlate the formations at Cerro de Cristo Rey with those in the Trinity River Valley generally in the same way as did Strain (1976).

Dry Cimarron Valley, New Mexico Section

The Dry Cimarron Valley in Union County, northeastern New Mexico and the panhandle of Oklahoma, exposes Albian-Cenomanian sandstones, mudstones, and shales of the Lytle Sandstone, the Glencairn Formation including the basal Long Canyon Sandstone Bed, the Mesa Rica Sandstone, the Pajarito Formation, and the Romeroville Sandstone overlain by the Graneros Shale (Kues and Lucas, 1987; Lucas, 1990). These lithostratigraphic units are separated into genetic sets by four regional unconformities that are well exposed at Tollgate Pass (Holbrook, 1992; Holbrook and Wright Dunbar, 1992; Holbrook and White, 1998). The basal unconformity sequence boundary (SB) 1 separates the Lower Cretaceous Lytle Formation from underlying Jurassic rocks, unconformity K-0 of Lucas et al. (1998). The Lytle ranges up to 20 m thick in this region and represents mainly fluviatile deposition. It is overlain by a regional unconformity designated SB 2 by Holbrook (1992) at the base of the Long Canyon Sandstone Member of the Glencairn, which is up to about 3 m thick. The Long Canyon is in the same stratal position as the Planiview Sandstone but is not physically continuous with it. The contact of the Long Canyon with the overlying Glencairn is a transgressive surface. The remaining Glencairn consists of paralic and nearshore marine shelf facies and is up to 20-25 m thick. The Glencairn here and in southeastern Colorado and its equivalent unit in east-central New Mexico, the Tucumcari Formation, yield age-diagnostic ammonites and bivalves (Scott, 1970; Cobban, 1985; Kues et al., 1985). Many of the taxa in the Tucumcari are also found in the Smeltertown Formation at Cristo Rey and in the Duck Creek Formation in north Texas, depositional cycle WA2.

In the Dry Cimarron Valley the Glencairn Formation is disconformably separated by SB 3 from overlying fluviatile strata of the Mesa Rica Sandstone, which represent the lowstand and transgressive deposits of a depositional cycle that includes the overlying Pajarito Shale. Westward at Romeroville SB3 merges with SB1 and SB2 where it is the K-0 unconformity of Lucas et al. (1998). Here the Mesa Rica is up to 10-12 m thick and the Pajarito is 12 m thick. In the Dry Cimarron Valley the Mesa Rica is unfossiliferous, however, southward in the Tucumcari Basin correlative Mesa Rica marine strata yield sparse ammonites, including *Mortoniceras equidistans*, which characterizes the Duck Creek Formation in north Texas. Also southern outcrops of the Pajarito yield the Upper Albian oyster, *Peilinia levicostata* (Kues, 1997). A closely related oyster, *Rastellum quadriplicata* (Shumard), is found in the Mesilla Valley Formation, depositional cycle WA4 at Cristo Rey, and in the Denton through Main Street formations in north Texas, depositional cycles WA3-WA5. The basal Pajarito is overlain disconformably at SB 4 by an unnamed upper Pajarito fluviatile sandstone up to about 8 m thick.

The overlying Romeroville Sandstone is up to 5 m thick and represents transgressive paralic deposition grading up into the Graneros Shale, which is up to 20-30 m thick in the region. In the upper part of the Graneros a thin limestone unit, Thatcher Limestone Member, contains Middle Cenomanian ammonites correlative with the uppermost part of the Woodbine Formation in north Texas (Cobban and Scott, 1972).

Huerfano Canyon, Colorado Section

The section at Huerfano Canyon is about 40 km southeast of Pueblo, Colorado, where the pre-Graneros section is well exposed (Long, 1966; Taylor, 1974; Holbrook and White, 1998). The section of the Graneros Shale, however, is a composite based on several measured sections of Cobban and Scott (1972) at the Rock Canyon area near Pueblo, the Old Hatchet Ranch and the Model anticline. The base of the Lytle is a regional unconformity (SB1) that locally cuts into red mudstone and white sandstone of the Morrison Formation below. The Lytle is 10-20 m thick. The base of the overlying Plainview Sandstone is a sharp surface of valley incision. The Plainview is about 8 m thick and dominantly estuarine fill. The Glencairn Formation is about 25 m thick and consists of interbedded sandstone, shale and mudstone. Besides shallow shelf trace fossils, *Inoceramus bellvuensis* is present. The Glencairn correlates with Washita Group depositional cycle WA2 in north Texas. The unconformity between the Lytle and Plainview is extremely widespread and marks fluvial incision. SB2 probably records a separate major regression.

The overlying Muddy Sandstone at Huerfano Canyon consists of a lower, thick-bedded, medium grained, cross-bedded sandstone about 15 m thick, the Dry Creek Canyon Member up to 10 m thick, an upper thick-bedded sandstone about 12 m thick, and a 3-m-thick sandstone with trace fossils indicating transgression. We place SB3 at the Glencairn-Muddy contact and SB4 between the Dry Creek Canyon and the upper Muddy units. We suggest that SB4 correlates with the same contact between the lower and upper Dakota units in the Amoco No. 1 Bounds core, and thus, is the Albian-Cenomanian boundary. The overlying Graneros Shale at Huerfano Canyon consists of a lower shale unit about 21 m thick, the Thatcher Limestone Member 0.5 m thick, and an upper shale unit 13 m thick. The 1-m-thick "X" Bentonite is at the top of this upper unit. The basal Graneros contact is a transgressive lag contact, and the upper contact with the Greenhorn Formation is abruptly gradational. In the Pueblo section, Eicher (1965) identified the *Trochammina mellariolum* Zone in the lower shale and the *Evolutinella apricarius* Zone at the top of the Graneros.

Amoco No. 1 Bounds, Kansas Section

The Amoco No. 1 Bounds core near the Kansas-Colorado border presents a complete section of the Plainview-Graneros interval (Hamilton, 1989, 1994; Scott et al., 1994; Scott et al.,1998). The Plainview Formation is the 20-m-thick basal sandstone-shale interval that disconformably overlies the Upper Jurassic Morrison Formation. It is a nearshore paralic facies with two transgressive contacts within it, and its top is a transgressive contact with the Glencairn Formation, which is 44 m thick. The maximum flooding interval in the Glencairn is about 1400-1410 ft depth, where *Adkinsites bravoensis* (Böse) and *Inoceramus comancheanus* Cragin occur. These fossils occur in north Texas in the Kiamichi-Duck Creek formations. The contact between the Glencairn and Dakota formations is an erosional disconformity (SB3).

The Dakota Formation consists of two intervals separated by a regional erosional disconformity, SB4. This contact separates nonmarine beds below from bioturbated transgressive sandstone above. The lower interval is 33 m thick, and the upper interval is 17 m thick. Graphic correlation postulates that the lower interval correlates with the Pawpaw-Mainstreet formations depositional cycle WA5 in north Texas, and the upper unit with either the Grayson-Buda formations or the basal Woodbine Formation, both of Cenomanian age (Scott et al., 1994).

The Graneros Shale is divided into a lower interval about 10 m thick capped by a very thin Thatcher Member, and an upper interval about 17 m thick. The "X" Bentonite is within 2.4 m of the contact between the Graneros and the Greenhorn Limestone. The *Evolutinella apricarius* Zone is represented by two species at a depth of 1050 ft in the upper Graneros interval about 2.4

m below the "X" Bentonite. The first nannofossils in the Graneros at a depth of 1084.75 ft and the first planktic foraminifers at 1080 ft indicate connection between the Caribbean and Boreal seas (Scott et al., 1998).

Interstate-70 Roadcut, Colorado Section

The Lower Cretaceous section exposed along the I-70 roadcut west of Denver is one of the most informative along the Front Range. It has been described carefully by Waage (1955) and Weimer and Land (1972), and our sampling of the Mowry has added a few details. The base of the section is the contact between the Jurassic Morrison Formation and the Lytle Formation, which is characterized by red mudstone below and tan sandstone above. The Lytle is 36.5 m thick and represents fluvial deposition. The basal South Platte Formation consists of the Plainview Sandstone, 10.1 m thick. The Glencairn Member is 18.9 m thick, and shallow water trace fossils occur in sandstone interbedded with mudstone. It is disconformably overlain by the Kassler Sandstone Member, which is 29.5 m thick and is characterized by trough cross bedding. The Van Bibber Shale Member is 8.8 m thick and did not yield foraminifers. The upper sandstone member is 20.9 m thick and consists of cross-bedded sandstone with clay clasts at the base, which is in sharp contact with bioturbated sandstone of the Van Bibber. Waage (1961) referred to this unit as the "First sandstone" of the South Platte Formation, and Weimer (1984) combines this unit with the Van Bibber and Kassler in the "J" sand unit. The sharp contact with the overlying Mowry is marked by a lamina of ferruginous sandstone, and it appears to be a transgressive disconformity. The Mowry is 5.6 m thick and yields foraminfers of the *Miliammina manitobensis* Zone, which extends into the overlying Graneros Shale at least 2 m. A thin sandstone bed marks the conformable Mowry/Graneros contact. Although bentonitic laminae are common in both units, the Clay Spur Bentonite is absent. Rojas (1980) measured more than 29 m of Graneros at this roadcut. Eicher measured about 33 m of Graneros up to the "X" Bentonite at the Alameda roadcut section about 3 km south of the I-70 roadcut, and that measurement is used here. His data also show that the Upper Albian-Lower Cenomanian *M. manitobensis* Zone spans the Mowry-Graneros contact; the Lower Cenomanian *Trochammina mellariolum* Zone is from 7.3 to 13.4 m above the base of the Graneros, and the mainly Middle Cenomanian *Evolutinella apricarius* Zone is from 13.4 to at least 31.7 m. The Thatcher bed is not recognized here.

Northern Front Range, Colorado Section

The thickest Mowry section along the Front Range, measured by Eicher (1965, section 11), is in Larimer County. The Mowry Formation is 34.8 m thick, and the Clay Spur Bentonite is about 0.9 m thick. The Graneros Shale is 62.5 m thick, and the 0.8-m-thick "X" Bentonite is at the top. The Thatcher Member is not identified here. The *M. manitobensis* Zone is present in the Mowry, the *T. mellariolum* Zone is in the lower part of the Graneros, and the *E. apricarius* Zone is in the upper part.

The lithologic succession and thicknesses of the pre-Graneros interval are based on cross sections by Eicher (1962), Dolson et al. (1991), and Dubois (1996). The Lytle Formation is estimated to be about 18 m thick; the Plainview Formation is about 16 m thick; the Skull Creek Formation is about 33 m thick; the Fort Collins Sandstone is about 19 m thick; and the Horsetooth Sandstone is about 10 m thick. *Inoceramus bellvuensis* occurs in the shale and sandstone interval of the Skull Creek (Reeside, 1923). This section needs to be re-examined for details about the contacts, but it is assumed that the features seen near Bellvue and at Spring Canyon Dam at Horsetooth Reservoir (MacKenzie, 1965) also characterize the contacts here.

CONCLUSIONS

Depositional Cycles

In the U.S. Western Interior Basin, the middle Cretaceous stratigraphic section consists of two long-term transgressive-regressive depositional cycles, the Kiowa-Skull Creek Cycle and the Greenhorn Cycle (Kauffman, 1977) and four short-term cycles. The Kiowa-Skull Creek Cycle as defined by Kauffman incorporates the Lytle, Plainview, and Glencairn formations and is bounded by SB1 at the base and SB3 at the top. Although the Lytle is part of the Kiowa-Skull Creek, the regional incision marked by SB2 suggests a major episode of regional incision between Lytle and Plainview. It seems more reasonable that the Lytle records an earlier event, older than the Plainview-Glencairn-Skull Creek. The Greenhorn Cycle includes the Muddy Sandstone and the Mesa Rica Sandstone/Pajarito Shale/Romeroville Sandstone succession at its base overlain by the Graneros and Greenhorn formations. Additional unconformities in this section bound short-term depositional sequences (Holbrook and Wright Dunbar, 1992). Sequence 1 is bounded by sequence boundary 1 (SB1) at the base of the Lytle and by SB2 at the top of the Lytle. Sequence 2 is bounded by SB2 at the base of the Plainview/Long Canyon Sandstone Bed/Campana Sandstone Bed and by SB3 at the top of the Glencairn. Sequence 3 is bounded by SB3 at the base of the Muddy Sandstone and by SB4 at the base of either the Romeroville Sandstone or local fluvial sandstone lenses filling channels incised into the upper Pajarito Formation. The Romeroville grades upward into Upper Cretaceous marine strata in the Dry Cimarron Valley (Lucas, 1990; Lucas et al., 1998; Holbrook, 1996). Westward of the Dry Cimarron SB1, SB2, and SB3 merge into a single unconformity, K-0, between the Morrison and Mesa Rica (Lucas et al., 1998). They label the Pajarito/Romeroville unconformity K-1, which we name SB4.

Sequence 1

The basal Cretaceous sequence 1 consists of the Lytle Formation in New Mexico and Colorado, and the localized incised channel fill of the Cheyenne Sandstone in southern Kansas. The Lytle has been correlated with basal Cretaceous sandstone units in Wyoming (Cloverly Formation), South Dakota (Lakota Formation), and Montana (Kootenai Formation). These basal sandstones consist of fluvial channel complexes that represent lowstand deposits, but, successive systems tracts of a complete sequence are absent. In Colorado, the Lytle is overlain by transgressive bioturbated facies of the Plainview, and the age of the Lytle is indeterminate because age-diagnostic fossils have not been recovered. The Lytle may be a truncated part of an older separate sequence, if the Lytle is coeval with a more complete sequence to the north. Alternatively the Lytle may be the lowstand facies tract of the Plainview-Glencairn sequence 2.

Sequence 2

At the base of this sequence in Colorado and western Kansas are transgressive paralic facies of the Plainview Sandstone. In northeastern New Mexico lenticular Campana and Long Canyon sandstones represent fluvial, estuarine and lagoonal facies in topographic depressions (Holbrook and Wright Dunbar, 1992). The Glencairn and Tucumcari shales and nearshore sandstones represent the deeper shelf deposits of maximum transgression of the early Late Albian Kiowa-Skull Creek seaway (Scott, 1970; Holbrook and Wright Dunbar, 1992). A localized shoreface sand at the top of the Glencairn on the northern Front Range is mapped as the Fort Collins Sandstone (MacKenzie, 1965; Dubois, 1996). This sequence correlates with depositional cycles WA1 and WA2 and part of WA3 at the base of the Washita Group in north Texas (Scott et al, 1978; Scott et al., 1994), which are also developed in the Cerro de Cristo Rey section. These cycles were renumbered Al 4, 5, and 6, and the bases of these cycles are dated at 104.00 Ma, 101.30 Ma, and 100.2 Ma (Scott et al., 2000).

Sequence 3

The regional unconformity within the Mesa Rica Sandstone between the lower marine facies and the upper fluviatile facies in northeastern New Mexico correlates into western Kansas with the basal unconformity of the Dakota Formation in the Bounds core. The lower Dakota here is dated as 98.2 to 96.0 Ma by correlating it with the north Texas depositional cycle WA5, Pawpaw and Main Street formations (Scott et al., 1994, 1998). Consequently, the Mesa Rica correlates with the Anapra Formation at Cerro de Cristo Rey. Southward in the Tucumcari Basin Holbrook and Wright Dunbar (1992) did not recognize SB3 below deltaic facies of the Mesa Rica. Northward the Mesa Rica correlates with the lower Muddy sandstone at Pueblo and in the Bounds core, and with the Kassler Sandstone at the I-70 road cut. The Van Bibber Shale is in the stratigraphic position of a transgressive unit above the Kassler as is the Pajarito above the Mesa Rica.

The Kassler Sandstone pinches out northward between the I-70 roadcut and the Horsetooth Reservoir, where the Horsetooth Sandstone, a localized incised valley-fill up to 38 km wide in a north-south direction (MacKenzie, 1965; Weimer, 1984; Mateer, 1987; Dolson et al., 1991; Dubois, 1996), disconformably overlies SB3 and the underlying Fort Collins Sandstone. The Fort Collins is a nearshore facies grading down into the Skull Creek. Because the Horsetooth is overlain by the Mowry Shale and the *M. manitobensis* Zone, it is older than the 97 Ma-old Clay Spur Bentonite Bed. Between the southern pinchout of the Horsetooth and the northern pinchout of the Muddy Sandstone/"J" sand, the Fort Collins underlies the basal Mowry transgressive disconformity (Weimer, 1984, fig. 15). The upper sandstone of Weimer's "J" sand in the I-70 section unconformably overlies the Van Bibber Shale and underlies the Mowry Formation at a transgressive contact and the *M. manitobensis* Zone. The Horsetooth Sandstone is correlated with the upper sandstone of the Muddy Sandstone in the I-70 section because they both are overlain by and in transgressive contact with the Mowry. The upper one meter of the Horsetooth has shallow-water burrows indicating encroaching marine conditions and is overlain by a lag pebble lamina of phosphate and shark teeth (MacKenzie, 1965). MacKenzie's isopach map indicates northeast transport into the Denver Basin. It is clear that both the Horsetooth and the upper Muddy sandstone at the I-70 cut are part of an incised valley cycle younger than Van Bibber and older than the top of the Mowry and the basal contact is amalgamated SB3 and SB4. Both sequence 3 and the Horsetooth sandstones are paralic facies of the Mowry basin to the north. The Muddy extends eastward and northeastward into the Denver Basin where it is overlain by the Huntsman Shale, which is correlated as a tongue of the Graneros Shale and overlain by the "D" sandstone. South of the I-70 section the Van Bibber and part of the Kassler are, in turn, cut out by the transgressive sandstones of sequence 4. The Pajarito Shale is also dominantly nonmarine except in one southernmost exposure where it has oysters.

If the environments in which the Van Bibber and Pajarito shales were deposited were connected, a limited, filtered exchange of southern and northern biota could have occurred during the transgressive cycle of sequence 3, thus permitting shallow-dwelling planktic foraminifers and perhaps larvae of engonocerid ammonites to enter Mowry waters. However the Van Bibber in the I-70 section yields non-marine spores and pollen and no benthic foraminifers.

Sequence 4

The base of sequence 4 is SB4 in New Mexico, and in some sections it is at the base an upper fluvial sandstone bed filling a channel incised into the Pajarito, and elsewhere it lies at the base of the Romeroville Sandstone. The sandstones above SB4 in places are lowstand incised fluvial sands. Generally these sands are overlain by marine sands of the Romeroville, which are nearshore transgressive facies tracts grading into the Graneros Shale in northeastern New Mexico, Colorado, and Kansas. In some sections, such as near Pueblo, Colorado, and in the Bounds core, the channel sandstone in the upper part of the Pajarito is missing and the transitional sandstone interval directly overlies SB4 amalgamated with TSE. Sequence 4 is correlated with the north Texas cycle WA6 of the Grayson and Buda formations (Scott et al., 1994, 1998). Thus, the strata above SB4 are younger than the Clay Spur Bentonite in Wyoming at the top of the Mowry Formation. Finally by Middle Cenomanian time when the Thatcher Member was deposited the Caribbean and Boreal seas were connected and biota freely exchanged.

ACKNOWLEDGMENTS

The research is based upon work supported by the National Science Foundation under Grant No. 9909601 to Southeast Missouri State University, the University of Missouri-Rolla, and the University of Tulsa. Any opinions, findings and conclusions or recommendations expressed in this research are those of the authors and do not necessarily reflect the views of the National Science Foundation.

We appreciate insightful ideas and encouragement by Dr. W.A. Cobban, Dr. E.A. Merewether, Dr. R.W. Tillman, and Dr. J.P. Bhattacharya. Careful reviews by Dr. S.G. Lucas and Dr. B.S. Kues significantly improved this manuscript.

REFERENCES

Caldwell, W.G.E., North, B.R., Stelck, C.R., and Wall, J.H., 1978, A foraminiferal zonal scheme for the Cretaceous System in the Interior Plains of Canada; *in* Stelck, C.R., and Chatterton, B.D.E., eds., Western Arctic Canadian Biostratigraphy: Waterloo, Geological Association of Canada Special Paper 18, p. 495-575.

Caldwell, W.G.E., Diner, R., Eicher, D.L., Fowler, S.P., North, B.R., Stelck, C.R., and von Holdt Wilhelm, L., 1993, Foraminiferal biostratigraphy of Cretaceous marine cyclothems; *in* W. G. E. Caldwell and E.G. Kauffman, eds., Evolution of the Western Interior Basin: St. John's, Geological Association of Canada Special Paper 39, p. 477-520.

Cobban, W.A., 1985, The Lower Cretaceous ammonite *Schloenbachia leonensis* Conrad var. *equidistans* Cragin: U.S. Geological Survey Bulletin 1641-A, 8 p.

Cobban, W.A., 1987a, An ammonoid fauna from the Glencairn Shale Member of the Lower Cretaceous Purgatorie Formation, Baca County, Colorado: New Mexico Geological Society, 38th Field Conference, Guidebook, p. 217-222.

Cobban, W.A., 1987b, Ammonite faunas of the Sarten Sandstone (Cretaceous), Luna County, New Mexico: U.S. Geological Survey, Bulletin 2641-B, 30 p.

Cobban, W.A., 1990, Ammonites and some characteristic bivalves from the Upper Cretaceous Frontier Formation, Natrona County, Wyoming: Washington, D.C., U.S. Geological Survey Bulletin 1917-B, 38 p.

Cobban, W. A. and Kennedy, W. J., 1989, The ammonite *Metengonoceras* Hyatt, 1903, from the Mowry Shale (Cretaceous) of Montana and Wyoming: Washington, D.C., U. S. Geological Survey Bulletin 1787, p. L1-L11.

Cobban, W.A., and Scott, G.R., 1972, Stratigraphy and ammonite fauna of the Graneros Shale and Greenhorn Limestone near Pueblo, Colorado: Washington, D.C., U.S. Geological Survey Professional Paper 645, 109 p., 39 pl.

Dolson, J.C., Muller, D., Evetts, M.J., and Stein, J.A., 1991, Regional paleotopographic trends and production, Muddy Sandstone (Lower Cretaceous), central and northern Rocky Mountains: American Association of Petroleum Geologists Bulletin, v. 75, p. 409-435.

Dubois, D., 1996, Eldorado Springs Member of the Skull Creek Shale in the Denver Basin, Colorado: The Mountain Geologist, v. 33, p. 85-93.

Eicher, D.L., 1962, Biostratigraphy of the Thermopolis, Muddy, and Shell Creek formations: Wyoming Geological Association, Seventeenth Annual Field Conference, Guidebook, p. 72-93.

Eicher, D.L., 1965, Foraminifera and biostratigaphy of the Graneros Shale: Journal of Paleontology, v. 39, p. 875-909.

Eicher, D.L., 1967, Foraminifera from Belle Fourche Shale and equivalents, Wyoming and Montana: Journal of Paleontology, v. 41, p. 167-188.

Hamilton, V.J., 1989, Stratigraphic sequences and hydrostratigraphic units in Lower Cretaceous strata, Kansas [M.S. thesis]: Golden, Colorado School of Mines, 165 p.

Hamilton, V.J., 1994, Sequence stratigraphy of Cretaceous Albian and Cenomanian strata of Kansas; *in* Shurr, G.W., Ludvigson, G.A., and Hammond, R.H., eds., Perspectives on the Eastern Margin of the Cretaceous Western Interior Basin: Boulder, Geological Society of America Special Paper 287, p. 79-96.

Hancock, J. M., Kennedy, W. J., and Cobban, W. A., 1993, A correlation of the Upper Albian to basal Coniacian sequences of northwest Europe, Texas and the United States Western Interior; *in* W. G. E. Caldwell and E.G. Kauffman, eds., Evolution of the Western Interior Basin: St. John's, Geological Association of Canada Special Paper 39, p. 453-476.

Hardenbol, J., Thierry, J., Farley, M.B., Jacquin, T., DeGraciansky, P.-C., and Vail, P.R., 1998, Mesozoic-Cenozoic sequence chronostratgraphic framework or European basins; *in* De Graciansky, P.-C., Hardenbol, J., Jacquin, T., and Vail, P.R., eds., Mesozoic and Cenozoic Sequence Stratigraphy of European Basins: Tulsa, SEPM (Society for Sedimentary Geology) Special Publication 60, p. 3-13.

Harland, W. B., Armstrong, R. L., Cox, A. V., Craig, L. E., Smith, A. G., and Smith, D. G., 1990, A geologic time scale 1989: Cambridge, Cambridge University Press, 263 p.

Holbrook, J.M., 1992, Developmental and sequence-stratigraphic analysis of Lower Cretaceous sedimentary systems in the southern part of the United States Western Interior: Interrelationships between eustasy, local tectonics, and depositional environments [Ph.D. dissertation]: Bloomington, Indiana University, 271 p.

Holbrook, J.M., 1996, Complex fluvial response to low gradients at maximum regression: A genetic link between smooth sequence-boundary morphology and architecture of overlying sheet sandstone: Journal of Sedimentary Research, v. 66, p. 713-722.

Holbrook, J.M. and Wright Dunbar, R., 1992, Depositional history of Lower Cretaceous strata in northeastern New Mexico: Implications for regional tectonics and depositional sequences: Geological Society of America Bulletin, v. 104, p. 802-813.

Holbrook, J.M., and White, D.C., 1998, Evidence for subtle uplift from lithofacies distribution and sequence architecture by low-relief intraplate uplift: Examples from the Lower Cretaceous of northeastern New Mexico; *in* Shanley, K.W., and McCabe, P.J., eds. Relative Role of Eustacy, Climate, and Tectonism in Continental Rocks: Tulsa, SEPM Special Publication 59, p. 123-132.

Kauffman, E.G., 1977, Geological and biological overview: Western Interior Cretaceous Basin: The Mountain Geologist, v. 14, p. 75-99.

Kauffman, E.g., 1984, Paleobiogeography and evolutionary response dynamic to the Cretaceous Western Interior Seaway of North America; *in* G. E.g. Westermann, ed., Jurassic-Cretaceous Biochronology and Paleogeography of North America: St. John's, Geological Association of Canada Special Paper 27, p. 273-306.

Kauffman, E.J., Sageman, B.B., Kirkland, J.I., Elder, W.P., Harries, P.J., Villamil, T., 1993, Molluscan biostratigraphy of the Cretaceous Western Interior Basin, North America; *in* W. G. E. Caldwell and E.g. Kauffman, eds., Evolution of the Western Interior Basin: St. John's, Geological Association of Canada, Special Paper 39, p. 397-434.

Kennedy, W.J., Gale, A.S., Hancock, J.M., Crampton, J.S., and Cobban, W.A., 1999, Ammonites and inoceramid bivalves from close to the Middle-Upper Albian boundary around Fort Worth, Texas: Journal of Paleontology, v. 73, p. 1101-1125.

Kues, B.S., 1986, Paleontology and correlation of a Lower Cretaceous (Albian) outlier in Roosevelt County, New Mexico: New Mexico Geology, v. 8, p. 88-94.

Kues, B.S., 1997, New bivalve taxa from the Tucumcari Formation (Cretaceous, Albian), New Mexico, and the biostratigraphic significance of the basal Tucumcari fauna: Journal of Paleontology, v. 71, p. 820-839.

Kues, B.S., and Lucas, S.G., 1987, Cretaceous stratigraphy and paleontology in the Dry Cimarron Valley, New Mexico, Colorado and Oklahoma: New Mexico Geological Society, 38th Field Conference, Guidebook, p. 167-198.

Kues, B.S., and Lucas, S.G., 1993, Stratigraphy, paleontology and correlation of Lower Cretaceous exposures in southeastern New Mexico: New Mexico Geological Society, 44th Field Conference, Guidebook, p. 245-260.

Kues, B.S., Lucas, S.G., Kietzke, K., and Mateer, N.J., 1985, Synopsis of Tucumcari Shale, Mesa Rica Sandstone and Pajarito Shale paleontology, Cretaceous of east-central New Mexico: New Mexico Geological Society, 36th Field Conference, Guidebook, p. 261-281.

Long, C.S., Jr., 1966, Basal Cretaceous strata, southern Colorado [Ph.D. dissertation]: Boulder, University of Colorado, 479 p.

Lovejoy, E.M.P., 1976, Geology of Cerro de Cristo Rey Uplift, Chihuahua and New Mexico: New Mexico Institute of Mining and Technology Memoir 31, 84 p.

Lucas, S.G., 1990, Type and reference sections of the Romeroville Sandstone (Dakota Group) Cretaceous of northeastern New Mexico: New Mexico Geological Society, 41st Field Conference, Guidebook, p. 323-326.

Lucas, S.G., 1991, Cretaceous Dakota Group outlier, Sacramento Mountains, Otero County, New Mexico: New Mexico Geological Society, 42nd Field Conference, Guidebook, p. 261-264.

Lucas, S.G., Anderson, O.J., and Estep, J.W., 1998, Stratigraphy and correlation of middle Cretaceous rocks (Albian-Cenomanian) from the Colorado Plateau to the southern High Plains, north-central New Mexico; *in* Lucas, S.G., Kirkland, J.I., and Estep, J.W., eds., Lower and Middle Cretaceous Terrestrial Ecosystems: New Mexico Museum of Natural History and Science Bulletin 14, p. 57-66.

Lucas, S.G., and Estep, J.W., 1998, The late Albian ammonite *Engonoceras* from southwest New Mexico: New Mexico Geology, v. 20, p. 78-82.

MacKenzie, D.B., 1965, Depositional environments of Muddy Sandstone, western Denver Basin, Colorado: American Association of Petroleum Geologists Bulletin, v. 49, p. 186-206.

Mateer, N. J., 1987, The Dakota Group of northeastern New Mexico and southern Colorado: New Mexico Geological Society, 38th Annual Field Conference, Guidebook, p. 223-236.

McNeil, D.H., and Caldwell, W.G.E., 1981, Cretaceous rocks and their Foraminifera in the Manitoba Escarpment: Waterloo, Geological Association of Canada, Special Paper 21, 440 p.

Nichols, D.J., Jacobson, S.R., and Tschudy, R.H., 1982, Cretaceous palynomorph biozones for the central and northern Rocky Mountain region of the United States; *in* Powers, R.B., ed., Geologic Studies of the Cordilleran Thrust Belt: Denver, Rocky Mountain Association of Geologists, p. 721-733.

Obradovich, J. D., 1993, A Cretaceous time scale; *in* W. G. E. Caldwell and E.g. Kauffman, eds., Evolution of the Western Interior Basin: St. John's, Geological Association of Canada Special Paper 39, p. 379-396.

Reeside, J.B., 1923, The fauna of the so-called Dakota Formation of northern central Colorado and its equivalent in southeastern Wyoming: Washington, D.C., U.S. Geological Survey Professional Paper 131-H, p. 199-205.

Rojas, I. de, 1980, Stratigraphy of the Mowry Shale (Cretaceous), western Denver Basin, Colorado [M.S. thesis]: Golden, Colorado School of Mines, 134 p.

Scott, R.W., 1970, Stratigraphy and sedimentary environments of Lower Cretaceous rocks, Southern Western Interior: American Association of Petroleum Geologists Bulletin, v. 54, p, 1225-1244.

Scott, R.W., Fouke, B.W., and Nederbragt, A.J., 1996, Defining global Cretaceous events by quantitative stratigraphic techniques: The Paleontological Society, Special Publication 8, Sixth North American Paleontological Convention, Abst., p. 350.

Scott, R. W., Franks, P. C., Stein, J. A., Bergen, J. A., Evetts, M. J., 1994, Graphic correlation tests the synchronous Mid-Cretaceous depositional cycles: Western Interior and Gulf Coast; *in* J. C. Dolson, ed., Unconformity-related hydrocarbons in sedimentary sequences: Denver, Rocky Mountain Association Geologists, p. 89-98.

Scott, R. W., Franks, P. C., Evetts, M. J., Bergen, J. A., and Stein, J. A., 1998, Timing of Mid-Cretaceous relative sea level changes in the Western Interior: Amoco No. 1 Bounds Core; *in* W. A. Dean and M. A. Arthur, ed., Stratigraphy and paleoenvironments of the Cretaceous Western Interior Seaway, USA: Tulsa, SEPM Concepts in Sedimentology and Paleontology 6, p. 11-34.

Scott, R.W., Schlager,W., Fouke, B., Nederbragt, S.A., 2000, Are mid-Cretaceous eustatic events recorded in Middle East carbonate platforms?; *in* A.S. Alsharhan and R. W. Scott, eds., Middle East models of Jurassic/Cretaceous carbonate systems: Tulsa, SEPM (Society for Sedimentary Geology) Special Publication 69, p. 73-84.

Scott, R.W. and Stein, J.A., 1995, Age of the Albian/Cenomanian boundary: a graphic correlation solution: Institut royal des Sciences Naturelle de Belgique, Brussels, Second International Symposium on Cretaceous Stage Boundaries, Abstract Volume, p. 112.

Strain, W.S., 1976, Appendix 2—New formation names in the Cretaceous at Cerro de Cristo Rey, Doña Ana County, New Mexico: New Mexico Institute of Mining and Technology Memoir 31, p. 77-82.

Taylor, A.M., 1974, Stratigraphy and depositional environments of Upper Jurassic and Cretaceous rocks in Bent, Las Animas, and Otero counties, Colorado [Ph.D. dissertation]: Golden, Colorado School of Mines, 302 p.

Waage, K.M., 1955, Dakota Group in northern Front Range Foothills, Colorado: U.S. Geological Survey Professional Paper 274-B, p. 15-51.

Waage, K.M., 1961, Stratigraphy and refractory clayrocks of the Dakota Group along the northern Front Range, Colorado: U.S. Geological Survey Bulletin 1102, 150 p.

Weimer, R.J., 1984, Relation of unconformities, tectonics and sea level changes, Cretaceous of Western Interior, United States of America; *in* Schlee, J.S., ed., Interregional Unconformities and Hydrocarbon Accumulation: Tulsa, American Association of Petroleum Geologists Memoir 36, p. 7-36.

Weimer, R.J., and Land, C.B., Jr., 1972, Field guide to Dakota Group (Cretaceous) stratigraphy, Golden-Morrison area, Colorado: The Mountain Geologist, v. 9, p. 241-267.

Young, K., 1986, Cretaceous, marine inundations of the San Marcos Platform, Texas: Cretaceous Research, v. 7, p. 117-140.

NEARSHORE FAUNA OF THE TUCUMCARI FORMATION (LOWER CRETACEOUS, ALBIAN), QUAY COUNTY, NEW MEXICO

BARRY S. KUES[1] AND SPENCER G. LUCAS[2]
[1]Department of Earth and Planetary Sciences, University of New Mexico, Albuquerque, NM 87131; and
[2]New Mexico Museum of Natural History, 1801 Mountain Road N.W., Albuquerque, NM 87104

Abstract.—The Lower Cretaceous Tucumcari Formation of Quay County, New Mexico, contains a diverse marine fauna of early late Albian age. In this paper, 49 taxa from relatively nearshore, sandy facies of the Tucumcari (predominantly a midshoreface environment) are briefly discussed and illustrated, many for the first time. The taxa include one foraminiferan, three coral, one annelid, one ctenostome bryozoan, 29 bivalve, seven gastropod, one scaphopod, one ammonite, three echinoid, and two fish species. Taxa from more offshore, open-bay environments, represented by dark gray mudstone and shale facies of the Tucumcari, are not considered here. The nearshore fauna is dominated by bivalves, which account for nearly 60% of the total diversity and more than 90% of the total abundance of specimens. Common to abundant bivalves include *Texigryphaea pitcheri, Neithea occidentalis, Plicatula incongrua, Peilinia levicostata, Scabrotrigonia emoryi,* and *Protocardia texana,* whereas the only other species approaching these bivalves in abundance at some localities is the gastropod *Turritella "seriatim-granulata"*. Rare taxa represent a significant percentage (39%) of the total number of species. Among these rare taxa are the first reported occurrences in the Tucumcari Formation of colonial corals, ctenostome bryozoa, scaphopods, regular and spatangoid echinoids, and shark and pycnodontid fish teeth.

INTRODUCTION

The Tucumcari Formation and the basal sandstone of the overlying Mesa Rica Sandstone contain a marine fauna of early late Albian age, coeval with that of the lower Washita Group of Texas and Oklahoma. These formations are well exposed and locally abundantly fossiliferous in east-central New Mexico, especially in Quay and eastern Guadalupe Counties. Although the Tucumcari fauna has not been studied monographically, it is reasonably well known taxonomically (e.g., Stanton, 1947; Scott, 1974; Kues et al., 1985) and from the perspectives of community structure and biostratigraphic and paleobiogeographic relationships (e.g., Scott, 1970b, 1974, 1975, 1986a).

Kues et al. (1985) provided a summary of Tucumcari paleontology, including a catalog of all species reported from the formation at that time, and illustrations of some common taxa. Since that time, additional study of some elements of the fauna and of the stratigraphy and regional correlation of the Tucumcari Formation have been accomplished. Scott (1986a), in considering the paleobiology of late Albian faunas in the Texas and southern Western Interior regions, included the Tucumcari fauna in a western endemic center of the Southern Western Interior faunal province, noting significant differences of this fauna from those of the Caribbean province to the south and east, the Boreal province to the north, and an eastern endemic center (Kiowa Formation of Kansas and western Oklahoma) to the northeast. Later, Scott (1986b) studied protocardiid bivalve taxa, including species present in the Tucumcari, a paper similar to his earlier study of Albian carditid bivalves (Scott, 1977), which was inadvertently omitted in the summary of Kues et al. (1985). Cobban (1985) and Lucas and Hunt (2000) reported on specimens of the important index fossil *Mortoniceras equidistans* from the basal Mesa Rica Sandstone, and Kietzke (1985) illustrated numerous species of foraminiferans and ostracods from the Tucumcari Formation at its type locality on Pyramid Mountain. Kues (1989) studied the most abundant fossil in the Tucumcari Formation, the oyster *Texigryphaea tucumcarii* (Marcou), and concluded that it is synonymous with *T. pitcher* (Morton), based on extensive collections of both species from New Mexico, Texas, and Oklahoma. Later, Kues (1997) studied the fauna of the basal Tucumcari at the San Jon hill locality, describing new species of *Plicatula* and *Gyrostrea*, and establishing a new genus and species, *Peilinia levicostata,* for the common arcuate ostreid that previously had been called *Lopha quadriplicata* (Shumard). Kues (1986) and Kues and Lucas (1993) reported on the faunas of small outliers of the Tucumcari Formation in Roosevelt and northern Lea Counties, south of the main Tucumcari outcrop belt, and Kues and Lucas (1987) studied the stratigraphy and paleontology of the Glencairn Formation, the unit correlative with the Tucumcari to the north. Kauffman et al. (1993, fig. 6), in discussing the biostratigraphy of Cretaceous strata of the Western Interior, provided range zones for some species occurring in the Tucumcari Formation.

Since 1985, additions to our knowledge of Tucumcari stratigraphy have also been published. Lucas and Kisucky (1988) designated the exposure on Pyramid Mountain (first described by Marcou, 1855, 1858, and relocated by Kues and Kietzke, 1985), as the type section of the Tucumcari Formation, described the San Jon hill exposure as a reference section, and designated type sections for the overlying Mesa Rica and Pajarito formations as well. Holbrook and Dunbar (1992) discussed the sedimentology and sedimentary tectonics of the Tucumcari and related formations in the Tucumcari and Dalhart basins. Dyman et al. (1994) portrayed the Glencairn (and by implication, the Tucumcari) Formation as an Albian unit separated from the overlying Mesa Rica Sandstone, which they believed to be of Cenomanian age, by a 5-million-year unconformity. Kues (1997) discussed the biostratigraphy of the Tucumcari and correlative formations and considered the Tucumcari, Mesa Rica and Pajarito formations to be Albian, with the Albian-Cenomanian boundary within the overlying Romeroville Sandstone (fig. 2) or possibly within the Pajarito Formation (p. 834). He also suggested, based on lithologic similarity, that thin outliers of Albian strata around numerous playa lakes in west-central Texas, called Kiamichi and Duck Creek by Brand (1953), might well be considered part of the Tucumcari

Formation, in some cases representing deposition preceding the earliest Tucumcari strata exposed in Quay and Guadalupe Counties. Lucas and Estep (2000) briefly summarized Tucumcari stratigraphy and age in an overview of Lower Cretaceous stratigraphy throughout New Mexico. Their view of the late Albian sequence in east-central New Mexico as including, in ascending order, the Tucumcari, Mesa Rica, and Pajarito formations, with the Pajarito overlain unconformably by the lower Cenomanian Romeroville Sandstone, accords best with the biostratigraphic and stratigraphic evidence, and is accepted here.

Scott (1974), in a pioneering study of the Tucumcari Formation, recognized several lithofacies and corresponding substrates within the Tucumcari Formation, including dark gray mudstone, gray-brown to dark gray clayey-silty sandstone, yellow-brown to gray brown thin-bedded sandstone, and shell conglomerate. A fifth lithofacies, yellow-brown to brown-gray bioturbated sandstone, said to occur in the uppermost Tucumcari, is included by us in the basal Mesa Rica Sandstone on lithological grounds. Scott (1974) also defined several faunal associations that were utilized to reconstruct open-bay, lower, mid-, and upper shoreface, and biostromal communities. Our observations of Tucumcari stratigraphy and faunas accord with Scott's conclusions.

In this paper, we summarize the faunal constituents of the more sandy, shallow marine Tucumcari facies — predominantly the biostromal *Texigryphaea-Lopha*, and mid-shoreface *Scabrotrigonia-Turritella* communities of Scott (1974) — based on collections made since 1984. Sampling and study of the more seaward communities, especially those of the open-bay, dark-gray mudstone and lower shoreface, clayey to silty sandstone or mudstone environments, is incomplete, and the taxa restricted to those environments are not discussed here. Illustrations and brief descriptions are provided of the shoreward marine taxa, many of which have not previously been illustrated from the Tucumcari Formation, and some of which have been first identified since the review of Kues et al. (1985). Specimens mentioned in the text are reposited in the University of New Mexico Department of Earth and Planetary Sciences paleontology collections (UNM), Albuquerque, the collections of the New Mexico Museum of Natural History and Science (NMMNH), Albuquerque, and in the Mesalands Dinosaur Museum (MDM), Tucumcari, NM.

LOCALITIES

For a complete list of Tucumcari localities from which fossils have been collected, see Kues et al. (1985). Specimens illustrated in this paper are from the following six localities, all in Quay County, New Mexico (Fig. 1).

1. Pyramid Mountain: SE¼SW¼ sec. 19, T9N, R29E.
2. Apache Canyon: exposures along NM-278, NE¼ sec. 8, T8N, R33E.
3. San Jon hill: roadcut along NM-469, 14.6 km south of San Jon, SE¼ sec. 20, T9N, R34E.
4. Mesa Quemado outlier: small outcrop a little SE of Mesa Quemado, on S side of unpaved road (Quay County road QR-54), 7.0 km W of NM-209, SE¼SE¼ sec. 1, T9N, R29E.
5. West of Tucumcari: roadcut on U.S.-66, 14.1 km W of Tucumcari, SE¼ sec. 4, T10N, R29E.
6. Palomas locality: exposures ENE of Palomas railroad siding, west of Tucumcari, N½ sec. 32, T11N, R29E.

STRATIGRAPHY

The principal outcrops of the Tucumcari Formation are in Quay and eastern Guadalupe Counties, New Mexico (Fig. 1). Here, the Tucumcari Formation is up to about 25 m thick and consists dominantly of olive-gray shale and mudstone. Minor lithotypes are yellowish-brown and yellowish-orange locally bioturbated sandstone and siltstone, and dark yellowish-brown fossiliferous limestone. The Tucumcari Formation disconformably overlies the Upper Jurassic Morrison Formation (e.g., Lucas and Kisucky, 1988; Holbrook and Dunbar, 1992) and is overlain conformably or with minor disconformity by the Lower Cretaceous Mesa Rica Sandstone (Fig. 1).

Marcou (1889) first used the term "Tucumcari beds" for the fossiliferous strata he considered Jurassic (see Kues, 1985, for a summary of the development of lithostratigraphic terminology in this region). Marcou's main reference section for his "Tucumcari beds" was Pyramid Mountain, southwest of Tucumcari (sec. 19, T9N, R29E). This is the type section for the Tucumcari Formation, where it is about 9 m thick (Lucas and Kisucky, 1988; Fig. 1). It is also the type locality of "*Gryphaea dilatata* var. *tucumcarii*" (now *Texigryphaea pitcheri*), which Marcou (1855, 1858) described from fossils collected at Pyramid Mountain, probably about 1 m above the base of the Tucumcari Formation.

A much thicker (about 25 m) section of the Tucumcari Formation about 14.6 km south of San Jon, at "San Jon hill" (sec. 20, T9N, R34E) is probably the most intensively studied section of the formation, and was designated a reference section by Lucas and Kisucky (1988; Fig. 1). This section yields nearshore marine megafossils mainly from four distinct stratigraphic levels (Fig. 1): 1) the basal 1.2 m of the formation contain a distinctive bivalve assemblage including *Ceratostreon texanum, Gyrostrea hinchada,* and *Plicatula quayensis* (Kues, 1997); 2) about 3 to 4 m above the base, a shale and ledge-forming limestone and sandstone interval (best exposed towards the top of the southern roadcut at San Jon hill) contains a *Texigryphaea* and *Peilinia* dominated assemblage; 3) a pair of thin, light-gray, fine-grained sandstone intervals a little above the middle of the section, within the thick gray shale/mudstone interval comprising most of the Tucumcari Formation, is rich in *Texigryphaea* and other bivalves; and 4) the basal part of the overlying Mesa Rica Sandstone also yields marine, mostly bivalve, taxa that are generally identical to those of the Tucumcari Formation (Kues, this volume). This last interval is disconformably overlain by Mesa Rica channel deposits, and the contact is the SB3 erosion surface of Holbrook and Dunbar (1992).

The distinctive assemblage at the base of the formation is older than the basal Tucumcari in other areas of its outcrop belt in Quay County (Kues, 1997). Its preservation here reflects either a slightly lower paleotopography or slight differential subsidence in this area as the advancing Tucumcari sea spread into this area from the south and east. Albian sediments older than are present

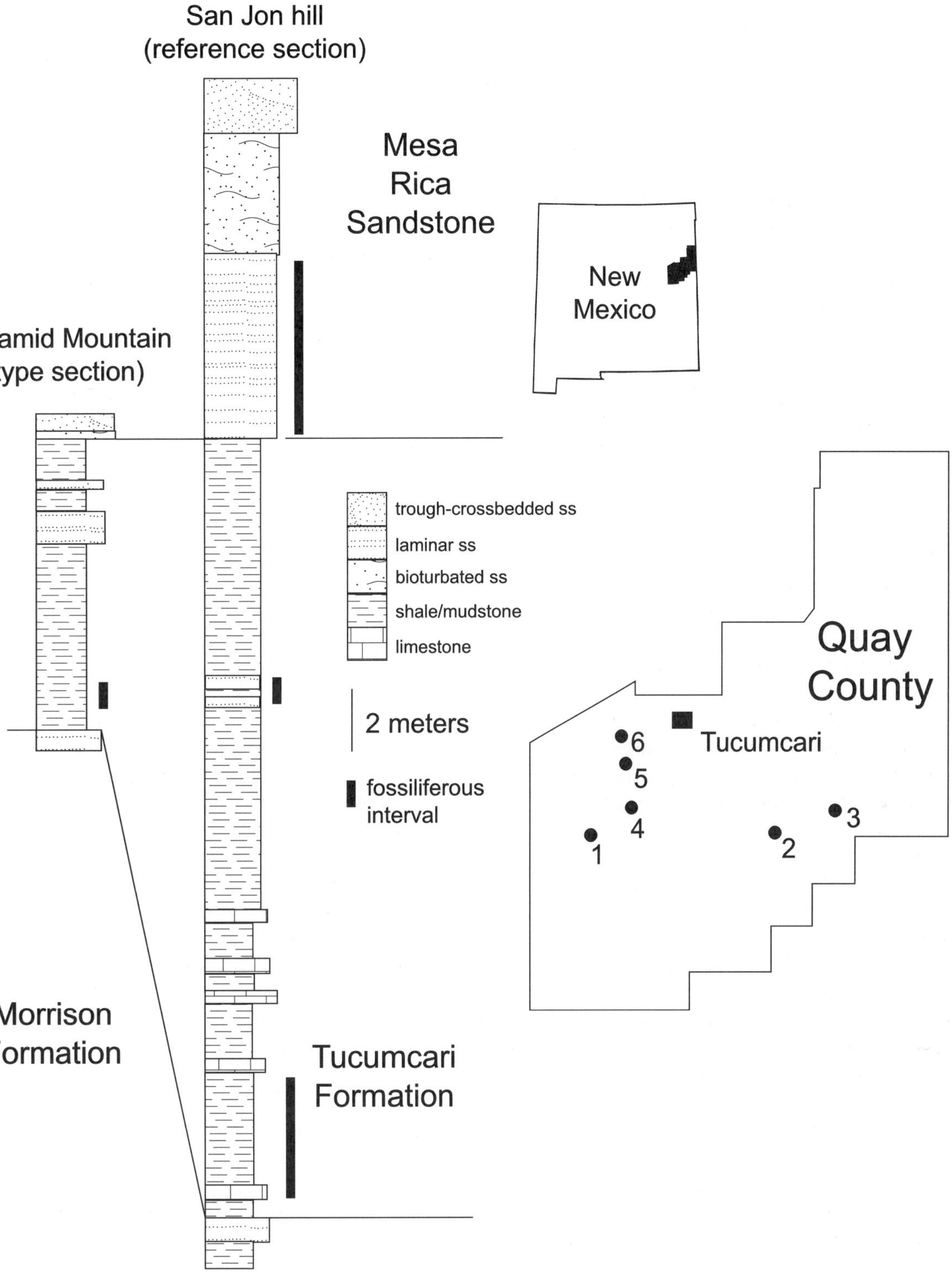

FIGURE 1. Map showing Quay County, localities of the Tucumcari Formation yielding fossils discussed in this report, and the type and reference sections of the Tucumcari Formation. Localities: 1, Pyramid Mountain; 2, Apache Canyon; 3, San Jon hill; 4, Mesa Quemado outlier; 5, roadcut west of Tucumcari; 6, Palomas locality. See text for exact coordinates of localities.

in the Quay County Tucumcari exposures are exposed in outliers to the south (e.g., in Roosevelt County; Kues, 1986) and in several restricted outcrops around playa lakes in west-central Texas, to the east (Brand, 1953; Kues, 1997).

THE NEARSHORE TUCUMCARI FAUNA

The nearshore, sandy facies of the Tucumcari Formation discussed here includes a total of 49 identified taxa: one large foraminiferan, three coral, one annelid worm, one bryozoan, 29 bivalve, seven gastropod, one ammonite, three echinoid, and two fish species (Table 1). Bivalves dominate this fauna, both in diversity and in abundance. As is typical of diverse marine assemblages, a few taxa are common to abundant, and a significant number are rare (defined as being limited to one or two specimens in our collections). The dominant elements of the nearshore Tucumcari fauna are the bivalves *Texigryphaea pitcheri, Neithea occidentalis, Plicatula incongrua, Peilinia levicostata, Scabrotrigonia emoryi, Protocardia texana,* and the gastropod *Turritella "seriatimgranulata"*. These taxa are characteristic of the *Scabrotrigonia-Turritella* and *Texigryphaea-Lopha* associations or communities discussed by Scott (1974). Rare taxa comprise 19 of the 49 species (39%) identified in this study. Among these rare taxa are the first reported occurrences in the Tucumcari Formation of a colonial coral, ctenostome bryozoan, scaphopod, regular and spatangoid echinoids, and shark and pycnodontid fish teeth.

Quality of preservation varies considerably among the nearshore Tucumcari faunal elements. The valves of oysters, the epizoans on them, and other pteriomorph bivalves (e.g., *Neithea, Plicatula, Lima*) are generally preserved well, although many are fragmented. Preservation of other bivalves ranges from weathered valves to steinkerns and molds that are rarely complete. Gastropod remains are typically weathered shells or steinkerns. Fragmentary steinkerns of some bivalves and gastropods are unidentifiable, suggesting that the total number of species in this nearshore fauna is probably greater than the number of species identified here, and that further collecting will probably yield additional taxa. Scott (1974) observed that the dark-gray mudstone and clayey-silty sandstone facies, representing offshore, open-bay and lower shoreface environments, respectively, have benthic communities that are taxonomically different from those of the nearer-shore, sandy lithologies. The lithofacies representing more offshore environments compose much of the total thickness of the Tucumcari Formation at most localities. We have not yet sampled these facies extensively, and description of the taxa preserved within them remains for the future. Similarly, detailed study of the stratigraphy, facies, and paleontology of most portions of the Tucumcari outcrop belt have not yet been done. Such studies would add much to our present knowledge of the taxonomic composition, geographic and stratigraphic distribution, and paleoecology of the Tucumcari faunas.

Foraminifera

Kietzke (1985) reported a richly diverse (66 taxa) foraminiferan fauna from the lower Tucumcari Formation at Pyramid Mountain, derived from offshore dark gray shale facies. Here, we document *Cribratina texana*, probably the only nonattached foraminifer large enough to be noted and collected routinely from outcrops, and which was a probable inhabitant of nearshore, sandy facies of the Tucumcari sea.

Cribratina texana (Conrad)

The test of these unusually large foraminiferans (Figs. 2A, B) is uniserial, consisting of a series of six or more convex chambers of gradually increasing diameter, and often attaining a length of more

TABLE 1. Fauna of the nearshore sandy facies of the Tucumcari Formation, Quay County, New Mexico, discussed in text. Abundance estimates are subjective, based on number of specimens and number of localities yielding specimens of each species. A = abundant; C = common; MC = moderately common; UC = uncommon; R = rare.

Foraminifera
Cribratina texana (Conrad) (MC)
Cnidaria
Rennensismilia stainbrooki (Wells) (MC)
Coelosmilia texana (Conrad) (UC)
Montastraea? sp. (R)
Annelida
Serpula (Cycloserpula) cragini Twenhofel (MC)
Bryozoa
Vinella sp. (R)
Bivalvia
Pinna guadalupae Böse (UC)
Phelopteria aff. *P. salinaensis* (White) (R)
"Inoceramus" bellvuensis Reeside? (R)
Neithea texana (Roemer) (MC)
Neithea occidentalis Conrad (C)
Plicatula incongrua Conrad (C)
Plicatula quayensis Kues (UC)
Lima wacoensis Roemer (UC)
Lima utahensis Stanton (UC)
Lima n. sp (MC)
Lima sp. (UC)
Texigryphaea pitcheri (Morton) (A)
Ceratostreon texanum (Roemer) (UC)
Gyrostrea hinchada Kues (UC)
Nicaisolopha subovata (Shumard) (UC)
Peilinia levicostata Kues (C)
Scabrotrigonia emoryi (Conrad) (C)
Parvilucina? sp. (R)
Ludbrookia belviderensis (Cragin) (R)
Crassatellina oblonga Meek (R)
Pleuriocardia kansasense (Meek) (UC)
Protocardia texana Conrad (C)
Protocardia filosa (Conrad) (MC)
Linearia (Liothyris)? aff. *L. (L.)*? *kansasensis* Twenhofel (R)
"Isocardia" aff. *"I." slatana* Stephenson (R)
Aphrodina? sp. (R)
Cyprimeria cf. *C. washitaensis* Adkins (MC)
Flaventia belviderensis (Cragin) (MC)
Pholadomya sp. (R)
Gastropoda
Turritella "seriatimgranulata" Roemer (C)
Turritella n. sp. (UC)
Amberleya mudgeana (Meek) (MC)
Calliostoma cragini Stanton (UC)
Drepanochilus aff. *D. mudgeana* (White) (UC)
Euspira cragini (Stanton) (R)
Hillites newberryi (Cragin) (UC)
Scaphopoda
Dentalium (Dentalium) sp. (R)
Cephalopoda - - Ammonoidea
Mortoniceras equidistans (Cragin) (R)
Echinoidea
Regular echinoid spines (two types) (R)
Washitaster bravoensis (Böse) (R)
Vertebrata
aff. *Cretodus* sp. (R)
Coelodus sp. (R)

FIGURE 2. Protista, corals, annelids, and bryozoans from the Tucumcari Formation, Quay County, New Mexico. **Protista.** A-B. *Cribratina texana* (Conrad), two tests, NMMNH P-33,451A and P-33,451B, x7. **Corals.** C-F. *Rennensismilia stainbrooki* (Wells). C-D. top and side view (with attached oyster), UNM 13,115, x1.25; E. convex side view, UNM 13,117, x1.25; F. side view, with attached oyster, UNM 13,116, x1.25. G-H. *Coelosmilia texana* (Conrad). G. side view, UNM 13,114, x 1.25; H. side view, UNM 13,113, x1.25. I. *Montastrea*? sp., small corallum with several corallites, top view, UNM 13,136, x6. **Annelid worms.** J-K. *Serpula (Cycloserpula) cragini* Twenhofel. J, large specimen attached to fragment of *T. pitcheri* left valve, UNM 13,009, x1.65; K. small specimen, attached to surface of small *T. pitcheri* left valve, UNM 13,100. **Ctenostome bryozoan.** L, *Vinella* sp., NMMNH P-33, 452, x5.75. Localities: A, B, L, San Jon hill; C-K, Apache Canyon.

than 5 mm. *Cribratina texana* is a good late Albian index fossil with a wide distribution in New Mexican strata of that age, ranging from Caribbean province faunas of extreme southern New Mexico, through the western endemic center as far north as the Glencairn Formation of northeastern New Mexico. Kues and Lucas (1987, fig. 11K), Lucas et al. (1988, fig. 8C), Lucas (1991, fig. 2C), Kues and Lucas (1993, fig. 8W) and Lucas and Estep (1998, figs. 5A, B) have documented the species from other areas in the state.

Cnidaria

Numerous Albian coral species have been reported from the Caribbean province, in Texas (e.g., Wells, 1933) and south-central New Mexico (at Cerro de Cristo Rey; Turnšek et al., 1983), but only two species have previously been reported from the Tucumcari Formation. These are *Coelosmilia* [?*Desmophyllum*] *texanum* and *Platycyathus scottianus* (Kues et al., 1985), both solitary forms. We have not observed the latter species in our collecting, but the first species is present, together with two others not previously reported from the Tucumcari Formation.

Rennensismilia stainbrooki (Wells)

These solitary corals (Figs. 2C-F) have a broadly conical, straight to somewhat arcuate corallite, that is nearly circular to oval in cross section. A typical large specimen (Figs. 2C, D) is 33 mm high, with a diameter of the oval corallite measuring 32 mm by 20 mm. The septa are thin and closely spaced; exact counts were not possible on the specimens at hand because of poor preservation and matrix within the calices, but the number of septa observed or extrapolated on 10 specimens from the Apache Hill locality ranges from 74 to 100. The septa converge towards the center of the calice, but no columella was observed on any specimen, probably because they are obscured by matrix. The calicular margin is broad and rounded, and the septa cross the margin to become conspicuous, sharp, strongly nodose costae on the external surface of the corallite. A peculiarity of this species, noted by Wells (1933) on the type specimens and observed on many of the Tucumcari specimens as well, is that the costae on the convex side of an arcuate corallite extend from the calicular margin to the base (apex), whereas costae on the concave side extend downward towards the base but intersect the first group of costae, typically at angles of 30° to 45°, before reaching it.

Wells (1933) initially described this species from Kiamichi-equivalent exposures around several of the playa lakes not far east of the New Mexico-Texas border, in west-central Texas, where they occur with *Platycyathus scottianus* Wells and *Coelosmilia texana* (Conrad), associated with *Texigryphaea pitcheri*. It also occurs in the Smeltertown and Mesilla Valley Formations at Cerro de Cristo Rey (Turnsek et al., 1983). In the Tucumcari Formation, *Rennensismilia stainbrooki* is moderately common at a few localities in silty-sandy lithologies, but is rare to absent at most outcrops. Epizoans, especially serpulids and juvenile oysters, are present on the external surfaces of many specimens.

***Coelosmilia texana* (Conrad)**

The corallite of *Coelosmilia texana* (Figs. 2G, H) is relatively small, narrowly conical, typically gently arcuate, and subcircular in cross section. The largest specimen in our collections is 21 mm high, with a maximum diameter of 9 mm by 12 mm. Although poor preservation prevents accurate determination of septa number on most specimens, 48 septa are present on the best preserved specimen (Fig. 2G), which is the constant number characteristic of the species (Wells, 1933). The septa cross the calice border and become an equal number of fine, sharp costae on the external corallite surface. These costae tend to be smooth to obscurely nodular. This species is readily distinguished from *Rennensismilia stainbrooki*, with which it may co-occur in the Tucumcari Formation, by its more narrowly conical shape, smaller size, and much fewer septa and costae. In north Texas, it ranges from the Walnut to Kiamichi Formation and at Cerro de Cristo Rey it occurs in the Del Norte Formation (Turnšek et al., 1983).

***Montastraea?* sp.**

A single small, worn specimen of a colonial coral, the first reported from the Tucumcari Formation, was recovered from the Apache Canyon locality. The specimen is an elongate, knob-like mass 11.5 mm long, 7 mm wide, and 6 mm high, consisting of seven closely packed circular, cerioid corallites (Fig. 2I). The calices are 2.0 to 2.5 mm in diameter; because they are filled with matrix, the presence of a columella cannot be ascertained. The septa number 24, in two or possibly three cycles, and continue as relatively coarse costae across the swollen calice margins. The costae are generally confluent with those of adjacent corallites.

Wells (1933) described numerous species of colonial corals from the upper Albian of Texas. Of these, the Tucumcari specimen most closely resembles several species of *Orbicella* that range from the Edwards Limestone to the Main Street Formation, although it is not obviously conspecific with any of them. *Orbicella* was considered a synonym of *Montastraea* by Wells (1956). Turnsek et al. (1983) reported 21 species of corals from the Albian at Cerro de Cristo Rey, most of them solitary; none of the colonial species resemble the Tucumcari specimen. Only a single colonial coral is known from the Kiowa Formation of Kansas (*Actinastraea nidiformis* (Cragin); Scott, 1970a), but it differs from *Montastraea*? sp. in having much smaller, polygonal rather than circular corallites, and thin septa that do not extend across the corallite margins as costae.

Annelida

***Serpula (Cycloserpula) cragini* Twenhofel**

This serpulid worm (Figs. 2J, K) is characterized by cylindrical, unornamented tubes, occurring as twisted masses cemented to the shells of other organisms, most often the valves of *Texigryphaea pitcheri*. The tubes are typically 1.5 to 2.25 mm in diameter, display fine growth lines and local discontinuities probably reflecting temporary cessation of growth, and may expand or constrict slightly along their length. The walls of the tubes are thick, up to 0.5 mm in a tube 2.25 mm wide, and are composed of at least three shell layers. In these features, the Tucumcari specimens agree closely with *Serpula (Cycloserpula) cragini,* described initially from the Kiowa Formation of Kansas (Twenhofel, 1924). No comparable Albian species of *Serpula* have been described from the Texas region, but some Late Cretaceous species, such as the late Cenomanian *S. (C.) intrica* White, are quite similar to *S. (C.) cragini* (Kirkland, 1996).

Serpula (Cycloserpula) cragini is one of the most common epizoans of the Tucumcari fauna, and its tubes sometimes cover several cm^2 of the shell surface of a large host.

Bryozoa

In addition to rare, small colonies of membraniporoid cheilostome bryozoan colonies encrusting valves of *Texigryphaea*, the following ctenostome bryozoan also occurs in the Tucumcari Formation.

***Vinella* sp.**

Two examples of ctenostome bryozoan, one on a left valve, and the other on a right valve, of *Texigryphaea pitcheri*, were recovered from the Tucumcari Formation at the San Jon hill locality. The colonies consist of branching, stoloniferous tubes about 0.1 to 0.15 mm in diameter, forming an interconnected network on the surface of the valves. The stolons are compressed locally along their length, forming "segments", and display widely-spaced, round pores along their upper surface. The best-preserved colony (Fig. 2L) also includes a cluster of radiating stolons extending from a central node. These specimens fit the definition of *Vinella*, as diagnosed by Condra and Elias (1944). The type species of *Vinella* is of Ordovician age, and the range of the genus is Ordovician to Cretaceous (Bassler, 1953).

Bivalvia

***Pinna guadalupae* Böse**

Most specimens of *Pinna* from the Tucumcari Formation are small fragments, but a few nearly complete, albeit poorly preserved, specimens are as much as 170 mm long. The best preserved specimen (Figs. 3A, C) expands from the apex (anterior end) at a 30° to 35° angle, and has a nearly circular cross section near the apex, which becomes broadly fusiform with later growth. Each valve has a low median longitudinal carina. Ornamentation consists of approximately 15 strong, sharp, widely-spaced longitudinal costae on each valve, which are crossed by fine, very

FIGURE 3. **Bivalves** from the Tucumcari Formation, Quay County, New Mexico. A and C. *Pinna guadalupae* Böse, nearly complete right valve, x0.75 (A) and posterior cross-sectional view, x0.85 (B), UNM 13,062. B. *Phelopteria* aff. *P. salinaensis* (White), partial right valve, UNM 9,093, x0.85. D. *"Inoceramus" bellvuensis* Reeside?, partial left valve, MDM 205, x0.50. E-G. *Neithea occidentalis* Conrad. E. Small right valve, UNM 13,086, x2; F. Large right valve, UNM 13,084, x1.4; G. Right valve, UNM 13,085, x1.25. H-I. *Neithea texana* (Roemer). H. Laterally compressed right valve, UNM 13,088, x1.33; I. Laterally compressed right valve, UNM 13,089, x1.4. J-K. *Plicatula incongrua* Conrad. J. Small right valve, with attachment scar, UNM 13,096, x2; K. Right valve, UNM 13,097, x2. L-M. *Plicatula quayensis* Kues. L. right valve, holotype, UNM 11,987, x1.5; M. Right valve, paratype, UNM 11,988, x1.5. N. *Lima wacoensis* Roemer, incomplete left valve, UNM 13,090, x2. O. *Lima utahensis* Stanton, right valve, UNM 13,095, x2. P, *Lima* n. sp., right valve, UNM 13,094, x2. Localities: A, C, D, L, M, O, San Jon hill; B, Mesa Quemado locality; E-K, N, P, Apache Canyon locality.

closely-spaced growth lines. The longitudinal costae begin at the apex and extend the entire length of the valve.

Kues and Lucas (1987, p. 182) discussed the uncertainties in identifying Albian species of *Pinna* from the Texas-southern Western Interior region, especially in distinguishing *P. comancheana* (Cragin, 1894) from *P. guadalupae* (Böse, 1910). *Pinna comancheana* has been reported in numerous faunal lists from the Tucumcari Formation (see Kues et al., 1985), but Cragin's original description was vague, the species was not illustrated, and the location of the type specimen was not given. Scott (1970a, p. 54) reported that the type specimens of species established by Cragin (1894) are lost, so it may never be possible to ascertain in detail the characteristics of *P. comancheana*. The type specimen was collected from the Kiowa Formation of Kansas (Cragin also noted its presence in the Tucumcari Formation), yet the species was not collected by either of the two workers (Twenhofel, 1924; Scott, 1970a) who have monographed the Kiowa fauna. Adkins (1928, pl. 18, fig. 6) illustrated a poorly preserved specimen from an unknown location having much coarser and more closely-spaced longitudinal costae than is suggested by the original description of *P. comancheana*.

In contrast, *Pinna guadalupae* was well described and illustrated by Böse (1910), from Albian strata (Del Norte and Mesilla Valley formations) at Cerro de Cristo Rey, south-central New Mexico. The Tucumcari pinna specimens are conspecific with *P. guadalupae*. It is uncertain whether *P. guadalupae* is the same species as *P. comancheana*, but because the characteristics of *P. comancheana* are so poorly known, and the type and topotypic specimens are unavailable, we are referring the Tucumcari specimens to *P. guadalupae*. *Pinna comancheana* is probably best deemed a *nomen dubium*.

***Phelopteria* aff. *P. salinaensis* (White)**

Phelopteria aff. *P. salinaensis* was first documented from the Tucumcari Formation by Kues et al. (1985, fig. 9A). Unfortunately, no additional specimens have been collected that would allow a better determination of the morphology and identity of this large pterioid. This specimen (Fig. 3B), a right valve, is incomplete, but displays a straight hingeline about 55 mm long, a small beak extending slightly above the hingeline, a small, subtriangular anterior auricle, and a very wide, flaring posterior auricle that extends from the hingeline to the posteroventral margin of the valve. The main body of the value and umbo are only gently inflated. The specimen agrees in its relatively large size with *P. salinaensis*, originally described from the Kiowa Formation (Twenhofel, 1924; Scott, 1970a), but differs in having a smaller anterior and wider posterior auricle, and in having less inflated valves.

Phelopteria belviderensis, also from the Kiowa, is a large, poorly known species that may be synonymous with or a variant of *P. salinaensis* (Twenhofel, 1924; Scott, 1970a). The only extant illustration of *P. belviderensis* (Twenhofel, 1924, pl. 19, fig. 10) is of a poorly preserved right valve that is vaguely similar to that of the Tucumcari specimen in apparently having a short anterior auricle and broad posterior auricle, but the axial portion of the valve was said to be "quite convex", in contrast to the low convexity of the Tucumcari specimen.

***"Inoceramus" bellvuensis* Reeside?**

Specimens of *"Inoceramus"* are rare in the Tucumcari Formation; even small, fibrous-prismatic fragments of their distinctive shell are uncommon. The most complete specimen available to us (courtesy of A. Hunt, Mesalands Dinosaur Museum, Tucumcari, NM) is a steinkern with some shell adhering of a partial left valve (Fig. 3D), collected from San Jon hill. The valve is about 190 mm in length along the growth axis, but was even larger when complete, as the ventral margin of the valve is not preserved. It is moderately inflated, bears a relatively small beak that projects slightly beyond the hingeline, and is ornamented with generally strong but unevenly spaced comarginal folds that vary considerably in size. Because the anterior part of the valve is missing, its exact shape cannot be observed, but the hingeline is relatively long and the posterior region broad. The angle between the hingeline and the principal valve axis changed with growth, beginning at nearly 90° and curving gradually to produce a final angle of about 60°.

Scott (1970a) noted the difficulties in distinguishing *"I." comancheanus* Cragin from *"I." bellvuensis* Reeside, the two species of *"Inoceramus"* known from upper Albian strata of the southern Western Interior. From published descriptions, *"I." bellvuensis* has a larger, broader and more subquadrate shell, less convex valves, a larger angle between the hingeline and anterior margin, and narrower, more widely spaced comarginal folds. However, Reeside's (1923) descriptions and illustrations of both species suggest significant overlap in the distinguishing characters, especially among smaller specimens, but no large assemblages of either species have been reported that would allow biometric analysis of their morphology and the degree of overlap in various features.

Positive assignment of the large Tucumcari specimen is hindered by its incomplete state. Valve convexity and ornamentation seem closer to that of *"I." comancheanus,* but extrapolated shell shape, low obliquity and very large size are more characteristic of *"I." bellvuensis,* to which we questionably assign this specimen. It is quite different from the inoceramids (species and morphs of *Actinoceramus*) reported recently from the Comanche Peak and Kiamichi formations of eastern Texas (Kennedy et al., 1999).

***Neithea texana* (Roemer)**

Neithea texana (Figs. 3H, I) is common in the Tucumcari Formation and represents a large species, attaining a valve height of 40 mm or more, although complete valves are rarely found. The moderately convex right valve is ornamented with simple, strong, high, rounded radial ribs that are of approximately equal size and wider than the intervening interspaces, which have the appearance of deep troughs between the ribs. Early ribs are fine, sharp, about of equal size, and display few bifurcations; splitting of ribs is restricted to early growth. The left valves are flat to very gently convex and bear simple, strong, relatively narrow ribs. The auricles are seldom preserved, but are relatively small and lack radial ornamentation. Sharp, closely-spaced comarginal growth lines are present on the valve body and auricles.

Stanton (1947) noted that some specimens have six slightly enlarged primary ribs on the right valve, with two slightly smaller secondary ribs between each primary pair, but this variation was not observed on the Tucumcari specimens.

***Neithea occidentalis* Conrad**

Neithea occidentalis (Figs. 3E-G) is more abundant in the Tucumcari Formation than *N. texana*. It is easily distinguished from that species by its generally smaller valves and especially by its distinctive ornamentation. Six primary ribs on the right valves are larger than the secondary ribs between each pair of primaries, and all of the ribs tend to be more sharply rounded than on *N. texana*. In addition, much finer, narrow, radiating lirae are present along one or both sides of each primary rib. These develop early in growth by bifurcation of the primary ribs and are never present on *N. texana*. Small specimens of *N. occidentalis* display a delicate array of fine ribs, with the primary ribs being more conspicuously enlarged relative to the secondary ribs than is the case of fully grown specimens.

***Plicatula incongrua* Conrad**

Plicatula incongrua (Figs. 3J, K) is a small (typically 10-12 mm high), strongly costate bivalve with subtriangular to suboval valves. The right valve is moderately convex and on some specimens bears an attachment scar near the beak, whereas the left valve is nearly flat. Both valves bear eight to 12 strong, closely-spaced radiating ribs that are nodose and may bear small hyote spines. *Plicatula incongrua* was first described by Conrad (1857) from upper Albian strata at Cerro de Cristo Rey, and Kues (1997) discussed the possible synonymy of *P. incongrua* and *P. subgurgitis*, described from the same or nearly the same locality and stratum by Böse (1910). Tucumcari specimens of *P. incongrua* are variable, and some display features supposedly characteristic of *P. subgurgitis*, such as a strong ventral bend in the right valve.

***Plicatula quayensis* Kues**

This species (Figs. 3L, M) was described by Kues (1997) from the basal Tucumcari Formation at the San Jon road cut locality. It differs from the far more common and widespread *P. incongrua* in being much larger (attaining a height of 20+ mm), having a subcircular rather than subtriangular valve outline, and in possessing as many as 30 radiating ribs on the valves, which tend to be more irregular and spinose than the ribs of *P. incongrua*.

***Lima wacoensis* Roemer**

Lima wacoensis (Fig. 3N) is distinguished by its strongly oblique, nearly subrectangular valves, having sharp, widely-spaced radiating ribs. The angle between the short hingeline and the principal valve axis is typically low, between 30° and 35°, and the auricles are very small. A typical specimen is about 18 mm long and 13 mm high. Stanton (1947) provided a detailed description of this species, and the Tucumcari specimens closely resemble his specimens. The species ranges widely through the Fredericksburg and Washita Groups of the Texas region, and was previously reported from the Tucumcari Formation by Stanton (1947). However, it is much less common than other species of *Lima* in the Tucumcari.

***Lima utahensis* Stanton**

Scott (1974, 1986a) first reported *Lima utahensis* from the Tucumcari Formation, an unusual occurrence of a species first described from the Cenomanian of the Western Interior (Stanton, 1893) appearing in the Albian of New Mexico. Earlier, White (1877) had reported Cenomanian specimens of this species as *L. wacoensis*. Most of our specimens are fragmentary, but a complete right valve (Fig. 3O) from the San Jon roadcut locality displays the relatively erect proportions and fine radiating ribs characteristic of *L. utahensis*.

***Lima* n. sp.**

The most abundant species of *Lima* in the Tucumcari Formation (Fig. 3P) is a relatively small (height to about 20 mm) form having erect valves, with the principal axis of growth diverging from the narrow hingeline at an angle of 50°+. The valves are more inflated than those of *L. wacoensis* and *L. utahensis,* and possess coarser, more closely spaced radial ribs. These are strong, sharp crested, number 25-30 on a valve, and are broadly triangular in cross section, creating a finely undulating commissure. Some specimens possess a fine radial thread at the bottom of interspaces between pairs of major ribs. Fine, closely-spaced growth lines cross the ribs and locally produce small nodes at the crest of the ribs. This species is about the same size as *L. wacoensis* and *L. utahensis*, and other Albian species from the region (e.g., from Cerro de Cristo Rey, see Böse, 1910), but differs sufficiently in valve convexity and coarseness of ribbing to suggest that it is an unnamed species.

***Lima* sp.**

A few molds and one nearly complete steinkern (Fig. 4A) of an articulated specimen, indicate the presence of an unusually large species of *Lima* in the Tucumcari Formation. The best specimen has fairly erect valves (50°-55°), and has a height of 58 mm and length of 46 mm. The radiating ribs are relatively fine, and are widely separated by interspaces two or three times the width of a rib. A total of about 26 ribs occur on the valves. This taxon is considerably larger than typical specimens of *L. wacoensis* and *L. utahensis.* Stanton (1947) mentioned that some Washita specimens of his variety *L. wacoensis quadrangularis* attain a length of 56 mm, but the valves of that form are strongly oblique and subrectangular in shape.

***Texigryphaea pitcheri* (Morton)**

Specimens of *Texigryphaea pitcheri* (Figs. 4B, E, G) from the Tucumcari Formation were originally described as *Gryphaea tucumcarii* by Marcou (1855), based on specimens collected near Pyramid Mountain. However, Kues (1989) showed that *T. tucumcarii* was based on unusually large, deep, elongate left valves of *T. pitcheri*, a morph that is completely gradational with typical morphs of *T. pitcheri* in Texas, Oklahoma, and in the Tucumcari Formation of east-central New Mexico. Thus, *T. tucumcarii* is a synonym of the earlier described species *T. pitcheri*. *Texigryphaea pitcheri* is the only species of the genus observed in the main outcrop belt of the Tucumcari Formation, where it is by far the most abundant invertebrate species, occurring in huge numbers, sometimes in biostromal accumulations, throughout the formation. The species has been extensively described and illustrated by Kues et al. (1985), Kues and Lucas (1987), and Kues (1987, 1989).

A
B
C
D
E
F
G
H
I
J
K
L

Outliers of Albian strata, representing deposition preceding that of the main Tucumcari outcrop belt, contain earlier species of *Texigryphaea*, such as *T. navia*, at localities in Roosevelt County, New Mexico (Kues, 1986), and several Kiamichi-equivalent exposures in west-central Texas (Brand, 1953; Kues, 1997).

Kues (1989) noted that valves of *T. pitcheri* from the Tucumcari Formation are typically colonized by a variety of epizoans, including small texigryphaeas and other oysters, serpulid worms (see above), clinonid sponge and acrothoracic barnacle borings, endolithic algae, and rare membraniporoid bryozoans. As noted above, rare ctenostome bryozoans also colonized the valves of *T. pitcheri.*

***Ceratostreon texanum* (Roemer)**

Ceratostreon texanum (Fig. 4F) is a distinctive large oyster characterized by its suboval to reniform valves, strongly coiled beak, and ornamentation of large, arcuate, rugose, irregularly branching, radiating ribs. The valves attain a height of up to about 75 mm. This species, which is widespread and abundant in strata of the Fredericksburg Group in Texas and Oklahoma, and in equivalent strata in southern New Mexico, is present only at the very base of the Tucumcari Formation in Quay County, where it has been observed at San Jon hill (see Kues, 1997). It became extinct prior to the onset of Tucumcari deposition in most areas of east-central New Mexico; its uppermost occurrence coincides, or nearly so, with the first appearance of *Peilinia levicostata*.

***Gyrostrea hinchada* Kues**

Kues (1997) described this medium-sized oyster (Figs. 4C, D) from the basal part of the Tucumcari Formation at the San Jon road cut. The species is characterized by a fairly deep, subcircular left valve with low, lamellate growth lines, and a smaller, flat to slightly convex right valve. Most specimens are 15 to 40 mm high, but the largest specimen observed is 57 mm high. Most specimens display conspicuous attachment scars (flattened areas or concave pits) on the beak of the left valve. Some of these are large, extending for most of the height of the valve, and indicate that the individual remained cemented to its host organism for much of its life. Isolated left valves cemented to the shell surfaces of other organisms, especially *Texigryphaea pitcheri*, are common in the Tucumcari Formation.

FIGURE 4. **Bivalves** from the Tucumcari Formation, Quay County, New Mexico. A. *Lima* sp., right valve steinkern, UNM 13,091, x1.4. B, E, G, *Texigryphaea pitcheri* (Morton). B. interior of left valve, UNM 9077, x1; E, exterior, left valve, UNM 9078, x1; G. Interior of left valve, relatively narrow specimen, UNM 9085, x1. C-D. *Gyrostrea hinchada* Kues. C. Right valve view of articulated specimen, paratype, NMMNH P-18774, x1; D. Left valve with prominent attachment scar at beak, paratype, UNM 12,526, x1. F. *Ceratostreon texanum* (Roemer), external surface of large left valve, with gypsiferous encrustations, UNM 13,133, x1. H-I. *Nicaisolopha subovata* (Shumard), external (H) and internal (I) views of partial left valve, MDM 206, x0.5. J-L. *Peilinia levicostata* Kues. J. External view, large left valve, UNM 13,082, x1.5; K. External view, left valve, UNM 13,083, x1.5; L. External view, juvenile left valve, x2. Localities: A, J-L, Apache Canyon locality; B, E, Pyramid Mountain; C, D, F, San Jon hill; G, West of Tucumcari locality; H, I, Palomas locality.

***Nicaisolopha subovata* (Shumard)**

Fragmentary valves of this large oyster are occasionally encountered in sandy facies of the Tucumcari Formation, but complete valves are rare. The largest complete valve in our collections (Kues et al., 1985, fig. 9H) is about 110 mm in height, but incomplete valves (e.g., Figs. 4H, I) suggest a maximum height of about 175 mm or more. The valves are suboval to subtrigonal in shape, with a relatively short hingeline, and overall are not very convex. Both valves bear up to 8 to 10 low, wide plications that produce a pronounced plicate commissure, and these are crossed by coarse, comarginal growth lamellae. As this species was among the largest organisms living on the Tucumcari seafloor, its valves invariably host epizoans, especially oysters, which on some specimens completely cover the external surface of the valves. The species has a wide geographic distribution in upper Albian strata, from the Kiowa Formation of Kansas, through east-central and southern New Mexico, to north-central Texas.

Malchus (1990) provided evidence to indicate that true *Lopha* originated in the Tertiary, and that Mesozoic species assigned to that genus are not closely related to *Lopha*. *"Lopha" subovata* agrees morphologically with *Nicaisolopha* Vyalov, as redefined by Malchus (1990, p. 171), and is here transferred to that genus.

***Peilinia levicostata* Kues**

Peilinia levicostata (Figs. 4J-L) was described in detail by Kues (1997), who established the new genus and species for Tucumcari Formation specimens that previously had been referred to *Lopha quadriplicata* (Shumard). *Peilinia levicostata* appears to be restricted to the western and possibly eastern endemic centers of the southern Western Interior province, whereas *P. quadriplicata* appeared somewhat later in the southern New Mexico and Texas-Oklahoma areas of the Caribbean province. Fully grown specimens of *P. levicostata* are characterized by moderately large (40+ mm long), arcuate valves. The left valve is relatively convex and bears strong growth lamellae and five or six prominent radial costae that often extend past the ventral margin of the valve, whereas the right valve is smaller, nearly flat, and lacks costae. The shape and ornamentation of this species is highly variable, especially in the earlier growth stages, which lack or may have only one or two small costae, and may have strongly arcuate to even L-shaped valves. The species is common in the Tucumcari Formation, and at some localities, such as San Jon hill, is present in large numbers in some horizons.

***Scabrotrigonia emoryi* (Conrad)**

This large distinctive bivalve (Figs. 5A-C) is abundant in sandy facies of the Tucumcari Formation. It attains a valve height of 45 mm or more, and its valves are strongly arcuate in shape, with prominent, strongly opisthogyrate beaks situated anterior of mid-length. The anterior margin is broadly convex, and the concave posterodorsal margin extends to a moderately elongate, rather sharply convex posterior margin. A long, depressed escutcheon occupies the posterodorsal margin along the hingeline. The valves are strikingly ornamented with 25 or more strong, widely spaced, comarginal costae, which are sharply rounded near the beak but display conspicuous, closely spaced nodes along their length later in growth, across most of the valve surface. On some specimens these nodes are developed into short, sharp, conical spines. The costae are deflected posteriorly across a pair of fine radial ridges bordering the escutcheon, and become smaller, nodose, chevron-shaped ribs across the escutcheon.

The Tucumcari specimens display no significant differences from Conrad's (1857) types or from Böse's (1910) specimens from the Muleros and Mesilla Valley formations of Cerro de Cristo Rey. *Scabrotrigonia emoryi* is widely distributed in correlative strata in the Texas-southern Western Interior region.

Parvilucina? sp.

One small left valve (Fig. 5D) from the San Jon hill locality is questionably assigned to *Parvilucina* based on valve shape and nature of the beak and umbonal areas; dentition and other internal valve features are unknown. The valve is 6.1 mm long and 5.4 mm high at the beak. It is subcircular in outline, with a small, slightly prosogyrate beak extending a little beyond the dorsal margin, has a broadly convex posterodorsal/posterior margin, and a straight anterodorsal margin that curves rather abruptly into the somewhat truncated anterior margin. The umbo is small and slightly inflated, and ornamentation is limited to fine, slightly irregular growth lines. The dorsal margin displays neither a lunule nor an escutcheon but appears to be slightly thickened; otherwise the valve is thin.

The small size of this valve suggests it may be a juvenile, but its shape is distinctive and closely resembles that of several small to moderately small Late Cretaceous species assigned to *Lucina* or *Parvilucina* (probably all belong to *Parvilucina*). *Parvilucina juvensis* (Stanton, 1893; see also Kirkland, 1996), *L. dentonana* (Stephenson, 1952) and *L. parvilineata* Shumard (Stephenson, 1941) are all species that appear to be related to the Tucumcari specimen, although lacking knowledge of the dentition, even its generic identification must remain tentative.

Ludbrookia belviderensis (Cragin)

Scott (1977) described *Ludbrookia belviderensis* (Fig. 5E) and illustrated specimens from the Tucumcari Formation at San Jon hill. This species is small, with subtrigonal to suborbicular valves that are higher than wide, a prominent, prosogyrate, curved beak, and numerous strong radial ribs. It is rare in our Tucumcari collections.

Crassatellina oblonga Meek

Crassatellina oblonga, described initially from the Kiowa Formation of Kansas, is a relatively small, elongate species; a typical specimen (Fig. 5F) measures 18 mm long and 11.5 mm high. The hingeline is long and slopes gently posteriorly, curving sharply ventrally to form a moderately acute, spatulate posterior margin. The posterior umbonal ridge is sharp and separates the moderately inflated valve body from a depressed posterodorsal region. The beaks are small and situated about one third of the distance from the evenly convex anterior margin.

Pleuriocardia kansasense (Meek)

Pleuriocardia kansasense (Fig. 5G) is characterized by relatively small, subcircular, moderately inflated valves with sharp, nearly central beaks. Ornamentation consists of fine, evenly spaced, radial riblets that develop small nodes where crossed by growth lines. The posterior margin of the valves is nearly straight and gently truncated, whereas the anterior margin is moderately convex. The riblets produce a finely crenulated commissure. A typical specimen is 28 mm high and 27 mm long.

This species was originally described from the Kiowa Formation of Kansas (e.g., Scott, 1970a), and was assigned to the new genus *Pleuriocardia* in Scott's (1978) study of Comanche cardiids. Scott (1986a) first reported it from the western endemic center (Tucumcari and related formations of New Mexico), where it is an uncommon species.

Protocardia texana Conrad

Protocardia texana (Figs. 5H-J) in the Tucumcari Formation is characterized by relatively large (up to 50 mm long) valves that are slightly longer than high, have a suboval, subcircular, to vaguely subtrigonal shape, and posses a prominent slightly prosogyrate beak situated subcentrally, slightly closer to the anterior than to the posterior margin. Ornamentation consists of numerous comarginal lirae across most of the valve, which are fine on the umbo but become coarser towards the ventral margin, and prominent, sharp, closely-spaced radial ribs, numbering from 12 to 15+, across the posterior part of the valve. Variation in the strength of the comarginal lirae is significant among individuals within an assemblage, ranging from relatively fine throughout growth, to conspicuous coarsening of the lirae towards the ventral margin of larger valves. These lirae are as wide or a little wider than the interspaces between them.

The Tucumcari specimens do not correspond exactly to any of the taxa described by Scott (1986b), but are clearly part of the *Protocardia texana* complex of forms. They differ from the type specimens of *P. texana texana* in having a subovate to subcircular rather than subquadrate shape. Shape and ornamentation of the more finely lirate specimens are similar to *P. texana salinaensis* Scott, from the Kiowa Formation of Kansas, but the Tucumcari specimens attain a larger size and develop stronger, more widely spaced comarginal lirae, both characters of *P. texana texana*. The Tucumcari specimens are either steinkerns or rather severely weathered valves, with the fine detail of the ornamentation obscured. Here, we assign these specimens to *P. texana*, without reference to subspecies. Of the taxa of *Protocardia* discussed by Scott (1986b), only *P. texana texana* was cited from eastern New Mexico.

Protocardia filosa (Conrad)

Protocardia filosa (Fig. 5K) is characterized generally by its high, subtriangular shell with prominent, sharp, nearly central prosogyrate beaks; its sharp, steeply sloping posterior umbonal ridge; a narrow, gently concave posterior area with numerous (20 to 25+) fine radiating ribs; and relatively coarse comarginal lirae across the remainder of the valve surface. These features allow *P. filosa* to be distinguished readily from *P. texana*, with which it may co-occur in the Tucumcari Formation.

Scott (1986b) showed that several Albian species are synonyms of *P. filosa*, including *Roudaria quadrans* Cragin, which was cited in some early reports from the Tucumcari Formation (see Kues et al., 1985, as *Veniella quadrans*). Some specimens from the Tucumcari are unusually large for the species, attaining a height of nearly 50 mm.

***Linearia (Liothyris)*? aff. *L. (L.)*? *kansasensis* Twenhofel**

The best preserved specimen of this rare species (Fig. 5L) is an incomplete, weathered right valve, missing the posterior margin and measuring 31 mm long and 24 mm high. The valve is suboval, only slightly inflated at the umbo, with a small, prosogyrate, slightly curved beak situated a little closer to the anterior than to the posterior end. The slightly convex anterodorsal margin slopes gently from the beak into a broadly convex anterior margin. A lunule and escutcheon appear to be lacking. The valve surface is characterized by somewhat irregular, raised, widely-spaced growth lines and by fine (5/mm) radiating lirae, visible on the posterodorsal and anterodorsal portions of the valve, but possibly extending across more of the valve.

Twenhofel (1924) described *Linearia kansasensis* from the Kiowa Formation of Kansas, emphasizing the radial lirae (which he termed striae) and noting the presence of even finer concentric striae as well. His illustrations (pl. 12, figs. 9, 10) are of steinkerns displaying internal molds, which show neither of these features. Scott (1970a) redescribed the species but provided no illustrations. The Tucumcari specimens appear to be slightly more elongate than those from the Kiowa, although exact proportions cannot be determined because the posterior margin is absent. The type species of *Linearia (Linearia)* is heavily costate, with a central beak, but gradation among species through forms with restricted fine lirae to smoothness and a somewhat elongate form (subgenus *Liothyris*) was noted by Stephenson (1952). We questionably assign this taxon to *Linearia* (*Liothyris*) on the basis of its subdued radial ornamentation and slightly elongate form.

"*Isocardia*" aff. "*I.*" *slatana* Stephenson

This taxon is represented by several small, poorly preserved specimens (Fig. 5M). The valves are distinctive among Tucumcari bivalves in being high-subtrigonal in shape, strongly inflated, and having a strongly prosogyrate, very incurved beak situated nearly centrally. The anterodorsal margin slopes steeply away from the beaks and is gently concave, and the posterodorsal margin slopes steeply into the broadly convex posterior margin. The shell surface on all specimens is strongly weathered, so nothing may be said about ornamentation. The best-preserved specimen is articulated, with some valve distortion and lateral slippage, but shows this species to be equivalved.

The Tucumcari specimens closely resemble the taxon named *Isocardia slatana* by Stephenson (1952), from the Cenomanian of Texas, but poor preservation precludes positive assignment to that species. The Treatise (Cox et al., 1969, p. 657) treats the type species of *Isocardia* as a synonym of the Cenozoic genus *Glossus*, and distributes other species originally assigned to *Isocardia* among numerous genera within three different families, based mainly on characters of the dentition. Because the teeth of the Tucumcari specimens are unknown, reassignment of this taxon to a different genus is not possible. Albian species assigned by early workers to *Isocardia* (e.g., *I. washita* Marcou; *I. medialus* Conrad) are far larger than the Tucumcari specimens, although generally similar in valve morphology.

***Aphrodina*? sp.**

Specimens assigned questionably to the veneroid genus *Aphrodina* (Fig. 5N) have small, subtrigonal valves, with a high, prominent, prosogyrate beak and broadly convex anterior and posterior margins. Both posterodorsal and anterodorsal margins slope steeply away from the beak; the former is moderately convex and the latter is distinctly concave. No lunule or escutcheon were observed along the hingeline and the nature of the dentition is unknown. Ornamentation is limited to fine (5-6/mm), closely-spaced, low, rounded, comarginal lirae.

This taxon does not closely resemble any species reported from the Kiowa Formation by Twenhofel (1924) or Scott (1970a). It most closely resembles species of *Callistina* reported from the Cenomanian of Texas (Stephenson, 1952); *Callistina* was considered a synonym of *Aphrodina* in the Treatise (Cox et al., 1969).

***Cyprimeria* cf. *C. washitaensis* Adkins**

Fragments of a large species of *Cyprimeria* (Figs. 5O, P) are locally common in sandy facies of the Tucumcari Formation. Maximum estimated valve length is about 85 mm. The valve outline is subcircular to slightly elongate-subovate, with length slightly exceeding height, and the valves are gently inflated and possess a small anterior beak.

Several species of *Cyprimeria* in this size range have been described from the Albian of the Texas and southern Western Interior regions, but typically on the basis of incomplete specimens or steinkerns, so adequate comparison of the species is difficult. The size and slightly ovate outline of the Tucumcari specimens closely resemble the characters of *C. washitaensis*, described by Adkins (1918) from the Weno Formation of eastern Texas. *Cyprimeria kiowana* Cragin appears to have a higher, more closely circular valve outline, and *C. texana* (Roemer), reported from the Tucumcari by Scott (1974), is a smaller species that was originally described on the basis of steinkerns (Adkins, 1928) from Fredericksburg Group strata. It is possible that with better specimens and more study of these species, differences in size and outline reported for these species may be found to be not especially significant, and that some are synonyms.

***Flaventia belviderensis* (Cragin)**

Flaventia belviderensis (Figs. 5Q, R) is a medium-sized bivalve (length up to about 40 mm) with high, sharp, strongly curved, prosogyrate beaks situated about 35 to 40% of the distance from the anterior to posterior valve margins. The valves are a little longer than high (length/height averages about 1.25), are moderately inflated (articulated width/length is about 0.45), and lack ornamentation except for growth lines, which typically become conspicuous comarginal folds near the moderately convex ventral margin. The posterodorsal margin is strongly convex, and the anterodorsal margin, just below the beaks, is straight to gently concave. The lunule and escutcheon along the hingeline are both narrow and poorly developed. Specimens are almost always preserved with the valves articulated.

A
B
C
D
E
F
G
H
I
J
K
L
M
N
O
P
Q
R
S

Flaventia belviderensis, initially described from the Kiowa Formation by Cragin (1894), is uncommon at most Tucumcari localities, but may be abundant locally in sandy to silty lithologies.

***Pholadomya* sp.**

A fragment of the posterior portion of a medium-sized *Pholadomya* (Fig. 5S) from Apache Canyon is 34 mm long and 28 mm high, indicating an elongate shell with a length estimated at 75+ mm when complete. The posterior margin is evenly and relatively sharply convex, and appears to have gaped slightly. The umbonal ridge is inflated and distinct, and sloped gently away from the (unpreserved) beak area to terminate at the posterior margin. Ornamentation consists of relatively strong, closely spaced growth lines that trend posteriorly near the dorsal margin and curve sharply anteriorly across the umbonal ridge. The growth lines are crossed by faint, finer radiating lirae, forming obscure nodes at their intersection with growth lines.

Although only the posterior one third or so of the valves are present, they closely resemble the typical morphology of this part of the shell of *Pholadomya*. Numerous species of *Pholadomya* and the closely related *Homomya* have been described from the Albian of Texas and Mexico (e.g., Adkins, 1928; Perkins, 1961), but too little of this Tucumcari specimen is preserved to assign it to a species. Scott (1974) reported *P.* sp. from the Tucumcari Formation, and Kues et al. (1985) added two unidentified species of *Homomya*. The present specimen belongs to *Pholadomya* based on the presence of radiating ornamentation, which *Homomya* lacks.

Gastropoda

***Turritella "seriatimgranulata"* Roemer**

These large, high-spired gastropods (Figs. 6A-E) attain a height of more than 70 mm and are the most abundant gastropods in the Tucumcari Formation. Ornamentation consists of four strong, nodose spiral lirae on each whorl. The upper one or two lirae often are conspicuously larger than those below, and single fine lirae may be present between pairs of major lirae. Five or more less-nodose lirae are present on the base of the body whorl. Stanton (1947, pl. 56, figs. 19, 21) illustrated typical specimens from Mesa Tucumcari (now Tucumcari Mountain). Large collections from several localities display relatively little variation in shell proportions or ornamentation; some specimens possess somewhat finer lirae and a few have a fifth lira exposed just above the lower suture.

It is not certain that this distinctive Tucumcari species is properly assigned to *Turritella seriatimgranulata*. Roemer (1852) characterized *T. seriatimgranulata* by a strongly noded spiral cord just below the upper suture and four finer cords, with additional finer lirae sometimes inserted at later growth stages. The species was based on a single specimen from the Fredericksburg Group (Walnut or Comanche Peak Formation) of east Texas (Roemer, 1849, 1852), which, according to Stanton (1947, p. 75) "left the impression of its outer surface on the beak of the lower valve" of the oyster *Ceratostreon texanum*. An artificial mold of the impression of the holotype (Stanton, 1947, pl. 56, fig. 7) indicates ornamentation of five, relatively fine, nodose spiral lirae of equal size per whorl, and a shell no more than about 25 mm high when complete. Stanton doubted that the Tucumcari forms warranted even varietal separation from *T. seriatimgranulata*, but his concept of the species seems excessively broad, including also specimens from Sonora, Mexico, that have much finer ornamentation and which we would regard as distinct (see below). The same may be true of the Tucumcari specimens Stanton assigned to *T. seriatimgranulata*, which not only differ significantly from the holotype, but are also considerably younger. Specimens conspecific with *T. "seriatimgranulata"* were described as *T. macropleura* by Stainbrook (1940), from the "Kiamichi" Shale at Guthrie Lake, west-central Texas. Kollman (written communication, R.W. Scott, 2001), noted that the ornamentation of *T. seriatimgranulata,* with a dominant spiral cord at the upper suture that is separated from the other spiral cords by a broad space, is characteristic of *Turritella (Torquesia) granulata* J. de C. Sowerby, a species abundant in the upper Albian of Europe, and suggested that the two species may be synonyms.

Tucumcari specimens of *T. "seriatimgranulata"* have been compared with collections of a related species, *T. belviderei* Cragin, from the Kiowa Formation of Kansas. The latter species attains nearly the size of the former, but is distinguished by more numerous (typically seven per whorl), finer, less coarsely nodose spiral lirae. Although *T. belviderei* has been reported from the Tucumcari Formation (Scott, 1974), we have encountered no specimens that can be unquestionably assigned to that species.

***Turritella* n. sp.**

A second species of *Turritella* (Figs. 6F-H) that locally occurs with *T. "seriatimgranulata"* has a smaller, more slender shell, with a flat rather than gently convex whorl profile, and distinctive ornamentation that differs considerably from that of *T. "seriatimgranulata"*. Based on fragments, this species attains a maxi-

FIGURE 5. **Bivalves** from the Tucumcari Formation, Quay County, New Mexico. A-C. *Scabrotrigonia emoryi* (Conrad). A. Incomplete left valve, showing typical ornamentation, UNM 13,074, x2; B. Beak area of small left valve, UNM 13,076, x2; C. Top view of articulated specimen showing beaks and early ornamentation, UNM 13,073, x 1.8. D. *Parvilucina*? sp., left valve on matrix, UNM 13,065, x2.5. E. *Ludbrookia belviderensis* (Cragin), incomplete, weathered right valve, UNM 13,071, x2. F. *Crassatellina oblonga* Meek, left valve, UNM 13,067, x2. G. *Pleuriocardia kansasense* (Meek), weathered left valve, in matrix, UNM 13,080, x2. H-J. *Protocardia texana* Conrad. H. left valve, UNM 13,118, x1; I. incomplete, rather coarsely lirate left valve, UNM 13,120, x1; J, weathered left valve, with strong ventral comarginal lirae, UNM 13,119, x 0.9. K. *Protocardia filosa* (Conrad), weathered left valve, UNM 13,121, x1.4. L. *Linearia (Liothyris)*? aff. *L. (L.)*? *kansasensis* Twenhofel, incomplete right valve, UNM 13,070, x1.8. M. *"Isocardia"* aff. *"I." slatana* Stephenson, oblique, anterodorsal view of articulated, slightly distorted specimen, UNM 13,072, x1.5. N. *Aphrodina*? sp., incomplete right valve, UNM 13,079, x2. O-P. *Cyprimeria* cf. *C. washitaensis* Adkins. O. incomplete left valve, UNM 13,123, x0.7; P. incomplete right valve, UNM 13,124, x0.7. Q-R. *Flaventia belviderensis* (Cragin). Q. right valve, UNM 13,125, x1.75; R. dorsal view of articulated specimen, UNM 13,126, x1.75. S. *Pholadomya* sp., fragment of left valve, UNM 13,066, x2. Localities: A, B, F, G, N, S, Apache Canyon locality; C, E, H-M, O-R, Mesa Quemado locality; D, San Jon hill.

mum height of about 50 mm, but is considerably narrower, with a lower spiral angle, than comparable specimens of *T. "seriatimgranulata"* of similar height. The ornamentation consists of a single prominent nodose lira just below the upper suture, followed ventrally by five to seven much finer lirae of approximately equal size that bear fine nodes or are subnodose. At least five sharp, non-nodose spiral lirae occupy the base of the body whorl. These specimen most closely resemble forms Stanton (1947, pl. 56, figs. 17, 18, 23, 24) described from the Potrero Formation of Sonora, Mexico, and which he assigned to *T. seriatimgranulata*. As noted above, however, we doubt that specimens of this kind are conspecific with the distinctive large Tucumcari turritellas we refer to *T. "seriatimgranulata."*

***Amberleya mudgeana* (Meek)**

Amberleya mudgeana (Figs. 6I-K) is a medium-sized, low-spired anomphalous turbiniform species of four or five whorls that locally is moderately common in the Tucumcari Formation. Smaller specimens are slightly wider than high, but with growth the shell becomes higher spired, so that height/width of the largest specimens (height = 27+ mm) is about 1.2 to 1.35. Three widely spaced, sharp to nodose, spiral carinae are present on the shell, the highest just below the upper suture, the second forming a sharp angulation between the upper and outer whorl faces, and the lowest at the angular intersection of the outer and basal surfaces of the whorl. Most specimens display a conspicuous medial lira on the outer whorl face, and some possess one or two additional fine lirae on the upper or outer whorl faces.

FIGURE 6. Gastropods, scaphopod, ammonite, and echinoids from the Tucumcari Formation, Quay County, New Mexico. **Gastropods.** A-E. *Turritella "seriatimgranulata"* Roemer. A. Apertural view, UNM 9090, x1.25; B. Lateral view, UNM 13,107, x1.25; C. Adapertural view, UNM 13,106, x1.25; D, adapertural view, UNM 13,108, x1.25; E. Adapertural view, early whorls, UNM 13,109, x1.25. F-H. *Turritella* n. sp. F. Apertural view, incomplete small specimen, UNM 13,112, x1.75; G. Apertural view, incomplete weathered specimen, UNM 13,110, x1.75; H. Adapertural view, incomplete weathered specimen, UNM 13,111, x1.75. I-K. *Amberleya mudgeana* (Meek). I. Adapertural view, specimen with nodose spiral carinae, UNM 13,104, x1.75; J. Adapertural view, specimen with relatively smooth carinae, UNM 13,105, x1.75; K. Adapertural view, large specimen, UNM 13,103, x1.75. L-N. *Drepanochilus* aff. *D. mudgeana* (White). L. Adapertural view of small specimen, UNM 13,101, x1.75; N. Adapertural view, incomplete larger specimen, UNM 13,102, x1.75. M. *Calliostoma cragini* Stanton, adapertural view, UNM 13,068, x1.75. O. *Euspira cragini* (Stanton), apertural view, UNM 13,069, x1.75. P-Q. *Hillites newberryi* (Cragin). P. Oblique adapertural view, large specimen, UNM 13,078, x1.75; Q. Apertural view, small specimen, UNM 13,077, x1.75. **Scaphopod.** R. *Dentalium (Dentalium)* sp., shell fragment, NMMNH P-33,449, x2. **Ammonite.** S-T. *Mortoniceras equidistans* (Cragin), side view of part of outer whorl (S) and view of venter (T), NMMNH P-33,447, x0.75. **Echinoids.** U-V. *Washitaster bravoensis* (Böse), dorsal view, x1.75 (U) and side view, x2 (V) of an incomplete specimen, NMMNH P-33,448. W. Small fragment of spine of regular echinoid, type 1, UNM 13,063, x2. X. Small fragment of spine of regular echinoid, type 2, UNM 13,064, x2. Localities: A-L, N-Q, Mesa Quemado locality; M, W, X, Apache Canyon locality; R, U, V, San Jon hill; S, T, Pyramid Mountain.

Variation in the shells of *A. mudgeana* in the Tucumcari Formation occurs mainly in the degree to which nodes have developed on the three major carinae. On one extreme are specimens having sharp, smooth carinae, whereas on the other extreme are specimens on which the carinae are strongly nodose. Cragin (1894) described the latter form as a separate species, *Margarita marcouana*, but Stanton (1947) considered the two species, both of which were originally described from the Kiowa Formation of Kansas, as synonyms based on the intergradational nature of the ornamentation.

***Calliostoma cragini* Stanton**

These Tucumcari specimens (Fig. 6M) agree closely with specimens of *Calliostoma cragini* described by Stanton (1947) from the Kiowa Formation in Kansas, and from Kiamichi-equivalent strata near Tahoka Lake, west-central Texas. The shell is relatively low spired and turbinate, small (less than 20 mm high), and consists of about seven whorls. The whorl profile is slightly convex, the sutures shallowly incised, and the surface of the shell is ornamented by about six coarsely nodose cords on the outer whorl face. The lowest one or two of these above the lower suture are a little larger and form the whorl periphery, with an additional 10 nodose cords on the flattened, anomphalous whorl base. The surface of the whorls is covered by very fine spiral lirae superimposed upon the nodose primary ornamentation.

Before Stanton's work, Kiowa and west-central Texas "Kiamichi" shells of this type had been assigned to *Trochus texanus* Roemer (e.g., Twenhofel, 1924; Stainbrook, 1940), a species somewhat more evenly conical in shape and with a tooth-like fold on the columella.

***Drepanochilus* aff. *D. mudgeana* (White)**

Three severely weathered, incomplete specimens appear to be closely related to *Drepanochilus mudgeana* (Figs. 6L, N). These fragments indicate a shell about 25 to 30 mm high when complete, with coarse, gently arcuate to nearly straight, opisthocline collabral ribs numbering 12 to 14 per whorl, crossed by much finer spiral lirae. The expanded, wing-like outer lip is not preserved on these specimens. Stanton (1947) noted the close similarity between *D. mudgeana* and *D. kiowana* (Cragin); the latter species has been reported from the Tucumcari Formation (Scott, 1974). *Drepanochilus kiowana* was characterized by Stanton (1947) as being considerably smaller and more slender, with finer and more numerous ribs, and a narrower, more carinate wing, as compared with *D. mudgeana*. He suggested that the relationship between the two species is sufficiently close to suggest that *D. kiowana* is no more than a variety of *D. mudgeana*.

The Tucumcari specimens were compared with well-preserved specimens of *D. kiowana* from the Kiowa Formation near Belvidere, Kansas, in the UNM collections, and were found to be larger, with fewer, coarser, collabral ribs and more pronounced spiral lirae. The Tucumcari specimens more closely resemble a specimen from the Kiowa Formation described as Aporrhaid gastropod indet. by Scott (1970a). Restudy of the type specimens of *D. mudgeana* and *D. kiowana*, and better preserved material from the Tucumcari Formation, are needed to ascertain accurately the relationships of these species.

Euspira cragini **(Stanton)**

The single specimen (Fig. 6O) available is small (height = 16.1 mm; width = 11.5 mm), rather soleniscoid in shape, and consists of five whorls. The shell is severely weathered, and lacks ornamentation except for fine, moderately prosocline growth lines across the body whorl. The aperture is obscured by matrix but appears to be suboval, higher than wide, and rather large. This specimen corresponds well with, but is somewhat smaller than, *Lunatia*? *cragini,* described by Stanton (1947) from the Kiowa Formation near Belvidere, Kansas. Sohl (1960) considered *Lunatia* a synonym of *Euspira.* His Late Cretaceous species of *Euspira* are generally very similar to the Albian Kiowa specimens and to the Tucumcari specimen.

Hillites newberryi **(Cragin)**

The shell of this species (Figs. 6P, Q) is low spired, with about four whorls, the last of which is greatly inflated. The largest specimen is about 22 mm high, and its aperture is large and subcircular, with its base extended into a curved siphonal canal. Ornamentation consists of two strong, rather widely separated spiral ribs on the upper whorl surface, one immediately below the upper suture and the other marking a rounded shoulder between the upper and outer whorl surfaces; and 10 to 13 strong spiral cords across the outer whorl surface and base of the body whorl. These cords are subnodose, especially where crossed by growth lines, and a much finer, sharp lira occupies the interspace between some pairs of cords.

The few available specimens, all of which are distorted, most closely resemble two poorly known species, originally described from the Kiowa Formation of Kansas, *Margarita newberryi* Cragin, 1894, and *M. ornata* Twenhofel, 1924. Neither Twenhofel nor Scott (1970a) obtained species of *M. newberryi* in their Kiowa faunal studies, but Stanton (1947) redescribed the species (as *Turbo*? *newberryi*) based in part on a specimen from the Tucumcari Formation at Tucumcari Mountain. Twenhofel (1924) based *M. ornata* on one incomplete specimen. Its relationship with *M. newberryi* is unclear; neither Twenhofel nor Stanton compared the two species. The descriptions and illustrations available suggest that they are closely related, if not conspecific. *Margarita ornata* appears to have fewer, coarser, spiral cords; its aperture is unknown so detailed comparison with *M. newberryi* is not possible.

Both species, as well as closely related forms described by Stanton (1947) as *Turbo*? *cookensis* and *T.*? *gainsvillensis* closely resemble *Hillites multilirata* Stephenson, the type species of *Hillites* (Stephenson, 1952). All of these species appear to best be placed within that genus, as Stephenson (1952, p. 81) and Scott (1970a, p. 57) suggested for some of them.

Scaphopoda

Dentalium (Dentalium) **sp.**

A single fragment of a small scaphopod shell (Fig. 6R) was recovered from the San Jon road cut locality. It is slender and gently arcuate in shape, and is 21 mm long. The fragment is probably part of a much longer shell, as it increases only slightly, from 4.0 to 4.6 mm, in diameter along its length. The internal cross section on both ends is circular, and the thin shell is ornamented by 22 strong, rounded, longitudinal ribs, separated by interspaces of equal width. At the posterior (narrower) end of the fragment, a smaller internal tube (pipe) projects slightly and bears ornamentation similar to that of the external shell surface. At the anterior (wider) end of the specimen, in cross-sectional view, there appear to be at least three nested pipes, probably the result of additional secretion of shell material internally towards the posterior end after full growth of the external shell had occurred. Because this fragment does not include the apical end of the shell, the presence of a notch or slit there could not be determined. The ornamentation and circular internal cross section suggest assignment to *Dentalium (Dentalium)*, following the classification of Palmer (1974).

Scaphopods are apparently quite rare in the Albian of the Texas-southern Western Interior region. One species of *Cadulus* is present in the Kiowa Formation (Scott, 1970a), but scaphopods are not known in the Albian sequence at Cerro de Cristo Rey (Böse, 1910, and collections by BSK), and are not cited in monographs and compilations of Albian taxa from Texas (e.g., Adkins, 1928; Brand, 1953; Perkins, 1960; Lokke, 1967), nor by any other author so far as we are aware. The only previous reports from the New Mexico Albian is a fragmentary specimen of *Dentalium* from the Mesilla Valley Shale at the Lee Ranch locality west of the Cornudas Mountains (Kues and Lucas, 1993, fig. 8X), and reports in faunal lists of *Dentalium* and *Cadulus* from the Mojado Formation, Big Hatchet and Animas Mountains, by Zeller (1965) and Zeller and Alper (1965). Although several species of *Dentalium* have been described from Cenomanian strata in Texas (Stephenson, 1952) and in the Western Interior (e.g., Kirkland, 1996), none of them appear closely related to the Tucumcari specimen.

Ammonoidea

Although several species of ammonites have been reported from the main outcrop belt of the Tucumcari Formation (Kues et al., 1985), none have been illustrated. We have collected specimens of only one species of ammonoid during our long-term studies of the Tucumcari, and it is discussed below.

Mortoniceras equidistans **(Cragin)**

The best specimen of *Mortoniceras equidistans*, collected from the Tucumcari type section at Pyramid Mountain, is a fragment of the outer whorl of a large individual (Figs. 6S, T). Its length is 107 mm, whorl height is 55 mm, and whorl width about 35 mm, all suggesting that the complete shell had a diameter of at least 150 mm. The fragment is weathered and heavily encrusted with juvenile oysters, and displays ornamentation on only one side. A low keel is present along the venter, and preserved suture lines are limited to a single lobe and saddle. The ornamentation consists of strong, straight, relatively closely spaced simple ribs, each of which bears a wide ventral-lateral, a narrower lateral, and a minor umbilical node.

Although assigned to *M. equidistans*, the specimen is somewhat atypical in having relatively coarse, closely-spaced ribs, seven occurring within a circumference distance of 100 mm, compared to five or six per 100 mm on the youngest part of the shell of Cragin's type specimen from the Duck Creek Formation of Texas (see Cobban, 1985), a somewhat smaller whorl than the Tucumcari specimen. Specimens of *M. equidistans* illustrated from the overlying basal Mesa Rica Sandstone (e.g., Kues et al., 1985; Lucas and Hunt, 2000) in Quay County likewise bear somewhat finer, more widely-spaced ribs.

Arthropoda

Although we observed no arthropods in the Tucumcari Formation, Kietzke (1985) reported 20 species of ostracods from the Pyramid Mountain locality and illustrated most of them.

Echinoidea

Echinoids are very rare in the Tucumcari Formation, and the taxa discussed here are the first reports of this group in the formation. Fragments of two types of regular echinoid spines, both having a diameter of 4.0 to 4.6 mm and probably measuring 40 to 50 mm in length when complete, are present. The ornamentation of the first type (Fig. 6W) consists of sharp, high, longitudinal ribs that are subtly nodose, with about 8 ribs across 5 mm of circumferance. The second type (Fig. 6X) displays longitudinal rows of isolated nodes, about 4 or 5 rows per 2 mm of circumferance. The third type of echinoid, a spatangoid represented by a single incomplete test, is described below.

***Washitaster bravoensis* (Böse)**

One incomplete specimen (Figs. 6U, V) from the San Jon road cut is the first record of an echinoid test from the Tucumcari Formation. The specimen is laterally crushed, and is missing the anterior, most of the ventral, and part of the right dorsal portions of the test, with much of the left dorsal region covered with matrix. Because of lateral compression, the right side of the posterior part of the test has overridden part of the left side, making the test appear narrower than it was originally, and obscuring the periproct, which was situated high on the posterior surface. The specimen is about 37 mm long, 28 mm wide, and 18 mm high.

Because of poor preservation, a complete description of the morphology of this specimen cannot be given. However, parts of the left anterior, left posterior, and right posterior petals are preserved, and those plates of the test that are visible are unweathered and preserve surface details excellently. In addition, portions of the lateral and peripetalous fascioles are well preserved.

The anterior petals are long, relatively narrow, and gently arcuate, whereas the posterior petals are straight, moderately wide, less than half the length of the anterior petals, and diverge at a 90° angle. The outer portions of the anterior petals consist of paired pores, with the external pore elongate and the internal pores circular, and this pattern is visible on the posterior petals as well. Ornamentation is limited to small round tubercles on the interambulacral plates, ranging from two or three per plate near the petals but increasing in number on the plates forming the lateral and posterior test surfaces. The lateral fasciole band is relatively wide and extends along the test a little above midheight.

The shape and proportions of the test, orientation of the petals and ambulacral pores, fascioles, and ornamentation all indicate assignment to *Washitaster bravoensis*, originally described from the Muleros and Mesilla Valley formations of Cerro de Cristo Rey by Böse (1910; see also Cooke, 1955). Most Albian spatangoids of similar size and characters with which this specimen might be confused (especially species of *Heteraster*) lack fascioles.

Vertebrata

Fossils of vertebrates are extremely rare in the Tucumcari Formation, never having been previously reported. Here, we document two taxa of fishes based on isolated teeth from the Apache Canyon locality.

aff. *Cretodus* sp.

A single tooth with an incomplete root base (Figs. 7A-C) clearly belongs to a lamniform selachian. The crown is dominated by

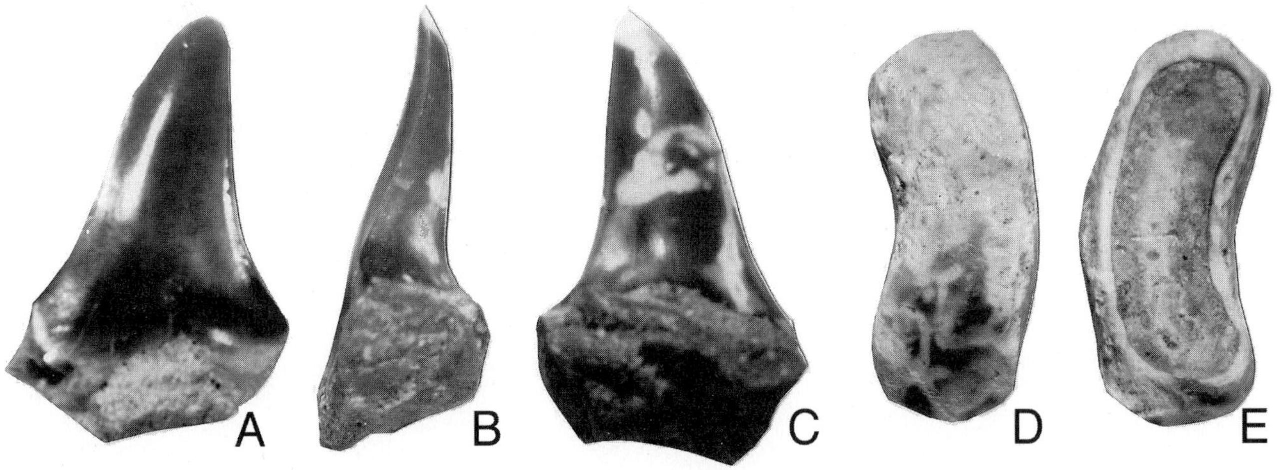

FIGURE 7. **Fish teeth** from the Tucumcari Formation, Quay County, New Mexico; all from the Apache Canyon locality. A-C. aff. *Cretodus* sp., lingual (A), lateral (B) and labial (C) views, UNM 13,134, x5. D-E. *Coelodus* sp., crown (D) and root base (E), UNM 13,135, x5.

a broad, smooth, slightly sigmoid, triangular cusp that is about 9 mm high. Breakage of the crown at its base makes it impossible to determine if the main cusp was flanked by two smaller accessory cusps, as is common among lamniforms (Welton and Farish, 1993). The root base appears to be broad and thick, but it is incomplete. Definite assignment of this tooth to a lamniform genus is not possible because of its incomplete crown base and root. However, greatest similarity is to *Cretodus* (Welton and Farish, 1993, p. 98), and we tentatively identify it as aff. *Cretodus* sp.

***Coelodus* sp.**

An isolated reniform tooth is readily recognized as that of a pycnodontid osteichthyan. This tooth (Figs. 7D-E) has a crown that measures about 11 by 4 mm, is slightly convex and covered with a smooth (though highly weathered), thick enamel cap. The root base of the tooth is a convex depression that underlies the entire crown. This tooth is identical to teeth of the pycnodontid *Coelodus*, from the Kiowa Formation of Kansas (Williston, 1900, p. 254-255, pl. 26, fig. 6), to which we assign it. However, an isolated tooth is not a sufficient basis for a species-level identification.

ACKNOWLEDGMENTS

We thank William A. Cobban and Robert W. Scott for reviewing an earlier version of this paper and offering suggestions that improved it. Kenneth Kietzke collected some of the specimens discussed here, including several of the rarer taxa, and kindly made them available to the authors for study. We also thank Adrian Hunt, Mesalands Dinosaur Museum, Tucumcari, for the loan of several specimens, and Mabel Chavez for word-processing the manuscript.

REFERENCES

Adkins, W.S., 1918, The Weno and Pawpaw Formations of the Texas Comanchean: University of Texas Bulletin no. 1856, p. 1-172.

Adkins, W.S., 1928, Handbook of Texas Cretaceous fossils: University of Texas, Bulletin no. 2838, 385 p.

Bassler, R.S., 1953, Part G, Bryozoa: Treatise on Invertebrate Paleontology, Geological Society of America and University of Kansas Press, 253 p.

Böse, E., 1910, Monografia geológica y paleontológica del Cerro de Muleros cerca de Ciudad Juarez, Estado de Chihuahua, y descripcion da la fauna cretácea de la Encantada, placer de Guadalupe, Estado de Chihuahua: Instituto Geológico de Mexico, Boletin 25, 193 p.

Brand, J.P., 1953, Cretaceous of Llano Estacado of Texas: Bureau of Economic Geology, University of Texas at Austin, Report of Investigations 20, 59 p.

Cobban, W.A., 1985, The Lower Cretaceous ammonite *Schloenbachia leonensis* Conrad var. *equidistans* Cragin: U.S. Geological Survey, Bulletin 1641-A, p. 1-4.

Condra, G.E. and Elias, M.K., 1944, Carboniferous and Permian ctenostomatous Bryozoa: Geological Society of America Bulletin, v. 55, p. 517-568.

Conrad, T.A., 1857, Description of Cretaceous and Tertiary fossils; *In* Emory, W.H., Report on the United States and Mexican boundary survey, volume I: 34th Congress, 1st Session, Executive Document 108, p. 141-174.

Cooke, C.W., 1955, Some Cretaceous echinoids from the Americas: U.S. Geological Survey, Professional Paper 264-E, p. 87-112.

Cox, L.R. et al., 1969, Treatise on Invertebrate Paleontology, Part N, Volumes 1, 2, Mollusca 6, Bivalvia: Geological Society of America and University of Kansas Press, p. N1-N952.

Cragin, F.W., 1894, New and little known Invertebrata from the Neocomian of Kansas: American Geologist, v. 14, p. 1-12.

Dyman, T.S., Merewether, C.M., Molenaar, C.M., Cobban, W.A., Obradovich, J.D., Weimer, R.J. and Bryant, W.A., 1994, Stratigraphic transects for Cretaceous rocks, Rocky Mountains and Great Plains regions; *In* Caputo, M.V., Peterson, J.A. and Franczyk, K.J. (eds.), Mesozoic Systems of the Rocky Mountain region, U.S.A.: Rocky Mountain Section, SEPM, p. 365-391.

Holbrook, J.M. and Dunbar, R.W., 1992, Depositional history of Lower Cretaceous strata in northeastern New Mexico: implications for regional tectonics and depositional sequences: Geological Society of America Bulletin, v. 104, p. 802-813.

Kauffman, E.G., Sageman, B.B., Kirkland, J.L., Elder, W.P., Harriess, P.J. and Villamil, T., 1993, Molluscan biostratigraphy of the Cretaceous Western Interior basin, North America; *In* Caldwell, W.G.E. and Kauffman, E.G. (eds.), Evolution of the Western Interior basin: Geological Association of Canada, Special Paper 39, p. 397-434.

Kennedy, W.J., Gale, A.S., Hancock, J.M., Crampton, J.S. and Cobban, W.A., 1999, Ammonites and inoceramid bivalves from close to the middle-upper Albian boundary around Fort Worth, Texas: Journal of Paleontology, v. 73, p. 1101-1125.

Kietzke, K.K., 1985, Microfauna of the Tucumcari Shale, Lower Cretaceous of east-central New Mexico: New Mexico Geological Society, Guidebook 36, p. 247-260.

Kirkland, J.I., 1996, Paleontology of the Greenhorn cyclothem (Cretaceous: late Cenomanian to middle Turonian) at Black Mesa, northeastern Arizona: New Mexico Museum of Natural History and Science, Bulletin 9, 131 p.

Kues, B.S., 1985, Stratigraphy of the Tucumcari area: a historical account: New Mexico Geological Society, Guidebook 38, p. 119-140.

Kues, B.S., 1986, Paleontology and correlation of a Lower Cretaceous (Albian) outlier in Roosevelt County, New Mexico: New Mexico Geology, v. 8, p. 88-94.

Kues, B.S., 1987, *Texigryphaea* in the Glencairn Formation near Two Buttes, Colorado, with notes on an assemblage of *Texigryphaea* from the Kiowa Formation of southern Kansas: New Mexico Geological Society, Guidebook 38, p. 207-215.

Kues, B.S., 1989, Taxonomy and variability of three *Texigryphaea* (Bivalvia) species from their Lower Cretaceous (Albian) type localities in New Mexico and Oklahoma: Journal of Paleontology, v. 63, p. 454-483.

Kues, B.S., 1997, New bivalve taxa from the Tucumcari Formation (Cretaceous, Albian), New Mexico, and the biostratigraphic significance of the basal Tucumcari fauna: Journal of Paleontology, v. 71, p. 820-839.

Kues, B.S. and Kietzke, K.K., 1985, Supplemental road log 3, Mesa Quemado and Pyramid Mountain: New Mexico Geological Society, Guidebook 36, p. 80-85.

Kues, B.S. and Lucas, S.G., 1987, Cretaceous stratigraphy and paleontology in the Dry Cimarron Valley, New Mexico, Colorado and Oklahoma: New Mexico Geological Society, Guidebook 38, p. 167-198.

Kues, B.S. and Lucas, S.G., 1993, Stratigraphy, paleontology and correlation of Lower Cretaceous exposures in southeastern New Mexico: New Mexico Geological Society, Guidebook 44, p. 245-260.

Kues, B.S., Lucas, S.G., Kietzke, K.K. and Mateer, N.J., 1985, Synopsis of Tucumcari Shale, Mesa Rica Sandstone and Pajarito Shale paleontology, Cretaceous of east-central New Mexico: New Mexico Geological Society, Guidebook 36, p. 261-281.

Lokke, D.H., 1967, Status and bibliographic review of Comanchean Series (Lower Cretaceous) paleontology in Texas; *In* Hendricks, L. (ed.), Comanchean (Lower Cretaceous) stratigraphy and paleontology of Texas: Permian Basin Section, SEPM, Pub. 67-8.

Lucas, S.G., 1991, Cretaceous Dakota Group outlier, Sacramento Mountains, Otero County, New Mexico: New Mexico Geological Society, Guidebook 42, p. 261-264.

Lucas, S.G. and Estep, J.W., 1998, The late Albian ammonite *Engonoceras* from southwest New Mexico: New Mexico Geology, v. 20, p. 78-82.

Lucas, S.G. and Estep, J.W., 2000, Lower Cretaceous stratigraphy, correlation and paleogeography of New Mexico: New Mexico Museum of Natural History and Science, Bulletin 16, p. 45-61.

Lucas, S.G. and Hunt, A.P., 2000, The Lower Cretaceous ammonite *Mortoniceras equidistans* (Cragin) from the Mesa Rica Sandstone, Quay County, New Mexico; New Mexico Museum of Natural History and Science, Bulletin 16, p. 97-100.

Lucas, S.G. and Kisucky, M.J., 1988, Type and reference section of the Tucumcari, Mesa Rica and Pajarito Formations, Cretaceous of east-central New

Mexico: New Mexico Geology, v. 10, p. 82-89.
Lucas, S.G., Kues, B.S., Hayden, S.N., Allen, B.D., Kietzke, K.K., Williamson, T.E., Sealey, P. and Pence, R., 1988, Cretaceous stratigraphy and biostratigraphy, Cooke's Range, Luna County, New Mexico: New Mexico Geological Society, Guidebook 39, p. 143-167.
Malchus, N., 1990, Revisions der Kreide-Austern (Bivalvia: Pteriomorpha) Ägyptens (Biostratigraphie, Systematik): Berliner geowissenschaftliche Abhandlungen, Reihe A, v. 125, 231 p.
Marcou, J., 1855, Résumé éxplicatif d'une carte geologique des Etats-Unis et des provinces anglaises de l'Amerique du Nord, avec un profil géologique allant de la vallée du Mississippi aux côtes du Pacifique, et une planche des fossiles: Société Géologique de France, Bulletin, series 2, v. 12, p. 813-936.
Marcou, J., 1858, Geology of North America, with two reports on the prairies of Arkansas and Texas, the Rocky Mountains of New Mexico, and the Sierra Nevada of California, originally made for the United States Government: Zurich, Zurcher and Furrer, 144 p.
Marcou, J., 1889, American Neocomian and chalk of Arkansas: American Geologist, v. 4, p. 357-367.
Palmer, C.P, 1974, A supraspecific classification of the scaphopod molluscs: Veliger, v. 17, p. 115-123.
Perkins, B.F., 1961, Biostratigraphic studies in the Comanche (Cretaceous) Series of northern Mexico and Texas: Geological Society of America, Memoir 83, 138 p.
Reeside, J.B. Jr., 1923, The fauna of the so-called Dakota Formation of northern central Colorado and its equivalent in southwestern Wyoming: U.S. Geological Survey, Professional Paper 131-H, p. 199-207.
Roemer, F., 1849, Texas.....: Adolph Marcus, Bonn, 464 p.
Roemer, F., 1852, Die Kreidabildungen von Texas und ihre organischen Einschlüsse: Adolph Mercus, Bonn, 100 p.
Scott, R.W., 1970a, Paleoecology and paleontology of the Lower Cretaceous Kiowa Formation, Kansas: The University of Kansas Paleontological Contributions, Article 52 (Cretaceous 1), 94 p.
Scott, R.W., 1970b, Stratigraphy and sedimentary environments of Lower Cretaceous rocks, Southern Western Interior: American Association of Petroleum Geologists Bulletin, v. 54, p. 1225-1244.
Scott, R.W., 1974, Bay and shoreface benthic communities in the Lower Cretaceous: Lethaia, v. 7, p. 315-330.
Scott, R.W., 1975, Patterns of Early Cretaceous molluscan diversity gradients in south-central United States: Lethaia, v. 8, p. 241-252.
Scott, R.W., 1977, Paleobiology of Lower Cretaceous carditid bivalves, North America: Journal of Paleontology, v. 51, p. 1150-1160.
Scott, R.W., 1978, Paleobiology of Comanchean (Cretaceous) cardiids (Cardiinae), North America: Journal of Paleontology, v. 52, p. 881-903.
Scott, R.W., 1986a, Biogeographic influences on Early Cretaceous paleocommunities, Western Interior: Journal of Paleontology, v. 60, p. 197-207.
Scott, R.W., 1986b, Paleobiology of Early Cretaceous protocardiids, Caribbean Province: Journal of Paleontology, v. 60, p. 1186-1211.
Sohl, N.F., 1960, Archeogastropoda, Mesogastropoda and stratigraphy of the Ripley, Owl Creek, and Prairie Bluff Formations: U.S. Geological Survey, Professional Paper 331-A, 151 p.
Stainbrook, M.A., 1940, Gastropoda of the Kiamichi Shale of the Texas panhandle: University of Texas, Publication 3945, p. 705-716.
Stanton, T.W., 1893, The Colorado Formation and its invertebrate fauna: U.S. Geological Survey, Bulletin 106, 288 p.
Stanton, T.W., 1947, Studies of some Comanche pelecypods and gastropods: U.S. Geological Survey, Professional Paper 211, 256 p.
Stephenson, L.W., 1941, The larger invertebrate fossils of the Navarro Group of Texas: University of Texas, Publication 4101, 641 p.
Stephenson, L.W., 1952, Larger invertebrate fossils of the Woodbine Formation (Cenomanian) of Texas: U.S. Geological Survey, Professional Paper 242, 226 p.
Turnšek, D., LeMone, D.V. and Scott, R.W., 1983, Corals from Cerro de Cristo Rey, Dona Ana County, New Mexico and Chihuahua, Mexico: Abstracts, Fourth International Symposium on Fossil Cnidaria (and Archeocyathids and Stromatoporoids), Washington, D.C., p. 44.
Twenhofel, W.J., 1924, The geology and invertebrate paleontology of the Comanchean and "Dakota" formation of Kansas: Kansas Geological Survey, Bulletin 9, 128 p.
Wells, J.W., 1933, Corals of the Cretaceous of the Atlantic and Gulf Coastal Plains and Western Interior of the United States: Bulletins of American Paleontology, v. 18, no. 67, p. 83-292.
Wells, J.W., 1956, Scleractinia: Treatise on Invertebrate Paleontology, Part F, Coelenterata, Geological Society of America and University of Kansas Press, p. F328-F477.
Welton, B.J. and Farish, R.F., 1993, The collector's guide to fossil sharks and rays from the Cretaceous of Texas: Lewisville, Before Time, 204 p.
White, C.A., 1877, Report upon the invertebrate fossils collected in portions of Nevada, Utah, Colorado, New Mexico, and Arizona, by parties of the expeditions of 1871, 1872, 1873, and 1874: U.S. Geographical Surveys West of the One Hundredth Meridian [Wheeler Survey], v. 4, pt. 1, 219 p.
Williston, S.W., 1900, Cretaceous fishes [of Kansas]. Selachians and pycnodonts: University Geological Survey of Kansas, v. 6, p. 237-256.
Zeller, R.A. Jr., 1965, Stratigraphy of the Big Hatchet Mountains area, New Mexico: New Mexico Bureau of Mines and Mineral Resources, Memoir 16, 128 p.
Zeller, R.A. Jr. and Alper, A.M., 1965, Geology of the Walnut Wells quadrangle, Hidalgo County, New Mexico: New Mexico Bureau of Mines and Mineral Resources, Bulletin 84, 105 p.

At Mesa Redonda in Quay County, the type section of the Redonda Formation forms red-bed slopes and ribbed cliffs underneath a locally thick, cliff-forming sandstone of the Middle Jurassic Entrada Sandstone. Mesa Redonda is close to the southern margin of the vast Middle Jurassic San Rafael basin that extended from here into Utah.

STRATIGRAPHY AND BIOSTRATIGRAPHY OF AN UPPER CRETACEOUS OUTLIER NEAR ROY, HARDING COUNTY, NEW MEXICO

SPENCER G. LUCAS[1] AND ANDREW B. HECKERT[2]

[1]New Mexico Museum of Natural History, 1801 Mountain Rd NW, Albuquerque, 87104;

[2]Department of Earth and Planetary Sciences, University of New Mexico, Albuquerque, NM 87131-1116

Abstract.—Upper Cretaceous strata exposed as outliers near Roy in Harding County, New Mexico, are assigned to the Graneros Shale and the overlying Greenhorn Formation. The Greenhorn Formation locally consists of three members (in ascending order): Lincoln, Hartland, and Bridge Creek. The Lincoln Member is 8.3 m thick and dominated by yellowish gray and very pale orange calcareous shale and characteristic persistent beds of calcarenite. The Hartland Member is 11.0 m thick and consists mostly of medium and brownish gray bentonitic shale. The Bridge Creek Member here is erosionally truncated by the Neogene Ogallala Formation 4.1 m above its base and consists of interbeds of ledge-forming limestone and calcareous shale. Fossils from the Lincoln Member include numerous specimens of the bivalve *Ostrea beloiti* Logan and of the selachian *Ptychodus anonymus* Williston. The Hartland Member lacks biostratigraphically significant fossils. The Bridge Creek Member produces the ammonites *Euomphaloceras* sp. and *Sciponoceras gracile* (Shumard) near its base and *Watinoceras coloradoense* (Henderson) and *Vascoceras birchbyi* Cobban and Scott near its top. These fossils can be assigned to the upper Cenomanian *Sciponoceras gracile* ammonite zone and the lower Turonian *Vascoceras birchbyi* ammonite zone. Younger ammonite zones normally present in the Bridge Creek Member are not present at Roy, having been stripped off by Neogene erosion.

INTRODUCTION

Wanek (1962) mapped Upper Cretaceous strata along the edge of the Canadian Escarpment west and northwest of Roy in T20N, R25-26E, Harding County, New Mexico (Fig. 1). He assigned these strata to the Graneros Shale ("chiefly black fissile marine shale containing thin beds of dark brown-gray limestone," no thickness given) and the Greenhorn Limestone ("thin interbeds of marine limestone and shale and is 35 feet thick").

Subsequent statewide maps by Dane and Bachman (1965) and Clemons (1982) have indicated that the Upper Cretaceous strata near Roy are continuous with Graneros and Greenhorn strata to the north. However, the outcrops near Roy are actually isolated outliers, largely covered in the subdued local topography (Fig. 2). As no data beyond those provided by Wanek (1962) are available, we describe the lithostratigraphy and biostratigraphy of the best exposed Upper Cretaceous outlier near Roy, which is in sec. 12, T20N, R25E (Figs. 1, 2). Here, NMMNH refers to the New Mexico Museum of Natural History and Science, Albuquerque.

LITHOSTRATIGRAPHY

Wanek (1962) correctly assigned Upper Cretaceous strata exposed near Roy to the Graneros Shale and Greenhorn Limestone. The Graneros Shale, however, is so poorly exposed in this area that little can be said about it other than: (1) it conformably (?) overlies ledge-forming quartzarenite sandstone of the Romeroville Sandstone of the Dakota Group; (2) its estimated thickness is about 20 m; and (3) it consists primarily of dark gray, light brownish gray and grayish orange shale with minor, thin (<0.3 m) interbeds of calcarenite, siltstone and airfall bentonitic ash. Near Roy, the Graneros Shale forms covered slopes and plains.

The Greenhorn Formation conformably overlies the Graneros Shale near Roy and consists of three members (ascending): Lincoln, Hartland and Bridge Creek (Figs. 1, 2). Total Greenhorn thickness is approximately 23 m, but this is not a complete thickness as the top of the Bridge Creek Member has been removed by Neogene erosion (Fig. 1).

The Lincoln Member is 8.3 m thick and consists of a few persistent ledges of very pale orange calcarenite and several thin bentonite beds in slopes of yellowish gray and very pale orange calcareous shale. We pick the base of the Lincoln Member at a prominent calcarenite ledge full of shell debris of inoceramid bivalves, numerous valves of the small oyster *Ostrea beloiti* Logan and occasional sharks' teeth. This bed is 1.6 m above a prominent bentonite (correlated to the "x bentonite") in the uppermost Graneros Shale. We identify the top of the Lincoln Member as the highest persistent calcarenite ledge below slope-forming, bentonitic shale of the Hartland Member (Fig. 1).

Near Roy, the Hartland Member is 11.0 m thick and consists mostly of medium gray and yellowish gray bentonitic shale that includes three distinctive bentonites. Each of these bentonites is mottled yellowish gray, dark yellowish orange and very pale orange.

The base of the Bridge Creek Member is the first persistent ledge of light gray to very pale orange dense fossiliferous limestone. At least 4.1 m of Bridge Creek Member are present near Roy, and these strata are interbeds of ledge-forming limestone and medium-gray calcareous shale. Calcrete of the Neogene Ogallala Formation unconformably overlies the Bridge Creek Member.

PALEONTOLOGY

Fossils are present in virtually all of the calcarenite and limestone beds of the Greenhorn Formation near Roy. However, we collected identifiable and biostratigraphically significant fossils from only three beds: (1) NMMNH locality 4618, in the basal calcarenite of the Lincoln Member at UTM 569352E, 3980960N (Zone 13, NAD 27); (2) NMMNH locality 4619, in the lower part of the Bridge Creek Member at UTM 569670E, 3980844N (Zone 13, NAD 27); and (3) NMMNH locality 4620, near the top of the preserved Bridge Creek section at UTM 569304E, 3980533N (Zone 13, NAD 27) (Fig. 1). Fossils from these localities are of bivalves, ammonites and selachian teeth.

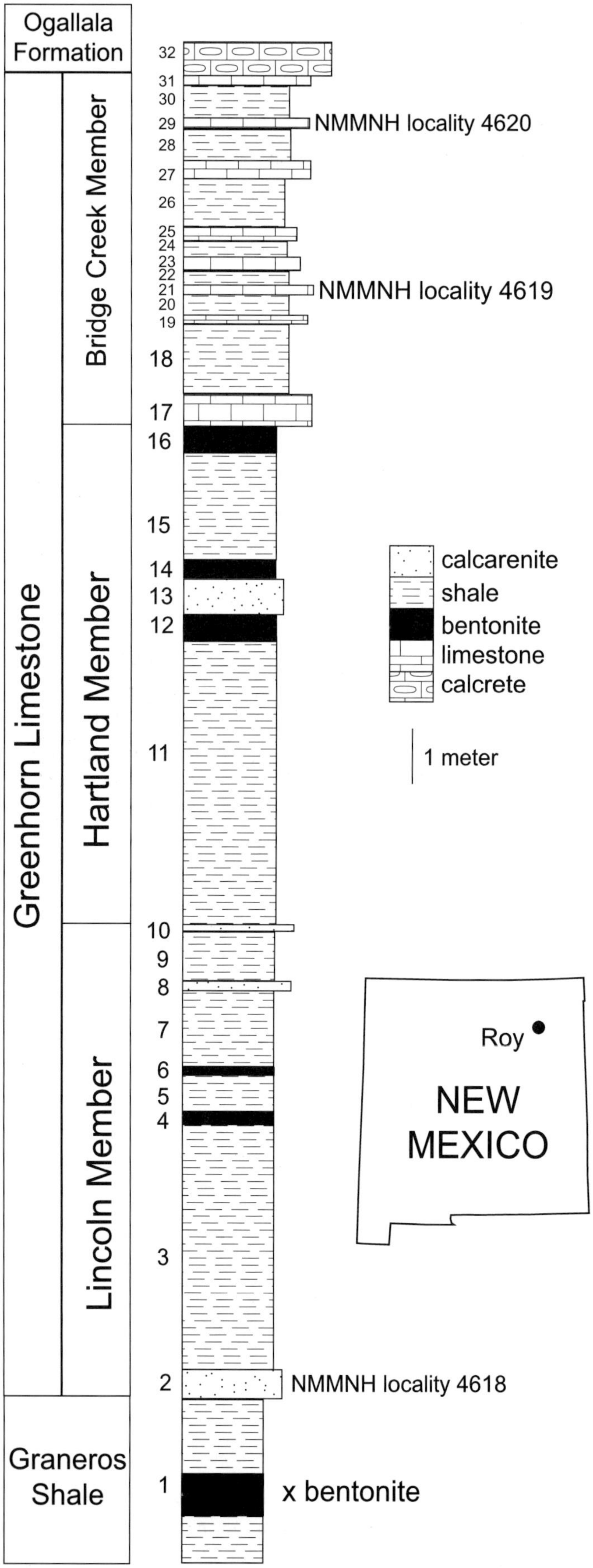

Bivalves

We collected both oysters and inoceramids from the Greenhorn strata near Roy. The oysters (Fig. 3A-B) are from the basal Lincoln Member calcarenite and have small, irregularly convex valves that often show linear, closely set denticles around the inner shell margin. They are readily assigned to *Ostrea beloiti* Logan (cf. Hattin, 1975, pl. 2b-e; Kauffman et al., 1977, p. 92, pl. 8; Lucas et al., 2000, figs. 4A-B).

Inoceramids are abundant in the upper part of the Bridge Creek Member (locality 4620) and are mostly *Mytiloides* (Fig. 3C-D). Most common are specimens of *M. puebloensis* Walaszczyk and Cobban. These specimens have a slightly convex growth axis, maximum inflation at the umbo, a very small posterior auricle and regular rugae with rounded edges, separated by evenly spaced microrugae (cf. Kennedy et al., 2000, p. 321, pls. 6-8, 10). Some small inoceramids from locality 4620 have single rugae in the juvenile stage (Fig. 3D), so we tentatively refer them to *Mytiloides goppelnensis* Badillet and Sornay (cf. Kennedy et al., 2000, p. 323, pls. 9-10).

Ammonites

Ammonites are abundant as chalky molds and casts at NMMNH localities 4619 and 4620 (Fig. 3). Numerous specimens from locality 4619 (e.g., Fig. 3E) are readily assigned to *Euomphaloceras* sp. (cf. Cobban and Scott, 1972, p. 70-71, pls. 4, 5). These fairly evolute ammonites have straight, rectiradiate to prorsiradiate ribs. They have two tubercles on each rib, and the ribs are closely spaced and alternate as primaries and secondaries. One strong tubercle is on the ventrolateral shoulder, and the second is at the umbilicus. All specimens we collected are small, inner whorls, so we refrain from assigning them to a species.

A second common ammonite taxon from locality 4619 is the heteromorph *Sciponoceras gracile* (Shumard) (Fig. 3F). This taxon has a straight shell with a subcircular cross section. The shell surface has prorsiradiate ribs, and no suture is preserved (cf. Cobban and Scott, 1972, p. 47, pl. 17; Hattin, 1975, pl. 6).

The most common ammonite at NMMNH locality 4620 can be assigned to *Vascoceras (Greenhornoceras) birchbyi* Cobban and Scott (Fig. 3G-H). These subglobose ammonites are moderately involute with a deep umbilicus and have an essentially smooth body chamber. There are strong ribs and umbilical tubercles on the later inner whorls (cf. Cobban and Scott, 1972, p. 85, pls. 22-25).

A less common ammonite from locality 4620 is *Watinoceras coloradoense* (Henderson) (Fig. 3I-J). These are relatively large (diameter ~80 mm), moderately involute with a steep umbilical wall and have strong, rectiradiate ribs, each with a bullate umbilical tubercle, a clavate lower ventrolateral tubercle and a clavate upper ventrolateral tubercle that is the largest of the three tubercles (cf. Cobban and Scott, 1972, p. 76, pls. 27-28).

FIGURE 1. Measured section of the Greenhorn Formation near Roy, Harding County, New Mexico. See the Appendix for description of the numbered lithologic units.

FIGURE 2. Selected Upper Cretaceous outcrops near Roy, Harding County. A. Overview of Graneros Shale (G) and Greenhorn Formation (GR) outcrop along border of secs. 12 and 13, T20N, R25E. B. Basal *Ostrea beloiti*-bearing calcarenite bed of Lincoln Member of Greenhorn Formation (L) above Graneros Shale (G) (bed 2 of measured section in Figure 1). C. Typical calcareous shale and thin calcarenite of Hartland Member of Greenhorn Formation (beds 11-12 of measured section in Figure 1). D. Characteristic limestone bed of Bridge Creek Member of Greenhorn Formation (bed 23 of measured section in Figure 1).

Selachian Teeth

Selachian teeth from the basal calcarenite of the Lincoln Member at NMMNH locality 4618 resemble those described by Lucas et al. (2000) from the same unit in Quay County, New Mexico. Near Roy, the most common teeth are of *Ptychodus anonymus* Williston. These teeth are relatively small, have a weakly conical high cusp, fine transverse ridges and a gentle transition between the cusp and granular marginal area. *P. anonymus* is a common Cenomanian and Turonian taxon (Welton and Farish, 1993, p. 57).

BIOSTRATIGRAPHY

Ammonites from the Bridge Creek Member of the Greenhorn Formation near Roy identify two of the ammonite zones characteristic of that unit elsewhere (e.g., Kennedy et al., 2000), the uppermost Cenomanian *Sciponoceras gracile* zone and the earliest Turonian *Vascoceras birchbyi* zone. Therefore, the Cenomanian-Turonian boundary is between localities 4619 and 4620 in the Bridge Creek Member near Roy. The younger ammonite zones normally present in the Bridge Creek Member—*Mammites nodosoides* and *Collignoniceras woollgari* zones—are not present at Roy, having been stripped off by Neogene erosion.

ACKNOWLEDGMENTS

We thank W. A. Cobban and J. I. Kirkland for comments on an earlier version of this manuscript. The NMMNH provided logistical support for this research.

REFERENCES

Clemons, R. E., 1982, New Mexico geological highway map: New Mexico Geological Society, scale 1:1,000,000.

Cobban, W. A., and Scott, G. R., 1972, Stratigraphy and ammonite fauna of the Graneros Shale and Greenhorn Limestone near Pueblo, Colorado: U.S. Geological Survey, Professional Paper, 101 p.

Dane, C. H., and Bachman, G. O., 1965, Geologic map of New Mexico: U.S.

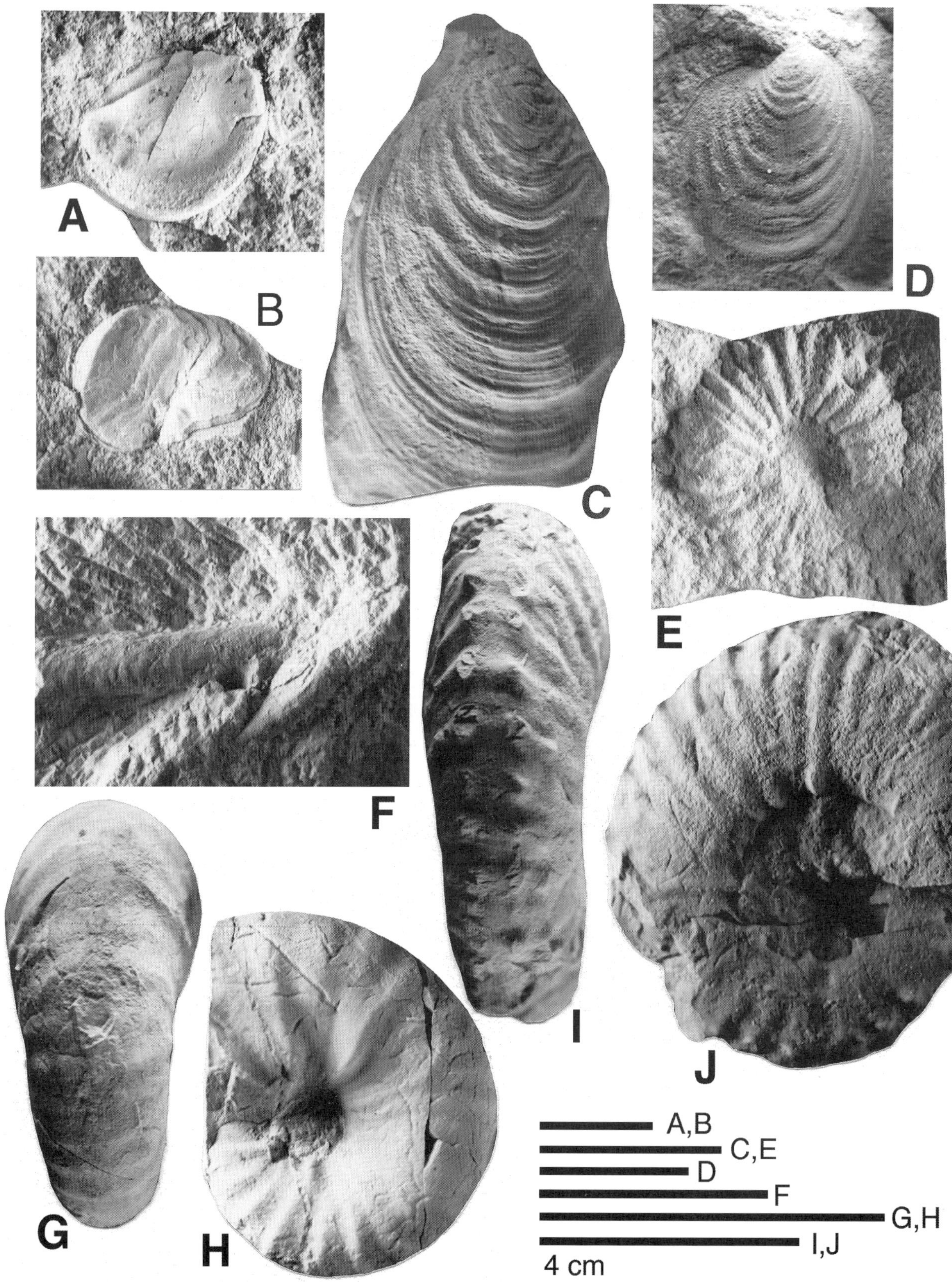

FIGURE 3. Selected fossils from the Greenhorn Formation near Roy, New Mexico. A-B. *Ostrea beloiti* Logan, NMMNH P-33150 from locality 4618. C. *Mytiloides puebloensis*, NMMNH P-33159 from locality 4620. D. *Mytiloides goppelnensis*, NMMNH P-33161 from locality 4620. E. *Euomphaloceras* sp., NMMNH P-33154 from locality 4619. F. *Sciponoceras gracile* (Shumard), NMMNH P-33152 from locality 4619. G-H. *Vascoceras* (*Greenhornoceras*) *birchbyi* Cobban and Scott, NMMNH P-33162 from locality 4620. I-J. *Watinoceras coloradoense*, NMMNH P-33157 from locality 4620.

Geological Survey, scale 1:500,000.

Goddard, E. N., Trask, P. D., DeFord, R. K., Rove, O. N., Singewald, G. T. J., and Overbeck, R. M., 1984, Rock Color Chart: Boulder, Geological Society of America.

Hattin, D. E., 1975, Stratigraphy and depositional environment of Greenhorn Limestone (Upper Cretaceous) of Kansas: Kansas Geological Survey Bulletin 209, 128 p.

Kauffman, E.g., Hattin, D. E., and Powell, J. D., 1977, Stratigraphic, paleontologic, and paleoenvironmental analysis of the Upper Cretaceous rocks of Cimarron County, northwestern Oklahoma, Geological Society of America Memoir, 150 p.

Kennedy, W. J., Walaszczyk, I. and Cobban, W. A., 2000, Pueblo, Colorado, USA, candidate global boundary stratotype section and point for the base of the Turonian Stage of the Cretaceous, and for the Middle Turonian substage, with a revision of the Inoceramidae (Bivalvia): Acta Geologica Polonica, v. 50, p. 295-334.

Lucas, S. G., Estep, J. W., Pence, R., and Hunt, A. P., 2000, Upper Cretaceous outlier in Quay County, New Mexico: Stratigraphy and biostratigraphy: New Mexico Museum of Natural History and Science, Bulletin 16, p. 121-127.

Wanek, A. A., 1962, Reconnaissance geologic map of parts of Harding, San Miguel, and Mora counties, New Mexico: U.S. Geological Survey, Oil and Gas Investigations Map OM-208, scale 1:96,000.

Welton, B. J., and Farish, R. F., 1993, The collector's guide to fossil sharks and rays from the Cretaceous of Texas. Lewisville, Texas, Before Time, 204 p.

APPENDIX—DESCRIPTION OF MEASURED SECTION

Stratigraphic section measured in two segments. Base of the first segment (unit 1) is at UTM 569352E, 3980960N, ending with unit 17 at UTM 569571E, 3980813N. The second segment begins with unit 18 at UTM 569257E, 3980494N and ends at 569300E, 3980495N. All UTM coordinates are in zone 13, NAD27. Rock colors follow Goddard et al. (1984). Section measured 8 July, 2000.

unit	lithology	thickness (m)
Ogallala Formation:		
32	White and red calcrete.	**Not measured**
unconformity		
Greenhorn Formation: Bridge Creek Member:		
31	Limestone, same color and lithology as unit 21.	**0.1**
30	Shale; calcareous; mostly covered.	**0.7**
29	Limestone, same color and lithology as unit 21; NMMNH locality 4620.	**0.1**
28	Shale; calcareous; mostly covered.	**0.6**
27	Limestone, same color and lithology as unit 21. Stratigraphic horizon of NMMNH locality 4620.	**0.2**
26	Shale; calcareous; mostly covered.	**0.9**
25	Limestone, same color and lithology as unit 21.	**0.3**
24	Shale; calcareous; mostly covered.	**0.3**
23	Limestone, same lithology as unit 21. Stratigraphic horizon of NMMNH locality 4619.	**0.1**
22	Shale, same lithology as unit 20.	**0.2**
21	Limestone; light greenish gray (5GY8/1) to light gray (N7); NMMNH locality 4619.	**0.1**
20	Shale; calcareous; mostly covered.	**0.4**
19	Limestone, bioclastic; yellowish gray (5Y8/1) to very pale orange (10YR8/2).	**0.1**
18	Shale and calcarenite; shale medium gray (N5) and calcareous; calcarenite is grayish orange (10YR7/4).	**1.2**
17	Limestone; light gray (N7) to very pale orange (10YR8/2); forms a ledge.	**0.4**
Greenhorn Formation: Hartland Member:		
16	Bentonite.	**0.5**
15	Silty shale; grayish orange (10YR7/4); calcareous; bentonitic.	**1.9**
14	Bentonite.	**0.2**
13	Calcarenite; pale yellowish brown (10YR6/2) and yellowish gray (5Y8/1); clasts are shell hash.	**0.7**
12	Bentonite.	**0.1**
11	Calcareous shale; mostly covered.	**6.0**
Greenhorn Formation: Lincoln Member:		
10	Calcarenite; same color and lithology as unit 8.	**0.1**
9	Shale, same colors and lithology as unit 5.	**0.8**
8	Calcarenite; very pale orange (10YR8/2); calcareous.	**0.8**
7	Shale, same colors and lithology as unit 5.	**1.5**
6	Bentonitic claystone; mottled yellowish gray (5Y8/1), dark yellowish orange (10YR6/6) and very pale orange (10R8/2); calcareous.	**0.1**
5	Shale; yellowish gray (5Y8/1) and very pale orange (10YR8/2); calcareous.	**0.7**
4	Bentonitic claystone; mottled yellowish gray (5Y8/1), dark yellowish orange (10YR6/6) and very pale orange (10R8/2); calcareous.	**0.1**
3	Shale; mostly covered.	**4.5**
2	Calcarenite; very pale orange (10R8/2); laminated; locally coquinoid; forms a ledge; NMMNH locality 4618.	**0.4**
Graneros Shale:		
1	Shale and siltstone, pale yellowish brown (10YR6/2); shale is very calcareous; includes 0.9-m-thick bentonite bed (x bentonite) about 1.6 m below top of unit.	**3.0+**

The spillway at Clayton Lake in Union County exposes sandstone and siltstone beds in the Lower Cretaceous Mesa Rica and Pajarito formations that are covered with dinosaur tracks. Here, volunteers from the New Mexico Museum of Natural History are working on a map of the tracksite.

PALEONTOLOGICAL RESOURCES OF LAKE MEREDITH NATIONAL RECREATION AREA AND ALIBATES FLINT QUARRIES NATIONAL MONUMENT, WEST TEXAS

ADRIAN P. HUNT[1] AND VINCENT L. SANTUCCI[2]

[1]Mesalands Dinosaur Museum, Mesa Technical College, 911 South Tenth Street, Tucumcari, NM 88401;
[2]Fossil Butte National Monument, P. O. Box 592, Kemmerer, WY 83101

Abstract. – Lake Meredith National Recreation Area and Alibates Flint Quarries National Monument are units of the National Park Service located northeast of Amarillo in the Panhandle of Texas. They preserve a sequence of sedimentary rock units that range from Late Permian to Holocene in age. These strata have yielded significant paleontological resources, but they have not been widely reported upon. The fossils are Triassic, Miocene-Pliocene, Pleistocene and Holocene in age and include trace fossils and plant, invertebrate and vertebrate specimens. The most important specimen found to date is a female skull of the giant Pleistocene bison, *Bison latifrons*.

INTRODUCTION

Lake Meredith National Recreation Area and Alibates Flint Quarries National Monument are located northeast of Amarillo in the Panhandle of Texas (Fig. 1). Lake Meredith was established primarily to provide a surface water supply for 11 municipalities in the Texas Panhandle experiencing increased urbanization and industrialization and a lowering water table. Construction of Sanford Dam, named for the adjacent community, was started in 1962 and the resultant reservoir was named Lake Meredith for A. A. Meredith, a civic leader who was instrumental in the completion of the project. Water is pumped to the area communities, including Amarillo and Lubbock, via 518 km of pipeline, 10 pumping plants and 3 regulating reservoirs. The Canadian River Municipal Water Authority operates and maintains the facilities. Since 1965, a recreation area of over 41,000 acres around the reservoir has been managed by the National Park Service, since 1972 as Lake Meredith National Recreation Area (Fig. 2).

Alibates Flint Quarries National Monument was established by Congress in 1965 to protect prehistoric chert quarries. This area has been quarried since Clovis time for the distinct and abundant chert. Flint is concentrated within the monument in a horseshoe-shaped area that includes 736 quarry pits that are mainly oblong in shape. The current monument includes 1,370.97 acres adjacent to the east-central portion of Lake Meredith National Recreation Area.

Lake Meredith National Recreation Area and Alibates Flint Quarries National Monument have yielded significant paleontological specimens, but they have been poorly documented. The purpose of this paper is to document what is currently known of the paleontological resources of the area for scientific reasons and to act as an aid to resource management, interpretation, and strategic planning for the National Park Service. This study is based on fieldwork at Lake Meredith National Recreation Area and Alibates Flint Quarries National Monument, study of museum collections and a literature review. WTSU refers to West Texas State University, now West Texas A and M University, Canyon and PPHM to Panhandle Plains Historical Museum, Canyon.

GEOLOGIC OVERVIEW

Lake Meredith National Recreation Area and Alibates Flint Quarries National Monument straddle the boundary between two major structural basins - the Palo Duro basin and the Anadarko basin. This boundary is formed by the Amarillo-Wichita uplift that is a large regional structure that extends southeast-northwest from New Mexico to central Oklahoma. Because of this structure, the stratigraphy of the southwestern end of Lake Meredith National Recreation area is distinct from the remainder of the recreation area and from Alibates Flint Quarries National Monument. Notably, Triassic strata are only preserved in the southwest portion of the study area (Palo Duro basin).

There is limited primary geologic literature on the local geology. The most important reference is the Master's thesis of Wilson (1988) which includes a detailed study and mapping of four U. S. Geological Survey 7.5 quadrangles (Alibates Ranch, Chunky, Berry Sand Draw and McDowell Creek). A useful older reference is the 1963 guidebook of the Panhandle Geological Society "Field trip, September 14, 1963, Alibates Flint Quarries, Alibates Indian Ruin, Santa Fe Trail, Sanford Dam." Summaries of the geology and paleontology of the area can be found in sections of the draft EIS for the area written by Lisa Norby and Vincent Santucci, respectively (Santucci, 2000).

Permian

There are four distinct Upper Permian (Ochoan) rock units exposed in the study area, but they have a confused stratigraphic nomenclature (Wilson, 1988)(Table 1). The medial unit is the 6-7 m-thick Alibates Dolomite or Alibates Formation, the latter name being utilized here because this unit contains lithologies other than dolostone. Above this unit are 13 m of red beds that have been referred to the Quartermaster Formation/Group or Dewey Lake Formation and below it are more red beds that have been referred to as the Artesia Group or Whitehorse Sandstone/Formation/Group or Salado-Tansill Formation. At the base of the lower red-

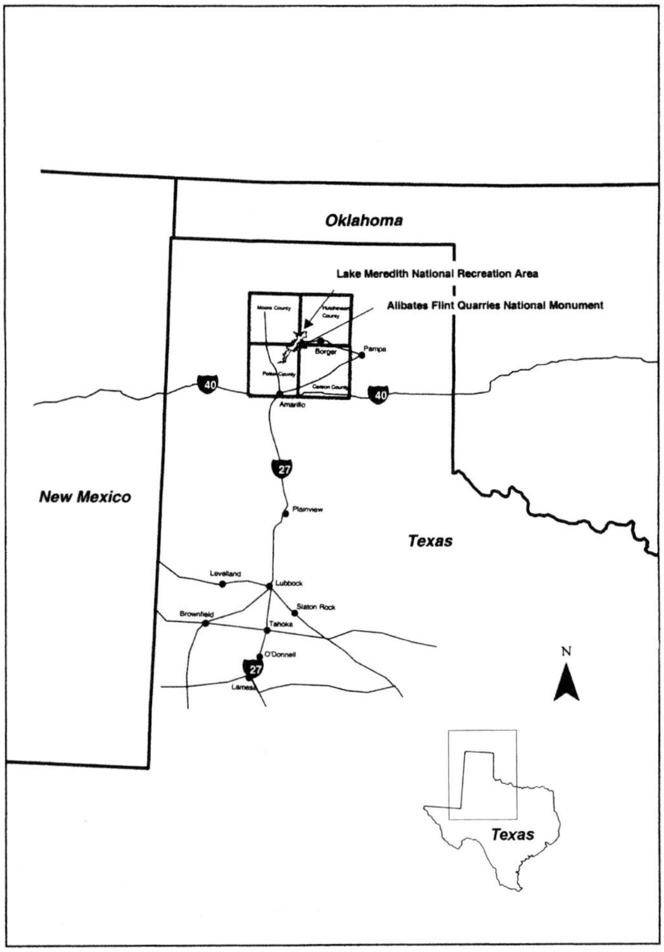

FIGURE 1. Location of Lake Meredith National Recreation Area and Alibates Flint Quarries National Monument, Panhandle of Texas.

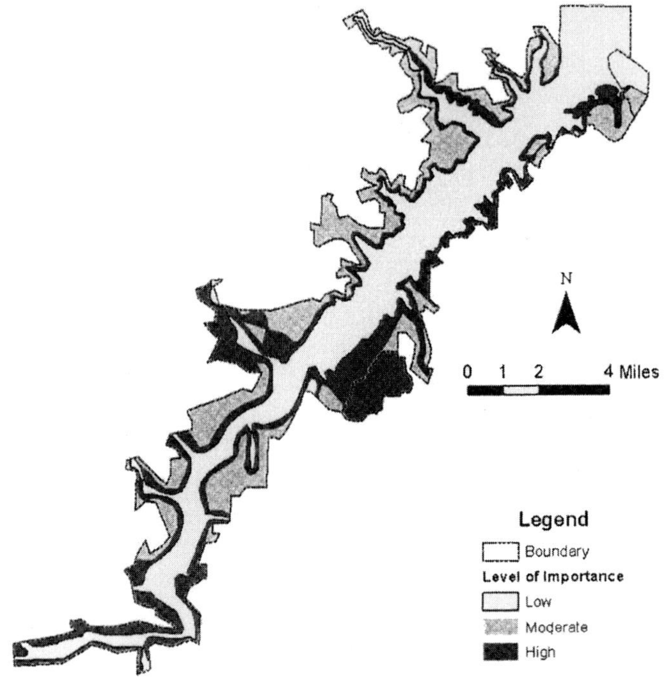

FIGURE 2. Map showing boundaries of Lake Meredith National Recreation Area and Alibates Flint Quarries National Monument and paleontological sensitivity ratings of different outcrop areas.

beds is a prominent gypsum originally called the Saddlehorse Gypsum and now referred to the Cloud Chief Gypsum. The stratigraphic nomenclature that we will use is, from youngest to oldest: Dewey Lake Formation, Alibates Formation, Whitehorse Formation and Cloud Chief Gypsum. These Permian strata were deposited in restricted marine and intertidal settings (Wilson, 1988).

Triassic

The Upper Triassic (upper Carnian) strata in the western portion of Lake Meredith National Recreation area are the Tecovas and Trujillo formations of the Chinle Group (sensu Lucas, 1993, 2001)(= Dockum Formation of older workers). The lowermost unit is the Tecovas Formation, which is up to 60 m thick and consists of mudstone and lesser amounts of sandstone and conglomerate of fluvial origin. The Trujillo is up to 13 m thick and is mostly fluvial sandstone with minor conglomerate and mudstone. Locally, basal conglomerates in the Tecovas can be assigned to the Camp Springs Conglomerate (Table 1).

TABLE 1. Overview of stratigraphy and paleontology of Lake Meredith National Recreation Area and Alibates Flint Quarries National Monument.

Age	Formation/Member	Environment of deposition	Paleontology
Holocene	Unnamed sands, gravels, soils	Fluvial, soil	Diverse mammals, reptiles, birds, fish
Pleistocene	Unnamed gravels, sands, conglomerate, loess	Fluvial, eolian, lacustrine	*Bison latrifrons*, mammoth, gastropods, Cretaceous oysters, plants, wood, trace fossils
Miocene-Pliocene	Ogallala Formation:		
	"Caprock"	Paleosol	Unfossiliferous
	Coetas Member	Fluvial, lacustrine	"Mastodon", mammals, fish, gastropods, plants, trace fossils
	Potter Creek Member	Fluvial	Bone scrap, Cretaceous oysters
Triassic	Trujillo Formation	Fluvial	Unfossiliferous
	Tecovas Formation	Fluvial	Amphibians, reptiles, logs
	Camp Springs Conglomerate	Fluvial	Unfossiliferous
Permian	Dewey Lake Formation	Restricted marine	Unfossiliferous
	Alibates Formation	Intertidal	Unfossiliferous
	Whitehorse Formation	Restricted marine	Unfossiliferous
	Cloud Chief Gypsum	Restricted marine	Unfossiliferous

Miocene-Pliocene

Vertebrate fossils date the Ogallala Formation (or Group) as Miocene-Pliocene (Wood et al., 1941; Schultz, 1977). In general, the Ogallala in the Panhandle consists of a basal conglomerate, medial sandy interval and an upper calcrete-rich interval. These units are all of fluvial origin except for the calcrete that is pedogenic.

The lower Potter Creek Member consists of conglomerate and conglomeratic sandstone that formed in low-sinuosity streams; it is locally 130 m thick. The medial sandy unit is the Coetas Member, which is about 55 m thick and consists of sandstone, mudstone and locally flaggy limestone. Wilson (1988) separated the lower 33.5 m of this unit as his LX member which was dominantly composed of sandstone and silty sandstone that he distinguished from an upper unit (his Coetas Member) that included flaggy limestone, sandy limestone and thinly laminated mudstone and siltstone. The name LX Member is still unpublished and thus is an informal name. The upper unit of the Ogallala is up to about 75 m thick and includes sandstone, gravel and calcrete, including a capping calcrete caprock that is up to 10 m thick.

Pleistocene and Holocene.

The study area contains significant Pleistocene deposits that include terrace gravels, fluvial sandstones and conglomerates, alluvial sandstone and loess. Holocene deposits in the study area include alluvium, eolian sand and soils.

PALEONTOLOGY

Permian

There are no known fossils from the Permian strata of Lake Meredith National Recreation area or Alibates Flint Quarries National Monument. The restricted marine environment represented by these rocks indicates a low chance of fossil preservation. However, rare footprints have been found in intertidal deposits associated with similar marine environments in the Permian Mesita Blanca Member of the Yeso Formation in central New Mexico (Robert Colpitts, personal commun., 1994). However, Late Permian fossils are rare worldwide, and any occurrences in the study area would be of great significance.

Triassic

The Panhandle of Texas in general, and the Canadian River Valley in particular, have yielded very significant Triassic vertebrate fossils as well as abundant petrified wood and smaller numbers of invertebrate fossils (Murry, 1986, 1989). These fossils include a variety of reptiles notably phytosaurs, aetosaurs and rauisuchians, and metoposaurid amphibians. Regionally, the Chinle of West Texas is one of the most important sources of Late Triassic vertebrate fossils in the world.

Wilson (1988) reported numerous fragments of Triassic amphibians and reptiles from the Tecovas Formation, although it is unclear how these localities relate to the boundaries of the recreation area.

The Triassic rocks of the Panhandle of Texas have also yielded significant plant fossils, including the best-preserved specimen of the enigmatic *Sanmiguelia*. Wilson (1988) reported fossil logs up to 4 m long and 1 m in diameter on the southwest rim of Corral Creek in the Tecovas Formation. Wes Phillips (personal commun., 2000) also indicated that significant petrified wood had been found near the west end of Lake Meredith National Recreation Area.

In conclusion, Lake Meredith National Recreation Area and Alibates Flint Quarries National Monument are located in a region where the Late Triassic rocks contain abundant fossils, and there is evidence that some fossils have been found in the study area, even though there has been no systematic paleontological exploration. The Triassic rocks thus have great potential for yielding significant specimens.

Miocene-Pliocene

There are at least six documented localities in the Ogallala Formation at Lake Meredith National Recreation Area and Alibates Flint Quarries National Monument:

1. A "mastodon" (probably gomphothere) tooth, now lost, was found at Cedar Canyon when a restroom pipe was placed. The tooth derived from the medial sandy interval of the Ogallala.

2. Wes Phillips (personal commun., 2000) reported a locality with *in situ* bone including turtle specimens in Alibates Flint Quarries National Monument.

3. Wes Phillips (personal commun., 2000) reported a bonebed with multiple elements in the Turkey Creek area.

4. Wilson (1988) reported root casts, silicified grass anthoecia, endocarps of *Celtis* sp. and insect burrows from the "LX Member. He also mentioned that vertebrate fossils of Clarendonian (late middle Miocene) age are found in the upper 6-7 m of the LX Member.' However, it is unclear what specimens were found within the boundaries of Lake Meredith Recreation Area or Alibates Flint Quarries National Monument.

5. Wilson (1988) reported molds of gastropods, imprints of two fish and Clarendonian (late middle Miocene) vertebrate fossils from thin, flaggy beds of the upper Coetas Member (above his LX Member). Again, it is unclear what specimens were found within the boundaries of Lake Meredith Recreation Area or Alibates Flint Quarries National Monument.

6. Wilson (1988) reported a large number of Cretaceous oysters (presumably *Texigryphaea*), clasts of Cretaceous fossiliferous limestone and bone scraps from the Potter Member. Again, it is unclear what specimens were found within the boundaries of Lake Meredith Recreation Area or Alibates Flint Quarries National Monument.

The last three Cenozoic land mammal "ages" (Clarendonian, Hemphillian, Blancan) were named for the abundant vertebrate faunas from the Texas Panhandle (Wood et al., 1941; Schultz, 1977; Lucas and Morgan, 2001). These ages are standards of reference for North America during the late Cenozoic. They were based on faunas from the Texas Panhandle because of the abundance and diversity of fossils of this age in this area. There are at least six known localities within the study area, and possibly

FIGURE 3. *Bison latifrons* skulls at Panhandle Plains Museum. A. WTSU unnumbered, female skull from Lake Meredith National Recreation Area, Potter County, Texas. B. PPHM 2315-1, male skull, Waters Ranch, Lipscomb County, Texas. Scale bars are 25 cm.

several more, depending on the details of Wilson's (1988) discoveries. The known paleontological potential of the region and the fact that several sites have been found within the study area, without any paleontological exploration, suggests a strong potential for finding significant fossils of this age in Lake Meredith National Recreation Area and Alibates Flint Quarries National Monument.

Pleistocene

There are 5 documented sites of Pleistocene age in Lake Meredith National Recreation Area and Alibates Flint Quarries National Monument.

1. The most significant, and best documented, site produced a bison skull and other specimens in the 1970's (Anderson, 1977; Dalquest and Schultz, 1992). The principal, and first specimen, is a partial skull of a large bison with both horn cores complete almost to their tips and is at the Panhandle Plains Historical Museum, Canyon (Fig. 3A). This specimen (WTSU unnumbered) has a single horn core length on the upper curve of 520 mm (measurements follow McDonald, 1981)), on the lower curve of 540 mm, with a basal circumference of 305 mm and vertical and transverse diameters of 110 mm (Anderson, 1977). This specimen is referable to the giant Pleistocene bison, *Bison latifrons,* on the basis of an upper curve length of the horn core of greater than 500 mm and to a female of the species on the basis of a basal horn core circumference of less than 400 mm (McDonald, 1981). Females of the species have more gracile horn cores than males (Fig. 3B). Complete specimens of female *Bison latifrons* are rare and McDonald (1981, table 19) lists only one other.

Anderson (1977) examined the site and found additional specimens, including a bison rib fragment and assorted snails, as well as a rodent burrow with charcoal that could be Holocene in age. The bones were found in a layer of fine sand and gravel about 2-m-thick. The gastropods and rodent burrow occur in another fine sand that is stratigraphically lower than the bone-bearing sand and is separated from it by a 1.5 m-thick layer of coarse gravel (Anderson, 1977). This locality is on the north side of Lake Meredith, east of the boat ramp near Plum Creek, in an inset terrace.

2. A mammoth humerus (?) was discovered, and removed prior to study, from just west of Alibates Flint Quarries National Monument. With information from Wes Phillips, Jim Rancier and APH rediscovered this locality. It is located on a cliff of alluvium adja-

cent to an oil well. The bone came from a 6.5 m-thick coarse, pebbly sandstone, that is overlain by 2 m of reddish brown conglomerate and underlain by 1 m of brown conglomerate that is poorly sorted and has clasts up to 10 cm in diameter. The conglomerates contain petrified wood and *Texigryphaea* shell fragments. This is part of an inset terrace that occurs over a fairly wide area.

3. In the area of South Canyon, there are thick exposures of the Lava Creek B (Pearlette type O) ash that dates to about 610,000 years ago. This was produced by the final eruptions of the Yellowstone caldera. One locality was investigated on a hill behind a well. The white ash bed is about 3-m-thick at this locality. The ash contains abundant trace fossils in the form of burrows. Also, the ash contains abundant plant debris and has the potential to produce leaves or other diagnostic megafossil remains as well as pollen. The lensoid geometry of the ash and its thickness at this locality suggests an aqueous pond deposit that probably contains microfossils of diatoms and ostracodes. Crayfish burrows have apparently been found in this ash (Wes Phillips, personal commun., 2000).

4. Petrified wood and *Texigryphaea* fragments were found in Pleistocene gravels near the Ogallala "mastodon" site mentioned previously.

5. Within the boundaries of Lake Meredith National Recreation Area and Alibates Flint Quarries National Monument there are several breccia pipes that are presumably associated with Pleistocene? dissolution. The Eagle's Nest was examined just west of Alibates Flint Quarries National Monument. This pipe extends through bedded Quartermaster strata. The infilling consists of chert and Permian sedimentary rocks and other lithologies. There is an outer rind consisting principally of rounded chert clasts. The only fossil found at this locality was a small piece of petrified wood. However, such pipes have the potential to be natural traps and to yield vertebrate fossils. These pipes may, in some cases, date back to the Miocene (Eck and Renfield, 1963).

In addition, it is important to mention two significant Pleistocene localities that lie within a few km of the northeast border of Lake Meredith National Recreation Area that demonstrate the potential of strata with the National Park Service lands.

i. The Sanford-Big Creek Local Fauna was collected from a roadcut on the north side of Farm-to-Market Road 687, 0.8 km west of Big Creek, approximately 3.2 km north of the town of Sanford in Section 39, Blk. 41, H and TC RR Co. Survey (Dalquest and Schultz, 1992). This diverse fauna is reposited at West Texas A & M University (Table 2).

ii. The Merchant Ranch locality lies 120 m south of Farm-to-Market Road 687 in the banks of a tributary of Big Creek, 1.6 km north of the Canadian River, 3.2 km north of Sanford in Section 39, Blk. 46, H and TC RR Co. Survey (Dalquest and Schultz, 1992). This fauna, reposited at the Panhandle Plains Historical Museum, includes a partial mammoth skeleton (Table 2).

Very significant vertebrate fossils have been found in the Pleistocene deposits of the study area, notably the *Bison latifrons* skull, without any systematic paleontological survey. These finds, together with the proximity of important localities to the study area indicates that the Pleistocene of Lake Meredith National Recreation Area and Alibates Flint Quarries National Monument has great potential to yield significant specimens in the future. It also appears that petrified wood and abraded *Texigryphaea* specimens are relatively common in conglomerates.

Holocene

Duffield (1964; 1970) reviewed the vertebrate material from most of the Panhandle Aspect archeological sites in the Lake Meredith National Recreation Area and Alibates Flint Quarries National Monument. These localities are Alibates 28, Pickett Ruin, Roper site, Sanford Ruin, Medford Ranch and Spring Canyon:

1. Alibates 28 was excavated by W.P.A. workers from 1938-1941 and is situated on the top edge of the bluffs overlooking Alibates Creek Canyon (Table 3).

2. In 1958, the Panhandle Plains Historical Museum and the Norpan Archeological Society excavated the Pickett Ruin site on the eastern side of the promontory that now forms the southern end of the Sanford Dam (Table 4).

3. The Roper site was excavated by the Panhandle Plains Historical Museum and the Norpan Archeological Society in 1957. This site is west-northwest of the town of Sanford on the west side of South Canyon (Table 4).

4. The Sanford Ruin site was excavated in 1953 by the Panhandle Plains Historical Museum. This site was on a promontory that now forms the southern end of Sanford Dam (Table 4).

5. The Medford Ranch site was excavated in 1961 by the Texas Archeological Salvage Project, a co-operative project of the University of Texas and the National Park Service to salvage sites in the vicinity of the proposed location of the Sanford Reservoir dam. This locality is on the west fork of Spring Canyon (Table 4).

TABLE 2. Pleistocene local faunas of the of the Lake Meredith Area, Hutchinson County, Texas (Dalquest and Schultz, 1992).

PLEISTOCENE LOCAL FAUNAS FROM NEAR LAKE MEREDITH NATIONAL RECREATION AREA
Sanford-Big Creek Local Fauna
Gastropoda (Snails - at least 12 species)
Fish
Frog
Snake
Castor canadensis (Beaver)
Geomys sp. (Pocket gopher)
Ondatra sp. (Muskrat)
Microtus pennsylvanicus (Meadow vole)
Canis cf. *latrans* (Coyote)
Canid sp. (Small fox)
Camelid sp. (Camel)
Odocoileus sp. (Deer)
Equus sp. (Horse)
Merchant Ranch Local Fauna
Castor canadensis (Beaver)
Geomyid (Pocket gopher)
Ondatra cf. *nebracensis* (Nebraska muskrat)
Mammuthus sp. (Mammoth)
Equus sp. (Horse)

6. The Spring Canyon site, on the eastern side of Spring Canyon, was excavated in 1961 by the Texas Archeological Salvage Project (Table 5).

It is clear that there are many paleontological specimens that have been found in an archeological context within Lake Meredith National Recreation Area and Alibates Flint Quarries National Monument. The collection housed at Lake Meredith National Recreation Area includes many vertebrate specimens, principally of *Bison* sp.

PALEONTOLOGICAL RESOURCES AND DEVELOPMENT OF OIL AND GAS RESOURCES

Lake Meredith National Recreation Area and Alibates Flint Quarries National Monument are unusual among National Park Service units in that they include active and extensive exploitation of oil and gas reserves (Santucci et al., in press). During preparation of an Oil and Gas Management Plan/Environmental Impact Statement (EIS) for Lake Meredith National Recreation Area and Alibates Flint Quarries National Monument, paleontological resource inventories were initiated for each of these National Park Service units. This document was written in 2000 in order to define a long-term management direction for existing and anticipated oil and gas operations. Specifically the document addressed the issues associated with the exercise of non-federal oil and gas interests underlying the parks and the activities undertaken in the existing transpark pipeline right-of-ways.

TABLE 3. Holocene vertebrate fauna from Alibates 28 archeological site, Alibates Flint Quarries National Monument (Duffield (1964; 1970).

VERTEBRATE FAUNA OF ALIBATES 28 ARCHEOLOGICAL SITE
Sylvilagus sp. (Cottontail rabbit)
Lepus californicus (Jack rabbit)
Cynomys ludovicianus (Prairie dog)
Geomys bursarius (Plains pocket gopher)
Pappogeomys castanops (Chestnut-faced pocket gopher)
Canis familiaris (Dog)
Canis latrans (Coyote)
Vulpes velox (Swift fox)
Taxidea taxus (Badger)
Canidae
Lynx rufus (Bobcat)
Odocoileus hemionus (Black-tailed deer)
Odocoileus sp. (Deer)
Deer and/or antelope
Antilocapra americana (Antelope)
Bison bison (Bison)
Bos taurus (Cow)
Olor buccinator(?) (Trumpeter swan)
Branta canadensis (Canada goose)
Anser albifrons (White-fronted goose)
Anas acuta (Pintail)
Anas discors (Blue-winged teal)
Spatula clypeata (Shoveler)
Aythya americana (Redhead)
Aythya vaslisineria (Canvasback)
Buteo jamaicensis (Red-tailed hawk)
Pediocetes phasianellus (Sharp-tailed grouse)
Bubo virginianus (Great horned owl)
Corvus sp. (Crow or raven)
Terrapene sp. (Box turtle)
Trionyx sp. (Softshell turtle)
Kinosternon flavescens (Mud turtle)
Pseudemys scripta elegans (Red-eared turtle)
Snake

TABLE 4. Holocene vertebrate fossils from archeological sites at Lake Meredith National Recreation Area (Duffield, 1964; 1970).

Archeological site	Vertebrate fauna
Pickett Ruin site	*Antilocapra americana* (antelope) *Bison bison* (Bison) *Terrapene* sp. (Box turtle)
Roper site	*Scalopus aquaticus* (Eastern mole) *Sylvilagus* sp. (Cottontail rabbit) *Lepus californicus* (Jack rabbit) *Geomys bursarius* (Plains pocket gopher) *Pappogeomys castanops* (Chestnut-faced pocket gopher) *Odocoileus* sp. (Deer) Deer and/or antelope *Antilocapra americana* (Antelope) *Bison bison* (Bison) Bird *Terrapene* sp. (Box turtle)
Sanford Ruin site	*Sylvilagus* sp. (Cottontail rabbit) *Spermophilus* sp. (Ground Squirrel) *Geomys bursarius* (Plains pocket gopher) *Pappogeomys castanops* (Chestnut-faced pocket gopher *Neotoma* sp. (Wood rat) Deer and/or antelope *Antilocapra americana* (Antelope) *Bison bison* (Bison) *Terrapene* sp. (Box turtle) Snake Frog and/or toad
Medford ranch site	*Sylvilagus* sp. (Cottontail rabbit) *Geomys bursarius* (Plains pocket gopher) *Odocoileus* sp. (Deer) *Antilocapra americana* (Antelope) *Bison bison* (Bison) Birds *Chelydra serpentina* (Snapping turtle) *Terrapene* sp. (Box turtle) *Trionyx mutica* (Spineless soft-shell turtle) *Trionyx spinifer*(?) (Spiney soft-shell turtle)

As part of the planning effort, the National Park Service developed a Reasonably Foreseeable Development Scenario (RFD) to project future oil and gas development in the parks and to provide a basis to measure potential environmental impacts. The RFD scenario estimated that over the next 15 to 20 years, in areas of the parks where drilling and production could be permitted, up to 85 new wells could be developed. Ground disturbing activities associated with oil and gas development can potentially damage or destroy non-renewable paleontological resources.

The Oil and Gas Management Planning Team identified the need to consider the paleontological resources in the planning process. A comprehensive paleontological resource inventory was conducted for Lake Meredith and Alibates Flint Quarries consisting of literature reviews, museum collections searches and field inventories (Hunt, 2000). A Paleontological Resource Sensitivity Map was developed identifying areas of high, moderate, and low probability for the occurrence of fossils (Fig. 2).

The Oil and Gas Management Plan/EIS identifies Standard Operating Procedures for Locating and Protecting Paleontological Resources (Santucci, 2000). These procedures outline circumstances when a paleontological survey should be required and how the survey should be implemented. The procedures also provide guidance in situations where there is an unanticipated discovery of fossils during approved operations or when damage occurs to fossils within previously identified paleontological localities.

Three alternative actions are identified in the EIS for paleontological resources. Alternatives B and C designate the entire park as a Special Management Area for protection of the paleontological resources. Additionally, Alternatives B and C indicate the application of the Standard Operating Procedures for Locating and Protecting Paleontological Resources and other park resources.

TABLE 5. Holocene vertebrate fauna from Spring Canyon archeological site, Lake Meredith National Recreation Area (Duffield, 1964; 1970).

Vertebrate fauna of Spring Canyon site
Sylvilagus sp. (Cottontail rabbit)
Lepus californicus (Jack rabbit)
Spermophilus sp. (Ground squirrel)
Cynomys ludovicianus (prairie dog)
Geomys bursarius (Plains pocket gopher)
Pappogeomys castanops (Chestnut-faced pocket gopher)
Neotoma cf. *albigula* (White-throated wood rat)
rodent
Canis familiaris (dog)
Taxidea taxus (Badger)
Odocoileus sp. (Deer)
Deer and/or antelope
Antilocapra americana (Antelope)
Bison bison (Bison)
Corvus sp. (Crow)
Duck
Passerine bird
Terrapene sp. (Box turtle)
Trionyx sp. (Soft-shell turtle)
Turtle
Snake
Fish
Cypridae
Ictalurus sp. (Catfish)

The Lake Meredith National Recreation Area and Alibates Flint Quarries National Monument Oil and Gas Management Plan and Environmental Impact Statement represents the first time that paleontological resources have been considered in this type of planning effort within a National Park Service area. The benefits resulting from the consideration of fossils in the planning process include: the completion of paleontological resource inventories for the sites, the development of new Standard Operating Procedures for locating and protecting fossils, and an increase in park management's ability to better protect the non-renewable paleontological resources at Lake Meredith and Alibates Flint Quarries (Santucci et al., in press).

CONCLUSIONS

Lake Meredith National Recreation Area and Alibates Flint Quarries National Monument are located in the Texas Panhandle, which has produced abundant and important collections of fossils of Triassic, Miocene-Pliocene, Pleistocene and Holocene age (Table 1). All of these ages of rocks are found within the study area. In addition, all these ages of rocks have produced important fossils within Lake Meredith National Recreation Area and Alibates Flint Quarries National Monument despite the fact that there has been no paleontological survey of any part of the area.

The fossils from Lake Meredith National Recreation Area and Alibates Flint Quarries National Monument are Triassic, Miocene-Pliocene, Pleistocene and Holocene in age and include trace fossils and plant, invertebrate and vertebrate specimens. This represents a remarkable range of fossils for such a restricted geographic area. It is evident that paleontology represents a significant resource at Lake Meredith National Recreation Area and Alibates Flint Quarries National Monument.

ACKNOWLEDGMENTS

This study would not have been possible without the help of Jim Rancier, Chief of Resource Management and Wes Phillips, retired Ranger, at Lake Meredith National Recreation Area and Alibates Flint Quarries National Monument and Lisa Norby of the National Park Service Geologic Resources Division. Spencer Lucas and Gary Morgan provided very useful reviews of this paper.

REFERENCES

Anderson, B. A., 1977, Overview of bison remains from the Plum Creek area, Lake Meredith Recreation Area, Texas: Southwest Cultural Resources Center, Santa Fe, unpublished report on file at Lake Meredith National Recreation Area, 10 p.

Dalquest, W. W. and Schultz, G. E., 1992, Ice Age mammals of northwestern Texas: Wichita Falls, Midwestern State University Press, 309 p.

Duffield, L. F., 1964, Three Panhandle Aspect sites at Sanford Reservoir, Hutchinson County, Texas: Bulletin of Texas Archeological Society, v. 35, p. 19-81.

Duffield, L. F., 1970, Some Panhandle Aspect sites in Texas: their vertebrates and paleoecology [Ph.D. dissertation]: Madison, University of Wisconsin, 505 p.

Eck, W. and Redfield, R. C., 1963, Geology of Sanford Dam, Borger, Texas, *in* Anthony, E. D., Jr., ed., Field trip, September 14, 1963 Alibates Flint Quarries, Alibates Indian Ruin, Santa Fe Trail, Sanford Dam: Amarillo, Panhandle Geological Society, Guidebook, p. 54-59.

Hunt, A. P., 2000, Preliminary assessment of the paleontological resources of Lake Meredith National Recreation Area and Alibates Flint Quarries National Monument: Unpublished National Park Service Report on file at Lake Meredith National Recreation Area, 17 p.

Lucas, S. G., 1993, The Chinle Group: revised stratigraphy and biochronology of Upper Triassic nonmarine strata in the western United States, in Morales, M., ed., Aspects of Mesozoic Geology and Paleontology of the Colorado Plateau: Flagstaff, Museum of Northern Arizona Bulletin, v. 59, p. 27-50.

Lucas, S. G., 2001, Abandon the term Dockum!: New Mexico Geological Society, 52nd Field Conference, Guidebook, this volume.

Lucas, S. G. and Morgan, G. S., 2001, Neogene land-mammal ages in the Texas Panhandle: New Mexico Geological Society, 52nd Field Conference, Guidebook, this volume.

McDonald, J. N., 1981, North American Bison: their Classification and Evolution: Berkeley, University of California Press, 315 p.

Murry, P. A., 1986, Vertebrate paleontology of the Dockum Group, western Texas and eastern New Mexico; *in* Padian, K., ed., The Beginning of the Age of Dinosaurs: Faunal Change Across the Triassic-Jurassic Boundary: Cambridge, Cambridge University Press, p. 109-137.

Murry, P. A., 1989, Geology and paleontology of the Dockum Formation (Upper Triassic), West Texas and eastern New Mexico; *in* Lucas, S. G. and Hunt, A. P., eds., The Dawn of the Age of Dinosaurs in the American Southwest: Albuquerque, New Mexico Museum of Natural History and Science, p. 102-144.

Phillips, J. W., 2000, Paleontology of the Lake Meredith area. Unpublished report on file at Lake Meredith National Recreation area, 4 p.

Santucci, V. L., 2000. Standard Operating Procedures for locating and protecting paleontological resources (Appendix F), *in* Lake Meredith National Recreation Area and Alibates Flint Quarries National Monument Oil and Gas Management Plan/ Environmental Impact Statement: Denver, National Park Service, p. F1-F2.

Santucci, V. L., Hunt, A. P. and Norby, L., in press, Oil/gas management planning and the protection of paleontological resources: Park Science.

Schultz, G. E., ed., 1977, Guidebook field conference on late Cenozoic biostratigraphy of the Texas Panhandle and adjacent Oklahoma, August 4-6, 1977: Canyon, West Texas State University, Kilgore Research Center, Department of Geology and Anthropology Special Publication, v. 1, 145 p.

Wilson, G. A., 1988, The effects of subsurface dissolution of Permian salt on the deposition, stratigraphy and structure of the Ogallala Formation (Late Miocene age), northeast Potter County, Texas [M.S. thesis]: Canyon, West Texas State University (dba West Texas A & M University), 254 p.

Wood, H. E., Chaney, R. W., Chaney, J., Colbert, E. H., Jepsen, G. L., Reeside, J. B. and Stock, C., 1941, Nomenclature and correlation of the North American continental Tertiary: Bulletin of the Geological Society of America, v. 51, p. 1-48.

A REVIEW OF PLEISTOCENE VERTEBRATE FAUNAS FROM NORTHEASTERN NEW MEXICO

GARY S. MORGAN[1], SPENCER G. LUCAS[1], PAUL L. SEALEY[1] AND ADRIAN P. HUNT[2]

[1]New Mexico Museum of Natural History, 1801 Mountain Rd., NW, Albuquerque, NM 87104, and [2]Mesalands Dinosaur Museum, Mesa Technical College, 911 S. 10th St. Tucumcari, NM 88401

Abstract.—There are 45 known Pleistocene vertebrate localities in northeastern New Mexico. The number of Pleistocene vertebrate sites (in parentheses) in each of the northeastern New Mexico counties is as follows: Colfax (2), Union (1), Mora (2) , Harding (2), San Miguel (2), Quay (6), Torrance (10), Guadalupe (2), De Baca (6), Curry (0), and Roosevelt (12). This region comprises part of the southwestern margin of the Southern High Plains. The western portions of several of these counties do include some of the higher mountain ranges in New Mexico, but only two fossil sites are located in this mountainous region. Twenty-eight sites in northeastern New Mexico contain the mammoth *Mammuthus*, a genus characteristic of North American Pleistocene (Irvingtonian and Rancholabrean) faunas. There is no definitive evidence for Irvingtonian faunas in this region. Twenty-one sites in northeastern New Mexico contain *Bison*, the diagnostic genus of Rancholabrean faunas. Even though the other 24 sites lack *Bison*, most of these are probably Rancholabrean as well. Several sites have associated radiocarbon dates, clearly establishing a late Pleistocene (late Rancholabrean) age. Eight latest Pleistocene sites in this region document the association of extinct megafauna with Paleoindian artifacts, including Blackwater Draw/Clovis, Folsom, Manzano Cave, Lucy, San Jon, Stolle, Milnesand, and Ted Williamson. These localities are best known as Paleoindian archaeological sites, but we stress their vertebrate faunas.

Most of the Pleistocene vertebrate faunas from northeastern New Mexico are low diversity assemblages. Only ten of the 45 sites have three or more species of mammals. Twenty sites contain a single mammal (13 sites with *Mammuthus*, four with *Bison*, and three with *Equus*), and ten sites contain two mammals (six sites with *Mammuthus* and *Equus*, three with *Equus* and *Camelops*, and one with *Mammuthus* and *Bison*). The faunas are dominated by large grazing mammals, including *Mammuthus*, *Bison*, *Equus*, *Camelops*, and *Paramylodon*, in decreasing order of abundance. The most diverse vertebrate fauna in northeastern New Mexico is Blackwater Draw Locality 1, also called the Clovis Site, which includes more than 60 species. The mammalian fauna from Blackwater Draw (combined Gray Sand and Brown Sand Wedge local faunas, ranging from 13,000-11,000 yr B. P.) consists of 39 species, including 13 members of the extinct Pleistocene megafauna. The abundance of large grazing mammals (*Mammuthus*, *Bison*, and *Equus*) at Blackwater Draw indicates extensive grassland or savanna habitats. A diverse fauna of small vertebrates also suggests the presence of permanent water and nearby forests. The Brown Sand Wedge fauna includes three extant small mammals now found farther north or at higher elevations (*Sorex cinereus*, *Microtus mexicanus*, and *M. pennsylvanicus*) and four species generally found in faunas farther south (*Hesperotestudo wilsoni*, *Terrapene carolina putnami*, *Dasypus bellus*, and *Sigmodon hispidus*). This is indicative of a late Pleistocene "disharmonius fauna" in which species co-occur that are now found in different habitats. The climatic conditions during the late Pleistocene in northeastern New Mexico, as typified by the Blackwater Draw fauna, included milder, frost-free winters, cooler summers, and more available moisture, with permanent streams, ponds, and small lakes.

INTRODUCTION

Northeastern New Mexico is not well known for its Pleistocene vertebrate record. In a recent review of Pleistocene vertebrate faunas from New Mexico, Harris (1993) listed 25 sites from this region of the state. We add 11 previously unpublished sites and nine other little known sites from the literature. Figure 1 shows the location of 37 Pleistocene vertebrate assemblages in northeastern New Mexico, several of which are composed of more than one site, bringing the total to 45 sites. The great majority of these sites represent the Rancholabrean land-mammal "age." More than half (28) of the sites contain only one or two mammals, most commonly *Mammuthus*, *Equus*, and/or *Bison*. The richest fauna in this region of New Mexico, Blackwater Draw Locality 1 (Gray Sand and Brown Sand Wedge local faunas, see below), also called the Clovis Site, has 39 species of mammals and over 60 total species of vertebrates (Lundelius, 1972; Slaughter, 1975; Katz, 1997). Blackwater Draw/Clovis and Folsom are perhaps the two best known late Pleistocene sites in northeastern New Mexico. However, both Clovis and Folsom are better known as archaeological sites, documenting the association of extinct Pleistocene megafauna with Paleoindian artifacts. Six other sites in this region also document the association of Pleistocene megafauna and Paleoindians, including: Manzano Cave (Hibben, 1941), Lucy site (Roosa, 1956), San Jon site (Roberts, 1942; Hill et al., 1995), Milnesand and Ted Williamson sites (Holliday, 1995), and possibly the Stolle mammoth site (Agogino and Sweetland, 1985). Because of the great interest in Paleoindian archaeology in this region of New Mexico, the archaeological literature on latest Pleistocene fossil sites is more extensive than the paleontological literature. However, we concentrate almost exclusively on the vertebrate paleontological aspects of the Paleoindian sites.

This paper is primarily a literature review of late Pleistocene vertebrate fossil sites in northeastern New Mexico. However, we document 11 new sites, and provide substantial additional information on several previously published sites. We consider northeastern New Mexico to comprise the eleven counties included in our review (from north to south): Colfax, Union, Mora, Harding, San Miguel, Quay, Torrance, Guadalupe, De Baca, Curry, and Roosevelt (Fig. 1). This region primarily encompasses the Southern High Plains, east of the southern Rocky Mountain ranges and the Basin and Range province. However, the western portions of Colfax, Mora, and San Miguel counties include some of the highest mountains in the Sangre de Cristo Range, reaching nearly 4000 m in elevation, and the westernmost part of Torrance County includes the Manzano Mountains, with elevations over 3000 m. Only two of the 45 fossil sites in northeastern New Mexico are located in the mountainous portion of this region.

Each site is numbered, corresponding to the numbers on the map in Figure 1. For each site we provide the following information (not all data are available for each site): site name; site

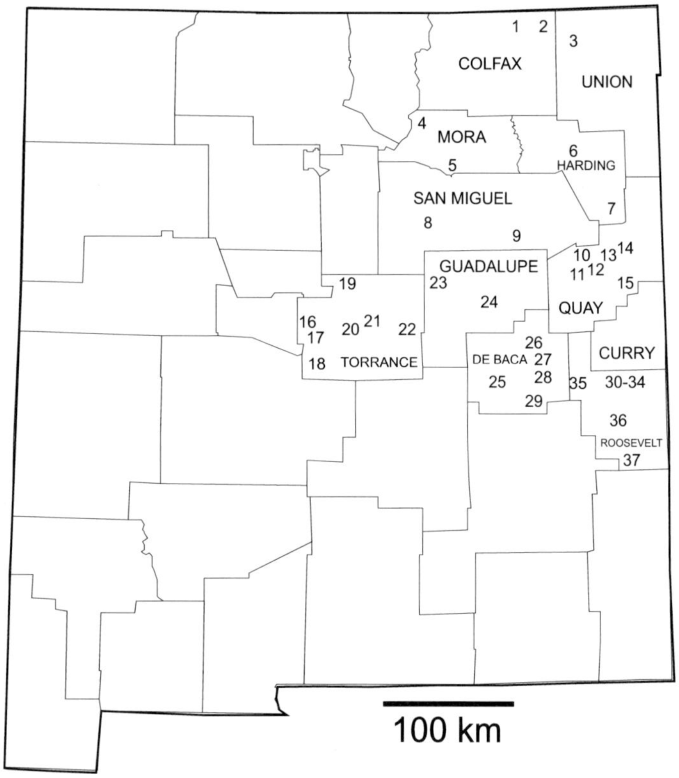

FIGURE 1. Map showing the location of Pleistocene vertebrate faunas in northeastern New Mexico. The numbers on the map correspond to sites discussed in the text. **Colfax County**: 1. Raton; 2. Folsom. **Union County**: 3. Wetherly. **Mora County**: 4. Guadalupita; 5. Dogie Jones Ranch. **Harding County**: 6. Albert; 7. Casados Ranch. **San Miguel County**: 8. Las Vegas; 9. Rio Conchas. **Quay County**: 10. Pajarito Creek; 11. Quay; 12. Mesa Redonda; 13. Badlands Ranch; 14. Plaza Larga Creek; 15. San Jon. **Torrance County**: 16. Manzano Cave; 17. Quarai and Punta de Agua; 18. Keers; 19. Moriarty; 20. Lake Estancia sites; 21. Lucy; 22. Encino. **Guadalupe County**: 23. Milagro; 24. Agua Negra. **De Baca County**: 25. Yeso Creek; 26. Fort Sumner; 27. Taiban Creek; 28. Eighteenmile Bend; 29. X-Bar-X Ranch. **Roosevelt County**: 30. Blackwater Draw Locality 1, including Gray Sand LF and Brown Sand Wedge LF; 31. *Arctodus* site; 32. McCullum Ranch; 33. Anderson Basin sites; 34. Circus Basin and Elephant Tusk Basin; 35. Floyd; 36. Stolle; 37. Milnesand and Ted Williamson.

number if known (in particular, sites collected by the New Mexico Museum of Natural History); general locality information; latitude and longitude; geologic and stratigraphic data; historical information relating to its discovery and collection; a brief discussion of the fauna; acronym(s) where fossils from the site are housed; age of the fauna, including any radiocarbon dates; and a list of references. Several of the numbered sites actually consist of two or more individual sites in the same general area (e.g., Lake Estancia) that have similar vertebrate faunas and stratigraphic occurrence. Detailed locality data (Township, Range, Section, UTM coordinates, etc.) are available for many of the sites in the original publications, or in the New Mexico Museum of Natural History vertebrate paleontology locality files (site name preceded by NMMNH site L-). We provide map coordinates for sites where these data have not been published before.

Many of the Pleistocene sites in northeastern New Mexico have been mentioned only briefly in the literature, and details regarding their exact location are unclear, as is information on stratigraphy, fauna, age, and disposition of the fossils. Very few of the sites occur in named stratigraphic units because most of them are late Pleistocene in age; instead, they are derived from what is usually termed Quaternary alluvium (QAL) on geologic maps. Blackwater Draw is the only one of the sites for which the fauna is large enough to be presented in tabular form. Although we list all species in the various faunas, our discussion stresses the mammals. Only sites that are clearly Pleistocene in age (i.e., older than 10,000 years Before Present=yr B. P.), and have extinct species of mammals are included in our survey. Sites that contain only the living bison, *Bison bison*, or sites in which the species of *Bison* is not identifiable, are excluded. We do not attempt to provide an exhaustive list of literature citations for the various sites. For the larger and better known sites, such as Blackwater Draw, we provide several citations that provide an entry into the literature.

The following are institutional abbrevations of museums known to possess Pleistocene vertebrate fossils from northeastern New Mexico. This is not an exhaustive list, as it is based primarily on specimens published in the literature.

AMNH: American Museum of Natural History, New York

ANSP: Academy of Natural Sciences, Philadelphia, Pennsylvania

BWDM: Blackwater Draw Museum, Portales, New Mexico

BWDS: Blackwater Draw Site, Portales, New Mexico

CIT/LACM: California Institute of Technology collection-now housed at the Los Angeles County Museum, Los Angeles, California

DMNH: Denver Museum of Nature and Science (formerly Denver Museum of Natural History and Colorado Museum of Natural History), Denver, Colorado

ENMU: Eastern New Mexico University, Portales, New Mexico

ISM: Illinois State Museum, Springfield, Illinois

MDM: Mesalands Dinosaur Museum, Tucumcari, New Mexico

MMA: Maxwell Museum of Anthropology, University of New Mexico, Albuquerque, New Mexico.

MNM: Laboratory of Anthropology, Museum of New Mexico, Santa Fe, New Mexico

NMMNH: New Mexico Museum of Natural History, Albuquerque, New Mexico

SMM: Sternberg Memorial Museum, Fort Hays State University, Hays, Kansas

SMP: Shuler Museum of Paleontology, Southern Methodist University, Dallas, Texas

TMM: Texas Memorial Museum, University of Texas, Austin, Texas

TTU: Texas Tech University, Lubbock, Texas

UASM: University of Arizona State Museum, Tucson, Arizona.

UMMP: University of Michigan, Museum of Paelontology, Ann Arbor, Michigan

UNM: University of New Mexico, Museum of Geology, Albuquerque, New Mexico

USNM: U. S. National Museum of Natural History, Smithsonian Institution, Washington, D. C.

PLEISTOCENE VERTEBRATE LOCALITIES IN NORTHEASTERN NEW MEXICO

Colfax County

1. Raton—McDonald (1981) reported a skull of the giant bison *Bison latifrons* from near Raton in Colfax County. No further information is available for this locality. The skull is housed in the AMNH vertebrate paleontology collection. The presence of *B. latifrons* suggests that this site is early Rancholabrean, although there is no associated fauna or stratigraphy to substantiate the age. The only other New Mexico record of *B. latifrons*, from the Los Duranes Formation near Bernalillo, Sandoval County in the middle Rio Grande Valley (Smartt et al., 1991), appears to be early Rancholabrean (Morgan and Lucas, 2000).

2. Folsom—The Folsom site is located on a branch of Dry Cimarron Creek about 15 km northwest of Folsom, 35 km east of Raton, and 12 km south of the Colorado border, in the northeastern corner of Colfax County (36°53'N, 104°04'W). Folsom is one of the most significant Paleoindian sites in North America (Hay and Cook, 1930). The site was excavated in the late 1920s by field crews from the Colorado Museum of Natural History (CMNH, now the Denver Museum of Nature and Science) and the American Museum of Natural History (AMNH). The Folsom site is located in a valley fill developed in the Upper Cretaceous Pierre Shale (site descriptions from Hay and Cook, 1930, p. 7). Bison fossils and associated artifacts were discovered when floods eroded a deep, narrow arroyo into the valley fill. The main part of the deposit, where a large number of skeletons of extinct bison were excavated, is about 2-4 m below the original land surface.

The original CMNH and AMNH excavations at the Folsom site yielded more than 30 skeletons of the extinct bison, *Bison antiquus*, originally described as a distinct species, *B. taylori* (Hay and Cook, 1928, 1930). The bison sample from Folsom, including the type skull of *B. taylori* (DMNH 1236), was thoroughly described, together with a series of measurements and illustrations (Hay and Cook, 1930, p. 26-29, pl. 8, figs. 1, 2; pl. 10, figs. 2, 3). McDonald (1981) referred the Folsom bison to *B. antiquus occidentalis*, an extinct form characteristic of latest Pleistocene and early Holocene faunas (between about 5,000 and 11,000 yr B. P.). Four other mammals were reported from the Folsom site, all of which are extant species (Hay and Cook, 1930): the black-tailed prairie dog *Cynomys ludovicianus*, the ground squirrel *Spermophilus* sp., the black-tailed jackrabbit *Lepus californicus*, and the mule deer *Odocoileus hemionus*. Associated with the *B. antiquus* skeletons were characteristic projectile points now known as Folsom points, which are associated with the Paleoindian Folsom culture and date to the latest Pleistocene and earliest Holocene. Bones of the extinct bison *B. antiquus* are often associated with Folsom sites, but fossils of other extinct members of the Pleistocene megafauna (mammoth, horse, camels, etc.) are generally absent. We were unable to locate published radiocarbon dates for the Folsom site, but it is probably latest Pleistocene in age (~10,000-11,000 yr B. P).

Union County

3. Wetherly—Effinger and Lucas (1990) reported a skull and numerous post-cranial elements of bison from a site about 1.6 km northwest of Wetherly Lake and 11 km east of Des Moines in northwestern Union County (36°45'N, 103°42'W). These fossils and associated notes are housed in the Geology Museum at UNM. A partial bison skull with a horn core from the Wetherly site was identified in the UNM collection as "*B. preoccidentalis*." Effinger and Lucas (1990) tentatively referred this skull to *B. priscus*. The presence of *B. priscus*, an extinct species, indicates a late Pleistocene (Rancholabrean) age.

Mora County

4. Guadalupita—Gillette et al. (1985b) reported several mammoth teeth and a horse tooth from the Guadalupita site (NMMNH site L-4625) in northwestern Mora County. The Guadalupita site is on the ranch of Jimmy Gallegos in Guadalupita Canyon, about 5 km south of Guadalupita and 14 km north-northeast of Mora (36°06'N, 105°14'W). These fossils were collected at an elevation of 2,280 m on the eastern side of the Rincon Mountains on the eastern flank of the Sangre de Cristo Mountains, and represent the highest elevation mammoth known from New Mexico (Gillette et al., 1985b). Mammoth teeth and bones were first discovered at the Guadalupita site in 1982 by Mr. Gallegos, who found the fossils in peaty sediments that he was excavating to drain a bog in his pasture. Dave Gillette and a field crew from the New Mexico Museum of Natural History visited the site in the summer of 1984. They dug several trenches to a depth of about 3 m to find additional fossils, and also to determine the stratigraphy. The fossils were recovered from a dark gray micaceous sand about 1 m below the surface.

Three mammoth specimens from Guadalupita may represent a single individual, including a right M2 (NMMNH 33276, Fig. 2E, and Gillette et al., 1985b, fig. 2.27), and a fragmentary tooth (NMMNH 33277), and two small tusk fragments (NMMNH 33278). Gillette et al. (1985b) reported a tooth fragment of the horse *Equus* from Guadalupita, but this fossil is not in the NMMNH collection. The right M2 (NMMNH 33276) of *Mammuthus* from Guadalupita has 14 plates and a lamellar frequency (plates/100 mm) of 9 (Table 1). The high lamellar frequency, which reflects both the thin enamel and narrow plates, is characteristic of *Mammuthus columbi*. The advanced characters of the *M. columbi* tooth from Guadalupita suggest a late Pleistocene (late Rancholabrean) age.

5. Dogie Jones Ranch—The Dogie Jones Ranch site (NMMNH site L-4324) is located 0.2 km north of the Sapello River and 4 km southwest of Watrous, in south-central Mora County (35°46'N, 105°01'W). Fossils were first discovered at this locality in 1993 by Mr. Dogie Jones, who found a partial mammoth tooth that he brought to the attention of the NMMNH. Gary Morgan and Pete Reser visited the Dogie Jones Ranch in September 1994. They excavated a mammoth M3 from a buff-colored silty sand about 1 m below the ground surface. This fossil and a partial mammoth tooth found by Mr. Jones were eroding from the side of a small, narrow arroyo that trends south and eventually joins the Sapello River.

Two mammoth teeth were found at the Dogie Jones Ranch site, a nearly complete right M3 (NMMNH 31522, Figs. 2F, 2G) and a partial right m3 (NMMNH 31523), possibly from the same individual. NMMNH 31522 has 20 plates, a lamellar frequency of 9 plates/100 mm, and thin enamel (Table 1). These dental charactacters are indicative of the progressive late Pleistocene (late Rancholabrean) mammoth species *Mammuthus columbi*.

Harding County

6. Albert—The Albert site is located about 27 km east of Roy on Carrizo Creek near Albert in north-central Harding County (Hay, 1924; Lucas and Effinger, 1991). Charles Gilmore of the USNM visited the Albert site in the summer of 1922. By the time he arrived, a local farmer had already removed a tusk and two teeth of a mammoth from the site. Gilmore excavated an area "10 by 20 feet to a depth of 4 or 5 feet," but found nothing else except for broken fragments of the skull (Hay, 1924, p. 108-109). Gilmore was not able to obtain the fossils from the collector, and thus it is unknown whether they belonged to *Mammuthus columbi* or *M. imperator*. No specimens from the Albert site apparently exist in museum collections.

7. Casados Ranch—The Casados Ranch site (NMMNH site L-4627) is 20 km northwest of Logan and 7 km north of New Mexio Highway 39, at an elevation of 1295 m in southeastern Harding County (35°34'N, 103°30'W). The site is located at UTM coordinates 636114E, 3937532N (Zone 13, NAD 27) where it is crossed by an all-weather, unpaved road. Fred Trauger discovered this site in 1969, and McMullen and Zakrzewski (1972) described the Casados Ranch vertebrate and invertebrate faunas. Spencer Lucas, Gary Morgan, Mark Gordon, and Paul Sealey visited the Casados Ranch site in December 2000, and collected additional samples of vertebrates and molluscs, took site photos, and recorded geological data.

The principal fossiliferous layer has abundant gastropods, and also contains bones and teeth, notably of salamander, horse, and mammoth. This layer is a 0.8-m-thick bed of light olive gray (5 Y 6/1) to light brownish gray (5 YR 6/1) blocky, very calcareous, silty quartz sandstone. It has an outcrop area of at least 1.4 hectares; erosion to the west has removed much of the bed, so its original extent was greater. Locally, at the top of the bed are 0.2-m-thick lenses of sandy mudstone that is mottled moderate brown (5 YR 4/4) and light olive gray (5 Y 6/1). The fossiliferous bed rests with sharp contact on grayish orange (10 YR 7/4) to light brown (5 YR 6/4) very fine to fine-grained, quartzose eolian sand. It is also sharply overlain by a similar eolian sand that is grayish orange (10 YR 7/4) to grayish orange pink (5 YR 7/2). A partial mammoth tusk was found at the base of this overlying eolian sand. Figure 3 is a stratigraphic section at Casados Ranch.

McMullen and Zakrzewski (1972, p. 134) described the geology of the Casados Ranch site as a "sand deposit." Barnes (1983) mapped the Casados Ranch fossiliferous layer as a playa deposit, described in the map legend as "clay and silt, sandy, light-gray, in shallow depressions (Wisconsinan), mostly covered by thin deposit of Holocene sediment." Certainly, the lithotypes and geometry of the Casados Ranch fossiliferous layer and the fossils themselves indicate lacustrine deposi-

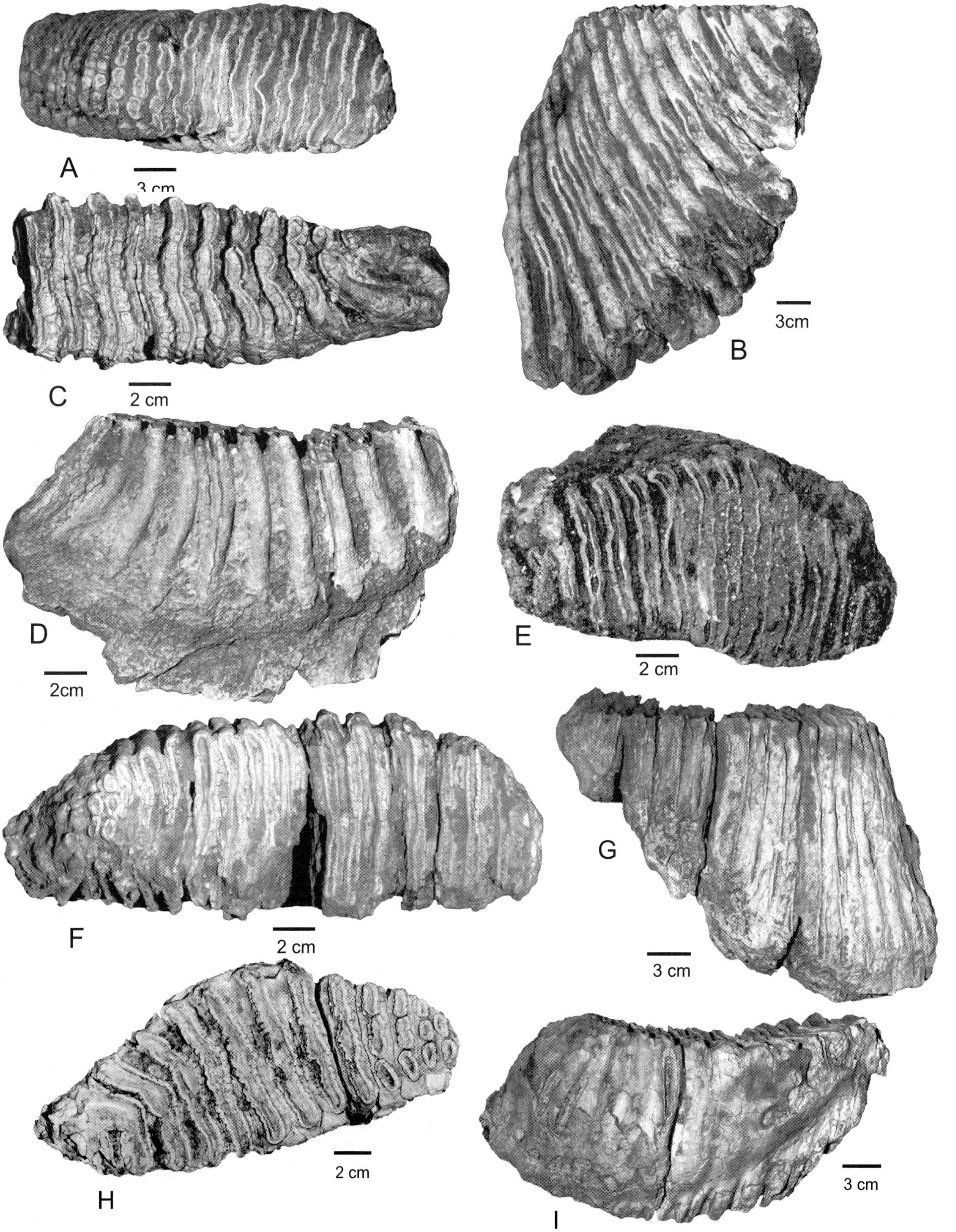
A
3 cm
B
3cm
C
2 cm
D
2cm
E
2 cm
F
2 cm
G
3 cm
H
2 cm
I
3 cm

tion in a small, playa lake. The bracketing eolian sand sheets indicate the fossiliferous bed was deposited during a wet episode between intervals of relative aridity. The dune field in which the Casados Ranch deposit formed is developed on a geomorphic surface no older than Kansan (Dolliver, 1985), and similar dune fields across the Llano Estacado are no older than late Pleistocene (Reeves, 1972).

The Casados Ranch Local Fauna (LF) consists of four species of vertebrates: the salamander *Ambystoma*, a small turtle, the horse *Equus*, and the mammoth *Mammuthus*. The original Casados Ranch fossil collection described by McMullen and Zakrzewski (1972) is in the Sternberg Memorial Museum (SMM 13579-13602) at Fort Hays State University in Hays, Kansas. A second sample from this site collected in December 2000 is in the NMMNH. McMullen and Zakrzewski (1972) reported the large horse *Equus niobrarensis* (now referred to the *Equus laurentius* group of Winans, 1989) from Casados Ranch. Although they did not specify which element of *E. niobrarensis* they had from the site, their identification of this horse to the species level strongly suggests they had a tooth. We collected a complete medial phalanx (NMMNH 33465, Fig. 4E) of a large horse from Casados Ranch that is also tentatively referred to the *E. laurentius* group. Measurements (in mm) of this horse toe are: proximal-distal length, 53; proximal width, 60; distal width, 53. A partial tooth (NMMNH 33469), fragments of a second tooth (NMMNH 33471), and a tusk fragment about 500 mm long (NMMNH 33470) are referred to the mammoth *Mammuthus columbi*. The partial tooth has a lamellar frequency (number of plates/100 mm) of 7, and the enamel is fairly thin, ranging from 2.3-3.0 mm in thickness, both characteristic of *M. columbi*. The Casados Ranch site also has an extensive freshwater and terrestrial molluscan fauna, composed of about 20 species of gastropods and bivalves (McMullen and Zakrzewski, 1972).

McMullen and Zakrzewski (1972) concluded that the fossiliferous layer at Casados Ranch represented a permanent, quiet-water environment surrounded by savannas and short grass uplands, based on the presence of a tiger salamander and the abundance of the gastropods *Gyraulus parvus* and *Vallonia gracilicosta*. Furthermore, the presence of several species of northern or high altitude gastropods led these authors to propose that deposition occurred during an episode of cool, humid climate, presumably a glacial period. McMullen and Zakrzewski (1972) favored an Illinoian (early Rancholabrean) age because the Casados Ranch molluscan fauna is more similar to Illinoian molluscan faunas from the High Plains (e.g., Miller, 1966) than it is to Wisconsinan molluscan faunas from the same region. The Casados Ranch molluscs are similar to molluscan faunas from the Doby Springs LF of Oklahoma and the Butler Spring LF of southwestern Kansas, both of which also have associated early Rancholabrean vertebrate faunas (McMullen and Zakrzewski, 1972). The vertebrate fauna from Casados Ranch is not particularly informative with regard to age, although most New Mexico records of *Mammuthus columbi* are from Wisconsinan (late Ranchcolabrean) sites. Barnes (1983) considered the Casados Ranch playa deposit to be Wisconsinan.

San Miguel County

8. Las Vegas—Hay (1924) reported a partial M3 (AMNH 15544) of a mammoth from a site near Las Vegas in San Miguel County, but more precise locality data are not available. The Las Vegas mammoth tooth consists of 10 plates from the anterior end of the tooth and has a lamellar frequency of 7 plates/100 mm, indicating referral to *Mammuthus columbi* (Hay, 1924; Lucas and Effinger, 1991).

9. Rio Conchas—Hay (1927) reported a mammoth tooth from about 90 km southeast of Las Vegas, New Mexico, on or near the "Rio Conchas" (Conchas River on modern maps), about 10 km west of Conchas in eastern San Miguel County (35°21'N, 104°16'W). The tooth was collected in 1925 by T. W. La Rue of Las Vegas, New Mexico, who found it on a mesa occupying sections 14 and 15, T13N, R25E (Bookout Ranch quadrangle). These coordinates place this locality on the south shore of what is now Conchas Lake, about 2 km east of the point where Cuervo Creek and Garita Creek enter the lake at its southwestern extremity. However, the mammoth tooth was collected in 1925, and Conchas Lake did not come into existence until after Conchas Dam was built by the U. S. Army Corps of Engineers in 1939. At the time the fossil was collected, this site was about 3-4 km south of the Conchas River.

The mammoth tooth from the Rio Conchas site is a right m3 (USNM 11648). The tooth is worn anteriorly, and several plates also appear to be missing from the posterior (Hay, 1927). The enamel is fairly thick and crenulated. As preserved, the tooth consists of 10 plates with a lamellar frequency of 5 plates/100 mm. On the basis of its thick enamel and low lamellar frequency, Hay (1927) identified this tooth as the imperial mammoth, *Mammuthus imperator*.

FIGURE 2. Columbian mammoth (*Mammuthus columbi*) teeth from Pleistocene sites in northeastern New Mexico. A. occlusal and B. medial view, NMMNH 33453, right M3, Fort Sumner (NMMNH site L-4633), De Baca County; C. occlusal view and D. lateral view, NMMNH 15018, right m2, Milagro (NMMNH site L-1496), Guadalupe County; E. occlusal view, NMMNH 33276, right M2, Guadalupita (NMMNH site L-4625), Mora County; F. occlusal view and G. lateral view, NMMNH 31522, right M3, Dogie Jones Ranch (NMMNH site L-4324), Mora County; H. occlusal view and I. medial view, MDM 202, left m3, Quay, Quay County.

Quay County

10. Pajarito Creek—A mammoth tooth was found near Pajarito Creek (W½ T11N, R30E, Liberty Mesa quadrangle) in northwestern Quay County. The tooth is a partial lower molar (MDM 203) tentatively referred to *Mammuthus columbi*.

11. Quay—A single mammoth tooth was collected near Quay (E½ T8N, R30E, Quay quadrangle) in western Quay County, just southwest of the Mesa Redonda mammoth site. This tooth, a left m3 (MDM 202; Figs. 2H, 2I), has 15 plates and a lamellar frequency (plates/100 mm) of 6 (Table 1). The measurements, lamellar frequency, and enamel thickness of this tooth suggest referral to the late Pleistocene mammoth *Mammuthus columbi*.

12. Mesa Redonda—The Mesa Redonda mammoth site (NMMNH site L-3736) is located along a small, unnamed arroyo that flows northeast from the base of Mesa Redonda, about 23 km south of Tucumcari in central Quay County (34°58'30"N, 103°37'30"W). Mammoth fossils were first found at this site in 1929 by Robert Abercrombie of Tucumcari, who contacted Harold Cook, a paleontologist at the Colorado Museum of Natural History in Denver (CMNH, now the DMNH). The Mesa Redonda site was excavated in 1930 by a field crew from the CMNH led by Nelson J. Vaughn. The CMNH crew mostly found isolated fossils of mammoths. Gillette et al. (1985a) briefly discussed the Mesa Redonda site, and Morgan et al. (this volume) review the site and its mammoths. Much of the Mesa Redonda mammoth sample was transferred from the DMNH to the NMMNH in Albuquerque in 1985, and was mounted as a composite skeleton now on display in the NMMNH (Morgan et al, this volume, fig. 4).

The bone-bearing stratum at the Mesa Redonda site is a 1.0- to 1.5-m thick layer of muddy sand that rests directly on Triassic bedrock of the Redonda Formation. The Quaternary sediments at the Mesa Redonda mammoth site are part of an alluvial fan system that flowed northeast from the mesa during part of the Pleistocene. The bone-bearing sand with its lenticular breccias probably was deposited in a low energy channel or interchannel portion of an alluvial fan. The stratigraphy and sedimentology of the Mesa Redonda mammoth site are similar to that of the Badlands Ranch late Pleistocene site (see below and Morgan et al., 2000), located about 20 km northeast, where a localized Pleistocene valley fill developed in Triassic bedrock. There is a strong possibility that these two sites are nearly contemporaneous.

The Mesa Redonda mammoths have been mentioned in the literature previously (Gillette et al., 1985a; Lucas and Effinger, 1991; Harris, 1993), and are studied in detail in Morgan et al. (this volume), who provide a list of catalogued specimens, descriptions of the teeth and mandibles, illustrations of the most complete specimens, tables of measurements, and a brief taxonomic discussion. They calculated a minimum number of five individuals of *Mammuthus columbi* from Mesa Redonda based on lower jaws. Three M3s of *Mammuthus* from Mesa Redonda (Table 1) range from: 241-257 mm in length, 92-101 mm in width, and 187-271 mm in crown height. These teeth have from 17-19 plates, a lamellar frequency of 6-8, and enamel thicknesses averaging 2.7 mm (Table 1). The size, plate numbers, lamellar frequency, and enamel thicknesses of the Mesa Redonda mammoths are within the range of *M. columbi*. The horse *Equus* is represented at Mesa Redonda by two associated incisors. Nelson Vaughn's correspondence about the Mesa Redonda site mentioned ground sloth teeth and a carnivore tooth, but the presence of these two mammals at Mesa Redonda is not documented by fossils in the DMNH or NMMNH collections.

The presence of *Mammuthus columbi* at Mesa Redonda strongly suggests a late Pleistocene (late Rancholabrean) age, although there are no radiocarbon dates or age-diagnostic mammals (e.g., *Bison*) from the site to corroborate this age. All other well-dated records of *M. columbi* from New Mexico and the southwestern United States (Madden, 1981) are late Rancholabrean. The similarity of the Mesa Redonda mammoths to the sample of *M. columbi* from the latest Pleistocene

TABLE 1. Measurements (in mm) of *Mammuthus* teeth from various Pleistocene sites in northeastern New Mexico. Presence of a "+" after a measurement or plate number indicates that a portion of the tooth has worn away, and thus these measurements are minimum values. All specimens were measured by the authors, except the M3 of *Mammuthus imperator* (UMMP 44424) from the Eighteenmile Bend site.

species, locality, and specimen number	tooth	length	maximum width	crown height	plate number	lamellar frequency[1]	enamel thickness[2]
Mammuthus imperator							
Eighteenmile Bend[3]							
UMMP 44424	m3	434	124	142	16	4	3.0
Mammuthus columbi							
Guadalupita							
NMMNH 33276	M2	181	91	128	14	9	2.5 (2.3-2.8)
Dogie Jones Ranch							
NMMNH 31522	M3	220	92	195	20	9	2.3 (2.0-2.7)
Dogie Jones Ranch[4]							
NMMNH 31523	m3	--	85	172	--	8	2.5 (2.2-2.7)
Quay							
MDM 202	m3	230	97	160	15	6	3.0 (2.4-3.4)
Mesa Redonda							
NMMNH 33255	M3	257	92	271	19	8	2.7 (2.5-3.0)
Mesa Redonda							
NMMNH 33256	M3	241	101	265	17	8	2.6 (2.3-3.0)
Mesa Redonda[4]							
DMNH 15136	M3	--	101	187	--	6	--[5]
Mesa Redonda							
DMNH 1639b	m3	220	73	160	18	8	2.4 (2.2-2.7)
Mesa Redonda[4]							
DMNH 1639a	m3	--	80	--[6]	19	8	2.4 (2.1-2.6)
Mesa Redonda							
NMMNH 33257	m2	215	83	--[6]	14	6	2.8 (2.6-3.1)
Mesa Redonda[4]							
NMMNH 33257	m3	--	--	--[6]	14+[7]	9[8]	--[5]
Milagro							
NMMNH 15018	m2	205	76	80+	11	7	2.2 (2.0-2.4)
Fort Sumner							
NMMNH 33453	M3	338	104	335	18	6	2.4 (2.0-2.6)
Blackwater Draw							
BWDS (Mammoth #1)	M3	215+	98	263	21	6.5	4.0 (3.6-4.4)
Blackwater Draw							
BWDM 1749	M3	220+	93	245	17	7.5	3.0 (2.2-3.6)
Blackwater Draw[4]							
BWDM 1517	M3	--	88	260	--	7	2.7 (2.3-3.4)
Blackwater Draw							
BWDM 1205	M3	190+	98	260	12+	7	3.8 (3.4-4.3)
Blackwater Draw							
BWDM 1948	m3	270	90	160+	17	6	4.2 (3.8-4.6)
Blackwater Draw							
BWDM 38	m3	240	90	110+	18	7	3.8 (2.8-4.8)

[1]Lamellar frequency is the number of plates/100 mm of tooth length.

[2]Enamel thickness measurements are the means of five measurements taken on each individual tooth (observed range in parentheses).

[3]Measurements of UMMP 44424 from Madden (1981).

[4]Incomplete tooth, length and/or total number of plates cannot be measured.

[5]Tooth not fully erupted, enamel thickness measurements could not be taken.

[6]Tooth in dentary, crown height could not be measured.

[7]Tooth not fully erupted, an undetermined number of plates were unerupted and could not be counted.

[8]Tooth not fully erupted; lamellar frequency is probably too high.

FIGURE 3. Casados Ranch locality (NMMNH site L-4627), Harding County, New Mexico. A. View of site looking southeast. B. Close-up view of fossiliferous calcareous sand layer (rock hammer for scale). C. Casados Ranch stratigraphic section.

Blackwater Draw Locality 1 (see Table 1) further supports a late Rancholabrean age for Mesa Redonda.

13. Badlands Ranch—The Badlands Ranch site is located about 3 km east of Barranca Creek and 15 km southeast of Tucumcari in central Quay County (35°06'N, 103°30'W). The Badlands Ranch vertebrate and invertebrate faunas were recently described (Morgan et al., 2000). The site is near the crest of the western face of a north-south trending terrace. Vertebrate fossils comprising the Badlands Ranch LF (NMMNH site L-2972) were derived from Quaternary sediments unconformably overlying red beds of the Triassic Bull Canyon Formation. The vertebrate fossils were collected from the base of a sand unit about 1 m thick, representing a stream deposit. A large sample of molluscs was found in an overlying sandy mudstone about 1 m thick, probably representing a relatively quiet freshwater depositional environment such as a pond. Morgan et al. (2000) interpreted the fossil site as a localized valley fill developed in Triassic bedrock during the Pleistocene.

The Badlands Ranch vertebrate fauna consists of six species of mammals (all fossils are housed in the NMMNH under catalogue numbers 25547-25557, 27719-27727): *Paramylodon harlani*, *Equus* sp., *Camelops* sp., *Bison* sp., an indeterminate artiodactyl, and *Mammuthus* sp. The ground sloth *Paramylodon harlani* is the best represented species in the Badlands Ranch, including more than ten postcranial elements that appear to belong to one individual (Morgan et al., 2000). This is one of only three Rancholabrean records of Harlan's ground sloth from New Mexico. The other records are from the Albuquerque Gravel Pits Fauna in Bernalillo County (Morgan and Lucas, 2000) and the Blackwater Draw Locality 1 in Roosevelt County (Harris, 1993; see below and Table 2). The Badlands Ranch sample includes two complete lower premolars of a small horse that are closest in size to either the *Equus alaskae* or *E. francisci* groups of Winans (1989), both of which are present in Blackwater Draw Locality 1. The camel, bison, and mammoth are represented by fragmentary teeth that are not identifiable to the species level. The molluscan fauna from Badlands Ranch is composed of five species of freshwater gastropods, two species of terrestrial gastropods, and three species of freshwater bivalves. The presence of *Mammuthus* establishes a Pleistocene age for the Badlands Ranch LF, and the occurrence of *Bison* further restricts the age to Rancholabrean.

14. Plaza Larga Creek—A fossil horse tooth was found along Plaza Larga Creek, near its confluence with Revuelto Creek (sec. 6, T11N, R33E, San Jon NW quadrangle) in east-central Quay County, just northeast of the Badlands Ranch site (35°12'N, 103°29'W). The tooth is a left M3 of *Equus* (MDM 204, Figs. 5C, 5D). Measurements (in mm) of this tooth are: length, 30.5; width, 26.6; mesostyle crown height, 71. This tooth is from a rather large species of *Equus* with fairly complicated fossettes, and is referred to the *E. laurentius* group of Winans (1989). The presence of *E. laurentius* (=*E. niobrarensis* of previous usage, e.g., Lundelius, 1972) suggests a Pleistocene age for the Plaza Larga Creek site, but no further refinement of the age is possible based on this single horse tooth.

15. San Jon—San Jon is a multicomponent archaeological/paleontological site (LA/MNM site 6437) located along the northwestern margin of the Llano Estacado, about 17 km south of San Jon in southeastern Quay County (34°58'N, 103°21'W). The site consists of an older basin about 1 km wide filled with late Pleistocene sediments containing mammoth fossils, and a smaller and younger basin about 350 km wide with latest Pleistocene and Holocene strata containing archaeological deposits (Roberts, 1942; Hill et al., 1995; Holliday, 1997). Both basins are dissected by deep arroyo tributaries of a canyon cut deeply into the Llano Estacado escarpment. The various localities that comprise the San Jon site are exposed as peninsulas of sediment between the deep arroyos eroded into the playa-basin fill. The San Jon site was first excavated by Frank Hibben and a group of University of New Mexico students in 1940. Hibben turned the site over to Frank Roberts of the Smithsonian Institution, who excavated four areas of the site during the summer of 1941 (Roberts, 1942). Matthew Hill and Vance Holliday of the University of Wisconsin conducted additional field work at the San Jon site between 1993 and 1995 (Hill et al., 1995; Holliday, 1997).

Late Rancholabrean vertebrate fossils occur in two stratigraphic levels at the San Jon site, an older late Pleistocene basin fill containing mammoth, and the oldest Paleoindian level (latest Pleistocene) containing extinct bison (Roberts, 1942; Hill et al., 1995). Although several younger layers at San Jon contain archaeological materials of Holocene age (Hill et al., 1995), only the strata with extinct Pleistocene megafauna are discussed here. Roberts (1942), Hill et al. (1995), and Holliday (1997) reviewed the Paleoindian and younger archaeology of the San Jon site. Roberts (1942, p. 8) mentioned mammoth bones, teeth, and tusk fragments from late Pleistocene lacustrine sediments at the San Jon site, but the location of these fossils is unknown. The fossils presumably represent the species *Mammuthus columbi*, based on the late Pleistocene (late Rancholabrean) age of these deposits. The overlying layer contains a bone bed of the extinct bison *Bison antiquus* in association with Paleoindian artifacts (Hill et al., 1995, figs. 3, 4). Hill et al. (1997) recorded a minimum of five individuals of *B. antiquus* from this bone bed, which are curated in the archaeological collections of the USNM. Roberts (1942) reported numerous examples of lower leg bones and articulated feet of *B. antiquus* in this bone bed, found in an upright position suggesting that the animals became mired in the mud and were then butchered. Roberts (1942) recorded several other mammals in the *Bison* bone bed, including deer and rabbit, but remains of these other taxa could not be located in the USNM collection (Hill et al., 1995).

Late Pleistocene lacustrine sediments from the San Jon site, containing unspecified mammoth fossils collected by Hibben in 1940, were radiocarbon dated at 13,145 ± 315 yr B. P. (Hill et al., 1995), and apparently predate Paleoindian occupation of this region. The overlying layer, containing the *Bison antiquus* bone bed and associated Paleoindian artifacts, has bracketing latest Pleistocene radiocarbon ages of 12,510 ± 230 yr B. P. and 11,450 ± 300 yr B. P. (Hill et al., 1995, figs. 3, 4). Hill et al. (1995) also obtained a younger, early Holocene radiocarbon date of 8,360 ± 210 yr B. P. associated with the *Bison antiquus* bone bed (San Jon site, Area II). Hill et al. (1995) considered *B. antiquus* as the early Holocene Southern High Plains "phenotype" of bison. Although Holocene bison records are often referred to the extant species *B. bison*, according to McDonald (1981) *B. bison* did not appear until the middle Holocene about 5,000 yr B. P. McDonald identified latest Pleistocene and early Holocene (~5,000-11,000 yr B. P.) bison as *B. antiquus occidentalis*, including the San Jon bison sample, and late Pleistocene bison (older than ~11,000 yr B. P.) as the subspecies *B. antiquus antiquus*, including specimens from the Clovis level at Blackwater Draw dated at 11,000-12,000 yr B. P. (see below). Therefore, according to McDonald (1981), the extinct bison *B. antiquus*

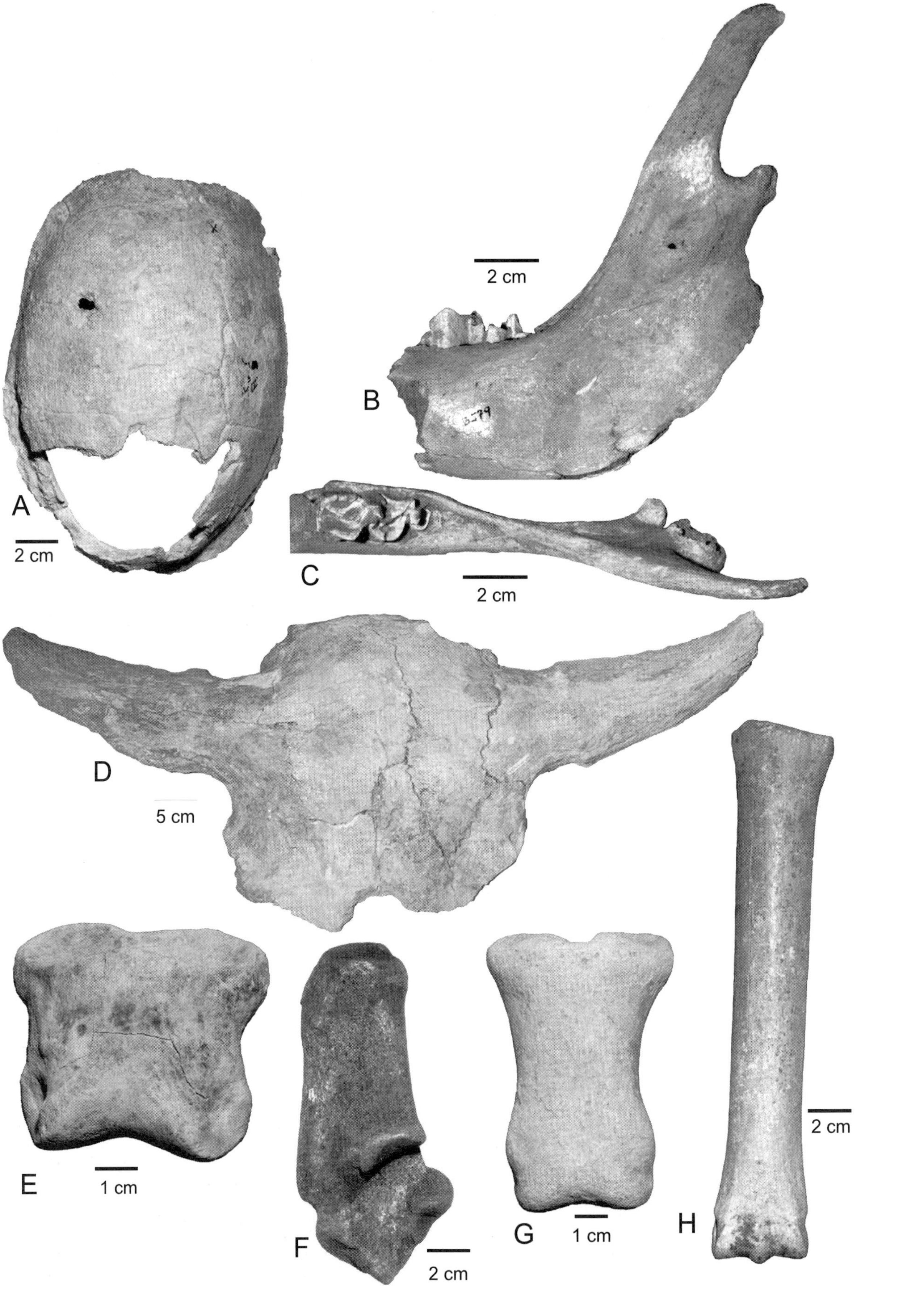
A
2 cm
B
2 cm
C
2 cm
D
5 cm
E
1 cm
F
2 cm
G
1 cm
H
2 cm

TABLE 2. Mammalian fauna from the latest Pleistocene (late Rancholabrean) Blackwater Draw Locality Number 1, Roosevelt County, New Mexico. The fauna is divided into the two stratigraphic levels that produce late Pleistocene vertebrates: the Gray Sand Local Fauna is 13,000-11,500 yr B. P. in age and the Brown Sand Wedge Local Fauna is 11,500-11,000 yr B. P. in age (Haynes, 1995)."X" indicates presence, "X?" indicates a tentative identification (cf. or ? in original publication), and "—" indicates absence. A scientific name preceded by "†" designates an extinct species. The Gray Sand faunal list is from Lundelius (1972) and the Brown Sand Wedge faunal list is from Slaughter (1975), with additions from Harris (1993).

Species	Gray Sand	Brown Sand Wedge
Marsupialia		
Didelphidae		
Didelphis virginiana	--	X
Xenarthra		
Dasypodidae		
†*Dasypus bellus*	--	X
Mylodontidae		
†*Paramylodon harlani*	--	X[1]
Insectivora		
Soricidae		
Sorex cinereus	--	X
Carnivora		
Canidae		
†*Canis dirus*	X	--
Canis latrans	X	X
Canis lupus	X	X
Urocyon cinereoargenteus	--	X[1]
Vulpes velox	X	X[1]
Vulpes vulpes	--	X?
Felidae		
†*Smilodon fatalis*	X	--
Mustelidae		
"small skunk"	--	X
Procyonidae		
Procyon lotor	--	X
Ursidae		
Ursus americanus	--	X
Lagomorpha		
Leporidae		
Lepus californicus	--	X[1]
Rodentia		
Sciuridae		
Cynomys ludovicianus	--	X
Sciurus carolinensis	--	X?
Geomyidae		
Geomys bursarius	--	X?
Rodentia, cont.		
Muridae		
Microtus mexicanus	--	X?
Microtus pennsylvanicus	--	X
Neotoma sp.	--	X
Ondatra zibethicus	X	X
Peromyscus leucopus	--	X?
Peromyscus truei	--	X?
Pitymys ochrogaster	--	X
Reithrodontomys megalotis	--	X?
Sigmodon hispidus	--	X
Perissodactyla		
Equidae[2]		
†*Equus alaskae*	X	--
†*Equus francisci*	X?	X[1]
†*Equus laurentius*	X	X
Artiodactyla		
Tayassuidae		
†*Platygonus compressus*	X	X[1]
Camelidae		
†*Camelops hesternus*	X	X
†*Hemiauchenia macrocephala*	X	X[1]
Antilocapridae		
†*Capromeryx* sp.	--	X
antilocaprid, gen. and sp. indet.	X	--
Cervidae		
Odocoileus hemionus	--	X
Odocoileus virginianus	--	X[1]
Bovidae		
†*Bison antiquus*	X	X
Proboscidea		
Elephantidae		
†*Mammuthus columbi*	X	X
TOTAL:	15 species	35 species

[1]Species from Blackwater Draw listed in Harris (1993) as "Blackwater Draw Locality 1, Clovis age", are placed in the Brown Sand Wedge LF, as Haynes (1995) has shown that the Gray Sand LF is older than the first appearance of Clovis people in Blackwater Draw.

[2]Species names in the genus *Equus* follow Winans (1989) and Harris (1993).

survived well into the Holocene, and the presence of this extinct species does not necessarily indicate a late Pleistocene (late Rancholabrean) age.

Torrance County

16. Manzano Cave—The cave is located in the limestone foothills of the Manzano Mountains, about 8 km west of Manzano in westernmost Torrance County. Like Sandia Cave in the Sandia Mountains, Manzano Cave supposedly contained Paleoindian artifacts in association with late Pleistocene mammals (Hibben, 1941). The artifacts and bones were found (Hibben, 1941, p. 35) "...embedded in rock fragments and limestone debris intermixed with disintegrated and amorphous dung." The vertebrate fossils reported from Manzano Cave (Hibben, 1941) included the ground sloth *Nothrotheriops* and the camel *Camelops*. We have not examined these fossils, but presumably they are housed in the Maxwell Museum of Anthropology at the University of New Mexico.

17. Quarai—Quarai actually represents two sites reported by Hibben (1941), located within just a few kilometers of one another in western Torrance County, including a site near the Quarai Pueblo ruins and a second site near Punta de

FIGURE 4. Vertebrate fossils from Pleistocene sites in northeastern New Mexico. A-D. Blackwater Draw Locality 1, Roosevelt County. A. dorsal view of box turtle shell, *Terrapene carolina putnami*, BWDS 313; B. lateral view and C. occlusal view, left mandible with m3, *Hemiauchenia macrocephala*, BWDS B579; D. dorsal view of partial skull with horncores, *Bison antiquus*, BWDS 139; E. medial phalanx, *Equus* sp. NMMNH 33465, Casados Ranch (NMMNH site L-4627), Harding County; F-H, Moriarty (NMMNH site L-4622), Torrance County. F. left calcaneum, *Camelops hesternus*, NMMNH 33265; G. proximal phalanx, *Equus* sp., NMMNH 33264; H. metacarpal 3, *Equus* sp., NMMNH 33263.

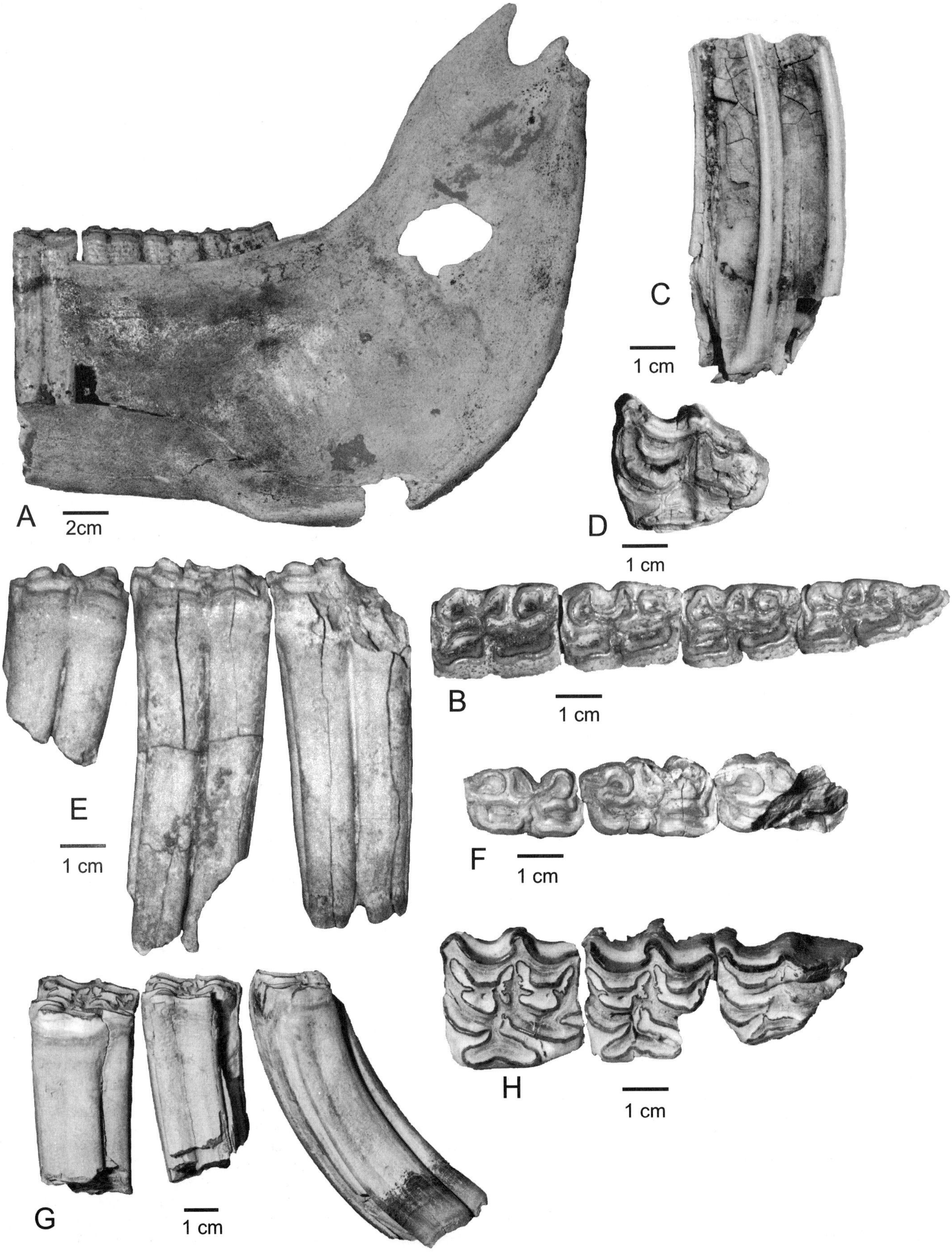
C
1 cm
A
2cm
D
1 cm
B
1 cm
E
1 cm
F
1 cm
H
1 cm
G
1 cm

Agua. Acccording to Hibben (1941, p. 6), "A complete but disarticulated mammoth was excavated by the expedition near the ruins of Quarai." Hibben's (1941, p. 6) description of the Punta de Agua site was as follows, "Pleistocene bones, especially mammoth, are common in the vicinity of Punta de Agua. They occur just beneath a layer of alluvium on top of a reddish clay." Nothing further is known about the Quarai or Punta de Agua sites, and the disposition of the mammoth fossils collected there is unknown.

18. Keers—This site (NMMNH site L-4489) is located about 22 km south of Mountainair and 18 km north of the Gran Quivira archaeological site in southwestern Torrance County (34°23'N, 106°08'W). The Keers Environmental Company buries environmenal waste at this site in deep man-made trenches. In March 2000, a piece of heavy machinery was excavating one of these trenches, and uncovered a section of proboscidean tusk about 5 m below the surface in a reddish-brown silty sand. The tusk fragment (NMMNH 31598), almost certainly *Mammuthus*, measures 70 cm in length and 8-10 cm in width.

19. Moriarty—The NMMNH Vertebrate Paleontology Collection has three fossils bearing the label "Moriarty, NM Gravel Pit, no other data." There are several gravel pits in the vicinity of Moriarty in northwestern Torrance County, and thus we are not able to provide any further locality data for this site (NMMNH site L-4622). The fossils are well preserved and heavily mineralized. They consist of a metacarpal 3 (NMMNH 33263, Fig. 4H) and proximal phalanx (NMMNH 33264, Fig. 4G) of the horse *Equus* and a calcaneum of the camelid *Camelops* (NMMNH 33265, Fig. 4F). Measurements (in mm) of the *Equus* metacarpal and phalanx, respectively, are: total length, 248, 82; proximal width, 48, 57; distal width, 45, 45. The *Camelops* calcaneum is 151 mm long. The presence of *Equus* and *Camelops* is not particularly age diagnostic, although a Pleistocene age is most likely.

20. Lake Estancia—The NMMNH Vertebrate Paleontology Collection has vertebrate fossils from three different sites in the vicinity of Willard in the Estancia basin in central Torrance County. These three sites occur in lacustrine sediments of late Wisconsinan Lake Estancia and have similar vertebrate faunas, and are thus combined as the Lake Estancia Fauna. Bachhuber (1989) reported two species of aquatic vertebrates from his Playa E-28 locality (NMMNH site L-3540), located just east of Laguna del Perro, about 15 km northeast of Willard (34°41'N, 105°57'W). The fossiliferous sediments at the Playa E-28 site occur in a section about 7 m thick, and consist of three units of thick-bedded, flint-gray clays deposited under freshwater lacustrine conditions (Bachhuber, 1989). These three freshwater units are separated by unfossiliferous layers of silty clay, silt, and gypsarenite that represent drier periods and water of high salinity. Vertebrate fossils occur in all three of Bachhuber's freshwater phases of his late Lake Estancia. The first and last of these freshwater units also contain abundant freshwater molluscs. The lowest of Bachhuber's freshwater phases, radiocarbon dated at 24,300 yr B. P., contains both the cutthroat trout *Oncorhynchus clarki* and the tiger salamander *Ambystoma tigrinum*. The middle and upper freshwater phases, dated at 20,040 yr B. P. and 11,740 yr B. P., respectively, contain only cutthroat trout (Bachhuber, 1989). Fish bones were first found in the Estancia basin in 1966 by Platt Bradbury, and hundreds of skull elements, vertebrae, and ribs, as well as several skeletons, have been found since. Surprisingly, the cutthroat trout *O. clarki* is the only species of fish represented in the Lake Estancia collection (Bachhuber, 1989). A partial skeleton of *O. clarki* with a body outline and scales is catalogued as NMMNH 32, and a representative collection of skeletal elements of this species is catalogued as NMMNH 33-35. The cutthroat trout no longer inhabits the Estancia Valley, but occurs in adjacent areas, including the Rio Grande and Pecos River.

The second site in the Lake Estancia Fauna is from Bachhuber's (1971) Playa E-9 locality (NMMNH site L-4623), located about 1 km south of Willard (34°35'N, 106°02'W). Ken Kietzke collected vertebrate fossils from the lowest gray lacustrine clays (unit E) at this site in September 1975. The same two freshwater species Bachhuber (1989) reported from the Playa E-28 site are present in Playa E-9, *Oncorhynchus clarki* and *Ambystoma tigrinum*. The collection from Playa E-9 includes four cranial elements and 11 vertebrae of *O. clarki* (NMMNH 33266) and a single vertebra of *A. tigrinum* (NMMNH 33267).

The third site in the Lake Estancia Fauna (NMMNH site L-4624) is located on the south side of Bachhuber's (1971) Playa E-12, about 10 km east of Willard along U. S. Route 60 (34°36'N, 105°55'W). Susan Block and Kate Zeigler collected vertebrate fossils from gray lacustrine clays at site L-4624 in September 1999. The fauna from Playa E-12 consists of: one vertebra of *Oncorhynchus clarki* (NMMNH 33273), a partial coracoid of a duck (NMMNH 33274), and a right M1-M3 of the horse *Equus* (NMMNH 33275, Figs. 5G, 5H). The upper molars of *Equus* have a fairly simple enamel pattern, a tiny pli caballin, and an elongated protocone with a lingual indentation. Measurements (in mm) of the *Equus* teeth are (M1, M2, M3, respectively): length, 30, 29, 29; width, 30, 27, 25; crown height, --, --, 87). The morphology and measurements of the Lake Estancia *Equus* teeth are similar to *Equus niobrarensis* (e.g., Lundelius, 1972; Harris and Porter, 1980), now generally included in the *Equus laurentius* group of Winans (1989).

The combined Lake Estancia vertebrate fauna from the three individual sites consists of four species: three freshwater taxa, including one fish (*Oncorhynchus clarki*), one salamander (*Ambystoma tigrinum*), and a duck (family Anatidae), and an extinct species of horse (*Equus laurentius* group). The Lake Estancia Fauna is late Wisconsinan (late Rancholabrean), ranging in age from 24,300-11,740 yr B. P.

21. Lucy—The Lucy site is located in the eastern portion of the Estancia basin, about 11 km south and 1.5 km west of Lucy, in the center of Torrance County. The site consists of a series of blowouts at the head of a broad valley, well above the eastern shoreline of Pleistocene Lake Estancia (Roosa, 1956). The Lucy site was discovered in 1954 by K. W. Kendall, and was excavated in the summer of 1954 by a University of New Mexico field party led by William Roosa under the direction of Frank Hibben. Fossils of mammoth (*Mammuthus* sp.), horse (*Equus* sp.) and extinct bison (*Bison antiquus*) were found at the Lucy site, in association with Paleoindian projectile points of both the Clovis and Sandia types (Roosa, 1956; Agogino and Stevens, 1972). According to Roosa (1956, p. 310), "A Sandia point was found in situ just below the surface alongside a portion of mammoth long bone which was eroding out of a layer of hard gray-white sand." There is a radiocarbon date of 14,330 yr B.P. from the Lucy site (Crane and Griffin, 1958; Agenbroad, 1984), but this date is probably too old. The association of extinct Pleistocene megafauna and Clovis-type Paleoindian artifacts in the Lucy site indicates an age range of ~11,000-12,000 yr B. P.

22. Encino Blowouts—This site consists of blowouts in a sand dune area along New Mexico Highway 3 north of Encino in eastern Torrance County. Fossils of mammoth (*Mammuthus* sp.), horse (*Equus* sp.), and bison (*Bison* sp.) were reported from the Encino Blowouts (Agogino and Stevens, 1972).

Guadalupe County

23. Milagro—A complete mammoth tooth was collected about 5 km southeast of Milagro in the northwestern corner of Guadalupe County (34°54'N, 105°12'W). Albert Gallegos found this tooth in 1977 near Milagro Canyon on the Gallegos Ranch, about 2.5 km east-southeast of the Gallegos Ranch headquarters (NMMNH site L-1496; Lucas and Effinger, 1991). The tooth is a right m2 of *Mammuthus columbi* (NMMNH 15018, Figs. 2C, 2D). The measurements, lamellar frequency (7 plates/100 mm), and thin enamel (2.2 mm) of NMMNH 15018 are indicative of *Mammuthus columbi* (Table 1). The presence of *M. columbi* suggests a late Pleistocene age for the Milagro site.

24. Agua Negra—Blair Salisbury collected a partial horse mandible with three teeth in 1993 on the Agua Negra Ranch, about 9 km south of Santa Rosa in central Guadalupe County (34°52'N, 104°41'30"W). This site is located on the east bank of a small arroyo that flows north into the Rio Agua Negra (NMMNH site L-4621). The fossil was found in gypsiferous sediment about 30 cm above the bottom of the arroyo. The horse jaw (*Equus*) consists of an associated right p3-m1 with fragments of the dentary (NMMNH 31599, Figs. 5E, 5F). The teeth have a fairly simple enamel pattern, a pli caballinid is present on p4 and m1, and the ectoflexid penetrates well into the isthmus on m1. Measurements (in mm) of the teeth are (p3, p4, m1, respectively): length, 29, 31, 27; width, 16, 17, 16; crown height, 76, 83, --). The morphology and measurements closely match the species *Equus niobrarensis* (e.g., Lundelius, 1972; Harris and Porter, 1980), now generally included in the *Equus laurentius* group of Winans (1989). The age of the Agua Negra site is almost certainly Pleistocene, but a more precise refinement is not possible pending the discovery of more age-diagnostic mammals.

FIGURE 5. Horse (*Equus*) dentary and teeth from Pleistocene sites in northeastern New Mexico. A. lateral view and B. occlusal view, left mandible with p4-m3, *Equus laurentius* group, BWDS B546, Blackwater Draw Locality 1, Roosevelt County; C. lateral view and D. occlusal view, left M3, *Equus laurentius* group, MDM 204, Plaza Larga Creek, Quay County; E. lateral view and F. occlusal view, associated right p3-m1, *Equus laurentius* group, NMMNH 31599, Agua Negra (NMMNH site L-4621), Guadalupe County; G. medial view and H. occlusal view, associated left M1-M3, *Equus laurentius* group, NMMNH 33275, Lake Estancia (NMMNH site L-4624), Torrance County.

De Baca County

25. Yeso Creek—The fossils from this site were collected in the walls of Yeso Creek, located about 12 km south of the town of Yeso in western De Baca County. This site was called Yeso Arroyo by Agogino and Stevens (1972) and Harris (1993), whereas Effinger and Lucas (1990) shortened the name to Yeso. The intermittent stream for which the site is named is called Yeso Creek on the Horney Ranch 7.5 minute quadrangle map and on all other New Mexico maps we have seen. Yeso Creek drains into the Pecos River about 30 km east of the fossil site. According to Agogino and Stevens (1972, p. 142), "The arroyo walls of this drainage are rich with fossil bone of Pleistocene age. The bones of mammoth, fossil bison, extinct ground sloth, and Pleistocene horse have been found along the drainage." George Agogino and Vance Haynes found three distinct layers in the walls of Yeso Creek that produced *Bison* fossils. The middle layer yielded the extinct bison, *B. antiquus*, and the upper layer produced *B. bison*. McDonald (1981) and Effinger and Lucas (1990) both mentioned a skull of *B. bison* (UASM Z-198) from Yeso Creek. We have not been able to locate any other fossils from Yeso Creek, but the presence of *Mammuthus* sp., *Bison antiquus*, *Equus*, and a ground sloth comprises a fairly typical eastern New Mexico late Pleistocene (Rancholabrean) fauna. All four of these taxa are present in the Badlands Ranch LF from Quay County (Morgan et al., 2000).

26. Fort Sumner—Two teeth and three postcranial elements of mammoths are known from the Fort Sumner site (NMMNH site L-4633), located in the vicinity of Fort Sumner in De Baca County. These specimens were part of the paleontology collection of the New Mexico Bureau of Mines and Mineral Resources (NMBMMR) that was transferred to the NMMNH in 1994. The only data associated with these fossils indicates they were found near Fort Sumner by Ned Slagel of the Roswell office of the U. S. Bureau of Land Management and donated to the NMBMMR collection. Attempts to locate additional information about the Fort Sumner site have been unsuccessful. The Fort Sumner *Mammuthus* sample consists of five fossils, including a complete right M3 (NMMNH 33453; Figs. 2A, 2B), a partial tooth (NMMNH 33454), a partial innominate (NMMNH 33455), a distal humerus (NMMNH 33456), and the head of a femur (NMMNH 33457). The M3 is from a very large mammoth. The length (338 mm), width (104 mm), and crown height (335 mm) of this tooth (NMMNH 33453) are larger than any other M3 of *Mammuthus columbi* measured from northeastern New Mexico (see Table 1). This tooth has 18 plates, and several are missing, suggesting that the total number of plates would have exceeded 20, the lamellar frequency is 6, and the enamel is thin (mean of five measurements is 2.4 mm). The large number of plates and thin enamel indicate that NMMNH 33453 is from a large individual of *Mammuthus columbi*, indicating a late Rancholabrean age. Several other sites along the Pecos River south of Fort Sumner contain the older, more primitive species of mammoth, *M. imperator*, suggesting an early Rancholabrean or late Irvingtonian age.

27. Taiban Creek—This locality is on the west bank of the Pecos River about 11 km south of Fort Sumner, 4 km south of the Fort Sumner State Monument, and 4 km north of the mouth of Taiban Creek, in central De Baca County (34°22'N, 104°12'W). Taiban Creek is one of three sites combined as the Blackdom Terrace locality in Lucas and Effinger (1991); the other two sites are Eighteenmile Bend and X-Bar-X Ranch (see below). The Taiban Creek assemblage includes two sites at the same elevation and stratigraphic level, located less than 1 km apart in sec. 27, T2N, R 26 E (Bonner Lake quadrangle; Jelinek, 1967). The first site, Locality U8 of Jelinek (1967), is in a small ravine trending southwest, in the SW¼ NE¼ sec. 27. Horse and camel teeth from Jelinek's Locality U8 were derived from a buff alluvium about 1 m thick, which overlies pond deposits of light yellowish-gray sands and interbedded clays. The second site, Locality P41 of Jelinek (1967), is about 0.5 km southeast of Locality U8, in the NE¼ SE¼ sec. 27. Locality P41 has a similar buff alluvium, with caliche pebbles, many of which contain bone fragments, including a partial horse jaw. Most of the bones and teeth from the buff alluvium are covered with a heavy incrustation of caliche (Jelinek, 1967).

The Taiban Creek fauna is composed of at least one species of the horse *Equus* and the large camel *Camelops* (Jelinek, 1967). The sample of fragmentary bones and teeth of *Equus* from Locality U8 includes a nearly complete right M2 measuring 27 mm in antero-posterior length and 24 mm in transverse width (Jelinek, 1967, pl. 1A). The size and characters of this tooth suggest referral to *Equus niobrarensis*, now generally placed in the *E. laurentius* species group of Winans (1989). A badly broken jaw fragment with two partial teeth of a horse from Locality P41 can be identified only as *Equus*. A fragmentary tooth of a large camel from Locality U8 is referred to *Camelops* (Jelinek, 1967). The presence of *E. laurentius* and *Camelops* sp. in Taiban Creek is not particularly age diagnostic, although the fauna is certainly Pleistocene.

The stratigraphy and geologic history of the terrace deposits along this portion of the Pecos River provide some information regarding the age of the Taiban Creek fauna. The fossils occur in the Blackdom terrace, which is the highest in elevation, and thus the oldest, of the three terraces found along the Pecos River. Jelinek (1967) correlated the fossiliferous unit of the Taiban Creek fauna, the buff alluvium, with the Yarmouthian interglacial period, roughly corresponding to the late Irvingtonian. He suggested that the caliche deposits in the buff alluvium indicated a prolonged arid interval characteristic of an interglacial. Jelinek thought the upper gravels of the Blackdom terrace overlying the fossil-bearing unit correlated with the next to last or Illinoian glaciation, and the upper surface of the Blackdom terrace correlated with the Sangamonian interglacial. The younger two terraces, the Orchard Park and Lakewood, are Wisconsinan and Holocene in age, respectively, according to Jelinek (1967). Pending the discovery of age-diagnostic mammals (e.g., *Bison* or *Mammuthus*), we regard the Taiban Creek fauna as Pleistocene, probably either late Irvingtonian or early Rancholabrean.

28. Eighteenmile Bend—This site is located on the west side of the Pecos River near Eighteenmile Bend, about 24 km south of Fort Sumner and 13 km south of the Taiban Creek site, in central De Baca County (34°15'N, 104°13'W). Jelinek (1967) reported a mammoth jaw, proboscidean limb bone fragments, and fragmentary *Equus* teeth from his Locality U6, here named the Eighteenmile Bend site, located in the SE¼ SW¼, sec. 4, T1S, R26E (Eighteenmile Hill quadrangle). These fossils were derived from water-laid sands near the top of the Blackdom terrace. Jelinek (1967) considered the upper Blackdom sands to be pre-Wisconsinan in age, probably either Yarmouthian or Illinoian, based on the substantial downcutting in the Pecos River valley adjacent to the site, indicating a prolonged interval following the deposition of the sands (see more detailed discussion of Blackdom terrace above under Taiban Creek site).

The most complete specimen from the Eighteenmile Bend site consisted of a heavily weathered pair of mammoth mandibles with both m3s in place; however, only the left m3 was collected (Jelinek, 1967, pl. 1D). According to Madden (1981), the Eighteenmile Bend tooth (UMMP 44424) is one of the largest known specimens of the imperial mammoth, *Mammuthus imperator* (Table 1). The length (434 mm) and width (124 mm) of UMMP 44424 are considerably larger than any m3 of *M. columbi* from northeastern New Mexico. Furthermore, the lamellar frequency of 4 plates/100 mm is lower than that of any tooth of *M. columbi* measured (all have a lamellar frequency of 6 or higher). The large size of UMMP 44424, low crown height, and low lamellar frequency are all characters of *M. imperator*. According to Madden (1981), the youngest records of *M. imperator* are early Illinoian (=early Rancholabrean). This age fits well with the age of the Blackdom terrace suggested by Jelinek (1967).

29. X-Bar-X Ranch—This site is on the X-Bar-X Ranch, located near the Pecos River about 40 km south of Fort Sumner in T2S, R25E (Conejo Creek East or La Espia Peak quadrangles) in southern De Baca County (Hay, 1927; Northrop, 1931). Harris (1993) listed both the X-Bar-X Ranch site of Northrop (1931) and the Dunlap site of (Hay, 1927). However, details in Hay's (1927) paper verify that the site he described, later called the Dunlap site by Harris (1993), is the same locality as the X-Bar-X Ranch site of Northrop (1931). The X-Bar-X Ranch site was discovered by Alfred and Wallace Moss in 1926 (Hay, 1927), and was excavated in December 1930 by Stuart Northrop and geology students from the University of New Mexico (Northrop, 1931). Northrop and his students collected a nearly complete tusk, a partial tusk, two teeth, and various postcranial elements of mammoths. Northrop (1931) provided a detailed description and photographs of the collection of the two mammoth tusks from X-Bar-X Ranch. The fossils were found near the surface in sand and gravel deposits of the ancestral Pecos River.

The nearly complete mammoth tusk from X-Bar-X Ranch was about seven feet long and eight inches in diameter, and the largest mammoth molar weighed 23 pounds (Northrop, 1931, p. 12-13). This is probably the same mammoth tooth mentioned earlier by Hay (1927, p. 281, quoting a letter of March 16, 1926 from Wallace Moss), who noted that "...the tooth was said to weigh 22 pounds..." and the "...length of the grinding surface was given as 12 inches." Most of the X-Bar-X Ranch fossils were deposited in the UNM Geology Museum. However, Northrop (1931, p. 13) noted that one of the two tusks and "An enormous leg bone, three and a half feet long..." (almost certainly a mammoth femur) were retained by the Moss family. Hay (1927) identified the largest mammoth tooth as *Mammuthus imperator* based on its dimensions (about 300 mm in length) and the presence of 13 thick plates. Northop (1931) and Lucas and Effinger (1991) also considered the X-Bar-X sample to be *M. imperator*. Although we have not examined the mammoth teeth from this site, we follow previous authors and refer these specimens to *M. imperator*. The age of the X-Bar-X Ranch site is unknown, although the presence of *M. imperator* indicates that, like the Taiban Creek and

Eighteenmile Bend sites on the Pecos River, the X-Bar-X Ranch site is probably older than late Pleistocene (early Rancholabrean or possibly late Irvingtonian).

Roosevelt County

30. Blackwater Draw Locality 1—Blackwater Draw Locality 1, also known as the Clovis site, is located in Blackwater Draw, 10 km north of Portales and 17 km southwest of Clovis, in northeastern Roosevelt County (34°17'N, 103°20'W). Blackwater Draw Locality 1 is the type site of the Clovis Culture, and is one of the most important Paleoindian archaeological sites in North America. Blackwater Draw Locality 1 also has two important latest Pleistocene (late Rancholabrean) vertebrate faunas. The Gray Sand LF (Lundelius, 1972) immediately pre-dates Paleoindian occupation of this region, and the Brown Sand Wedge LF (Slaughter, 1975) is contemporaneous with the occupation of this site by Clovis people (Haynes, 1995). Higher stratigraphic layers at Blackwater Draw contain remains of extinct bison, *Bison antiquus*, and Paleoindian artifacts that date to the younger Folsom and Agate Basin cultures (Haynes, 1995; Holliday, 1997). Only the Gray Sand and Brown Sand Wedge vertebrate faunas are discussed here. Several other smaller late Pleistocene sites also located in Blackwater Draw, including McCullum Ranch and Anderson Basin, are discussed below.

The Blackwater Draw site has a long and complex history, and only a brief review is provided here, with citations to the pertinent literature. We concentrate on the vertebrate fossils, whereas the majority of research and publications on this site have focused on the Paleoindian archaeology. Gravel mining operations led to the initial discovery of vertebrate fossils and associated Paleoindian artifacts at Blackwater Draw in 1932. The site was an active gravel pit in the 1930s and again from 1952 to the late 1960s. Vertebrate fossils were first reported from Blackwater Draw by Stock and Bode (1936), who based their work on collections made from 1932-1936 and now housed in the Academy of Natural Sciences in Philadelphia (ANSP). They also reported fossils from several other nearby basins in Blackwater Draw (e.g., Anderson Basin, see below). Lundelius (1972) reviewed the Gray Sand LF based on fossils in the Texas Memorial Museum (TMM) at the University of Texas, collected in 1949-1950 and 1953-1957 by field crews led by E. H. Sellards. Slaughter (1975) reviewed the Brown Sand Wedge LF based on fossils he collected in 1962, and now housed in the Shuler Museum of Paleontology (SMP) at Southern Methodist University. Extensive salvage operations were conducted at Blackwater Draw between 1962 and 1964 to save fossils and artifacts from destruction by gravel mining operations. Although large samples of fossils were collected during this time period, few of these have been studied. Collections from the 1962-1964 excavations are housed at the Blackwater Draw Museum (BWDM), Blackwater Draw Site (BWDS), Eastern New Mexico University (ENMU), Museum of New Mexico (MNM), and Texas Tech University (TTU). In 1962 and 1963, a joint field team from the MNM and the El Llano Archaeological Society (ELAS) excavated five mammoth skeletons associated with more than 100 Clovis artifacts (Hester, 1972). Lundelius (1972) provided descriptions and measurements of the mammoths uncovered in the MNM/ELAS excavation. Many of the mammoth and other mammalian fossils from the MNM/ELAS excavations are now curated at the BWDS and BWDM, some of which are illustrated (Figs. 4A-D, 5A, 5B, 8A-H) and measured here (Table 1). Geochronological studies of Blackwater Draw provide detailed stratigraphic and chronologic data (Haynes and Agogino; 1966; Haynes, 1995; Holliday, 1997). Hester (1972) presented the most detailed review of the Blackwater Draw site.

Blackwater Draw Locality 1 (Fig. 6) is located on the Llano Estacado, and consists of a late Pleistocene basin connected to the upper portion of Blackwater Draw by a small channel called the "outlet channel" or "spring draw" (Haynes, 1995; Holliday, 1997). This basin is inset into two much older, underlying units. The oldest unit consists of sands and gravels of the Ogallala Formation and is overlain by the Blackwater Draw Formation. The Ogallala Formation at Blackwater Draw Locality 1 has produced a mandible of the shovel-tusked gomphothere *Amebelodon*, indicating a late Miocene (Hemphillian) age (see Morgan and Lucas, this volume). The Gray Sand (Unit B of Haynes, 1995) occurs above an unconformity at the top of the Blackwater Draw Formation (Fig. 7), and produces a late Pleistocene (late Rancholabrean) vertebrate fauna (Gray Sand LF of Lundelius, 1972) that apparently lacks any association with Paleoindian artifacts (Haynes, 1995). The Brown Sand Wedge (Unit C of Haynes, 1995) occurs above the Gray Sand and has a late Pleistocene (late Rancholabrean) vertebrate fauna associated with Paleoindian artifacts of the Clovis type (Brown Sand Wedge LF of Slaughter, 1975). Figure 6 presents a series of photographs of Blackwater Draw Locality 1, including an overview of the site and various views of vertebrate fossils in place, particularly mammoths.

The Gray Sand LF (Lundelius, 1972) has 15 species of mammals, and the Brown Sand Wedge LF (Slaughter, 1975) has 35 species of mammals (Table 2). The primary difference between these two faunas is the significant number of small mammals (15 species) in the Brown Sand Wedge LF and the rarity of small mammals in the Gray Sand LF (Table 2), although this could be related to differences in sampling. Slaughter (1975, p. 179) washed ten tons of sediment from the Brown Sand Wedge, whereas there is no evidence that sediments were screen-washed from the Gray Sand. Futhermore, the Brown Sand Wedge contains seven species of lower vertebrates, including a frog, two turtles, and four snakes. The only lower vertebrate in the Gray Sand LF is the box turtle *Terrapene*. The two faunas are similar in their large mammal component, and both are considerably more diverse than any other Rancholabrean faunas in this region. Like most other late Pleistocene faunas from northeastern New Mexico, and the Southern High Plains in general, the Blackwater Draw faunas are dominated by large grazing mammals, in particular, mammoth, bison, horse, and camel.

Most of the fossils from the two Blackwater Draw faunas were found as isolated elements, with the exception of mammoths. Parts of five associated or articulated individuals of *Mammuthus columbi*, together with Clovis artifacts, were found in the Gray Sand according to Lundelius (1972). However, it is more likely from the stratigraphic studies of Haynes (1995) that these skeletons occurred in sediments equivalent in age to the younger Brown Sand Wedge. We measured six *Mammuthus* teeth from Blackwater Draw Locality 1 in the BWDM and BWDS collections (Table 1; Figs. 8A-H), most of which were collected by the El Llano Archaeological Society in 1962-1963. These probably include some of the teeth examined by Lundelius (1972). Four of the Blackwater Draw mammoth teeth we measured have unusally thick enamel (ranging from 3.8-4.2 mm) for late Pleistocene *Mammuthus*. Otherwise, the length, width, total number of plates on M3/m3 (17-21), and the lamellar frequency (6-7.5 plates/100 mm) are well within the range of *M. columbi*, and we confidently refer the Blackwater Draw mammoths to that species.

Slaughter (1975) discussed the paleoecological implications of the Brown Sand Wedge LF. This fauna contains numerous species of small vertebrates that no longer occur in the immediate vicinity of Blackwater Draw. Four categories of extralimital vertebrates occur in the Brown Sand Wedge LF: (1), species characteristic of cooler climates, now generally found north of Blackwater Draw or at higher elevations (e.g., the masked shrew *Sorex cinereus*, the meadow vole *Microtus pennsylvanicus*, and the Mexican vole *M. mexicanus*); (2), species characteristic of warmer climates, generally found south of Blackwater Draw (e.g., the extinct land tortoise *Hesperotestudo wilsoni*, the extinct large subspecies of box turtle *Terrapene carolina putnami*, the extinct armadillo *Dasypus bellus*, and the cotton rat *Sigmodon hispidus*); (3), species characteristic of permanent water (e.g., the frog *Rana*, the ribbon snake *Thamnophis sauritus*, and the muskrat *Ondatra zibethicus*); and (4), species typical of forested conditions (e.g., the opossum *Didelphis virginiana*, the black bear *Ursus americanus*, and the tree squirrel *Sciurus carolinenis*). Three of the four species typical of warmer climates are extinct; however, extinct land tortoises of the genus *Hesperotestudo* are usually thought to indicate frost-free winters (Hibbard, 1960), and the large extinct armadillo *D. bellus* is most common in faunas from southern North America, in particular Florida. The living armadillo, *Dasypus novemcinctus*, is susceptible to cold temperatures, and is currently unknown from New Mexico (Findley et al., 1975). Although found in the general vicinity of Blackwater Draw, *Sigmodon hispidus* is typically a south temperate species. The occurrence of these warmer climate species with the two voles and the shrew that are typical of cooler climates provides an example of a "disharmonius fauna" (Graham and Lundelius, 1984), in which species that are found in different habitats at the present time (i.e., allopatric) co-occur in late Pleistocene faunas. The presence of the frog, ribbon snake, and muskrat indicate that it was wetter during the late Pleistocene in the Blackwater Draw area, with the presence of permanent water (streams, springs, ponds, or small lakes). The tree squirrel, black bear, and opossum are indicative of nearby forested areas.

The most common mammals in the vicinity of Blackwater Draw at the present time are two desert rodents that inhabit sandy areas in grasslands habitats, the kangaroo rat *Dipodomys ordii* and the grasshopper mouse *Onychomys leucogaster* (Findley et al. 1975; Slaughter, 1975). These two rodents are absent from the Brown Sand Wedge LF, further indicating moister conditions in the late Pleistocene. In general, the climate in the southwestern United States at the end of the late Wisconsinan glacial was cooler, especially the summers, and moister than at present without marked seasonal extremes (Harris, 1985; Graham, 1987). The winters apparently were somewhat milder and generally frost-free, as indicated by the presence of the land tortoise and armadillo. In the latest Pleistocene, when the Brown Sand Wedge fauna inhabited Blackwater Draw, there were cooler summers, milder winters, and considerably more moisture was available, including permanent bodies of water.

FIGURE 6. Photographs of Blackwater Draw Locality 1, Roosevelt County, New Mexico. A. Eastern New Mexico University excavation, 1964-1968, south pit. B. Bones of Mammoth #2, woman pointing to large unifacial blade, El Llano Archaeological Society (ELAS) excavation, December 1962. C. Bones of Mammoth #1, ELAS excavation, December 1962. D. Preparing Mammoth #2 for removal, ELAS excavation, February 1963. E. Close-up view of Mammoth #1, ELAS excavation. F. Antler of deer (*Odocoileus*) from gray sand, ELAS excavation, north bank. (Photographs courtesy of Joanne Dickenson, Blackwater Draw Archaeological Site).

Previous authors do not agree on the stratigraphy and age of the Gray Sand and Brown Sand Wedge Local Faunas at Blackwater Draw Locality 1. We follow Haynes (1995) for the ages of the strata that produced these two faunas. Both faunas are late Rancholabrean, but they differ slightly in age, and particularly in their association with Paleoindians, or lack thereof. Haynes (1995) regarded the Gray Sand layer (his Unit B) as older than the appearance of Clovis people in this region (between 13,000 and 11,500 yr B. P.), whereas the Brown Sand Wedge layer (Haynes' Unit C) is contemporaneous with occupation of this region by Clovis people (11,500-11,000 yr B. P.). Although Clovis artifacts have been found in the Gray Sand layer, Haynes considered these to be intrusive or to occur on an occupation surface on top of the Gray Sand. We follow Haynes (1995) and consider the Gray Sand LF as a late Pleistocene paleontological fauna with no Paleoindian association. Other authors (e.g., Green, 1992) consider the Gray Sand level to be of Clovis age (~12,000-11,000 yr B. P.). For purposes of our discussion, these disagreements on the stratigraphy and age of the two superposed late Rancholabrean faunas at Blackwater Draw are not critical. The vertebrate faunas from the Gray Sand and the slightly younger Brown Sand Wedge are very similar, and therefore of essentially the same age within our level of biostratigraphic resolution.

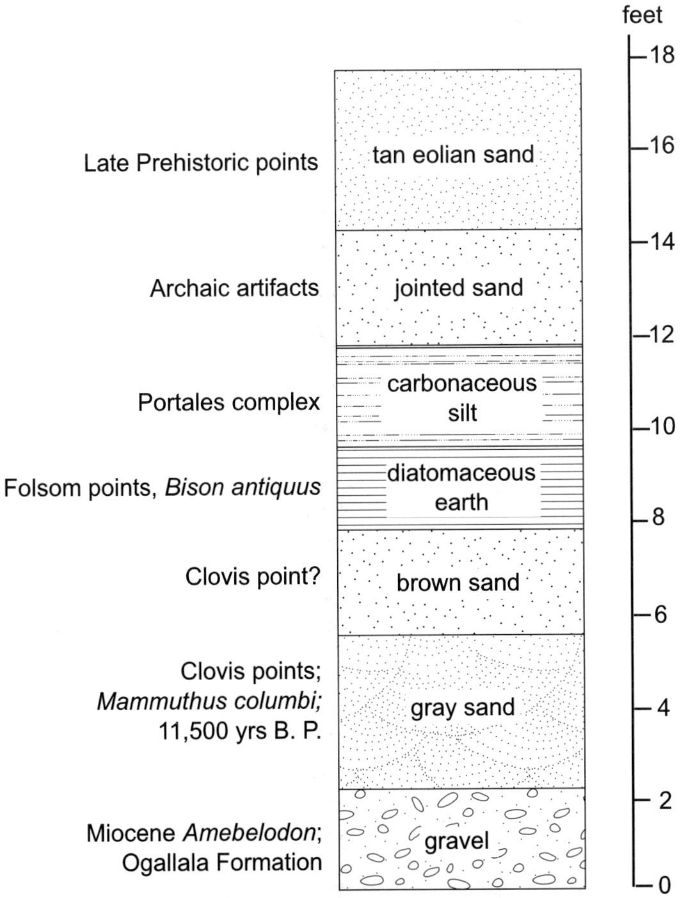

FIGURE 7. Stratigraphic section of Blackwater Draw Locality 1, Roosevelt County, New Mexico. Late Pleistocene fossil vertebrates were derived from the gray sand (Gray Sand Local Fauna of Lundelius, 1972) and brown sand (Brown Sand Wedge Local Fauna of Slaughter, 1975).

31. *Arctodus* Site—The *Arctodus* site is located in Blackwater Draw, 6 km southwest of Blackwater Draw Locality 1 and 8 km northwest of Portales, in northeastern Roosevelt County (34°16'N, 103°24'W). The only reference to this site is Haynes (1975), who reported fossil remains of the giant bear *Arctodus*, presumably *A. simus*, from a red eolian sand (Unit A) of pre-late Wisconsinan age. Haynes (1975, p. 89) also identified "bones of camel and mammoth" from Unit A at the *Arctodus* site, as well as "bones of mammoth and bison" from Unit C, a sand apparently correlative with the Brown Sand Wedge at Blackwater Draw Locality 1. The *Arctodus* site has two superposed Pleistocene faunas, an older pre-Wisconsinan fauna containing *Arctodus*, camel, and *Mammuthus*, and a younger latest Wisconsinan (late Rancholabrean) fauna with *Bison antiquus* and *Mammuthus columbi*.

32. McCullum Ranch—The McCullum Ranch site is located 5 km southeast of Blackwater Draw Locality 1 and 10 km northeast of Portales in northeastern Roosevelt County (34°16'N, 103°16'W). This site is a large blowout on the north side of Blackwater Draw (Haynes, 1975). Vertebrate fossils from McCullum Ranch occur in a green sand deposited in a small pond. A radiocarbon date of 15,750 ± 760 yr B. P. was obtained on a black carbonaceous zone just above the fossiliferous layer (Haynes, 1975). The only reference that mentions the McCullum Ranch vertebrate fauna (Haynes, 1975) lists ground sloth, horse, camel, bison, and mammoth, but scientific names and the fossils represented are not provided. On the basis of the late Pleistocene age of this site, it is most likely that the bison is *Bison antiquus* and the mammoth is *Mammuthus columbi*.

33. Anderson Basin—Anderson Basin is located 18 km south of Clovis and 12 km northeast of Portales in northern Roosevelt County (34°14'N, 103°12'W). Anderson Basin is on the north side of Blackwater Draw about 10 km southeast (downstream) of Blackwater Draw Locality 1. There are actually two different sites in Anderson Basin about 1 km apart, Anderson Basin 1 and Anderson Basin 2 (Haynes, 1975; Holliday, 1995). Both sites are blowouts in the Muleshoe Dunes, in which deflation has removed much of the valley fill, exposing late Pleistocene lake sediments, especially around the margins of the blowouts. Vertebrate fossils were first reported from Anderson Basin by Stock and Bode (1936), who found horse, bison, camel, and mammoth in a unit they called the "Blue Sand." In a stratigraphic section measured in Anderson Basin 2, Haynes (1975, p. 93), reported "Numerous horse, mammoth, and bison bones" from a speckled gray sand (Unit B-1), that he considered correlative with the Gray Sand at Blackwater Draw Locality 1. Holliday (1995) reported fossils of late Pleistocene megafauna in indurated lake carbonates exposed in the floor of Anderson Basin 2. He found no evidence that the fossil vertebrates in Anderson Basin 2 were associated with human activity.

There is some confusion regarding the occurrence of fossil vertebrates in the two Anderson Basin sites. Stock and Bode (1936) listed specimens from their sites Anderson Basin Number 1 and Anderson Basin Number 2, whereas Haynes (1975, p. 71) stated that Anderson Basin 2 was the "...main area of investigation in Anderson Basin in 1936..." Anderson Basin Numbers 1 and 2 as recognized by Stock and Bode (1936) have very similar vertebrate faunas, and are combined here as the Anderson Basin Fauna. The vertebrate fauna from Anderson Basin is composed of six species (from Stock and Bode, 1936): the box turtle *Terrapene ornata* and five mammals, including the skunk *Mephitis mephitis*; a horse of the *Equus laurentius* group (=*E. cf. E. excelsus* of Stock and Bode, 1936); the camel *Camelops* cf. *C. hesternus*; the extinct bison *Bison antiquus*; and the mammoth *Mammuthus columbi*. Stock and Bode (1936) described and illustrated several fossils from Anderson Basin, including: the anterior portion of a carapace of *Terrapene ornata* (ANSP 13780; pl. 8, fig. 1); a skull and jaws of *Mephitis mephitis* (ANSP 13733); a distal metapodial (ANSP 13728, fig. 5A) and proximal phalanx (ANSP 13727, fig. 5B) of *Camelops* cf. *C. hesternus*; two teeth of the *Equus laurentius* group (ANSP 13674, 13686; figs. 4D, 4J); and an upper molar of *Mammuthus columbi* (ANSP 13738, pl. 10). Like the Gray Sand LF, the Anderson Basin vertebrate fauna is latest Pleistocene (late Rancholabrean) in age, but pre-dates Paleoindian occupation of Blackwater Draw.

34. Circus Basin/Elephant Tusk Basin—The Circus Basin and Elephant Tusk Basin sites, both named by Stock and Bode (1936), are located in Blackwater Draw, east of Blackwater Draw Locality 1 and northeast of Portales, in northeastern Roosevelt County. However, we have been unable to determine the exact location of these two basins. Stock and Bode (1936) are the only paleontologists who described fossils from the Circus and Elephant Tusk basins. They identified a horse of the *Equus laurentius* group from both Circus Basin and Elephant Tusk Basin, and *Mammuthus columbi* from Circus Basin. Even though they did not specifically mention the presence of mammoths in Elephant Tusk Basin, this basin was obviously named for the abundance of "elephant" tusks found there, presumably belonging to *M. columbi*. Stock and Bode (1936) described and illustrated a right mandible with p2-m3 (ANSP 13722, fig. 4A) and an upper tooth (ANSP 13689, fig. 4B) of *Equus* cf. *E. excelsus* (=*E. laurentius* group of Winans, 1989) and an upper molar of *M. columbi* (ANSP 13740, pl. 9) from the Circus Basin, and a lower tooth *Equus* cf. *E. excelsus* (ANSP 13688, fig. 4G) from the Elephant Tusk Basin. The Circus Basin and Elephant Tusk Basin vertebrate faunas are similar in age to the late Pleistocene (late Rancholabrean) Gray Sand LF from Blackwater Draw Locality 1.

35. Floyd—The Floyd site (NMMNH site L-4634) is located about 30 km west of Floyd, just southwest of The Mesa and the Melrose Bombing Range, in northwestern Roosevelt County. The site was discovered in 1986 by Ron Egan of Carlsbad, New Mexico, who collected horse and camel fossils in a large blow-out surrounded by sand hills and grassland. Although large bones, presumably proboscidean, supposedly were present at the Floyd site about ten years before, Egan found only fossils of *Equus* and *Camelops*. The *Equus* sample consists of six associated incisors (NMMNH 33458), a proximal phalanx (NMMNH 33459), and a medial phalanx (NMMNH 33460). None of the horse fossils are diagnostic at the species level. The *Camelops* fossils include a partial right mandible with m2-m3 (NMMNH 33462), a distal tibia (NMMNH 33463), and a distal metapodial (NMMNH 33464). Measurements (in mm) of the *Camelops* teeth (NMMNH 33462) are: m2 length, 38.9; m2 width, 20.5; m3 length, 57.9; m3 width, 18.6. The distal width of the *Camelops* tibia is 91. These measurements are within the range of variation for *Camelops hesternus* from Rancho la Brea in southern California (Webb, 1965), and the Floyd fossils are referred to that species. Neither *Equus* nor *Camelops hesternus* are particularly age diagnostic. The Floyd site is almost certainly Pleistocene, but its age can not be further delimited based on the known fauna.

36. Stolle—The Stolle Mammoth site is located approximately 27 km south of Portales and 8 km north of Dora in the center of Roosevelt County. Fossils were

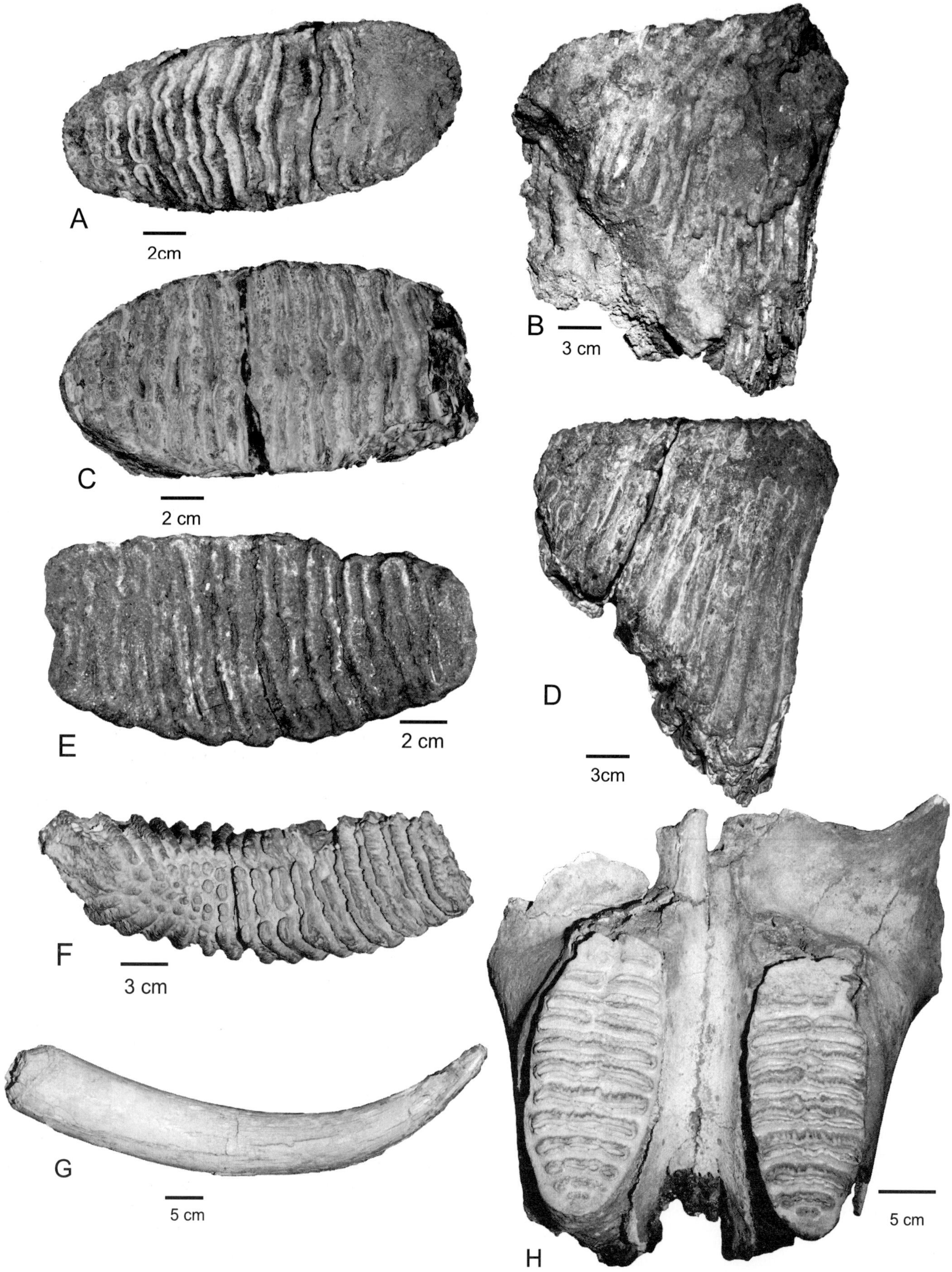
A
2cm
B
3 cm
C
2 cm
D
3cm
E
2 cm
F
3 cm
G
5 cm
H
5 cm

first discovered at the Stolle site in the spring of 1977 by Lester Shirley of Eastern New Mexico University (Agogino and Sweetland, 1985). Shirley found the fossils on the property of A. W. Stolle during the excavation for a cattle tank located close to the center of a playa about 1 km wide. The site was excavated in the summer and fall of 1977 by a field team from ENMU led by Bill Sweetland. The fossils from the Stolle Mammoth site consist of a partial skeleton of a young mammoth that supposedly became mired in a small pond (Agogino and Sweetland, 1985). The front limbs of the mammoth were found in an upright or vertical position in a light gray sandy silt, overlain by a thick black organic layer, overlain by a sandy layer possibly representing a sand bar. The articulated forefeet were preserved in the sandy silt layer (Agogino and Sweetland, 1985, fig. 1). Two large caliche rocks found near the skull were thought to have been used by Clovis hunters to smash open the braincase of the mammoth. However, no diagnostic Paleoindian tools were discovered at the Stolle Mammoth site. The evidence for cultural association was based on the smashed skull, the positioning of the bones, and the absence of the tusks and the left hind leg, which were supposedly removed from the skeleton and butchered elsewhere (Agogino and Sweetland, 1985).

The Stolle mammoth skeleton has not been examined, but presumably it is in the collections of ENMU. It is not clear how much of the skeleton was recovered (Agogino and Sweetland, 1985). The skeleton is from a young or adolescent individual, and consisted of a broken skull lacking the tusks, the lower jaws, both front limbs, and the right hind limb. No mention was made of vertebrae, ribs, and limb girdles. The authors did not provide any morphological information about the mammoth skull, teeth, or bones, and thus we can only identify the specimen as *Mammuthus* sp. The Stolle Mammoth site is a probable Clovis kill site, which would establish a very late Pleistocene age (between 11,000 and 12,000 yr B. P.), and would indicate referral of the mammoth to *M. columbi*.

37. Milnesand—This Paleoindian site is located 6 km northeast of Milnesand in southern Roosevelt County (33°41'N, 103°18'W). A second Paleoindian site, called the Ted Williamson site, is located about 0.5 km from the Milnesand site. Both sites occur in small blowouts in the Lea-Yoakum dunes, near a narrow constriction of Sulphur Draw (Holliday, 1997). Because the Milnesand and Ted Williamson sites are in close proximity, occur in a similar stratigraphic context, and expose bone beds of the extinct bison, *Bison antiquus*, they are combined as one paleontological site. Both sites were exposed by deflation of blowouts in the early 1950s, and were investigated by E. H. Sellards and Jim Warnica (Holliday, 1997). The Milnesand site was excavated by Sellards in 1953, whereas little is known about the Ted Williamson site. The two sites were completely destroyed by erosion and weathering following their exposure in the 1950s (Holliday, 1997).

Both the Milnesand and Ted Williamson sites contained extensive bone beds of *Bison antiquus*, with the Milnesand bone bed covering at least 300 m^2 (Holliday, 1997). Although the recovery and curation of bison bones from Milnesand was highly selective, more than 30 individuals of *B. antiquus* were found at this site (Holliday, 1997). The bison bone bed at Milnesand also produced a large sample of Paleoindian projectile points, called Milnesand points. The age of the Milnesand site is probably latest Pleistocene (late Rancholabrean) based on the association of *B. antiquus* with Paleoindian artifacts. However, it is possible that this site is early Holocene in age, as *B. antiquus* survived until about 5,000 yr B. P. (McDonald, 1981).

OVERVIEW OF PLEISTOCENE VERTEBRATE FAUNAS FROM NORTHEASTERN NEW MEXICO

Forty-five Pleistocene vertebrate localities are known from northeastern New Mexico. We have attempted to determine the ages of these sites based on biochronology, radiocarbon dates, or other geochronologic data. The mammoth *Mammuthus* occurs in 28 of the sites. The presence of *Mammuthus* is characteristic of North American Pleistocene faunas (1.8 Ma-10 ka), including both the Irvingtonian and Rancholabrean land-mammal "ages." There is no definitive evidence that any vertebrate faunas from northeastern New Mexico are Irvingtonian, which includes the entire early Pleistocene and much of the middle Pleistocene (1.8-0.3 Ma). Most of the faunas for which a more specific age can be determined are Rancholabrean, characterized by the presence of *Bison*, which arrived in North America from Eurasia at about 0.3 Ma (Lundelius et al., 1987). Of the 45 Pleistocene sites in northeastern New Mexico, 21 sites contain *Bison* and are clearly Rancholabrean in age, whereas 24 sites lack *Bison*, and thus a Rancholabrean age cannot be firmly established without other associated geochronologic data. Nevertheless, we suspect that most of the faunas lacking *Bison* are Rancholabrean as well, and the absence of *Bison* is related to factors other than age. A few of these sites have radiocarbon dates that establish a Rancholabrean age, but for most of the sites lacking *Bison* we cannot be more specific than Pleistocene for their age.

The Rancholabrean is subdivided into the early and late Rancholabrean, with the boundary between these two subages corresponding to the beginning of the last, or Sangamonian interglacial (130 ka), which is also the boundary between the middle and late Pleistocene. Discriminating early from late Rancholabrean mammalian faunas is difficult (e.g. Morgan and Hulbert, 1995), particularly with regard to the low-diversity Pleistocene faunas from northeastern New Mexico. The species of *Bison* have been used to subdivide the Rancholabrean, with the giant bison *B. latifrons* generally found in older early Rancholabrean faunas, and *B. antiquus* characteristic of late Rancholabrean faunas. The only specimen of *B. latifrons* from northeastern New Mexico, from near Raton in Colfax County (McDonald, 1981), has no associated fauna or stratigraphy, and thus there is no additional information regarding its age. The only other New Mexico record of *B. latifrons*, a horn core and upper dentition from a gravel pit near Bernalillo in Sandoval County in the central Rio Grande Valley (Smartt et al., 1991), is derived from the Los Duranes Formation, which appears to be early Rancholabrean in age (older than 130 ka) based on stratigraphic data (S. Connell, oral. comm., 2000). The presence of the extinct species *B. antiquus* is often used to distinguish late Pleistocene faunas from Holocene faunas, which generally contain the living species *B. bison*. However, McDonald (1981) has shown that the extinct bison *B. antiquus occidentalis* survived until the middle Holocene.

Several sites from the Blackdom terrace along the Pecos River in De Baca County, including Taiban Creek, Eighteenmile Bend, and X-Bar-X Ranch, may be early Rancholabrean or perhaps late Irvingtonian. According to Jelinek (1967), the stratigraphy and geologic history of the terraces along the Pecos River indicate that the Blackdom terrace corresponds in age to the Yarmouthian interglacial and Illinoian glacial, roughly equivalent to the early Rancholabrean, whereas the younger two terraces, the Orchard Park and Lakewood, are Wisconsinan and Holocene in age, respectively. It should be stressed that older publications (e.g., Jelinek, 1967; McMullen and Zakrzewski, 1972) often referred to the ages of terraces and vertebrate faunas in terms of their

FIGURE 8. Columbian mammoth (*Mammuthus columbi*) teeth and tusk from Blackwater Draw Locality 1, Roosevelt County, New Mexico. A. occlusal view, right m3, BWDM 1948. B. lateral view and C. occlusal view, left M3, BWDM 1749. D. lateral view and E. occlusal view, M3, BWDM 1205. F. occlusal view, left m3, BWDM 38. G. worked tusk, BWDS B503. H. occlusal view, palate with right and left M3, BWDS Mammoth #1.

relationship with the four classic North American glacial and interglacial ages (i.e., from oldest to youngest, the Nebraskan, Kansan, Illinoian, and Wisconsinan glacials, and the Aftonian, Yarmouthian, and Sangamonian interglacials). However, more recent studies suggest there may have been as many as 25 glacial/interglacial cycles in the late Pliocene and Pleistocene (e.g., Richmond and Fullerton, 1986). Thus, most of the old glacial/interglacial terms are outdated, and the ages associated with them are suspect. Geochronologic data provide accurate dates for the last two glacial/interglacial cycles, from oldest to youngest (Richmond and Fullerton, 1986): penultimate interglacial (Yarmouthian?), 250-200 ka; penultimate glacial (Illinoian), 200-130 ka; last interglacial (Sangamonian), 130-120 ka; last glacial (Wisconsinan), 120-10 ka.

The imperial mammoth, *Mammuthus imperator*, found in several northeastern New Mexico faunas, appears to be characteristic of early Rancholabrean and older Pleistocene faunas. According to Madden (1981), *M. imperator* is characteristic of late Irvingtonian and early Rancholabrean faunas, and *M. columbi* occurs only in the late Rancholabrean. *M. imperator* also occurs in several New Mexico early Irvingtonian faunas (Lucas et al., 1999; Morgan and Lucas, 2000). Although *M. imperator* has been reported from Rio Conchas (Hay, 1927), X-Bar-X Ranch (Northrop, 1931; Lucas and Effinger, 1991), and Eighteenmile Bend (Jelinek, 1967; Madden, 1981) in northeastern New Mexico, only the Eighteenmile Bend record is unquestionably *M. imperator*. Madden (1981) noted that a huge m3 from Eighteenmile Bend was the largest tooth of this species known from North America. The presence of *M. imperator* in the Eighteenmile Bend site indicates an early Rancholabrean or perhaps older age, supporting the pre-Wisconsinan age suggested by Jelinek (1967) from the terrace stratigraphy. Two other sites from northeastern New Mexico have been regarded as pre-Wisconsinan in age (i.e., early Rancholabrean or older), Casados Ranch (McMullen and Zakrezewski, 1972) and the *Arctodus* site (Haynes, 1975). The Illinoian (=early Rancholabrean) age of Casados Ranch was primarily based on similarities of its terrestrial and freshwater molluscan fauna to early Rancholabrean molluscan faunas from Kansas and Oklahoma (Miller, 1966; McMullen and Zakrezewski, 1972). However, geologic data suggest this site may be Wisconsinan. The sparse vertebrate fauna from Casados Ranch provides limited biochronologic information. The pre-Wisconsinan age of the *Arctodus* site in Blackwater Draw, which includes the giant bear *Arctodus*, camel, and mammoth, was based on the occurrence of these fossils in a stratigraphic unit underlying the late Wisconsinan Gray Sand (Haynes, 1975). The stratigraphy suggests the *Arctodus* site is pre-late Wisconsinan, but the age of the site also could be earlier in the Wisconsinan. The vertebrate fauna provides no clues to further limit the age of the *Arctodus* site.

The majority of vertebrate faunas from northeastern New Mexico are late Pleistocene (late Rancholabrean) in age. Many of these sites have associated radiocarbon dates that confirm a late Pleistocene age. The two best biostratigraphic indicators of a late Rancholabrean age in these faunas are the extinct bison *Bison antiquus* and the Columbian mammoth *Mammuthus columbi*, although as noted above, *B. antiquus* survived into the early Holocene. Seven of the vertebrate faunas from northeastern New Mexico are associated with Paleoindian artifacts of the Clovis or Folsom cultures, indicating a latest Rancholabrean age, between 12,000 and 10,000 yr B.P. (Holliday, 1995). The only late Pleistocene fauna in northeastern New Mexico with radiocarbon dates that is not associated with Paleoindians is Lake Estancia. Three radiocarbon dates ranging in age from 24,300-11,740 yr B.P. clearly establish a late Wisconsinan age (late Rancholabrean) for the Lake Estancia Fauna (Bachhuber, 1989). A radiocarbon date of 14,330 yr B.P. from the Lucy site (Crane and Griffin, 1958; Agenbroad, 1984) is probably too old, as the association of Pleistocene megafauna and Paleoindian artifacts indicates an age of ~12,000-11,000 B.P.

Blackwater Draw Locality 1 is the best known and most thoroughly dated late Pleistocene site in northeastern New Mexico. A series of radiocarbon dates from various stratigraphic layers at the Blackwater Draw site has established a very detailed chronology (Haynes, 1995; Holliday, 1997). Only the two strata that contain late Rancholabrean vertebrates, the Gray Sand and Brown Sand Wedge, are discussed here. Haynes (1995) regarded the Gray Sand as pre-Clovis (radiocarbon dates between 13,000 and 11,500 yr B.P.), whereas the Brown Sand Wedge is contemporaneous with the Clovis culture (radiocarbon dates between 11,500-11,000 yr B.P.). The combined late Rancholabrean mammalian fauna from the Gray Sand and Brown Sand Wedge LFs contains 13 members of the extinct Pleistocene megafauna. Four of the Paleoindian sites from northeastern New Mexico, Folsom, San Jon, Milnesand, and Ted Williamson, represent the Folsom or similar cultures, and lack Pleistocene megafauna, with the exception *of Bison antiquus*. All four of these sites have bone beds of *B. antiquus occidentalis* associated with Paleoindian artifacts, and thus could be either latest Pleistocene or early Holocene in age (Holliday, 1995). The San Jon *B. antiquus* bone bed has conflicting radiocarbon dates, with two dates of 12,510 yr B.P. and 11,450 yr B.P. indicating a late Pleistocene age, and a younger early Holocene date of 8,360 yr B.P. (Hill et al., 1995). A date of 13,145 yr B.P. from San Jon is on pre-Clovis late Pleistocene lake sediments containing mammoth (Hill et al., 1995). The other three Paleoindian sites lack a radiocarbon chronology. The extinction of the Pleistocene megafauna occurred sometime between 11,000 and 10,000 yr B.P. (Mead and Meltzer, 1984). Thus, it is possible that some sites contemporaneous with the Folsom Culture and lacking megafaunal species are latest Pleistocene in age.

The Pleistocene vertebrate record from northeastern New Mexico consists primarily of low-diversity faunal assemblages. Only ten of the 45 sites have three or more species of mammals, and only two of these, the Gray Sand LF and Brown Sand Wedge LF, both from Blackwater Draw Locality 1, have more than 10 species of mammals. Twenty sites contain a single species of mammal (13 sites with *Mammuthus*, four sites with *Bison*, and three sites with *Equus*), and ten sites contain two species of mammals (six sites with *Mammuthus* and *Equus*, three sites with *Equus* and *Camelops*, and one site with *Mammuthus* and *Bison*). Overall, the faunas are dominated by large grazing mammals, in particular, *Mammuthus*, *Bison*, *Equus*, *Camelops*, and the ground sloth *Paramylodon*, in decreasing order of abundance. Low-diversity faunal assemblages consisting primarily of large grazing mammals are typical of late Pleistocene sites throughout New Mexico derived from fluvial or alluvial depositional environments (e.g., Lucas and Morgan, 1996; Morgan and Lucas, 2000). More diverse late Pleistocene faunas are primarily restricted to caves (Harris, 1993) and quiet-water depositional situations, such as the spring deposits at Blackwater Draw (Lundelius, 1972; Slaughter, 1975).

Only five of the vertebrate faunas from northeastern New Mexico, including Lake Estancia, Casados Ranch, Folsom, Gray Sand, and Brown Sand Wedge, contain small vertebrates. The Lake Estancia Fauna is dominated by freshwater species, including the cutthroat trout *Oncorhynchus clarki*, the tiger salamander *Ambystoma tigrinum*, and a duck. The Lake Estancia Fauna is consistent with the geologic evidence for a large lake in the Estancia basin during the Wisconsinan (Bachhuber, 1989). The presence of abundant freshwater molluscs and a salamander from Casados Ranch indicates the fossils were deposited in a small lake or pond.

The diverse mammalian faunas from the Gray Sand and Brown Sand Wedge at Blackwater Draw Locality 1 provide the most detailed picture of late Pleistocene paleoenvironmental conditions in northeastern New Mexico (Slaughter, 1975; Graham, 1987). The primary difference between these two faunas, which range in age from about 13,000-11,000 yr B.P., is the virtual absence of small mammals from the Gray Sand. Otherwise, the two faunas have very similar large mammal assemblages. For purposes of this discussion, we have combined the mammalian faunas from the Gray Sand and Brown Sand Wedge. The two Blackwater Draw faunas have a total of 39 species of mammals, including 13 members of the extinct Pleistocene megafauna. The most common members of the fauna are the same species of large grazing mammals found in late Pleistocene sites throughout northeastern New Mexico: *Mammuthus columbi*, *Bison antiquus*, *Equus* (three species), and *Camelops hesternus*. The fauna also includes two large extinct carnivores, the dire wolf *Canis dirus* and the sabretooth cat *Smilodon fatalis*. The fauna of large grazing mammals, and the carnivores that preyed upon them, are indicative of extensive grassland and savanna habitats in this region. The diverse fauna of small mammals and other small vertebrates from Blackwater Draw provide additional information suggesting the presence of permanent streams, ponds, or small lakes, as well as trees, probably occurring as riparian forests along permanent streams (Slaughter, 1975; Graham, 1987). Several extralimital species of living small mammals from Blackwater Draw now occur farther north or at higher elevations, including *Sorex cinereus*, *Microtus mexicanus*, and *M. pennsylvanicus*. Four species from Blackwater Draw are generally found in faunas farther south, including the extinct land tortoise *Hesperotestudo wilsoni* and the extinct armadillo *Dasypus bellus*, indicating that winters were not marked by extreme cold spells (Hibbard, 1960; Graham, 1987). Blackwater Draw is a late Pleistocene "disharmonius fauna" (Graham and Lundelius, 1984) in which species co-occur that are now allopatric in New Mexico (e.g., masked shrew and cotton rat). The climatic conditions during the late Wisconsinan in northeastern New Mexico, as typified by the Blackwater Draw vertebrate fauna (Graham, 1987), included milder winters, cooler summers, and more available moisture with permanent water in the form of streams, ponds, and small lakes.

ACKNOWLEDGMENTS

Frederick Bachhuber, Susan Block, Ron Egan, Albert Gallegos, Jimmy Gallegos, Junior Jaramillo, Dogie Jones, Ken Kietzke, Blair Salisbury, Paul Sealey, Ned Slagel, and Kate Zeigler generously donated Pleistocene vertebrate fossils from northeastern New Mexico to the New Mexico Museum of Natural History. Howard Abercrombie, Robert Abercrombie, Jr., Mark Gordon, and Paul Sealey helped in the field. Matthew Hillsman and Joanne Dickenson allowed us to examine and photograph fossils in the Blackwater Draw Museum and Blackwater Draw Site, respectively. Joanne Dickenson loaned us photos of Blackwater Draw and gave us permission to reproduce those photos. Barry S. Kues and Michael O'Neill provided helpful comments on an earlier draft of the manuscript.

REFERENCES

Agenbroad, L. D., 1984, New World mammoth distribution; in Martin, P. S. and Klein, R. G., ed., Quaternary Extinctions: Tucson, University of Arizona Press, p. 90-108.

Agogino, G. A and Sweetland, B., 1985, The Stolle mammoth: A probable Clovis kill site: National Geographic Society, Research Reports, 1977 projects, v. 18, p. 89-93.

Agogino, G. A. and Stevens, D. E., 1972, Tours of archaeological interest in east-central New Mexico: New Mexico Geological Society, 23rd Field Conference Guidebook, p. 141-143.

Bachhuber, F. W., 1971, Paleolimnology of Lake Estancia and the Quaternary history of the Estancia Valley, central New Mexico [PhD thesis]: Albuquerque, New Mexico, University of New Mexico, 238 p.

Bachhuber, F. W., 1989, The occurrence and paleolimnologic significance of cutthroat trout (*Oncorhynchus clarki*) in pluvial lakes of the Estancia Valley, central New Mexico: Geological Society of America Bulletin, v. 101, p. 1543-1551.

Barnes, V. E., 1983, Geologic atlas of Texas Tucumcari sheet: Austin, Bureau of Economic Geology, The University of Texas at Austin, scale 1:250,000.

Crane, H. R. and Griffin, J. B., 1958, University of Michigan radiocarbon dates II. Science, v. 127, p. 1098-1105.

Dolliver, P. N., 1985, The Plio-Pleistocene Canadian breaks of New Mexico: A profile: New Mexico Geological Society, 36th Field Conference Guidebook, p. 315-318.

Effinger, J. A. and Lucas, S. G., 1990, Fossil *Bison* in New Mexico: New Mexico Journal of Science, v. 30, p. 7-15.

Findley, J. S, Harris, A. H., Wilson, D. E., and Jones, C., 1975, Mammals of New Mexico: Albuquerque, University of New Mexico Press, 360 p.

Gillette, D. D., Gillette, J. L., and Murphy, J., 1985a, The Tucumcari mammoths excavated by the Denver Museum of Natural History in 1930: New Mexico Geological Society, 36th Field Conference Guidebook, p. 14-17.

Gillette, D. D., Haynes, C. V., Saunders, J. J., and Stanford, D, 1985b, Preliminary report on the occurrence of *Mammuthus* sp. in Mora County, New Mexico: New Mexico Geological Society, 36th Field Conference Guidebook, p. 48-50.

Graham, R. W., 1987, Late Quaternary mammalian faunas and paleoenvironments of the southwestern plains of the United States; *in* Graham, R. W., Semken, H. A., Jr. and Graham, M. A., eds.: Late Quaternary mammalian biogeography and environments of the Great Plains and prairies. Illinois State Museum, Scientific Papers, v. 22, p. 24-86.

Graham, R. W. and Lundelius, E. L., Jr., 1984, Coevolutionary disequilibrium and Pleistocene extinctions, *in* Martin, P. S. and Klein, R. G. eds.: Quaternary extinctions, a prehistoric revolution. Tucson, University of Arizona Press, p. 223-249.

Green, F. E., 1992, Comments on the report of a worked mammoth tusk from the Clovis site: American Antiquity, v. 57: 331-337.

Harris, A. H., 1985, Late Pleistocene vertebrate paleoecology of the West: Austin, University of Texas Press, 293 p.

Harris, A. H., 1993, Quaternary vertebrates of New Mexico, *in* Lucas, S. G.

and Zidek, J., eds., Vertebrate paleontology in New Mexico: New Mexico Museum of Natural History and Science Bulletin 2, p. 179-197.

Harris, A. H. and Porter, L. S. W., 1980, Late Pleistocene horses of Dry Cave, Eddy County, New Mexico: Journal of Mammalogy, v. 61, p. 46-65.

Hay, O. P., 1924, The Pleistocene of the middle region of North America and its vertebrated animals: Carnegie Institution of Washington, Publication No. 322 A, 385 p.

Hay, O. P., 1927, The Pleistocene of the western region of North America and its vertebrated animals: Carnegie Institution of Washington, Publication No. 322 B, 346 p.

Hay, O. P. and Cook, H. J., 1928, Preliminary descriptions of fossil mammals recently discovered in Oklahoma, Texas and New Mexico: Proceedings of the Colorado Museum of Natural History, v. 8, no. 2, p.33.

Hay, O. P. and Cook, H. J., 1930, Fossil vertebrates collected near, or in association with, human artifacts at localities near Colorado, Texas; Frederick, Oklahoma; and Folsom, New Mexico: Proceedings of the Colorado Museum of Natural History, v. 9, no. 2, p.1-40.

Haynes, C. V., 1975, Pleistocene and recent stratigraphy; *in* Wendorf, F. and Hester, J. J., eds., Late Pleistocene environments of the Southern High Plains: Fort Burgwin Research Center, Southern Methodist University, Publication No. 9, p.57-96.

Haynes, C. V., 1995, Geochronology of paleoenvironmental change, Clovis type site, Blackwater Draw, New Mexico: Geoarchaeology, v. 10, p. 317-388.

Haynes, C. V. and Agogino, G. A., 1966, Prehistoric springs and geochronology of the Clovis site, New Mexico: American Antiquity, v. 31, p. 812-821.

Hester, J. J., 1972, Blackwater Draw Locality No. 1. A stratified, Early Man site in eastern New Mexico: Fort Burgwin Research Center, Southern Methodist University, Publication No. 8, 239 p.

Hibbard, C. W., 1960, An interpretation of Pliocene and Pleistocene climates in North America: Michigan Academy of Science, Arts and Letters, 62nd Annual Report, p. 5-30.

Hibben, F. C., 1941, Evidences of early occupation in Sandia Cave, New Mexico, and other sites in the Sandia-Manzano region: Smithsonian Miscellaneous Collections, v. 99, n. 23, p. 1-44.

Hill, M.G., Holliday, V. T., and Stanford, D. J., 1995, A further evaluation of the San Jon site, New Mexico: Plains Anthropologist, v. 40 (154), p. 369-390.

Holliday, V. T., 1997, Paleoindian geoarchaeology of the Southern High Plains: Austin, University of Texas Press, 297 p.

Jelinek, A. J., 1967, A prehistoric sequence in the middle Pecos Valley, New Mexico: Museum of Anthropology, University of Michigan, Anthropological Papers, No. 31, 175 p.

Katz, L., 1997, Late Pleistocene fauna at Blackwater Draw Locality #1: An archaeological perspective; *in* Lucas, S. G., Estep, J. W., Williamson, T. E., and Morgan G. S. eds., New Mexico's fossil record 1: New Mexico Museum of Natural History and Science Bulletin 11, p. 135-143.

Lucas, S. G. and Effinger, J. E., 1991, *Mammuthus* from Lincoln County and a review of the mammoths from the Pleistocene of New Mexico: New Mexico Geological Society, 42nd Field Conference Guidebook, p. 277-282.

Lucas, S. G. and Morgan, G. S., 1996, Pleistocene vertebrates from the Pecos River Valley near Roswell, Chaves County, New Mexico: New Mexico Geology, v. 18, p. 93-96.

Lucas, S. G., Morgan, G. S., Estep, J. W., Mack, G. H. and Hawley, J. W., 1999, Co-occurrence of the proboscideans *Cuvieronius*, *Stegomastodon*, and *Mammuthus* in the lower Pleistocene of southern New Mexico: Journal of Vertebrate Paleontology, v. 19, p. 595-597.

Lundelius, E. L. Jr., 1972, Vertebrate remains from the Gray Sand, *in* Hester, J. J., ed., Blackwater Draw Locality No. 1. A stratified early man site in eastern New Mexico: Fort Burgwin Research Center, Southern Methodist University, Publication No. 8, p. 148-163.

Lundelius, E. L. Jr., Churcher, C. S, Downs, T., Harington, C. R., Lindsay, E. H., Schultz, G. E., Semken, H. A., Webb, S. D., and Zakrzewski, R. J., 1987, The North American Quaternary sequence, *in* M. O. Woodburne (ed.), Cenozoic mammals of North America, geochronology and biostratigraphy: Berkeley, University of California Press, p. 211-235.

Madden, C. T., 1981, Mammoths of North America: PhD Dissertation, University of Colorado, Boulder, 271 p.

McDonald, J. N., 1981, North American *Bison*: Their classification and evolution: Berkeley, University of California Press, 316 p.

McMullen, T. L. and Zakrzewski, R. J., 1972, A new late Pleistocene fauna from northeastern New Mexico: New Mexico Geological Society, 23rd Field Conference Guidebook, p. 134-136.

Mead, J. I. and Meltzer, D. J., 1984, North American Late Quaternary extinctions and the radiocarbon record, *in* Martin, P. S. and Klein, R. G. (eds.): Quaternary extinctions, a prehistoric revolution. Tucson, University of Arizona Press, p. 440-450.

Miller, B. B., 1966, Five Illinoian molluscan faunas from the southern Great Plains: Malacologia, v. 4, p. 173-260.

Morgan, G. S., and Hulbert, R. C., Jr., 1995, Overview of the geology and vertebrate paleontology of the Leisey Shell Pit Local Fauna, Hillsborough County, Florida, *in* R. C. Hulbert, Jr., G. S. Morgan, and S. D. Webb, eds., Paleontology and geology of the Leisey Shell Pits, early Pleistocene of Florida: Bulletin of the Florida Museum of Natural History, v. 37, part I, p. 1-92.

Morgan, G. S. and Lucas, S. G., 2000, Pliocene and Pleistocene vertebrate faunas from the Albuquerque basin, New Mexico, *in* Lucas, S. G., ed., New Mexico's Fossil Record 2: New Mexico Museum of Natural History and Science Bulletin 16, p. 217-240.

Morgan, G. S., Lucas, S. G., Sealey, P. L., McDonald, H. G., and Gordon, M. E., 2000, Pleistocene (Rancholabrean) vertebrate and invertebrate fossils from Badlands Ranch, Quay County, New Mexico, *in* Lucas, S. G., ed., New Mexico's Fossil Record 2: New Mexico Museum of Natural History and Science Bulletin 16, p. 241-251.

Northrop, S. A., 1931, Geologists collect tusks: New Mexico Alumnus, v. 4, n. 5, p. 11-13.

Reeves, C. C., Jr., 1972, Tertiary-Quaternary stratigraphy and geomorphology of West Texas and southeastern New Mexico: New Mexico Geological Society, 23rd Field Conference Guidebook, p. 108-117.

Richmond, G. M. and Fullerton, D. S., 1986, Summation of Quaternary glaciations in the United States of America, in Sibrava, V., Bowen, D. Q., and Richmond, D. S. (eds.), Quaternary glaciations in the Northern Hemisphere: Quaternary Science Reviews, v. 5, p. 183-296.

Roberts, F. H. H., Jr., 1942, Archaeological and geological investigations in the San Jon district, eastern New Mexico: Smithsonian Miscellaneous Collections, v. 103, no. 4, 30 p.

Roosa, W. B., 1956, The Lucy site in central New Mexico: American Antiquity, v. 21, p. 310.

Slaughter, B. H., 1975, Ecological interpretation of the Brown Sand Wedge local fauna, *in* Wendorf, F. and Hester, J. J., eds., Late Pleistocene environments of the Southern High Plains: Fort Burgwin Research Center, Southern Methodist University, Publication No. 9, p. 179-192.

Smartt, R. A., Lucas, S. G., and Hafner, D. J., 1991, The giant bison (*Bison latifrons*) from the middle Rio Grande Valley of New Mexico: Southwestern Naturalist, v. 36, p. 136-137.

Stock, C. and Bode, F. D, 1936, The occurrence of flints and extinct animals in pluvial deposits near Clovis, New Mexico. Part III,--Geology and vertebrate paleontology of the Late Quaternary near Clovis New Mexico: Proceedings of the Academy of Natural Sciences of Philadelphia, vol. 88, p. 219-241.

Webb, S. D., 1965, The osteology of *Camelops*: Bulletin of the Los Angeles County Museum, Science, no. 1, 54 p.

Winans, M. C., 1989, A quantitative study of North American fossil species of the genus *Equus*, *in* Prothero, D. R. and Schoch, R. M., eds., The Evolution of the Perissodactyls: New York, Oxford University Press, p. 262-297.

LATE PLEISTOCENE MAMMOTHS (*MAMMUTHUS COLUMBI*) FROM MESA REDONDA, QUAY COUNTY, EAST-CENTRAL NEW MEXICO

GARY S. MORGAN, SPENCER G. LUCAS AND MARK E. GORDON
New Mexico Museum of Natural History, 1801 Mountain Road NW, Albuquerque, NM 87104

Abstract.—In 1929, Robert Abercrombie discovered parts of at least five mammoth skeletons near Mesa Redonda, Quay County, New Mexico. Nelson Vaughn of the Colorado Museum of Natural History (now the Denver Museum of Nature and Science) supervised the excavation of these mammoths in 1930. Many of the Mesa Redonda mammoth fossils were transferred to the New Mexico Museum of Natural History and Science in 1985 where they were used to construct a composite mounted skeleton now on display. The mammoth fossils were recovered from an alluvial fan deposit of probable late Pleistocene age. Measurements of three *Mammuthus* M3s from Mesa Redonda result in ranges of: length, 241-257 mm; width, 92-101 mm; crown height, 187-271 mm; enamel thickness, 2.3-3.0 mm; 17-19 plates; and lamellar frequency (plates/100 mm) of 6-8. The measurements and morphology of the Mesa Redonda mammoth teeth support their assignment to *Mammuthus columbi*. The presence of *M. columbi* and the associated fauna of extant terrestrial gastropods indicates a late Pleistocene (late Rancholabrean) age for the Mesa Redonda fauna. The gastropod assemblage suggests a relatively mesic, riparian habitat of grasses and woody, deciduous vegetation.

INTRODUCTION

In 1929, Robert Abercrombie discovered mammoth remains in a dry stream bed northeast of Mesa Redonda, south of Tucumcari, Quay County, northeastern New Mexico (Fig. 1). Abercrombie's find generated considerable interest because of the discovery just a few years earlier of Paleoindian artifacts associated with extinct Pleistocene *Bison* at Folsom, New Mexico. Several museums expressed an interest in the mammoths, including the Colorado Museum of Natural History (CMNH), the Museum of New Mexico, and the Frick Laboratory of the American Museum of Natural History in New York. The mammoths were subsequently excavated in 1930 by the CMNH. Gillette et al. (1985) briefly discussed the history of the Mesa Redonda mammoths. A portion of the Mesa Redonda mammoth sample was transferred to the New Mexico Museum of Natural History (NMMNH) in Albuquerque in 1985, and was eventually mounted as a composite skeleton now on display in the NMMNH. We review the history of the Mesa Redonda site and the excavation of the mammoths, describe the locality and geology, provide descriptions, measurements, and photographs of mammoth teeth, and discuss the associated molluscan fauna.

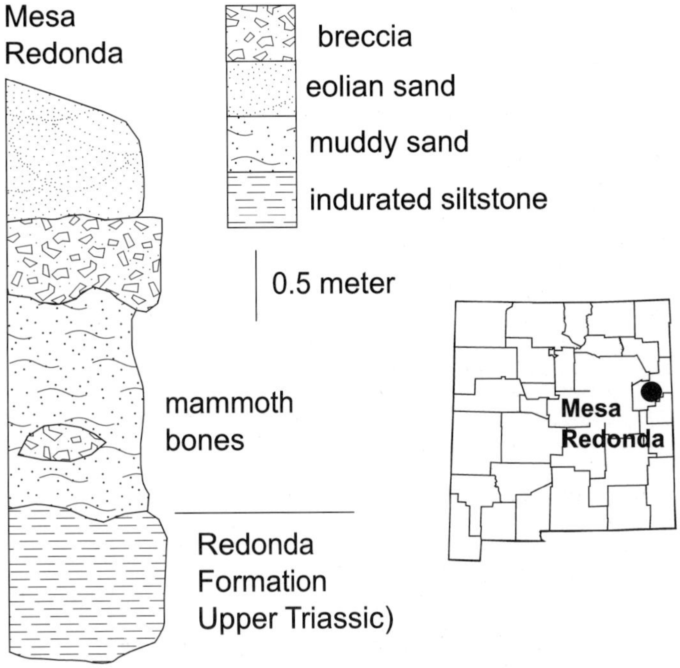

FIGURE 1. Stratigraphic section of Mesa Redonda site and map of New Mexico showing location of Mesa Redonda site (NMMNH site L-3736) in Quay County.

MATERIALS AND METHODS

The authors visited the Mesa Rendonda mammoth site in December 2000 with Howard Abercrombie and Robert Abercrombie, Jr., the son and grandson, respectively, of the discoverer of the site, Robert Abercrombie, Sr. We made geological observations, took site photographs, collected samples for lithological analysis, and collected sediments to screenwash for microvertebrates and mollusks. The vertebrate fossils from the Mesa Redonda site are housed in the Denver Museum of Nature and Science in Denver, Colorado (formerly known as the Denver Museum of Natural History-DMNH) and the New Mexico Museum of Natural History and Science in Albuquerque (NMMNH). The mollusk samples are housed in the NMNMH. When the Mesa Redonda mammoth site was discovered and excavated in 1929 and 1930, the DMNH was called the Colorado Museum of Natural History (CMNH). To keep the account of the site in its proper historical context, the museum in Denver is referred to as the Colorado Museum of Natural History or CMNH when discussing the events surrounding the excavation of the mammoths. However, all specimens housed in Denver have the acronym DMNH. Measurements of fossils are in mm, and were taken with either digital calipers or a metric tape measure.

HISTORY OF THE SITE

Much of the history regarding the discovery and excavation of the Mesa Redonda mammoth site is preserved in the correspondence of Harold J. Cook, Honorary Curator of Paleontology at the CMNH, who was instrumental in determining the importance of the site, and Nelson J. Vaughn, who conducted the actual excavation of the mammoths for the CMNH. The following historical account is mostly derived from their letters, now curated in the archives of the DMNH (copies in NMMNH).

The Mesa Redonda mammoth site was discovered in 1929 by Robert Abercrombie, Sr., a rancher and amateur fossil collector from Tucumcari, New Mexico. Abercrombie contacted Dr. Hewett, the director of the State Museum in Santa Fe, about the Mesa Redonda mammoth find. Hewett did not have the personnel to assist Abercrombie in the excavation and recommended that he contact Harold Cook. At that time, Cook was investigating late Pleistocene sites in eastern New Mexico in search of localities documenting the association between Paleoindians and extinct Pleistocene mammals. In the summer of 1929, Abercrombie showed the Mesa Redonda site to Cook, who expressed a strong interest in excavating the site. The earliest reference to the mammoth site was in a letter dated 24 July 1929, from Harold J. Cook to Walter C. Mead. The following is an excerpt from that letter referring to the Mesa Redonda mammoths:

> "I believe we have located...perhaps the most remarkable deposit of Elephant bones yet known in this country...There are at least six mammoths present, and probably more, in one spot not forty yards across, in gravels and sand deposits of Pleistocene age...Parts of eight tusks are exposed as it is, and parts of several jaws and skulls, and many limb bones, ribs, vertebrae, etc. They seem to be scattered and disarticulated, for the most part..."

A letter dated just five days later, 29 July 1929, to Childs Frick from Charles Falkenbach, a collector for the Frick Laboratory at the American Museum of Natural History, indicated that the Frick Lab also was interested in the Mesa Redonda mammoths. Cook decided that the CMNH would excavate the Mesa Redonda mammoth site, and sent Nelson Vaughn to Tucumari in September 1929 to make the arrangements.

The CMNH leased the parcel of land containing the Mesa Redonda mammoth site from the landowner, Charles J. Ford, for $100 on 26 March 1930, for the specific purpose of recovering the mammoth fossils. The excavation at the Mesa Redonda mammoth site lasted from March to June 1930, and was under the direction of Nelson Vaughn.. A considerable amount of information pertaining to the site is contained in an extensive correspondence between Vaughn and J. D. Figgins, then Director of the CMNH. Vaughn drew a rough map of the Mesa Redonda site (redrawn here as Fig. 3) showing a segment of the dry wash approximately 150 ft long, with the main bone deposit at the northern end of the site and a smaller concentration of bones about 75-100 ft to the south. Vaughn's letters indicate that the main bone concentration contained at least four mammoth tusks; an associated mammoth with a maxilla with one molar, pair of lower jaws, scapula, humerus, sacrum, pelvis, six vertebrae, 11 ribs, and two foot bones; and another mammoth with 14 articulated vertebrae. The southern site had a partial skull with a tusk, a pair of lower jaws, humerus, pelvis, vertebrae, and ribs. The CMNH excavations at the Mesa Redonda site ended in June 1930, and to our knowledge no one has excavated there since.

LOCALITY AND GEOLOGY

The Mesa Redonda mammoth site (NMMNH site L-3736; DMNH site 1006) is located in and along the eastern bank of a small, unnamed arroyo that flows northeast from the base of Mesa Redonda, about 23 km south of Tucumcari, Quay County, New Mexico (Figs. 1-3). The site is in the NE¼ SW¼ sec. 26, T9N, R31E (Norton, NM, USGS 7.5' quadrangle), and the UTM coordinates are 3870956N, 625678E, (Zone 13, NAD 27). The bone-bearing stratum at the site is a 1.0- to 1.5-m thick layer of friable, light brown (5 YR 5/6) to moderate reddish-brown (10 R 4/6), very fine-grained to fine-grained, muddy sand (Figs. 1, 2). This bed rests directly on Upper Triassic bedrock of light brown (5 YR 3/4) to moderate reddish-brown (10 R 6/6) calcareous siltstone that is blocky and locally bioturbated of the Redonda Formation of the Chinle Group. Above the bone layer is a 0.3- to 0.5-m thick sedimentary breccia with imbricated clasts of Mesozoic sandstone and limestone that indicate paleoflow to the northeast. Lenses of such breccia also occur in the bone-bearing layer. Pedogenesis at the top of the bone-bearing sand locally has produced a 0.2-m thick layer of pale yellowish-brown (10 YR 6/2) sandy clay. The top of the section at the mammoth site is an unconsolidated, light brown (5 YR 5/6) eolian sand of probable Holocene age (Fig. 1).

The geometry and lithotypes of the Quaternary sediments at the Mesa Redonda mammoth site clearly were part of an alluvial fan system that flowed N/NE from the mesa during part of the Pleistocene. Imbricated fan breccias thicken just east of the mammoth site and are indicative of a stream channel subject to high, although possibly periodic, current velocities. While the bone-bearing sand represents deposition of weathered Triassic bedrock, the presence of the Cretaceous oyster, *Texigraphaea pitcheri*, demonstrates that the breccias were derived at least in part from fluvial redistribution of colluvial deposits at the base of Mesa Redonda. The lenticular nature of the breccias reflects channel migration due to stream meander and/or braided channels. These processes contributed to the subsequent redeposition of finer-grained sediment over these channels (see Malanson, 1993) and thus created the fan structure. As such, the bone-bearing sand probably represents inter-channel portions of the fan and bank deposition related to lower energy discharge and/or stream meander.

There has been little study of Quaternary deposits in the Tucumcari area, including those at the Mesa Redonda mammoth site. Regional mapping (Barnes, 1977) assigns these deposits to Quaternary pediment deposits (Qpd), which are described as "silt, sand, and coarser debris derived chiefly from Triassic rocks." The Mesa Redonda mammoth site is similar to the late Pleistocene (Rancholabrean) Badlands Ranch Local Fauna described by Morgan et al. (2000), located near Barranca Creek about 20 km to the northeast. Badlands Ranch is a localized Pleistocene valley fill developed in Triassic bedrock. Geological and faunal features at Badlands Ranch also are similar to those at Mesa Redonda.

PALEONTOLOGY

The Mesa Redonda mammoths have been mentioned previously in the literature (Gillette et al., 1985; Lucas and Effinger,

FIGURE 2. Photographs of Mesa Redonda site (NMMNH site L-3736). A. Overview of site looking toward Mesa Redonda to the southwest. People are standing on the main CMNH excavation. B. Close-up view of bone-bearing muddy sandstone. Note bone to right of hammer and alluvial fan breccia above bone-bearing stratum.

1991; Harris, 1993), but the sample was not studied in detail. Gillette et al. (1985, fig. 1.24) illustrated a complete pair of lower jaws, and Lucas and Effinger (1991, fig. 4) illustrated the mounted skeleton in the NMMNH (see Fig. 4). We provide a list of all catalogued specimens of mammoths from the Mesa Redonda site, tables of measurements (Tables 1-2), descriptions of the teeth and mandibles, illustrations of the most complete specimens (Figs. 5-6), and a brief taxonomic discussion.

In addition to mammoths, Vaughn's letters state that he collected ground sloth teeth, horse teeth, and an indeterminate carnivore tooth from Mesa Redonda. However, in our inventory of the Mesa Redonda collection at the DMNH, the only non-mammoth specimen we located consisted of two associated incisors of the horse *Equus* (DMNH 15950). We screenwashed a small sample of the fine-grained sand containing the mammoth fossils. The only microvertebrate fossil we recovered was an edentulous maxilla of a small indeterminate species of rodent (NMMNH 34000). This sediment sample also yielded six species of terrestrial gastropods, all presently extant (see Table 3).

Class Mammalia
Order Proboscidea
Family Elephantidae
***Mammuthus columbi* (Falconer, 1857)**

Referred specimens

DMNH 1639a (DMNH 1639 includes mandibles from two individuals; we have given these specimens the suffixes, a and b, so they can be distinguished in Table 1 and Figs. 5-6), right dentary with m3 (Figs. 6C, D); DMNH 1639b, associated dentaries with complete left m3 and partial right m3 (Figs. 5B-E); DMNH 15136, partial left M3; NMMNH 33255, palate with right and

TABLE 1. Measurements (in mm) of teeth of *Mammuthus columbi* from Mesa Redonda (NMMNH site L-3736), Quay County, New Mexico.

specimen number	tooth	length	maximum width	crown height	plate number	lamellar frequency[1]	enamel thickness[2]
DMNH 15136	LM3	--	101	187	--	6	--[3]
NMMNH 33255	LM3	257	92	271	19	8	2.7 (2.5-3.0)
NMMNH 33256	LM3	241	101	265	17	8	2.6 (2.3-3.0)
DMNH 1639a	Rm3	--	80	--[4]	19	8	2.4 (2.1-2.6)
DMNH 1639b	Lm3	220	73	160	18	8	2.4 (2.2-2.7)
NMMNH 33257	Lm2	215	83	--[4]	14	6	2.8 (2.6-3.1)
NMMNH 33257	Rm3	--	--	--[4]	14[5]	9[6]	--[3]

[1]Lamellar frequency is the number of plates/100 mm of tooth length.
[2]Enamel thickness measurements are the means of five measurements taken on each individual tooth (observed range in parentheses).
[3]Tooth not fully erupted, enamel thickness measurements could not be taken.
[4]Tooth in dentary, crown height could not be measured.
[5]Tooth not fully erupted, an undetermined number of plates were unerupted and could not be counted.
[6]Tooth not fully erupted; lamellar frequency is probably too high.

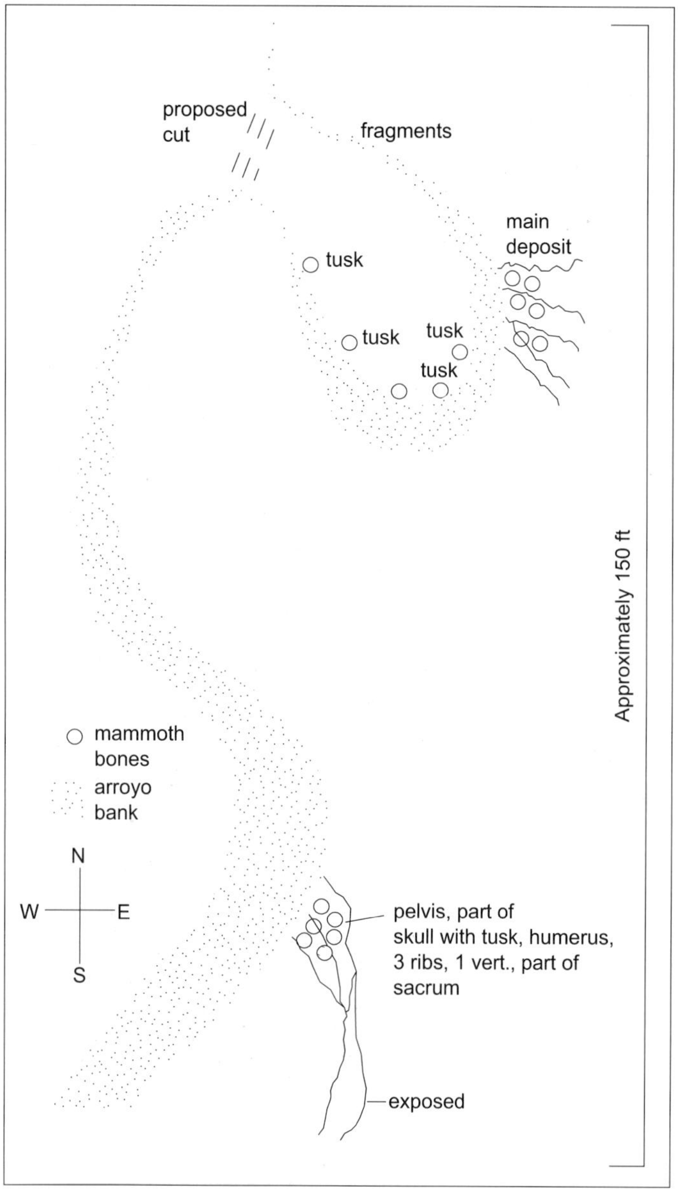

FIGURE 3. Map of Mesa Redonda site (NMMNH site L-3736). The primary bone concentration (main deposit on map) is at the northern end of the site, with a secondary concentration about 100 feet farther south. This map is redrawn from the original site map hand-drawn by Nelson Vaughn in 1930 and housed in the archives of the Denver Museum of Nature and Science.

TABLE 2. Measurements (im mm) of a complete pair of right and left dentaries of *Mammuthus columbi* (NMMNH 33257) from Mesa Redonda (NMMNH site L-3736), Quay County, New Mexico.

Total length of mandible from tip of spout to posterior edge of ascending ramus	728
Height from ascending ramus to tip of coronoid process	372
Height of horizontal ramus below m2	215
Maximum width across horizontal rami	635
Length of mandibular spout	80
Width of mandibular spout	37

FIGURE 4. Composite mounted skeleton of *Mammuthus columbi* (NMMNH 27234) from the Mesa Redonda site (NMMNH site L-3736), Quay County, New Mexico. This skeleton is on display in the New Mexico Museum of Natural History in Albuquerque.

left M3 (Fig. 5A); NMMNH 33256, maxillary fragment with left M3 (Fig. 5F); NMMNH 33257 (original number DMNH 1639), right and left dentaries with m2s and partially erupted m3s (Figs. 6A, B); NMMNH 33258, reconstructed scapula; NMMNH 33259, 33260, two partial innominates; NMMNH 33261, thoracic vertebra; NMMNH 33262, shaft of femur; DMNH 15317, carpal. There are two additional mammoth mandibles from the Mesa Redonda site in the DMNH collection that we did not examine: DMNH

TABLE 3. Terrestrial gastropods recovered from alluvium at Mesa Redonda (NMMNH site L-3736), QuayCounty, New Mexico.

PHYLUM MOLLUSCA

Class Gastropoda
- Subclass Pulmonata
  - Order Stylommatophora
    - Family Pupillidae
      - *Gastrocopta cristata* (Pilsbry and Vanatta, 1900)
        - *Pupilla blandi* Morse, 1865
    - Family Succineidae
      - *Catinella vermata* (Say, 1829)
      - *Succinea concordialis* Gould, 1848
      - *Succinea grosvenori* Lea, 1864
      - *Succinea vaginacontorta* Lee, 1951

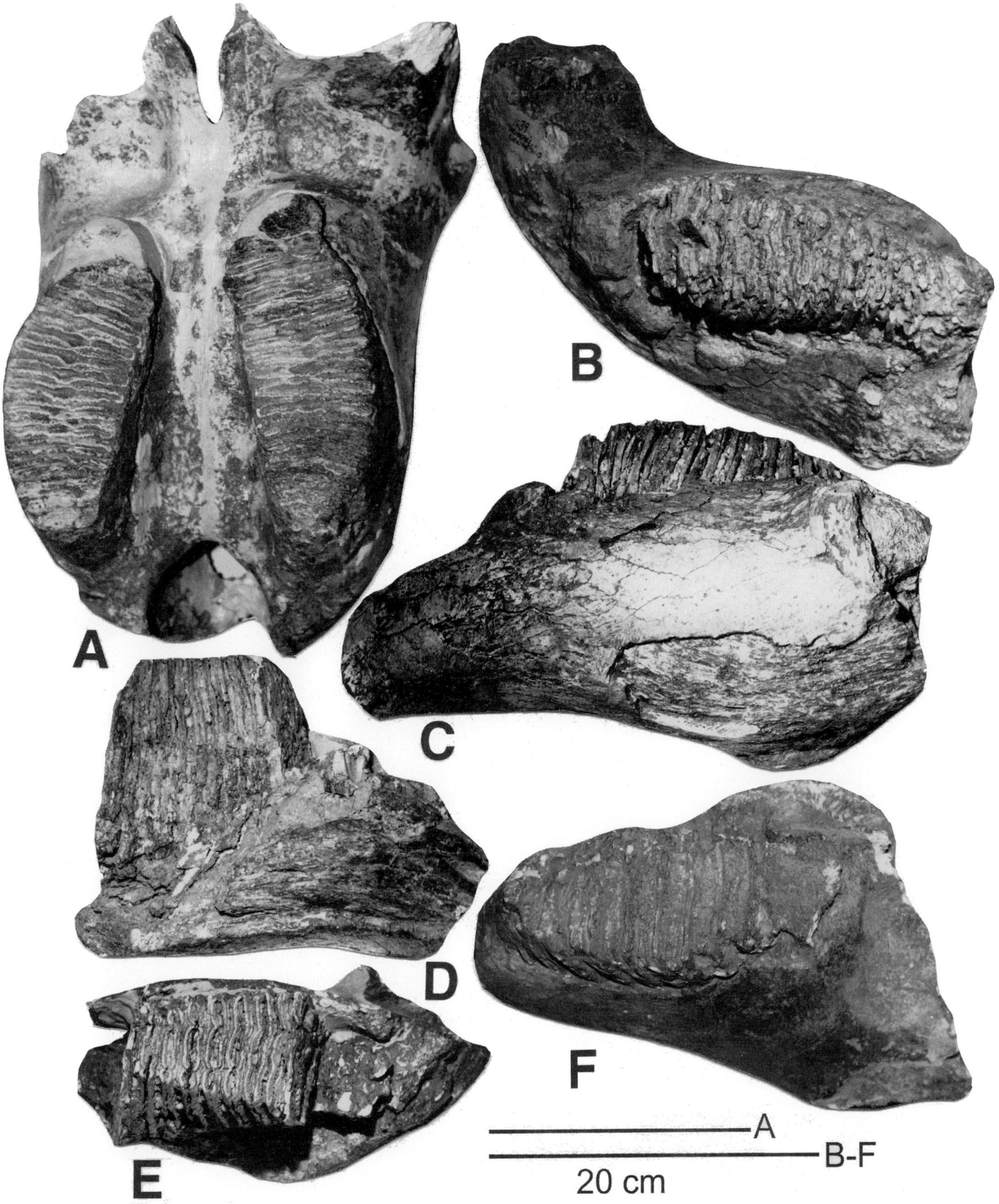

FIGURE 5. Maxillae and dentaries of *Mammuthus columbi* from the Mesa Redonda site (NMMNH site L-3736), Quay County, New Mexico. A. Occlusal view, palate with left and right M3, NMMNH 33255; B. occlusal view and C. lateral view, left dentary with m3, DMNH 1639b; D. lateral view and E. occlusal view, right dentary with partial m3, DMNH 1639b; F. occlusal view, maxillary fragment with LM3, NMMNH 33256. Scale bars are 20 cm.

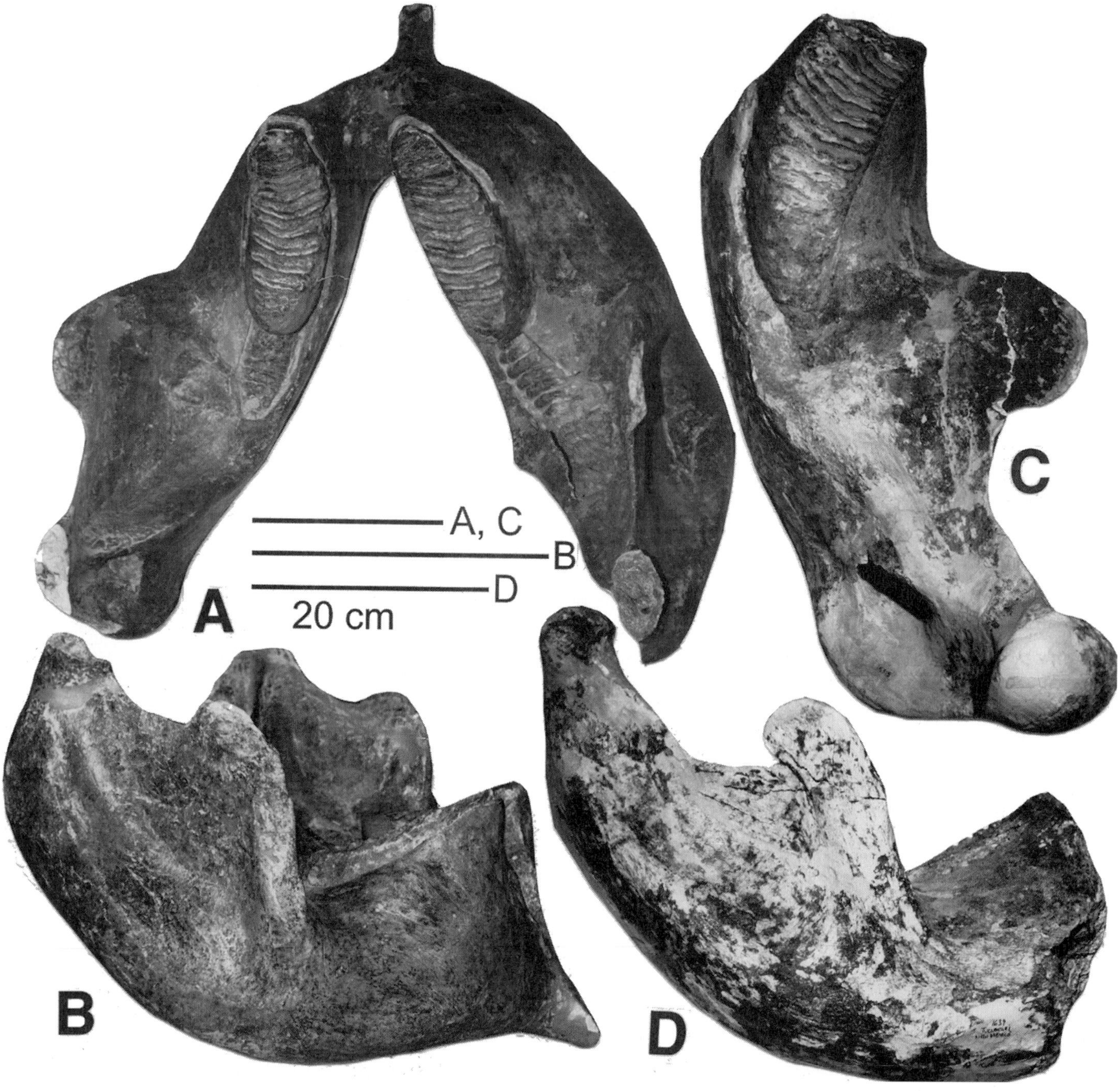

FIGURE 6. Dentaries of *Mammuthus columbi* from the Mesa Redonda site (NMMNH site L-3736), Quay County, New Mexico. A. occlusal view and B. right lateral view of right and left dentaries with m2 and unerupted m3, NMMNH 33257; C. occlusal view and D. lateral view of right dentary with m3, DMNH 1639a. Scale bars are 20 cm.

1485, partial dentary (on exhibit in the DMNH, apparently used in the reconstruction of another mammoth from Milo, Kansas); DMNH 1638, right and left dentaries (on exhibit in the DMNH).

NMMNH 27234 is a composite mounted skeleton on display in the New Mexico Museum of Natural History and Science (Fig. 4). An examination of this skeleton, as well as an inventory of the mammoth fossils from Mesa Redonda received from the DMNH, reveals that this mount is composed of about 30% fossil material. Actual mammoth fossils from Mesa Redonda in the NMMNH mounted skeleton include: right humerus, radius, and ulna; right and left scapulae (reconstructed); right and left innominates (reconstructed); six cervical vertebrae; 15 thoracic vertebrae (found as an articulated series), and two lumbar vertebrae. The remainder of the skeleton consists of casts.

Descriptions and comparisons

Four or five species of mammoths (*Mammuthus*) are known from North American Pleistocene faunas (Maglio, 1973; Kurtén and Anderson, 1980; Madden, 1981; Graham, 1986; Agenbroad, 1994). Three species of mammoths occur in New Mexico (Lucas and Effinger, 1991): *M. meridionalis*, *M. imperator*, and *M. columbi*. *M. meridionalis* (referred to *M. hayi* by Webb and Dudley, 1995) is a primitive species from the early Pleistocene (early Irvingonian) that differs from the Mesa Redonda teeth in having fewer plates with much thicker enamel. *M. imperator* from the Irvingtonian and early Rancholabrean also has fewer plates and thicker enamel than the Mesa Redonda *Mammuthus* teeth. *M. columbi* is found primarily in late Rancholabrean sites and is distinguished from *M. imperator* by its relatively short, broad molars, larger number of plates, and thinner enamel. Morphological characters and measurements of the Mesa Redonda mammoth teeth compare most closely with *M. columbi*. Three M3s from Mesa Redonda (Table 1) range from: 241-257 mm in length, 92-101 mm in width, and 187-271 mm in crown height. These teeth have from 17-19 plates, a lamellar frequency (number of plates/100 mm) of 6-8, and enamel thicknesses of 2.3-3.0 mm (mean 2.7 mm). The size, plate number, lamellar frequency, and enamel thickness of the Mesa Redonda mammoth teeth compare well with the morphometric data on the large sample of *M. columbi* from Hot Springs, South Dakota (Dutrow, 1980; Agenbroad, 1994).

Remarks

Correspondence and various historical accounts suggest that as many as six individual mammoths, including eight tusks, were found at the Mesa Redonda site. Our complete inventory of the Mesa Redonda mammoth samples housed in the DMNH and NMMNH reveals a smaller number of specimens (see above), and not a single tusk or even tusk fragment. We calculated a minimum number of five individuals of *Mammuthus columbi* from Mesa Redonda based on lower jaws (DMNH 1485, 1638, 1639a, b and NMMNH 33257). With the exception of a single carpal (DMNH 15317), the current DMNH sample consists only of jaws and isolated teeth, because the postcranial sample was transferred to the NMMNH in 1985 and subsequently incorporated into a composite mounted skeleton (NMMNH 27234; Fig. 4).

DISCUSSION

Mammuthus columbi is the most common mammoth in New Mexico, where it is known from more than 20 sites (Lucas and Effinger, 1991). All mammoth records from late Pleistocene (late Rancholabrean) sites in New Mexico and elswhere in the southwestern United States appear to be *M. columbi* (Madden, 1981). The presence of *M. columbi* suggests that the Mesa Redonda site is late Pleistocene (late Rancholabrean), although there are no radiocarbon dates or other age-diagnostic mammals (e.g., *Bison*) to confirm this age.

Besides Mesa Redonda, the only other large sample of *Mammuthus columbi* from New Mexico is from the Blackwater Draw/Clovis site in Roosevelt County, where Paleoindians of the Clovis culture apparently killed and butchered mammoths. Limited morphometric data have been published on the *Mammuthus* sample from Blackwater Draw. Examination and measurements of a small sample of mammoth teeth in the Blackwater Draw Museum and the Blackwater Draw Site (see Morgan et al., this volume) substantiate the identification of the Blackwater Draw mammoths as *M. columbi*. The mammoth teeth from Mesa Redonda and Blackwater Draw are very similar in size and other morphological characters (see comparative measurements in Morgan et al., this volume), confirming the presence of *M. columbi* in both sites. Radiocarbon dates from Blackwater Draw (between 13,000 and 11,000 yr B.P.) establish a latest Pleistocene age for *M. columbi* from this site (Haynes, 1995).

The presence of a gastropod fauna composed of six currently extant taxa further supports a late Pleistocene (late Rancholabrean) age for the Mesa Redonda site. All species recovered from the alluvial deposits were terrestrial taxa. This lack of an aquatic fauna suggests that the stream flows through the alluvial fan may have been periodic; however, subsurface flow apparently was sufficient to support riparian vegetation. Ecological preferences for these gastropod species (see Pilsbry, 1948; Metcalf and Smartt, 1997) indicate a riparian community with muddy stream banks, grasses, and deciduous trees or shrubs.

ACKNOWLEDGMENTS

We thank Howard Abercrombie and Robert Abercrombie, Jr. for allowing access to the Mesa Redonda mammoth site, and for providing information regarding the discovery of the mammoths by their father/grandfather, Robert Abercrombie, Sr. Paul Sealey helped us in the field. The Denver Museum of Nature and Science archives provided copies to the NMMNH of all relevant letters and other documents relating to the Mesa Redonda mammoth site. Suzanne Cowan assisted us in examining these documents. We thank Logan D. Ivy and Russell W. Graham for allowing access to the Mesa Redonda mammoth specimens housed in the DMNH vertebrate paleontology collection. Russell W. Graham and Michael O'Neill provided helpful comments on an earlier draft of the manuscript.

REFERENCES

Agenbroad, L. D., 1994, Taxonomy of North American *Mammuthus* and biometrics of the Hot Springs mammoths; in Agenbroad, L. D. and Mead, J. I., eds., The Hot Springs mammoth site: The Mammoth Site of South Dakota, Hot Springs, South Dakota, p. 158-207.

Barnes, V. E., 1977, Geologic atlas of Texas, Clovis sheet: Bureau of Economic Geology, The University of Texas at Austin, scale 1;250,000.

Dutrow, B. L., 1980, Metric analysis of a late Pleistocene mammoth assemblage, Hot Springs, South Dakota: MS Thesis, Dallas, Texas, Southern Methodist University, 165 p.

Gillette, D. D., Gillette, J. L., and Murphy, J., 1985, The Tucumcari mammoths excavated by the Denver Museum of Natural History in 1930: New Mexico Geological Society, Guidebook 36, p. 14-17.

Graham, R. W., 1986, Taxonomy of North American mammoths; *in* Frison, G. C. and Todd, L. C., eds., The Colby Mammoth Site-Taphonomy and archaeology of a Clovis kill in northern Wyoming: Albuquerque, University of New Mexico Press, p. 165-229.

Harris, A. H., 1993, Quaternary vertebrates of New Mexico; *in* Lucas, S. G. and Zidek, J., eds., Vertebrate paleontology in New Mexico: New Mexico Museum of Natural History and Science Bulletin 2, p. 179-197.

Haynes, C. V., 1995, Geochronology of paleoenvironmental change, Clovis type site, Blackwater Draw, New Mexico: Geoarchaeology, v. 10, p. 317-388.

Kurtén, B., and Anderson, E., 1980, The Pleistocene mammals of North America: New York, Columbia University Press, 442 p.

Lucas, S. G. and Effinger, J. E., 1991, *Mammuthus* from Lincoln County and a review of the mammoths from the Pleistocene of New Mexico: New Mexico Geological Society, Guidebook 42, p. 277-282.

Madden, C. T., 1981, Mammoths of North America: PhD Dissertation, University of Colorado, Boulder, 271 p.

Maglio, V. J., 1973, Origin and evolution of the Elephantidae: Transactions of the American Philosophical Society, v. 63, part 3, p. 1-149.

Malanson, G. P., 1993, Riparian landscapes: Cambridge, Cambridge University Press, 296 p.

Metcalf, A. L., and Smartt, R. A., 1997, Land snails of New Mexico: a systematic review; *in* Metcalf, A. L., and Smartt, R. A., eds., Land snails of New Mexico: New Mexico Museum of Natural History and Science Bulletin 10, 69 p.

Morgan, G. S., Lucas, S. G., Sealey, P. L., McDonald, H. G., and Gordon, M. E., 2000, Pleistocene (Rancholabrean) vertebrate and invertebrate fossils from Badlands Ranch, Quay County, New Mexico; *in* Lucas, S. G., ed., New Mexico's Fossil Record 2: New Mexico Museum of Natural History and Science Bulletin 16, p. 241-251.

Pilsbry, H. A., 1948, Land Mollusca of North America (north of Mexico), volume II, part 2: The Academy of Natural Sciences of Philadelphia Monographs 3, p. 521-1113.

Webb, S. D. and Dudley, J. P. 1995, Proboscidea from Leisey Shell Pits, Hillsborough County, Florida; *in* R. C. Hulbert, Jr., G. S. Morgan, and S. D. Webb, eds., Paleontology and geology of the Leisey Shell Pits, early Pleistocene of Florida: Bulletin of the Florida Museum of Natural History, v. 37, part II, p. 645-660.

A NEW LATE NEOGENE PALEO-FAUNA FROM THE KIOWA/RITA BLANCA NATIONAL GRASSLAND, NORTHEASTERN NEW MEXICO

BRUCE A. SCHUMACHER
USDA Forest Service, 1420 E. 3rd St., La Junta, CO 81050

Abstract.—The Perico Creek local paleo-fauna was informally documented by USDA Forest Service archaeologists in 1990 when nearly 350 partial bones and bone fragments of large mammals were recovered from the surface of an actively eroding cut-bank along Perico Creek, Union County, New Mexico. The bony elements are dominated by *Camelops* sp., with paired elements of similar size perhaps belonging to a single individual. Also present are a partial skeleton of *Platygonus*, remains of *Equus*, a large bone fragment referable to Proboscidea, and a possible cervid antler base. The bones are eroding from thin beds of unconsolidated sand and clay and are associated with (minimally) two species of freshwater gastropods (*Lymnaea* cf. *L. caperata*, *Physa* sp.), two species of terrestrial gastropods (unidentified taxon, *Succinea* cf. *S. grosvenori*), and two species of freshwater bivalves (*Uniomerus tetralasmus*, *Sphaerium* cf. *S. striatinum*). Future work will likely increase the known diversity of the assemblage and produce age-diagnostic taxa to further constrain the relative age of the deposit. The first excavation and detailed analysis of the stratigraphic interval will be undertaken this year.

INTRODUCTION

In 1990, Dr. Joseph Tainter, an archaeologist with the USDA Forest Service, undertook an investigation of paleo-faunal remains recovered from an actively eroding cut-bank along Perico Creek, just west of the Texas-New Mexico border on the Kiowa/Rita Blanca National Grassland (Fig. 1). Tainter originally suspected that the collection might represent a Paleoindian site on the basis of apparent burning and anomalous breakage patterns of the bony materials. Laboratory examination revealed, however, that the "burned" bone merely possessed patchy black surface staining due to manganese mineralization. In the end Tainter determined that the site showed no evidence of cultural affiliation and labeled it a paleontological occurrence.

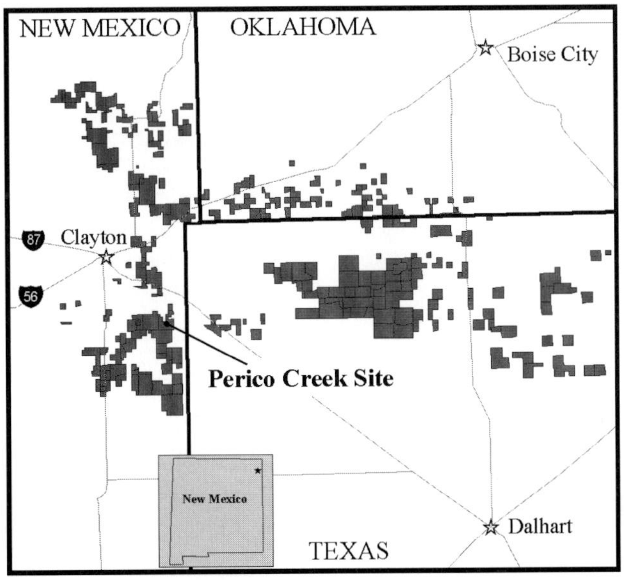

FIGURE 1. Location of the Perico Creek Site (New Mexico Museum of Natural History Locality 4638) in northeastern New Mexico, shaded areas denote a portion of the Kiowa/Rita Blanca National Grassland.

The site was re-visited in 1997 by Greg Liggett of the Sternberg Museum of Natural History, Hays, Kansas. In the same general area, Liggett located and salvaged a partial *Platygonus* skeleton on the surface, but limited exploratory excavation revealed no further *in situ* material. In October of 2000, the author visited the site and on the surface located abundant small molluscs, a complete *Camelops* phalanx, and a shattered *Equus* molar. Exploratory excavation in several locations uncovered only small fragments of *in situ* bone. So, although the Perico Creek locality has produced numerous vertebrate remains over the past decade, the context of recovered specimens with respect to the exposed sediments remains in question (Fig. 2).

The materials collected by Tainter have until now been stored at the Cibola National Forest office in Albuquerque. This year the Tainter specimens, together with those collected by Liggett, will be transferred and curated into the collections at the New Mexico Museum of Natural History, Albuquerque.

FIGURE 2. Overview of the Perico Creek Site, Kiowa/Rita Blanca National Grassland, northeastern New Mexico.

TABLE 1. Late Negoene paleo-fauna from the Perico Creek Site, Kiowa/ Rita Blanca National Grasslands, northeastern New Mexico.

Invertebrates
- Phylum Mollusca
  - Class Gastropoda
    - Order Basommatophora
      - Family Lymnaeidae
        - *Lymnaea* cf. *L. caperata*
      - Family Physidae
        - *Physa* sp.
      - Family Succineidae
        - *Succinea* cf. *S. grosvenori*
      - Family Planorbidae
        - unidentified terrestrial form
  - Class Bivalvia
    - *Uniomerus tetralasmus*
    - Order Veneroida
      - Family Sphaeriidae
        - *Sphaerium* cf. *S. striatinum*

Vertebrates
- Class Mammalia
  - Order Artiodactyla
    - Family Camelidae
      - *Camelops* sp.
        - partial right scapula, proximal one-half
        - partial left scapula, proximal one-half
        - head of right humerus
        - head of left humerus lacking epiphysis
        - partial shaft of humerus
        - distal one-third of metapodial
        - complete astragalus
        - complete proximal phalange
    - Family Tayassuidae
      - *Platygonus* sp.
        - partial juvenile skeleton including anterior one-half of mandible (with incisors and canines), partial premaxilla with 2 incisors, lower M1, right and left humeri and scapulae, 3 thoracic vertebrae, and numerous vertebral and rib fragments
  - Order Perrisodactyla
    - Family Equiidae
      - *Equus* sp.
        - five splintered cheek teeth
        - partial axis vertebrae
        - humeral shaft
    - ?Family Cervidae
        - base of antler
  - Order Proboscidea
        - large chunk of bone not attributable to other taxa based upon size

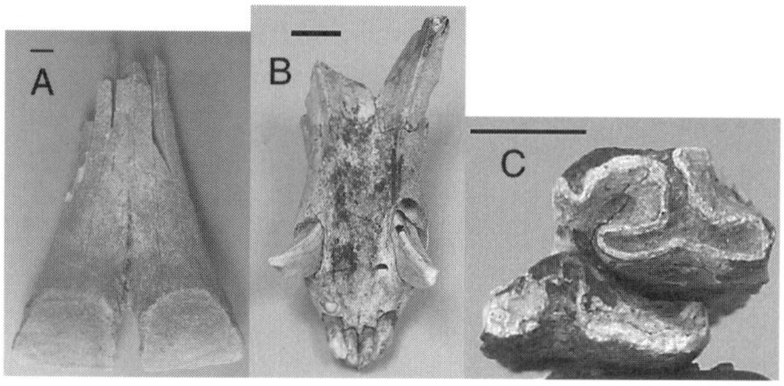

FIGURE 3. Select vertebrate specimens from the Perico Creek Site, bar is 1 cm. A) *Camelops* distal metapodial; B) *Platygonus* mandibular symphysis; C) *Equus* partial lower tooth.

FIGURE 4. Invertebrate fauna of the Perico Creek Site, bar is 1 cm. A) *Uniomerus tetrelasmus*; B) *Sphaerium* cf. *S. striatinum*; C) unidentified terrestrial form; D) *Physa* sp.; E) *Succinia* cf. *S. grosvenori*; F) *Lymnaea* cf. *L. caperata*.

PALEO-FAUNAL LIST

Diagnostic specimens from the Perico Creek Site are listed in Table 1. The skeletal remains are dominated by *Camelops* sp. (Fig. 3a), and the majority of preserved elements may comprise a single individual due to the presence of paired scapulae, humeri, and various podial elements of similar size. A *Platygonus* partial skeleton is represented by cranial (Fig. 3B), axial, and appendicular elements. Also present are several teeth (Fig. 3C), a partial humerus, and a partial axis vertebrae attributable to *Equus,* a possible cervid antler base, and a large chunk of bone referable to Proboscidea based upon size.

There are minimally two bivalve and four gastropod taxa present (Fig. 4), including: two species of freshwater gastropods (*Lymnaea* cf. *L. caperata*, *Physa* sp.), two species of terrestrial gastropods (unidentified taxon, *Succinea* cf. *S. grosvenori*), and two species of freshwater bivalves (*Uniomerus tetralasmus*, *Sphaerium* cf. *S. striatinum*). Small mollusc shells are abundant on the surface of the site, and those few variants noted here resulted from brief surface inspection of the site. A vigorous search specifically for these small shells will likely produce higher diversity.

AGE OF THE DEPOSIT

Until such time that the Perico Creek paleo-fauna can be investigated in more detail, the age of the deposit can only be constrained to somewhere within the Blancan through Rancholabrean land mammal "ages". Three large vertebrate taxa are well represented at the site, but *Camelops*, *Equus*, and *Platygonus* all have their first appearances in the Blancan and range through the end of the Pleistocene (Lundelius et al., 1987). The lack of *Bison* or *Mammuthus* material does not permit restriction to Pleistocene or younger age. However, the nature of the deposit (unconsolidated sands just below ground surface) suggests a relatively young age, more likely Rancholabrean than Blancan. A Rancholabrean age is supported by the fact that the illustrated invertebrate paleo-fauna includes largely extant forms, and some of these taxa are present at two other late Pleistocene sites in New Mexico including the Casados Ranch local fauna (McMullen and Zakrzewski, 1972) and the Badlands Ranch local fauna (Morgan et al., 2000).

CONCLUSIONS

The known Perico Creek paleo-fauna is composed of well preserved invertebrate and vertebrate remains including (minimally) four gastropods, two bivalves, a camel, a horse, a peccary, and possibly a cervid and a proboscidean. Further work is needed to fully elucidate the composition of the Perico Creek local paleo-fauna. To date, only minor exploratory excavation has been undertaken, and the USDA Forest Service is planning an intensive excavation effort in the near future in conjunction with the New Mexico Museum of Natural History. This work will likely increase the number of vertebrate and invertebrate taxa present at the site. Invertebrate paleo-ecology, distribution of the vertebrate remains, and analysis of the sedimentary record will help to unravel the taphonomic history of the site. Discovery of additional taxa may permit the assignment of a more closely constrained relative age to the assemblage.

ACKNOWLEDGMENTS

Mark Gordon kindly provided identifications of invertebrate specimens. Gary Morgan confirmed vertebrate identifications and provided helpful comments on the manuscript.

REFERENCES

Lundelius, E. L. Jr., Churcher, C. S., Downs, T., Harington, C. R., Lindsay, E. H., Schultz, G. E., Semken, H. A., Webb, S. D., and Zakrzewski, R. J., 1987, The North American Quaternary sequence, *in* Woodburne, M. O., ed., Cenozoic Mammals of North America, Geochronology and Biostratigraphy: Berkeley, University of California Press; p. 211-235.

McMullen, T. L., and Zakrzewski, R. J., 1972, A new late Pleistocene fauna from northeastern New Mexico: New Mexico Geological Society, 23rd Field Conference, Guidebook, p. 134-136.

Morgan, G. S., Lucas, S. G., Sealey, P. L., McDonald, H. G., and Gordon, M. E., 2000, Pleistocene (Rancholabrean) vertebrate and invertebrate fossils from Badlands Ranch,Quay County, New Mexico, *in* Lucas, S. G., ed., New Mexico's Fossil Record: New Mexico Museum of Natural History and Science, Bulletin 16, p. 241-251.

About 800 dinosaur tracks are exposed in the dam spillway at Clayton Lake in Union County. Most of these Early Cretaceous tracks are like this one, and represent some sort of ornithopod (iguanodontid or hadrosaurid?). Clayton Lake and similar Early Cretaceous tracksites in eastern New Mexico indicate an abundance of such ornithopods in local shoreline facies during the latest Albian.

QUANTIFICATION OF SPATIALLY VARYING HYDROGEOLOGIC PROPERTIES FOR A PREDICTIVE MODEL OF GROUNDWATER FLOW IN THE OGALLALA AQUIFER, NORTHERN TEXAS PANHANDLE

ALAN R. DUTTON[1], ROBERT E. MACE[2] AND ROBERT C. REEDY[1]

[1]Bureau of Economic Geology, The University of Texas at Austin, University Station Box X, Austin, TX 78713-8924 alan.dutton@beg.utexas.edu

[2]Texas Water Development Board, P.O. Box 13231, Austin, TX 78711-3231 robert.mace@twdb.state.tx.us

Abstract.—The Ogallala aquifer, an extensive unconfined aquifer, is the main source of agricultural and drinking water across the Texas Panhandle and the U.S. High Plains. After 50 yrs of water production at rates in excess of recharge, less than half of the original saturated thickness remains in parts of the aquifer in the northern Texas Panhandle. Numerical models of groundwater flow in the Ogallala are tools with which to manage the aquifer; at least 15 numerical groundwater-flow models have been developed for different parts of the Ogallala aquifer in Texas. Regional models of groundwater flow in the heterogeneous Ogallala aquifer should draw on as many spatially varying data as possible. Besides the basic geometry of the aquifer (top and bottom and lateral boundaries), model input for hydraulic conductivity and recharge can be delimited using spatial attributes. Contouring hydraulic conductivity on the basis of trends in depositional systems and percentage sand and gravel in the aquifer represents both geological heterogeneity and a degree of lateral continuity not captured in simple variogram geostatistics. Recharge rates can be predicted on the basis of long-term precipitation trends and locally adjusted where soils are more or less permeable than average. The combination of mapping hydraulic conductivity using depositional systems and recharge using precipitation and soil factors minimized model calibration error to less than 5 % of the hydraulic-head change across the study area. Reaching a low model calibration error with negligible parameter adjustment raises confidence in the applicability of a groundwater-flow model to aquifer management.

INTRODUCTION

The Ogallala aquifer, which makes up the main part of the High Plains aquifer, along with adjacent and hydraulically interconnected older and younger formations, is the main source of agricultural and public-water supply in much of the Texas Panhandle (Fig. 1). Prediction of the amount of remaining groundwater in the Ogallala aquifer is an important part of managing the aquifer's resource and of developing regional plans to meet future water needs. Because of the complexity of the aquifer, due to heterogeneous geological materials as well as numerous points of groundwater withdrawal, it is necessary to use a numerical or computer model to make predictions accurate and detailed enough to be used in aquifer management.

Few regional aquifers have been as extensively studied as the Ogallala aquifer (for example, Gutentag and others, 1984; Knowles and others, 1984; Luckey and others, 1986; Nativ, 1988; Weeks and Gutentag, 1988; Dutton, 1995; Gustavson, 1996; Mullican and others, 1997; Scanlon and Goldsmith, 1997). Numerical models of groundwater flow have been important tools for managing the groundwater resource and evaluating future changes in water level and saturated thickness. At least 15 numerical groundwater-flow models have been developed for different parts of the Ogallala aquifer in Texas (Fig. 1). Numerical models integrate much of the known information on an aquifer, allow consideration of how the water-level response to pumping is influenced by aquifer properties, and help identify what information and conceptual understanding needs additional development. Each of the previous Ogallala models has had a specific purpose and carried associated strengths and weaknesses (Mace and Dutton, 1998).

This paper focuses on how spatially variable properties for hydraulic conductivity, storativity, and recharge were characterized for input to a predictive model of groundwater flow in the northern part of the Texas Panhandle (Fig. 1f). Previous studies have used fewer data on hydraulic conductivity. This study draws extensively on specific-capacity well-test data to estimate hydraulic conductivity and contours hydraulic conductivity guided by depositional facies and percentage sand-and-gravel maps. Previous studies have also used either constant recharge rates or varied recharge in order to improve model calibration. This study assigned recharge on the basis of precipitation and spatially varying soil properties. We found that these combined maps of hydraulic conductivity and recharge were important for minimizing model calibration error with negligible parameter adjustment.

MODEL INFORMATION

The study area lies north of the Prairie Dog Town Fork of the Red River and the Canadian River. The model area, however, extends westward and eastward to the limits of the Ogallala Formation, and northward to the Cimarron River, so that the model perimeter is defined by natural or hydrologic boundaries (Fig. 1f).

Hydrostratigraphy

Gutentag and others (1984) advocated referring to the groundwater system in the study area as the High Plains aquifer for two main reasons. First, groundwater can move between the Ogallala Formation and adjacent Permian, Mesozoic, and Quaternary formations, so the term Ogallala aquifer is inadequate to refer to the whole aquifer system. Second, not all of the Ogallala Formation is saturated. The term "High Plains aquifer" addresses these issues and avoids use of a formational name that is also an aquifer name. Because the focus of this study is on groundwater in the Ogallala Formation, however, the term "Ogallala aquifer" is used in this report, following local usage.

The Ogallala Formation in the study area consists of Tertiary-age alluvial-fan, fluvial, lacustrine, and eolian deposits derived

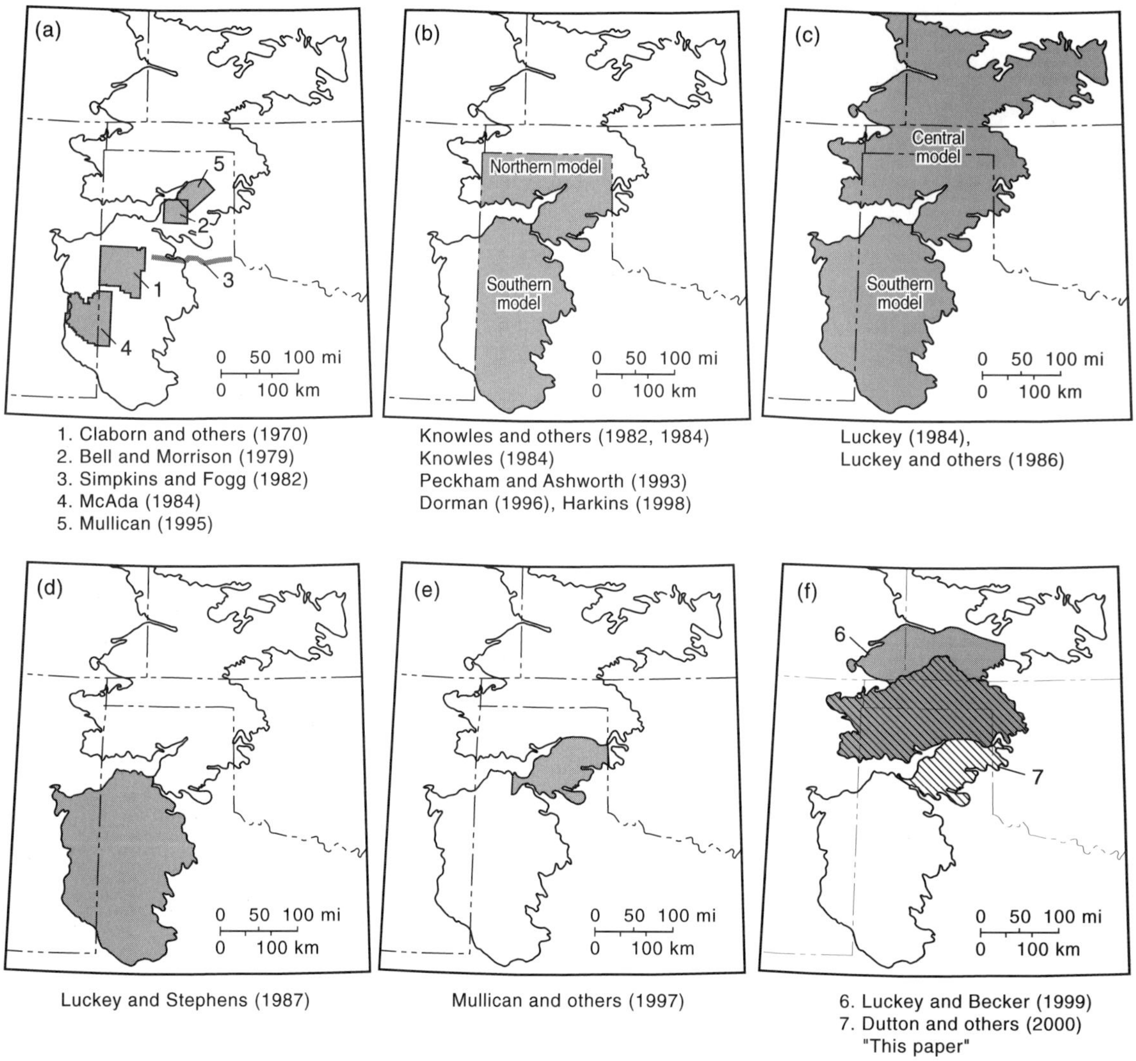

FIGURE 1. Location and area of coverage of models of the Ogallala aquifer in Texas.

from erosion of the Rocky Mountains (Seni, 1980; Gustavson and Winkler, 1988). The Ogallala Formation in the study area unconformably overlies Permian, Triassic, and other Mesozoic formations (Gutentag and others, 1984) and in turn may be covered by Quaternary deposits (Fig. 2). Ogallala sediments filled paleovalleys eroded into the pre-Ogallala surface (Seni, 1980; Gustavson and Winkler, 1988). Deposition of the Ogallala Formation in some areas was contemporaneous with dissolution of underlying Permian salt beds, resulting in additional ground-surface subsidence and increased accumulation of Ogallala sediment (Gustavson and Finley, 1985). At the northwestern limit of the study area in northeastern New Mexico, the Ogallala Formation is also interbedded and locally covered with Tertiary-age volcanic deposits.

Seni (1980) interpreted the Ogallala to have accumulated in alluvial-fan depositional environments with overlapping fan lobes (Fig. 3). According to Seni (1980), much of the study area includes distal-fan facies with narrow channel systems separated by broad, low net-sand areas. Seni (1980) did not everywhere break out the overlying Blackwater Draw Formation. Gustavson (1996) distinguished between alluvial deposits filling paleoval-

AGE	GEOLOGIC UNIT	
Quaternary	Blackwater Draw Formation	Tahoka Formation Double Lakes Formation Tule Formation
Tertiary	Blanco Formation	Cita Canyon lake beds
	Ogallala Formation	
Cretaceous	Edwards Group	
Triassic	Dockum Group	
Permian	Ochoan Series Guadalupian Series Leonardian Series	

FIGURE 2. Stratigraphic nomenclature of Permian and younger strata, including the Ogallala Formation, in the study area.

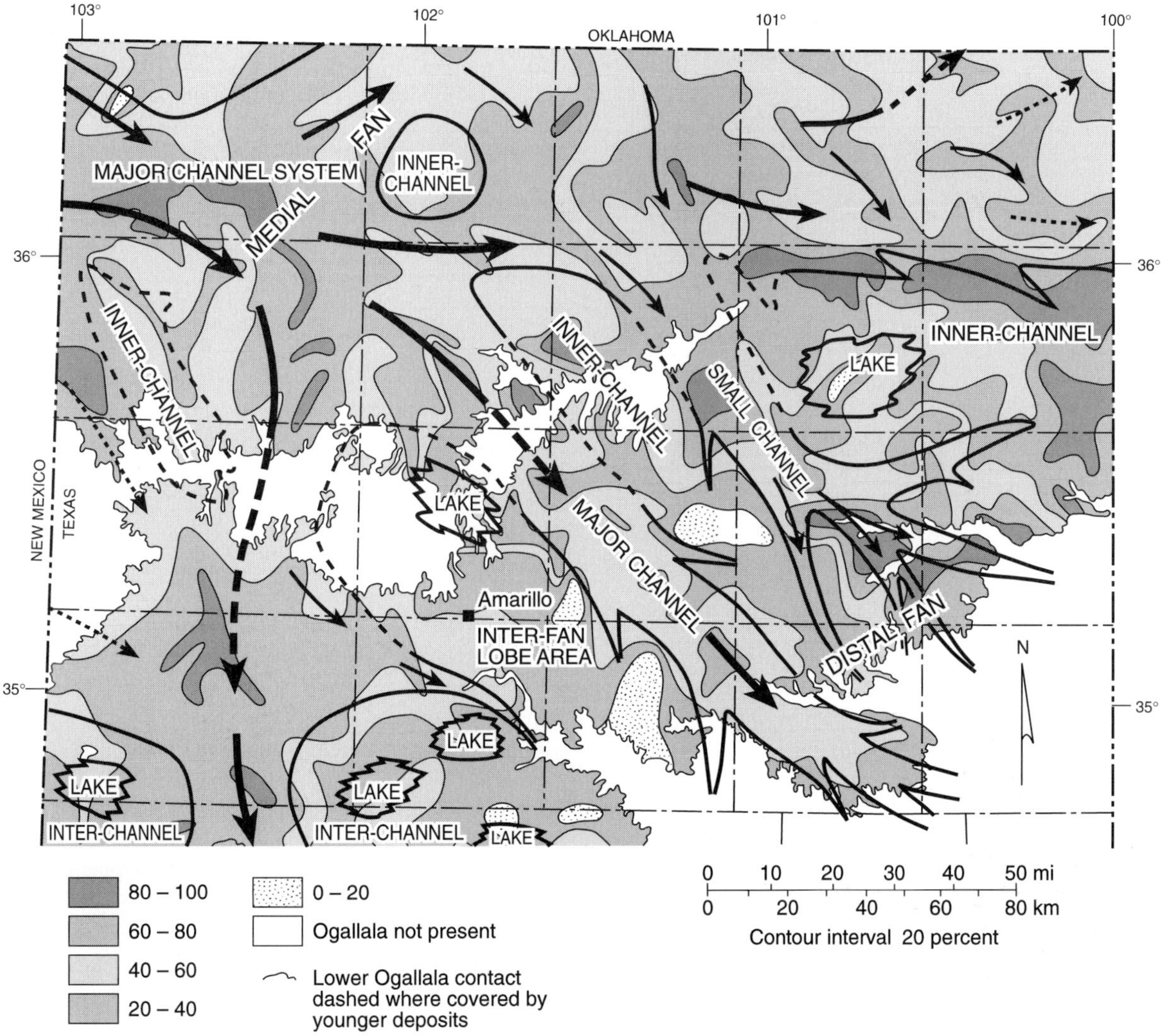

FIGURE 3. Percentage sand and gravel in the Ogallala Formation and inferred depositional facies (modified from Seni, 1980, figs. 10 and 11).

leys and widespread eolian sediments capping the fluvial sections and paleouplands and mapped the top of the Ogallala at an erosion-resistant caprock comprised of a calicified soil layer.

Both depositional models define an aquifer framework or architecture having appreciable lateral and vertical heterogeneity. Aquifer heterogeneity is the spatial variability in properties that control the occurrence and movement of groundwater, such as hydraulic conductivity and specific yield, and is largely related to geologic features. Areas of the aquifer with a greater amount of sand and gravel have greater hydraulic conductivity. The lower part of the formation in paleovalley-fill alluvium tends to have more coarse-grained sediment and greater hydraulic conductivity than the upper part. Within any section, sediment bedding may slightly impede the vertical circulation of groundwater.

The Ogallala Formation and overlying Quaternary Blackwater Draw Formation underlie the High Plains. Retreat of the edge of the High Plains surface has left a steep escarpment in most areas, which is held up in part by the caprock caliche (Gustavson and Simpkins, 1989; Gustavson, 1996). The other main physiographic feature in the study area is the Canadian River Breaks, consisting of dissected erosional drainage bordering the Canadian River.

Aquifer geometry

Aquifer geometry (ground surface, aquifer base, water table) is probably the best characterized of all the input data. Nonetheless, the water table and base of the aquifer are not perfectly known, and data input to the model still requires some simplification and approximation.

Structure of the bottom of the aquifer is defined by numerous wells. The base of the aquifer for the model was mapped locally at the base of "red beds," Permian and Mesozoic rocks bearing groundwater in continuity with Ogallala groundwater. Mapping the base of the aquifer using tools in ArcInfo/ArcView (triangulated irregular networks [TIN's], gridding the TIN surfaces, and assigning values to the model grid) resulted in a reasonable representation of regional trends (Fig. 4). The map might not accurately depict local features, however, especially where data are sparse.

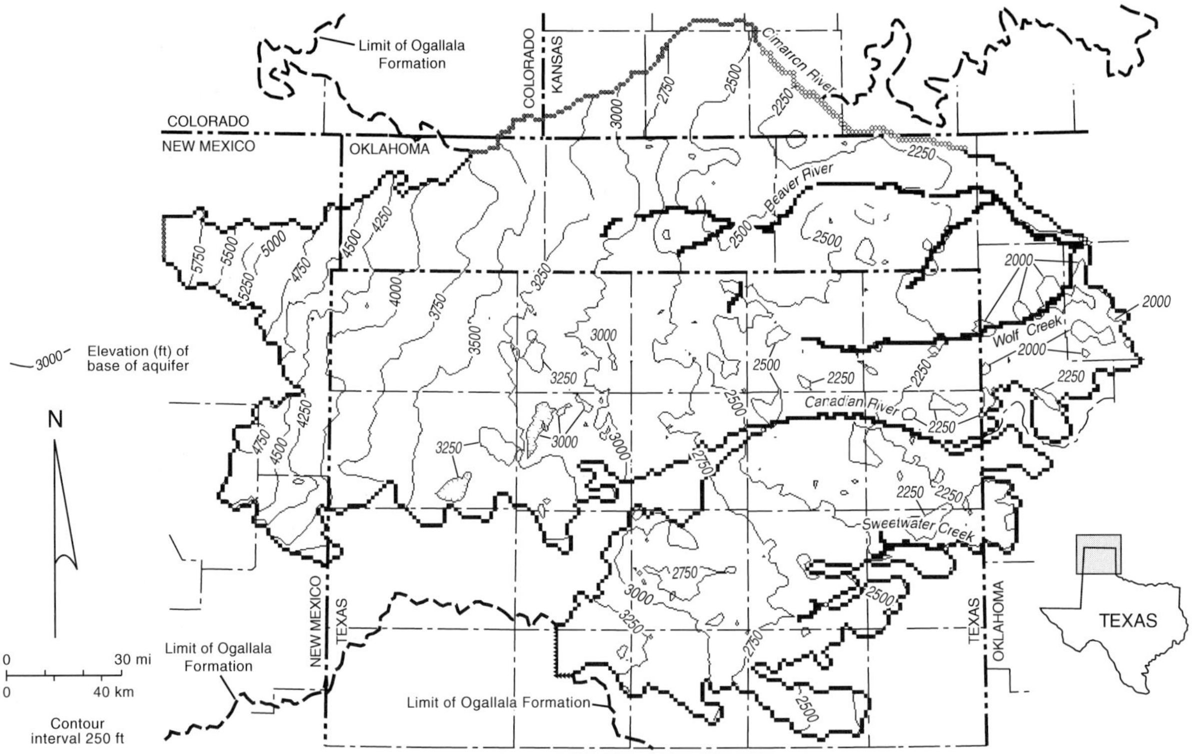

FIGURE 4. Structural elevation of the base of the Ogallala Formation as used in the numerical model. Elevation contours range from ~610 to ~1750 m (2000 to 5750 ft) above mean sea level.

Ground surface was input using digital elevation model data; it is most precisely known. The elevation of the "predevelopment" water-table surface (Fig. 5) was based on the earliest measured water levels. Hydraulic head is an expression of potential energy per unit weight of water; a map of hydraulic-head elevations represents the water table in an unconfined aquifer. This picture of the "predevelopment" water table (Fig. 5) is imperfect because (a) data were composited from a wide range of years to include the first recorded measurements in different areas of the model; (b) some amount of groundwater was already being withdrawn in each area of the model when the earliest water levels were being reported; and (c) early data are sparse, and elevation of the water table is extrapolated partly on the basis of the shape of ground-surface topography. The major features of the estimated predevelopment water table (Fig. 5) resemble those depicted by Knowles and others (1984) and Luckey and others (1986); each study used a common set of data.

Saturated thickness of groundwater in the Ogallala aquifer in the study area was more than 210 m (>690 ft) in southwestern Kansas and the Oklahoma Panhandle, but it was generally less than 90 m (<295 ft) in Texas under predevelopment conditions. Given that the top of the saturated section is fairly smooth, much of the variation in saturated thickness is due to relief on the base of the Ogallala. In Carson County, the thick accumulation of Ogallala sediments reflects continued Tertiary-age deposition contemporaneous with ground-surface subsidence above salt-dissolution zones (Gustavson and Finley, 1985). The thinnest saturated sections of the Ogallala aquifer were in eastern New Mexico and around the perimeter or limit of the aquifer.

The elevation of the water table has changed more than 46 m (>150 ft) in the study area since the 1950's owing to withdrawal of groundwater at rates greatly exceeding the recharge rate. The drawdown of water levels in well fields such as the Amarillo well field in Carson County locally changes the direction of regional flow paths. Water levels in the aquifer in the northern part of the Texas Panhandle declined an average of about 2 m/yr (5.5 ft/yr) during 1960–80 (Knowles and others, 1984), although there also was comparable water-level recovery in parts of the aquifer beneath the southern High Plains south of the Canadian River.

Aquifer parameters

To estimate hydraulic properties for the study area in Texas and expand upon previous studies, we (1) compiled available information on aquifer properties or tests from published reports and well records, (2) used specific-capacity information to estimate transmissivity and hydraulic conductivity, (3) used statistics to summarize results, and (4) used geological maps to "condition," or map, values of hydraulic conductivity. Hydraulic conductivity expresses the ease with which water moves through a unit cross section of the aquifer under a unit gradient in hydraulic head. A major improvement to hydraulic properties over previous studies is the inclusion of specific-capacity information, which significantly increases the number of measurement points for the aquifer.

We took information from Mullican and others (1997) and compiled test results from a groundwater database (Texas Water Development Board [TWDB], 1999). Mullican and others (1997) had information on 70 aquifer tests, which included high-quality

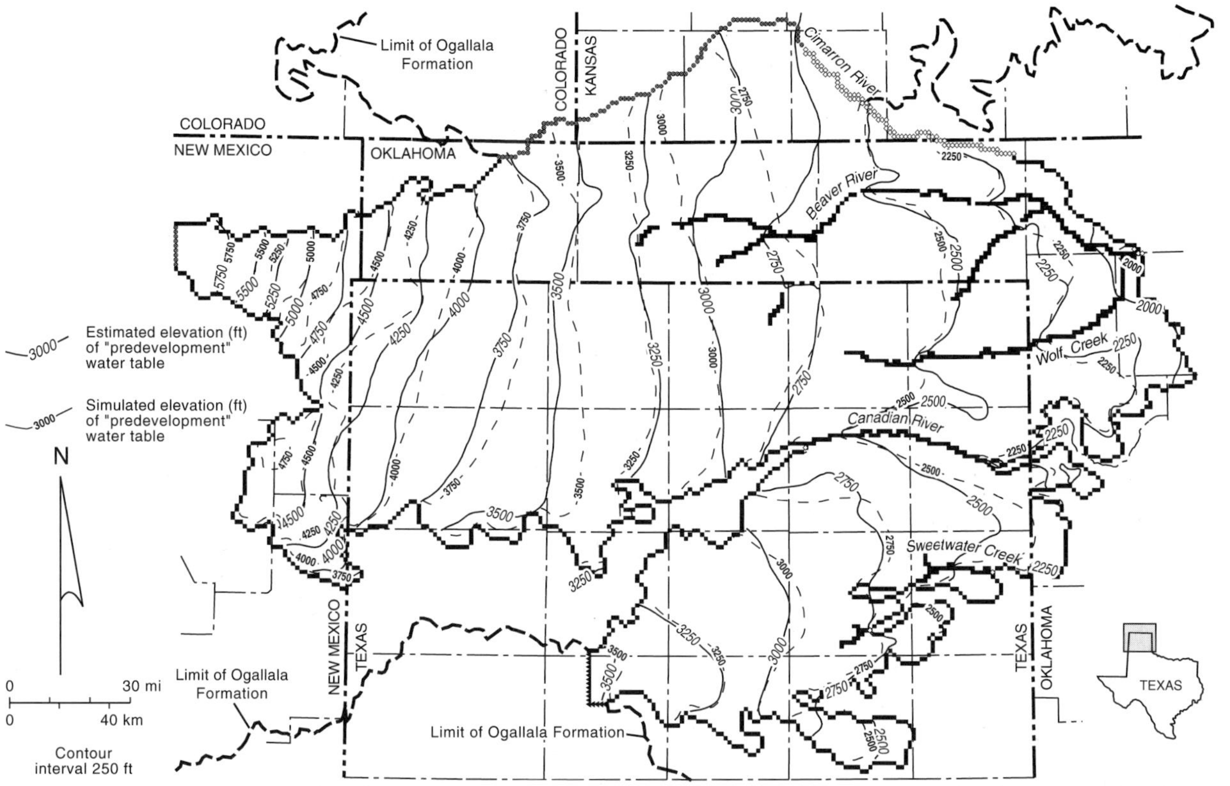

FIGURE 5. Comparison of estimated and simulated elevations of "predevelopment" water table. Contours of water-table elevation range from ~610 to ~1750 m (2000 to 5750 ft) above mean sea level.

specific-capacity tests. We were able to cull data from an additional 1270 specific-capacity tests in the TWDB groundwater database. To estimate transmissivity and hydraulic conductivity from specific capacity, we used an analytical technique developed by Theis (1963) and described in Mace (in press). Hydraulic conductivity was determined by dividing transmissivity by the saturated thickness exposed to the well bore. Locations of tested wells are recorded with an accuracy generally less than ±1 km (<±3200 ft). Few records include quality-measured spatial coordinates.

Based on results from the data compilation and specific-capacity analysis, we found that hydraulic conductivity for all the tests in the Ogallala aquifer appears to be lognormally distributed (Fig. 6) with a geometric mean of about 4.5 m/d (14.8 ft/d) and a standard deviation that spans from 1.5 to 13.4 m/d (5 to 44 ft/d). A lognormal distribution means that the logarithms of the values are normally distributed, and a geometric mean is the antilogarithm of the mean of the logarithms of the values.

Semivariograms (see Clark, 1979; McCuen and Snyder, 1986) show that hydraulic conductivity in the Ogallala aquifer is spatially correlated. Spatial correlation implies that points that are closer together are more similar to each other than points that are farther apart. Fitting a spherical theoretical semivariogram to the experimental semivariogram resulted in a nugget of 0.12 [log(m/day)]^2, a sill of 0.22 [log(m/day)]^2, and a range of ~44 km (~26 mi). The range suggests that hydraulic conductivity is spatially correlated within that distance in the Ogallala aquifer.

Hydraulic conductivity in the Ogallala aquifer was contoured on the basis of posted data and the Seni (1980) maps of inter-

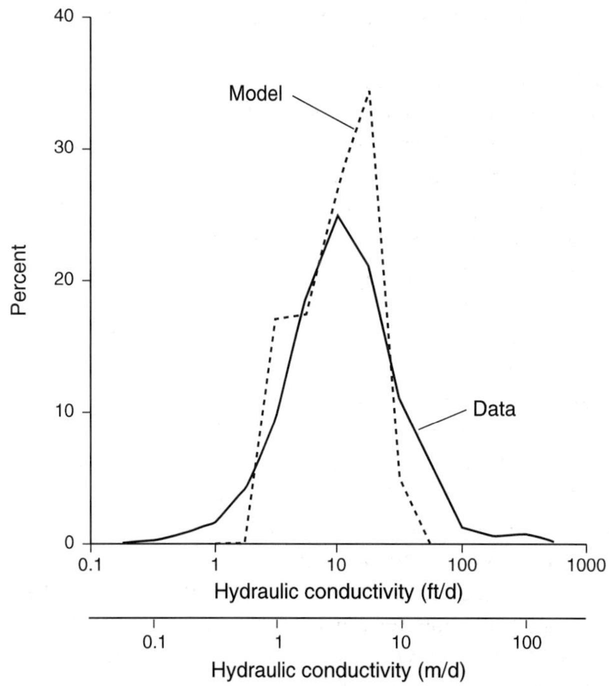

FIGURE 6. Comparison of measured and calibrated values of hydraulic conductivity used in the Texas part of the model.

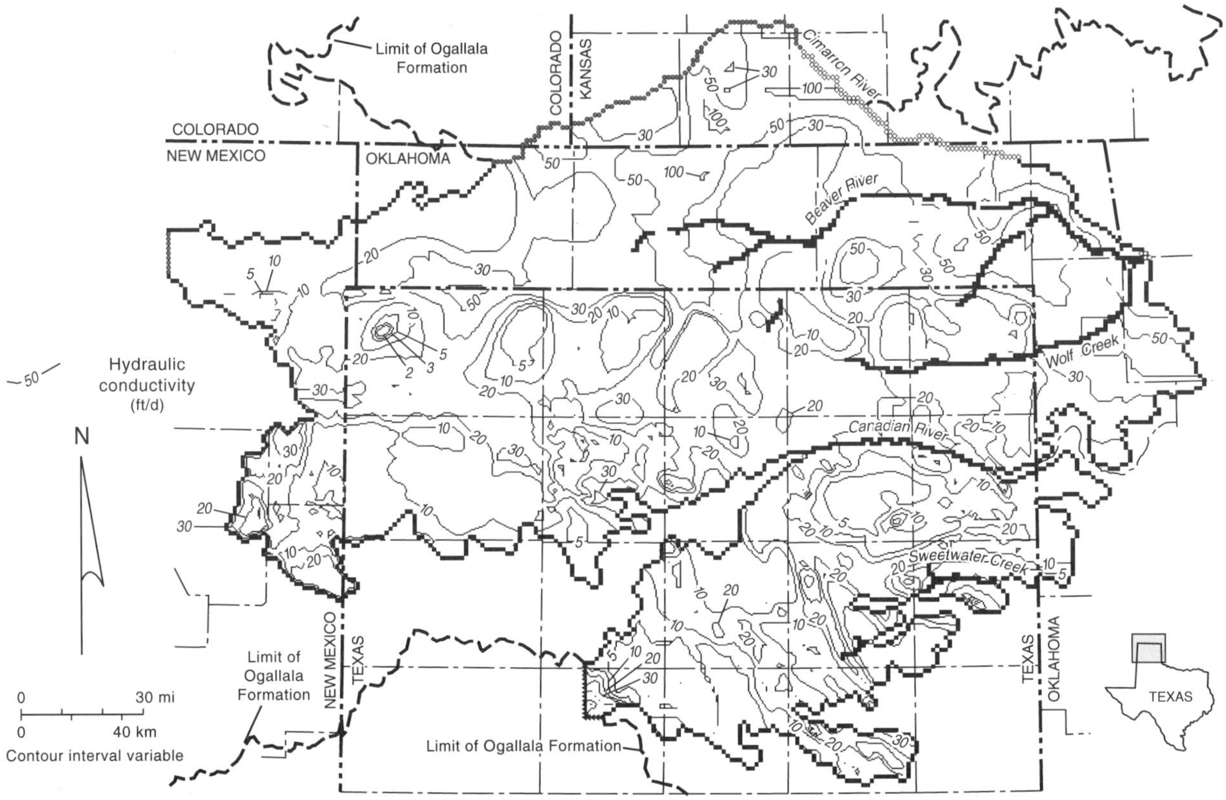

FIGURE 7. Hydraulic conductivity of the Ogallala aquifer used for historical and predictive simulations. Contours of hydraulic conductivity range from ~0.6 to ~30 m/d (2 to 100 ft/d).

preted depositional facies and percent of sand and gravel (Fig. 3). Contours and trend lines from the geologic maps were then used as a guide to contour the hydraulic-conductivity data (Fig. 7). We assumed that the elongate trends at the base of the aquifer interpreted by Seni (1980) as channel systems reasonably approximate laterally continuous flow units in the aquifer. Sand-percent maps were used because the Ogallala Formation is eroded in the Canadian River Valley. Hydraulic-conductivity data are consistent with Seni's (1980) channel axes and bifurcating lobes. Details of depositional channels of high hydraulic conductivity can be resolved in data-rich areas corresponding to channels within various fan lobes. For other depositional axes, data are too sparse to constrain the precise location of narrow channels of high hydraulic conductivity. Depositional channels of high hydraulic conductivity probably follow narrow and meandering bifurcating paths within those broadly defined trends. Statistical distributions of measured and final calibrated hydraulic conductivity are similar for the Texas part of the model (Fig. 6), with only minor changes to hydraulic conductivity made during model calibration.

Hydraulic conductivity of the Ogallala was previously mapped by Knowles and others (1984; see also Nativ, 1988). Without geologic conditioning, the Knowles and others (1984) map suggests no regional continuity in hydraulic conductivity. For comparison, we also mapped the hydraulic conductivity of the Ogallala aquifer by kriging, using geostatistical parameters from the semivariogram analysis. Because the kriged and geologically conditioned maps share the same hydrologic data, they have some resemblance, with northwest-southeast trends of high hydraulic conductivity corresponding to the depositional axes. The kriged map, however, is more like the Knowles and others (1984) map in suggesting a much lower degree of regional continuity in "high-permeability" channels. This study's map of hydraulic conductivity (Fig. 7) assumes a greater degree of lateral continuity because it is drawn on the basis of the interpreted map of depositional systems (Fig. 3).

Contemporaneous and postdepositional salt dissolution and subsidence (Gustavson and Finley, 1985; Gustavson and Winkler, 1988; Gustavson, 1996) may have affected hydraulic conductivity of the Ogallala. Additional analysis is needed to determine which, if either, process might have contributed to several apparent disruptions in continuity of the permeability channels.

Maps of specific yield were taken from Knowles and others (1984) and merged with cell values used by Luckey and Becker (1999) for the non-Texas part of the model. Storativity is a bulk or composite hydrologic property that expresses the volume of water released from a vertical column of an aquifer per unit surface area and unit decline in hydraulic head. It is a function of porosity, compressibility of water, and elasticity of the aquifer matrix. For an unconfined aquifer, storativity is represented by specific yield. Specific yield is generally less than porosity because some water held in the matrix cannot be removed by gravity drainage. Additional work is needed to determine whether the depositional systems maps of aquifer fabric can be used to map or predict specific yield in the Ogallala aquifer.

Recharge

The study area has a dry continental climate with moderate precipitation, low humidity, and high evaporation. Precipitation decreases from east to west across the Texas Panhandle from more than 56 cm/yr (>22 in/yr) to less than 40 cm/yr (<16 in/yr), whereas potential evapotranspiration increases (Larkin and Bomar, 1983).

The Ogallala aquifer is recharged by downward percolation of water from the surface of the High Plains. The distribution of recharge is poorly known; estimates of recharge rates range from 0.25 to 15 cm/yr (0.01 to 6 in/yr) (Mullican and others, 1997). In much of the study area, runoff of surface water is not well integrated in streams, and much of the runoff collects in playa basins. Playas can focus recharge to the aquifer (Wood and Petraitis, 1984; Mullican and others, 1997). Estimates of regional recharge rates are weighted averages of the higher rates beneath playas and lower rates beneath interplaya settings (Mullican and others, 1997). Regional and local recharge rates may vary with the characteristics of the soils that underlie playa and interplaya areas.

Recharge rates (Fig. 8) were set as a function of precipitation and soil types (Table 1). Data on long-term average (1950 to 1990) precipitation were compiled from U.S. National Oceanic and Atmospheric Administration (2000). These data were contoured and interpolated for the cells in the model area. Recharge was assumed initially to vary linearly from 0.25 to 1.27 cm/yr (0.1 to 0.5 in/yr) where precipitation ranged from 42 to 57 cm/yr (16.5 to 22.5 in/yr), respectively. During calibration the straight-line relationship between recharge and precipitation was changed so that (1) recharge is a greater percentage of precipitation on the wetter, eastern side of the study area than to the west and (2) recharge is at a minimum wherever precipitation is less than 48 cm/yr (<19 in/yr) in the central to western part of the study area. Further research on the relation of recharge to precipitation is needed.

TABLE 1. Soil-based weighting factors for recharge rates. Recharge rates were assigned in the model on the basis of long-term average precipitation and locally adjusted on the basis of weighting factors derived from soil textures. Soil data compiled from U.S. Department of Agriculture (2000).

Soil group	Soil textures	Area in model (km^2)	Soil permeability (cm/hr)	Weighting factor
1	Loam-Silt loam	17,956	2.5	1.0
2	Loamy sand-Sandy loam	21,445	37.1	1.0
3	Sandy loam-Clayey loam-Silty clay loam	5,840	11.2	1.0
4	Silty clay loam-Silty clay	13,755	0.3	0.67
5	Silt loam-Clayey loam	1,339	1.3	0.67
6	Clay loam-Clay	883	0.8	0.67
7	Sandy loam-Loam-Clay loam	321	11.2	0.67
8	Sand	2,479	75.4	2.77

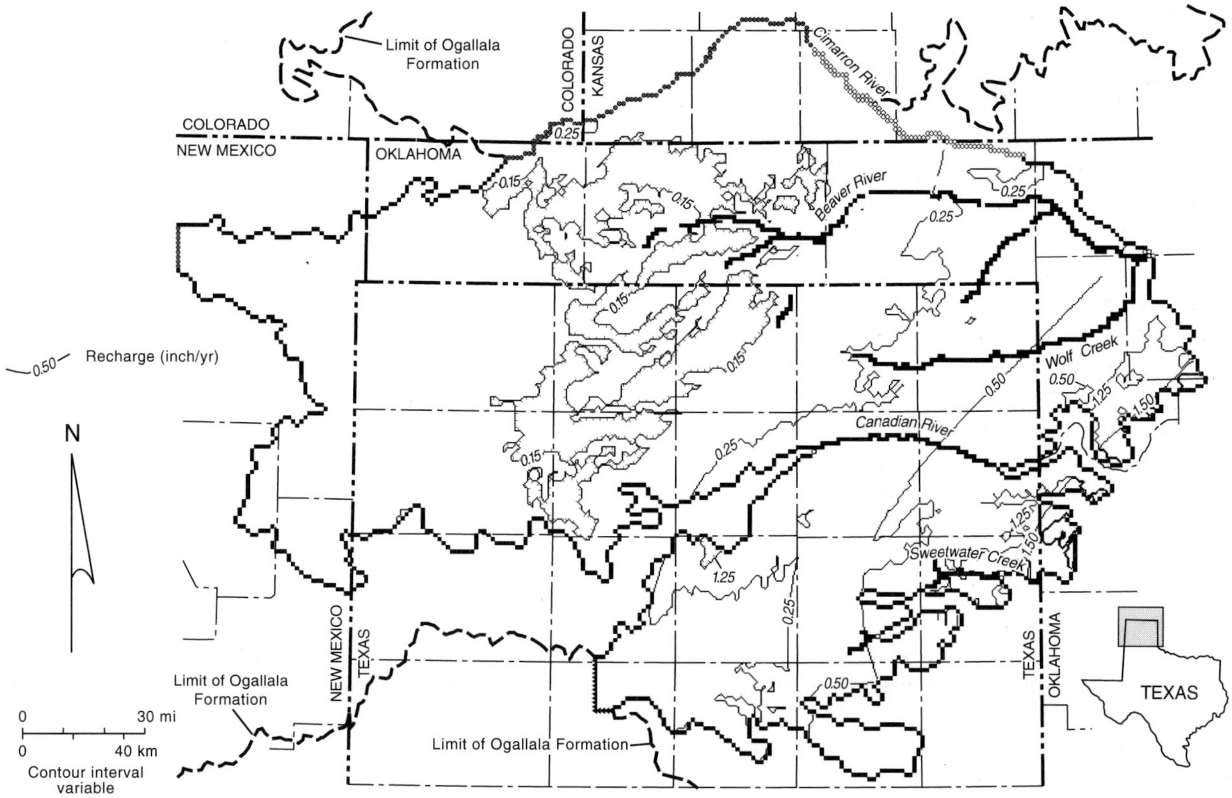

FIGURE 8. Recharge used in the numerical model assigned on the basis of precipitation and soil texture. Recharge contours range from ~0.06 to 0.6 cm/yr (0.15 to 1.5 in./yr).

Recharge estimates were furthered adjusted by soil type. GIS polygons of soil types were downloaded from U.S. Department of Agriculture (2000) and classified into eight groups (Table 1). Groups 1 to 3 have mainly loamy surface and subsurface soils, whereas Groups 4 to 7 have loamy surface but clayey subsurface soils. Groups 1 and 2 roughly correspond to the extent of the Ogallala Formation outcrop, especially south of the Canadian River. Group 4 south of the Canadian River corresponds to soils developed on the Blackwater Draw Formation (Table 1). Group 8 is made up of windblown sands that are younger deposits than the Blackwater Draw Formation. Recharge estimated from precipitation was not changed (weighting factor of 1.0) for "Ogallala" soils. Recharge was decreased for "Blackwater Draw" soils and increased for sandy Group 8 soils (Table 1).

Groundwater recharge determined through model calibration was less than 1% of annual precipitation across about 72% of the model area. The other 99% of precipitation is assumed to have returned to the atmosphere by evapotranspiration or run off as surface water. Groundwater recharge was set at less than 2% of precipitation across 92% of the model area but was between 5 and 6% of precipitation in 3% of the area. The higher recharge rates were on sandy soils on the eastern, wetter side of the study area.

Return flow

Between 1950 and 1998, approximately 68 km^3 (55 million acre-feet) of groundwater was pumped from the Ogallala aquifer for irrigation in Texas. Average annual withdrawal for irrigation was greatest during the 1980's at approximately 1.9 km^3/yr (1.5 million acre-feet per year) (Fig. 9a). During the 1990's the total rate of irrigation withdrawal decreased to about 1.5 km^3/yr (1.2 million acre-feet per year). Irrigation water in 1998 made up, on average, 86% of total groundwater production from the Ogallala aquifer but ranged from 59 to 98%. Irrigation application rates ranged from about 5 to 16 cm/yr (0.17 to 0.52 acre-foot per acre per year) during 1960 to 1998.

Irrigation withdrawal in the Texas part of the model was distributed on the basis of results of a 1994 survey obtained in GIS format from the Texas Natural Resources Information System (TNRIS). That database identified polygons with irrigated acreage and specified the percentage of the polygon area under irrigation in 1994. We assumed that the same pattern of irrigated acreage applied to the entire modeling period. This assumption is reasonable because irrigation rate changed but location of irrigated fields did not change. Total county withdrawal of groundwater for irrigation for a given year was proportionately distributed across the model grid to those cells with irrigated acreage. Decadal estimates of irrigation withdrawal by county were made on the basis of rainfall and irrigation efficiencies (S. Amosson, personal communication, 2000).

An irrigation efficiency (e_i) of 100% indicates all applied irrigation is used by plants or is evapotranspired. Irrigation inefficiency ($1 - e_i$) expresses the generally unknown proportion of irrigation water that passes below the root depth and out of the reach of evapotranspiration. Return flow is the amount of irrigation water that ends up recharging the water table. Some water originally may have gone to increasing moisture content of the unsaturated zone and may not have reached the water table. Irrigation inefficiency probably was high during the 1940's and 1950's but decreased during the past few decades. Luckey and Becker (1999) assumed that irrigation inefficiency decreased from 24% during the 1940's and 1950's to less than 4% by the 1980's (Fig. 9b).

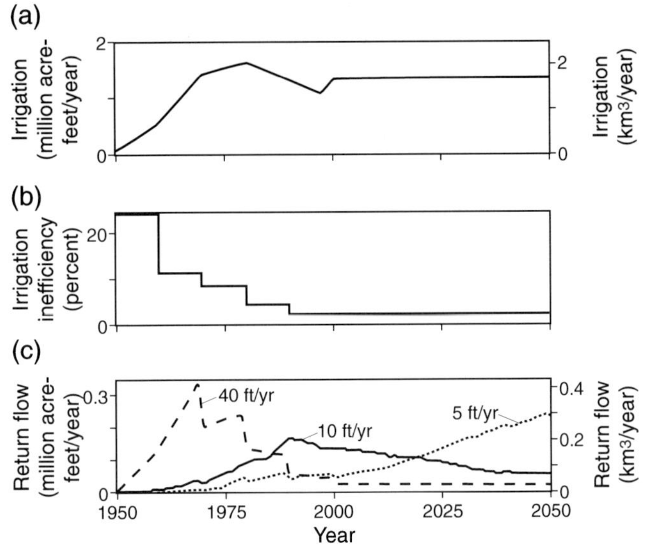

FIGURE 9. Estimation of return flow (c) from irrigation rates (a), inefficiency (b), and velocity at which water moves through the unsaturated zone. Changing depth to water and soil types were also taken into account. Irrigation inefficiency is 1.0 – irrigation efficiency. There is no lag in return flow (c) at high velocity (e.g., ~12 m/d [40 ft/d]). At lower velocity (e.g., 1.5 and 3 m/d [5 and 10 ft/d]), return flow is increasingly delayed from catching the falling water table.

Return flow to the water table was estimated for the groundwater flow model on the basis of irrigation rate, inefficiency, soil type, depth to water, and velocity or rate of downward movement of water from the root zone to the water table. Irrigation inefficiency was initially set equal to 24% for the 1950's and decreased to 2% since the 1990's (Fig. 9b); simulations were also made with other loss rates. The same soil-weighting factors were applied to return flow as to recharge from precipitation (Table 1). Less return flow was predicted from irrigation on Blackwater Draw soils than on Ogallala-derived soils as if the former intrinsically have higher irrigation efficiency. Depth to water was approximated using preliminary model results without return flow. Depth to water increased through time at most model cells, as water levels were drawn down by pumping, increasing the travel time for water to move from the root zone to the water table. Accordingly, return flow may recharge the water table later than the year in which irrigation was applied, and the delay or lag may increase through time as depth to water increases. Finally, velocity of water through the unsaturated zone was assumed to lie between 1.5 and 12.2 m/yr (5 and 40 ft/yr).

Velocity of water moving downward through the unsaturated zone is an important but poorly constrained variable. Velocity through the unsaturated zone was varied to see the impact of transit time on simulated water levels. If that velocity is much greater than the rate of water-level decline (~2 m/yr [~5.5 ft/yr]), return flow quickly reaches the water table (Fig. 9c). If that downward velocity is similar to the rate of water-level decline, much of the return flow may be significantly delayed in reaching the water table, leaving more water in storage in the unsaturated zone. Figure 9c shows that at a low velocity, return flow may catch up to the water table in the future as rate of water-level decline slows down.

The magnitude and effect of return flow remain poorly known. The difference between maximum rate of return flow and no return flow accounts for less than 6 m (~20 ft) of drawdown between 1950 and 1998, and not much more than 6 m (~20 ft) by 2050. Other model uncertainties associated with hydraulic properties and pumping rate account for at least this much error. Comparison of observed and simulated hydrographs, therefore, does not very well constrain the return-flow rate. Return flow may be important to future water budgets in areas that had high irrigation rates and low irrigation efficiency.

DISCUSSION

Contours of the simulated water table reasonably match the estimated 1950 water table at the start of the calibration period (Fig. 5). The estimated water table includes appreciable extrapolation error itself. Comparison of measured and simulated water levels for specific wells for 1998 at the end of the calibration period yields a model calibration error (root mean square error) of about 22.5 m (74 ft) (Fig. 10). This error has no evident bias and is less than 4% of the head drop across the Texas part of the model, whereas a typical calibration goal is 10 % for a numerical model. The calibration error includes uncertainties owing to all model parameters and inputs in the "predevelopment" calibration, including hydraulic conductivity, specific yield, pumping rates, recharge rates, return flow, and boundary conditions. This very good model calibration was obtained with negligible adjustment of parameters after data input. Ease of calibration partly reflects the use of a well-posed conceptual model and a geologically reasonable distribution of parameter values.

Early attempts at model calibration with recharge assigned uniformly across the study area and hydraulic conductivity mapped using kriging solely on the basis of variogram parameters yielded poorer calibration with root mean square error of more than 36.6 m (>120 ft). Considerable and somewhat arbitrary parameter adjustment was needed to decrease the numerical value of the calibration error. Although the calculated calibration errors of the preliminary model and the final model reported here were similar, the real calibration of the final model is better because it more closely honors original data and was obtained with negligible parameter adjustment.

Calculating model calibration error on the basis of measured and simulated water levels does not fully characterize a model. The root mean square error (RMSE) is slightly sensitive to whether hydraulic conductivity is assigned on the basis of depositional systems (RMSE = 22.5 m [74 ft]), kriged geostatistics (RMSE = 24.4 m [80 ft]), or average hydraulic conductivity (RMSE = 26.8 m [88 ft]). Depending on the purpose and scope of a model, ignoring heterogeneity may give satisfactory results. For example, a model predicting the drawdown of water levels in a well field in a small geographic area may be fairly insensitive to locally small variations in hydraulic conductivity. On the other hand, prediction of particle paths for contaminant transport calculations or groundwater-age modeling may be more sensitive at any scale to heterogeneities. Estimates of regional, long-term drawdown of water levels in extensive aquifers may also be sensitive enough to justify accounting for heterogeneity.

There is a direct relation between hydraulic conductivity and recharge rate in model calibration, everything else being held constant. If recharge rate were set higher in all or part of the model, hydraulic conductivity would have to be increased to compensate and keep model-calibration error unchanged. It would take a higher hydraulic conductivity to move the greater volume of water recharging the aquifer and keep simulated water level the same. This pattern was documented in sensitivity analyses by Luckey and Becker (1999, p. 52). The combination of mapping hydraulic conductivity using depositional systems and recharge using precipitation and soil factors was important to minimizing model error without arbitrary parameter adjustment.

Geological conditioning of maps of hydrologic properties has advantages and limitations. Chief among the advantages is that the depositional fabric or architecture of the aquifer that controls hydrologic properties is taken into account and not ignored (Koltermann and Gorelick, 1996; Fogg and others, 1998) This should result in a more realistic numerical model of groundwater flow. A key difference between geologically and nongeologically contoured maps of hydrologic properties, however, is the apparent extent of regional continuity. Examples in this paper show extensive lateral continuity of permeability channels enveloped in volumes of lower permeability areas because the depositional model implied a high degree of lateral continuity. Depositional systems

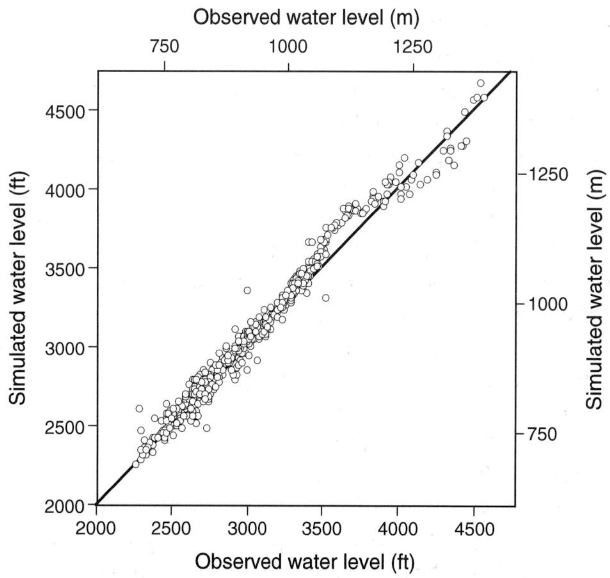

FIGURE 10. Calibration results for 1998 water-level measurements. The mean square error between observed and simulated water levels is 22.5 m (74 ft).

maps help constrain the architecture of the aquifer and distribution of hydraulic conductivity. We are also looking at whether the depositional systems maps of aquifer fabric can be used to map or predict specific yield.

Having sufficient data density to resolve fine-scale permeability features may limit the application of this geological conditioning approach. Key to our approach was using specific-capacity data to accumulate enough measurements of hydraulic conductivity to allow this detail of mapping. As previously stated, it is likely that the fine-scale details such as high-permeability channels revealed in data-rich areas probably occur also in data-poor areas. Discernment and judgment in drawing meandering trends of high hydraulic conductivity within channel lobes defined by net-sand or sand-percent maps is balanced against the uncertainty of where the "channel" axis is located. Another limitation may be the level of discernment or artistic license in a geological map. There may be multiple ways to contour and honor the hydrologic data while still reflecting the underlying geological fabric. In addition, the degree of lateral continuity implied by various depositional-systems interpretations may vary.

CONCLUSIONS

Use of appropriate data sets of spatially varying hydrogeologic parameters can facilitate obtaining well-calibrated groundwater-flow models. In this example, hydraulic conductivity was mapped consistently with trends in depositional facies and recharge was assigned on the basis of precipitation and soil type. Other parameters such as base of aquifer and predevelopment water table were mapped using extensive data. Reaching a low model-calibration error with negligible parameter adjustment raises confidence in applicability of a groundwater-flow model to aquifer management. Other factors besides calculated values of root mean square error of hydraulic head can determine how well a model may perform.

ACKNOWLEDGMENTS

This study was sponsored by the Panhandle Water Planning Group (PWPG) of the Panhandle Regional Planning Commission under interagency contract number UTA99-0230. A Modeling Committee of the PWPG provided helpful discussion, data for use in the model, and helpful reviews of the draft report. Liying Xu assisted in data compilation and digitization. Thet Naing assisted with model simulations. Illustrations were prepared by William Bergquist under the direction of Joel L. Lardon, Graphics Manager. Text was edited by Lana Dieterich under direction of Susann Doenges. Publication was authorized by the Director.

REFERENCES

Bell, A. E., and Morrison, S., 1979, Analytical study of the Ogallala aquifer in Carson County, Texas, projections of saturated thickness, volume of water in storage, pumpage rates, pumping lifts, and well yields: Texas Department of Water Resources Report 242, 64 p.

Claborn, B. J., Austin, T. A., and Wells, D. M., 1970, Numerical model of the Ogallala as a management tool, *in* Mattox, R. B., and Miller, W. D., eds., Ogallala Aquifer Symposium: Texas Tech University, International Center for Arid and Semi-Arid Land Studies, Special Report Number 39, p. 89-110.

Clark, Isobel, 1979, Practical geostatistics: London, Applied Science Publishers, Limited, 129 p.

Dorman, T. M., 1996, The Texas High Plains Aquifer System- modeling and projections for the southern region [Master's thesis]: Lubbock, Texas, Texas Tech University.

Dutton, A. R., 1995, Groundwater isotopic evidence for paleorecharge in U.S. High Plains aquifers: Quaternary Research v. 43, p. 221–231.

Dutton, A. R., Reedy, R. C., and Mace, R. E., 2000, Saturated thickness in the Ogallala aquifer in the Panhandle Water Planning Area—Simulation of 2000 through 2050 withdrawal projections: The University of Texas at Austin, Bureau of Economic Geology, Topical Report prepared for the Region A Panhandle Water Planning Group, Panhandle Regional Planning Commission under Contract number UTA99-0230, December 2000, 75 p.

Fogg, G. E., Noyes, C. D., and Carle, S. F., 1998, Geologically based model of heterogeneous hydraulic conductivity in an alluvial setting: Hydrogeology Journal v. 6, p.131–143.

Gustavson, T. C., 1996, Fluvial and eolian depositional systems, paleosols, and paleoclimate of the upper Cenozoic Ogallala and Blackwater Draw Formations, Southern High Plains, Texas and New Mexico: The University of Texas at Austin, Bureau of Economic Geology Report of Investigations No. 239, 62 p.

Gustavson, T. C., and Finley, R. J., 1985, Late Cenozoic geomorphologic evolution of the Texas Panhandle and northeastern New Mexico—case studies of structural controls on regional drainage development: The University of Texas at Austin, Bureau of Economic Geology Report of Investigations No. 148, 42 p.

Gustavson, T. C., and Simpkins, W. W., 1989, Geomorphic processes and rates of retreat affecting the Caprock Escarpment: The University of Texas at Austin, Bureau of Economic Geology Report of Investigations No. 180, 49 p.

Gustavson, T. C., and Winkler, D. A., 1988, Depositional facies of the Miocene-Pliocene Ogallala Formation, northwestern Texas and eastern New Mexico: Geology, v. 16, p. 203–206.

Gutentag, E. D., Heimes, F. J., Krothe, N. C., Luckey, R. R., and Weeks, J. B., 1984, Geohydrology of the High Plains aquifer in parts of Colorado, Kansas, Nebraska, New Mexico, Oklahoma, South Dakota, Texas, and Wyoming: U.S. Geological Survey Professional Paper 1400-B, 57 p.

Harkins, D., 1998, The future of the Texas High Plains aquifer system—modeling and projections [Ph.D. dissertation]: Lubbock, Texas, Texas Tech University, 278 p.

Knowles, T. R., 1984, Assessment of the ground-water resources of the Texas High Plains, *in* Whetstone, G. A., ed., Proceedings, Ogallala Aquifer Symposium II: Texas Tech University Water Resources Center, p. 217–237

Knowles, T. R., Nordstrom, Phillip, and Klemt, W. B., 1982, Evaluating the ground-water resources of the High Plains of Texas: Texas Department of Water Resources, Final Report LP-173, v. 1, 174 p.

Knowles, Tommy, Nordstrom, Phillip, and Klemt, W. B., 1984, Evaluating the groundwater resources of the High Plains of Texas: Austin, Texas, Department of Water Resources Report 288, volumes 1 to 3.

Koltermann, C. E., and Gorelick, S. M., 1996, Heterogeneity in sedimentary deposits: a review of structure-imitating, process-imitating, and descriptive approaches: Water Resources Research, v. 32, no. 9, p. 2617–2658.

Larkin, T. J., and Bomar, G. W., 1983, Climatic atlas of Texas: Austin, Texas, Department of Water Resources, Report LP-192, 151 p.

Luckey, R. R., 1984, The High Plains regional aquifer—flow system simulation of the Central and Northern High Plains, *in* Whetstone, G. A., ed., Proceedings, Ogallala Aquifer Symposium II: Texas Tech University Water Resources Center, p. 48–66.

Luckey, R. R., and Becker, M. F., 1999, Hydrogeology, water use, and simulation of flow in the High Plains aquifer in northwestern Oklahoma, southeastern Colorado, southwestern Kansas, northeastern New Mexico, and northwestern Texas: U.S. Geological Survey Water-Resources Investigations Report 99-4104, 68 p.

Luckey, R. R., Gutentag, E. D., Heimes, F. J., and Weeks, J. B., 1986, Digital simulation of groundwater flow in the High Plains aquifer in parts of Colorado, Kansas, Nebraska, New Mexico, Oklahoma, South Dakota, Texas, and Wyoming: U.S. Geological Survey Professional Paper 1400-D, 57 p.

Luckey, R. R., and Stephens, D. M., 1987, Effect of grid size on digital simulation of ground-water flow in the southern High Plains of Texas and New Mexico:

U.S. Geological Survey Water-Resources Investigations Report 87-4085, 32 p.

Mace, R. E., *in press*, Estimating transmissivity using specific-capacity data: The University of Texas at Austin, Bureau of Economic Geology Geological Circular.

Mace, R. E., and Dutton, A. R., 1998, Numerical modeling of groundwater flow in the Ogallala aquifer in Texas, *in* Castellanos, J.Z., Carrillo, J.J., and Yanez, C.H., eds., Memoria del Simposio Internacional de Aguas Subterraneas: Sociedad Mexicana de la Ciencia del Suelo, p. 98–109.

McAda, D. P., 1984, Projected water-level declines in the Ogallala aquifer in Lea County, New Mexico: U.S. Geological Survey Water-Resources Investigations Report 84-4062, 84 p.

McCuen, R. H., and Snyder, W. M., 1986, Hydrologic modeling: statistical methods and applications: Englewood Cliffs, New Jersey, Prentice Hall, 568 p.

Mullican, W. F., III, 1995,. A technical review of the Canadian River Municipal Water Authority application for permit to transport water from Roberts and Hutchinson Counties: Panhandle Ground Water Conservation District, variously paginated.

Mullican, W. F., III, Johns, N. D., and Fryar, A. E., 1997, Playas and recharge of the Ogallala aquifer on the southern High Plains of Texas—an examination using numerical techniques: The University of Texas at Austin, Bureau of Economic Geology Report of Investigations No. 242, 72 p.

Nativ, Ronit, 1988, Hydrogeology and hydrochemistry of the Ogallala aquifer, Southern High Plains, Texas Panhandle and eastern New Mexico: The University of Texas at Austin, Bureau of Economic Geology Report of Investigations No. 177, 64 p.

Peckham, D. S., and Ashworth, J. B., 1993, The High Plains aquifer system of Texas; 1980 to 1990 overview and projections: Texas Water Development Board 341, p. 1–13.

Scanlon, B. R., and Goldsmith, R. S., 1997, Field study of spatial variability in unsaturated flow beneath and adjacent to playas: Water Resources Research, v. 33, no. 10, p. 2239–2252.

Senger, R. K., and Fogg, G. E., 1987, Regional underpressuring in deep brine aquifers, Palo Duro Basin, Texas. 1. Effects of hydrostratigraphy and topography: Water Resources Research v. 23, p. 1481–1493.

Seni, S. J., 1980, Sand-body geometry and depositional systems, Ogallala Formation, Texas: The University of Texas at Austin, Bureau of Economic Geology Report of Investigations No. 105, 36 p.

Simpkins, W. W., and Fogg, G. E., 1982, Preliminary modeling of groundwater flow near salt-dissolution zones, Texas Panhandle, *in* Gustavson, T. C., and others, Geology and geohydrology of the Palo Duro Basin, Texas: The University of Texas at Austin, Bureau of Economic Geology Geological Circular 82-7, p. 130–137.

Texas Water Development Board, 1999, Ground water database: http://www.twdb.state.tx.us.

Theis, C. V., 1963, Estimating the transmissivity of a water-table aquifer from the specific capacity of a well: U.S. Geological Survey Water-Supply Paper 1536-I, p. 332–336.

U.S. Department of Agriculture, 2000, USDA-NRCS National STATSGO database: http://www.ftw.nrcs.usda.gov/stat_data:html.

U.S. National Oceanic and Atmospheric Administration, 2000, National Weather Service data and products database: http://www.nws.noaa.gov/dataprod.html.

Weeks, J. B., and Gutentag, E. D., 1988, High Plains, Region 17, *in* Back, W., Rosenshein, J. S. and Seaber, P. R., eds., Hydrogeology: Geological Society of America, The Geology of North America, v. O-2, p. 157–164.

Wood, W. W., and Petraitis, M. J., 1984, Origin and distribution of carbon dioxide in the unsaturated zone of the Southern High Plains: Water Resources Research, v. 20, no. 9, p. 1193–1208.

This fluvial sandstone is in the Upper Triassic Bull Canyon Formation at Bull Canyon in eastern Guadalupe County. Sandstone bodies like this are common in the Bull Canyon Formation, and make it difficult to accept the interpretation that the unit is primarily of lacustrine origin. Instead, it represents a vast riverine floodplain deposit.

GEOPHYSICAL AND HYDROCHEMICAL ANALYSIS OF THE WHITE RIVER ALLUVIAL AQUIFER, CROSBY COUNTY, TEXAS

BARRY J. HIBBS

Department of Geological Sciences, California State University-Los Angeles, Los Angeles, CA 90032, bhibbs@calstatela.edu

Abstract.—A hydrogeologic and electrical resistivity study was conducted on White River alluvial aquifer in order to identify deposits that could provide water of sufficient quantity and quality to help augment municipal water supply for local communities. Fifty-four Schlumberger vertical electrical soundings were performed on topographically high terraces and recent alluvial deposits adjacent to White River, an ephemeral stream. Groundwater samples were collected for water quality analysis from a few livestock and irrigation wells in the alluvium. Water levels were measured in several water wells. These data, along with drillers logs and geotechnical logs, were used to constrain and interpret electrical resistivity sounding data. Test hole logs indicate that thicknesses of the alluvial deposits vary from about 11 to 25 m. Alluvial deposits are underlain in most areas by the Permian Quartermaster Group (Permian "redbeds"). Depth to groundwater is usually less than 4.5 m in recent deposits near White River, and usually exceeds 7.5 m on topographically high terraces at distances from the river. Groundwater quality analyses indicate that total dissolved solids (TDS) exceed 1,500 mg/L at low lying areas near the river and are most often less than 1,100 TDS on high terrace deposits. Electrical resistivity sounding curves were processed and correlated to these ground-truth data. The electrical resistivity sounding data also indicate that groundwater is relatively poor quality at low-lying areas near the river, and is relatively good quality on high terraces. Different ranges of TDS are likely due to evaporative concentration of groundwater at low-lying areas and less or negligible evaporation of groundwater on high terraces where depth to groundwater is greater.

INTRODUCTION

Statement of the problem

White River Reservoir is located in Crosby County, Texas, approximately 110 km southeast of Lubbock (Fig.1). The region was profoundly affected by the 1994 to 1996 drought. As of late April 1996, the reservoir contained 10,250,400 m^3 of water, or 26 percent of its 39,287,000 m^3 of storage. White River Reservoir is the primary water supply for the communities of Crosbyton, Post, Ralls, and Spur, Texas. In 1996, reservoir managers were investigating alternative ways to supplement the lake supply, including the addition of groundwater to the reservoir distribution system. Hydrogeologic investigations were conducted on the White River alluvial aquifer to identify likely areas of groundwater occurrence that meet both quality and quantity requirements. Methods of investigation included standard hydrogeologic techniques, surface geophysical surveys, and collection and analysis of water quality data.

Hydrogeologic characteristics

The study area is located along the bottomlands below the escarpment that separates the "caprock" region from the erosional bottomlands. The pre-erosional section in the study area include the basal Quartermaster Formation, which consists of shale, sandstone, gypsum, and dolomite, commonly referred to as Permian "redbeds" by locals in the study area (Fig. 2). Above the Quartermaster Formation is the Dockum Group, a lake-deposited mudstone and fluvial sandstone of Triassic age. Dockum Group deposits consist of interbedded sandstones, clay, shale, and conglomerate. Capping the Dockum is the Tertiary-age Ogallala Formation, a fluvial deposit consisting of interbedded sand, silt, clay, and gravel.

After the caprock was eroded away, the Quartermaster Formation was exposed along the bottomlands. Fluviatile terrace

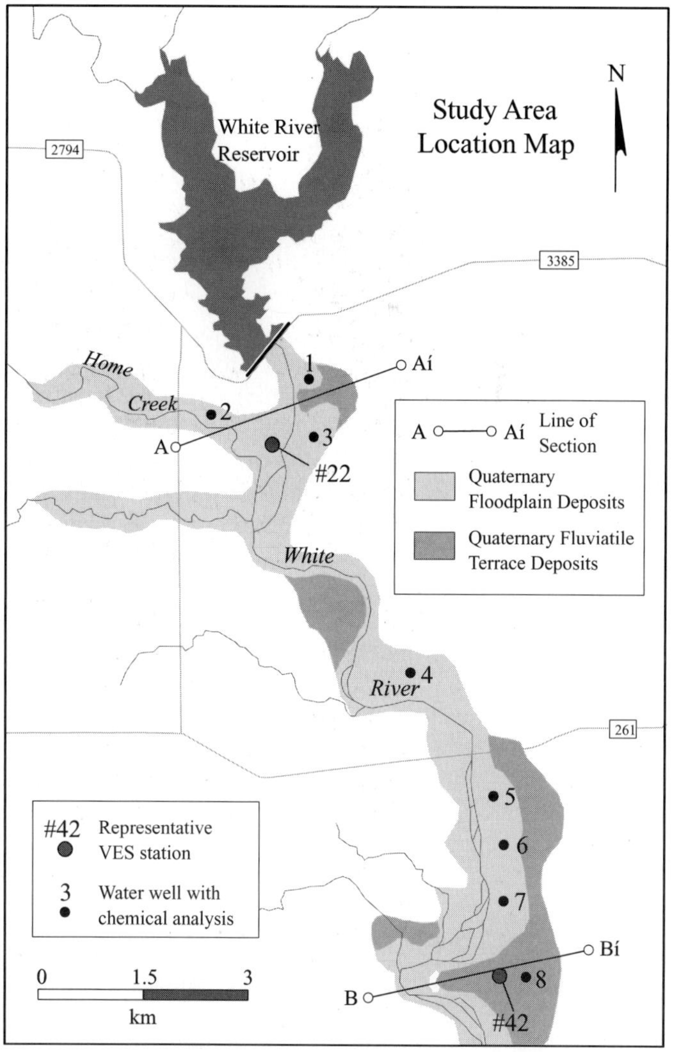

FIGURE 1. Study area location map showing the White River alluvium, lines of section for hydrogeologic cross sections, and water wells that were sampled for water quality analysis.

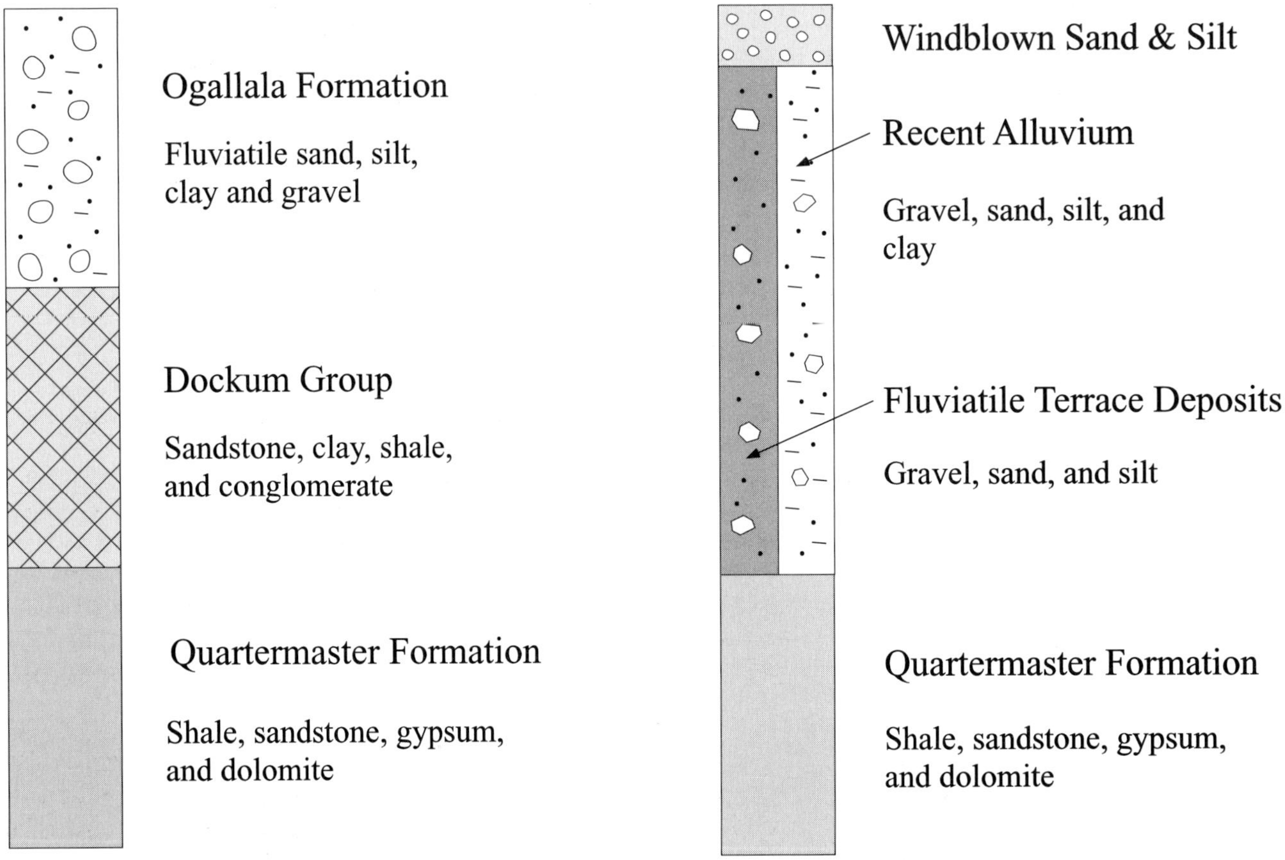

FIGURE 2. Pre-erosional and erosional stratigraphic sections for the study area.

deposits and recent alluvial deposits began to form along drainages of the ancestral and modern White River. The Quaternary-age terrace deposits are coarse textured, consisting mostly of gravel, sand and silt (Fig. 2). The recent alluvium is coarse to fine textured, consisting of interbedded gravels, sands, silts, and clays. In some areas, windblown sand and silt mantle the alluvial terraces and remnant deposits of the Quartermaster Formation.

White River is mostly an ephemeral channel above and below White River Reservoir. In some areas, the water table intersects the streambed of White River where marshy areas form. Flow in the White River channel does not persist for more than a few tens of meters below these marshes.

The principal water supply aquifer for local residents is the White River alluvial aquifer. The population is approximately 25 people who engage in irrigated agriculture and cattle ranching. Several domestic and livestock watering wells supply the needs of the local population and livestock industry. A few irrigation wells also exist within the study area.

METHODS OF INVESTIGATION

The following steps were taken to determine possible sources of potable water in the White River alluvial aquifer.

* Test hole drilling and water level measurements
* Surface electrical resistivity investigations
* Groundwater sampling and analysis

Test hole drilling and water level measurements

Test holes were drilled to determine depth to the basal Permian Quartermaster deposits, to determine texture and mineralogy of alluvial deposits, and to ascertain potential aquifer yields. Test holes also provided a means to calibrate and interpret earth resistivity data. The rotary drilling method was employed for drilling seven test holes. Test holes were supplemented by other drillers logs, which were available in the central files of the Texas Water Development Board. High quality geotechnical logs were also available from studies carried out when the White River Reservoir dam was designed and constructed (Freese, Nichols and Endress Consulting Engineers, 1962).

To characterize local groundwater gradients, water levels in monitoring wells were determined with electric lines and steel tape and chalk. Water-level measurements were measured in twenty-nine wells located in the White River alluvial aquifer. Measurements were collected after pumping wells were shut off and allowed to recover.

Surface electrical resistivity investigations

As an initial step in the investigation, a resistivity survey was selected as a convenient and economical means of potable water delineation in White River alluvial aquifer. Numerous studies have demonstrated the usefulness of surface electrical resistivity as a means to detect fresh and saline groundwaters (Lehr, 1969; Stollar and Roux, 1975; Cartwright et al., 1977; Urish, 1983; Murphy et al., 1988). The resistivity method employs two current electrodes, placed within the soil mantle, and the injection and migration of current from the positive to the negative electrode. The voltage difference, measured between two internally placed voltage electrodes, is a function of the intrinsic resistance to electrical current flow of the subsurface media and the ionic content of pore waters contained within the media. Because high levels of total dissolved solids increase the electrical conductance within geologic materials, resistivity readings often are much lower when taken above environments with high salinity. Alternatively, resistivity readings are high in aquifers with low total dissolved solids (TDS).

When performing a resistivity survey, increasing the distance between the two external current electrodes increases the depth to which subsurface materials are energized. Because of the layering within subsurface environments, an electrical current which penetrates two or more layer boundaries will create a resistivity reading which is an integrated weighted average of the individual layer resistivities. The resultant reading is known as the "apparent resistivity."

Apparent resistivity readings often are taken with one specific separation of the current electrodes at each designated resistivity station. This method is generally employed for rapid reconnaissance and is known as profiling. Alternatively, a vertical electrical sounding (VES) is generated if the current electrodes are symmetrically expanded at each station, such that apparent resistivity readings are obtained at each electrode spacing. This method produces a field curve which permits the quantitative interpretation of layer properties and depth.

The Schlumberger configuration was used in this investigation to collect VES data. In the Schlumberger configuration, the two inner potential electrodes have a spacing that remains fixed as the current electrodes are symmetrically expanded. The Schlumberger configuration is the preferable method for conducting VES as it permits easier resolution of subsurface lateral heterogeneities when soundings are quantitatively interpreted (Koefoed, 1979). Using the Schlumberger electrode configuration, fifty-four soundings were performed (Fig. 3). Soundings were obtained on the recent alluvium near White River and on the older alluvial terraces that stand above White River floodplain. Electrode spacings were extended from 1.5 m to 68 m and apparent resistivity readings were taken at eleven incrementally greater electrode spacings at each station. This provided a sufficient number of measurements to define VES curves. In performing resistivity surveys, care was taken to avoid potential interferences such as buried underground electrical lines and conduits, fences, and overhead power lines. A Sting Earth Resistivity meter was used to collect VES data.

Groundwater sampling and analysis

Groundwater sampling for chemical analysis was employed as a necessary step to resolve spatial and temporal variations in water quality and as a means to interpret resistivity sounding data. To accommodate sampling and to remove stagnant water, water wells were either evacuated of several casing volumes of water, or were sampled after they had been pumped continuously for long periods of time. Eight samples were collected in new polyethylene containers to be analyzed, respectively, for anions and cations (Fig. 1). These samples were iced down immediately after collection. Once in the laboratory the samples were filtered and refrigerated until chemical analysis was performed. Chemical analysis was performed by A&L Plains Agricultural Laboratories of Lubbock, Texas.

RESULTS

Test hole drilling and water level measurements

Test hole drilling indicated that alluvial thicknesses are as much as 20 m thick in the recent deposits along White River. Alluvial thicknesses are as much as 26 m thick in the older alluvial terraces. Permian redbeds were encountered at all drilling sites at the base of the alluvium. In most test holes, materials

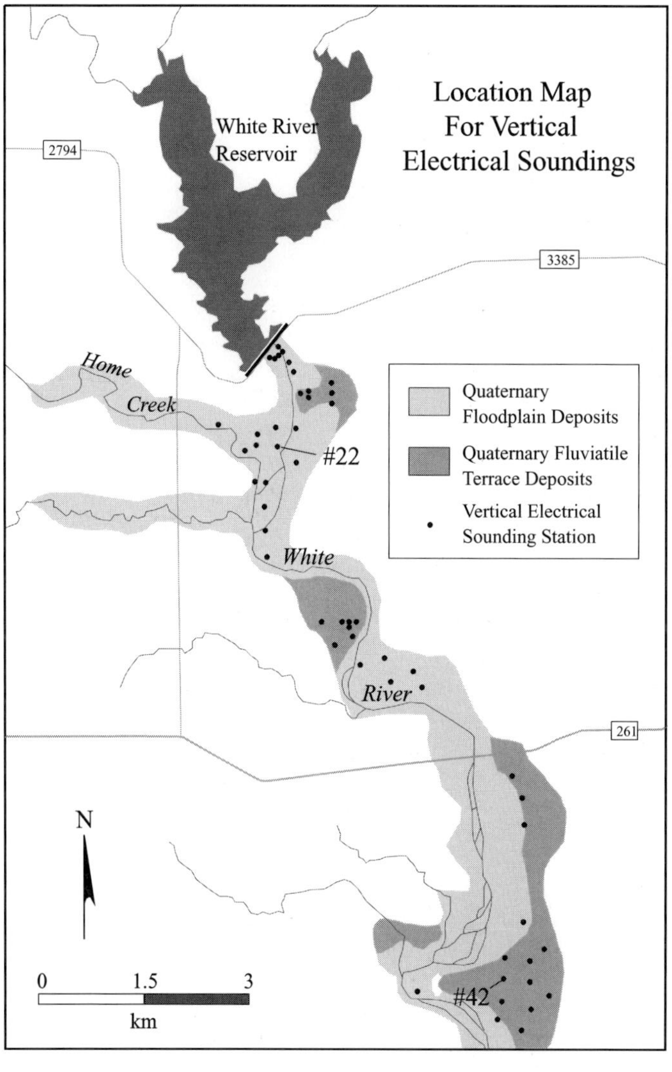

FIGURE 3. Location map for vertical electrical sounding stations.

encountered during drilling included thick sequences of sand and gravel, with intervening clay stringers and occasional clay beds (Fig. 4). At most locations, the lithologic sections were dominated by fine, medium, and coarse sand deposits with some gravel. Drilling sites near Home Creek indicated thicker and more abundant clay layers and lenses (Fig. 1 and Fig. 4). Test hole cuttings from the alluvial terraces included coarse sediments derived from weathering of Ogallala Formation and Dockum Group deposits. Recent alluvial deposits near White River had more clay stringers and lenses derived from weathering of Permian Quartermaster deposits.

Water level measurements in the alluvial terraces encountered the water table at depths that varied from about 6 to 11 m below land surface. The hydraulic gradient on the alluvial terraces is oriented perpendicular to sub-perpendicular to White River channel. Measurements in recent alluvial floodplain deposits varied from 0 to 7 m below land surface, with most values less than 4.5 m. The hydraulic gradient is oriented parallel to sub-parallel to the White River channel in the recent alluvial floodplain deposits. Groundwater flow in the recent alluvial deposits is down the axis of the floodplain.

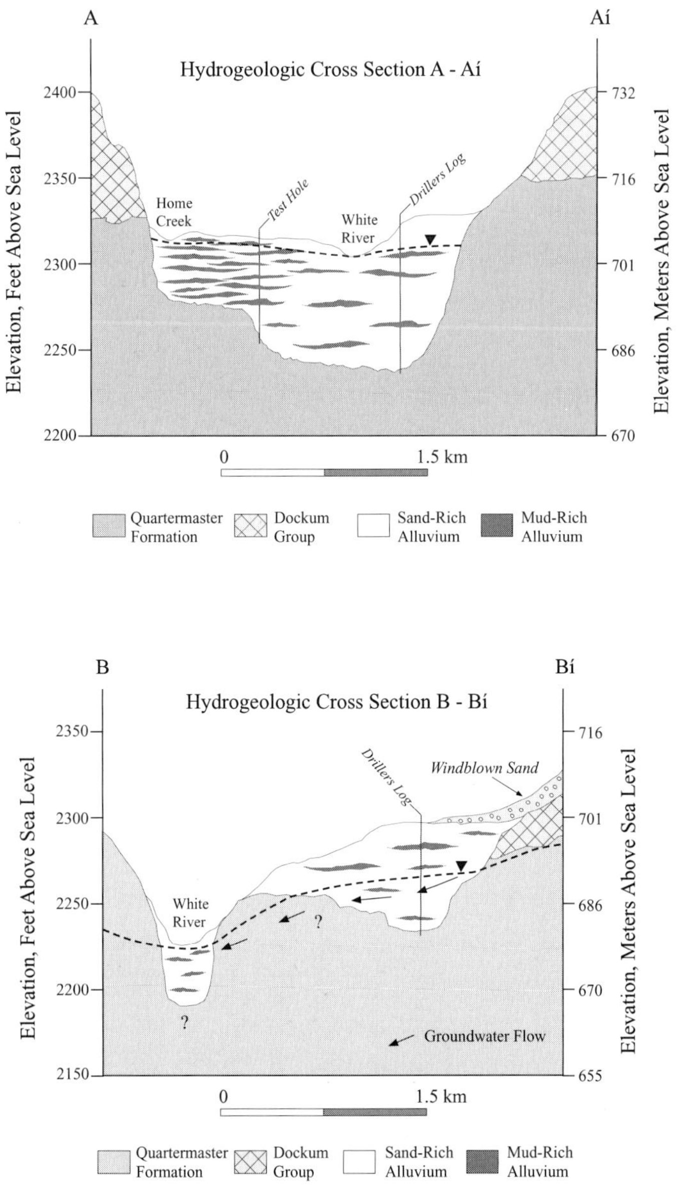

FIGURE 4. Hydrogeologic cross sections for the White River alluvial aquifer (lines of section shown in Figure 1).

Surface electrical resistivity investigations

VES data were interpreted with the method of steepest descent (Koefoed, 1979). Program DESCENT uses field VES data and a preliminary layered trial model supplied by the user as input. The program adjusts the trial model parameters to produce a theoretical VES curve which is compatible with the new model and within a limited root-mean-square error of the field VES curve. By minimizing error between field and theoretical curves, a subsurface interpretation of geoelectric layers is obtained.

To provide a means to interpret the data, VES data obtained next to the test holes were calibrated with lithologic and water level data. Drilling information and water level date were used to constrain the model depth-layer boundaries so that layer resistivities in the vicinity of the test holes could be determined. Inter-test hole models were then constrained by water level information extrapolated from the local potentiometric map. DESCENT was applied, with estimated layer resistivities input as unconstrained trial model data.

The program adjusts layer boundaries and resistivity values to provide an interpretation of geoelectric layers. Computer generated interpretations were considered to be reasonable when at most a 3% root-mean-square error existed between field and computed VES curves. Forty-nine of fifty-four VES curves were within the 3% root-mean-square error criterion.

Figure 5 presents representative VES data and interpretations obtained from program DESCENT. VES station 22 was located on the recent alluvial sediments near White River (Fig. 1 and Fig. 3). Water level measurements indicated that depth to groundwater is about 3.6 m near station 22. Drillers logs indicated that bedrock depth is about 18 m near the station. The resistivity sounding data suggests that the unsaturated zone is very dry and resistive at station 22 (resistivity equals 253 to 541 ohm-m). Below the water table, the resistivity is much less, decreasing from 46 ohm-m near the water table to 9.0 ohm-m near the base of the aquifer. The basal Permian redbeds are interpreted to have a resistivity of 23.1 ohm-m.

Drillers logs and water level measurements near station 42 indicated that depth to groundwater is about 9 m, and depth to bedrock is about 21 m. The resistivity sounding data indicate that the unsaturated zone is moderately to highly resistive at station 42 (50.6 to 315.5 ohm-m). In the saturated zone, the resistivity is moderately high, ranging from 168.2 ohm-m to 263.9 ohm-m. The basal formation is interpreted to have a resistivity of 37.3 ohm-m.

In general, the resistivity of the saturated zone at station 42 is more than an order of magnitude higher than the resistivity of the saturated zone at station 22. The sharply contrasting curve shapes support this interpretation. VES curve 42 ascends and

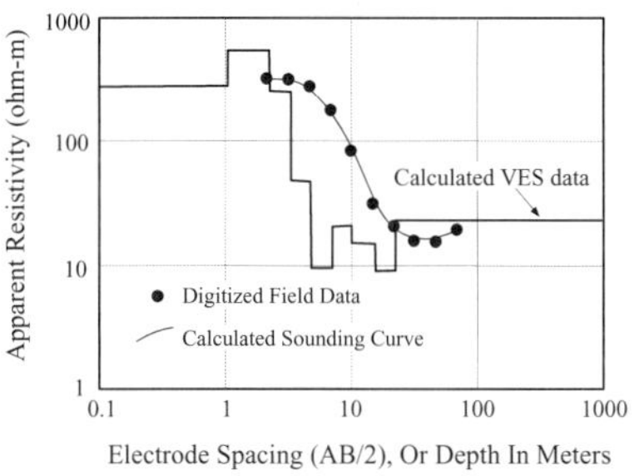

Calculated VES Data

Depth (m)		Resistivity (ohm-m)
1.0		253.2
2.3	*Vadose Zone*	541.1
3.3		253.1
4.9		46.21
7.1		9.5
10.5	*Aquifer*	20.3
15.4		14.7
19.8		9.0
Infinite	*Bedrock*	23.1

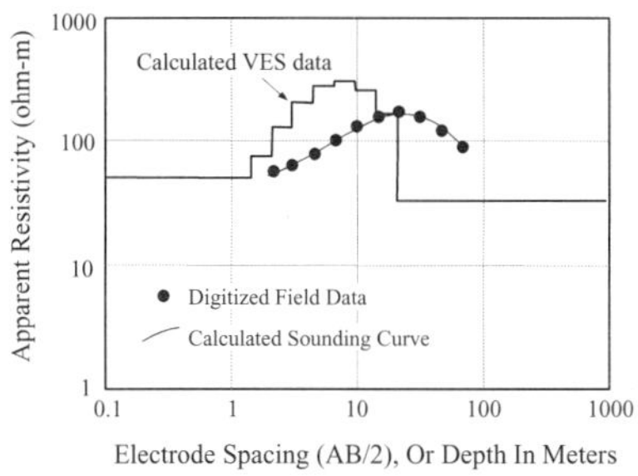

Calculated VES Data

Depth (m)		Resistivity (ohm-m)
1.4		50.6
2.0		76.8
2.9	*Vadose Zone*	129.6
4.4		208.7
6.4		287.9
9.4		315.5
13.8	*Aquifer*	263.9
20.3		168.2
Infinite	*Bedrock*	37.3

FIGURE 5. Representative vertical electrical sounding curves showing digitized field data, calculated sounding curves, and interpreted layer resistivities (see Figure 1 and Figure 3 for station locations).

then descends as the current electrode separation increases. VES curve 22 ascends slightly, descends, and then re-ascends as the electrode separation increases. This is because apparent resistivity values are higher in the resistive saturated zone materials at station 42 and lower in the relatively conductive saturated zone materials at station 22. Insofar as the lithologies of the saturated zone at stations 22 and 42 are dominated by sands and gravels with subordinate silts and clays, the contrasting resistivity values are attributed to substantially different total dissolved solids in the aquifer. On the basis of the geophysical assessment, the water quality is assumed to be poor at station 22 and relatively good at station 42 (e.g., TDS is relatively high at station 22 and relatively low at station 42). Similar interpretations of the other VES curves allowed us to develop the apriori water quality map for White River alluvial aquifer (Fig. 6). This map shows zones that are predicted to have good to very poor water quality. Good water quality zones are limited in extent.

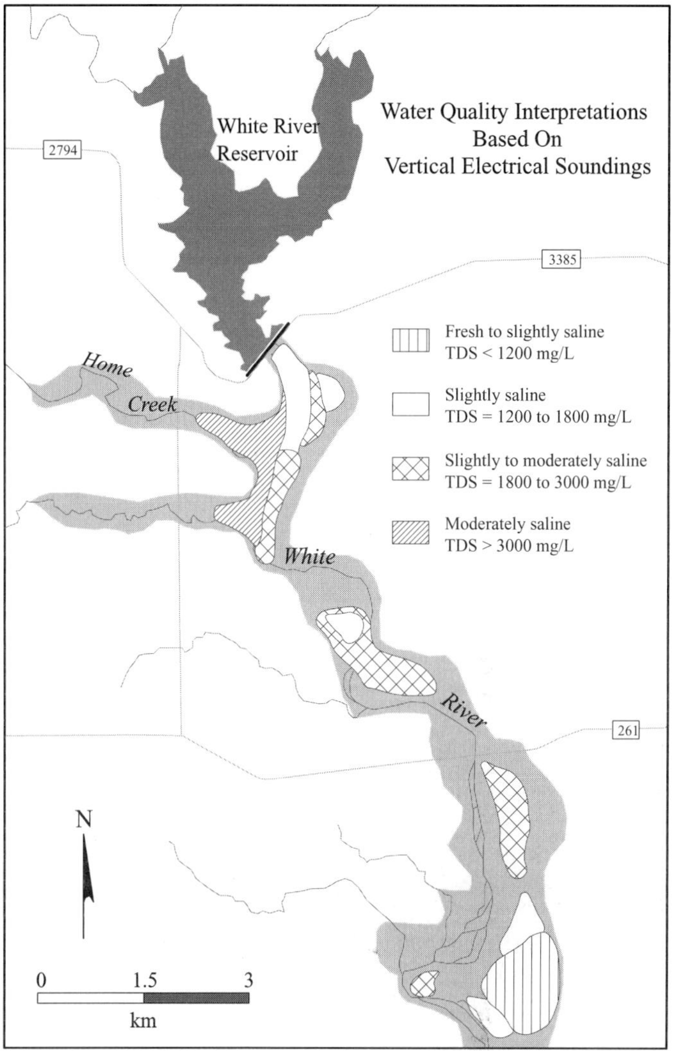

FIGURE 6. Water quality map showing aquifer salinities based on vertical electrical sounding interpretations.

Groundwater sampling and analysis

Water quality data collected at eight water wells are in excellent agreement with the water quality map interpreted by VES. Water quality is quite poor in most of the alluvial aquifer (Fig. 1 and Table 1). Chloride concentrations vary from 265 mg/L to 1900 mg/L, with all but one value above 400 mg/L. Sulfate concentrations vary from 179 to 775 mg/L, with all but one value over 300 mg/L. Sodium values vary from 190 mg/L to 775 mg/L. Only one sodium value is less than 300 mg/L. Many of these samples exceed the drinking water standards for chloride (250 mg/L), sulfate (250 mg/L), and sodium (250 mg/L). These are USEPA National Secondary Drinking-Water Standards, also known as recommended maximum concentration levels. These are maximum desirable concentrations for potable waters,

TABLE 1. Selected water quality parameters in White River alluvial aquifer.

Well Number	Chloride (mg/L)	Sulfate (mg/L)	Sodium (mg/L)	Nitrate (mg/L)	Dissolved Solids (mg/L)
1-Anderson	435	775	500	<4.43	1873
2-Harper	1900	508	1010	<4.43	3608
3-Slack	415	488	370	7	1601
4-Arnett	590	888	590	<4.43	2238
5-Short	405	300	400	4	1406
6-Gregory	560	588	340	11	1986
7-McArthur	1030	703	540	2	2906
8-Sparling	265	179	190	54	1002

although most humans can tolerate higher concentrations of these inorganic constituents.

Total dissolved solids are above 1000 mg/L in all wells. All but one value are greater than 1400 mg/L (Fig. 1 and Table 1). The recommended maximum concentration level for TDS of 500 mg/L is seldom achieved in many regions in West Texas. Even so, most of the groundwater in White River alluvial aquifer is too saline for domestic consumption.

Nitrate values vary from non-detect concentrations to concentrations as high as 54 mg/L (Table 1). The USEPA maximum contaminant level (MCL) for nitrate is 45 mg/L. The MCL is the concentration that must not be exceeded in drinking water due to health concerns. Only one sample exceeds this drinking standard (8-Sparling). This same sample was the most dilute with respect to standard inorganic constituents.

ORIGIN OF SALINITY IN WHITE RIVER ALLUVIAL AQUIFER

Groundwater quality in the White River alluvial aquifer, characterized by VES and water quality sampling, is slightly to moderately saline. The highest salinity concentrations are found in the recent alluvial deposits, where depth to groundwater is usually less than 4.5 m. The most dilute groundwater is found in the high alluvial terraces that flank the White River floodplain (e.g., 8-Sparling) (Fig. 1 and Table 1).

Recent alluvial deposits are interpreted to have relatively high salinity concentrations for the following reasons:

1. Shallow depth to groundwater intensifies salinization of groundwater due to evaporation and evaporative concentration of salts by phreatophytes.
2. Lithologies in the recent alluvium contain more lithic fragments from the Quartermaster Formation. These are rich in gypsum and other soluble minerals. The alluvial terraces contain smaller amounts of sediments from the Quartermaster Formation and greater amounts of lithic fragments from the Dockum Group and Ogallala Formation. These have less soluble mineralogies.
3. The direction of groundwater flow in the terraces is oriented sub-perpendicular to White River. Groundwater residence times are shorter in the terraces. Groundwater in the recent alluvial deposits moves axially down the White River Valley and have greater residence times.

When depth to groundwater is more than 9 m, discharge by evaporation is considered to be a small component of groundwater loss (Davis and DeWiest, 1966; Bouwer, 1978; Mifflin, 1988). Soil gas humidities nearly reach 100% at these depths (Stephens, 1996). Groundwater evaporation is slow at these humidities. At shallower depths, venting of soil gases by the atmosphere reduces soil humidity and accelerates groundwater evaporation. This has the tendency to concentrate salts in solution (Fig. 7).

Many phreatophytes also concentrate salts in groundwater by extracting dilute water from solution through root osmosis, leaving behind a more saline solution. Insofar as depth to groundwater is often less than 4.5 m in the recent alluvial deposits, it is likely that phreatophytes help to increase groundwater salinity in these low-lying deposits. Phreatophytes are less abundant, less important agents of groundwater salinization in the high-standing terrace deposits.

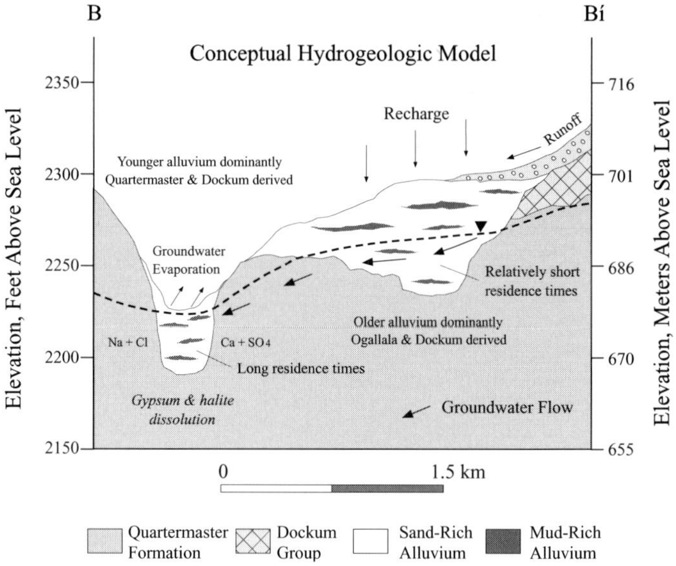

FIGURE 7. Conceptual hydrogeologic model showing mechanisms of salinization in the White River alluvial aquifer.

Lithic fragments from the Quartermaster Formation exert important controls on aquifer salinization. Quartermaster deposits contain gypsum and other soluble minerals that contribute to the dissolved load of the recent alluvial deposits. Soluble minerals from the Quartermaster Formation are not as abundant in the alluvial terrace deposits, and do not have the same controls on aquifer salinity.

Groundwater residence times in the alluvial terraces are also relatively short. Recharge waters move down gradient from the alluvial terraces into the recent alluvial deposits, where they mix with groundwaters already moving down the axis of White River floodplain (Fig. 7). There, groundwater becomes even more saline by evaporation and mineral dissolution as it travels down the axis of the floodplain. Occasional flooding in White River apparently does not dilute shallow groundwater significantly.

CONCLUSIONS

Due to the relatively poor quality groundwaters in White River alluvial aquifer, the Reservoir Watermaster was advised not to install high capacity wells in the alluvium to augment water supply. However, this study did identify relatively good quality groundwater in parts of the alluvial aquifer. The study also provided explanations for contrasting salinities in the aquifer.

REFERENCES

Bouwer, H., 1978, Groundwater Hydrology, McGraw-Hill Book Co., New York, 480 p.

Cartwright, K., Griffin, R.A., and Gilkerson, R.H., 1977, Migration of landfill leachate through glacial tills, Ground Water, v. 15, no. 4, p. 294-305.

Davis, S.N., and DeWeist, R.J.M., 1966, Hydrogeology, John Wiley & Sons, Inc., New York, 463 p.

Freese, Nichols and Endress Consulting Engineers, 1962, White River Municipal Water District, Contract I, White River Dam, Blueprints and Geotechnical Plans, various pagination.

Koefoed, O., 1979, Geosounding Principles 1, Resistivity Sounding Measurements, Elsevier Book Co., New York, 276 p.

Lehr, J.H., 1969, A study of groundwater contamination due to saline water disposal in the Morrow County oil fields, Office of Water Resources Research, United States Department of the Interior, 78 p.

Mifflin, M.D., 1988, Region 5, Great Basin, in Hydrogeology, the Geology of North America, Geological Society of America DNAG volume, v.0-2, p. 69-78.

Murphy, E.C., Kehew, A.E., Groenewold, G.H., and Beal, W.A., 1988, Leachate generated by an oil-and-gas brine pond site in North Dakota, Ground Water, v. 26, no. 1, p. 31-38.

Stephens, D.B., 1996, Vadose Zone Hydrology, Lewis Publishers, Boca Raton, 347 p.

Stollar, R.L., and Roux, P., 1975, Earth resistivity surveys, a method for defining ground water contamination, Ground Water, v. 13, no. 2, p. 145-150.

Urish, D.W., 1983, The practical application of surface electrical resistivity to detection of ground water pollution, Ground Water, v. 21, no. 2, p. 144-152.

In east-central New Mexico, the youngest Mesozoic strata are erosional outliers of the Cenomanian-Turonian Greenhorn Formation. This is an outstanding and very fossiliferous Greenhorn outcrop in the Bonita fault zone southeast of Tucumcari.

THE ENDEE PIPELINE LEAK — IMPLICATIONS FOR TREATING GROUND-WATER CONTAMINATION BENEATH THE GREAT PLAINS, EASTERN NEW MEXICO

WILLIAM J. STONE
Los Alamos National Laboratory, P.O. Box 1663, MS-F649, Los Alamos, NM 87545 wstone@lanl.gov

Abstract.—In November 1992, a major petroleum pipeline between Amarillo, TX and Albuquerque, NM ruptured, spraying thousands of gallons of jet fuel onto the ground at a site 27 mi east of Tucumcari, NM. The so-called "Endee site" is characterized by Triassic redbeds (Chinle Formation) overlain by Quaternary eolian sand. Monitoring wells installed at the site revealed up to 8 ft of free product floating on a body of perched ground water within the eolian cover sand, approximately 16 to 36 ft below the site. Although yields are low and quality is poor, the redbeds are the major source of water for area ranches. However, this ground water is vulnerable to contamination due to its shallow depth. The overlying eolian sand is porous and readily takes up spills. Such contamination may first show up in local perched water bodies in the sand, where it is easier to clean up than if it reaches the redbeds, which are fairly tight. Where perched saturation is present, it serves as a buffer for the deeper groundwater, but clean-up should be prompt before contamination reaches the regional system.

INTRODUCTION

A major high-pressure petroleum pipeline passes through the Tucumcari area. It was built in 1958 and extends from Amarillo, Texas to Albuquerque, New Mexico. In the east-central portion of Quay County it parallels State Route 392 on the south side. Different products flow through the line at different times.

At 5:00 AM on the morning of 15 November 1992 (a Sunday) the pipeline company in Amarillo detected a pressure drop in the line. The leak was located by 7:20 AM and block valves straddling the portion of the pipeline where the rupture occurred (spaced every10 mi) were closed by 7:35 AM, halting flow through the pipeline. It is estimated that 2,088 barrels (87,695 gal) were released. By 9:40 PM the leak was repaired and flow was resumed.

The 34-yr-old, 6-in. pipeline had ruptured approximately 27 mi east of Tucumcari, spraying jet fuel high into the air. More specifically, the leak was located at a point 7 mi north and 5.4 mi east of San Jon, New Mexico. Although it is 4.4 mi east of the village of Porter, the pipeline operator assigned the name "Endee" to the leak, after a village farther to the southeast. The leak was also detected on-site, as the resulting fountain of fuel was spectacular. It was said to be clearly visible and audible from both the road and a nearby ranch house.

Because the spill occurred on a Sunday, it was reported to the New Mexico Environment Department (NMED) emergency hotline. First thing Monday morning, the incident was reported to the Chief of the Ground Water Protection and Remediation Bureau, in keeping with state regulations. Both the state and the pipeline operator's consultant initiated investigations. I was with NMED's Ground-Water Protection and Remediation Bureau (Albuquerque) and overseeing this case at the time.

SITE CHARACTERIZATION

Steps were taken to characterize the leak site. This included defining the local geology, the area hydrology, as well as the nature and extent of the contamination. Of particular concern was the threat of contamination posed by the leak to ground water supplying two adjacent ranch houses, one located ¼ mi east of the leak and the other 1 mi to the northeast.

Geology

The Endee site lies in the Great Plains physiographic province. The area is characterized by a gently rolling surface that is drained toward the northeast. The site lies on the northern slope of a branch of Rana Canyon, a small, ephemeral tributary of the Canadian River.

The surface geology of this portion of Quay County is characterized by Triassic redbeds (Chinle Formation) overlain by Quaternary eolian sand (Berkstresser and Mourant, 1966; Anonymous, 1982). The Chinle consists mainly of red shale and red siltstone, but also includes sandstone, clay and limestone. The thickness of the redbeds is on the order of 250 ft at the site. The eolian sand is very fine to medium grained and uncemented. Thickness of the dune sand is <50 ft throughout the area. Hand augering for samples of the sand at the site revealed the presence of caliche nodules at depths of approximately 4 ft and an interbedded layer of light green clay (at least 19 in. thick) at a depth of 5 ft.

Hydrology

The regional water table occurs at a shallow depth in the Chinle redbeds (Berkstresser and Mourant, 1966). Regional ground water flows northeasterly toward the Canadian River, where it discharges. Perched ground water has been reported in the San Jon and Porter areas (Berkstresser and Mourant, 1966). The hydraulic conductivity of the redbeds is low in contrast to that of the sand. As more suitable aquifers are not available, the Chinle Formation is an important source of stock and domestic water for area ranches. However, the water is of poor quality: total-dissolved-solids often exceed 1000 ppm. In most of the area the water is not used for drinking or cooking. In some places, it is not even usable for stock. The well at the closest ranch house is reportedly 80 ft deep and yields poor quality water from the redbeds at a rate of 8-9 gpm.

To define the nature and extent of the contamination, six monitoring wells were installed in and around the impacted area (Fig. 1). More specifically, five wells were located within the area of the surface plume (area of discolored soil) and one was located outside of that to the west. These wells revealed that there is a shallow body of saturation (16 – 36 ft below ground surface). This water appears to be perched on a well-cemented pebbly zone near the top of the redbeds (Stone, 1999). Water-level measurements suggest flow is southeasterly toward local drainage (Fig. 1).

CONTAMINANT PLUMES

The initial site visit revealed that the southeasterly wind at the time of the leak had spread the fuel over an elongate area, 600 ft long and 15 –150 ft wide. The maximum width occurred next to the source of the leak and the minimum occurred at the toe of the surface plume.

The location of the leak on a valley margin facilitates the movement of contaminated sand and water downslope during runoff events. Such movement is enhanced by the fact that much of the porosity of the sand is taken up by the fuel. Before the leak, most precipitation would have soaked into the sand; after the leak, most runs off. During one heavy rainstorm, fuel-contaminated sand and water flowed as far as the yard of the closest ranch house (Stone, 1999).

Characterization of the subsurface plume was made possible by the six monitoring wells. Three of the wells located within the surface plume encountered up to 8 ft of free product floating on the perched water. The other two wells detected major fuel components (benzene, toluene, ethylbenzene and xylene or BTEX) at 11–311 ug/L in the perched ground water, but no free product. The extent of the ground-water plume was found to generally mirror that of the surface plume.

REMEDIATION

Three remediation targets were recognized by the state: the free product, the potential spread of contaminated soil and water by runoff and the in-place contaminated soil. Thus, a three-phase remediation plan was implemented.

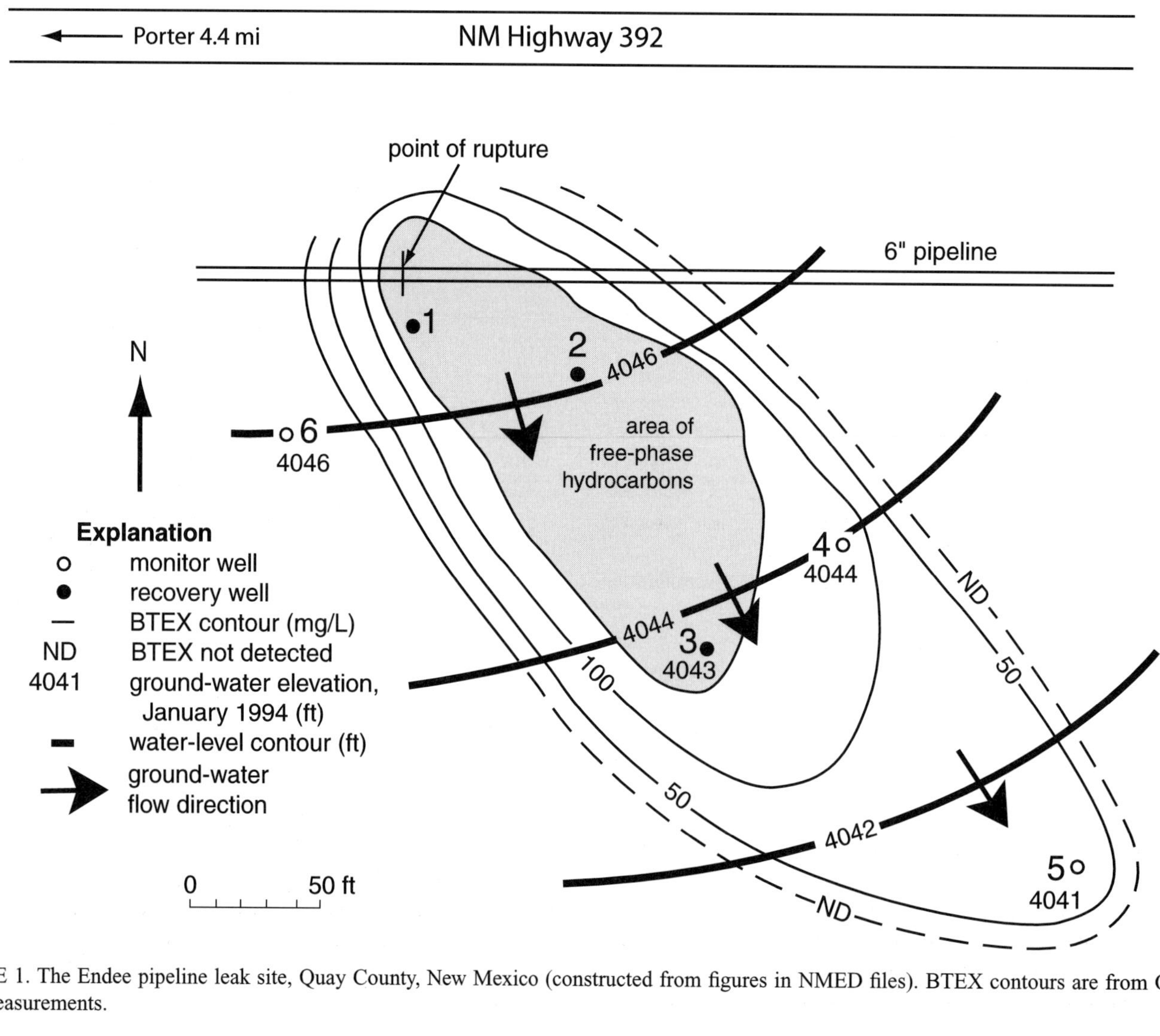

FIGURE 1. The Endee pipeline leak site, Quay County, New Mexico (constructed from figures in NMED files). BTEX contours are from October, 1996 measurements.

TABLE 1. Early product-recovery results at the Endee pipeline leak site, Quay County, New Mexico (compiled from data in NMED files).

Date	Removed From Tank (gal) Product	Water	Cumulative Product Recovered (gal)
1993			
05/25	400	0	400
08/04	300	0	700
10/26	300	0	1000
12/22	650[1]	0	1650
12/24	200	0	1850
12/29	300	0	2150
1994			
01/17	500	0	2650
02/07	250[2]	0	2900
03/02	200	200	3100
03/15	80	170	3180
03/29	120	100	3300
04/18	225	0	3525
05/02	160	0	3685
06/07	200	300	3885
06/29	250	250	4135
09/22	300	100	4435
10/03	350	25	4485
12/20	250	100	5035
1995			
01/12	380	0	5415

[1] pumps replaced in wells 1 and 2 on 12/13/93 (see Figure 1)
[2] pump installed in well 3 on 02/07/94

Product remediation was undertaken by means of a solar-powered recovery system. This involved outfitting initially two and eventually all three of the wells that penetrated free-phase hydrocarbons with pumps (Fig. 1). The pumps are connected to a storage tank that is emptied periodically. Recovered product is taken to a plant in Tucumcari for oil/water separation and appropriate reclamation. Until March 1994, only free product was collected by the recovery system (Table 1). Thereafter, quantities of water, sometimes exceeding those of product, were also produced. By 1995, more than 5,000 gal of fuel had been removed from the ground. Product-recovery efforts continue today.

The spread of contamination by runoff was minimized by installing impediments to flow and sediment transport along the slope below the leak. These included earthen diversion berms and fabric silt dikes.

The leak contaminated some 17,000 yds^3 of cropland. Natural attenuation was proposed by the pipeline operator as an approach to cleaning up such a large volume of in-place contaminated soil. This has been rejected by NMED. The final plan for soil remediation has not yet been determined.

Organic contaminants were never found in periodic analyses of water from the two ranch house wells. However, the rancher on whose land the leak occurred suffered a loss of production from the soil that was directly impacted and sued the pipeline operator for damages. He was awarded a yearly payment until the site is cleaned up. Furthermore, it was agreed that a new well would be provided if water in the current one is ever found to be contaminated.

IMPLICATIONS FOR OTHER SITES

Aging pipelines are not the only potential source of ground-water contamination on the Great Plains of eastern New Mexico. Accidental spills from tanker trucks and railroad tank cars, as well as seepage from feedlots, dairies, landfills and septic tanks all pose threats.

Findings at the Endee site have several implications for dealing with ground-water contamination in similar settings on the Great Plains of eastern New Mexico:

1. The regional ground water system in the redbeds of the Chinle Formation is vulnerable to contamination due to its shallow depth.
2. Although the redbeds yield small amounts of poor quality water, that water is all that is available in some areas and must be protected for local users.
3. The overlying eolian sand is very porous and permeable. Liquid spilled or leaked in areas where it lies at the surface is quickly taken up.
4. The first ground water such contamination encounters may be that perched in the eolian sand.
5. Unless remediated, contamination in the perched water may move down to the regional system.
6. It is easier to recover contaminated water from the sand than from the underlying redbeds, which are fairly tight.

In other words, where present, the perched water serves as a buffer for the regional system. However, protecting ground-water supplies requires prompt and effective action to clean up spills before contamination leaves the soil and perched saturation.

ACKNOWLEDGMENTS

Michael Space (now with Terradigm, Inc.) took the state's initial soil samples at the site. Charles Ferguson, (NMED/Tucumcari and Clovis Field Offices) was a valuable on-site resource as I was based in Albuquerque. William McDonald (now with NMED's Hazardous Materials Bureau) assisted me on various site visits. The cooperation of Jack Johnson (landowner), in allowing us to sample his well and that of his son Jimmy, as well as keeping me informed of activities at the site during my involvement with the case, is gratefully acknowledged. Mike Space and Patrick Longmire (Los Alamos National Laboratory) provided useful peer reviews of the paper. Annie Loweree and Randi Moore (Los Alamos National Laboratory) prepared the figure.

REFERENCES

Anonymous, 1982, New Mexico highway geologic map: New Mexico Geological Society

Berkstresser, C.F., Jr., and Mourant, W.A., 1966, Ground-water resources and geology of Quay County, New Mexico: New Mexico Bureau of Mines and Mineral Resources, Ground-Water Report 9, 115 p.

Stone, W.J., 1999, Hydrogeology in practice – a guide to characterizing ground-water systems: Prentice Hall, Upper Saddle River, New Jersey, 248 p.

Red Peak southeast of Tucumcari is an erosional outlier developed in the Upper Triassic Bull Canyon and Redonda formations of the Chinle Group. Fossil vertebrates have been collected here, and copper was also mined here. (The fossil collecting was much better, by all accounts, than the copper mining.)

LONG-TERM MONITORING OF THE GEOCHEMISTRY OF SURFACE WATER AND STREAM-SEDIMENT SAMPLES FROM THE UPPER PECOS RIVER, FROM THE SOUTHERN PECOS WILDERNESS TO BRANTLEY DAM, NORTH OF CARLSBAD, EASTERN NEW MEXICO: YEAR NINE

VIRGINIA T. MCLEMORE, LYNN A. BRANDVOLD AND ELIZABETH A. CAREY
New Mexico Bureau of Geology and Mineral Resources, New Mexico Institute of Mining and Technology, Socorro, New Mexico, USA 87801

Abstract.—Since reclamation of the Pecos mine waste piles, area roads and campgrounds, and the Alamitos Canyon mill area began in 1990-1991, there has been continued monitoring of base metals and trace element concentrations in stream sediments along the Pecos River, extending from the upper Pecos River to below Brantley Dam, north of Carlsbad. In 1992, 1996, and 2000, water and stream sediments were sampled and analyzed for the entire reach. Between 1992-1996, the upper reach of the river between the Pecos Wilderness and Villanueva was sampled on eight occasions as part of a multi-disciplinary study. Five areas of concern included the Pecos mine waste pile (a volcanogenic massive-sulfide deposit containing Pb, Zn, Cu, Ag, and Au), the Alamitos Canyon mill tailings (El Molino site 18 km south of the mine), roads and campgrounds north of Terrero (mine waste material from the Pecos mine waste pile was used as fill), and the Lisboa Springs Fish Hatchery (where several fish kills in 1991-1995 have been attributed to drainage from the mine site and contaminated campground sites). Geochemical analyses of surface-water samples indicate that drainage from the Pecos mine is not significantly affecting the composition of the surface water in the area, except in the immediate vicinity of the mine and mill site. Elevated concentrations of Cu, Pb, Zn, and Cd occurred in stream sediments below both the Pecos mine and Alamitos Canyon mill sites, before reclamation began. Collectively, multi-disciplinary studies suggest that Cu, Pb, Zn, and other metals were eroded and leached from the Pecos mine waste pile and the tailings piles in Alamitos Canyon. These geochemical trends confirm a decrease in concentrations with time since reclamation began, especially in the immediate vicinity of the Pecos mine (where Cu levels decreased from 310 to 92 ppm; Cd from 17 to 4.7 ppm; Pb from 300 to 160 ppm; and Zn from 3100 to 2080) and below the confluence of Alamitos Creek (Cu decreased from 270 to 30 ppm; Pb from 590 to 93 ppm; and Zn from 990 to 287 ppm). The overall metal concentrations dramatically decrease in stream sediments below Pecos Village, mostly due to dilution of sediment derived from the red bed sedimentary units.

INTRODUCTION

Water quality of the Pecos River has regional and international concerns as the population in eastern New Mexico increases (Fig. 1). The Pecos mine and Alamitos Canyon mill in the upper Pecos River area had been identified by state and federal agencies as point sources for metals contamination for Pb, Zn, Cu, Al, Se, Cd, and Cr (Fig. 2, 3; Johnson and Deeds, 1995a, b). Material from the mine waste pile had been used as fill in the campgrounds and roads north of Terrero and these areas had been also identified as point sources for metal contamination. Furthermore, the Lisboa Springs Fish Hatchery (New Mexico Game and Fish Department) experienced several fish kills from March 1991 through 1995 that may be attributed to contamination from the Pecos mine. Reclamation of the campgrounds, roads, and Alamitos Canyon mill site began in the early 1990s and is nearly completed. Reclamation of the Pecos mine began in 1999 and is on going. The Lisboa Springs Fish Hatchery closed, temporarily, in 2000 because the fish were diagnosed with Whirling disease, which is not related to the metals contamination from the Pecos mine.

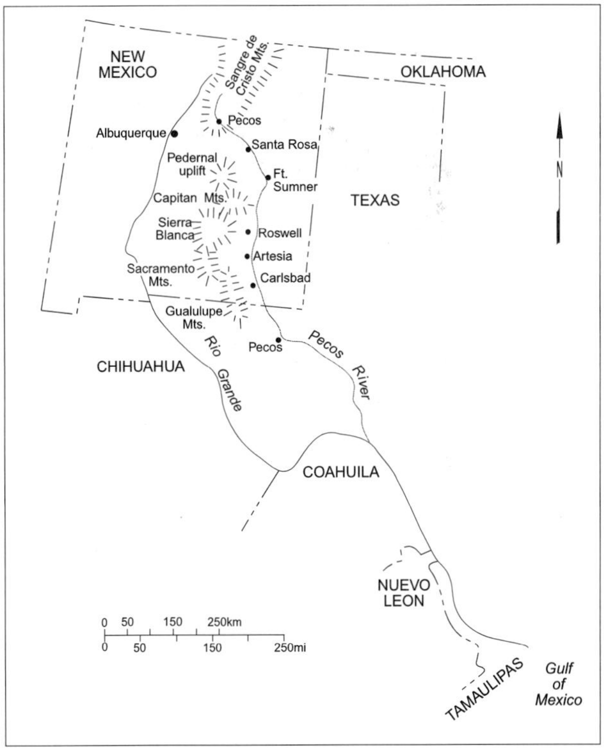

FIGURE 1. Pecos River drainage basin, eastern New Mexico and western Texas.

In September 1992, the New Mexico Bureau of Geology and Mineral Resources (NMBGMR) personnel sampled and analyzed surface water and stream sediments from the Pecos River at sites near the Pecos Wilderness boundary southward to below Brantley Lake, north of Carlsbad (McLemore et al., 1993). Subsequently, the U.S. Bureau of Reclamation, in cooperation with the New Mexico Institute of Mining and Technology (NMIMT) and NMBGMR, initiated a three-year multi-disciplinary study of the headwaters of the Pecos River (Pecos Wilderness boundary to Villanueva) in order to identify and prioritize point and non-point sources of contamination so that measures could be taken to protect water supplies for municipal, irrigation, and recreational use and to protect wildlife habitat (Brandvold et

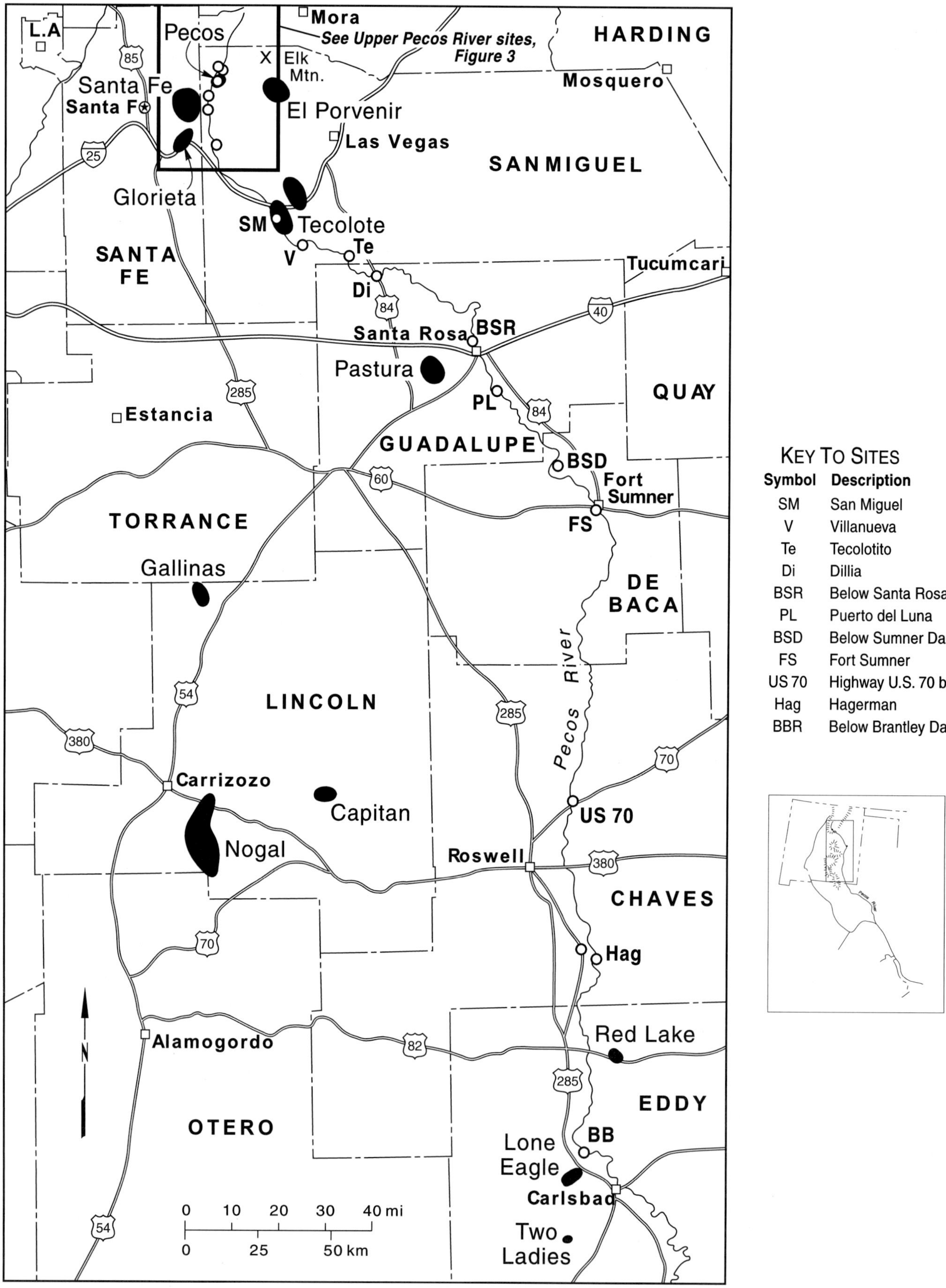

FIGURE 2. Location of sample sites and mining districts along the Pecos River, eastern New Mexico.

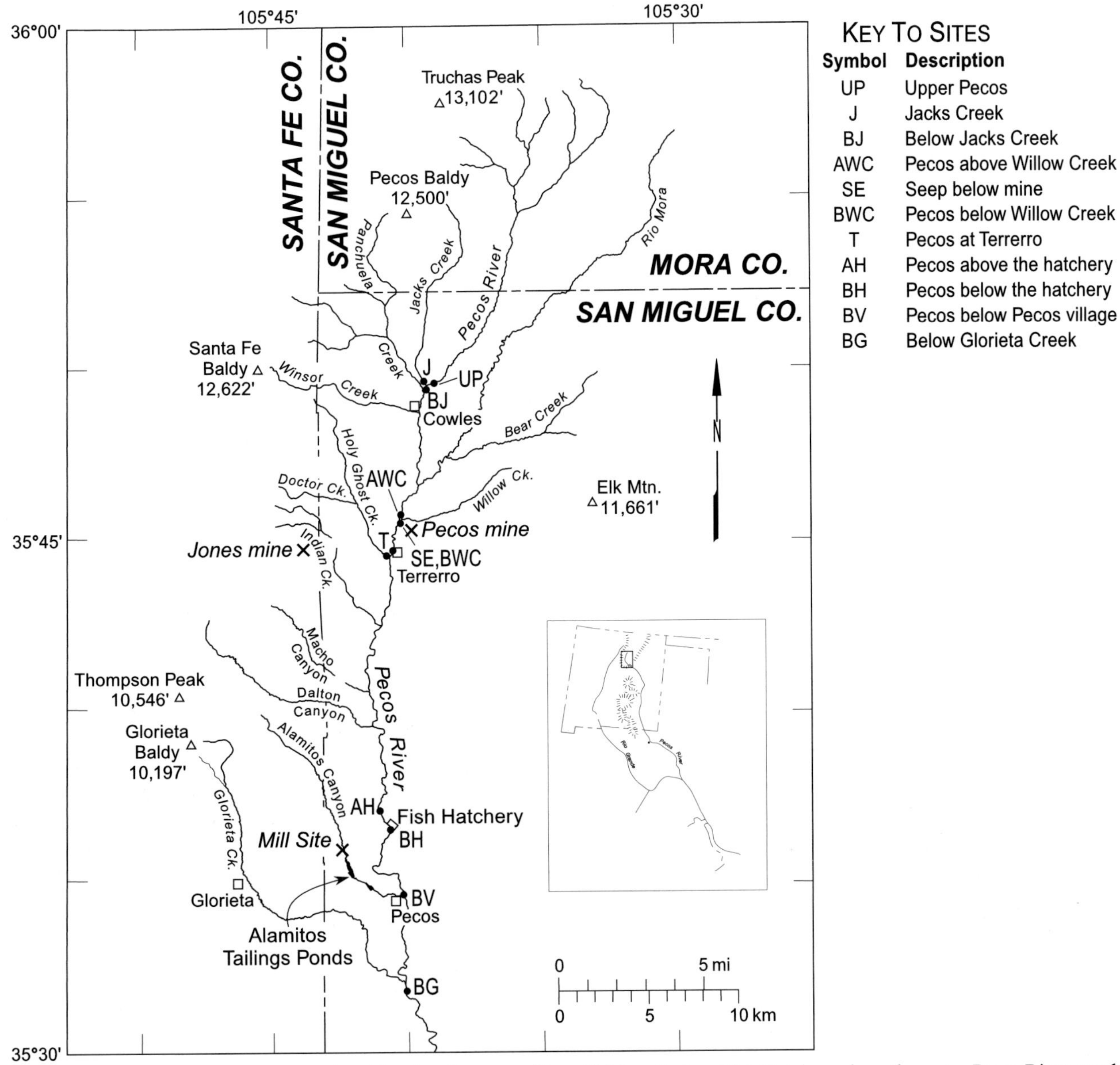

FIGURE 3. Location of sample sites, Pecos mine, Alamitos Canyon mill, and Lisboa Springs Fish Hatchery along the upper Pecos River, northeastern New Mexico.

al., 1995; Brandvold and McLemore, 1999; McLemore et al., 1995a, b; Popp et al., 1996). In September 1992, 1996, and June 2000, the NMBGMR sampled water and stream sediments from the entire Pecos River from the Pecos Wilderness southward to below Brantley Lake. This report presents a summary of the first nine sampling trips in 1992-1995 (McLemore et al., 1993; Popp et al., 1996) and presents new data from the sampling trips in 1996 and 2000 (Fig. 2, 3).

STUDY AREA

The study area covers the southern Pecos Wilderness to south of Brantley Lake and includes many tributaries and the main stem of the Pecos River. The headwaters of the Pecos River are in the southern Sangre de Cristo Mountains, with additional drainages from the Pedernal uplift, Capitan Mountains, Sierra Blanca, and Guadalupe Mountains (Fig. 1). Elevations range from 1000 to over 3600 m. The river flows southward through the towns and cities of Pecos, Santa Rosa, Fort Sumner, Roswell, Artesia, and Carlsbad into west Texas and finally enters the Rio Grande at the international border with Coahuila, Mexico (Fig. 1). The Pecos River flows through the Santa Fe National Forest and the Pecos National Historical Park (U.S. National Park Service) and four state parks. Water from the Pecos River is stored in three main reservoirs: Santa Rosa, Sumner, and Brantley Lake. Eastern New Mexico is predominantly a rural agricultural area, but the larger cities have some light industry. Petroleum and potash production is important in southeastern New Mexico near Artesia and Carlsbad. There is currently no metals mining within the Pecos drainage basin.

REGIONAL GEOLOGY

The geology of the Pecos River drainage basin is complex with rocks ranging in age from Proterozoic to recent. Lithologies are likewise diverse, ranging from metamorphic volcanic rocks to granites to syenites to shales, limestones, and sandstones (Anderson et al., 1994). Proterozoic rocks crop out along the upper Pecos River and several tributaries north of Pecos. The oldest rocks are mafic metamorphic and volcaniclastic rocks that comprise the Pecos greenstone belt. The most abundant Proterozoic rocks are the plutonic rocks that consist of granite, tonalite-trondjemite, gabbro, diabase, and ultramafic rocks. Overlying sedimentary rocks consist of Mississippian limestone, sandstone, and shale of the Arroyo Penasco Group (Espiritu Santo and Terrero Formations) and unconformably overlie the Proterozoic rocks. Pennsylvanian siltstones, sandstones, shales, thin coals, and limestones overlie the Mississippian and, locally Proterozoic rocks in the Pecos River area. The Pennsylvanian-Permian rocks consist of the Magdalena Group and the Sangre de Cristo Formation. Permian siltstones, limestones, and sandstones overlie the Pennsylvanian-Permian rocks and consist of the Yeso, San Andres, and Bernal Formations and Artesia Group. Triassic sandstones and shales crop out in the area south of Pecos and consist of the Chinle Group. Permian evaporites that comprise the Permian Basin section crop out south of Roswell. Quaternary rocks are found throughout the area and consist of Pleistocene to recent alluvial, terrace, and floodplain deposits.

Mining began in the late 1880s in the southern Sangre de Cristo Mountains (Harley, 1940; McLemore, 1995). The largest of these deposits is the volcanogenic massive-sulfide deposit containing Pb, Zn, Cu, Ag and Au at the Pecos mine in the Willow Creek district, which produced during 1902-1904, 1927-1939, and 1943-1944. The Pecos mine was the largest lead and zinc producer in New Mexico from 1927 to 1939 and was one of the top ten past producers of lead and zinc in New Mexico. Production from other deposits in the southern Sangre de Cristo Mountains is minor and insignificant (McLemore, 1995).

TABLE 1. Mineralogy of the ore deposits at the Pecos mine and vicinity, in order of perceived abundance (Stott, 1931; Krieger, 1932; Northrop, 1959; Fulp, 1982; Fulp and Rensaw, 1985; Popp et al., 1996). * Found on mine waste piles at Pecos mine (Popp et al., 1996).

Major Ore Minerals
*sphalerite ZnS
*galena PbS
*chalcopyrite CuFeS2
Minor Ore Occurrences
*bornite Cu5FeS4
tetrahedrite (Cu,Fe)12Sb4S13
native gold Au
argentite Ag2S
proustite Ag3AsS3
Oxidized Minerals
*malachite Cu2CO3(OH)2
*azurite Cu3(CO3)2(OH)2
*angelsite PbSO4
*smithsonite ZnCO3
*chalcocite Cu2S
cerussite PbCO3
*chalcanthite CuSO4 _ 5H2O
Selected Gangue Minerals
*pyrite FeS2
*pyrrhotite Fe1-xS
*quartz SiO2
*chlorite (complex silicate)
*actinolite Ca2(Mg,Fe)5Si8O22(OH)2
*tourmaline (complex silicate)
*epidote Ca2(Al,Fe)2(Al,Si3,O12)(OH)
*Fe and Mn oxides
*arsenopyrite FeAsS
*biotite K (Mg,Fe)3AlSi3O10(OH)2
*fluorite CaF2
*kaolinite Al2Si2O5(OH)4
*sericite KAl2(AlSi3O10)(OH)2
*calcite CaCO3
Additional Minerals in the Jones Hill, Dalton Canyon and Macho Canyon Areas
molybdenite MoS2
scheelite CaWO4

DESCRIPTION OF PECOS MINE WASTE PILE AND ALAMITOS CANYON TAILINGS PILE

The waste rock generated during mining activities was piled at the Pecos mine site and consists of several mine waste piles within an area of approximately 7.7 ha. Estimates of the volume of waste rock vary from 66,151 to 72,630 m^3 of material and it varies in thickness from 3.1 to 13.7 m. The waste pile is underlain by a layer of colluvial materials less than 5 m thick and consists of clay and sedimentary rock fragments (limestone, shale, sandstones; Johnson and Deeds, 1995 a, b). Mineralogically and chemically the mine waste pile is very heterogeneous (Tables 1, 2). The Pecos River flows less than 152 m west of the mine site and Willow Creek flows west along the northern edge of the mine waste pile.

Numerous discrete and diffuse seeps occur along the base of the mine waste pile. The number of seeps and amount of water flow varies according to storm events and snowmelt as well as yearly fluctuations in precipitation. During August 1994 many seeps were dry and the total amount of flow was lower than on previous visits in 1992 and 1993. White and brown precipitates or froths are present along the course of the seeps that drain into Willow Creek. X-ray diffraction and chemical analyses (Table 3; Popp et al., 1996) of the precipitates from the seeps indicated that they are a mixture of Fe, Cu, and Zn sulfates, and Fe and Al oxides and hydroxides including goslarite ($ZnSO_4 \cdot H_2O$). Chemical analyses of water samples from the seeps indicate a low pH and high total metals (Table 4; Berger et al., 2000). The main controls on pH of the water is controlled by reaction between the water from the mine waste piles and local bedrock (i.e. limestone) and mixing with water from Willow Creek (Berger et al., 2000). Neutral pH waters were collected from the Jones Hill mines (Table 4).

The ore was crushed at the mine and then transported by aerial tramway to the Alamitos Canyon mill (El Molino site) 18 km south

TABLE 2. Chemical analyses of waste rock at the Pecos mine in ppm (* from S. M. Stoller Corp., 1993; — no data).

Element	*Average	*Standard deviation	*Maximum	*Minimum	*Number of samples	NMBMMR sample of Pecos mine piles (Pecos)
Cu	1,840	1,960	7,140	40	25	11,810
Pb	9,380	10,400	33,100	41.8	25	238
Zn	17,330	30,600	150,000	204	25	31,350
As	25	23	101	1	20	—
Cd	68	136	607	0.65	20	7.5
Hg	0.65	0.81	2.49	0.033	10	<0.10
Ag	10	11	32.6	0	20	—
Al	13,740	9,880	27,000	23	5	—
Fe	52,650	35,120	129,150	10,960	12	5.3
Co	18	14	59.5	3.2	13	—
Cr	18	13	52.4	2.4	20	—
Ni	12	6	23.3	1.7	13	29

TABLE 3. Chemical analyses of sediment and precipitate from seeps along the base of the mine waste pile at the Pecos mine. Samples were digested using agua regia and analyzed by FAAS (NMBMMR Chemistry laboratory). (— no analyses)

Lab No.	Sample name	Date collected	Cu ppm	Pb ppm	Zn ppm	Cd ppm	Fe%
—	Seep 1 white	9/92	4,700	1,500	12,700	—	4.09
—	Seep 1 brown	9/92	1,000	600	12,000	—	4.9
360	Seep 1 (at pipe)	5/94	1,279	461	3,455	16	—
361	Seep 2 (at confluence with Willow Creek)	5/94	1,933	1,340	5,900	15	—
2523	Seep	6/93	1,630	490	38,000	31	—
—	Seep	7/7/93	680	310	70,000	40	—

TABLE 4. Chemical analyses of water samples from seeps at the Pecos and Jones Hill mines (in mg/L). Analyses by NMBMMR Chemical Laboratory. (— no analyses)

	SEEP 1 5/3/94	SEEP 2 5/3/94	SEEP 1 7/1/94	SEEP 2 7/1/94	UPPER JONES ADIT 5/3/94	LOWER JONES ADIT 5/3/94
pH	3.7	4.1	3.6	—	7.0	6.9
TDS	14,405	8,614	14,148	—	79	223
conductivity (umhos/cm)	11,000	7.7	10,300	—	139	312
HCO3	0	11	0	—	53	75
Cl	23	25	23	—	3.1	1.1
SO4	13,000	7,660	13,000	—	17	113
Na	11	25	9.8	—	1.1	5.4
K	13	11	5.1	—	2.1	2.5
Mg	950	475	800	—	7.2	14.5
Ca	380	400	310	—	14	40
Al	630	8.1	—	—	<1.0	1.1
Cd	16.5	8.6	15	0.06	<0.03	<0.03
Co	2.1	1.2	1.5	0.39	<0.03	<0.50
Cu	80	35	65	0.07	<0.10	0.14
Fe	1.2	0.38	2.7	0.24	0.13	0.7
Pb	1.3	0.04	0.8	<0.002	0.022	0.017
Mn	27.8	18.4	45	1	<0.10	<0.10
Hg (ppb)	1.5	<1.0	—		1.5	<1.0
Si	28	12	—	10	5.4	8.1
Zn	19,500	2,000	3,460	26	0.17	1.2

of the mine. The sulfide grains entering the mill were as large as 3 mm. A typical analysis of the mill feed was reported at 4.9% Pb, 15.4% Zn, 0.8% Cu, 113 ppm Ag, and 3.4 ppm Au (Bemis, 1932). Mill recoveries were good by the standard of the day although variable (Bemis, 1932). The rejected waste material was conveyed to the tailings pond in Alamitos Canyon and held in place by two earthen dams, 0.7 km and 1.5 km downstream of the mill site.

Today the tailings piles occupy approximately 14 ha in the bottom of Alamitos Canyon (Sidle et al., 1991) and have been covered with an impermeable cover. The depth varies from 0 to 20 m and averages 9-12 m. The tailings consist of predominantly sand size material (Table 5; Sidle et al., 1991). More silt is found in the southern, lower end of the pile than the north end. Higher concentrations of metals are found in the lower end as well. Discharge from the tailings pile enters Alamitos Creek, was acidic before reclamation, and contained elevated levels of metals (Table 6; Sidle et al., 1991; Brandvold et al., 1995). Wet meadows with pockets of water covered some of the tailings. Reclamation consisted of construction of a stream channel through the tailings pile and stabilization of the tailings by covering with clay, gravel, and soil (Johnson and Deeds, 1995a, b).

From 1950 through 1980, various state and federal agencies used the mine waste material from the Pecos mine as fill in various campgrounds (Jacks Creek campground and trailhead, Panchuela Creek campground), Winsor Trailhead parking area and the access roads to these sites from the Pecos mine. Mine waste rock from the Pecos mine was used as fill in these areas and the U.S. Forest Service had determined that erosion from these areas are point sources of Cu, Pb, Zn, and other metals (Table 7; Koch and Barkmann, 1995; Science Applications International Corp., 1995). All of these areas, except for Panchuela campground, have been reclaimed by removing the material to the Pecos mine waste pile.

TABLE 5. Selected average physical and chemical properties of the tailings pile (from Sidle et al., 1991).

Property	Upper, northern portion of pile	Lower, southern portion of pile
% Sand	65.7	37.7
% Silt	22.9	46.5
% Clay	11.3	15.8
% Organic matter	0.91	1.61
pH	3.41	3.71
Cu mg/kg (total)	586.9	1,873.0
Pb mg/kg (total)	4,154.4	7,225.4
Zn mg/kg (total)	245.6	1,572.2
Cd mg/kg (total)	0.6	5.9

TABLE 6. Summary of chemical analyses (mg/L) of water samples from the Alamitos Canyon mill site (El Molino)(from Johnson and Deeds, 1995b). (— no analyses)

Element (mg/l)	Range of Ground Water Analyses	Range of Surface Water Analyses
Cd	<0.0001-3.4	<0.1-8.4
Cu	<0.001-0.13	<1.0-205
Fe	<0.02-13,000	0.042-7.3
Mn	<0.01-150	—
Pb	<0.001-0.85	<1.0-18
Ag	<0.0001-0.015	<0.1-3.0
Zn	<0.01-3,500	0.016-1.8
SO4	33-90,200	—
pH	9.62-4.2	8.36-4.95

SUMMARY OF PREVIOUS RESULTS

Results from the first sampling trip in September 1992 (McLemore et al., 1993) found that samples below the Pecos mine waste pile were elevated in Hg, Pb, Cu, and Zn. Multi-disciplinary studies reported by Popp et al. (1996) and McLemore et al. (1995b) confirmed these results.

Chemical analyses of six sediment size-fractions from six sites suggested that the metals were predominantly traveling both as suspended and/or absorbed material and as larger minerals or other grains weathered from the mine waste pile and the tailings pile (McLemore et al., 1995a). Statistical analyses of XRF data indicated that Cu, Pb, and Zn had a high correlation with Fe and Mg (Popp et al., 1996), suggesting that these metals were associated with iron-bearing and ferro-magnesium minerals such as magnetite, pyrite, biotite, pyroxene, hornblende, etc. In addition, Cu, Pb, and Zn in the small size fraction (< 63 microns) had a high correlation with Al, suggesting that these metals were also associated with clay minerals. The weight percent of the size fraction had a high positive correlation with SiO_2 and Na_2O and a high negative correlation with TiO_2 and Al_2O_3, indicating that SiO_2 is concentrated in quartz in the large-size fraction and that TiO_2 and Al_2O_3 are concentrated in the small-size fraction, probably as clay minerals. Cu, Pb, and Zn also had a high correlation

TABLE 7. Geochemical analyses (FAAS) in mg/kg of background soils and soils from the contaminated campgrounds and roads that used mine waste pile material as fill (from Science Applications International Corp., 1995).

Element	Background Soils Range	Background Soils Mean	Environmental Site Range	Environmental Site Mean
Ag	0.5-5.0	1.3	1.0-38	4.4
Al	10,850-32,050	18,015	7,400-28,350	15,178
As	1.3-4.7	2.5	0.9-9.3	3.2
Be	0.5-0.9	0.7	0.4-0.9	0.6
Ca	1,575-10,15	3,518	427-100,000	3,491
Cd	0.1-1.2	0.2	0.7-124	8.3
Cr	20-63	37	12-66	26
Cu	10.5-26.5	14.7	25-1,311	245
Fe	20,550-31,750	25,227	11,800-41,050	25,222
Hg	<0.01	<0.01	0.02-0.35	0.05
K	1,135-7,500	2,337	805-4,275	2,206
Mg	1,945-5,660	2,831	1,825-23,250	5,779
Mn	271-960	590	251-1,060	494
Na	30-111	53	29-225	68
Pb	10-41	18	37-10,600	702
Se	0.9-1.0	0.9	0.8-6.7	1.9
Si	1,375-3,070	2,052	1,110-3,365	2,034
V	25-70	46	25-85	42
Zn	60-129	96	485-33,550	2,417

with each other and with Ga and V, and Pb had a high correlation with Ni. These elements are probably elevated in the ore and waste material at the Pecos mine.

McLemore and Brandvold (1997) also concluded that their studies suggested that Cu, Pb, Zn, and other metals are eroded and leached from the Pecos mine waste tailings and the tailings piles in Alamitos Canyon. However, the total metal addition from these sources is insignificant compared to sediment input from the Wilderness Area and drainages south of Pecos Village, especially Glorieta Creek. Zinc concentrations were elevated in samples collected from the dam above the Lisboa Springs Fish Hatchery, especially in the <63 micron size-fraction, and may have contributed to fish kills at the hatchery.

Partial dissolution techniques were employed to determine the following metal forms: (1) exchangeable, (2) organic, and (3) adsorbed to Fe and Mn oxides (Brandvold et al., 1995). The smallest size fraction (<63 micron), which constituted 3% or less of the sample weight, contained the highest concentration of metals. Partial dissolution techniques suggested that for the sites above the mine waste tailings the metals are concentrated in the crystalline or mineral phase as opposed to adsorption to metal hydroxide, organic, clay or dissolved phases. Below the mine waste pile and below the mill tailings there is a marked increase in the amount of Zn adsorbed to clay and hydroxides. This phenomenon changed downstream in both cases to the initial conditions. Copper, Pb, and Zn were higher in samples collected during the Fall of 1992 (low flow conditions) compared to samples collected during the Spring and Summer of 1993 (high flow conditions). This difference, in part, may be due to scouring of the river channel and dispersion of sediment during spring runoff. Partial dissolution techniques designed to estimate exchangeable, organic, and oxide-bound forms indicate that the largest amounts of Cu and Pb were in forms solubilized by *aqua regia* but not by three extraction techniques (Brandvold et al., 1995). The largest amount of Zn was associated with Fe and Mn oxides.

Brandvold and McLemore (1999) used ANOVA statistics on duplicate analyses of stream-sediment samples to study Cu, Pb, and Zn variations. Analytical error was found to be the greatest contributor to total variance. Differences in the adjacent site means were most apparent in the smallest size fraction (<63 micron). But even these differences were shown to be mostly due to random error.

METHODS OF STUDY

Sample sites along the Pecos River and its tributaries were determined prior to collection on the basis of well-maintained roads, easy access by vehicle, and proximity to known mines (Table 8, Figs. 2, 3). Additional sites were added as needed. Samples were collected from the Pecos River during eleven separate trips from September 1992 through June 2000. Not all sites were sampled during each trip. The entire Pecos River reach was sampled only in September 1992 and 1996 and June 2000; the remaining sample trips were in the upper Pecos River from the Pecos Wilderness southward to Villanueva (Fig. 3). Stream-sediment and water samples were taken at each site during each trip. Lake sediments were collected only in 1992 and 1996, but data for lake sediments are not discussed in this report. Outcrops, mine waste piles, and tailings piles were examined and sampled for chemical analyses at various times from 1992 to 1995 (Popp et al., 1996). Water samples were collected at each site and stored in polypropylene bottles. A portion of the water was acidified with trace pure nitric acid for trace metal analysis.

Stream-sediment samples were collected from each site along the Pecos River and adjacent tributaries. The river was typically shallow, enabling the sample collector to sample near the center of the river. During the spring runoff, the river was deeper and safety ropes were used to enable the collector to sample the stream sediments. Samples were stored in plastic bags and air dried and sieved at the laboratory (Fig. 4). Mineralogy is described in Table 9. These samples were homogenized using the cone and quarter method, and then screened in order to obtain different size fractions. Only the <63 micron fraction was analyzed for samples collected in 2000. A portion of the sample was dissolved in acid (*aqua regia*) for analysis using flame atomic absorption spectroscopy (FAAS) (Carey et al., 2001). Pressed powder briquettes were prepared for selected samples for analysis of trace elements by X-ray fluorescence (XRF).

Stream-sediment and rock samples were analyzed for metal concentrations using FAAS; water samples were analyzed using FAAS and graphite-furnace atomic absorption spectroscopy (GAAS). Selected rock, ore, and stream-sediment samples were also analyzed for major and trace elements by XRF and instrumental neutron activation (INAA) (Popp et al., 1996). FAAS and XRF analytical procedures and quality control data are described by McLemore et al. (1993, 1995a), Brandvold et al (1995), Popp et al. (1996), and Carey et al. (2001).

RESULTS AND DISCUSSION

The pH values of Pecos water samples collected in 1996 and 2000 ranged between 7.4 and 8.5 and were similar to samples collected in 1992 (Table 10; McLemore et al., 1993). The metal-content of water samples from the Pecos River is low, in many cases below detection limits (Table 10). Collectively, these data indicate that drainage from the Pecos mine is not significantly affecting the composition of the surface water in the area, except in the immediate vicinity of the Pecos mine site (Popp et al., 1996).

FAAS, XRF, and INAA analyses of samples collected from 1992 through 1994 confirmed that elevated concentrations of Cu, Pb, Zn, Cd, Cs, As, Ag, Au, W, Br, Se, Sb, Co, Al, and Fe occurred in stream sediments below both the Pecos mine (BWC) and Alamitos Canyon mill (BV) sites (Tables 11, 12; Popp et al., 1996; McLemore et al., 1995b). Samples collected in 1996 and analyzed by FAAS for Cu, Pb, Zn, and Cd showed similar results below the Pecos mine (Table 11). Samples collected in 2000 were still elevated in Cu (310 ppm in 1992, 420 ppm in 1996, and 92 ppm in 2000), Pb (300 ppm, 280 ppm, and 160 ppm), Zn (3100 ppm, 2900 ppm, and 2080 ppm), and Cd (17 ppm, 14 ppm, and 4.7 ppm) below the Pecos mine, but not as high as in earlier years (Table 10, 11). This is consistent with the fact that reclamation didn't really begin in this area until after 1996. Metal concentrations are no

TABLE 8. Description of sample sites (collected by NMBMMR). *Sample site consists of predominantly boulders. Sand-sized sediment collected from natural sand traps or bends in river.

Symbol	Name	Latitude	Longitude	
UP	Upper Pecos	35°49′30"	105°39′5"	*Near Pecos Wilderness boundary, along of main stream of Pecos River.
J	Jacks Creek	35°49′30"	105°39′7"	*Along Jacks Creek upstream of the confluence with the Pecos River approximately 1 m wide.
BJ	Below Jacks Creek	35°49′28"	105°39′14"	*West side of Pecos River approximately 50 m downstream of the confluence of Jacks Creek (upstream of bridge). Pecos River is approximately 4 m wide.
AWC	Pecos above Willow Creek	35°45′43"	105°40′13"	*Along Pecos River on east side at Willow Creek Campground approximately 200 m upstream of Pecos Mine. Pecos River is approximately 9 m wide.
SE	Seep, below mine	35°45′33"	105°40′7"	From seep along drainage downstream of Pecos mine. White and brown iron and aluminum oxide and hydroxide precipitates containing Cu, Pb, and Zn occur along the course of the seep.
BWC	Pecos below Willow Creek	35°45′27"	105°40′10"	*From Pecos River. Sample collected from sand bar or bank (depending upon water flow) approximately 10 m downstream of the confluence with Willow Creek.
T	Pecos at Terrerro campground	35°44′35"	105°40′30"	*From west side of Pecos River, 30 m upstream of Terrerro campground.
AH	Pecos above hatchery	35°37′00"	105°41′00"	*From Pecos River at small diversion dam approximately 1.0 km upstream of hatchery. Sample from west bank upstream of dam.
BH	Pecos below Hatchery	35°36′30"	105°41′00"	*From Pecos River approximately 10 m below discharge pipes from hatchery.
BV	Pecos below village	35°34′32"	105°40′10"	*From Pecos River at bridge where Highway 223 crosses the river. Pecos River is approximately 12 m wide.
BG	Pecos Below Glorita River	35° 31' 59"	105° 40' 00"	From Pecos River, at confluence with Glorieta River.
SM	Pecos at San Miguel	35°22′00"	105°26′45"	From Pecos River at large dam south of San Miguel. Samples taken upstream of dam on west side of river.
V	Pecos at Villanueva	35°15′30"	105°20′15"	From Pecos River south end of Villanueva State Park campground. Samples taken on north bank or middle of river (depending upon water flow).
Te	Tecolitito	35°20' 20"	105°14' 20"	*From Pecos River at Tecolitito.
Di	Dillia	35°26' 15"	105°11' 25"	From Pecos River at Dillia.
BSR	Below Santa Rosa dam	35° 1' 35"	104°40' 00"	Pecos River below Santa Rosa Dam.
PL	Puerto del Luna	34°54' 00"	104°37' 30"	Pecos River at Puerto del Luna.
BSD	Below Sumner Dam	34°35' 30"	104°23' 00"	Pecos River below Sumner Dam.
FS	Ft. Sumner bridge	34°29' 00"	104°15' 03"	Pecos River at Ft. Sumner bridge.
US70	U.S. 70	33°38' 45"	104°22' 30"	Pecos River at U.S. 70 bridge.
Hag	Hagerman	33° 7' 2"	104°16' 45"	Pecos River at Hagerman.
BB	Below Brantley dam	32°32' 25"	104°21' 00"	Pecos River below Brantley Dam.

longer elevated below Alamitos Canyon (Tables 11, 12). Elevated Cu and Pb concentrations were detected for the first time below the fish hatchery (Table 13), possibly as a result of the cleaning of the raceways and discharging the water into the Pecos River.

XRF analyses are typically higher than FAAS (Table 14). XRF analysis measures the entire solid sample, whereas FAAS measures the amount of dissolved material. Therefore, not all of the sample was dissolved by *aqua regia* dissolution methods used in order to obtain FAAS analyses. However, FAAS probably reflects a better measurement of how much material is available for dissolution and mobility in surface and ground water. XRF is a measurement of the total metal concentrations in the solid.

The metal concentrations of most samples reflect the metal concentrations of the predominant lithology in this area. Stream-sediment samples collected above the Pecos mine contain similar metal concentrations as the Precambrian rocks (Popp et al., 1996). Stream-sediment samples collected below Pecos Village contain similar metal concentrations as the Paleozoic and Mesozoic rocks. These results further confirm that the Pecos mine and Alamitos Canyon mill were point sources of metals contamination.

Zinc is elevated in sediments collected at a dam above the fish hatchery compared to that in sites above the Pecos mine (Fig. 6; Tables 11, 13). Several fish kills have occurred at the hatchery 1991-1995 that have been attributed to drainage from the mine site and the contaminated campground sites. Sediment samples from the dam above the fish hatchery contained elevated concentrations of Zn in 1992 (1300 ppm) and 1996 (944 ppm), which suggested that precipitates formed from the leachate and suspended material eroded from the Pecos mine waste piles could be deposited at the diversion dam above the fish hatchery. During sporadic periods of turbulent flow in the fall, when the Pecos River is at low flow (i.e. low volume), suspended material elevated in Zn could be entering the fish hatchery and contributing to the recent fish kills. After 1995, the fish hatchery personnel established new procedures that reduced the incidence of fish kills. The fish hatchery personnel removed stream sediment above the dam to reduce the amount of Zn entering the hatchery. The fish hatchery personnel also lined the culvert and pumps from the dam to the hatchery with lime.

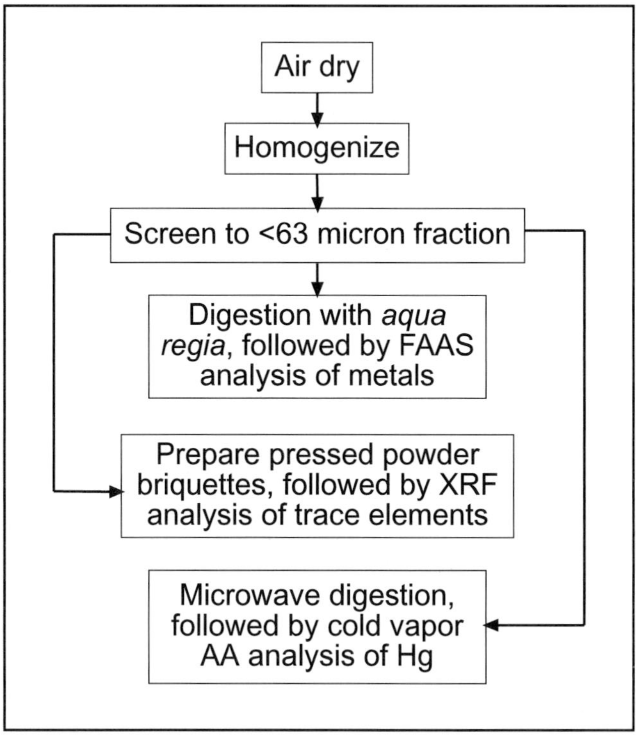

FIGURE 4. Sampling procedures.

TABLE 9. Mineralogy of stream-sediment samples from the upper Pecos River.

Major Constituents (>10% in most samples)
Quartz SiO_2
Lithic or rock fragments
Feldspars
K-feldspar $K(AlSi_3O_8)$
Plagiocalse $(Na,Ca)(AlSi_3O_8)$
Pyroxene (complex silicate mineral)
Hornblende $Ca_2Na(Mg,Fe^{+2})_4(Al,Fe^{+3},Ti)_3Si_8O_{22}(O,OH)_2$
Minor Constituents (<10% in most samples)
Biotite $K(Mg,Fe)_3(AlSi_3O10)(OH)_2$
Chlorite $Mg_3(Si_4O10)(OH)_2Mg_3(OH)_6$
Muscovite $K\,Al_2(AlSi_3O_{10})(OH)_2$
Magnetite Fe_3O_4
Zircon $ZrSiO_4$
Calcite $CaCO_3$
Titanite $CaTiO(SiO_4)$
Apatite $Ca_5(F,Cl,OH)(PO_4)_3$
Clay minerals
Kaolinite $Al_4(Si_4O10)(OH)_8$
Illite (K silicate)
Smectite (complex silicate)
Fe and Mn oxides
Trace Constituents (<1% in some samples)
Garnet (complex silicate)
Monazite $(Ce,La,Y,Th)PO_4$
Rutile, anatase TiO_2
Pyrite FeS_2
Fluorite CaF_2

Finally, the fish hatchery personnel shuts off river water during periods of rain or snow melt and utilizes spring water. Results from sampling in 1996 indicated that Zn was still elevated in stream sediments above the dam, even though the fish hatchery personnel removed material earlier in the year. By 2000, Zn was only slightly elevated in sediments (325 ppm)(Table 13).

During the remediation efforts at Jacks Creek Campground, an erosional event occurred (Richard Koch, personal communication, Sept. 28, 1995) that coincided with an increase in metal concentrations in stream sediments below Jacks Creek (Table 15). This event, in summer 1994, contaminated Jacks Creek with mine waste material, resulting in elevated metal concentrations (Popp et al., 1996). This suggests that remediation efforts can release additional metals to the Pecos River. Reclamation at the Alamitos Canyon mill site is evident in the stream-sediment sampling; metal concentrations decreased during and after reclamation (Tables 11, 14; Popp et al., 1996; Carey et al., 2001). Copper values decreased from 270 ppm in 1992 to 30 ppm in 2000. Lead values decreased from 590 ppm in 1992 to 30 in 2000. Zinc values decreased from 990 ppm in 1992 to 290 ppm in 2000.

SUMMARY

Elevated concentrations of Cu, Pb, Zn, Cd and other metals occurred in stream sediments below both the Pecos mine and Alamitos Canyon mill sites before reclamation began. Collectively, multi-disciplinary studies suggest that Cu, Pb, Zn, and other metals were eroded and leached from the Pecos mine waste pile and Alamitos Canyon mill tailings. These studies confirm a decrease in metal concentrations with time since reclamation began, especially in the immediate vicinity of the Pecos mine and Alamitos Canyon. Reclamation efforts at the Alamitos Canyon mill site are successful in that the metal concentrations in both surface water and stream sediments have decreased since reclamation. The overall metal concentrations dramatically decreased in stream sediments below Pecos village in all sampling trips, mostly due to dilution of sediment derived from the red bed sedimentary units. Therefore, the total metals addition from the Pecos mine waste pile and Alamitos Canyon mill tailings is insignificant compared to sediment input from the Pecos Wilderness and drainages south of Pecos village, except in the immediate vicinity of the Pecos mine and mill site. In 2000, a sharp increase was noted in Cu and Pb levels in stream sediments immediately below the fish hatchery. More samples should be taken from this site to confirm the finding.

ACKNOWLEDGMENTS

We would like to thank personnel from the New Mexico Game and Fish Department, Lisboa Springs Hatchery, U.S. National Park Service, and the U.S. Forest Service, Pecos District, for their assistance in collecting samples. We also acknowledge laboratory assistance since 1992 from Kelly Donahue, David Bright,

TABLE 10. Chemical analyses of water samples collected in 1992, 1996, and 2000. Values in ppm except for conductivity, pH and where noted. Major and trace element analyses by flame atomic absorption spectrometry (FAAS) and graphite-furnace absorption spectrometry (GAAS). (— no analyses)

Site	Date	Temp. C°	Cond. uS	pH	HCO_3	Cl	SO_4	NO_3	F	Na	K	Mg	Ca	As ppb	Hg ppb	Pb ppb	Cu ppb	Zn ppb
UP	9/92	15.0	200	7.60	170	—	—	—	—	—	—	—	—	—	<0.2	<1	3	<50
	9/96	9.6	230	8.27	100	1.4	20	0.7	<0.2	2.8	0.6	4.2	37	<10	—	—	—	—
	6/00	15.3	240	8.10	120	2.0	15	<0.2	0.1	1.4	0.6	4.1	38	<1	—	<5	15	<50
BJ	9/92	12.0	170	8.20	140	—	—	—	—	—	—	—	—	—	<0.2	2	2	<50
	9/96	9.0	270	8.33	146	2.2	22	0.7	<0.2	2.6	0.5	4.5	46	<10	—	—	—	—
J	6/00	13.8	280	8.28	148	2.1	14	<0.2	<0.1	1.4	0.5	3.9	48	<1	—	<5	5	<50
AW	9/96	12.0	140	8.22	73	1.2	8.2	0.4	<0.2	2.5	0.6	2.4	20	<10	—	—	—	—
	6/00	18.2	160	8.13	82	2.0	9.3	<0.2	0.1	1.5	0.5	2.4	25	<1	—	<5	6	<50
BW	9/92	21.5	160	9.10	120	—	—	—	—	—	—	—	—	—	<0.2	6	3	410
	9/96	14.0	130	7.44	73	1.3	8.6	0.9	<0.2	2.6	0.6	2.4	22	<10	—	—	—	—
	6/00	16.1	160	8.16	88	2.0	8.5	0.1	0.1	1.6	0.6	2.5	26	<1	—	<5	5	<50
T	9/96	15.2	130	7.68	73	1.1	8.6	0.5	<0.2	2.5	0.5	2.4	21	<10	—	—	—	—
	6/00	17.5	170	8.02	83	2.0	9.3	<0.2	0.1	1.7	0.6	2.4	25	<1	—	<5	5	<50
AH	9/92	18.0	180	8.30	140	—	—	—	—	—	—	—	—	—	<0.2	2	3	60
	9/96	16.9	160	8.01	91	1.3	9.9	0.4	<0.2	3.0	0.6	2.7	25	<10	—	—	—	—
	6/00	20.6	190	8.12	97	2.3	10	0.1	0.1	2.0	0.7	2.7	30	<1	—	<5	6	<50
BH	9/92	16.5	180	8.05	140	—	—	—	—	—	—	—	—	—	<0.2	2	1	60
	9/96	15.8	190	8.08	100	1.4	9.8	0.6	<0.2	3.0	0.6	2.8	27	<10	—	—	—	—
	6/00	22.6	200	8.18	100	2.2	11	<0.2	0.2	2.1	0.6	2.8	32	<1	—	<5	5	<50
BV	9/92	18.0	200	8.35	140	—	—	—	—	—	—	—	—	—	<0.2	<1	<1	30
	9/96	17.9	180	7.78	100	1.5	11	0.7	<0.2	3.2	0.6	3	30	<10	—	—	—	—
	6/00	22.0	240	8.08	125	2.3	12	0.1	0.2	2.4	0.6	3.2	38	<1	—	<5	4	<50
BG	9/96	18.4	230	8.47	110	5.5	15	0.5	<0.2	6.6	0.8	3.9	34	<10	—	—	—	—
	6/00	25.9	250	8.41	130	2.7	14	<0.2	0.2	3.3	0.7	3.4	40	<1	—	<5	3	<50
SM	9/92	21.5	220	7.95	140	—	—	—	—	—	—	—	—	—	<0.2	9	4	60
	9/96	19.9	220	8.49	120	2.6	13	0.4	<0.2	5.0	0.8	3.8	35	<10	—	—	—	—
	6/00	26.1	290	8.03	140	3.4	19	0.3	0.2	5.3	1.1	4.2	44	<1	—	<5	7	<50
V	9/92	23.0	260	8.20	170	—	—	—	—	—	—	—	—	—	<0.2	<1	2	<50
	9/96	21.8	250	8.46	150	3.1	14	0.6	<0.2	5.5	1.0	4.8	40	<10	—	—	—	—
	6/00	24.5	370	8.21	200	4.0	19	<0.2	0.2	7.3	1.5	7.8	55	<1	—	<5	4	<50
Te	9/92	20.5	250	8.10	170	—	—	—	—	—	—	—	—	—	<0.2	1	2	<50
	9/96	22.4	250	8.45	140	3.7	15	0.9	<0.2	6.4	1.1	5.0	39	<10	—	—	—	—
	6/00	22.6	350	7.90	190	3.9	17	<0.2	0.2	6.8	1.6	7.2	51	<1	—	<5	6	<50
Di	9/92	26.0	310	8.60	220	—	—	—	—	—	—	—	—	—	<0.2	3	3	<50
	9/96	22.8	280	8.47	150	3.6	16	0.4	<0.2	6.0	1.5	6.0	44	<10	—	—	—	—
	6/00	22.8	350	8.08	170	5.2	28	0.2	0.2	8.7	1.4	9.6	45	<1	—	<5	6	<50
BSR	9/92	24.5	590	8.05	170	—	—	—	—	—	—	—	—	—	—	—	—	—
	9/96	26.9	2100	7.68	160	21.1	1320	0.4	0.4	51	14	140	250	<10	—	—	—	—
	6/00	24.6	1550	7.84	200	15	650	0.4	0.2	31	4.3	75	185	<1		<5	16	<50
PL	9/92	24.0	2480	8.00	190	—	—	—	—	—	—	—	—	—	<0.2	<1	1	14
	9/96	23.4	2710	8.05	155	135	1320	0.6	0.4	89	2.4	61	460	<10	—	—	—	—
	6/00	25.1	3060	8.06	140	130	1540	<0.2	0.4	100	2.4	68	550	<1	—	<5	15	<50
BSL	9/92	21.0	1250	8.40	140	—	—	—	—	—	—	—	—	—	—	—	—	—
	9/96	17.7	1960	8.02	100	67	1000	0.9	0.2	62	3.1	41	290	<10	—	—	—	—
	6/00	25.0	1110	8.12	140	30	400	0.2	0.2	28	2.6	21	180	<1	—	<5	6	<50
FS	9/92	20.0	1250	8.10	150	—	—	—	—	—	—	—	—	—	—	—	—	—
	9/96	17.8	1860	7.65	100	69	875	0.5	0.2	65	3.1	44	260	<10	—	—	—	—
	6/00	32.8	1610	8.00	170	70	510	<0.2	0.3	64	2.5	40	210	<1	—	<5	12	<50
US70	9/92	29.0	2000	8.00	120	—	—	—	—	—	—	—	—	—	<0.2	3	4	10
	9/96	18.6	2190	8.05	90	300	610	2.0	<0.2	203	3.8	48	170	<10	—	—	—	—
	6/00	33.5	1780	8.02	110	140	560	3.7	0.4	156	6.0	34	170	<1	—	<5	18	<50
H	9/92	27.0	3490	8.00	150	—	—	—	—	—	—	—	—	—	<0.2	2	<1	8
	9/96	19.3	3150	7.98	130	430	880		0.3	295	4.5	77	250	<10	—	—	—	—
	6/00	31.5	6380	7.88	130	1300	1420	1.8	0.4	790	7.1	130	480	<1	—	<5	18	<50
BBR	9/92	22.0	3250	8.30	120	—	—	—	—	—	—	—	—	—	—	—	—	—
	9/96	22.3	2860	7.70	100	400	770	0.8	<0.2	275	4.9	66	225	<10	—	—	—	—
	6/00	26.5	3880	7.42	110	530	1040	0.2	0.3	345	5.0	85	350	<1	—	<5	<1	<50

TABLE 11. Chemical analyses of stream-sediment samples collected in 1992, 1996, and 2000, as determined by FAAS (<63-micron size fraction). Entire data set can be obtained from the authors.

Site	Cu ppm			Pb ppm			Zn ppm			Cd ppm			Mn ppm			Fe %		
	1992	1996	2000	1992	1996	2000	1992	1996	2000	1992	1996	2000	1992	1996	2000	1992	1996	2000
UP	18	23	20	30	31	17	110	86	88	1.2	1.3	1.2	410		585	2.20		2.28
BJ	40	42	14	110	99	21	330	325	97	1.5	1.0	1.2	400	440	440	2.40	1.70	1.72
AWC	70	44	23	95	65	32	300	250	160	1.7	1.5	1.1	330	370	370	2.20	2.20	2.22
BWC	310	420	92	300	280	160	3100	2900	2080	17	14	4.9	360	425	425	2.05	2.90	2.58
T	92	78	48	125	84	80	1160	1000	755	2.2	5.0	4.7		555	830	1.90	2.00	2.42
AH	55	44	24	94	57	37	1300	945	325	2.2		1.7	310		277	2.00		2.17
BH	35	39	130	70	54	545	830	490	375	2.7		2.0	280		270	1.85		2.49
BV	270	38	30	590	38	93	990	320	290	2.5		2.2	380		320	1.88		2.61
BG		24	22		39	44		175	215			2.0			445			1.93
SM		15	8	29	28	13	51	33	35	0.8		1.1	310		310	1.65		1.89
V	14	16	4	44	27	15	75	27	160	0.7		1.0	310		300	1.80		1.96
Te		14	8	18	20	31	45	25	49	0.7		0.9	280		770	4.00		1.74
Di	15	10	10	21	20	22	32	20	62	0.7		1.0	210		450	1.50		1.71
BSR		12	4		25	32		34	38			1.0			795			2.26
PL		12	6	21	24	17	37	30	59			0.6			275			1.89
BSD		18	14		68	31		36	39			0.9			665			2.83
FS		37	9	21	32	28	27	28	69			0.9			710			2.12
US70		36	17	114	27	26	66	31	145			1.4			410			3.18
Hag		33	9	19	21	16	18	25	60			0.9			380			3.01
BBR	9	32	5	30	24	18	24	27	41			1.0			285			1.50

TABLE 12. Comparison of chemical analyses of stream-sediment samples (<63-micron size fraction, FAAS) Below Pecos Village (BV) before and after reclamation of the Alamitos Canyon mill site.

DATE	Cu ppm	Pb ppm	Zn ppm
9/92	270	590	990
5/94	25	40	290
9/94	22	28	211
9/96	28	38	319
6/00	30	93	287

TABLE 13. Comparison of chemical analyses of stream-sediment samples (<63-micron size fraction, FAAS) above and below the fish hatchery.

DATE	Cu ppm	Pb ppm	Zn ppm
Pecos Above Fish Hatchery (AH)			
9/92	55	94	1300
5/94	33	46	580
9/94	38	77	697
9/96	44	57	944
6/00	24	37	325
Pecos Below Fish Hatchery (BH)			
9/92	35	67	830
5/94	41	74	550
9/94	31	53	373
9/96	39	54	488
6/00	130	543	374

Doug Jones, Roger Duncan, Patrick Martinez, Forest Hawman, Tim Looper, Scott Salvas, Catherine O'Connor, Victoria Romero, Mathew Hind, Jessica McElhiney, Tim Pease, Terry Thomas, and Daniel Hack. Special thanks to Jeffery Todd (AMAX) for providing us with data on remedial programs at the Pecos mine. Chris McKee (NMBGMR) analyzed some samples by X-ray-fluorescence using the Phillips PW 2400 instrument purchased with funds from NSF grant EAR-9316467. We would like to thank Donald Brandvold and Frank Kottlowski for reviewing this manuscript; their comments were beneficial. Finally we acknowledge financial support for this project since 1992 from the NMBGMR (Peter Scholle, Director and State Geologist (1999-present); Charles Chapin, former Director and State Geologist (1991-1999)), U.S. Bureau of Reclamation (cooperative agreement no. 3-FC-40-13830), and the Chemistry Department of NMIMT.

REFERENCES

Anderson, O. J., Jones, G. E., and Green, G. N., 1994, Geologic map of New Mexico: New Mexico Bureau of Mines and Mineral Resources, Open-file report 408A, scale 1:1:500,000.

Bemis, H. D., 1932, Milling methods and costs at the Pecos concentrator of the American Metal Co., Terrero, New Mexico: U.S. Bureau of Mines, Information Circular 6605, 23 p.

Berger, A. C., Bethke, C. M., and Krumhansl, J. L., 2000, A process model of natural attenuation in drainage from a historic mining district: Applied Geochemistry, p. 655-666.

Brandvold, L. A. and McLemore, V. T., 1999, A study of the analytical variation of sampling and analysis of stream sediments from mining and milling contaminated areas: Journal of Geochemical Exploration, v. 64, pp. 185-196.

Brandvold, L. A., McLemore, V. T., Brandvold, D. K., and O'Conner, C., 1995, Distribution and partitioning of copper, lead, and zinc in stream-sediments above and below an abandoned mining and milling area near Pecos, New Mexico, U.S. A.: The Analyst, v. 120, p. 1485-1495.

TABLE 14. Comparison of chemical analyses by FAAS and X-ray fluorescence (XRF) methods for samples collected in 2000 (<63-micron size fraction). As was determined only by XRF methods. (— no analyses).

Site	Cu		Pb		Zn		Cr		Mn		Fe		As
	(FAAS)	(XRF)	(FAAS)	(XRF)	(FAAS)	(XRF)	(FAAS)	(XRF)	(FAAS)	(XRF)	(FAAS)	(XRF)	(XRF)
UP	20	20	17	24	88	106	35	75	585	770	2.27	2.94	5
J	14	21	21	33	97	117	32	63	438	540	1.70	2.21	4
AWC	23	29	32	45	156	180	32	75	369	620	2.20	3.15	4
BWC	92	92	157	143	2080	1600	33	81	424	620	2.91	3.11	3
T	48	58	80	97	755	915	36	78	827	1160	2.41	3.21	4
AH	24	28	37	45	325	432	33	70	284	540	3.44	2.97	3
BH	130	146	543	680	374	599	28	70	283	700	3.81	4.01	<2
BV	30	43	93	103	287	368	42	69	318	460	3.83	3.29	3
BG	22	28	44	51	213	306	36	64	434	620	2.88	2.69	3
SM	8	18	13	13	35	48	34	73	309	390	3.15	2.38	3
V	4	14	15	15	157	57	32	66	305	390	2.90	2.39	3
Tec	8	—	31	—	49	—	27	—	764	—	1.66	—	—
Di	10	—	22	—	62	—	69	—	442	—	1.62	—	—
BSR	4	19	32	32	38	49	46	92	793	930	3.31	2.65	5
PL	6	16	17	17	59	47	31	145	304	390	2.33	2.28	3
BSD	14	—	31	—	39	—	35	—	658	—	2.75	—	—
FS	9	—	28	—	69	—	59	—	701	—	2.03	—	—
US70	17	—	26	—	146	—	47	—	424	—	4.97	—	—
Hag	9	20	16	21	60	54	50	257	370	700	4.80	4.36	7
BBR	5	14	18	18	41	44	38	99	290	390	2.27	1.83	3

TABLE 15. Comparison of chemical analyses of stream-sediment samples (<63-micron size fraction, FAAS) Below Jacks Creek.

DATE	Cu ppm	Pb ppm	Zn ppm
9/92	40	110	330
5/93	23	45	130
5/94	33	100	170
9/94	580	1681	481
9/96	42	99	325

Carey, E. A., Brandvold, L. A., and McLemore, V. T., in press, Base metal and trace element concentrations in stream sediments along the Pecos River, from the southern Pecos Wilderness to Brantley Dam, north of Carlsbad, eastern New Mexico: American Society for Surface Mining and Reclamation, 18th annual meeting, June 2001.

Fulp, M. S., 1982, Precambrian geology and mineralization of the Dalton Canyon volcanic center, Santa Fe County, New Mexico [MS thesis]: University of New Mexico, Albuquerque, 199 p.

Fulp, M. S. and Renshaw, J. L., 1985, Volcanogenic-exhalative tungsten mineralization of Proterozoic age near Santa Fe, New Mexico, and implications for exploration: Geology, v. 13, p. 66-69.

Harley, G. T., 1940, The geology and ore deposits of northeastern New Mexico (exclusive of Colfax County): New Mexico Bureau of Mines and Mineral Resources, Bulletin 15, 104 p.

Koch, R. J., and Barkmann, G., 1995, Summary of environmental sampling at Forest Service roads and campgrounds near Cowles, New Mexico: New Mexico Geological Society, Guidebook 46, p. 43-44

Krieger, P., 1932, Geology of the zinc-lead deposits at Pecos, New Mexico, Part II: Economic Geology, v. 27, p. 450-470.

Johnson, P. S., and Deeds, J. L., 1995a, Summary of environmental issues at El Molino mill, north-central New Mexico: New Mexico Geological Society, Guidebook 46, p. 319-322.

Johnson, P. S. and Deeds, J. L., 1995b, A site conceptual model of environmental issues at the Pecos mine: New Mexico Geological Society, Guidebook 46, p. 41-43.

McLemore, V. T., 1995, Pecos mine and Alamitos Canyon mill: New Mexico Geological Society, Guidebook 46, p. 40-41.

McLemore, V. T. and Brandvold, L. A., 1997, Transport of metals from the Pecos mine and Alamitos Canyon mill site in the Pecos River, northeastern New Mexico; *in* WERC-HSRC, 97 Joint conference on the environment: WERC and HSRC/S&SW, Proceedings Volume, p. 189-193.

McLemore, V. T., Brandvold, L. A., and Brandvold D. K., 1993, A reconnaissance study of mercury and base metal concentrations in water, stream- and lake-sediment samples along the Pecos River in eastern New Mexico: New Mexico Geological Society, Guidebook 44, p. 339-351.

McLemore, V. T., Brandvold, L. A., and Pease, T. C., 1995a, The effect of particle size distribution on the geochemistry of stream sediments from the upper Pecos River, San Miguel County, New Mexico: New Mexico Geological Society, Guidebook 46, p. 323-329.

McLemore, V. T., Brandvold, L. A., Brandvold D. K., Kirk, K., Popp, C., Hanson, S., Radtke, R., Kyle, P. R., and Hossain, A. M., 1995b, A preliminary summary of multidisciplinary studies in the upper Pecos River area, Santa Fe and San Miguel Counties, New Mexico: New Mexico Geological Society, Guidebook 46, p. 331-338.

Northrop, S. A., 1959, Minerals of New Mexico: University of New Mexico Press, Albuquerque, 665 p.

Popp, C. J., Brandvold, D. K., Kirk, K., Brandvold, L. A., McLemore, V. T., Hansen, S., Radtke, R., and Kyle, P. R., 1996, Reconnaissance investigation of trace metal sources, sinks, and transport in the upper Pecos River Basin, New Mexico: U.S. Bureau of Reclamation, Report No. 3-FC-4D-13830, 224 p.

Science Applications International Corp., 1995, Santa Fe National Forest—Pecos Roads and campgrounds environmental site investigation report—phase 1: U.S. Forest Service report, contract no. 53-8371-2-68, 87 p.

Sidle, R. C., Chambers, J. C. and Amacher, M. C., 1991, Fate of heavy metals in an abandoned lead-zinc tailings pond: II sediment: Journal of Environmental Quality, v. 20, p. 752-758.

S. M. Stoller Corp., 1993, Tererro mine site, Pecos mine operable unit, San Miguel County, New Mexico: Background report prepared for AMAX Resource Conservation Co., 117 p.

Stott, C. E., 1931, Geology of the Pecos mine: Engineering and Mining Journal, v. 131, p. 270-275.

A GEOLOGIC MEMBRANE—MICROBIAL METABOLISM MECHANISM FOR THE ORIGIN OF THE SEDIMENTARY COPPER DEPOSITS IN THE PASTURA DISTRICT, GUADALUPE COUNTY, NEW MEXICO

V. W. LUETH[1] AND T. M. WHITWORTH[2]

[1]New Mexico Bureau of Geology and Mineral Resources, New Mexico Institute of Mining & Technology, Socorro, NM 87801, vwlueth@nmt.edu

[2]Department of Geological & Petroleum Engineering, University of Missouri-Rolla, Rolla, MO 65409, mikew@umr.edu

Abstract.—Examination of copper deposits in Pastura district shows that mineralization occurred at the sand/shale interface. Mineralization was greatest adjacent to the thinnest shales and decreased with increasing shale thickness, becoming non-existent adjacent to the thickest shales. This pattern of mineralization is consistent with geologic membrane effects that can occur in shales and clays in the subsurface. We propose a conceptual model invoking solute-sieving effects to help explain the observed mineralization in the copper deposits in the Pastura district. We suspect that the role of the membrane-functioning shales was to concentrate the copper and the dissolved nutrients necessary for metabolic-functioning of sulfate reducing bacteria in "layers of increased solute concentration," usually called the concentration polarization layer (CPL), adjacent to the shales. The sulfate-reducing bacteria, as a part of their metabolic process, produce H_2S, which precipitated the copper sulfides. Experimental studies have demonstrated that the CPL can range from several to more than 100 times that of background concentrations within the aquifer. Flux magnitude through shale is inversely proportional to its thickness and strongly affects CPL development. For any given solute, there is a threshold flux value that must be exceeded before rejected solute becomes unable to effectively back diffuse into the reservoir. Therefore, thick shales (with correspondingly low flux rates) typically do not form a CPL. This physical fact corresponds with the field observation that the sands adjacent to the thicker shales in the Pastura district are not mineralized. Our model explains why copper mineralization is localized in proximity to thin shale units and is not ubiquitous in the host rock.

INTRODUCTION

Becker (1892) appears to be the first to suggest the possibility of the natural concentration of metals by dialysis (membrane processes). A later paper by Brown (1971) also briefly suggested the possibility that sedimentary copper deposits may be formed by salt-sieving processes. Whitworth and DeRosa (1997) performed a series of clay membrane experiments in which they passed undersaturated copper carbonate, copper chloride, cobalt chloride, and lead chloride solutions through clay membranes. In each case, the solutions became sufficiently concentrated due to retardation of solute by the membrane to precipitate the mineral salts. These experiments offer solid experimental evidence that metals can be concentrated by geologic membrane-functioning materials.

The purpose of this paper is to suggest a geologic membrane-related mechanism for the origin of the copper deposits in the Pastura district, using field observations of copper mineral occurrence and stratigraphic unit thickness as a basis for our study. These field observations are 1) copper mineralization is always most extensive in those portions of sandstone adjacent to shale units, 2) sandstones adjacent to the thinnest shales exhibit greater mineralization, and 3) sandstones adjacent to the thickest shales contain only a trace or lack mineralization. Any model proposed to explain the distribution of copper mineralization in these deposits should also be able to explain the copper sulfide precipitation mechanism and, very importantly, why that mineralization is restricted instead of extending uniformly throughout the more permeable and laterally continuous sandstones.

Several sulfide precipitation mechanisms for sandstone-hosted copper occurrences have been proposed. Kirkham (1989) suggests that sulfide precipitation may be engendered by the mixing of groundwaters and ore-carrying brines. Garven and Freeze (1984) pointed out, however, that conditions which could lead to the mixing of two different fluids in the subsurface are rare. Eugster (1989) suggested that sulfide precipitation occurs at oxidation/reduction fronts within the strata. However, LaPointe (1989) points out that certain deposits have redox haloes surrounding them, which suggests the redox conditions may be related to mineralization, rather than the cause of mineralization. Most genetic models of sedimentary copper deposits, including ours, postulate that sulfide was produced as a product of bacterial reduction of sulfate (Hoy and Ohmoto, 1989). Our model explains why the postulated bacteriological-initiated redox conditions necessary for sulfide precipitation were confined to the locations at which mineralization occurs in the Pastura district.

GEOLOGY OF THE DEPOSITS

Pastura District

The Pastura district is located in east-central New Mexico southwest of Santa Rosa (Fig. 1), and hosts mineralization in two different formations. The Pintada mine exploited copper mineralization from the Permian-age Grayburg-Queen Formation. The Guadalupe mine (also known as the Stauber mine) is hosted by the Triassic-age Santa Rosa Formation (Anderson, 1957). The beds at both deposits are relatively flat-lying and dip angles do not exceed 4°, although both deposits are found in structural depressions of uncertain origin; these structures appear to be associated with mineralization (McLemore and North, 1985). Previous studies concerning the geology of the district (Kelly, 1972) and nature of mineralization in are provided by Soulé (1956), Holmquist (1947), Sandusky and Kaufman (1972), and LaPointe (1974, 1976, 1979).

Pintada Mine

The Grayburg-Queen Formation hosts all of the mineralization in the Pintada mine (Fig. 1). The stratigraphic relationship of

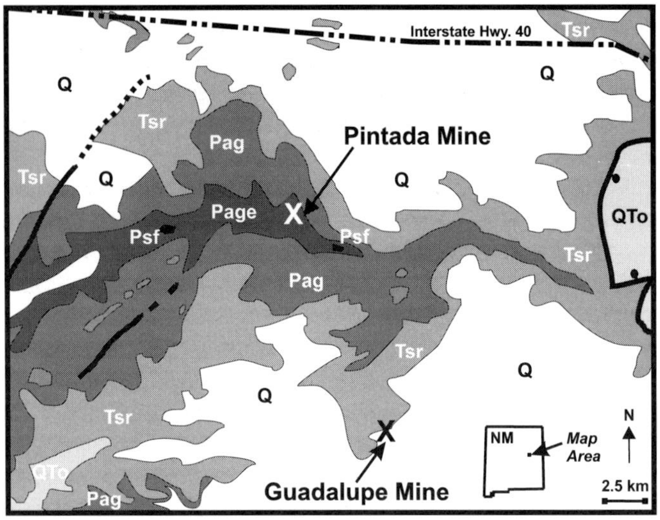

FIGURE 1. Location and generalized geologic map of the Pastura District. Symbols: Q – surficial deposits, QTo – older alluvium, Tsr – Santa Rosa Formation, Pag – upper Grayburg-Queen Formation, Page – lower Grayburg-Queen Formation, Psf – San Andres Formation. Map modified from Kelly, 1972.

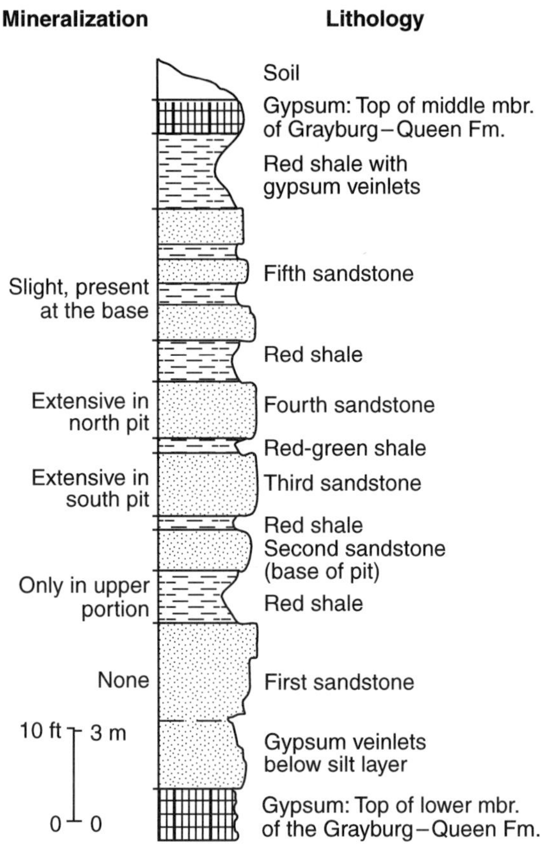

FIGURE 2. Stratigraphic column of the middle member, "lower sandstone unit," of the Grayburg-Queen Formation at the Pintada Mine in the Pastura District, New Mexico.

the basal contact of the Grayburg-Queen Formation and the San Andres Limestone (Permian) is obscure in the mine area although Kelly (1972) suggested a minor disconformity is present. The Santa Rosa Sandstone disconformably overlies the Grayburg-Queen Formation. Previous workers have divided the Grayburg-Queen Formation into three distinct units (Sandusky and Kaufman, 1972). The lower "gypsiferous" unit consists of massive gypsum which grades upward into the "lower sandstone" unit which hosts the mineralization. The "lower sandstone" consists of five separate, light gray to medium blue gray, fine to very fine-grained sandstone units separated from each other by thin green to red shale/siltstone interbeds (Fig. 2). The "upper sandstone", which contains no mineralization, is composed of fine grained, red-brown, thin bedded sandstones interbedded with thin layers of red-brown siltstone, mudstone, and shale. The contact between the two sandstone units is marked by a 0.3 meter gypsum layer at the top of the "lower sandstone" (Sandusky and Kaufman, 1972).

Mineralization at the Pintada mine is confined to the fine-grained quartz arenite units of the "lower sandstone". Chalcocite is the dominant primary sulfide mineral. Pyrite is present in small amounts and is irregularly distributed. Malachite and azurite are present and are commonly thought to have formed from *in situ* chalcocite oxidation due to subaerial exposure. Secondary carbonate minerals are found mainly on fractures, shale contacts and on surfaces of recently mined exposures although some appear to occupy original porosity. Organic material (hydrocarbon?) is found as wispy disseminations in some of the sandstones; importantly, the organic matter is not always mineralized.

Copper is most concentrated at the contact of the sandstones and the confining shale/siltstone units (Fig. 3.). At these contacts, mineralization is parallel to layering and can completely fill pores. Removed from the sandstone-shale contacts, the sandstone layers are characterized by wispy to disseminated mineralization, with the wispy mineralization apparently following primary bedding planes. Minor copper mineralization is observed in the silty units which separating the sandstones yet is absent in the clay-rich shales.

The amount of mineralization varies along the strike of the units as noted in the pit face (Fig 4). In the northern portion of this pit, mineralization is most well developed in the fourth sandstone unit (refer to Fig. 2). In the southern portion, the third sandstone is more intensely mineralized. The grade and thickness of copper mineralization in each sandstone unit also correlates inversely to the thickness of the bounding shale/siltstone such that copper grade and thickness are greater where confining shales are thinnest.

Guadalupe (Stauber) Mine

The Guadalupe mine was the largest copper producer in northeastern New Mexico (Anderson, 1957). Copper mineralization is confined to the lower member of the Santa Rosa Sandstone, which rests on an erosional unconformity on the Grayburg-Queen Formation (Kelly, 1972) in the mine area. In the vicinity of the mine, no other rocks cover the Santa Rosa Sandstone, which is also the case at the type section 27 km northeast from the mine (Kelly, 1972).

Mineralization at the Guadalupe mine is confined to a single sandstone layer characterized by a thickness varying from one to six meters that is confined by low permeability shale units.

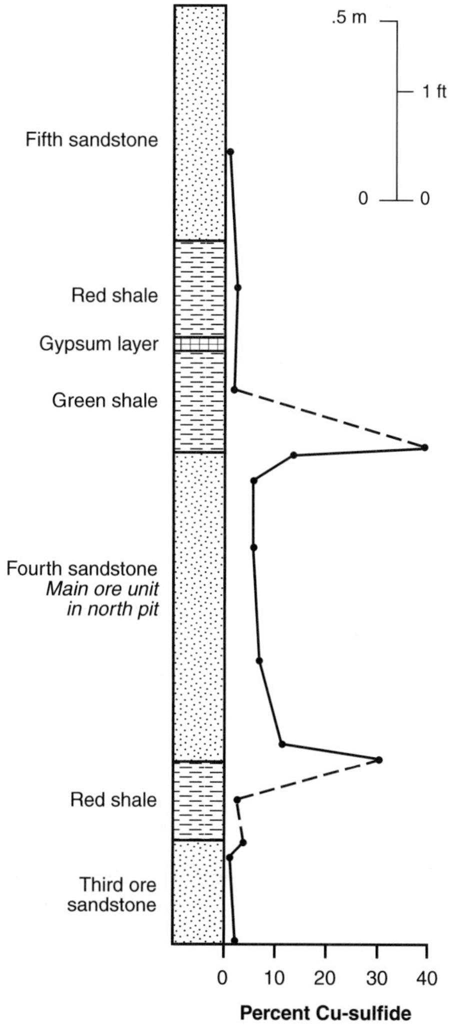

FIGURE 3. Variation in lithology and volume percent copper sulfide concentration at the Pintada Mine in the 4th sandstone unit in the north pit area. Dashed line indicates sampling was not continuous in that interval due to the friable nature of the rock.

Chalcocite is the main primary sulfide and has been replaced by supergene malachite and azurite. Mineralization shows greatest development at the base and top of the sandstone unit and weakly extends into the bounding shales for distances of up to twenty centimeters. Mineralization within the sandstone is a maximum of 4 m thick, although the majority of ore-grade copper is confined to the lower portions of the bed. The overlying shale is 12 m thick and the lower shale, described as plastic clay by Harley (1940), is 6 m thick. The lower shale unit is undulating with greatest development of mineralization in proximity to anticlinal structures in this lower shale. These structures are probably the result of soft sediment deformation or diagenetic compaction.

GEOLOGIC MEMBRANE PROCESSES

A semi-permeable membrane is defined as a material which will permit the passage of some molecules but not others (Noggle, 1984). As defined, such a membrane would be perfect, *i.e.*, it would absolutely preclude the passage of certain molecular species. Because such perfect membranes probably do not exist in nature (Fritz, 1986), a better working definition for a semi-permeable geological membrane would comprise any lithology that allows one solution component to pass through more easily than another.

Many studies have experimentally confirmed the ability of clays, which are a major component of shales, to act as semi-permeable membranes (Fritz and Whitworth, 1994). For example, Young and Low (1965) conducted experiments in which they passed saline solutions through actual discs of shale and siltstone, demonstrating that shales and siltstone are capable of acting as semi-permeable membranes. In addition to laboratory studies, many field studies have suggested the existence of geologic membranes. For example, the Milk River aquifer in western Canada has been cited as an example of a rock unit exhibiting membrane processes (Berry, 1960, Phillips, *et al.*, 1986). More recently, Neuzil (2000) presented field evidence of membrane-functioning shales.

Hyperfiltration (reverse osmosis) occurs when sufficient pressure is applied to the high-concentration membrane face to reverse the direction of water flow through the membrane. Under hyperfiltration conditions, both solute and water flow across the membrane from high to low solute concentration. Both osmosis and hyperfiltration can occur in the subsurface in conjunction with geologic membranes; however, osmotic processes are dissipative, meaning that they do not increase solute concentrations. Therefore, the role of osmosis in sediments is not considered in this paper.

We can use a simple diagram (Fig. 5) to describe what happens if groundwater is forced through a membrane-functioning lithology. Theoretically this situation begins with identical solute concentrations, c_i, on both sides of the membrane at time t = 0 (Fig. 5a). Because water passes more easily through the membrane than the solute, as time passes the solute begins to accumulate, c_o, at the high-pressure membrane face (Fig. 5b). The zone of increased concentration at the high-pressure membrane face is called the concentration polarization layer (CPL). As the CPL grows, ever more solute is available at the high-pressure membrane face to enter the membrane. Hence, membrane efficiency decreases and the effluent concentration tends to increase concurrently with growth of the CPL (Fritz and Marine, 1983). The concentration at the high-pressure membrane face continues to increase with time as does the width and amount of solute in the CPL (Fig. 5c). After a sufficient amount of time has passed, an equilibrium is reached (Fig. 5d) in which, if no precipitation is occurring within the CPL, the effluent concentration stabilizes at that of the original input concentration, c_i. However, if chemical reactions and/or mineral precipitation are occurring within the CPL, then the effluent concentration will always be less than the input concentration.

DISCUSSION

We will describe a geologic membrane-related model for the origin of the sedimentary-hosted copper deposits at the Pintada and Stauber mines. LaPoint (1974) discusses possible sources of copper in northern New Mexico, specifically the Pedernal uplift region. Specifically, these oxidized, copper-bearing solutions flowed along an aquifer bounded by confining shales. When groundwater passes

FIGURE 4. Photograph illustrating lithologic and mineralogic variation along strike at the Pintada Mine. Throughout the mine copper mineralization is most well developed where the bounding shale is the thinnest and least developed where the bounding shale is thickest. Note the variation in shale thickness from A to B on the photograph. Mineralization becomes higher grade and thickens in the area of B (shale thickness = 0.23 m) compared to A (shale thickness = 0.45 m)

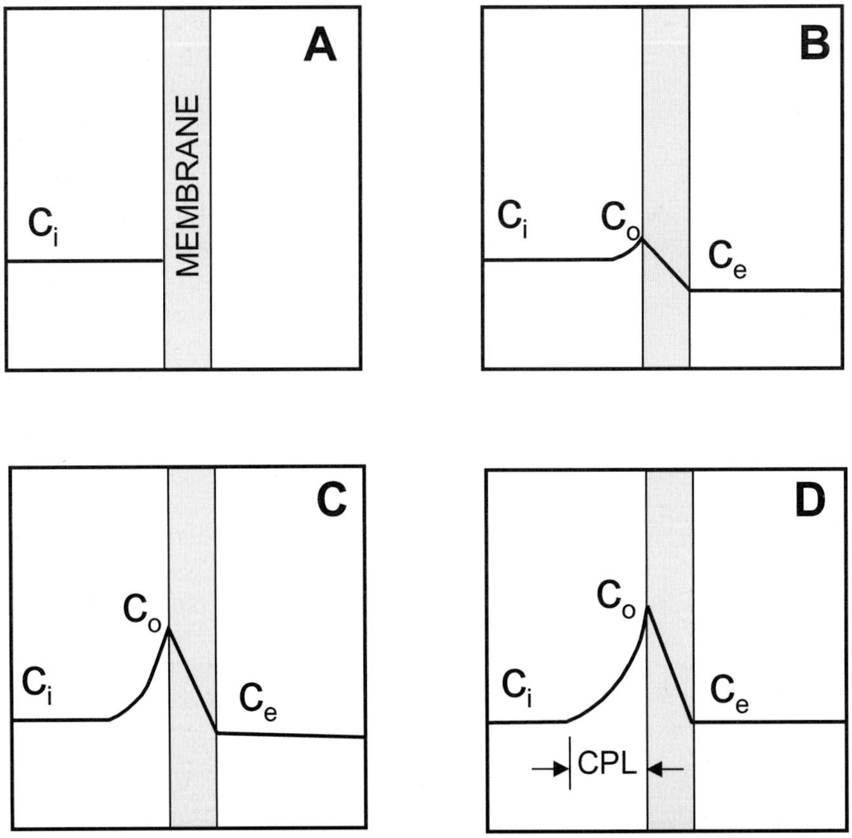

FIGURE 5. Conceptual development of the concentration polarization layer in a static cell. The initial conditions (A) are such that the solute is all on the high pressure side of the membrane and the pore fluids within the membrane contain no solute. Some time after solute flux through the membrane begins (B) the concentration at the high pressure membrane face C_o increases because some of the solute is rejected by the membrane. Some solute also begins to pass through the membrane so that the effluent now contains some solute as well. Even later (C) C_o has increased further as has the effluent concentration C_e. At steady state (D) the input concentration C_i is now equal to C_e and the value of C_o is constant. (Redrawn from Fritz and Marine, 1983).

through the shale-membrane, a portion of the solute is rejected, beginning a dynamic concentration of metals. This process continues as fluid flow occurs through a membrane-functioning unit. Solutes, while perhaps not always sufficiently concentrated to cause mineralization, are continually "swept" along the flow path.

If the volume flux of groundwater toward the membrane is sufficient to prevent the rejected solute from completely diffusing back into the aquifer, a CPL forms. Even though the component of flow parallel to the membrane interface will tend to diminish the CPL by dispersing a portion of the rejected solute back into the aquifer, groundwater flows are typically slow enough so that the CPL is not likely to be significantly diminished (Whitworth, 1998).

It is possible that metal concentrations within the CPL, near the membrane, would reach (super) saturation thus encouraging copper mineral precipitation in the sandstone adjacent to the bounding shale units. Notably, malachite and azurite are present in the copper deposits we examined; laboratory experiments (Whitworth and DeRosa, 1997) have demonstrated that copper carbonate can be precipitated by geologic membranes via the saturation of cupric copper and carbonate. Copper carbonate also forms from the weathering of copper sulfide minerals. Our observations indicate that no particular mechanism was solely responsible for the presence of copper carbonate mineralization in the Pastura district.

Most genetic models of sedimentary copper deposits, including ours, postulate that sulfide was produced by the bacterial reduction of sulfate (Hoy and Ohmoto, 1989). Certain anaerobic, sulfate-reducing bacteria (e.g. *Desulfovibro sp.*) exist at depth within groundwater systems and different varieties of bacteria can exist over a wide range of chemical and pressure conditions in the subsurface. For example, different varieties of bacteria can live in pH conditions of from 1 to 10, under fluid pressures of up to many hundreds of bars, and can survive temperatures of up to 75°C (Oppenheimer, 1963). Rose (1976) states that mineralization in red bed copper deposits seems to have occurred at temperatures of less than 75°C. Therefore, bacterial mediation of copper mineralization is not ruled out on the grounds of temperature. Furthermore, Rose (1976) reported that values of $\delta^{34}S$ in sulfides in these deposits range from -20 to -50‰ (Jensen, 1971) and are thus consistent with bacterial origin, although these values do not constitute proof of bacterial origin (Ohmoto, 1972). The lack of chalcocite twinning in these deposits indicates chalcocite precipitation temperatures were less than 115°C.

We suspect that the main role of the membrane-functioning shales in the formation of the copper deposits in the Pastura district is in concentrating both the dissolved copper and nutrients for use by sulfate reducing and other bacteria. Sulfate reducing bacteria are dependent on fermentative bacteria for the simple organic compounds their metabolism exploits (Chapelle, 1992). Organic matter is present in groundwater in colloidal form. In a study of colloids in groundwater at the Nevada Test Site, Kingston, *et al.* (1988) determined that organic colloids were present in the form of CH_2 and CH_3. We suggest that it is possible that one nutrient source for *Desulfovibro sp.* and the fermentative bacteria which supply their food source ultimately derive some, if not most, of their nutrients from groundwater colloids.

Hyperfiltration (to be technically correct in the case of colloidal materials—ultrafiltration) of groundwater colloids by geologic membranes would tend to concentrate a bacterial nutrient source within the CPL. Other dissolved nutrients such as phosphates and nitrates would also be concentrated in the CPL. This effect, occurring in a natural system, is similar to that described by Cheryan and Mehaia (1986) for membrane bioreactors, which use synthetic membranes to achieve microbial concentration for improving fermentation efficiencies in commercial fermentation processes.

In many sedimentary copper deposits, fossil wood and plant material have been a source of the carbon necessary for bacteriological metabolism or direct reduction/replacement, as evidenced by their mineralization. However, it is not always the presence of such organic material that causes sulfide precipitation. We observed unmineralized organic matter in the center of some of the sandstone strata in the Pastura district, while similar organic matter near the sandstone/shale boundary in the same unit were mineralized. This suggests that carbon may not be the limiting factor for bacterial growth and, therefore, generation of a sulfate-reducing environment at these deposits. Rather, we suggest that the limiting nutrient may be a solute, such as nitrogen or phosphorous, that was concentrated by the membrane-functioning shales. Because both nutrient and copper concentrations would be greatest at the membrane interface and diminishing with distance into the aquifer, mineralization related to microbial vigor and metals concentration in solution should be a maximum closest to the membrane. This is exactly the pattern of mineralization we observe in the Pastura district (Fig. 3).

When mass flux toward the membrane is insufficient to prevent the rejected solute from diffusing back into the aquifer, no CPL will form. Therefore, in cases in which flux toward and through the membrane is very low, one would expect no or only incipient and erratic copper mineralization at the membrane interface. Mineralization should only occur where the CPL develops to the extent that the solute concentration at the higher-pressure membrane face is greater than the solubility of a mineral species. This would explain why copper mineralization is not ubiquitous throughout the host rock and why mineralization favors the thinner shales where, for a given set of conditions, the solution flux through the shale is highest. In the Pintada and Guadalupe mines, the intensity of mineralization is inversely proportional to the thickness of the adjacent shale. We believe this occurs because volume flux through the shale is inversely proportional to shale thickness, as predicted by Darcy's law.

The concept of a "redox front" encountered at some distance downdip in the aquifer that encourages mineral precipitation, as is commonly proposed as a mineralization mechanism at many sedimentary copper deposits, does not appear to be reasonable in the Pastura district. We did not observe color changes, attributable to variable oxidation states of iron, cutting across sandstone bodies; instead, the more permeable sandstones tended to be gray, irrespective of mineralization, and the shales and less permeable silts and sands tend to be a reddish color. We suggest the color of the sediments may indeed be the result of redox phenomena, but not necessarily redox phenomena tied to mineralization.

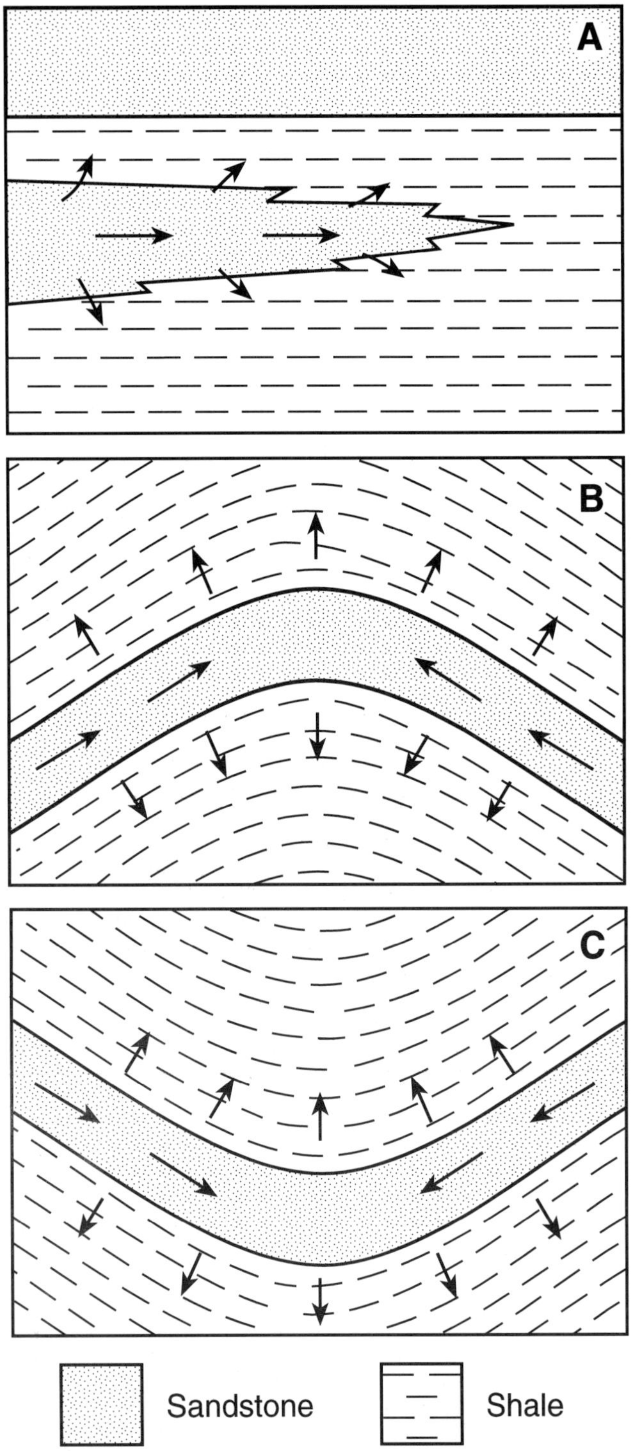

FIGURE 6. Conceptual diagram of focusing structures. A focusing structure is one which encourages flow through geologic membranes rather than through the aquifer. For example a facies change from sandstone to shale (A) may act as a focusing structure. Anticlines (B) and Synclines (C) may also direct flow through otherwise lower permeability rocks.

Fluid flow plays a major role in the formation of sedimentary copper deposits (and other similar deposits). Classically, groundwater scientists have typically considered fluid flow only within the aquifer based on the belief that fluid flow through confining beds is negligible (Freeze and Cherry, 1988). This approach does not adequately describe fluid flow in typical sedimentary depositional systems. Given a flow path of sufficient length, a facies change, pinch out, unconformity, fold, or fault will always be encountered.

When a facies change occurs in an aquifer, such that a permeability decrease is encountered, the effect is much like a partially closed valve on an open pipe. The valve, by reducing the cross-sectional area available for flow, effectively reduces the velocity of the fluid in the open pipe. So too does a permeability decrease slow the velocity of the fluid farther updip within the aquifer. Due to permeability decreases, some pinchouts, faults, unconformities, and folds have the capability to act as subsurface valves. The valve effect, then, by reducing flow velocity within the aquifer, results in a significantly lower flux than would be predicted by Darcy's Law using the hydraulic conductivity of the "good" aquifer matrix. As a result, the flux through surrounding shale membranes (aquitards) assumes a greater proportion of the flow. Thus, we expect subsurface flow through shales and other relatively low permeability rocks to be of significantly greater importance in most subsurface depositional systems than simple analysis using Darcy's Law would suggest.

Structures and stratigraphic transitions have been noted to aid in the development of larger ore deposits (Fig. 6). Hydrologic flow is focused by or along structures or stratigraphic facies changes, such as in synclines or basins, where fluid flow is directed toward the center of the structure. Because fluid flow occurring through a membrane-functioning unit results in some concentration of solutes, as fluid continues to flow through the membrane-functioning unit even greater solute concentration occurs until the solute concentration and bacterial activity leads to the precipitation of ore minerals. In the Pastura district, field observations suggest this mineralization was most likely to occur at such focusing structures. At Pastura, focusing structures are those stratigraphic sections where thinning in a membrane-functioning unit allows a significantly greater throughput of groundwater, thus concentrating or pooling most of the dissolved copper and microbial nutrients at a locus.

Features of fluid focusing are present at both the Guadalupe and Pintada mines. Mineralization is most well-developed in basinal or synclinal structures (Fig. 6B) yet is also focused in minor anticlines (Fig. 6C) where fluid flow is updip or is concentrated in a membrane structure which changes fluid flow characteristics. The latter situation is represented by more intense mineralization in proximity to the "humps" in the flooring shale described by Harley (1940) at the Guadalupe mine. At Pintada, mineralization is found in the center of a synclinal structure, a situation that encourages over-pressured, confined aquifers due to basin-edge recharge. In addition to a structural depression at the Guadalupe mine, stratigraphic pinchouts (Fig. 6A) may also have served as an additional mechanism to focus mineralization in the mine area (Holmquist, 1947). Stratigraphic pinchouts at the margins of basins are a particularly good place for the pooling of fluids and subsequent enhancement of the membrane effect. This feature is also noted at other deposits (Lange *et al.,* 1989) where minor structures or stratigraphic transitions within larger basins localize mineral deposition.

SUMMARY

Examination of two sedimentary copper deposits in the Pastura district shows that mineralization occurs in sandstones immediately adjacent to the confining shales. At each deposit, mineralization displays greatest development in copper concentrations and thickness adjacent to the thinnest shales and, conversely, with increasing shale thickness, becoming non-existent adjacent to the thickest shales. This pattern of mineralization is compatible with a geologic membrane-related process.

The fact that clays and shales can act as membranes has been firmly established by experimental and field observations. When solute is rejected by a shale membrane, and if flux through the membrane is sufficient to prevent effective back diffusion of the solute away from the membrane into an aquifer, a concentration polarization layer (CPL) is formed. Calculations demonstrate that, for reasonable hydrogeologic conditions, solute concentrations within the CPL can range from several to more than 100 times that of the background concentration within the aquifer (Whitworth, 1998). It is also reasonable to expect that shale membranes may also concentrate nutrients necessary for bacteriological metabolization of sulfate. Hence, we expect the CPL may be favorable for subsurface microbial activity with resulting H_2S production by sulfate-reducing bacteria producing sulfide mineral precipitation within the CPL.

Our mechanism also explains why copper mineralization is localized and not ubiquitous in the host rock. Because the magnitude of flux through a shale is inversely proportional to shale thickness, there is a threshold value of flux for any given solute through the membrane which must be exceeded before the solute rejected by the membrane becomes unable to effectively diffuse back into the reservoir. Therefore, one would expect that thick shales inhibit the formation of a CPL, while thinner shales would provide an environment appropriate for mature CPL development. Our observations indicate that the presence of a CPL is necessary for mineralization.

Finally, features of fluid focusing are present at both the Guadalupe and Pintada mines. At Pintada, mineralization is found in the center of a synclinal structure in an area of facies interfingering. In addition to a structural depression at the Guadalupe mine, a stratigraphic pinchout is also present and probably served as an additional mechanism to focus mineralization in the mine area. Stratigraphic pinchouts at the margins of basins are a particularly good place for "pooling" of fluids and subsequent enhancement of geologic membrane-microbial interactions, with related mineralization.

ACKNOWLEDGMENTS

This research was supported by the New Mexico Bureau of Geology and Mineral Resources, Peter Scholle, Director and State Geologist. Leo Gabaldon (NMBGMR) prepared a number of the figures for this paper. We appreciate the constructive reviews by William X. Chavez, Jr. (NMIMT) and Robert M. North (Phelps-Dodge, Chino).

REFERENCES

Anderson, E.C., 1957, The metal resources of New Mexico and their economic features through 1954: New Mexico Bureau of Mines and Mineral Resources, Bulletin 39, 183 p.

Becker, G. F., 1892, Quicksilver ore deposits, *in* Day, D. D., ed., Mineral Resources of the United States for 1892: U.S. Geological Survey Report, p. 139-168.

Berry, F. A. F., 1966, Proposed origin of subsurface thermal brines, Imperial Valley, California (abs.): American Society of Petroleum Geologists Bulletin, v. 50, p. 644-645.

Brown, A. C., 1971, Zoning in White Pine copper deposit, Ontonagon county, Michigan: Economic Geology, v. 66, p. 543-573.

Chapelle, F. H., 1992, Ground-water Microbiology and Geochemistry: New York, John Wiley & Sons, Inc., 424 p.

Chervyan, M. and Mehaia, M. A., 1985, Membrane bioreactors for high-performance fermentations, *in* S. Sourirajan and T. Matsuura, eds., Reverse Osmosis and Ultrafiltration: Washington D.C., American Chemical Society, p. 231-245.

Eugster, H.P., 1989, Geochemical environments of sediment-hosted Cu-Pb-Zn deposits, *in* Boyle, R.W., Brown, A.C., Jefferson, C.W., Jowett, E.C., and Kirkham, R.V. eds., Sediment-hosted Stratiform Copper Deposits: Geological Association of Canada, Special Paper 36, p. 111-126.

Freeze, R. A. and Cherry, J. A., 1988, Groundwater: Englewood Cliffs, NJ, Prentice-Hall, 604 p.

Fritz, S. J., 1986, Ideality of clay membranes in osmotic processes: a review: Clays and Clay Minerals, v. 34, p. 214-223.

Fritz, S. J. and Marine, I. W., 1983, Experimental support for a predictive osmotic model of clay membranes: Geochimica et Cosmochimica Acta, v 47, p. 1515-1522.

Fritz, S. J. and Whitworth, T. M., 1994, Hyperfiltration-induced fractionation of lithium isotopes: ramifications relating to representativeness of aquifer sampling: Water Resources Research, v. 30, p. 225-235.

Garvin, G. and Freeze, R. A., 1984, Theoretical analysis of the role of groundwater flow in the genesis of stratabound ore deposits. 2. quantitative results: American Journal of Science, v. 284, p. 1125-1174.

Harley, G. T., 1940, The geology and ore deposits of northeastern New Mexico: New Mexico School of Mines Bulletin No. 15, p. 91-97.

Holmquist, R. J., 1947, Stauber copper mine, Guadalupe county, New Mexico: U. S. Bureau of Mines, Report of Investigations 4026, 7 p.

Hoy, L. D. and Ohmoto, H., 1989, Constraints for the genesis of red-bed associated stratiform Cu deposits from sulfur and carbon mass-balance relations, *in* Boyle, R.W., Brown, A.C., Jefferson, C.W., Jowett, E.C., and Kirkham, R.V. eds., Sediment-hosted Stratiform Copper Deposits: Geological Association of Canada, Special Paper 36, p. 135-149.

Jensen, M. L., 1971, Sulfur isotopes of stratabound sulfide deposits: Society of Mining Geologists Japan, Special Issue 3, p. 300-303 (Proc. IMA-IAGOD).

Kelly, V.C., 1972, Geology of the Fort Sumner sheet, New Mexico: New Mexico Bureau of Mines and Mineral Resources, Bulletin 98, 55 p.

Kingston, L., Jacobson, R., and Whitbeck, M., 1988, Sampling and analysis of colloids in groundwater at the Nevada Test Site, in Mobility of colloidal particles in the subsurface: chemistry and hydrology of colloid-aquifer interactions, October 4-6, 1988, DOE/ER-0425, p. 35-39.

Kirkham, R. V., 1989, Distribution, settings and genesis of sediment-hosted stratiform copper deposits, *in* Boyle, R.W., Brown, A.C., Jefferson, C.W., Jowett, E.C., and Kirkham, R.V. eds., Sediment-hosted Stratiform Copper Deposits: Geological Association of Canada, Special Paper 36, 710 p.

Lange, I.M., Herberger, D., Whipple, J.W., and Krouse, H.R., 1989, Stratbound Cu-Ag and Pb-Zn mineralization in the Spokane and Helena Formations in the eastern part of the Belt Basin, Montana: A common origin?, *in* Boyle, R.W., Brown, A.C., Jefferson, C.W., Jowett, E.C., and Kirkham, R.V. eds., Sediment-hosted Stratiform Copper Deposits: Geological Association of Canada, Special Paper 36, p. 287-304.

LaPointe, D.J., 1989, A model for the diagenetic formation of sandstone copper deposits in sedimentary rocks of Permian and Triassic age in New Mexico, U.S.A., *in* Boyle, R.W., Brown, A.C., Jefferson, C.W., Jowett, E.C., and Kirkham, R.V. eds., Sediment-hosted Stratiform Copper Deposits: Geological Association of Canada, Special Paper 36, p. 357-370.

LaPointe, D.J., 1979, Geology, geochemistry, and petrology of sandstone copper deposits in New Mexico: unpublished Ph.D. dissertation, University of Colorado, Boulder, 333 p.

LaPointe, D.J., 1976, Comparison of selected sandstone copper deposits in New Mexico: Oklahoma Geological Survey Circular 77, p. 80-96.

LaPointe, D.J., 1974, Possible source areas for sandstone copper deposits in northern New Mexico: New Mexico Geological Society Guidebook, 25th Field Conference, p. 305-308.

McLemore, V. T. and North, R. M., 1985, Copper and uranium mineralization in east-central New Mexico: New Mexico Geological Society Guidebook, 36th Field Conference, p. 289-299.

Mitsoyannis, E. and Saravacos, G. D., 1977, Precipitation of calcium carbonate on reverse osmosis membranes: Desalination, v. 21, p. 235-240.

Neuzil, C. E., 2000, Osmotic generation of anomalous fluid pressures in geological environments: Nature, 403, p. 182-184.

Noggle, J. H., 1984, Physical Chemistry: Little, Brown & Company Limited, 953 p.

Ohmoto, H., 1972, Systematics of sulfur and carbon isotopes in hydrothermal ore deposits: Economic Geology, v. 67, p. 551-578.

Oppenheimer, C. H., 1963, Editor's preface to Introduction to geological microbiology, by Kuznetsov, S. V., Ivanov, M. V., and Lyalikova, N. N., New York, McGraw-Hill Book Company, 251 p.

Phillips,F. M.; Bentley, H. M.; Davis, S. N.; Elmnore, D.; and Swanick, G. H., 1986, Chlorine 36 dating of very old groundwater 2. Milk River Aquifer, Alberta, Canada: Water Resources Research, v. 22, p. 2003-2016.

Rose, A. W., 1976, The effect of cuprous chloride complexes in the origin of red-bed copper and related deposits: Economic Geology, v. 71, p. 1036-1048.

Sandusky, C. L. and Kaufman, W. H., 1972, Geology of the Pintada mine: New Mexico Geological Society Guidebook, 23rd Field Conference, p. 175-177.

Soulé, J. H., 1956, Reconnaissance of the "Red Bed" copper deposits in southeastern Colorado and New Mexico: U.S. Bureau of Mines Information Circular 7740, 74 p.

Young, A. and Low, P. F., 1965, Osmosis in argillaceous rocks: American Association of Petroleum Geologists Bulletin, v. 49, p. 1004-1008.

Whitworth, T. M., 1998, Steady-state mathematical modeling of geologic membrane processes in aquifer systems: WERC/WRHSRC/NMHWMS Joint Conference on the Environment, Proceedings, Albuquerque, NM, March 31—April 2, 1998, p. 37-41.

Whitworth, T. M. and DeRosa, G., 1997, Geologic membrane controls on saturated zone heavy metal transport. New Mexico Water Resources Research Institute Report No. 303, Las Cruces, New Mexico, 88 p.

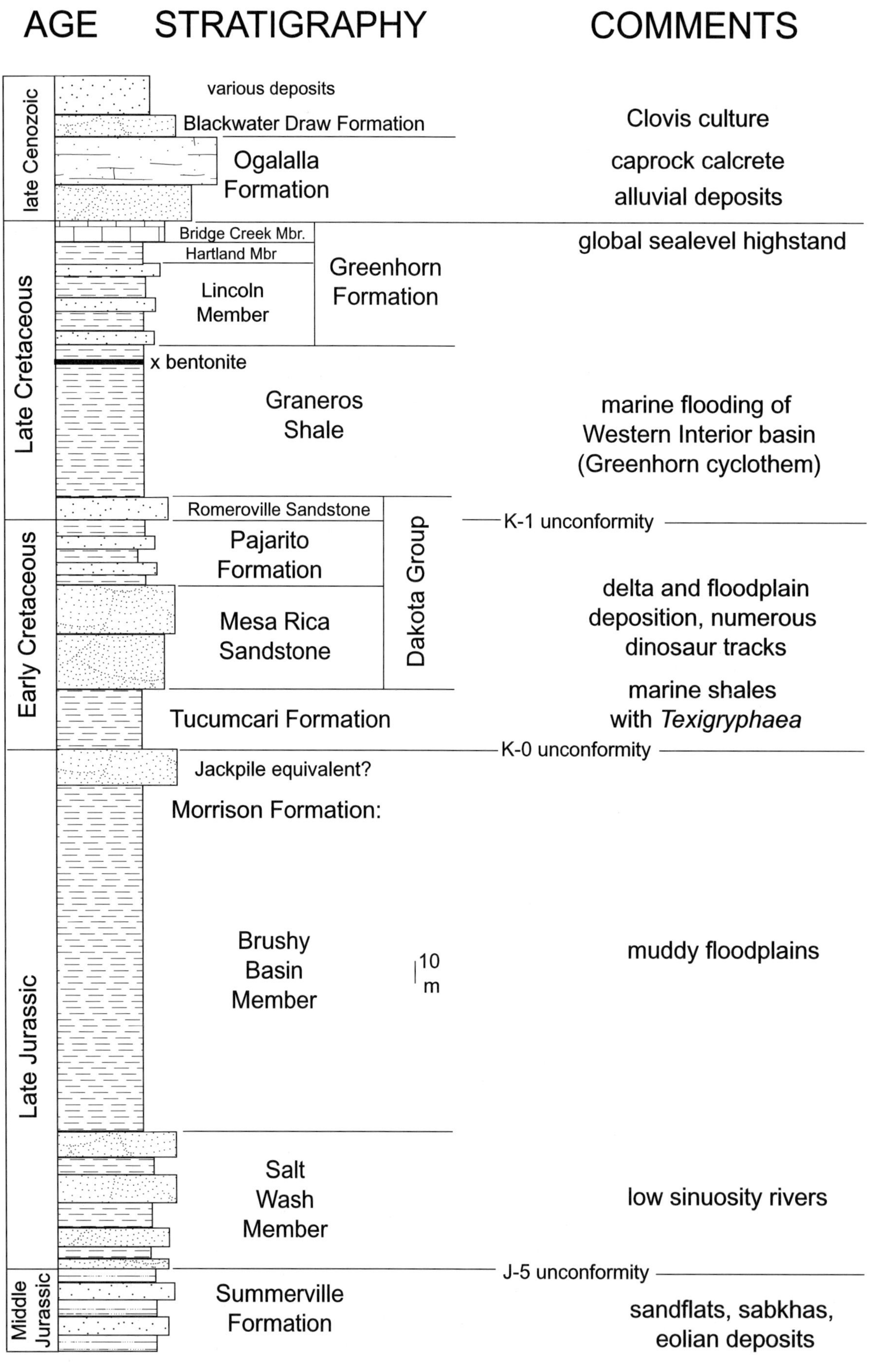
AGE
STRATIGRAPHY
COMMENTS
late Cenozoic
various deposits
Blackwater Draw Formation
Clovis culture
Ogalalla Formation
caprock calcrete
alluvial deposits
Late Cretaceous
Bridge Creek Mbr.
Hartland Mbr
Lincoln Member
Greenhorn Formation
global sealevel highstand
x bentonite
Graneros Shale
marine flooding of Western Interior basin (Greenhorn cyclothem)
Early Cretaceous
Romeroville Sandstone
K-1 unconformity
Pajarito Formation
Mesa Rica Sandstone
Dakota Group
delta and floodplain deposition, numerous dinosaur tracks
Tucumcari Formation
marine shales with *Texigryphaea*
K-0 unconformity
Late Jurassic
Jackpile equivalent?
Morrison Formation:
Brushy Basin Member
10 m
muddy floodplains
Salt Wash Member
low sinuosity rivers
J-5 unconformity
Middle Jurassic
Summerville Formation
sandflats, sabkhas, eolian deposits